Lecture Notes in Computer Science 4110

Commenced Publication in 1973
Founding and Former Series Editors:
Gerhard Goos, Juris Hartmanis, and Jan van Leeuwen

Josep Díaz   Klaus Jansen
José D.P. Rolim   Uri Zwick (Eds.)

Approximation, Randomization and Combinatorial Optimization

Algorithms and Techniques

9th International Workshop on Approximation Algorithms
for Combinatorial Optimization Problems, APPROX 2006
and 10th International Workshop
on Randomization and Computation, RANDOM 2006
Barcelona, Spain, August 28-30 2006
Proceedings

 Springer

Volume Editors

Josep Díaz
Universitat Politecnica de Catalunya
Departament de Llenguatges i Sistemes Informatics
08034 Barcelona, Spain
E-mail: diaz@lsi.upc.edu

Klaus Jansen
University of Kiel
Institute for Computer Science
24098 Kiel, Germany
E-mail: kj@informatik.uni-kiel.de

José D.P. Rolim
Centre Universitaire d'Informatique
1227 Carouge, Geneva, Switzerland
E-mail: Jose.Rolim@cui.unige.ch

Uri Zwick
Tel-Aviv University
School of Computer Science
Tel-Aviv 69978, Israel
E-mail: zwick@tau.ac.il

Library of Congress Control Number: 2006931401

CR Subject Classification (1998): F.2, G.2, G.1

LNCS Sublibrary: SL 1 – Theoretical Computer Science and General Issues

ISSN   0302-9743
ISBN-10   3-540-38044-2 Springer Berlin Heidelberg New York
ISBN-13   978-3-540-38044-3 Springer Berlin Heidelberg New York

Springer is a part of Springer Science+Business Media

springer.com

© Springer-Verlag Berlin Heidelberg 2006

Typesetting: Camera-ready by author, data conversion by Scientific Publishing Services, Chennai, India
Printed on acid-free paper   SPIN: 11830924   06/3142   5 4 3 2 1 0

Preface

This volume contains the papers presented at the 9th International Workshop on Approximation Algorithms for Combinatorial Optimization Problems (APPROX 2006) and the 10th International Workshop on Randomization and Computation (RANDOM 2006), which took place concurrently at the Universitat Politècnica de Catalunya, on August 28–30, 2006. APPROX focuses on algorithmic and complexity issues surrounding the development of efficient approximate solutions to computationally hard problems, and was the ninth in the series after Aalborg (1998), Berkeley (1999), Saarbrücken (2000), Berkeley (2001), Rome (2002), Princeton (2003), Cambridge (2004), and Berkeley (2005). RANDOM is concerned with applications of randomness to computational and combinatorial problems, and was the tenth workshop in the series following Bologna (1997), Barcelona (1998), Berkeley (1999), Geneva (2000), Berkeley (2001), Harvard (2002), Princeton (2003), Cambridge (2004), and Berkeley (2005).

Topics of interest for APPROX and RANDOM are: design and analysis of approximation algorithms, hardness of approximation, small space and data streaming algorithms, sub-linear time algorithms, embeddings and metric space methods, mathematical programming methods, coloring and partitioning, cuts and connectivity, geometric problems, game theory and applications, network design and routing, packing and covering, scheduling, design and analysis of randomized algorithms, randomized complexity theory, pseudorandomness and derandomization, random combinatorial structures, random walks/Markov chains, expander graphs and randomness extractors, probabilistic proof systems, random projections and embeddings, error-correcting codes, average-case analysis, property testing, computational learning theory, and other applications of approximation and randomness.

The volume contains 2 papers as invited lectures, 22 contributed papers, selected by the APPROX Program Committee out of 56 submissions, and 22 contributed papers, selected by the RANDOM Program Committee out of 49 submissions.

We would like to thank all of the authors who submitted papers, the members of the Program Committees:

APPROX 2006

Jittat Fakcharoenphol, Kasetsart University, Bangkok
Uriel Feige, Microsoft Research & Weizmann Institute
Anupam Gupta, Carnegie Mellon University, Pittsburgh
Magnús M. Halldórsson, University of Iceland
Johan Håstad, KTH, Stockholm
Amit Kumar, IIT, New Delhi
James R. Lee, IAS, Princeton
Mohammad Mahdian, Microsoft Research
Jiří Sgall, Academy of Sciences of the Czech Republic
Vijay Vazirani, Georgia Institute of Technology
Gerhard Woeginger, Eindhoven University of Technology
Uri Zwick, Tel Aviv University (Chair)

RANDOM 2006

Dimitris Achkioptas, University of California Santa Cruz
Andris Ambanis, University of Waterloo
Eli Ben-Sasson, Technion
Amin Coja-Oghlan, Humboldt-Universität zu Berlin
Colin Cooper, Kings College, London
Josep Díaz, Technological University of Catalunya (Chair)
Ravi Kannan, Yale University
Colin McDiarmid, University of Oxford
Rémi Monasson, CNRS Paris
Alessandro Panconesi, Università degli Studi di Roma La Sapienza
Vijaya Ramachandran, University of Texas at Austin
Vishal Sanwalani, University of Waterloo
Pavlos Spirakis, University of Patras
Madhu Saudan, MIT

We would also like to thank the external subreferees: Mikhail Alekhnovich, Christoph Ambühl, Albert Atserias, Per Austrin, Amitabha Bagchi, Maria-Florina Balcan, Markus Bläser, Josh Benaloh, Manuel Bodirsky, Andrej Bogdanov, Ioannis Caragiannis, Deeparnab Chakrabarty, Hubert Chan, Chandra Chekuri, Marek Chrobak, Artur Czumaj, Atish Das Sarma, Nikhil Devanur, Martin Dyer, Michael Elkin, Leah Epstein, Henrik Eriksson, Thomas Erlebach, Eldar Fischer, Martin Furer, Nicola Galesi, Naveen Garg, Ricard Gavalda, Stefanie Gerke, Ernst-Günter Giessmann, Gagan Goel, Andreas Goerdt, Leslie Ann Goldberg, Mordecai Golin, Daniel Golovin, Vineet Goyal, Sudipto Guha, Sam Gutmann, MohammadTaghi Hajiaghayi, Bjarni V. Halldórsson, Rafi Hassin, Nicole Immorlica, Robert W. Irving, Tejas Iyer, Dominik Janzing, Mark Jerrum, Raja Jothi, Ragnar Karlsson, Iordanis Kerenidis, Samir Khuller, Lefteris Kirousis, Johannes Köbler, Jochen Könemann, S. Kontogiannis, Guy Kortsarz, Michal Koucky, Annamaria Kovacs, Sven Krumke, Oded Lachish, Andre Lanka,

Erik van Leeuwen, Asaf Levin, Azarakhsh Malekian, Daniel Marx, Conrado Martinez, Frank McSherry, Aranyak Mehta, Michael Mitzenmacher, Cris Moore, Viswanath Nagarajan, Danupon Nanongkai, Assaf Naor, Tim Nieberg, Krzysztof Onak, Rasmus Pagh, Martin Pál, Anna Palbom, David Peleg, Kirk Pruhs, Ramamoorthi Ravi, Oded Regev, Omer Reingold, Tim Roughgarden, Dana Ron, Ronitt Rubinfeld, Atri Rudra, Jared Saia, Rishi Saket, Mathias Schacht, Elad Schiller, Hadas Shachnai, Ronen Shaltiel, Rene Sitters, Michiel Smid, Christian Sohler, Frits Spieksma, Aravind Srinivasan, Rob van Stee, Maxim Sviridenko, Kunal Talwar, Anusch Taraz, Luca Trevisan, Danny Vilenchik, Emanuele Viola, Maxwell Young, Neal E. Young, Raphael Yuster, Riccardo Zecchina, Alexander Zelikovsky, and Yan Zhang

We gratefully acknowledge the support from the Spanish Ministerio Educación i Ciencia, The Universitat Politécnica de Catalunya, the Department of LSI at the the UPC, Yahoo! Research Barcelona, the Institute of Computer Science of the Christian-Albrechts-Universität zu Kiel and the Department of Computer Science of the University of Geneva.

August 2006   Uri Zwick and Josep Díaz, Program Chairs
                        Klaus Jansen and José D. P. Rolim, Workshop Chairs

Table of Contents

Invited Talks

Contributed Talks of APPROX

Contributed Talks of RANDOM

On Nontrivial Approximation of CSPs

Johan Håstad

KTH– Royal Institute of Technology, Sweden
johanh@nada.kth.se

Abstract. Constraint satisfaction problems, more simply called CSPs are central in computer science, the most famous probably being Satisfiability, SAT, the basic NP-complete problem. In this talk we survey some results about the optimization version of CSPs where we want to satisfy as many constraints as possible.

One very simple approach to a CSP is to give random values to the variables. It turns out that for some CSPs, one example being Max-3Sat, unless P=NP, there is no polynomial time algorithm that can achieve a an approximation ratio that is superior to what is obtained by this trivial strategy. Some other CSPs, Max-Cut being a prime example, do allow very interesting non-trivial approximation algorithms which do give an approximation ratio that is substantially better than that obtained by a random assignment.

These results hint at a general classification problem of determining which CSPs do admit a polynomial time approximation algorithm that beats the random assignment by a constant factor. Positive results giving such algorithms tend to be based on semi-definite programming while the PCP theorem is the central tool for proving negative result.

We describe many of the known results in the area and also discuss some of the open problems.

J. Diaz et al. (Eds.): APPROX and RANDOM 2006, LNCS 4110, p. 1, 2006.

Analysis of Algorithms on the Cores of Random Graphs

Nick Wormald

Dept. Combinatorics and Optimization, University of Waterloo
nwormald@uwaterloo.ca

Abstract. The *k-core* of a graph is the largest subgraph of minimum degree at least k. It can be found by successively deleting all vertices of degree less than k.

The threshold of appearance of the k-core in a random graph was originally determined by Pittel, Spencer and the speaker. The original derivation used approximation of the vertex deletion process by differential equations. Many other papers have recently given alternative derivations.

A pseudograph model of random graphs introduced by Bollobás and Frieze, and also Chvátal, is useful for simplifying the original derivation. This model is especially useful for analysing algorothms on the k-core of a sparse random graphs, when the average degree is roughly constant. It was used recently to rederive the threshold of appearance of the k-core (with J. Cain). In addition, the following have recently been obtained concerning either of the random graphs $G = G(n, c/n)$, $c > 1$, or $G = G(n, m)$, $m = cn/2$.

(i) Analysis of a fast algorithm for off-line load balancing when each item has a choice of two servers. This enabled us to determine the threshold of appearance of a subgraph with average degree at least $2k$ in the random graph (with P. Sanders and J. Cain),

(ii) Bounds on the mixing time for the giant component of a random graph. We show that with high probability the random graph has a subgraph H with "good" expansion properties and such that $G - H$ has only "small" components with "not many" such components attached to any vertex of H. Amongst other things, this implies that the mixing time of the random walk on G is $\Theta(\log^2 n)$ (obtained recently and independently by Fountoulakis and Reed). This work is joint with I. Benjamini and G. Kozma. The subgraph is found by successively deleting the undesired vertices from the 2-core of the random graph.

(iii)Lower bounds on longest cycle lengths in the random graph. These depend on the expected average degree c and improve the existing results that apply to small $c > 1$ (by Ajtai, Komlós and Szemerédi, Fernandez de la Vega, and Suen). The new bounds arise from analysis of random greedy algorithms. Suen's bounds for induced cycles are also improved using similar random greedy algorithms. This is joint work with J.H. Kim.

In all cases the analysis is by use of differential equations approximating relevant random variables during the course of the algorithm. Typically, this determines the performance of the algorithms accurately, even if the best bounds are not necessarily achieved by these algorithms.

J. Diaz et al. (Eds.): APPROX and RANDOM 2006, LNCS 4110, p. 2, 2006.
© Springer-Verlag Berlin Heidelberg 2006

Constant-Factor Approximation for Minimum-Weight (Connected) Dominating Sets in Unit Disk Graphs

Christoph Ambühl[1], Thomas Erlebach[2], Matúš Mihaľák[2],
and Marc Nunkesser[3]

[1] Department of Computer Science, University of Liverpool
christoph@csc.liv.ac.uk
[2] Department of Computer Science, University of Leicester
{te17, mm215}@mcs.le.ac.uk
[3] Institute of Theoretical Computer Science, ETH Zürich
mnunkess@inf.ethz.ch

Abstract. For a given graph with weighted vertices, the goal of the minimum-weight dominating set problem is to compute a vertex subset of smallest weight such that each vertex of the graph is contained in the subset or has a neighbor in the subset. A unit disk graph is a graph in which each vertex corresponds to a unit disk in the plane and two vertices are adjacent if and only if their disks have a non-empty intersection. We present the first constant-factor approximation algorithm for the minimum-weight dominating set problem in unit disk graphs, a problem motivated by applications in wireless ad-hoc networks. The algorithm is obtained in two steps: First, the problem is reduced to the problem of covering a set of points located in a small square using a minimum-weight set of unit disks. Then, a constant-factor approximation algorithm for the latter problem is obtained using enumeration and dynamic programming techniques exploiting the geometry of unit disks. Furthermore, we also show how to obtain a constant-factor approximation algorithm for the minimum-weight connected dominating set problem in unit disk graphs.

1 Introduction

The dominating set problem is a classical optimization problem on graphs. For a given undirected graph $G = (V, E)$, a subset $D \subseteq V$ of its vertices is called a *dominating set* if every vertex in V is contained in D or has a neighbor in D. A vertex in D is called a *dominator*. A dominator *dominates* itself and all its neighbors. The goal of the *minimum dominating set problem* (MDS) is to compute a dominating set of smallest size. In the weighted version, the *minimum-weight dominating set problem* (MWDS), each vertex of the input graph is associated with a weight, and the goal is to compute a dominating set of minimum weight.

A dominating set $D \subseteq V$ is called a *connected dominating set* in the graph $G = (V, E)$ if the subgraph induced by D is connected. The minimum connected

J. Diaz et al. (Eds.): APPROX and RANDOM 2006, LNCS 4110, pp. 3–14, 2006.

dominating set problem (MCDS) and minimum-weight connected dominating set problem (MWCDS) are defined in the obvious way.

For general graphs, MDS (and therefore MWDS) is $\mathcal{NP}$-hard [9]. Furthermore, MDS for general graphs is known to be equivalent to the *set cover* problem, implying that it can be approximated within a factor of $O(\log n)$ for graphs with n vertices using a greedy algorithm (see, e.g., [16]), but no better unless all problems in $\mathcal{NP}$ can be solved in $n^{O(\log \log n)}$ time [8]. Approximation ratio $O(\log n)$ can also be achieved for the weighted set cover problem and thus for MWDS. The best known approximation ratio for MWCDS in general graphs is $O(\log n)$ as well [10].

In this paper, we are concerned with MWDS and MWCDS in a special class of graphs: *unit disk graphs*. A unit disk graph is a graph in which each vertex is associated with a (topologically closed) unit disk in the plane and two vertices are adjacent if and only if the corresponding disks have a non-empty intersection. We are interested in efficient approximation algorithms. An algorithm for MDS (or MWDS) is called a *ρ-approximation algorithm*, and has *approximation ratio ρ*, if it runs in polynomial time and always outputs a dominating set whose size (or total weight) is at most a factor of ρ larger than the size (or total weight) of the optimal solution. The definitions for MCDS and MWCDS are analogous. A *polynomial-time approximation scheme* (PTAS) is a family of approximation algorithms with ratio $1 + \varepsilon$ for every constant $\varepsilon > 0$.

A major motivation for studying (connected) dominating sets in unit disk graphs arises from routing in wireless ad-hoc networks, where dominating sets have been proposed for the construction of routing backbones (see, e.g., [1]). Each node of the graph models a wireless device, and two nodes are connected by an edge if they are close enough to receive each other's transmissions. A message that is broadcast by all nodes of a dominating set will be received by all nodes of the network. Therefore, a small connected dominating set is an energy-efficient routing backbone. Recent work has emphasized that ad-hoc networks are often heterogeneous as different nodes have different capabilities. Thus, it is meaningful to assign weights to the nodes (giving small weight to nodes that have a large remaining battery life, for example) and aim to determine a (connected) dominating set of small weight [17]. Therefore, one arrives at the MWDS and MWCDS problems in unit disk graphs.

Clark et al. [7] have proved that MDS is $\mathcal{NP}$-hard for unit disk graphs. Lichtenstein [13] has shown that MCDS is $\mathcal{NP}$-hard for unit disk graphs. Constant-factor approximation algorithms for MDS and MCDS in unit disk graphs were given by Marathe et al. [14]. For MDS in unit disk graphs, a PTAS was presented by Hunt et al. [12], based on the shifting strategy [3, 11]. These algorithms, however, do not extend to the weighted version. In particular, the PTAS is heavily based on the fact that the optimal dominating set for unit disks in a $k \times k$ square has size at most $O(k^2)$ and can thus be found in polynomial time using complete enumeration if k is a constant. In the weighted case, there is no such bound on the size of an optimal (or near-optimal) solution, as an optimal solution may consist of a large number of disks with tiny weight. For MCDS in unit disk graphs, a PTAS was presented in [6]. For the special case of unit disk graphs with bounded density, asymptotic fully polynomial-time approximation schemes (with running time polynomial in $\frac{1}{\varepsilon}$ and in the size of the input, but achieving ratio $1 + \varepsilon$ only for large enough inputs) were presented for MDS and MCDS in [15].

Wang and Li [17] give distributed algorithms for MWDS and MWCDS in unit disk graphs that achieve approximation ratio $O(\min\{\log \Delta, \sigma\})$, where Δ is the maximum degree of the graph and σ is the ratio of the maximum weight to the minimum weight of a disk. Note that these approximation ratios are not better than the known ratios for general graphs in the worst case.

Our Results. In this paper, we present the first constant-factor approximation algorithms for MWDS and MWCDS in unit disk graphs. Our algorithm for MWDS solves the problem in two steps. First, we reduce MWDS in unit disk graphs to the problem of covering a set of points that are located in a small square using a minimum-weight set of unit disks. In the reduction we lose only a constant factor in the approximation ratio. Then, we present a constant-factor approximation algorithm for the latter problem using enumeration and dynamic programming techniques exploiting the geometry of unit disks. To solve the MWCDS problem, we first compute an $O(1)$-approximation for the MWDS problem and then use an approach based on a minimum spanning tree calculation to add disks to the solution in order to make the dominating set connected. It remains an interesting open problem whether MDS and MWDS admit approximation algorithms with constant ratio also for arbitrary disk graphs.

The remainder of the paper is structured as follows. Our top-level approach to solving MWDS, which consists of breaking the problem into subproblems in small squares, is presented in Section 2. In Section 3, we show how the subproblem can be reduced to a special disk cover problem and give a constant-factor approximation algorithm for the latter problem. Section 4 shows how we can make a dominating set connected while incurring a cost that is bounded by a constant factor times the cost of the optimal connected dominating set. Proofs omitted due to space restrictions can be found in [2].

2 Algorithm for Minimum-Weight Dominating Sets

Let an instance of MWDS in unit disk graphs be given by a set $\mathcal{D}$ of weighted unit disks in the plane. The weight of disk $d \in \mathcal{D}$ is denoted by $w_d \geq 0$. Each disk has radius 1 and is specified by the coordinates of its center. For $U \subseteq \mathcal{D}$, we write $w(U)$ for $\sum_{d \in U} w_d$.

Our algorithm uses a parameter $\mu < 1$; we can set $\mu = 0.999$. We partition the plane into squares of side length μ. The square S_{ij}, for $i, j \in \mathbb{Z}$, contains all points (x, y) with $i\mu \leq x < (i + 1)\mu$ and $j\mu \leq y < (j + 1)\mu$.

For a square S_{ij} that contains at least one disk center, let $\mathcal{D}_{ij}$ be the set of disks in $\mathcal{D}$ whose center is in S_{ij}. Let $N(\mathcal{D}_{ij})$ denote the set of all disks in $\mathcal{D} \setminus \mathcal{D}_{ij}$ that intersect a disk in $\mathcal{D}_{ij}$. We consider a subproblem to be solved for each square S_{ij} that can be stated as follows: Find a minimum-weight set of disks in $\mathcal{D}_{ij} \cup N(\mathcal{D}_{ij})$ that dominates all disks in $\mathcal{D}_{ij}$. Let OPT_{ij} denote an optimal solution to the subproblem for square S_{ij}. In Section 3, we will present an algorithm that outputs a solution U_{ij} for the subproblem satisfying $w(U_{ij}) \leq 2 \cdot w(\mathrm{OPT}_{ij})$. In the end, we output the union of all sets U_{ij} that we have computed. It is clear that this yields a dominating set.

Theorem 1. *There is a constant-factor approximation algorithm for the minimum-weight dominating set problem in unit disk graphs.*

Proof. The algorithm described above outputs a dominating set U of weight at most $\sum w(U_{ij})$. Here and in the following, the summation is over all squares S_{ij} that contain at least one disk center. As we will present a 2-approximation algorithm to solve each subproblem in Section 3, we have $w(U_{ij}) \leq 2 \cdot w(\mathrm{OPT}_{ij})$. Let OPT denote an optimal dominating set for the whole instance. Let $\mathrm{OPT}[S_{ij}] = \mathrm{OPT} \cap (\mathcal{D}_{ij} \cup N(\mathcal{D}_{ij}))$. Note that $\mathrm{OPT}[S_{ij}]$ is a feasible solution to the subproblem for square S_{ij} and therefore we have $w(\mathrm{OPT}_{ij}) \leq w(\mathrm{OPT}[S_{ij}])$.

We get $w(U) \leq \sum w(U_{ij}) \leq 2 \sum w(\mathrm{OPT}_{ij}) \leq 2 \sum w(\mathrm{OPT}[S_{ij}])$. The sum $\sum w(\mathrm{OPT}[S_{ij}])$ adds the costs of solutions $\mathrm{OPT}[S_{ij}]$ for all squares S_{ij} that contain at least one disk center. Note that a disk d in OPT can be in $\mathrm{OPT}[S_{ij}]$ only if its center is in S_{ij} or it intersects a disk with center in S_{ij}. Therefore, the distance between the center of d and the square S_{ij} is at most 2. Consequently, there are only $O(1/\mu^2)$ squares S_{ij} such that d can be in $\mathrm{OPT}[S_{ij}]$. More precisely, all such squares must be fully contained in a disk of radius $2 + \sqrt{2}\mu$ around the center of d, and for $\mu = 0.999$ that disk can contain at most $\lfloor (2 + \sqrt{2}\mu)^2 \pi / \mu^2 \rfloor = 36$ such squares. This means that the number of times each disk in OPT contributes its weight to $\sum w(\mathrm{OPT}[S_{ij}])$ is bounded by 36. We get $\sum w(\mathrm{OPT}[S_{ij}]) \leq 36 \cdot w(\mathrm{OPT})$ and, thus, $w(U) \leq 72 \cdot w(\mathrm{OPT})$.   □

3   Solving the Subproblem for a Small Square

In this section we present a 2-approximation algorithm for the following problem: Given a $\mu \times \mu$ square S_{ij}, where $\mu < 1$, and the set of disks $\mathcal{D}_{ij} \cup N(\mathcal{D}_{ij})$, compute a minimum-weight set of disks that dominates all disks in $\mathcal{D}_{ij}$.

Let OPT_{ij} denote the set of disks in an optimal solution for the problem. In the following, we will often write that the algorithm "guesses" certain properties of OPT_{ij}. Such guesses are to be interpreted as follows: The algorithm tries all possible choices for the guess (there will be a polynomial number of such choices) and computes a solution for each choice. In the end, the algorithm outputs the solution of minimum weight among all solutions found in this way. Some guesses may not lead to feasible solutions; such guesses are discarded. In the analysis, we concentrate on the solution in which the algorithm makes the right guess about OPT_{ij}. It then suffices to show that the solution the algorithm finds for that guess is a constant-factor approximation of the optimum, because the solution output by the algorithm in the end will be at least as good as the one it finds for that guess. Unfortunately, the running-time of the algorithm, although polynomial, is quite large, since the algorithm must try all possibilities for the many guesses it makes about the optimal solution.

First, the algorithm guesses the largest weight w of a disk in OPT_{ij}. Note that there are at most n possible values for this guess (where n is the number of disks in the instance). If there is a disk of weight at most w in $\mathcal{D}_{ij}$, the algorithm simply outputs that disk as the solution (note that the disk has its center in S_{ij} and therefore dominates all other disks in $\mathcal{D}_{ij}$), and this solution is optimal. If there is no disk of weight at most w in $\mathcal{D}_{ij}$, we know that OPT_{ij} consists entirely of disks in $N(\mathcal{D}_{ij})$ of weight at most w. In this case, we first discard all disks from $N(\mathcal{D}_{ij})$ that have weight larger than w and arrive at the following problem: Find a set of disks of minimum weight from $N(\mathcal{D}_{ij})$ that dominates all disks in $\mathcal{D}_{ij}$. A disk d_1 from $N(\mathcal{D}_{ij})$ dominates a disk d_2 from $\mathcal{D}_{ij}$ if and only if the distance of the centers of d_1 and d_2 is at most 2. Therefore, we can increase the radius of the disks in $N(\mathcal{D}_{ij})$ from 1 to 2 and reduce the radius of the disks in $\mathcal{D}_{ij}$ from 1 to 0 and obtain an equivalent problem: If $\mathcal{D}'$ denotes the set containing the enlarged version of the disks in $N(\mathcal{D}_{ij})$ and $\mathcal{P}$ denotes the set of centers of the disks in $\mathcal{D}_{ij}$, we need to find a minimum-weight subset of the disks in $\mathcal{D}'$ that covers all points in $\mathcal{P}$. Furthermore, we can renormalize the setting so that the disks in $\mathcal{D}'$ have radius 1. The renormalized square S is now a $\delta \times \delta$ square, with $\delta = \mu/2 < 1/2$. Therefore, the problem to be solved can be stated as follows:

> **Disk cover in a small square:** Given a set $\mathcal{P}$ of points in a $\delta \times \delta$ square S, where $\delta < 1/2$, and a set $\mathcal{D}'$ of weighted unit disks, find a minimum-weight subset of $\mathcal{D}'$ that covers all points in $\mathcal{P}$.

In the following subsection, we will present a 2-approximation algorithm for this problem. In view of the discussion above, this implies that we have a 2-approximation algorithm for the problem of computing a minimum-weight set of disks that dominates all disks in $\mathcal{D}_{ij}$ for a given $\mu \times \mu$ square S_{ij}, and this is the ingredient that we needed in the previous section to obtain the constant-factor approximation algorithm for MWDS in unit disk graphs.

3.1 Algorithm for Disk Cover in a Small Square

We are given a set $\mathcal{P}$ of points in a $\delta \times \delta$ square S and a set $\mathcal{D}'$ of n weighted unit disks, and we want to find a minimum-weight subset of $\mathcal{D}'$ that covers all points in $\mathcal{P}$. Let OPT' denote a set of disks constituting an optimal solution to this problem.

Let C be the area covered by the union of the disks in OPT'. A *hole* of OPT' is defined to be a topological component of $S \setminus C$. Intuitively, if S was a glass window and the disks in OPT' were to cover parts of this window, the holes would be the connected regions where one can still see through the window.

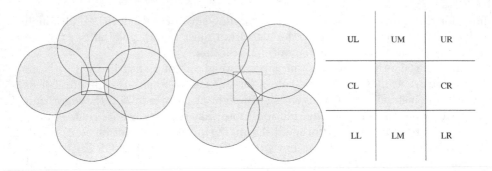

Fig. 1. One-hole solution (left), many-hole solution (middle), naming of regions (right)

Definition 1. OPT′ *is a* one-hole solution *if it has exactly one hole and each disk in* OPT′ *forms part of the boundary of that hole (and that part consists of more than 1 point).* OPT′ *is a* many-hole solution *if it has at least two holes.*

Definition 1 is illustrated in Fig. 1. If OPT′ is neither a one-hole solution nor a many-hole solution, it must be of one of the following types: Either OPT′ has no hole at all, or it has one hole but not all disks in OPT′ form part of the boundary of the hole. If OPT′ does not have a hole, we can delete one disk d from OPT′ (and remove all points in d from $\mathcal{P}$) to obtain a solution with at least one hole. If OPT′ has one hole but not all disks are on the boundary of the hole, let d' be a disk that is not on the boundary of the hole. If we delete d' from OPT′ (and the corresponding points from $\mathcal{P}$), we have at least two holes and arrive at a many-hole solution. Therefore, OPT′ can always be converted into a one-hole or many-hole solution by deleting at most two disks.

The algorithm guesses whether OPT′ is a one-hole solution or a many-hole solution. If OPT′ is neither of these, the algorithm also guesses this and additionally guesses the one or two disks that need to be removed from OPT′ (and added to the solution computed by the algorithm) in order to obtain a one-hole or many-hole solution. Hence, we can assume that OPT′ is a one-hole or many-hole solution and that the algorithm has guessed correctly which of the two is the case. In each of the two cases, we will encounter subproblems that can be solved efficiently by dynamic programming, as stated in the following lemma.

Lemma 1. *Let* $\mathcal{P}$ *be a set of points located in a strip between the horizontal lines* $y = y_1$ *and* $y = y_2$ *for some* $y_1 < y_2$. *Let* $\mathcal{D}$ *be a set of weighted unit disks with centers above the line* $y = y_2$ *(upper disks) or below the line* $y = y_1$ *(lower disks). Furthermore, assume that the union of the disks in* $\mathcal{D}$ *contains all points in* $\mathcal{P}$. *Then a minimum-weight subset of* $\mathcal{D}$ *that covers all points in* $\mathcal{P}$ *can be computed in polynomial time.*

In the following, we show how to deal with the one-hole case and the many-hole case.

One-Hole Solutions. Assume that OPT′ is a one-hole solution. The boundary of the hole is formed by disks from OPT′ and, potentially, some parts from sides of the square S (we view the latter as special kinds of disks with weight 0 and infinite radius, i.e., halfplanes, and do not treat them explicitly in the following). All disks in OPT′ have their centers outside S. Using the lines that are the extensions of the sides of S, we can partition the plane outside S into 8 regions in the natural way (see also Fig. 1): upper left region (UL), upper middle region (UM), upper right region (UR), central right region (CR), lower right region (LR), lower middle region (LM), lower left region (LL), and central left region (CL). The upper region (U) is the union of UL, UM and UR, and similarly for the lower region (L).

If we follow the boundary of the hole in counterclockwise direction, we will encounter disks with center in CL, then disks with center in L, then disks with center in CR, then disks with center in U. The points on the boundary that are in the intersection of two consecutive disks on the boundary are called *corners*. Each corner is *determined* by two disks (the disks on whose boundaries it lies).

Among all corners that are determined by at least one disk whose center is in CL, let p_ℓ denote the one with the smallest y-coordinate and let p_u denote the one with the largest y-coordinate. Let p'_ℓ and p'_u be defined analogously with respect to CR. (The case where no part of the boundary of the hole is created by disks with center in CL or CR is easier and is not treated in detail here.) The algorithm guesses the corners p_ℓ, p_u, p'_ℓ and p'_u and the pairs of disks determining them. As there are only $O(n^2)$ pairs of disks, the number of potential guesses is polynomial.

Let d_L be the unit disk that has p_ℓ and p_u on the boundary and has its center to the left of the line $\overline{p_\ell p_u}$. Note that in general d_L is not a disk that is part of the input of the problem. Let d_ℓ and d_u be the disks from OPT′ that have their center in CL and contain p_ℓ and p_u, respectively, on the boundary. Let x be the intersection point of the boundaries of d_ℓ and d_u that is closer to S. Let $\mathcal{L}$ be the connected region that is delineated by the boundary of d_L between p_u and p_ℓ, and by the boundary of d_ℓ between x and p_ℓ, and by the boundary of d_u between p_u and x. See Fig. 2 (left) for an illustration.

Lemma 2. *The only disks in* OPT′ *that intersect* $\mathcal{L}$ *have their center in CL or in the union of UR, CR and LR. Furthermore, no disk from* OPT′ *with center in CL can cover a point outside* $\mathcal{L}$ *that is not already covered by* d_u *or* d_ℓ.

Proof. As p_u and p_ℓ are on the boundary of the hole, no disk in OPT′ can contain p_u or p_ℓ in its interior. Hence, any disk d from OPT′ that intersects $\mathcal{L}$ must either have its center to the left of the line $\overline{p_\ell p_u}$ and intersect the parts of the boundaries of d_ℓ and d_u that define $\mathcal{L}$, or it must have its center to the right of the line $\overline{p_\ell p_u}$ and intersect the boundary of $\mathcal{L}$ twice on the part that is also a boundary of d_L. In the former case, the y-coordinate of the center of d must lie between the y-coordinates of the centers of d_ℓ and d_u, and hence d must have its center in CL. (To see this, consider the disk d' that is obtained from d by shifting it horizontally to the right until it first contains p_u or p_ℓ on its boundary; observe that the disk d_u can be rotated around p_u until it becomes identical to d', with its center continuously moving downward; the same argument can be applied

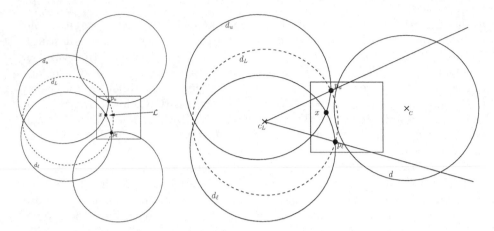

Fig. 2. The region $\mathcal{L}$ is defined by parts of the boundaries of disk d_L, drawn dashed, and disks d_u and d_ℓ (left). A disk d with center not in CL from OPT$'$ intersecting $\mathcal{L}$ must have its center in the cone of two halflines starting at the center c_L of d_L and passing through p_u and p_ℓ, respectively (right).

to the disk d_ℓ and shows that the center of d' must have larger y-coordinate than the center of d_ℓ. By the same argument, we also have that c_L must lie in CL.) In the latter case, the center c of d must lie in the cone of points between the halflines starting at the center c_L of d_L and passing through p_ℓ and p_u, respectively, see Fig. 2 (right). We want to show that c cannot be in UM or LM. Assume for a contradiction that c is in UM (the case for LM is similar). The slope of the line connecting c_L and p_u is at most $\delta/\sqrt{1-\delta^2}$. Therefore, the largest y-coordinate of a point in the intersection of the cone and UM is bounded by $y_{p_u} + \delta^2/\sqrt{1-\delta^2}$, so the distance between p_u and any point in that intersection is at most $\delta/\sqrt{1-\delta^2}$. Hence, for $\delta < \sqrt{2}/2$ (and we even have $\delta < 1/2$), a unit disk with center in that intersection must contain p_u. Thus, c cannot be in UM, as d would then contain p_u in its interior. Similarly, we get that c cannot be in LM. Furthermore, c clearly cannot be in UL or LL, as it must be to the right of p_u. Hence, we have shown that c must be in the union of UR, CR and LR.

We have shown that the only disks in OPT$'$ that intersect $\mathcal{L}$ have their center in CL or in the union of UR, CR and LR. It remains to show that no disk from OPT$'$ with center in CL can cover a point outside $\mathcal{L}$ that is not already covered by d_u or d_ℓ. Let d' be a disk from OPT$'$ with center in CL. All disks from OPT$'$ are on the boundary of the hole, and p_u and p_ℓ are the topmost and lowest corners, respectively, that are determined by at least one disk with center in CL. Therefore, d' must appear on the boundary of the hole between p_u and p_ℓ. This implies that $d' \setminus (d_u \cup d_\ell)$ consists of one region that is contained in $\mathcal{L}$ and a second region that is outside the square S (and thus cannot contain any points from $\mathcal{P}$). This establishes the claim.   □

Similar to $\mathcal{L}$, we can define a region $\mathcal{R}$ with respect to CR, p'_ℓ and p'_u, and the analogue of Lemma 2 holds for $\mathcal{R}$.

Let $\mathcal{P}'$ be the set of points that is obtained from $\mathcal{P}$ by removing the points that are contained in one of the disks defining the four corner points guessed by the algorithm. For the points in $\mathcal{P}' \cap (\mathcal{L} \cup \mathcal{R})$, we can compute an optimal disk cover using Lemma 1 (rotated by 90°), since the points are contained in the vertical strip containing S and the only disks that need to be considered for covering them have their center to the left or to the right of the strip. The remaining points in $\mathcal{P}'$ can only be covered by disks with center in U or in L by OPT', hence we can again compute an optimal disk cover for them using Lemma 1. If we output the union of the two disk covers, we have computed a 2-approximation to the overall disk cover problem in this square.

Many-Hole Solutions. Now we consider the case that OPT' is a many-hole solution. There must be two disks $d_1, d_2 \in$ OPT' such that $S \setminus (d_1 \cup d_2)$ consists of two disjoint regions and each of these two regions contains a hole of OPT'. (As a special case, we could also have a single disk from OPT' that intersects the square in such a way that two holes are created.) We use a new coordinate system in which the y-axis contains the centers c_1 and c_2 of d_1 and d_2, respectively, and the intersection points of the boundaries of d_1 and d_2 are on the x-axis. Let b be the smallest axis-parallel square containing the (rotated) square S. Let δ' be the side length of b. Note that $\delta' \leq \delta\sqrt{2} < \sqrt{2}/2$. As for the one-hole case, we partition the plane outside b into regions UL, UM, UR, CR, LR, LM, LL, CL, and we define regions U and L as before.

The disks d_1 and d_2 create two holes in S; we refer to the left hole as LH, and to the right hole as RH. Because OPT' is a superset of $\{d_1, d_2\}$, OPT' may contain more than two holes, but all the holes in OPT' are contained in either LH or RH. We can show the following lemma.

Lemma 3. *In OPT', no disk with center in the union of UR, CR and LR (in the union of UL, CL and LL) can intersect LH (RH).*

Due to space restrictions, we give only an outline of our solution for the weighted disk cover problem in the many-hole case: We can show that LH contains a region $\mathcal{L}$ such that points in $\mathcal{L}$ can be covered only by disks with center in CL by OPT'. Let $\mathcal{P}' \subseteq \mathcal{P}$ be the points in LH that are not in $\mathcal{L}$ and are not already covered by the disks the algorithm guesses to define $\mathcal{L}$. We can then show that points in $\mathcal{P}'$ can only be covered by disks with center in U or L. The same approach is applied to RH. This breaks the problem into two independent subproblems: covering points in $\mathcal{L}$ and in the corresponding region of RH using disks with center in CL or CR, and covering the remaining points using disks with center in U or L. Each of the two subproblems can be solved optimally by dynamic programming (Lemma 1). Since the subproblems are independent, the union of their optimal solutions gives an optimal solution to the disk cover problem in the many-hole case.

In summary, we have shown that in both the one-hole case and the many-hole case we can obtain a 2-approximation (in the many-hole case, even an optimal solution) of the minimum-weight disk cover for the given $\delta \times \delta$ square S. Furthermore, all other cases (no holes, or one hole with not all disks on the boundary of the hole) can be reduced to one of these cases by guessing one or two disks in the optimal solution. Therefore, we obtain a 2-approximation algorithm for the problem of computing a minimum-weight disk cover in a small square.

We remark that this result on disk cover in a small square also implies a constant-factor approximation algorithm for the general weighted disk cover problem with unit disks (i.e., given a set of points and a set of weighted unit disks, find a minimum-weight set of disks that covers all the points): We can simply partition the plane into $\delta \times \delta$ squares and compute an approximate disk cover for each square. Then we output the union of all computed disk covers as the solution. As a disk from the optimal solution can be used to cover points in at most $O(1/\delta^2)$ different $\delta \times \delta$ squares, we lose only a factor of $O(1/\delta^2)$ in the approximation ratio by solving the problem for each square separately. Previously, constant-factor approximation algorithms were known only for the unweighted case of the disk cover problem [4, 5].

4   Connecting the Dominating Set

In this section we consider the problem of adding disks to a given dominating set in order to produce a connected dominating set. We present an algorithm that solves this problem by adding disks of total weight at most $O(w^*)$, where w^* denotes the optimal weight of a connected dominating set for the given set of weighted disks. Note that the problem of connecting up a dominating set is a special case of the node-weighted Steiner tree problem; for general graphs, the best known approximation ratio for the latter problem is logarithmic in the size of the graph [10].

Let $\mathcal{D}$ be a set of weighted unit disks, and let $U \subseteq \mathcal{D}$ be a dominating set. Let G denote the unit disk graph corresponding to the disks in $\mathcal{D}$, and assume that G is connected (otherwise, G cannot have a connected dominating set). The vertex set of a connected component of $G[U]$ (the subgraph of G induced by U) is called a *cluster* of U. We create an auxiliary graph H. The vertices of H correspond to the clusters of U. For every path of length at most 3 in G that connects a vertex in one cluster c_1 of U to a vertex in another cluster c_2 of U and whose one or two internal vertices are not in U, we add an edge between c_1 and c_2 to H. The weight of the edge is the sum of the weights of the disks corresponding to the one or two internal vertices of the path. Note that H can have parallel edges. Next, we compute a minimum spanning tree T in H. (The proof of the theorem below shows that H is a connected graph.) Finally, we connect the dominating set U by adding all disks that correspond to internal vertices of the paths in G that correspond to the edges of T.

Theorem 2. *Let $\mathcal{D}$ be a set of weighted disks and U be a dominating set. Let w^* be the weight of a minimum-weight connected dominating set for $\mathcal{D}$. There is an efficient algorithm that computes a set U' of disks such that $U \cup U'$ is a connected dominating set and $w(U') \leq 17w^*$.*

Proof. We show that the auxiliary graph H contains a spanning tree T' of weight at most $17w^*$. This implies that H is connected. Furthermore, the weight of the set U' of disks that the algorithm adds to U is at most the weight of the minimum spanning tree, and the weight of the minimum spanning tree is upper bounded by the weight of T'. Therefore, we get $w(U') \leq 17w^*$.

It remains to show how to construct a spanning tree T' of H with weight at most $17w^*$. Let U^* be an optimal connected dominating set, $w(U^*) = w^*$. Let C be an arbitrary non-empty set of clusters of U, but not the set of all clusters of U. Let $\bar{C}$ be the set of the remaining clusters of U. We claim that G must contain a path π from a vertex in some cluster in C to a vertex in some cluster in $\bar{C}$ such that π contains at most two internal vertices and has the property that all its internal vertices are in $U^* \setminus U$. (Note that such a path π corresponds to an edge in H.) To prove the claim, we argue as follows. Let x be an arbitrary vertex in a cluster in C, and y an arbitrary vertex in a cluster in $\bar{C}$. As U^* is a connected dominating set, there must be a path p in G from x to y all of whose internal vertices are in U^*. Let x' be the last vertex on p that is not in U and that is dominated by a vertex x'' in a cluster in C. Note that such a vertex x' must exist. Furthermore, x' or the vertex y' after x' on p must be dominated by a vertex y'' in a cluster in $\bar{C}$. Therefore, we obtain the desired path as x'', x', y'' or x'', x', y', y''.

Now we can create a spanning tree of H as follows. We start with a tree consisting of a single vertex of H (corresponding to some cluster of U) and grow the tree by repeatedly finding a path π in G that connects a vertex from a cluster in the tree to a vertex in a cluster not in the tree and has the properties discussed above. The claim above shows that such a path must exist. We can thus grow the tree by adding the edge in H that corresponds to the path π. This is repeated until we have a spanning tree T'.

The weight of each edge in the spanning tree T' corresponds to the weight of the internal vertices (which are in U^*) of a path of length at most 3 that connects different clusters of U. Furthermore, a vertex (disk) d of U^* can contribute to at most 17 edges of H: Whenever d contributes to the weight of an edge, it is an internal vertex of a path that connects two clusters of U whose closest disks have (graph-theoretic) distance at most 2 from it. However, the set of disks at distance at most 2 from d can contain at most 18 disjoint disks (see e.g. [17]) and therefore at most 18 disks from different clusters of U. As the spanning tree can contain at most 17 edges between these 18 clusters, we obtain that d contributes its weight to at most 17 edges of the spanning tree T'. Consequently, $w(T') \leq 17w^*$. □

Together with Theorem 1, we obtain the following corollary.

Corollary 1. *There is a constant-factor approximation algorithm for the minimum-weight connected dominating set problem in unit disk graphs.*

The approximation ratio in Corollary 1 is at most $72 + 17 = 89$.

References

1. K. Alzoubi, P.-J. Wan, and O. Frieder. Message-optimal connected dominating sets in mobile ad hoc networks. In *Proceedings of the 3rd ACM International Symposium on Mobile Ad Hoc Networking and Computing (MobiHoc 2002)*, pages 157–164, 2002.
2. C. Ambühl, T. Erlebach, M. Mihaľák, and M. Nunkesser. Constant-factor approximation for minimum-weight (connected) dominating sets in unit disk graphs. Research Report CS-06-008, Department of Computer Science, University of Leicester, June 2006.
3. B. S. Baker. Approximation algorithms for NP-complete problems on planar graphs. *Journal of the ACM*, 41(1):153–180, 1994. Extended abstract published in the proceedings of FOCS'83, pp. 265–273, 1983.
4. H. Brönnimann and M. T. Goodrich. Almost optimal set covers in finite VC-dimension. *Discrete & Computational Geometry*, 14(4):463–479, 1995.
5. G. Calinescu, I. Mandoiu, P.-J. Wan, and A. Zelikovsky. Selecting forwarding neighbors in wireless ad hoc networks. *Mobile Networks and Applications*, 9(2):101–111, 2004.
6. X. Cheng, X. Huang, D. Li, W. Wu, and D.-Z. Du. A polynomial-time approximation scheme for the minimum-connected dominating set in ad hoc wireless networks. *Networks*, 42(4):202–208, 2003.
7. B. N. Clark, C. J. Colbourn, and D. S. Johnson. Unit disk graphs. *Discrete Mathematics*, 86:165–177, 1990.
8. U. Feige. A threshold of ln n for approximating set cover. In *Proceedings of the 28th Annual ACM Symposium on Theory of Computing (STOC'96)*, pages 314–318, 1996.
9. M. R. Garey and D. S. Johnson. *Computers and Intractability. A Guide to the Theory of NP-Completeness*. W. H. Freeman and Company, New York-San Francisco, 1979.
10. S. Guha and S. Khuller. Improved methods for approximating node weighted Steiner trees and connected dominating sets. *Information and Computation*, 150(1):57–74, 1999.
11. D. S. Hochbaum and W. Maass. Approximation schemes for covering and packing problems in image processing and VLSI. *Journal of the ACM*, 32(1):130–136, 1985.
12. H. B. Hunt III, M. V. Marathe, V. Radhakrishnan, S. S. Ravi, D. J. Rosenkrantz, and R. E. Stearns. NC-Approximation schemes for NP- and PSPACE-hard problems for geometric graphs. *Journal of Algorithms*, 26(2):238–274, 1998.
13. D. Lichtenstein. Planar formulae and their uses. *SIAM Journal on Computing*, 11(2):329–343, 1982.
14. M. V. Marathe, H. Breu, H. B. Hunt III, S. S. Ravi, and D. J. Rosenkrantz. Simple heuristics for unit disk graphs. *Networks*, 25:59–68, 1995.
15. E. J. van Leeuwen. Approximation algorithms for unit disk graphs. In *Proceedings of the 31st International Workshop on Graph-Theoretic Concepts in Computer Science (WG'05)*, LNCS 3787, pages 351–361, 2005.
16. V. V. Vazirani. *Approximation Algorithms*. Springer, 2001.
17. Y. Wang and X.-Y. Li. Distributed low-cost backbone formation for wireless ad hoc networks. In *Proceedings of the 6th ACM International Symposium on Mobile Ad Hoc Networking and Computing (MobiHoc 2005)*, pages 2–13, 2005.

Approximating Precedence-Constrained Single Machine Scheduling by Coloring

Christoph Ambühl[1], Monaldo Mastrolilli[2], and Ola Svensson[2]

[1] University of Liverpool - Great Britain
christoph@csc.liv.ac.uk
[2] IDSIA- Switzerland
{monaldo, ola}@idsia.ch

Abstract. This paper investigates the relationship between the dimension theory of partial orders and the problem of scheduling precedence-constrained jobs on a single machine to minimize the weighted completion time. Surprisingly, we show that the vertex cover graph associated to the scheduling problem is exactly the graph of incomparable pairs defined in dimension theory. This equivalence gives new insights on the structure of the problem and allows us to benefit from known results in dimension theory. In particular, the vertex cover graph associated to the scheduling problem can be colored efficiently with at most k colors whenever the associated poset admits a polynomial time computable k-realizer. Based on this approach, we derive new and better approximation algorithms for special classes of precedence constraints, including *convex bipartite* and *semi-orders*, for which we give $(1+\frac{1}{3})$-approximation algorithms. Our technique also generalizes to a richer class of posets obtained by lexicographic sum.

1 Introduction

We consider the problem of scheduling a set $N = \{1, \ldots, n\}$ of n jobs on a single machine, which can process at most one job at a time. Each job j has a processing time p_j and a weight w_j, where p_j and w_j are nonnegative integers. We only consider *non-preemptive* schedules, in which all p_j units of job j must be scheduled consecutively. A *partially ordered set* (or *poset*) is a structure $\mathbf{P} = (X, P)$ consisting of a *ground set* X and a *partial order*, i.e. a reflexive, antisymmetric, and transitive binary relation P on X. Jobs have precedence constraints between them that are specified in the form of a poset $\mathbf{P} = (N, P)$, where $(i, j) \in P$ $(i \neq j)$ implies that job i must be completed before job j can be started. The goal is to find a schedule which minimizes the sum $\sum_{j=1}^{n} w_j C_j$, where C_j is the time at which job j completes in the given schedule. In standard scheduling notation (see e.g. Graham et al. [9]), this problem is known as $1|prec|\sum_j w_j C_j$.

The general version of $1|prec|\sum_j w_j C_j$ was shown to be strongly NP-hard by Lawler [13] and Lenstra & Rinnooy Kan [14]. While currently no inapproximability result is known (other than that the problem does not admit a fully polynomial time approximation scheme), there are several polynomial

J. Diaz et al. (Eds.): APPROX and RANDOM 2006, LNCS 4110, pp. 15–26, 2006.

time 2-approximation algorithms [17, 20, 10, 3, 2, 15, 1]. For the general version of $1|prec| \sum_j w_j C_j$, closing the approximability gap is considered a longstanding open problem in scheduling theory (see e.g. [21]).

Due to this difficulty, more attention has recently been given to special classes [24, 12, 4, 1]. With this aim, it is worth mentioning that Woeginger [24] proved that the general case of $1|prec| \sum_j w_j C_j$ is not harder to approximate than many fairly restricted special cases, among them the case where all job weights are one. However, for a few relevant special posets with "nice" structural properties, one can obtain better approximation ratios than 2. For the special cases of *interval order* and *convex bipartite* precedence constraints, Woeginger [24] developed polynomial time approximation algorithms with worst case performance guarantee arbitrarily close to the golden ration $\frac{1}{2}(1 + \sqrt{5}) \approx 1.61803$. Recently, Ambühl & Mastrolilli [1] settled an open problem first raised by Chudak & Hochbaum [3] and whose answer was conjectured by Correa & Schulz [4]. The results in [1, 4] imply the existence of an exact polynomial time algorithm for the special case of two-dimensional partial orders, improving on previously known approximation algorithms [12, 4], and generalizing Lawler's exact algorithm [13] for series-parallel orders.

Moreover, the most significant implication in [1] is that problem $1|prec| \sum_j w_j C_j$ is a special case of the weighted vertex cover problem in an undirected graph $G_{CS}(\mathbf{P})$ (see [1, 4] and Section 2) that has a node for each ordered pair (i, j) of jobs $i, j \in N$ with $(i, j) \notin P$ and $(j, i) \notin P$ (we say i and j are *incomparable* and write $i \parallel j$ in P). By using this relationship several previous results for the scheduling problem can be explained, and in some cases improved, by means of the vertex cover theory.

Dimension is one of the most heavily studied parameters of partial orders, and many beautiful results have been obtained (see e.g. [22]). Dushnik & Miller [5] introduced dimension as a parameter of partial orders in 1941. Since that time, many theorems have been developed. The *dimension* of a partial order P is the minimum number of linear extensions which yield P as their intersection. More precisely, if P and Q are two partial orders on the same ground set, we say Q is an *extension* of P if $P \subseteq Q$, and we call Q a *linear extension* of P if Q is a linear order and an extension of P. A *realizer* $\mathcal{R}$ of P is a family of linear extensions of P such that $P = \cap \mathcal{R}$, i.e., for all $x, y \in X$, $(x, y) \in P$ if and only if $(x, y) \in L$ for every $L \in \mathcal{R}$. The *dimension* of $\mathbf{P}$, denoted by $dim(\mathbf{P})$ or $dim(X, P)$, is the smallest k such that there exists a realizer $\mathcal{R}$ of P with cardinality k, i.e., $|\mathcal{R}| = k$ ($\mathcal{R}$ is said to be a k-*realizer*). Obviously, $dim(X, P) = 1$ if and only if P is a linear order. With any finite poset $\mathbf{P}$, we can associate a hypergraph $\mathcal{H}_\mathbf{P}$ so that the dimension of $\mathbf{P}$ is equal to the chromatic number of $\mathcal{H}_\mathbf{P}$ [7, 22]. The vertices of $\mathcal{H}_\mathbf{P}$ are the incomparable pairs in P, and this hypergraph is called the *hypergraph of incomparable pairs*. The edges of size 2 in $\mathcal{H}_\mathbf{P}$ determine an ordinary graph $G_\mathbf{P}$, which is called the *graph of incomparable pairs*. Trotter [22] is a good source for further results involving dimension.

In this paper we continue to investigate the structure of problem $1|prec| \sum_j w_j C_j$. We point out an interesting relationship between the dimension theory of partial orders and problem $1|prec|\sum_j w_j C_j$. More specifically, in Section 3 we show that the vertex cover graph $G_{CS}(\mathbf{P})$ associated to $1|prec|\sum_j w_j C_j$ is exactly the graph of incomparable pairs $G_{\mathbf{P}}$ in dimension theory [7, 22]. This equivalence allows us to benefit from dimension theory. In particular, the chromatic number of $G_{CS}(\mathbf{P})$ is at most k, whenever the dimension of the poset at hand is (at most) k. Hochbaum [11] showed that if a given graph for the vertex cover problem can be colored by using k colors in polynomial time, then there exists a $(2 - 2/k)$-approximation algorithm for the corresponding weighted vertex cover problem. It follows that there exists a $(2 - 2/k)$-approximation algorithm for $1|prec|\sum_j w_j C_j$ for all those special classes of precedence constraints that admit a polynomial time computable k-realizer.

By following this general approach, we obtain approximation algorithms for relevant special classes of precedence constraints, such as[1] *convex bipartite* precedence constraints (Sections 4) and *semi-orders* (Section 5), for which we exhibit $(1 + \frac{1}{3})$-approximation algorithms that improve previous results by Woeginger [24]. However, the technique in [24] also extends to the case of interval order precedence constraints, for which we prove that our approach cannot yield a better approximation ratio (Section 5).

Our technique also generalizes to a richer class of posets obtained by lexicographic sum. Indeed we show, in Section 6, that the number of colors needed to color the graph of incomparable pairs does not increase under the lexicographic sum. In Section 7 we end up by discussing further posets and pointing out some related interesting open problems.

2 Preliminaries

Problem $1|prec|\sum_j w_j C_j$ was recently proved [1] to be a special case of MINIMUM WEIGHTED VERTEX COVER: Given a graph $G = (V, E)$ with weights w_i on the vertices, find a subset $V' \subseteq V$, minimizing the objective function $\sum_{i \in V'} w_i$, such that for each edge $(u, v) \in E$, at least one of u and v belongs to V'.

This result was achieved by investigating the relationship between several different linear programming formulations and relaxations [18, 3, 4] of $1|prec|\sum_j w_j C_j$, using linear ordering variables δ_{ij}. The variable δ_{ij} has value 1 if job i precedes job j in the corresponding schedule, and 0 otherwise. Correa & Schulz [4] proposed the following relaxation of $1|prec|\sum_j w_j C_j$:

$$[\text{CS-IP}] \quad \min \quad \sum_{i\|j} \delta_{ij} p_i w_j + \sum_{j \in N} p_j w_j + \sum_{(i,j) \in P} p_i w_j$$

$$\text{s.t.} \quad \delta_{ij} + \delta_{ji} \geq 1 \quad i \| j, \tag{1}$$

$$\delta_{ik} + \delta_{kj} \geq 1 \quad (i, j) \in P, i \| k \text{ and } k \| j, \tag{2}$$

$$\delta_{i\ell} + \delta_{kj} \geq 1 \quad (i, j), (k, \ell) \in P, i \| \ell \text{ and } j \| k, \tag{3}$$

$$\delta_{ij} \in \{0, 1\} \quad i \| j.$$

---

[1] Further special classes of posets can be found in [16, 22].

Note that [CS-IP] can be interpreted as the minimum weighted vertex cover in an undirected graph $G_{CS}(\mathbf{P})$, that has a node for each incomparable pair (i,j) of jobs. Two nodes (i,j) and (k,ℓ) are adjacent if either $j = k$ and $i = \ell$, or $j = k$ and $(i,\ell) \in P$, or $(i,\ell),(k,j) \in P$.

Correa & Schulz [4] conjectured that an optimal solution to $1|prec|\sum_j w_j C_j$ gives an optimal solution to [CS-IP] as well. The conjecture in [4] was recently solved by Ambühl & Mastrolilli [1], who proved that any feasible solution to [CS-IP] can be turned in polynomial time into a feasible solution to $1|prec|\sum_j w_j C_j$ without deteriorating the objective value. It follows that problem $1|prec|\sum_j w_j C_j$ is a special case of the weighted vertex cover problem in the graph $G_{CS}(\mathbf{P})$. We refer the interested reader to [1,4] for a more comprehensive discussion.

We already mentioned that Hochbaum [11] gave a $(2 - 2/k)$-approximation algorithm for the weighted vertex cover problem, whenever the vertex cover graph is k-colorable in polynomial time. Putting everything together we come up with the following result.

Theorem 1. *[1, 4, 11] Problem $1|prec|\sum_j w_j C_j$, for which the graph $G_{CS}(\mathbf{P})$ is k-colorable in polynomial time, has a polynomial time $(2-2/k)$-approximation algorithm.*

3   Posets: Dimension and Coloring

The aim of this section is to point out the connection between $1|prec|\sum_j w_j C_j$ and the dimension theory of partial orders. For this purpose, we need some preliminary definitions.

Let $\mathbf{P} = (N, P)$ be a poset. We say that the partial order $P^d = \{(x,y) : (y,x) \in P\}$ is the *dual* of P. An *alternating cycle* in (N, P) is a collection of incomparable pairs $\{(x_1,y_1),(x_2,y_2),\ldots,(x_k,y_k)\}$ such that $(y_i, x_{i+1}) \in P$ for all i (modulo k). We associate with $\mathbf{P}$ a hypergraph $\mathcal{H}_{\mathbf{P}} = (V, \mathcal{E})$ defined as follows. The vertex set V of $\mathcal{H}_{\mathbf{P}}$ is the set of incomparable pairs $inc(\mathbf{P}) = \{(x,y) \in X \times X : x\|y \text{ in } P\}$, and the edge set $\mathcal{E}$ consists of those subsets of V whose duals form alternating cycles. Let $G_{\mathbf{P}}$ denote the ordinary graph determined by all edges of size 2 in $\mathcal{H}_{\mathbf{P}}$. In the literature [22,7], $\mathcal{H}_{\mathbf{P}}$ and $G_{\mathbf{P}}$ are referred to as the *hypergraph* and the *graph of incomparable pairs*, respectively, and they play an important role in the understanding of dimension. We recall that the *chromatic number* of a hypergraph $\mathcal{H} = (V, \mathcal{E})$, denoted $\chi(\mathcal{H})$, is the least positive integer t for which there is a function $f : V \to [t]$ so that there is no $\alpha \in [t]$ for which there is an edge $E \in \mathcal{E}$ with $f(x) = \alpha$ for every $x \in E$. The following result associates a poset $\mathbf{P}$ to $\mathcal{H}_{\mathbf{P}}$ so that the dimension of $\mathbf{P}$ is the chromatic number of $\mathcal{H}_{\mathbf{P}}$.

Proposition 1 ([22,7]). *Let $\mathbf{P} = (N, P)$ be a poset, that is not a linear order. Then $dim(\mathbf{P}) = \chi(\mathcal{H}_{\mathbf{P}}) \geq \chi(G_{\mathbf{P}})$.*

Given a k-realizer $\mathcal{R} = \{L_1, L_2, \ldots, L_k\}$ of $\mathbf{P}$, we can easily color $\mathcal{H}_{\mathbf{P}}$ (and $G_{\mathbf{P}}$) with k colors: color vertex (i,j) with some color c whenever $(j,i) \in L_c$. Observe that if all nodes of a hyperedge are colored by the same color c then the linear extension L_c contains an alternating cycle, which is impossible.

The following proposition is immediate and it can be easily checked. It establishes a strong relationship between the dimension theory and $1|prec|\sum_j w_j C_j$.

Proposition 2. *The vertex cover graph $G_{CS}(\mathbf{P})$ associated to $1|prec|\sum_j w_j C_j$ and the graph of incomparable pairs $G_{\mathbf{P}}$ coincide.*

A large amount of combinatorial theory exists for posets. Tapping this source can help in designing approximation algorithms.

Theorem 2 ([22, 7]). *Let $\mathbf{P} = (N, P)$ be a poset, that is not a linear order. Then the graph $G_{\mathbf{P}}$ is bipartite if and only if $dim(\mathbf{P}) = 2$.*

Theorem 2 is a well-known result in dimension theory. Correa & Schulz [4] rediscovered it for the vertex cover graph $G_{CS}(\mathbf{P})$, unaware of the connection pointed out by Proposition 2. What is more, the following theorem follows easily from Theorem 1 and Propositions 2 and 1.

Theorem 3. *Problem $1|prec|\sum_j w_j C_j$, whenever precedence constraints are given by a k-realizer, has a polynomial time $(2 - \frac{2}{k})$-approximation algorithm.*

A natural question is for which posets one can construct a k-realizer in polynomial time. In the general case, Yannakakis [25] proved that determining whether the dimension of a poset is at most k is NP-complete for every $k \geq 3$. However, for several special cases, including convex bipartite orders (Section 4) and semi-orders (Section 5), a minimal realizer can be computed in polynomial time.

Finally, by Proposition 1, we remark that $dim(\mathbf{P})$ and $\chi(G_{\mathbf{P}})$ are, in general, not the same (see [7] for an example where $dim(\mathbf{P})$ is exponentially larger than $\chi(G_{\mathbf{P}})$). However, it is an immediate consequence of Theorem 2 that $dim(\mathbf{P}) = \chi(G_{\mathbf{P}})$ when $dim(\mathbf{P}) = 3$. Therefore, a 3-realizer for a 3-dimensional partial order P (as in Sections 4 and 5) immediately gives an optimal coloring for $G_{\mathbf{P}}$.

4   Convex Bipartite Precedence Constraints

In this section we consider $1|prec|\sum_j w_j C_j$ for which the precedence constraints form a so called convex bipartite order. For this class of partial orders, we show how to construct a realizer of size 3 in polynomial time. By Theorem 3, this gives a $(1 + \frac{1}{3})$-approximation algorithm.

A *convex bipartite order* $\mathbf{P} = (N = J^- \cup J^+, P)$ is defined as follows.

1. *The set of jobs are divided into two disjoint sets $J^- = \{j_1, \ldots j_a\}$ and $J^+ = \{j_{a+1}, \ldots, j_n\}$, the* minus-jobs *and* plus-jobs, *respectively.*
2. *For every $k = a+1, \ldots, n$ there are two indices $l(k)$ and $r(k)$ with $1 \leq l(k) \leq r(k) \leq a$ such that $(j_i, j_k) \in P$ if and only if $l(k) \leq i \leq r(k)$ (bipartiteness and convexity).*

It is not hard to check that convex bipartite orders can be recognized in polynomial time. Moreover, the class of convex bipartite orders forms a proper subset of the class of general bipartite orders, and a proper superset of the class of strong bipartite orders [16]. Lemma 3 states that the class of convex bipartite orders has dimension ≤ 3. This is indeed a tight bound, since a bipartite order $\mathbf{P}$ is 2-dimensional if and only if it is a strong bipartite order [16]. Finally, we observe that $1|prec| \sum_j w_j C_j$ with strong bipartite orders is solvable in polynomial time [1, 4, 16].

In the subsequent, we sometimes stress that a job j_i is a plus- or minus-job by writing j_i^+ and j_i^-, respectively. We also assume, without loss of generality, that the plus-jobs are numbered such that $i < j$ if and only if $l(i) \leq l(j)$ (breaking ties arbitrarily), where $j_i, j_j \in J^+$.

Given a convex bipartite poset $\mathbf{P} = (N, P)$, we partition its incomparable pairs into three sets E_1, E_2, and E_3 (also depicted in Fig. 1). A pair of incomparable jobs $(j_i, j_j) \in inc(\mathbf{P})$ is a member of

E_1 if $i > j$ and $j_i, j_j \in J^-$; **else if** $i < j$ and $j_i, j_j \in J^+$; **else if** $j_i \in J^-$ and $j_j \in J^+$.

E_2 if $i < j$ and $j_i, j_j \in J^-$; **else if** $j_i \in J^+, j_j \in J^-$ and there exists a $k > i$ such that $(j_j, j_k) \in P$.

E_3 if $i > j$ and $j_i, j_j \in J^+$; **else if** $j_i \in J^+, j_j \in J^-$ and $(j_j, j_k) \notin P$ for all $k > i$.

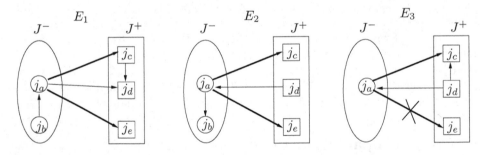

Fig. 1. The round and square nodes correspond to minus-jobs and plus-jobs, respectively. Bold edges correspond to precedence constrains, whereas the other edges are between incomparable jobs. In this example we assume that $a < b$ and $c < d < e$.

The following lemma is a direct consequence of the definition of E_1, E_2, and E_3.

Lemma 1. *Let* $\mathbf{P}$ *be a convex bipartite order then*

1. *The sets* E_1, E_2, *and* E_3 *form a partition of* $inc(\mathbf{P})$;
2. *For every* $(i, j) \in inc(\mathbf{P})$, *if* $(i, j) \in E_k$ *then* $(j, i) \notin E_k$, *where* $k \in \{1, 2, 3\}$.

Lemma 2. *Let* $\bar{E}_1 = E_1 \cup P, \bar{E}_2 = E_2 \cup P$, *and* $\bar{E}_3 = E_3 \cup P$. *Then* $\bar{E}_1, \bar{E}_2$, *and* $\bar{E}_3$ *are extensions of* P.

Proof. By the definition of $\bar{E}_i$, it follows that if $(j_i, j_j) \in P$ then $(j_i, j_j) \in \bar{E}_i$, where $i = 1, 2, 3$. Moreover, it is easy to see (Fig. 1) that the sets $\bar{E}_1$ and $\bar{E}_3$ do not contain cycles, i.e., are extensions of P.

Now suppose $\bar{E}_2$ contains an alternating cycle C, i.e., it is a non valid extension. By the definition of E_2 we have $C \cap P \neq \emptyset$ and thus $C \cap (J^+ \times J^-) \neq \emptyset$. Let $j_i^- \in J^-$ be the minus-job with largest index in the cycle, i.e., there does not exist a $k > i$ such that $j_k \in J^-$ is part of the cycle. Then $(j_i^-, j_j^+) \in P \cap C$ and $(j_j^+, j_m^-) \in C$ for some jobs $j_j \in J^+$ and $j_m \in J^-$, where $m < i$. However, this implies that there exists an $n > j$ such that $(j_m^-, j_n^+) \in P$ (recall the definition of E_2). Together with convexity and the numbering of plus-jobs this implies $(j_m^-, j_j^+) \in P$, which contradicts the existence of $(j_j^+, j_m^-) \in C$. □

Let L_1, L_2, and L_3 be any linear extensions of $\bar{E}_1, \bar{E}_2$, and $\bar{E}_3$, respectively. That $\mathcal{R} = \{L_1, L_2, L_3\}$ is a realizer follows from the facts that all incomparable pairs are reversed (Lemma 1), and that $\bar{E}_1, \bar{E}_2$, and $\bar{E}_3$ are valid extensions of $\mathbf{P}$ (Lemma 2). Furthermore, all steps involved in creating $\mathcal{R}$ can clearly be accomplished in polynomial time.

Lemma 3. *Given a convex bipartite order* $\mathbf{P} = (N, P)$, *a realizer of size three can be computed in polynomial time.*

Theorem 3 and Lemma 3 give us the following result.

Theorem 4. *Problem* $1|prec|\sum_j w_j C_j$ *for which the precedence constraints form a convex bipartite order has a polynomial time $(1 + 1/3)$-approximation algorithm.*

5   Interval Orders

A poset $\mathbf{P} = (N, P)$ is an *interval order* [16, 22, 23] if there is a function I assigning to each point $x \in N$ a closed interval $I(x) = I_x = [a_x, b_x]$ of the real line $\mathbb{R}$ so that $(x, y) \in P, x \neq y$ if and only if $b_x < a_y$ in $\mathbb{R}$. The function I is called an *interval representation* of the poset $\mathbf{P}$. Interval orders can be recognized in polynomial time and an interval representation can be computed in $O(n^2)$ time [16].

The best known approximation algorithm for $1|prec|\sum_j w_j C_j$ with interval order precedence constraints is due to Woeginger [24], who gave an (≈ 1.61803)-approximation algorithm. We observe that this ratio can be improved to $(1 + \frac{1}{3})$ in the special case of semi-order precedence constraints. Unfortunately, we show that our techniques do not generalize to interval orders.

5.1   Approximating Semi-orders

A *semi-order*, also called *unit* interval order, has a similar definition as interval orders, but the function I is restricted to only assign unit intervals, i.e., $I(x) = [a_x, a_x + 1]$. Semi-orders can be recognized in $O(n^2)$ time [16, 22]. Moreover, Rabinowitz proved, by constructing a realizer, that the dimension of *semi-orders* is at most three [19, 22]. The constructive proof can easily be turned into a polynomial algorithm and together with Theorem 3, we have the following theorem.

Theorem 5. *Problem* $1|prec|\sum_j w_j C_j$ *for which the precedence constraints form a semi-order has a polynomial time (1+ 1/3)-approximation algorithm.*

5.2   Coloring Interval Orders

For $1|prec|\sum_j w_j C_j$ with interval precedence constraints, one cannot obtain a better than 2-approximation by using our techniques. Indeed we exhibit interval orders where the associated graphs of incomparable pairs have arbitrarily large chromatic number. To prove this, we introduce the *canonical* interval orders. For an integer $n \geq 2$, let $\mathbf{I}_n$ denote the interval order determined by the set of all closed intervals with distinct integer end points from $[n]$. We will find it convenient to view the elements of $\mathbf{I}_n$ as 2-element subsets of $[n]$ with $(\{i_1, i_2\}, \{i_3, i_4\})$ in $\mathbf{I}_n$ if and only if $i_2 < i_3$ in $\mathbb{R}$ or $\{i_1, i_2\} = \{i_3, i_4\}$. The family $\{\mathbf{I}_n : n \geq 2\}$ is called the *canonical* interval orders [23].

Theorem 6. *For any integer k, there exists an integer n_0 so that if $n \geq n_0$, then the chromatic number $\chi(G_{\mathbf{I}_n})$ is larger than k.*

Proof. The chromatic number $\chi(G_{\mathbf{I}_n})$ is clearly a non-decreasing function of n. We assume that $\chi(G_{\mathbf{I}_n}) \leq k$ for all $n \geq 2$ and obtain a contradiction when n is sufficiently large.

Let the map $\varphi : \binom{[n]}{3} \rightarrow \{1, 2, \ldots, k\}$ denote a coloring of the 3-element subsets of $[n]$. Note that any coloring of $G_{\mathbf{I}_n}$, defines the map φ, by letting $\varphi(\{i, j, k\})$ equal the coloring of the vertex $(\{i, j\}, \{j, k\})$ [2] in $G_{\mathbf{I}_n}$.

Let n_0 equal the Ramsey number $R(3 : h_1, h_2, h_3 \ldots, h_k)$, where $h_1 = h_2 = \cdots = h_k = 4$. Now pick n to be greater or equal to n_0 and hence $|[n]| \geq n_0$. Consider any coloring of $G_{\mathbf{I}_n}$ and the corresponding map φ. By Ramsey's Theorem [22], there exists a subset H of $[n]$ with $|H| \geq 4$ so that $\varphi(A) = c$ for every 3-element subset A of H. Consider $\{i, j, k, l\} \subseteq H$, where $i < j < k < l$. We know that $\varphi(\{i, j, k\}) = c$ and $\varphi(\{j, k, l\}) = c$. However, this implies that the adjacent vertices $(\{i, j\}, \{j, k\})$ and $(\{j, k\}, \{k, l\})$ are colored with the same color. The vertices are adjacent because $\{(\{j, k\}, \{i, j\}), (\{k, l\}, \{j, k\})\}$ forms an alternating cycle.

Thus, for *any* k-coloring, we have two adjacent nodes in $G_{\mathbf{I}_n}$, which are colored by the same color. This contradicts the existence of a valid k-coloring for $G_{\mathbf{I}_n}$ when $n \geq n_0$.   □

6   Coloring Lexicographic Sums

So far, we have dealt with some classes of ordered sets and obtained approximation algorithms by coloring. In this section we will ask ourselves how we can use

---
[2] Note that we can assume without loss of generality that $i < j < k$.

existing posets to build new ordered sets for which the graph of incomparable pairs is still easily colorable. The construction we use here, lexicographic sums, comes from a very simple pictorial idea (see [22] for a more comprehensive discussion). Take a poset $\mathbf{P} = (X, P)$ and replace each of its points $x \in X$ with an ordered set $\mathbf{Q_x}$, the *module*, such that the points in the module have the same relation to points outside it. A more formal definition follows.

Let $\mathbf{P} = (X, P)$ be a poset, and let $\mathcal{F} = \{\mathbf{Q_x} = (Y_x, Q_x) : x \in X\}$ be a family of posets indexed by the elements of X. Define the **lexicographic sum** *of $\mathcal{F}$ over $\mathbf{P}$, denoted $\sum_{x \in \mathbf{P}} \mathcal{F}$, as the poset $\mathbf{S} = (Z, S)$ where $Z = \{z_{xy} : x \in X, y \in Y_x\}$ and $(z_{x_1 y_1}, z_{x_2 y_2}) \in S$ if and only if both $x_1 = x_2$ and $(y_1, y_2) \in Q_{x_1}$, or $(x_1, x_2) \in P$ (where $x_1 \neq x_2$).*

We observe that the resulting class of posets will be a new, larger class than its modules. For example, even if $\mathbf{P}$ and all posets in $\mathcal{F}$ are semi-orders, the lexicographic sum $\sum_{x \in \mathbf{P}} \mathcal{F}$ need not be an interval order: the two-element chain and the two-element antichain both carry semi-orders; Yet the lexicographic sum of two two-element chains over a two-element antichain is the forbidden poset for interval orders [22]. As another example, the lexicographic sum of any 3-irreducible convex bipartite poset and any non-bipartite semiorder poset over a two-element antichain is a poset that is none of the poset previously considered.

A natural question to ask is of course how approximation behaves under lexicographic constructions. With this aim, we prove that the number of colors needed to color the graph of incomparable pairs does not increase under the lexicographic sum. We remark that Hiraguchi (see e.g. [22]) proved that the dimension is "preserved" during lexicographic sum, i.e. $dim\left(\sum_{x \in \mathbf{P}} \mathcal{F}\right) = \max\{dim(\mathbf{P}), \max\{dim(\mathbf{Q_x}) : x \in X\}\}$. However, by Proposition 1 we know that $dim(\mathbf{P})$ and $\chi(G_\mathbf{P})$ are, in general, not the same. This motivates the following result.

Theorem 7. *Let $\mathbf{P} = (X, P)$ be a poset and let $\mathcal{F} = \{\mathbf{Q_x} = (Y_x, Q_x) : x \in X\}$ be a family of posets. Assume that for each $i \in \mathcal{P}$, where $\mathcal{P} = \{\mathbf{P}\} \cup \mathcal{F}$, the graph of incomparable pairs G_i can be colored with k_i colors. Then the graph of incomparable pairs $G_\mathbf{S}$ of the lexicographic sum $\mathbf{S} = \sum_{x \in \mathbf{P}} \mathcal{F}$ can be colored with $\max_{i \in \mathcal{P}} \{k_i\}$ colors.*

Proof. For every $i \in \mathcal{P}$, let C_i be a valid vertex coloring of graph $G_i = (V_i, E_i)$ that uses k_i colors, i.e. a map $C_i : V_i \rightarrow \{1, \ldots, k_i\}$ such that $C_i(u) \neq C_i(w)$ whenever u and w are adjacent. Let (z_{ai}, z_{bj}) be any incomparable pair of $G_\mathbf{S}$ and consider the following vertex coloring of $G_\mathbf{S}$:

$$C(z_{ai}, z_{bj}) := \begin{cases} C_\mathbf{P}(a, b) & \text{if } a \neq b; \\ C_{\mathbf{Q_a}}(i, j) & \text{otherwise}; \end{cases} \quad \text{for all } (z_{ai}, z_{bj}) \in inc(\mathbf{S}). \quad (4)$$

The claim follows by showing that (4) is a valid coloring of $G_\mathbf{S}$. With this aim it is sufficient to show that for any two adjacent incomparable pairs, namely (z_{ai}, z_{bj}) and $(z_{ck}, z_{d\ell})$, we always have $C(z_{ai}, z_{bj}) \neq C(z_{ck}, z_{d\ell})$. Note that $(z_{ai}, z_{d\ell}) \in P$ and $(z_{ck}, z_{bj}) \in P$, since (z_{ai}, z_{bj}) and $(z_{ck}, z_{d\ell})$ are assumed to be adjacent. We will consider two alternative cases: either we have $a = d$ and $b = c$, or at least one of the previous two conditions is not satisfied, say $a \neq d$, without loss of generality.

(i) $(a = d$ **and** $b = c)$ If $a = b$ then (i, j) and (k, ℓ) are adjacent in $G_{\mathbf{Q_a}}$, and $C(z_{ai}, z_{bj}) = C_{\mathbf{Q_a}}(i, j)$ and $C(z_{ck}, z_{d\ell}) = C_{\mathbf{Q_a}}(k, \ell)$. Otherwise $a \neq b$, and $C(z_{ai}, z_{bj}) = C_{\mathbf{P}}(a, b)$ and $C(z_{ck}, z_{d\ell}) = C_{\mathbf{P}}(b, a)$. The claim follows since $C_{\mathbf{Q_a}}$ and $C_{\mathbf{P}}$ are a valid vertex coloring of $G_{\mathbf{Q_a}}$ and $G_{\mathbf{P}}$, respectively.

(ii) $(a \neq d)$ We start observing that $b \notin \{a, d\}$ by the lexicographic construction. Indeed, by contradiction, if $a = b$ then $(z_{bj}, z_{d\ell}) \in P$ and this, together with $(z_{ck}, z_{bj}) \in P$, implies $(z_{ck}, z_{d\ell}) \in P$; a contradiction since $(z_{ck}, z_{d\ell}) \in inc(\mathbf{S})$. Moreover, if $b = d$ then $(z_{ai}, z_{bj}) \in P$, again a contradiction since $(z_{ai}, z_{bj}) \in inc(\mathbf{S})$. Similarly, we can prove that $c \notin \{a, d\}$. It follows that $C(z_{ai}, z_{bj}) = C_{\mathbf{P}}(a, b)$ and $C(z_{ck}, z_{d\ell}) = C_{\mathbf{P}}(c, d)$. Moreover, since $a \neq d$ we have $(a, d) \in P$. Finally, observe that either $b = c$ or $(c, b) \in P$ and in both cases $C_{\mathbf{P}}(a, b) \neq C_{\mathbf{P}}(c, d)$, and the claim follows.   □

A lexicographic sum $\sum_{x \in \mathbf{P}} \mathcal{F}$ is *trivial* if either $\mathbf{P}$ has only one point, or every poset in $\mathcal{F}$ is a one point poset; otherwise the sum is *non-trivial*. A poset is *decomposable* if it is isomorphic to a non trivial lexicographic sum; otherwise it is *indecomposable*. A poset can be decomposed into indecomposable posets in $O(n^2)$ time [16] and by Theorem 7, when coloring, we can restrict our attention on indecomposable posets.

7   Discussion and Open Problems

Semi-Order Dimension. The semi-order dimension of a poset $\mathbf{P} = (X, P)$, denoted $dim_S(\mathbf{P})$, is the smallest k such that there exists k semi-order extensions of P which realize P [6]. Since a linear extension is a semi-order and every semi-order has at most dimension 3 it follows that $dim_S(\mathbf{P}) \leq dim(\mathbf{P}) \leq 3 \cdot dim_S(\mathbf{P})$.

Proposition 3. *Problem* $1|prec| \sum_j w_j C_j$, *where precedence constraints are given as a semi-order realizer of size k, has a polynomial time* $(2 - \frac{2}{3k})$*-approximation algorithm.*

Recognizing posets with *interval* dimension 2 can be computed in time complexity $O(n^2)$ [22]. The complexity of recognizing posets with *semi-order* dimension 2 is not known. A polynomial constructive algorithm (constructs the semi-order realizer) would imply a $(1 + \frac{2}{3})$-approximation algorithm for $1|prec| \sum_j w_j C_j$ when precedence constraints form a poset with semi-order dimension at most 2. The class of semi-order dimension 2 posets is a proper superclass of the class of semi-orders and it is not contained in the class of interval orders.

Planar Posets. A poset is planar if its Hasse diagram [22] can be drawn without edge crossings. Our interest in planar posets stems from the fact that a planar poset $\mathbf{P} = (X, P)$ with a greatest or least element has at most dimension 3 [22]. Even though it is NP-complete to recognize if a given partial order is planar [8], we can construct a realizer of size 3 of P in polynomial time if the planar Hasse diagram is given as input [22].

Proposition 4. *Problem* $1|prec| \sum_j w_j C_j$, *where precedence constraints are given as a planar Hasse diagram with* a greatest or least element, *has a polynomial time (1+1/3)-approximation algorithm.*

We also note that planar posets with a greatest *and* least element have at most dimension two. As a consequence they can be recognized in polynomial time and $1|prec| \sum_j w_j C_j$ with precedence constraints of this type can be solved in polynomial time [1, 4]. The situation for planar posets without greatest or least element is more complex, because they can possess arbitrary high dimension [22].

Dimension Approximation. Finally, we remark that the complexity of computing a realizer of a poset is crucial for our approach. At the time being it is an open problem if there is a constant c such that for any partial order of dimension $k \geq 3$, it is possible to construct a realizer of size at most $c \cdot k$ in polynomial time. Any results on this problem would be interesting for the scheduling problem as well as for the dimension theory.

Acknowledgements

The authors thank Nikos Mutsanas for useful comments. This research is supported by Swiss National Science Foundation project 200021-104017/1, "Power Aware Computing", and by the Swiss National Science Foundation project 200020-109854, "Approximation Algorithms for Machine scheduling Through Theory and Experiments II".

References

1. C. Ambühl and M. Mastrolilli. Single machine precedence constrained scheduling is a vertex cover problem. In *Proceedings of the 14th Annual European Symposium on Algorithms (ESA)*, to appear, 2006.
2. C. Chekuri and R. Motwani. Precedence constrained scheduling to minimize sum of weighted completion times on a single machine. *Discrete Applied Mathematics*, 98(1-2):29–38, 1999.
3. F. A. Chudak and D. S. Hochbaum. A half-integral linear programming relaxation for scheduling precedence-constrained jobs on a single machine. *Operations Research Letters*, 25:199–204, 1999.
4. J. R. Correa and A. S. Schulz. Single machine scheduling with precedence constraints. *Mathematics of Operations Research*, 30(4):1005–1021, 2005. Extended abstract in Proceedings of the 10th Conference on Integer Programming and Combinatorial Optimization (IPCO 2004), pages 283–297.
5. B. Dushnik and E. Miller. Partially ordered sets. *American Journal of Mathematics*, 63:600–610, 1941.
6. S. Felsner and R. Möhring. Semi order dimension two is a comparability invariant. *Order*, (15):385–390, 1998.
7. S. Felsner and W. T. Trotter. Dimension, graph and hypergraph coloring. *Order*, 17(2):167–177, 2000.

8. A. Garg and R. Tamassia. On the computational complexity of upward and recti-linear planarity testing. *SIAM Journal on Computing*, 31(2):601–625, 2002.
9. R. Graham, E. Lawler, J. K. Lenstra, and A. H. G. Rinnooy Kan. Optimization and approximation in deterministic sequencing and scheduling: A survey. In *Annals of Discrete Mathematics*, volume 5, pages 287–326. North–Holland, 1979.
10. L. A. Hall, A. S. Schulz, D. B. Shmoys, and J. Wein. Scheduling to minimize average completion time: off-line and on-line algorithms. *Mathematics of Operations Research*, 22:513–544, 1997.
11. D. S. Hochbaum. Efficient bounds for the stable set, vertex cover and set packing problems. *Discrete Applied Mathematics*, 6:243–254, 1983.
12. S. G. Kolliopoulos and G. Steiner. Partially-ordered knapsack and applications to scheduling. In *Proceedings of the 10th Annual European Symposium on Algorithms (ESA)*, pages 612–624, 2002.
13. E. L. Lawler. Sequencing jobs to minimize total weighted completion time subject to precedence constraints. *Annals of Discrete Mathematics*, 2:75–90, 1978.
14. J. K. Lenstra and A. H. G. Rinnooy Kan. The complexity of scheduling under precedence constraints. *Operations Research*, 26:22–35, 1978.
15. F. Margot, M. Queyranne, and Y. Wang. Decompositions, network flows and a precedence constrained single machine scheduling problem. *Operations Research*, 51(6):981–992, 2003.
16. R. H. Möhring. Computationally tractable classes of ordered sets. In I. Rival, editor, *Algorithms and Order*, pages 105–193. Kluwer Academic, 1989.
17. N. N. Pisaruk. A fully combinatorial 2-approximation algorithm for precedence-constrained scheduling a single machine to minimize average weighted completion time. *Discrete Applied Mathematics*, 131(3):655–663, 2003.
18. C. N. Potts. An algorithm for the single machine sequencing problem with precedence constraints. *Mathematical Programming Study*, 13:78–87, 1980.
19. I. Rabinovitch. The dimension of semiorders. *Journal of Combinatorial Theory*, Series A(25):50–61, 1978.
20. A. S. Schulz. Scheduling to minimize total weighted completion time: Performance guarantees of LP-based heuristics and lower bounds. In *Proceedings of the 5th Conference on Integer Programming and Combinatorial Optimization (IPCO)*, pages 301–315, 1996.
21. P. Schuurman and G. J. Woeginger. Polynomial time approximation algorithms for machine scheduling: ten open problems. *Journal of Scheduling*, 2(5):203–213, 1999.
22. W. T. Trotter. *Combinatorics and Partially Ordered Sets: Dimension Theory*. Johns Hopkins Series in the Mathematical Sciences. The Johns Hopkins University Press, 1992.
23. W. T. Trotter. New perspectives on interval orders and interval graphs. In R. A. Bailey, editor, *Surveys in Combinatorics*, number 241 in Mathematical Society Lecture Note Series, pages 237–286, London, 1997.
24. G. J. Woeginger. On the approximability of average completion time scheduling under precedence constraints. *Discrete Applied Mathematics*, 131(1):237–252, 2003.
25. M. Yannakakis. On the complexity of partial order dimension problem. *SIAM Journal on Algebraic and Discrete Methods*, 22(3):351–358, 1982.

Minimizing Setup and Beam-On Times in Radiation Therapy

Nikhil Bansal[1], Don Coppersmith[2,*], and Baruch Schieber[1]

[1] IBM T.J. Watson Research Center, P.O. Box 218, Yorktown Heights, NY 10598
{nikhil, sbar}@us.ibm.com
[2] IDA Center for Communications Research, 805 Bunn Drive, Princeton, NJ 08540
don.coppersmith@idaccr.org

Abstract. Radiation therapy is one of the commonly used cancer therapies. The radiation treatment poses a tuning problem: it needs to be effective enough to destroy the tumor, but it should maintain the functionality of the organs close to the tumor. Towards this goal the design of a radiation treatment has to be customized for each patient. Part of this design are intensity matrices that define the radiation dosage in a discretization of the beam head. To minimize the treatment time of a patient the beam-on time and the setup time need to be minimized. For a given row of the intensity matrix, the minimum beam-on time is equivalent to the minimum number of binary vectors with the consecutive "1"s property that sum to this row, and the minimum setup time is equivalent to the minimum number of *distinct* vectors in a set of binary vectors with the consecutive "1"s property that sum to this row. We give a simple linear time algorithm to compute the minimum beam-on time. We prove that the minimum setup time problem is APX-hard and give approximation algorithms for it using a duality property. For the general case, we give a $\frac{24}{13}$ approximation algorithm. For unimodal rows, we give a $\frac{9}{7}$ approximation algorithm. We also consider other variants for which better approximation ratios exist.

1 Introduction

Radiation therapy is one of the commonly used cancer therapies. It has been shown to be effective, especially in cases where the tumor is localized and metastases have not yet started to form. The radiation treatment poses a tuning problem: the radiation needs to be effective enough to destroy the tumor, but it should maintain the functionality of the organs close to the tumor (organs at risk). Towards this goal the design of a radiation treatment has to be customized for each patient. This design is done using computer tomography that detects the exact location of the tumor and the organs at risk.

The radiation is done using a linear accelerator positioned in a beam head that is positioned in a gantry that can be rotated around the patient. (See

---

* This work was done while at IBM T.J. Watson Research Center.

J. Diaz et al. (Eds.): APPROX and RANDOM 2006, LNCS 4110, pp. 27–38, 2006.

Fig. 1(a).) The treatment design specifies the angles of the radiation and its intensity. The angles are specified by a set of positions at which the gantry stops to release radiation. (In most models these are a subset of 36 possible positions.) The desired intensity of the beam head at each gantry position is defined using an $m \times n$ *intensity matrix*, denoted $\mathcal{I}$, which corresponds to a discretization of the beam head into an $m \times n$ rectangular grid, with each of its entries called a *bixel*. The intensity matrix contains positive integral entries that determine the desired radiation dosage in each bixel.

The radiation generated by the accelerator is uniform. Thus, in order to achieve the varying intensity this radiation needs to be modulated. An emerging device for modulation is the *multileaf collimator (MLC)*. (See Fig. 1(b).) This device consists of a pair of leafs, a left leaf and a right leaf, for each row of the intensity matrix (often referred to as a *channel*). Consider a time t. If the left leaf of channel i is positioned at ℓ, for $1 \leq \ell \leq n$, and the right leaf is positioned at r, for $2 \leq r \leq n + 1$, where $\ell < r$, then the radiation is blocked in bixels $(i, 1), \ldots, (i, \ell - 1)$ and $(i, r), \ldots, (i, n)$, and a uniform amount of radiation is delivered by bixels $(i, \ell), \ldots, (i, r - 1)$. To achieve the desired intensity $\mathcal{I}(i, j)$, radiation needs to be delivered by bixel (i, j) for $\mathcal{I}(i, j)$ time units.

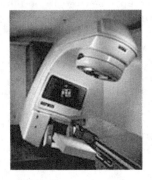

Fig. 1. (a) a linear accelerator in a gantry (b) multileaf collimator both made by Varian

The positions of the multileaf collimator in the m channels at time t define a 0-1 $m \times n$ *shape matrix* S_t. A "0" entry in S_t indicates a blocked bixel and a "1" entry indicates an active bixel. Note that the multileaf collimator function implies that each row in S_t satisfies the *consecutive 1's property*, that is, all the 1's in a row are consecutive.

To achieve the desired radiation intensity in T time units we need to find T shape matrices $S_1, \ldots, S_T$ such that $\sum_{i=1}^{T} S_i = \mathcal{I}$ where the summation is done entry-wise. The *beam-on* time of a treatment is determined by T, the number of shape matrices used. The *setup* time is determined by the time it takes to calibrate the multileaf collimator. The setup time varies according to the technology used. In some models the leafs can be calibrated simultaneously and thus the setup time is determined by the number of *distinct* shape matrices (once the shape matrix is set up, it can be used any number of times without any overhead). We call this setup time the *multileaf setup* time. In other models each pair of leafs needs to be calibrated separately, and thus the setup time depends on the number of movements of each pair. We call this setup time the *leaf setup* time. Observe that the leaf setup time is determined by the number of distinct rows in the shape matrices. To minimize the treatment time for a patient a linear combination of the beam-on time and setup time needs to be minimized. (We note that in some models the setup time is not constant but depends on the specific "from" and "to" shape matrices, in which case a more complex term is needed to estimate the setup time accurately.)

In this paper we consider the minimization of beam-on and setup times separately. In many cases one of these terms is dominant and thus minimizing each one separately leads in practice to treatment times that are close to optimal. Our results assume *unconstrained* multileaf collimator whose position in each channel is independent. Note that in the unconstrained case each channel can be considered separately.

From a combinatorial point of view, our problem is that of decomposing a matrix or a vector into a small set of matrices or vectors of a special kind. In the beam-on problem the intensity matrix needs to be decomposed into a small set of 0-1 shape matrices, while in the leaf setup problem, each channel in the intensity matrix needs to be decomposed into a small set of *interval vectors*. An interval vector is a 0-a vector that, for some $a > 0$, satisfies the *consecutive a's property*.

Results. First, we note that the beam-on time minimization problem can be solved in linear time using a simple algorithm. Most of the paper deals with the leaf setup time minimization problem. We note that setup time seems to dominate the treatment time in many cases [9]. First, we prove a "duality" relation between the setup time minimization problem and maximum partitioning of certain type of vectors. We use this "duality" to prove that the leaf setup time minimization problem is APX-Hard, and then give several approximation algorithms for this problem depending on the structure of the channels in the intensity matrix. As we shall see, there is a trivial greedy algorithm that achieves an approximation guarantee of 2. Our main contribution is to give algorithms that achieve a guarantee better than 2. For a "unimodal" channel, i.e., a row in the intensity matrix whose only local minima are at its two ends, we give a $\frac{9}{7}$ approximation algorithm. For the general case we give a $\frac{24}{13}$ approximation algorithm. We also give two variants of our general algorithm which are shown to have better performance in specific cases depending on the number of local minima in the channel.

Prior work. Radiation treatment design was considered extensively in the medical physics literature. (See, e.g., [6,5,9] and references therein.) Previous algorithmic analysis of the setup and beam-on minimization problems is quite limited. Boland, Hamacher and Lenzen [7] gave a polynomial time algorithm for minimizing beam-on time. Their algorithm is based on an integer programming formulation that is shown to be solved using network flow in a graph of quadratic size. An algorithm with improved running time for the beam-on time minimization problem was given by Ahuja and Hamacher [1]. Their algorithm is also based on network flow but in a graph of linear size. Our algorithm is much simpler and more efficient. To the best of our knowledge this paper is the first to algorithmically analyze the setup time minimization problem. A heuristic for a constrained variant of the (multileaf) setup time minimization problem was given in [8].

The rest of the paper is organized as follows. In Section 2 we show the connection between the leaf setup time minimization problem and maximum partition of *Prefix Positive Zero Sum* vectors (to be defined later). In Section 3 we prove the hardness of the leaf setup time minimization problem. In Section 4 we present the approximation algorithms for this problem (some proofs are deferred to the full version due to space constraints). Finally, in Section 5 we conclude with some open problems. The results for beam-on time minimization are deferred to the full version due to space constraints.

2   Setup Time and Vector Partition

In this section we consider the setup time minimization problem. To formulate the problem we first define *interval vectors*. For an integer $b > 0$ we call a 0-b vector $V = (v_1, \ldots, v_n)$ an *interval vector of height b* if all the b's are in consecutive positions. We use the triple (ℓ, r, b) to denote such a vector, where $v_\ell = \ldots = v_{r-1} = b$ and $v_i = 0$ everywhere else. We say that the vector *begins* at ℓ and *ends* at r. Recall that the (leaf) setup time minimization problem can be formulated as follows. Given a vector of non-negative integers $A = (a_1, \ldots, a_n)$ find a minimum set of *interval vectors* (ℓ_i, r_i, b_i) that sum to the input vector.

Consider an input vector $A = (a_1, \ldots, a_n)$. For notational convenience add one entry (a_0) to the head of A and another entry (a_{n+1}) to the tail of A and let $a_0 = a_{n+1} = 0$. Define the *difference vector* of A, denoted Δ^A, to be the vector $\Delta^A = (a_1 - a_0, a_2 - a_1, \ldots, a_{n+1} - a_n)$. From the definition of Δ^A it follows that for $i \in [1..n]$, the prefix sums $\sum_{j=1}^{i} \Delta^A(i) = a_i \geq 0$, and the sum $\sum_{j=1}^{n+1} \Delta^A(i) = a_{n+1} = 0$. We call such a vector *Prefix Positive Zero Sum (PPZS)* vector. We say that two vectors are "disjoint" if the index sets of their nonzero entries are disjoint. Consider the following maximum PPZS vector partitioning problem. Given a PPZS vector Δ find a maximum set of disjoint PPZS vectors that sum to Δ. In this section we prove the "duality" between the above two problems as follows.

Theorem 1. *The minimum setup time for a vector A is t if and only if the maximum partition of Δ^A is of size $z - t$, where z is the number of nonzero entries in Δ^A.*

To prove the theorem we need first to prove some properties of the setup time minimization problem. For an input vector A, we say that position i is an *uptick* if $\Delta^A(i) > 0$ (i.e., $a_i - a_{i-1} > 0$). Similarly, position i is a *downtick* if $\Delta^A(i) < 0$.

Lemma 1. *There exists an optimal solution to the setup time minimization problem that consists only of interval vectors that begin at an uptick and end at a downtick.*

Proof. We show that any arbitrary solution can be transformed to one that has the required property without increasing the number of interval vectors used. Given a solution $\mathcal{S}$, suppose that $\mathcal{S}$ contains interval vectors that do not begin at an uptick. Consider such an interval vector $V = (\ell, r, b)$ with the minimum ℓ. Thus $a_\ell \leq a_{\ell-1}$ and hence there must be at least one other interval vector $V' = (\ell', r', b')$ in $\mathcal{S}$ that ends at $r' = \ell$. As $\ell' < \ell$, by minimality of ℓ, V' must begin at an uptick. If $b = b'$ we could obtain a better solution by replacing V and V' by (ℓ', r, b). If $b < b'$, we replace V and V' by $(\ell', r', b' - b)$ and (ℓ', r, b), both of which begin at an uptick. If $b > b'$, we replace V and V' by (ℓ', r, b') and $\tilde{V} = (\ell, r, b - b')$. Note that the height of $\tilde{V}$ is strictly less than that of V and hence applying this transformation repeatedly, we get a solution where all interval vectors begin at upticks. An identical argument implies that all interval vectors end at downticks. $\square$

We call a solution that consists only of interval vectors that begin at upticks and end at downticks a *canonical* solution. We now show how to view any canonical solution to the setup problem as a graph, which will allow us to characterize the value of an optimum solution exactly. Let I_u denote the index set of the upticks and I_d denote the index set of the downticks.

Consider a canonical solution $\mathcal{S} = \{V_1, \ldots, V_t\}$, where $V_i = (\ell_i, r_i, b_i)$ are interval vectors. Since all interval vectors begin at upticks and end at downticks if some position j is an uptick, then the sum of heights of all interval vectors that begin at position j is exactly equal to $\Delta^A(j)$. Similarly, if j is a downtick then the sum of heights of all interval vectors that end at j is exactly equal to $|\Delta^A(j)| = -\Delta^A(j)$.

Associate a weighted bipartite graph $G(\mathcal{S}) = (I_u, I_d, E)$ with a canonical solution $\mathcal{S}$ as follows. The vertex sets are I_u and I_d, a vertex i (corresponding to position i) has a positive weight equal to $|\Delta^A(i)|$. For each interval vector $V_j \in \mathcal{S}$, we have an edge $e_j = (\ell_j, r_j)$ with weight b_j.

We will show that each connected component of $G(\mathcal{S})$ corresponds to a PPZS vector. Observe that $G(\mathcal{S})$ has t edges and z vertices. As $\mathcal{S}$ is canonical, the total weight of edges incident at a vertex is equal to the weight of the vertex. Consider a connected component (C_u, C_d, E_C) of $G(\mathcal{S})$. Notice that all the edges that are adjacent to vertices in $C = C_u \cup C_d$ are in E_C. It follows that every interval vector that begins at any of the upticks in C_u must end at a downtick in C_d, and vice versa.

Consider the vector Δ defined by the entries of the difference vector Δ^A in the set of positions C; that is, $\Delta(i) = \Delta^A(i)$, for $i \in C$, and $\Delta(i) = 0$ otherwise. We claim that Δ is a PPZS vector. To see this consider any prefix sum $\sum_{j=1}^{i} \Delta(j)$. It is not difficult to see that the value of this sum is exactly the total height of all interval vectors that begin at an uptick in positions $C_u \cap [1..i]$ and end at a downtick in positions $C_d \cap [i+1..n+1]$. Thus, $\sum_{j=1}^{i} \Delta(j) \geq 0$, for $i \in [1..n]$, and $\sum_{j=1}^{n+1} \Delta(j) = 0$. It follows that a canonical solution $\mathcal{S}$ to the setup problem for vector A induces a set of size p of disjoint PPZS vectors that sum to Δ^A, where p equals the number of connected components in $G(\mathcal{S})$. Since $G(\mathcal{S})$ has z vertices and p connected components, the number of interval vectors in $\mathcal{S}$ (which equals the number of edges in $G(\mathcal{S})$) is at least $z - p$. We conclude that a setup time t for vector A implies a partition of Δ^A into at least $z - t$ disjoint PPZS vectors.

To complete the proof of Theorem 1 we show that a partition of Δ^A into p disjoint PPZS vectors implies setup time at most $z - p$ for the vector A. Consider a PPZS vector Δ in the partition. Let S^Δ be the prefix sum vector of Δ; that is $A^\Delta(0) = 0$, and for $i \in [1..n+1]$, $S^\Delta(i) = \sum_{j=1}^{i} \Delta(j)$. Note that the sum of all the prefix sum vectors S^Δ, for all PPZS vectors Δ in the partition, is exactly A. Thus, it is sufficient to prove the following lemma.

Lemma 2. *Let Δ be a PPZS vector with z nonzero entries and let S^Δ be its prefix sums vector. There are $z - 1$ interval vectors that sum to S^Δ.*

Proof. We prove by induction on z. The base case is $z = 2$. In this case Δ consists of two nonzero entries at positions $\ell < r$. From the definition of a PPZS vector it follows that $\Delta(\ell) > 0$ and $\Delta(r) = -\Delta(\ell)$. Hence, the vector S^Δ is exactly the interval vector $(\ell, r, \Delta(\ell))$.

For the inductive step consider $z > 2$ assume that the lemma holds for $z' < z$. We simply show how to generate a single interval vector V such that the difference vector Δ' of the vector $S^\Delta - V$ has no more than $z - 1$ nonzero entries. Consider a nonzero entry in Δ with the minimum absolute value. Let the index of this entry be j. If $\Delta(j) > 0$, then generate the interval vector $(j = \ell, r, \Delta(j))$, where $r > j$ is the minimum index such that $\Delta(r) < 0$. Similarly, if $\Delta(j) < 0$, then generate the interval vector $(\ell, j = r, -\Delta(j))$, where $\ell < j$ is the maximum index such that $\Delta(\ell) > 0$. Note that $S^\Delta(i) \geq |\Delta(j)|$, for $i \in [\ell, r-1]$. Thus V is a valid interval vector. Also the difference vector of $S^\Delta - V$ has zero in position j and also has zeroes in all positions Δ has zeroes. $\qquad\square$

3   APX Hardness of the Setup Minimization Problem

We sketch the proof of APX-Hardness of setup minimization. The details can be found in the full version.

Theorem 2. *The setup minimization problem is APX-Hard even for vectors with entries polynomially bounded in n.*

Proof. The proof is by showing a gap preserving reduction from the 3-partition problem. The 3-partition problem is defined as follows. Given a threshold B and $3m$ integers $p_1, \ldots, p_{3m}$ such that $\sum_{i=1}^{3m} p_i = mB$ and $B/4 < p_i < B/2$ for each i, is there a partition of the $3m$ integers into m triples each of which sums exactly to B. Petrank [14] showed that unless P=NP, there exists an $\epsilon > 0$ such that it is impossible to distinguish in polynomial time whether there are exactly m or whether no more than $(1 - \epsilon)m$ disjoint triples that sum to exactly B. This is true even for instances where B is polynomially bounded in m.

Given an instance of the 3-partition problem we define an instance of the setup minimization problem. The input vector A consists of $4m - 1$ entries $A = (s_1, s_2, \ldots, s_{3m}, (m - 1)B, (m - 2)B, \ldots, B)$, where $s_i = \sum_{j=1}^{i} p_j$. Note that this instance is unimodal and each uptick has value p_i for some i and each downtick has value B. The corresponding difference vector with $4m$ entries is $\Delta^A = (p_1, \ldots, p_{3m}, -B, \ldots, -B)$. The proof idea is to use the duality in Theorem 1 to show that if a polynomial time algorithm can distinguish whether the setup time for vector A is exactly $3m$ or at least $3m(1 + \epsilon)$ for some $\epsilon > 0$, then it is possible to distinguish whether the 3-partition instance has exactly m or at least $m(1 - \epsilon')$ disjoint partitions, where $\epsilon' = O(\epsilon)$. $\square$

4   Approximating the Minimum Setup Time

In this section we describe several approximation algorithms for the minimum setup time for an input vector A. First, note that Lemma 2 implies a simple algorithm that finds $z - 1$ interval vectors that sum to A, where z is the number of nonzero entries in Δ^A. This is since Δ^A is a PPZS vector. On the other hand, the maximum size of any partition of Δ^A into disjoint PPZS vectors is at most $\lfloor \frac{z}{2} \rfloor$, since each such vector must have at least two nonzero entries, implying that the minimum setup time is at least $z - \lfloor \frac{z}{2} \rfloor = \lceil \frac{z}{2} \rceil$. Thus a factor two approximation is trivial. Below, we show approximation algorithms with better ratios.

The basic idea for our algorithms will be the following. Note that a PPZS vector may be in a partition of Δ^A if its nonzero entries are a subset of the nonzero entries of Δ^A. We call such a vector a *part* of Δ^A. We compute the collection of all possible PPZS vectors with at most y nonzero entries that are parts of Δ^A. This can be done in time $O(n^y)$ and hence is polynomial for constant y. We then find a large (close to optimum) cardinality subset S of disjoint PPZS vectors from the collection such that either S is a partition of Δ^A or in case it is not a partition of Δ^A, the vector V consisting of all the nonzero entries of Δ^A that are not entries of vectors in S is a PPZS vector. In this case $S \cup \{V\}$ is a disjoint partition of Δ^A. Our algorithms depends on the "shape" of the vector A. Roughly speaking, if the vector A has too many local minima, this complicates the task of finding the right set of PPZS vectors S such that the remaining vector V is also PPZS. We begin by considering the (simplest) unimodal case in which the only local minima of A are at its two ends. Then, we consider the general case and some of its variants.

4.1   Unimodal Input Vectors

Consider a unimodal vector A and let Δ^A be its corresponding difference vector. Note that Δ^A consists of a block of nonnegative entries followed by a block of non-positive entries. A useful property of such difference vectors is the following.

Lemma 3. *Consider a PPZS vector Δ that consists of a block of nonnegative entries followed by a block of non-positive entries. Let S be any set of disjoint PPZS vectors, each of which is a part of Δ^A. The vector V consisting of all the nonzero entries of Δ^A that are not entries of vectors in S is a PPZS vector.*

Proof. Note that the vector V also consists of a block of nonnegative entries followed by a block of non-positive entries. Also, the sum of all these entries is zero since the sum of all entries of the vectors in S is zero. Hence, each prefix sum must be nonnegative and V is a PPZS vector. □

It follows that in order to find a good approximation we need to find a large cardinality set of disjoint PPZS vectors, each of which is a part of Δ^A. We will find such a set consisting only of PPZS vectors with at most y nonzero entries. Identify each such vector with the set of indices of its nonzero entries. Then the problem is reduced to the following set packing problem: Given a collection $C = \{S_1, S_2, \ldots\}$ of sets where each set S_i has size at most y, find a maximum cardinality sub-collection $C' \subseteq C$ of disjoint sets.

The best known approximation algorithm for this set packing problem is an elegant local search based algorithm due to Hurkens and Schrijver [11] which achieves an approximation ratio of $y/2$.

Our algorithm for unimodal vectors is the following: Compute the set $\mathcal{S}_y$ of all PPZS vectors with at most y nonzero entries that are part of Δ^A, for $y = 2, 3$ and 4. Apply the algorithm of [11] to find a subset $\mathcal{C}_y \subseteq \mathcal{S}_y$ of disjoint vectors, for each $y = 2, 3$ and 4 and choose the subset $\mathcal{C}_y$ with the maximum cardinality.

Theorem 3. *The algorithm described above is a $\frac{9}{7}$ approximation algorithm for the minimum setup time problem for unimodal input vectors.*

Proof. Let Opt denote the optimum setup time for input vector A. By Theorem 1, Δ^A can be partitioned into $z - Opt$ PPZS vectors, where z is the number of nonzero entries in Δ^A. Let n_k denote the number of PPZS vectors with k nonzero entries in the partition defined by Opt. By definition n_i satisfies

$$\sum_{i>1} i \cdot n_i = z \qquad \text{and} \qquad Opt = \sum_{i>1}(i-1) \cdot n_i \qquad (1)$$

For a fixed choice of y, the algorithm of [11] guarantees that we can find at least $\sum_{i=2}^{y} \frac{2}{y} n_i$ disjoint vectors in $\mathcal{S}_y$, and hence a solution with setup cost at most $\sum_{i=2}^{y}(i-2/y) \cdot n_i + \sum_{i>y} i \cdot n_i$. To show that our algorithm is a $\frac{9}{7}$ approximation, it suffices to show that for any choice of $n_2, n_3, \ldots$ that satisfies condition (1), the inequality

$$\sum_{i=2}^{y}\left(i - \frac{2}{y}\right) \cdot n_i + \sum_{i>y} i \cdot n_i \leq \frac{9}{7}\sum_{i \geq 2}(i-1)n_i$$

holds for at least one of $y = 2$, $y = 3$ or $y = 4$.

Without loss of generality, we can assume that $n_i = 0$ for $i \geq 5$, because Opt incurs a setup time of at least $i - 1$ for any PPZS vector with i nonzero entries while our algorithm incurs a setup time of at most i. As $\frac{5}{4} \leq \frac{9}{7}$, an instance where $n_i > 0$ for $i \geq 5$ can only improve the approximation ratio. Thus it suffices to show that one of the following inequalities holds. (The first corresponds to $y = 2$, the second to $y = 3$ and the third to $y = 4$.)

$$n_2 + 3n_3 + 4n_4 \leq \frac{9}{7}(n_2 + 2n_3 + 3n_4)$$

$$\frac{4}{3}n_2 + \frac{7}{3}n_3 + 4n_4 \leq \frac{9}{7}(n_2 + 2n_3 + 3n_4)$$

$$\frac{3}{2}n_2 + \frac{5}{2}n_3 + \frac{7}{2}n_4 \leq \frac{9}{7}(n_2 + 2n_3 + 3n_4)$$

Multiplying the first inequality by 2, second by 3 and third by 2 and summing each side of the resulting inequalities gives an identity which implies that one of these inequalities always holds for any n_2, n_3 and n_4. □

It is easily seen that the analysis for the above algorithm is tight by considering an instance where the optimum solution has $n_2 = 2(n+1)/11$, $n_3 = (n+1)/11$ and $n_4 = (n+1)/11$. As $2n_2 + 3n_2 + 4n_4 = n+1$, this is a valid choice of n_i and the optimum setup time is $n+1 - n_2 - n_3 - n_4 = 7(n+1)/11$. Our algorithm on the other hand, finds $n_2 = 2(n+1)/11$ vectors for $y = 2$, $2/3 \cdot (n_2 + n_3) = 2(n+1)/11$ vectors for $y = 3$, and $(n_2 + n_3 + n_4)/2 = 2(n+1)/11$ vectors for $y = 4$. In either case, our solution has setup time equal to $(n+1) - 2(n+1)/11 = 9(n+1)/11$.

4.2   Arbitrary Input Vectors

For unimodal input vectors we used the key fact (Lemma 3) that any collection of disjoint PPZS vectors that are parts of Δ^A can be used in the solution. However, this is not true in general. Consider for example when $A = (10, 5, 10)$ and hence $\Delta^A = (10, -5, 5, -10)$. In this case, the PPZS vector $S = (10, 0, 0, -10)$ is a valid part of Δ^A. However we cannot choose this in the solution, because $V = \Delta^A - S = (0, -5, 5, 0)$ which is not a valid PPZS vector (the second prefix sum is negative).

Given an input vector A to the setup minimization problem, let z denote the number of non-zero entries in Δ^A. We say that position i is a *local minimum* if $\Delta^A(i) < 0$ and $\Delta^A(j) > 0$ where j is the smallest index greater than i such that $\Delta^A(j)$ is non-zero. Note that the number of local minima can be no more than $z/2$. We will show the following three results:

1. We give a $3/(2 - \epsilon)$ approximation when the number of local minima is no more than ϵz.
2. We improve this guarantee slightly when ϵ is $o(1/\log z)$. In particular, we give a $(11 + 9e^{-2} - 2e^{-3})/(8 + 6e^{-2} - 2e^{-3}) \approx 1.391$ approximation when the number of local minima is $o(z/\log z)$.
3. Finally, we give a $24/13 \approx 1.846$ approximation in the general case in which the number of local minima could be arbitrary.

Recall that our goal is to find a large cardinality set of PPZS vectors that are part of Δ^A, such that (1) no two vectors share a nonzero position, and (2) the vector given by subtracting all the prefix sum vectors of these vectors from A does not have any negative entries. We call property (1) *the independence property* and property (2) *the packing property*. Note that such a set of cardinality p induces a solution of size $z - p - 1$, where z is the number of nonzero entries in Δ^A.

Our first and the second algorithms above are based on rounding the solution of a certain linear program (LP) that models the properties above. We also show certain structural properties of these linear programs which are essential to our rounding scheme. For lack of space we defer the description of these algorithms to the full version of the paper.

We describe here the algorithm for the general case, where the number of local minima could be arbitrary.

The Algorithm for the General Case

Our main idea is the following. We will only be interested in PPZS vectors of size 2. Note that if some fixed optimum solution Opt does not use size 2 vectors, then the setup time is at least $2z/3$ and hence we trivially have a $3/2$ approximation. Thus we will focus on the case when Opt uses at lot of size 2 vectors (in particular it chooses close to $z/2$ such vectors). Our goal then will be to obtain a large subset of such vectors that simultaneously satisfies the independence and the packing properties.

We do not know how to find such a set of size 2 PPZS vectors directly and hence adopt an indirect approach. Recall that each such vector Δ with non-zeroes in positions ℓ and r (for $\ell < r$) corresponds to an interval vector $(\ell, r, \Delta(\ell))$ in the solution. Call such interval vectors *candidate* interval vectors. We show that there exists a certain "minimal" set of independent interval vectors $\mathcal{R}_2$, such that any feasible collection of candidate interval vectors can be transformed (without any loss) into one that only uses vectors from $\mathcal{R}_2$. Since $\mathcal{R}_2$ only contains independent vectors, $|\mathcal{R}_2| \leq z/2$. Now, if Opt uses close to $z/2$ vectors in its solution, then it must have discarded very few vectors from $\mathcal{R}_2$. This allows us to use the known results for the generalized caching problem considered by [2,3], and hence find a solution where the number of discarded intervals is no more than a constant times the number of intervals discarded by Opt.

We now describe the details. The following lemma describes the properties of the set $\mathcal{R}_2$. Due to space constraints its proof is deferred to the full version.

Lemma 4. *Given a vector A, there exists a set $\mathcal{R}_2$ of interval vectors that satisfies the following properties:*

1. *No two interval vectors in $\mathcal{R}_2$ begin or end at the same position (hence they are independent).*
2. *For any set S of candidate interval vectors that satisfies the independence and packing properties, there exists another set $S' \subseteq \mathcal{R}_2$, such that $|S'| \geq |S|$ and S' satisfies the packing property. (Since $S' \subseteq \mathcal{R}_2$ it also satisfies the independence property.)*

The set $\mathcal{R}_2$ is of linear size and can be obtained in linear time.

To find a large subset of $\mathcal{R}_2^h$ that satisfies the independence and packing properties we use known results for the general caching problem defined below.

In the general caching problem (with unit profit) we are given a vector $A = (a_1, \ldots, a_n)$ where a_i denotes the amount of cache available at time i. There is collection of tasks $\mathcal{T} = \{T_1, \ldots, T_m\}$, where each task T_i, specified by (ℓ_i, r_i, h_i), requires h_i units of cache during the interval $[\ell_i, r_i - 1]$. A set of tasks is feasible if the required cache size for these tasks does not exceed the available cache size at any time. The goal is to find a feasible collection of tasks such that the total number of tasks not included in the collection is minimized.

This problem was first considered by [2] who gave a logarithmic approximation for the problem. Later [3] gave an algorithm with an approximation ratio of 4. Note that the approximation ratio is for the number of tasks excluded rather than included in the collection.

Our algorithm is as follows. We construct an instance of the general caching problem where $\mathcal{T} = \mathcal{R}_2$. We apply the algorithm of [3] to this instance and obtain a collection S of interval vectors that satisfy the independence and packing properties. We use this to construct a solution with setup time at most $z - |S| - 1$, where z is the number of nonzero entries in Δ^A.

Theorem 4. *The algorithm stated above is a $\frac{24}{13} \approx 1.846$ approximation algorithm for the setup minimization problem.*

Proof. Consider an optimal canonical solution $\mathcal{S}$ for the setup problem. Let $\mathcal{S}_2$ be the set of interval vectors in the solution that begin at an uptick and end at a downtick of the same value. By the definition of n_i, we have $n_2 = |\mathcal{S}_2|$. By Lemma 4 the set $\mathcal{R}_2$ contains a subset of size n_2 that satisfies the independence and packing properties. Thus, the solution returned by the algorithm of [3] for the general caching problem contains at least $\max\{0, |\mathcal{R}_2| - 4(|\mathcal{R}_2| - n_2)\}$ interval vectors. Moreover, since $\sum_{i \geq 2} i \cdot n_i = z$, we have that $\sum_{i \geq 3} n_i \leq (z - 2n_2)/3$. Since $Opt = \sum_{i \geq 2}(i - 1) \cdot n_i$, we have $opt = \sum_{i \geq 2} i \cdot n_i - n_2 - \sum_{i \geq 3} n_i \geq z - n_2 - (z - 2n_2)/3 = (2z - n_2)/3$.

We consider two cases based on the magnitude of $\alpha = 4((|\mathcal{R}_2| - n_2)$. If $\alpha \geq |\mathcal{R}_2|$, or equivalently, $n_2 \leq 3|\mathcal{R}_2|/4 \leq 3z/8$, the approximation ratio is most

$$\frac{z}{Opt} \leq \frac{z}{(2z - n_2)/3} \leq \frac{3z}{2z - 3z/8} = \frac{24}{13}.$$

If $\alpha < |\mathcal{R}_2|$, let β denote $|\mathcal{R}_2| - \alpha$. Then, the approximation ratio is at most

$$\frac{z - \beta}{(2z - n_2)/3} = \frac{3(z - \beta)}{2z - (\mathcal{R}_2 - \alpha/4)} = \frac{3(z - \beta)}{2z - \beta/4 - 3\mathcal{R}_2/4}$$

As $|\mathcal{R}_2| \leq z/2$ this is at most

$$\frac{24(z - \beta)}{13z - 2\beta}$$

which is clearly maximized when $\beta = 0$ and hence is at most $\frac{24}{13}$. □

5 Conclusions and Open Problems

In this paper we considered the beam-on time and setup time minimization problems in radiation therapy. We presented an efficient linear time algorithm for the beam-on time minimization problem. We proved that the setup time minimization problem is APX Hard, and gave approximation algorithms for the problems that are better than the naive 2 approximation for the problem.

The area still has a lot of open problems, such as maximizing the combination of beam-on and setup times, considering multileaf rather than leaf setup time, considering constrained shape matrices and more.

References

1. R.K. Ahuja and H.W. Hamacher. A network flow algorithm to minimize beam-on time for unconstrained multileaf collimator problems in cancer radiation therapy. *Networks*, 44 (2005), 36–41.
2. S. Albers, S. Arora and S. Khanna. Page replacement for general caching problems. *Proc. 10th ACM-SIAM Symp. on Discrete Algorithms*, 31–40, 1999.
3. A. Bar-Noy, R. Bar-Yehuda, A. Freund, J. Naor and B. Schieber. A unified approach to approximating resource allocation and scheduling. *Journal of the ACM*, 48 (2001), 1069–1090.
4. S. Bernstein. Theory of Probability. Moscow, 1927.
5. A.L. Boyer. Use of MLC for intensity modulation. *Medical Physics*, 21 (1994), 1007.
6. T.R. Bortfeld, D.L. Kahler, T.J. Waldron and A.L. Boyer. X-ray field compensation with multileaf collimators. *Int. Journal of radiation Oncology, Biology, Physics*, 28 (1994), 723–730.
7. N.H. Boland, H.W. Hamacher and F. Lenzen. Minimizing beam-on time in cancer radiation therapy using multileaf collimators. *Networks*, 43 (2004), 226–240.
8. D.Z. Chen, X.S. Hu, S. Luan, C. Wang and X. Wu. Mountain reduction, block matching, and applications in intensity modulation radiation therapy. *Proc. 21st ACM Symp. on Computational Geometry*, 35–44, 2005.
9. J. Dai and Y. Zhu. Minimizing the number of segments in a delivery sequence for intensity modulated radiation therapy with multileaf collimator. *Medical Physics*, 28 (2001), 2113–2120.
10. H.N. Gabow, J.L. Bentley and R.E. Tarjan. Scaling and related techniques for geometry problems. *Proc. 16th ACM Symp. on Theory of Computing*, 135–143, 1984.
11. C.A.J. Hurkens and A. Schrijver. On the size of systems of sets every *t* of which have an SDR, with an application to the worst-case ratio of heuristics for packing problems. *SIAM Journal on Discrete Mathematics*, 2 (1989), 68–72.
12. S. Kamath, S. Sahni, J. Palta and S. Ranka. Algorithms for optimal sequencing of dynamic multileaf collimators. *Physics in Medicine and Biology*, 49 (2004), 33–54.
13. S. Kamath, S. Sahni, J. Palta, S. Ranka and J. Li. Optimal leaf sequencing with elimination of tongue-and-groove underdosage. *Physics in Medicine and Biology*, 49 (2004), N7–N19.
14. E. Petrank. The hardness of approximation: gap location. *Computational Complexity*, 4 (1994), 133–157.
15. J. Vuillemin. A Unifying Look at Data Structures. *Comm ACM*, 23 (1980), 229–239.

On the Value of Preemption in Scheduling

Yair Bartal[1,*], Stefano Leonardi[2], Gil Shallom[1], and Rene Sitters[3]

[1] Department of Computer Science, Hebrew University, Jerusalem, Israel
[2] Dipartimento di Informatica e Sistemistica, Universit di Roma La Sapienza, Rome, Italy
[3] Max-Planck-Insitut für Informatik, Saarbrücken, Germany

Abstract. It is well known that on-line preemptive scheduling algorithms can achieve efficient performance. A classic example is the Shortest Remaining Processing Time (SRPT) algorithm which is optimal for flow time scheduling, assuming preemption is costless. In real systems, however, preemption has significant overhead. In this paper we suggest a new model where preemption is costly. This introduces new considerations for preemptive scheduling algorithms and inherently calls for new scheduling strategies. We present a simple on-line algorithm and present lower bounds for on-line as well as efficient off-line algorithms which show that our algorithm performs close to optimal.

1 Introduction

Job scheduling is a common task in many computer systems. As a new job request arrives it is often necessary to preempt the current job in order to process the new one. It is evident that this operation involves significant overhead for the system, mostly due to context switching and extra paging (see e.g [4, 5, 11]) and therefore in some cases dramatically degrades the system's performance. In this paper we provide a theoretical model which incorporates this inherent cost of preemption.

The new model provides a provable explanation for the failure of some basic preemptive algorithms (Shortest Remaining Processing Time) in practice and hopefully leads to new approaches that may have more practical appeal. We focus on a classic problem in this context of flow time scheduling; Given a sequence of jobs we must schedule them for execution on a single or identical parallel machines, with the goal to minimize total flow time, i.e., the total time jobs spend in the system since arrival until they are run to completion. This includes the delay of waiting for service as well as the actual service time.

Formally, we are given a set J of n jobs and a set of m identical machines. Each job j is assigned with release time r_j and processing time p_j. The scheduling algorithm decides which of the jobs should be executed at each time. At any moment a machine can process at most one job and each job is processed on at most one machine. Moreover, a job cannot be processed before its release time. In the preemptive model a job that is running can be preempted and continued on an arbitrary moment on any machine. For a given schedule let C_j be the completion time of job j in this schedule. The flow time of job j in the schedule is $F_j = C_j - r_j$. The objective of the classical flow time problem

---

* Supported in part by a grant from the Israeli Science Foundation (195/02).

J. Diaz et al. (Eds.): APPROX and RANDOM 2006, LNCS 4110, pp. 39–48, 2006.

to minimize the total flow time $\sum_{j \in J} F_j$. We consider the problem for which the cost to make one preemption is K and the objective is to minimize $\sum_{j \in J} F_j + K \cdot M$, where M is the number of preemptions performed by the algorithm. This model naturally generalizes the two basic extreme cases – if $K = 0$ we are back in the costless preemption model, and when K is large enough we are essentially back to the non-preemptive case. Note that no preemption penalty is encountered if a job starts or terminates.

Preemption plays a crucial role in the context of flow time scheduling as well as many other scheduling problems. In the classical non-preemptive flow-time problem it is impossible to achieve any 'reasonable' approximation. Specifically, even for one machine there is no efficient $O(n^{\frac{1}{2}-\epsilon})$ approximation algorithm, unless $P = NP$ [7]. Moreover, no on-line algorithms can achieve better than the trivial $\Omega(n)$ bound. In contrast to these negative results we know that when preemption is allowed, the Shortest Remaining Processing Time algorithm achieves the optimal preemptive scheduling for a single machine [1]. This algorithm schedules at any time those jobs that are closest to completion. Hence, preemptions happen exactly at these moments that a newly released job has processing time smaller than the remaining processing time of a job on execution. The SRPT-algorithm is an on-line algorithm since at any time scheduling decisions are taken only on the basis of the knowledge of the jobs released so far. It has also been shown that the algorithm approximates the optimal preemptive flow time on parallel machines up to a logarithmic factor and that this is also the best competitive ratio that can be achieved by an on-line algorithm [8].

Whereas these theoretical results indicate that SRPT is a good heuristic, it may perform rather poorly in practice. The cause for this is provided by a number of recent experimental studies of real systems [4, 5, 11] showing that job preemption creates very significant overhead for the system, thus incurring an additional cost for the scheduler. We model this by associating a cost K with each preemption. It turns out that the right quantity to express the performance of on-line algorithms is not in K but in terms of κ, defined as the ratio of K and the length ϵ of the smallest job. However, the dependence of the performance ratio on κ is not linear as one may first expect. It turns out it is somewhat more delicate and depends on $\sqrt{\kappa}$. To be more exact, we denote $\alpha = \min\{\sqrt{\kappa}, \rho\}$, where ρ is the ratio of the length P_{max} of the largest job and the length ϵ of the smallest job. Our main result is an on-line algorithm with competitive ratio $(\alpha + 1)$ on a single machine and $O(C(\alpha + 1))$ for identical parallel machines, where $C = O(\log(\min\{\frac{n}{m}, \rho\}))$ is the performance ratio of the SRPT algorithm in the standard (costless) preemptive model on m machines. These results are complemented with an $\Omega(\alpha)$ lower bound on the competitive ratio of (randomized) on-line algorithms. Notice that the gap $C \leq \log \rho$ that remains is relatively small compared to the lower bound α.

We also provide an inapproximability result implying that the approximation ratio of any polynomial time algorithms is at least $\Omega(\alpha^{1/3-\delta})$, for any $\delta > 0$, unless $P = NP$.

Our new model may be compared with previous attempts to incorporate limitations on preemption in other theoretical studies. Such considerations have been imposed in the past in [9] and [3] were the algorithm must cope with an overall upper bound on the total number of preemptions, and in [12] in which the authors give a lower bound on the number of preemptions needed for competitiveness of any on-line algorithms aimed to minimize weighted sum of completion times. This paper introduces the cost of preemption as part of the objective function thus providing a natural motivation for minimizing the number of preemptions.

1.1 Performance of the Shortest Remaining Processing Time Algorithm

The shortest remaining processing time algorithm preempts the current job whenever a new job arrives whose processing time is smaller than the remaining processing time of the currently executing job. It is a well-known that this heuristic achieves the optimal flow time in the standard costless preemption model. Figure 1(a) shows why the SRPT algorithm performs poorly in the costly preemption model. The algorithm is given a

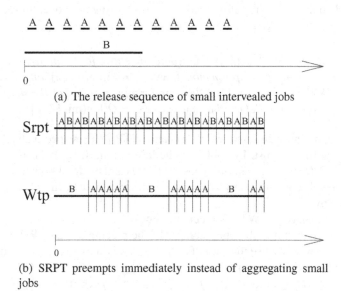

(a) The release sequence of small intervealed jobs

(b) SRPT preempts immediately instead of aggregating small jobs

Fig. 1. SRPT is presented with intervealed small jobs

job sequence in which a long job is released and then small jobs are released in non-consecutive intervals. The SRPT algorithm chooses to preempt the long job every time a short job is released. It is evident in this case that aggregating several small jobs before preempting is essential, if preemption incurs a cost. We note that the competitive ratio of SRPT can be easily shown to be $\Omega(\kappa) = \Omega(K/\epsilon)$. In the next section we show that aggregating a number of small jobs before actually preempting is indeed the preferred decision.

2   An On-Line Algorithm

We present an on-line algorithm for minimizing flow time with preemption cost for single and parallel machines. It is interesting to note that the algorithm considers the use of preemption every time some fixed amount of work is done, acting in a rather similar way to the actions of "real" operating system schedulers, which consider switching jobs every fixed interval of time (different fixed times for multi-level feedback schedulers). It turns out that this intuitive way in which "real" systems choose to act is competitive in our on-line model. We now turn to the description of the algorithm. We assume that the size ϵ of the smallest job is known to the algorithm.

Algorithm **Wait To Preempt (WTP):**

(i) If there is an idle machine and a job is available, then start processing the job with the shortest remaining processing time among the available jobs.

(ii) Every moment t that an additional $m\sqrt{K\epsilon}$ units (summed over all machines) are processed do the following: If the maximum processing time over all jobs released so far is more than $\sqrt{K\epsilon}$ preempt all jobs; Otherwise, do nothing.

Theorem 1. *Algorithm* WTP *is* $(C(\alpha+1)+\alpha)$-*competitive for minimizing flow time on m-machines, in the costly-preemption model, where* $\alpha = \min\{\rho, \sqrt{\kappa}\}$, *and C is the competitive ratio of SRPT for standard (costless) preemption model. For the single machine we have* $C = 1$ *and for parallel machines we have* $C = O(\log(\min\{\frac{n}{m}, \rho\}))$.

Proof. Consider an instance I and let $\mathrm{OPT}(I)$ and $\mathrm{WTP}(I)$ be, respectively, the optimal cost and the algorithm's cost. Let σ be a schedule produced by WTP for instance I. We partition $\mathrm{WTP}(I)$ in the preemption cost $\mathrm{WTP}_P(I)$ and the total flow time $\mathrm{WTP}_F(I)$. Denote the schedule constructed by WTP on I by σ and denote the number of processes in the schedule by n.

We begin with analyzing WTP 's preemption cost.

If $P_{max} \leq \sqrt{K\epsilon}$ then no job gets preempted. The preemption cost $\mathrm{WTP}_P(I)$ is zero in this case. Otherwise we use the sum of all processing times as a lower bound for the optimum. The number of preemptions is at most $m\lfloor\sum_j p_j/(m\sqrt{K\epsilon})\rfloor \leq \mathrm{OPT}(I)/(\sqrt{K\epsilon})$. The preemption cost is at most $K\cdot\mathrm{OPT}(I)/(\sqrt{K\epsilon}) = \mathrm{OPT}(I)\sqrt{\kappa}$. Further, the assumption $P_{max} > \sqrt{K\epsilon}$ equals $\rho > \sqrt{\kappa}$ which implies $\alpha = \sqrt{\kappa}$. The preemption cost is at most $\alpha\mathrm{OPT}(I)$.

To bound WTP's flow time we define an instance I' from I by modifying the release dates. Let S_j be the start time of job j in σ. We define $r'_j = \min\{S_j, r_j + \alpha\epsilon\}$. In lemma 1 we prove that the schedule σ is an SRPT schedule for instance I', implying

$$\mathrm{WTP}_F(I) \leq \mathrm{SRPT}(I') + \sum_j (r'_j - r_j) \leq C \cdot \mathrm{OPT}(I') + \sum_j (r'_j - r_j), \qquad (1)$$

where $(r'_j - r_j)$ is a bound on the additional flow time for a job incurred by $\mathrm{WTP}(I)$ beyond its flow time in $\mathrm{SRPT}(I')$.

Take any optimal schedule for I and shift it forward in time over $\alpha\epsilon$ time units. The resulting schedule is feasible for I' implying

$$\mathrm{OPT}(I') \leq \mathrm{OPT}(I) + \sum_j (\alpha\epsilon + r_j - r'_j). \qquad (2)$$

Combining (1) and (2) and using $\mathrm{OPT}(I) \geq n\epsilon$ for the second inequality, we obtain

$$\text{WTP}_F(I) \leq C \left(\text{OPT}(I) + n\alpha\epsilon\right) + (C - 1) \sum_j (r_j - r'_j)$$

$$\leq C(\alpha + 1)\text{OPT}(I).$$

The total cost becomes $\text{WTP} \leq (\alpha C + \alpha + C)\text{OPT}(I)$. ☐

Lemma 1. *Schedule σ satisfies the shortest remaining processing time rule for instance I'.*

Proof. We have to show that at any moment t the SRPT-rule is satisfied for schedule σ with respect to instance I'. Assume that at time t job j is processed and job k is available in instance I' but is not processed. We have to show the following property: At time t the remaining processing time of job j is at most the remaining processing time of job k.

We distinguish between the case $\alpha = \rho$ (i.e., $\rho \leq \sqrt{\kappa}$), and $\alpha = \sqrt{\kappa}$ (i.e., $\rho \geq \sqrt{\kappa}$).

If $\alpha = \rho$ then $P_{max} = \epsilon\rho \leq \epsilon\sqrt{\kappa} = \sqrt{K\epsilon}$ and no job gets preempted. In particular job k does not get preempted whence it is not processed before time t. Further, we assumed it is available in I' at time t but does not start at time t. Thus $r'_k < S_k$, the start time of job k. Now, using the definition of I', we must have $r'_k = \min\{S_k, r_k + \alpha\epsilon\} = r_k + \alpha\epsilon$. On the other hand, the start time of job j is at least $t - p_j \geq t - P_{max} = t - \rho\epsilon = t - \alpha\epsilon \geq r'_k - \alpha\epsilon \geq r_k$. Since the algorithm started j while k was available we have $p_k \geq p_j$ which is obviously at least the remaining processing time of job j at time t. Hence, the property holds in this case.

Now assume $\alpha = \sqrt{\kappa}$.

If job k was processed before time t then it must have been preempted. Consider the last time before t that job k was preempted. When a job gets preempted then all jobs get preempted at that time. Therefore, the segment of job j that is processed at time t started after job k was preempted. The property now follows directly from the algorithm (point (ii)).

If job k is not processed before time t then we have $r'_k < S_k$, since it is available in I' at time t but does not start at t. By definition of I' we must have $r'_k = r_k + \alpha\epsilon$, implying $r_k \leq r'_k - \alpha\epsilon \leq t - \alpha\epsilon$. If the segment of job j (that is processed at time t) started later than r_k then the claim obviously holds since the algorithm started (or resumed) the processing of j while job k was available. So assume the segment started before time r_k. We show that this gives a contradiction. Clearly, there can be no idle time between r_k and t since the algorithm has the choice to start job k. Since there is no idle time, at least $m(t - r_k) \geq m\alpha\epsilon = m\sqrt{K\epsilon}$ units are processed in the interval $[r_k, t]$ and it will contain at least one preemption point. At this point all jobs get preempted since job j is processed and $p_j \geq t - r_k \geq \alpha\epsilon = \sqrt{\kappa}\epsilon = \sqrt{K\epsilon}$. ☐

3 An Almost Tight Lower Bound

We will prove that the algorithm of Section 2 is best possible for the single machine model and only a factor C off from optimal for parallel machines, where C is the performance of SRPT in the standard model. Our proof is rather complicated and to facilitate this section we first present a bound for a more restricted model in which no machine is allowed to remain idle if any job is available for processing.

Lemma 2. *In the restricted model in which no idle time is allowed, any on-line algorithm has competitive ratio $\Omega(\alpha)$, where $\alpha = \min\{\rho, \sqrt{\kappa}\}$, independent of the number of machines available. Moreover, randomization does not improve upon this bound.*

Proof. The basic idea of the proof is to create a gap in the completion times of jobs being executed at a certain time by the on-line algorithm and the optimal solution. This is done by releasing two types of jobs: A-jobs and B-jobs of sizes x and $2x$ respectively. We set $x = \min\left\{\sqrt{K}\,\epsilon, \frac{\epsilon\rho}{2}\right\}$.

At time zero m jobs of type A and m jobs of type B are released. Since idle time is not allowed the start time of any job is a multiple of x. Moreover, no job can start at time $3x$ or later since in that case some machine must be idle before time $3x$. Let m_1, m_2, m_3 be the expected number of jobs of type B that starts respectively, at time 0, x and $2x$. We must have $m_k \geq m/3$ for some $k \in \{1, 2, 3\}$. The adversary starts releasing a sequence of $N \cdot m$ small jobs of size ϵ each, beginning at time kx, where a batch of m jobs is released at time $kx + i\epsilon$, for $0 \leq i \leq N - 1$ where $N = \frac{x}{\epsilon}$. See illustration in Figure 2.

The adversary does not preempt any job and executes all small jobs at their release date. If $k = 1$ then the adversary completes all A-jobs at time x and all B-jobs at time $4x$. The total flow time is at most $6xm$ in this case. Similarly, the total flow time is $7xm$ for $k = 2$ and $5xm$ for $k = 3$.

Consider a schedule σ given by the randomized algorithm A. Let S be the set of B-jobs that start processing at time $(k - 1)x$. At time kx the remaining processing time of these jobs is x. Let b be the number of jobs that are preempted. Since at least $|S| - b$ machines are not available for processing the small jobs, the total flow time of small jobs is more than $(|S| - b)Nx/2$. The total cost is at least $bK + (|S| - b)Nx/2 = bK + (|S| - b)x^2/(2\epsilon) \geq bK + (|S| - b)K/2 \geq |S|K/2$. Since $E|S| \geq m/3$ the expected cost is at least $mK/6$. We conclude that the expected competitive ratio is at least $mK/(42mx)$. Substituting $K \geq x^2/\epsilon$ yields the lower bound $\Omega(x/\epsilon) = \Omega(\min\{\sqrt{\kappa}, \rho\})$. □

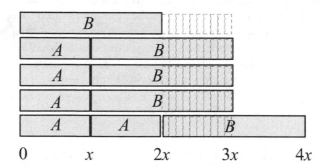

Fig. 2. The algorithm's schedule if no small jobs would be released. At time $2x$ there are still three B jobs with a remaining processing time of x. The adversary releases many small jobs (dotted in figure) at time $2x$.

Theorem 2. *An on-line algorithm cannot achieve a competitive ratio better than* $\Omega(\min\{\sqrt{\kappa},\rho\})$ *in the costly-preemption model with idle-time, independent of the number of machines available. Moreover, randomization does not improve upon this bound.*

Proof. First we prove the deterministic case and then sketch how to extend it for randomized algorithms. Let A be any deterministic algorithm and define $N = \frac{1}{2}\min\{\sqrt{\kappa},\rho\}$ and let $x = \epsilon N$. We define an instance with mN large jobs of length $2x$ and a number of batches of small jobs. A set of m large jobs is released at each time point $0, 4x, 8x, \ldots, (N-1)4x$. We refer to the interval $[(i-1)4x, i4x[$ as segment i.

Depending on the algorithm the adversary may decide to release a set of mN^2 small jobs, of length ϵ each, shortly after a set of large jobs. Consider the situation at time point $t = (i-1)4x$, i.e., at the beginning of segment i, for some $(i \in \{1,\ldots,N\})$. Let σ be the schedule that A constructs if no more jobs are released after time t. Let S be the set of large jobs in σ that start processing in segment i and are scheduled without preemption. If $|S| \geq m/2$ we say that the segment is of type 1. If $|S| < m/2$ we say the segment is of type 2. If $|S| \geq m/2$ then there is a $j \in \{1,2,3,4\}$ such that at least $|S|/4 \geq m/8$ jobs from S start processing in interval $[t+(j-1)x, t+jx]$. The adversary will release mN small jobs starting from time $t + ix$. More precisely, m jobs of length ϵ each are released at every time point $t + ix + k\epsilon$ for all $k \in \{0,\ldots,N-1\}$. If $|S| < m/2$ then no small jobs are released after the large jobs.

We construct a schedule with small cost in order to lower bound the optimal schedule. All small jobs are processed without preemption and start at their release time. The flow time of one small job is ϵ and the number of small jobs is at most mN^2 giving a total flow time of at most $mN^2\epsilon$. Next we add the large jobs without preemption to the schedule. We can do this such that the flow time of any large job is at most $5x$. (In any interval $[(i-1)4x + x, i4x + x]$ there is at most one batch of small jobs which block this interval for a time x.) The total flow time of this schedule is $mN^2\epsilon + mN5x = 6mNx$. Denote this value by z^*. Notice that the contribution of the large jobs is at most 5 times the contribution of the small jobs.

In order to be better than $\Omega(N)$ competitive the algorithm must complete almost all large jobs before time $4Nx$. More precisely, if αmN large jobs are completed after time $4Nx$ (for some $\alpha < 1$), then the average flow time for these jobs is at least $4Nx\alpha/2$. The total flow time for these jobs will be at least αmN times $4Nx\alpha/2$ is $2\alpha^2 mN^2 x = z^*N\alpha^2/3$.

Similarly, in order to be better than $\Omega(N)$ competitive the algorithm must schedule almost all large jobs without preemption. More precisely, if αmN large jobs are preempted ($\alpha < 1$), then the total cost is at least $\alpha mNK = \alpha mN\epsilon\kappa \geq \alpha mN\epsilon 4N^2 = 4\alpha mN^2 x = \frac{2}{3}\alpha z^* N$.

We conclude that in order to be better than $\Omega(N)$ competitive the algorithm must process almost all large jobs without preemption and before time $4xN$. Let U be the set of all large jobs with this property. By definition at most $m/2$ jobs from U start in any segment of type 2. In general, at most $4m$ jobs from U can start in a segment since no more jobs fit in. Let N_1 be the number of segments of type 1. Then $(N-N_1)m/2+N_14m \geq |U|$, implying $N_1 \geq (2|U|/m - N)/7$. For example, if $|U| \geq (17/20)mN$ then $N_1 \geq N/10$. Hence, we now assume that at least ten percent of the segments is of type 1.

Consider a segment of type 1. At the moment that the adversary starts releasing the small jobs there are at least $m/8$ large jobs which each have at least x processing time remaining. Denote this set by T. If the algorithm preempts at least half of them this adds $Km/16 = \epsilon \kappa m/16 \geq \epsilon 4x^2 m/16 = 2z^*/3$ to the total cost. In the other case we see that $m/16$ machines block the processing of small jobs for at least x time units. The total flow time of the small jobs is at least $Nm/16$ time $x/2$ is $Nmx/32 = 3z^*/16$.

Since we assumed that at least $N/10$ segments are of type 1, the total cost is at least $N/10$ times $3z^*/16$ is $3z^*N/160$. We conclude that there is a constant c such that the cost of any schedule is at least cNz^*.

The lower bound extends easily to randomized algorithms. We use Yao's min-max principle, i.e., again we consider a determinist algorithm but now we randomize the input. The set of large jobs remains the same but now a batch of small jobs is released for every segment i at time t_i, where t_i is uniformly taken from the set $\{(i-1)4x + jx \mid j = 1, 2, 3, 4\}$. The optimal schedule remains at most z^*. Again we can argue that, to be better than $\Omega(N)$-competitive, the algorithm must process almost all jobs without preemption and before time $4xN$. But in that case the expected flow time of the small jobs will be $\Omega(N)z^*$.   □

4 Inapproximability

In this section we sketch a proof for a lower bound on the approximation ratio achievable in polynomial time, assuming $P \neq NP$. Surprisingly, the result achieved shows that the competitive ratio achieved in the on-line scenario of the problem is not far from that achievable in polynomial time in the offline case. The proof is based on the inapproximability result for nonpreemptive flow time scheduling by Kellerer et al. [7].

Theorem 3. *There is no polynomial time algorithm which achieves an approximation $O(\kappa^{\frac{1}{4}-\delta})$ (assuming κ to be part of the input), or an approximation $O(\rho^{\frac{1}{3}-\delta})$, for any $\delta > 0$, in the single machine costly-preemption model with idle-time, unless $P = NP$.*

Proof. Kellerer et al. [7] showed a lower bound of $\Omega(\sqrt{n})$ on the approximation ratio of polynomial-time approximation algorithms for the problem of minimizing flow time in a single-machine non-preemptive environment. In our case preemption may be used but incurs a cost K. Therefore, we would use their proof in order to achieve an inapproximability result which is a function of the parameters of our problem, that is, ρ and κ.

We now turn to describe some details of the inapproximability proof in [7] so that we can describe how it can be used to achieve the desired inapproximability result. The authors reduce from the following strongly NP-complete version of numerical three-dimensional matching. (See Garey and Johnson [6] for the hardness proof.)

Problem: Numerical three-dimensional matching (N3DM).
Instance: Positive integers a_i, b_i and c_i, $1 \leq i \leq \ell$, with $\sum_{i=1}^{\ell}(a_i + b_i + c_i) = \ell D$.
Question: Do there exist permutations π, ψ such that $a_i + b_{\pi(i)} + c_{\psi(i)} = D$ holds for all i?

Given an arbitrary N3DM instance and some real number $0 < \gamma < \frac{1}{2}$, Kellerer et al. define the numbers

$$n = \left\lceil (20\ell)^{\frac{4}{\gamma}} D^{\frac{2}{\gamma}} \right\rceil, \qquad r = \left\lceil 2Dn^{\frac{1-\gamma}{2}} \right\rceil, \qquad g = 100r\ell^2.$$

The number of jobs in their instance is n. Further, the smallest job has processing time $1/(rg)$ and the largest has processing time at most $8r + D < 9r$. They show that the total flow time is smaller than $200r\ell^2$ if there is a solution to the N3DM problem and at least $100r^2\ell^2$ if no such solution exists. This results in a lower bound of $\frac{1}{2}r$ on the approximation ratio achievable.

In our model, preemption is allowed but if $K \geq 100r^2\ell^2$, then the optimal solution is non-preemptive. Thus, the lower bound of $\frac{1}{2}r$ on the approximability applies directly to our problem if we set $K = 100r^2\ell^2$.

We would like to express this lower bound in terms of ρ and κ:

$$\rho = \frac{P_{max}}{\epsilon} < \frac{9r}{1/(rg)} = 900r^3\ell^2,$$

$$\kappa = \frac{K}{\epsilon} = \frac{K}{1/(rg)} = 10000r^4\ell^4.$$

We simplify the parameters of the N3DM problem and use $\ell \leq r^{\frac{2\gamma}{1-\gamma}}$ in order to get:

$$\rho = O(r^{3+\frac{2\gamma}{1-\gamma}}) \Longrightarrow r = \Omega\left(\rho^{\frac{1}{3+\frac{2\gamma}{1-\gamma}}}\right),$$

$$\kappa = O(r^{4+\frac{4\gamma}{1-\gamma}}) \Longrightarrow r = \Omega\left(\kappa^{\frac{1}{4+\frac{4\gamma}{1-\gamma}}}\right).$$

We now recall that $0 < \gamma < \frac{1}{2}$ can be chosen arbitrarily. Given any $\delta > 0$, we choose a small enough γ in order to maintain the following relations:

$$\frac{1}{2}r = \Omega\left(\rho^{\frac{1}{3}-\delta}\right) \quad \text{and} \quad \frac{1}{2}r = \Omega\left(\kappa^{\frac{1}{4}-\delta}\right).$$

$\square$

5   Extensions

This paper introduces a more realistic scheduling model where preemption is costly. We study the model in the context of flow time scheduling and provide efficient on-line algorithms for this setting. Our work makes the first step in studying costly preemption models in the context of various scheduling problems, e.g. minimizing completion time, weighted flow time, or stretch [10]. It may also be possible to combine the costly preemption model with semi-clairvoyance [2].

Another important direction for further research is to consider a more sophisticated modelling of the preemption cost, e.g. the cost of preemption may be dependent on the processing time of the jobs involved, or more generally on the state of the system.

References

1. K.R. Baker. *Introduction to sequencing and scheduling*. Wiley, New York, 1974.
2. L. Becchetti, S. Leonardi, A. Marchetti-Spaccamela, and K. Pruhs. Semi-clairvoyant scheduling. In *Proc. 11th European Symp. on Algorithms (ESA)*, volume 2832 of *Lecture Notes in Comput. Sci.*, pages 67–77. Springer, 2003.
3. O Braun and G. Schmidt. Parallel processor scheduling with limited number of preemptions. *SIAM J. Comput.*, 32:671–680, 2003.
4. R.T. Dimpsey and R.K. Iyer. Performance degradation due to multiprogramming and system overheads in real workloads: Case study on a shared memory multiprocessor. In *Intnl. Conf. Supercomputing*, pages 227–238, 1990.
5. Y. Etsion, D. Tsafrir, and D.G. Feitelson. Effects of clock resolution on the scheduling of interactive and soft real-time processes. In *SIGMETRICS Conf. Measurement & Modeling of Comput. Syst*, pages 172–183, 2003.
6. M.R. Garey and D.S. Johnson. *Computers and Intractability: A Guide to the theory of NP-Completeness*. Freeman and Company, San Francisco, 1979.
7. T. Tautenhahn H. Kellerer and G.J. Woeginger. Approximability and nonapproximability results for minimizing total flow time on a single machine. *SIAM J. Comput.*, 28:1155–1166, 1999.
8. S. Leonardi and D. Raz. Approximating total flow time on parallel machines. In *Proc. 29th Symp. Theory of Computing (STOC)*, pages 110–119. ACM, 1997.
9. R. Motwani, S. Phillips, and E. Torng. Non-clairvoyant scheduling. *Theoret. Comput. Sci.*, 130:17–47, 1994.
10. S. Muthukrishnan, R. Rajaraman, A. Shaheen, and J. E. Gehrke. Online scheduling to minimize avarage strech. In *Proc. 40th Symp. Foundations of Computer Science (FOCS)*, pages 433–443. IEEE, 1999.
11. C. Natarajan, S. Sharma, and R.K. Iyer. Measurement-based characterization of global memory and network contention, operating system and parallelization overheads: Case study on a shared-memory multiprocessor. In *Ann. Intl. Symp. Computer Architecture*, volume 21, pages 71–80, 1994.
12. U. Schwiegelshohn. Preemptive weighted completion time scheduling of parallel jobs. In *Proc. 4th European Symp. on Algorithms (ESA)*, volume 1136 of *Lecture Notes in Comput. Sci.*, pages 39–51. Springer, 1996.

An Improved Analysis for a Greedy Remote-Clique Algorithm Using Factor-Revealing LPs*

Benjamin E. Birnbaum[1] and Kenneth J. Goldman[2]

[1]Department of Computer Science and Engineering
University of Washington, Seattle
Seattle WA 98195, USA
bbirnb@u.washington.edu
[2]Department of Computer Science and Engineering
Washington University in St. Louis
St. Louis MO 63130, USA
kjg@cse.wustl.edu

Abstract. Given a positive integer p and a complete graph with non-negative edge weights that satisfy the triangle inequality, the *remote-clique* problem is to find a subset of p vertices having a maximum-weight induced subgraph. A greedy algorithm for the problem has been shown to have an approximation ratio of 4, but this analysis was not shown to be tight. In this paper, we present an algorithm called d-GREEDY AUGMENT that generalizes this greedy algorithm (they are equivalent when $d = 1$). We use the technique of *factor-revealing linear programs* to prove that d-GREEDY AUGMENT, which has a running time of $O(pdn^d)$, achieves an approximation ratio of $(2p - 2)/(p + d - 2)$. Thus, when $d = 1$, d-GREEDY AUGMENT achieves an approximation ratio of 2 and runs in time $O(pn)$, making it the fastest known 2-approximation for the remote-clique problem. The usefulness of factor-revealing LPs in the analysis of d-GREEDY AUGMENT suggests possible applicability of this technique to the study of other approximation algorithms.

1 Introduction

Let $G = (V, E)$ be a complete graph with the weight for edge $\{v_1, v_2\} \in E$ given by $w(v_1, v_2)$. (Define $w(v, v) = 0$ for all $v \in V$.) The edge weights are nonnegative and satisfy the triangle inequality: for all $v_1, v_2, v_3 \in V$, $w(v_1, v_2) + w(v_2, v_3) \geq w(v_1, v_3)$. For a given integer parameter p, such that $1 \leq p \leq |V|$, the *remote-clique problem* is to find a subset $V' \subseteq V$ such that $|V'| = p$ and the average edge weight in V', $2/(p(p-1)) \cdot \sum_{\{v_1, v_2\} \in E \,:\, v_1, v_2 \in V'} w(v_1, v_2)$, is maximized. The remote-clique problem (also called *maxisum dispersion* [1] and *max-avg facility*

* This research was supported in part by the National Science Foundation under grant 0305954. It was performed while the first author was at Washington University in St. Louis.

J. Diaz et al. (Eds.): APPROX and RANDOM 2006, LNCS 4110, pp. 49–60, 2006.
© Springer-Verlag Berlin Heidelberg 2006

dispersion [2]) is one the so-called dispersion problems, which involve finding subsets of vertices that are in some way as distant from each other as possible. These problems are motivated by a number of applications in computer science and operations research (see [3], for example). The remote-clique problem can be shown to be NP-hard by an easy reduction from CLIQUE.

It has been shown that a simple greedy algorithm with running time $O(n^2)$ achieves an approximation ratio of 4 (i.e., the weight of the optimal solution is never more than four times the weight of the solution found by the greedy algorithm) [2]. An example is provided in which the optimal solution weighs twice as much as the algorithm's solution, but the question of whether a tighter bound for the algorithm can be proved remained open. In another paper, it is proved that a more complicated algorithm with running time $O(n^2 + p^2 \log p)$ achieves an approximation ratio of 2 [4]. However, a tight approximation ratio for the simple greedy algorithm has never been proved.

In this paper, we provide an algorithm parameterized by an integer d called d-GREEDY AUGMENT. When $d = 1$, the algorithm is the same as the greedy algorithm analyzed in [2], and when $d = p$, the algorithm amounts to examining all subsets of p vertices and choosing the one with maximum edge weight. (Clearly, this will return an optimal solution, but it does not run in polynomial time unless p is a constant.) We will show that d-GREEDY AUGMENT has an approximation ratio of $(2p - 2)/(p + d - 2)$ and has a running time of $O(pdn^d)$. Therefore if $d = 1$, our algorithm guarantees an approximation ratio of 2 and has a running time of $O(pn)$. This algorithm, then, is the fastest known (and easiest to implement) 2-approximation for the remote-clique problem. Our proof also answers an open question from [2] by showing that the greedy algorithm does indeed obtain an approximation ratio of 2.

Because the running time of d-GREEDY AUGMENT is exponential in d, it only runs in polynomial time for constant values of d. Furthermore, the remote-clique problem can be shown to be NP-hard only for non-constant values of p. (If p is a constant, then the problem is solvable in polynomial time by examining every subset of p vertices and choosing the maximum weight subset). Therefore, increasing the value of d does not affect the approximation ratio asymptotically. However, we include the analysis for all values of d both for completeness and because it could have some practical benefit for small values of p. For example, if $p = 4$, then the naive exact algorithm would take quartic time, whereas if we were willing to spend quadratic time, we could run d-GREEDY AUGMENT for $d = 2$ to guarantee finding a solution at least $\frac{2}{3}$ the value of the optimal solution.

To prove an approximation ratio of $(2p - 2)/(p + d - 2)$, we use the technique of *factor-revealing linear programs* [5, 6], which is a simple generalization of a method often used to provide bounds for approximation algorithms. Consider a maximization (resp., minimization) problem P. A typical analysis of an approximation algorithm ALG for P proceeds by using the behavior of ALG and the structure of P to generate a number of inequalities. These inequalities are then combined to provide a bound on the ratio of the value of the solution obtained by ALG to that of an optimal solution. Often, this can be done by straightforward manipulation, but not always. A more general way of obtaining a bound is to view the process as an optimization problem Q in its own right, in which an adversary tries to minimize (resp., maximize) the value of ALG's solution to P subject to the constraints given by the generated inequalities. The optimal solution to Q is then a bound on the performance of ALG. If Q can be formulated as a linear program, then this is a *factor-revealing LP*, which can be solved using duality. The simplicity of this technique makes it applicable to many problems, but in most cases it does not seem to be the easiest way to provide a bound. However, there are some algorithms, including the greedy algorithm examined here, in which it is the only known technique to provide a tight bound.

Before we proceed with the analysis of d-GREEDY AUGMENT, we define some notation that will be used in this paper:

- For any natural number n, let $[n] = \{1, \ldots, n\}$.
- For any set S and integer k, let $\binom{S}{k} = \{S' \subseteq S : |S'| = k\}$. If $k < 0$ or $k > |S|$, then $\binom{S}{k} = \emptyset$.

We will also need the following easy to prove combinatorial identities.

Lemma 1. *Let n be a positive integer and let i and j be two nonnegative integers such that $j \leq i \leq n$. Then if f is some function defined on the domain $\binom{[n]}{j}$,*

$$\sum_{I \in \binom{[n]}{i}} \sum_{J \in \binom{I}{j}} f(J) = \binom{n-j}{i-j} \sum_{J \in \binom{[n]}{j}} f(J) \ .$$

Proof. Omitted for brevity.[1]   □

Lemma 2. *For all integers k, d, and s, such that $0 \leq k \leq d \leq s$,*

$$\binom{s}{d}\binom{d}{k} = \binom{s}{k}\binom{s-k}{d-k} \ .$$

Proof. Omitted for brevity.   □

2 Analysis of d-Greedy Augment

The algorithm we analyze, called d-GREEDY AUGMENT, maintains a set of vertices T that starts empty. At each step in the algorithm, it augments T by the set of d vertices that will add the most weight to T. When $|T| = p$, d-GREEDY AUGMENT terminates and returns the set T. (Throughout this paper, we assume for simplicity that d divides p evenly.) To be more precise, we define the following notation. For any subset of vertices $V' \subseteq V$ that is disjoint from T, let

$$\text{aug}_T(V') = \sum_{v' \in V'} \sum_{v \in T} w(v', v) + \sum_{\{v'_1, v'_2\} \in \binom{V'}{2}} w(v'_1, v'_2) \ .$$

---

[1] The proofs omitted in this paper can be found in the thesis upon which this paper is based [7].

In other words, $\text{aug}_T(V')$ is the amount of edge weight that the vertices in V' add to T if T is augmented by V'. At each step in the algorithm, d-GREEDY AUGMENT chooses a set of d vertices V' that maximizes $\text{aug}_T(V')$ and adds them to T. For the first step, this means that d-GREEDY AUGMENT chooses a group of d vertices with the heaviest edge weights (breaking ties arbitrarily).[2] By storing and incrementally updating the value of $\text{aug}_T(V')$ for each set V' of size d, d-GREEDY AUGMENT can be be implemented to run in time $O(pdn^d)$.

Before we begin with the proof that d-GREEDY AUGMENT obtains an approximation ratio of $(2p-2)/(p+d-2)$, we observe in the following theorem that this ratio is a tight bound on the performance of d-GREEDY AUGMENT.

Theorem 1. *There exist infinitely many remote-clique problem instances in which the ratio of the average edge weight in an optimal solution to the average edge weight in the solution returned by d-GREEDY AUGMENT is $(2p-2)/(p+d-2)$.*

Proof. Consider the following problem instance $(G = (V,E), p)$, in which $|V| = 2p$ and V is partitioned into p/d subsets of d vertices called $V_1, \ldots, V_{p/d}$ and one group of p vertices called O. The edge weights are determined as follows:

$$w(v_1, v_2) = \begin{cases} 2 & \text{if } v_1, v_2 \in V_i \text{ for some } i \text{ or } v_1, v_2 \in O \\ 1 & \text{otherwise} \end{cases} .$$

This construction is illustrated in Fig. 1. The edges in this problem instance satisfy the triangle inequality since every edge either has weight 1 or has weight 2. It is clear that d-GREEDY AUGMENT could (if ties were broken badly) return the solution $T = \bigcup_i V_i$, whereas the optimal solution is O. If this happens, the total edge weight in T is

$$\frac{p}{d}\binom{d}{2} \cdot 2 + \left(\binom{p}{2} - \frac{p}{d}\binom{d}{2}\right) \cdot 1 = \frac{1}{2}p(p+d-2) ,$$

and since the total edge weight in O is $2\binom{p}{2}$, the ratio of the performance of an optimal algorithm to that of d-GREEDY AUGMENT is $(2p-2)/(p+d-2)$. To see that there are infinitely many such problem instances, note that without affecting the relative performance of d-GREEDY AUGMENT, we can add arbitrarily many vertices to this construction in which every edge incident to these new vertices has weight 1. □

We now continue with a proof of our approximation ratio. For an instance of the remote-clique problem, let OPT be the average edge weight in an optimal

---

[2] Note that if $d = 1$, then during the first step $\text{aug}_T(V') = 0$ for all $V' \subseteq V$ such that $|V'| = d$. Thus for $d = 1$, d-GREEDY AUGMENT just starts with an arbitrary vertex. This is in fact the only difference between d-GREEDY AUGMENT for $d = 1$ and the algorithm analyzed in [2]; instead of initializing T with an arbitrary vertex, the algorithm in [2] initializes T with two vertices that are endpoints of a maximum weight edge.

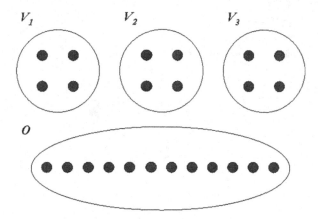

Fig. 1. An example (for $p = 12$ and $d = 4$) showing that the approximation ratio of $(2p - 2)/(p + d - 2)$ is a tight bound on the worst-case performance of d-GREEDY AUGMENT. Edges that are contained within the same circle have weight 2, and all other edges have weight 1.

solution. To prove that d-GREEDY AUGMENT achieves an approximation ratio of $(2p - 2)/(p + d - 2)$, we will prove that at each augmenting step, there exists a group of d vertices V' for which $\mathrm{aug}_T(V')$ is sufficiently high.

Lemma 3. *Before each augmenting step in the algorithm, there exists a group of d vertices $V' \subseteq V$ such that V' is disjoint from T and $\mathrm{aug}_T(V') \geq \frac{1}{2}d(|T| + d - 1)OPT$.*

Proof. We defer the proof of the lemma when $|T| > 0$ to the remainder of this section. When $|T| = 0$, the statement of the lemma is that there exists at least one group of vertices $V' \subseteq V$ of size d and total weight at least $\binom{d}{2}OPT$. We omit the details for brevity, but it is easy to show using an averaging argument that such a group of vertices exists inside an optimal solution. This is because the average weight of the edges in an optimal solution is the same as the average weight required of the edges in V'.   □

With this fact, it is straightforward to prove that d-GREEDY AUGMENT achieves an approximation ratio of $(2p - 2)/(p + d - 2)$.

Theorem 2. *The average weight of the edges in the solution returned by d-GREEDY AUGMENT is at least $(p + d - 2)/(2p - 2) \cdot OPT$.*

Proof. Since d-GREEDY AUGMENT adds d vertices to T during each augmenting step, we have that $|T| = (k - 1)d$ before the kth augmenting step. Thus by Lemma 3, there exists a set of d vertices $V' \subseteq V$ that can be added to T such that $\mathrm{aug}_T(V') = \frac{1}{2}d(dk - 1)OPT$. We know therefore that the weight of the edges added by d-GREEDY AUGMENT on the kth augmenting step is at least

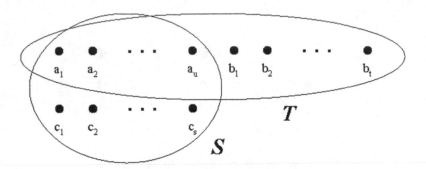

Fig. 2. An intermediate state of d-Greedy Augment

$\frac{1}{2}d(dk-1)OPT$, since d-Greedy Augment chooses the set of d vertices V' that maximizes $\text{aug}_T(V')$. Thus after k steps, the weight of the edges in T is at least

$$\sum_{j=1}^{k} \frac{1}{2}d(dj-1)OPT = \frac{1}{4}dk(dk+d-2)OPT .$$

Since the algorithm terminates after p/d steps, the final total weight of the edges in T is at least $\frac{1}{4}p(p+d-2)OPT$. Since there are $p(p-1)/2$ edges in this set of vertices, the average weight of the edges in the solution returned by d-Greedy Augment is at least $(p+d-2)/(2p-2) \cdot OPT$.   □

To prove Lemma 3 when $|T| > 0$, we begin with some notation. Consider an intermediate state of d-Greedy Augment. The set of vertices chosen so far is T, and let S be the set of vertices in an optimal solution. Let $u = |S \cap T|$, $t = |T - S|$, and $s = |S - T|$. As shown in Fig. 2, arbitrarily label the vertices in $S \cap T$ as $a_1, a_2, \ldots, a_u$, the vertices in $T - S$ as $b_1, b_2, \ldots, b_t$, and the vertices in $S - T$ as $c_1, c_2, \ldots, c_s$. Note that one of u or t may be equal to zero, but since we are in an intermediate state of d-Greedy Augment, we know that $d \le |T| \le |S| - d$, and hence $u + t \ge d$ and $s \ge t + d$.

We break the proof of Lemma 3 when $|T| > 0$ into five cases based on the values of u and t and state the proof for each case as its own lemma. For each of these cases, we show that a group of d vertices satisfying the condition of Lemma 3 can be found in the set $S - T$. We start with the case when S and T are disjoint, i.e., when $u = 0$ and $t \ge 1$. This case permits a direct analysis without the use of linear programming. It is instructive to analyze this case first, since it will become more clear why imitating this analysis does not seem to be possible for the other cases. This should help motivate the use of the more general technique of factor-revealing linear programs for the other cases.

Lemma 4. *Lemma 3 holds when $u = 0$ and $t \ge 1$. Specifically, there exists a set of indices $L \in \binom{[s]}{d}$ such that*

$$\sum_{\ell \in L} \sum_{j \in [t]} w(b_j, c_\ell) + \sum_{\{\ell, m\} \in \binom{L}{2}} w(c_\ell, c_m) \geq \frac{1}{2} d(d + t - 1) OPT .[3]$$

Proof. The key observation is that because of the triangle inequality, edges adjacent to the high-weight edges in S must also have high weight on average. By the triangle inequality,

$$w(b_j, c_\ell) + w(b_j, c_m) \geq w(c_\ell, c_m) \quad j \in [t], \{\ell, m\} \in \binom{[s]}{2} .$$

Summing over all j, this becomes

$$\sum_{j \in [t]} w(b_j, c_\ell) + \sum_{j \in [t]} w(b_j, c_m) \geq t w(c_\ell, c_m) \quad \{\ell, m\} \in \binom{[s]}{2} .$$

Now, summing over all $\{\ell, m\}$ yields

$$\sum_{\{\ell, m\} \in \binom{[s]}{2}} \left(\sum_{j \in [t]} w(b_j, c_\ell) + \sum_{j \in [t]} w(b_j, c_m) \right) \geq t \sum_{\{\ell, m\} \in \binom{[s]}{2}} w(c_\ell, c_m) = t \binom{s}{2} OPT ,$$

$$(1)$$

where the equality follows from the optimality of S. By applying Lemma 1 to the left-hand side of (1), we can rewrite it as

$$(s - 1) \sum_{\ell \in [s]} \sum_{j \in [t]} w(b_j, c_\ell) \geq t \binom{s}{2} OPT ,$$

which can be simplified to

$$\sum_{\ell \in [s]} \sum_{j \in [t]} w(b_j, c_\ell) \geq \frac{st}{2} OPT .$$

From this fact, along with the optimality of S, it follows that[4]

$$\binom{s-1}{d-1} \sum_{\ell \in [s]} \sum_{j \in [t]} w(b_j, c_\ell) + \binom{s-2}{d-2} \sum_{\{\ell, m\} \in \binom{[s]}{2}} w(c_\ell, c_m)$$

$$\geq \left(\binom{s-1}{d-1} \frac{st}{2} + \binom{s-2}{d-2} \binom{s}{2} \right) OPT$$

$$= \frac{1}{2} \binom{s}{d} d(d + t - 1) OPT ,$$

$$(2)$$

---

[3] Note that if $d = 1$, then the second sum in this inequality is empty. In general, there will be a number of formulas in this section that simplify significantly if $d = 1$.

[4] Note that if $d = 1$, then $\binom{s-2}{d-2} = 0$.

where the equality is obtained from some simplification using Lemma 2. We can now apply Lemma 1 to the left-hand side of this inequality to obtain

$$\sum_{L \in \binom{[s]}{d}} \left(\sum_{\ell \in L} \sum_{j \in [t]} w(b_j, c_\ell) + \sum_{\{\ell,m\} \in \binom{L}{2}} w(c_\ell, c_m) \right) \geq \frac{1}{2} \binom{s}{d} d(d+t-1)OPT \ .$$

But this implies that there must exist at least one $L \in \binom{[s]}{d}$ such that

$$\sum_{\ell \in L} \sum_{j \in [t]} w(b_j, c_\ell) + \sum_{\{\ell,m\} \in \binom{L}{2}} w(c_\ell, c_m) \geq \frac{1}{2} d(d+t-1)OPT \ .$$

$\square$

Now that we have proved Lemma 3 for the case when T is disjoint from S, we turn to proving Lemma 3 when some number of vertices in T are also in S. Intuitively, the algorithm should do no worse when T is not disjoint from S. If d-GREEDY AUGMENT has already found some of the optimal solution, then that should not hurt its performance. However, we will see that this case actually seems harder to analyze, and we will need to use factor-revealing linear programs for this case. We start with the most general non-disjoint case that we examine, when $u \geq 2$ and $t \geq 1$. (Our application of the triangle inequality leads to certain boundary cases that arise for smaller values of u and t, which we handle separately.)

Lemma 5. *Lemma 3 holds when $u \geq 2$ and $t \geq 1$. Specifically, there exists a set of indices $L \in \binom{[s]}{d}$ such that*

$$\sum_{\ell \in L} \left(\sum_{h \in [u]} w(a_h, c_\ell) + \sum_{j \in [t]} w(b_j, c_\ell) \right) + \sum_{\{\ell,m\} \in \binom{L}{2}} w(c_\ell, c_m) \geq \frac{1}{2} d(d+t+u-1)OPT \ .$$

Proof. It is sufficient to show that

$$\binom{s-1}{d-1} \sum_{\ell \in [s]} \left(\sum_{h \in [u]} w(a_h, c_\ell) + \sum_{j \in [t]} w(b_j, c_\ell) \right) + \binom{s-2}{d-2} \sum_{\{\ell,m\} \in \binom{[s]}{2}} w(c_\ell, c_m)$$

$$\geq \frac{1}{2} \binom{s}{d} d(d+t+u-1)OPT \qquad (3)$$

since we can then proceed to prove this lemma as we proved Lemma 4 from (2). To prove (3), we follow the same strategy that we used in Lemma 4; we use the triangle inequality and the optimality of S. By the triangle inequality, we have

$$w(a_h, c_\ell) + w(a_h, c_m) - w(c_\ell, c_m) \geq 0 \quad h \in [u], \{\ell, m\} \in \binom{[s]}{2} \qquad (4)$$

$$w(b_j, c_\ell) + w(b_j, c_m) - w(c_\ell, c_m) \geq 0 \quad j \in [t], \{\ell, m\} \in \binom{[s]}{2} \qquad (5)$$

$$w(a_h, c_\ell) + w(a_i, c_\ell) - w(a_h, a_i) \geq 0 \quad \{h, i\} \in \binom{[u]}{2}, \ell \in [s] \ . \qquad (6)$$

By the optimality of S, we have

$$\sum_{\{h,i\}\in\binom{[u]}{2}} w(a_h,a_i) + \sum_{\{\ell,m\}\in\binom{[s]}{2}} w(c_\ell,c_m) + \sum_{h\in[u]}\sum_{\ell\in[s]} w(a_h,c_\ell) \geq \binom{s+u}{2}OPT .$$

(7)

At the corresponding point in the proof of Lemma 4, it was possible to combine the inequalities expressing the triangle inequality and the optimality of S to yield (2) and finish the proof. In this lemma, however, the inequalities have a much more complicated form because of the overlap of S and T, and it does not seem possible to combine them directly to yield (3). To prove (3), we instead consider an adversary trying to minimize

$$\binom{s-1}{d-1}\sum_{\ell\in[s]}\left(\sum_{h\in[u]} w(a_h,c_\ell) + \sum_{j\in[t]} w(b_j,c_\ell)\right) + \binom{s-2}{d-2}\sum_{\{\ell,m\}\in\binom{[s]}{2}} w(c_\ell,c_m)$$

subject to the constraints given by (4), (5), (6), and (7). If we can show that the optimal value of this factor-revealing linear program (where the variables are the weights of the edges) is at least $\frac{1}{2}\binom{s}{d}d(d+t+u-1)OPT$, then we will have proved (3). Since the value of any feasible dual solution is a lower bound for the optimal value of the primal, we can prove (3) by finding a feasible dual solution with value $\frac{1}{2}\binom{s}{d}d(d+t+u-1)OPT$. The dual linear program is

maximize

$$\binom{s+u}{2}OPT\cdot z$$

subject to

$$-\sum_{\ell\in[s]} y_{\{h,i\},\ell} + z \leq 0 \quad \{h,i\}\in\binom{[u]}{2}$$

$$-\sum_{h\in[u]} w_{h,\{\ell,m\}} - \sum_{j\in[t]} x_{j,\{\ell,m\}} + z \leq \binom{s-2}{d-2} \quad \{\ell,m\}\in\binom{[s]}{2} \quad (8)$$

$$\sum_{m\in[s]-\{\ell\}} w_{h,\{\ell,m\}} + \sum_{i\in[u]-\{h\}} y_{\{h,i\},\ell} + z \leq \binom{s-1}{d-1} \quad h\in[u],\ell\in[s]$$

$$\sum_{m\in[s]-\{\ell\}} x_{j,\{\ell,m\}} \leq \binom{s-1}{d-1} \quad j\in[t],\ell\in[s]$$

where $w_{h,\{\ell,m\}}$ corresponds to (4), $x_{j,\{\ell,m\}}$ corresponds to (5), $y_{\{h,i\},\ell}$ corresponds to (6), and z corresponds to (7). It can be easily verified that the following dual solution is feasible.

$$w'_{h,\{\ell,m\}} = \frac{s-d-t+1}{(u+s)(s-1)}\binom{s-1}{d-1} \qquad h \in [u], \{\ell,m\} \in \binom{[s]}{2}$$

$$x'_{j,\{\ell,m\}} = \frac{1}{s-1}\binom{s-1}{d-1} \qquad j \in [t], \{\ell,m\} \in \binom{[s]}{2}$$

$$y'_{\{h,i\},\ell} = \frac{d+t+u-1}{(u+s)(u+s-1)}\binom{s-1}{d-1} \qquad \{h,i\} \in \binom{[u]}{2}, \ell \in [s]$$

$$z' = \frac{d(d+t+u-1)}{(u+s)(u+s-1)}\binom{s}{d}$$

The only constraint that is not trivial to verify is (8), but some straightforward manipulation shows that if $d = 1$, then the left-hand side is equal to

$$-\frac{su(u+t)}{(u+s)(u+s-1)(s-1)},$$

which is no greater than 0 since $s \geq 2$ when $t \geq 1$. Similarly, if $d > 1$, then the left-hand side can be written as

$$\left(1 - \frac{su(d+t+u-1)}{(u+s)(u+s-1)(d-1)}\right)\binom{s-2}{d-2},$$

which is clearly no greater than $\binom{s-2}{d-2}$. We conclude the proof by noting that the value of this dual solution is

$$\binom{s+u}{2}\binom{s}{d}\frac{d(d+t+u-1)}{(u+s)(u+s-1)}OPT = \frac{1}{2}\binom{s}{d}d(d+t+u-1)OPT,$$

which implies that the optimal value of the primal is no less than $\frac{1}{2}\binom{s}{d}d(d+t+u-1)OPT$ and hence implies (3), thus proving the lemma. $\qquad \square$

We have now proved the most general (and hardest) case of Lemma 3. It remains only to prove three boundary cases. We state the lemmas corresponding to the next two of these cases without proof. The paradigm for proving them is the same as was used to prove Lemma 5. First, the primal linear program is constructed with inequalities based on the triangle inequality and the optimality of S. Then the dual linear program is found, and a feasible solution is constructed which yields an appropriate lower bound on the primal. This bound on the primal is then used to prove the statement of Lemma 3. The next two cases we consider are when $u = 1$ and $t \geq 1$ (Lemma 6) and when $u \geq 2$ and $t = 0$ (Lemma 7). The former needs to be considered separately because we do not have constraint (6) (or dual variables of the form $y_{\{h,i\},\ell}$), and the latter needs to be considered separately because we do not have constraint (5) (or dual variables of the form $x_{j,\{\ell,m\}}$).

Lemma 6. *Lemma 3 holds when $u = 1$ and $t \geq 1$. Specifically, there exists a set of indices $L \in \binom{[s]}{d}$ such that*

$$\sum_{\ell \in L}\left(w(a_1,c_\ell) + \sum_{j\in[t]}w(b_j,c_\ell)\right) + \sum_{\{\ell,m\}\in\binom{L}{2}}w(c_\ell,c_m) \geq \frac{1}{2}d(d+t)OPT.$$

Proof. Omitted for brevity.   □

Lemma 7. *Lemma 3 holds when* $u \geq 2$ *and* $t = 0$. *Specifically, there exists a set of indices* $L \in \binom{[s]}{d}$ *such that*

$$\sum_{\ell \in L} \sum_{h \in [u]} w(a_h, c_\ell) + \sum_{\{\ell, m\} \in \binom{L}{2}} w(c_\ell, c_m) \geq \frac{1}{2} d(d + u - 1)OPT .$$

Proof. Omitted for brevity.   □

The final case yet to be covered is when $u = 1$ and $t = 0$. If this is true, then it must be the case that $d = 1$, since $u + t \geq d$. The proof of this case follows from a simple contradiction argument.

Lemma 8. *Lemma 3 holds when* $u = 1$ *and* $t = 0$ *(and hence* $d = 1$*). Specifically, there exists an* $\ell \in [s]$ *such that* $w(a_1, c_\ell) \geq \frac{1}{2} OPT$.

Proof. Suppose by way of contradiction that no such ℓ exists. Then by the triangle inequality, $w(c_\ell, c_m) < OPT$ for all $\{\ell, m\} \in \binom{[s]}{2}$. But this contradicts the optimality of S, since it implies that every edge in S has weight strictly less than OPT. Thus we conclude that the set $S - T$ does indeed contain a vertex c_ℓ satisfying the statement of the lemma.   □

Lemmas 4, 5, 6, 7, and 8 together imply Lemma 3, which in turn implies Theorem 2, stating that d-GREEDY AUGMENT has approximation ratio $(2p - 2)/(p + d - 2)$.

3   Conclusion

For the remote-clique problem, we have shown that the algorithm d-GREEDY AUGMENT achieves an approximation ratio of $(2p - 2)/(p + d - 2)$ and that this ratio is a tight bound on the worst-case performance of the algorithm. When $d = 1$, the algorithm is equivalent to the greedy algorithm analyzed in [2], in which it is proved that the algorithm has an approximation ratio of 4. By using factor-revealing linear programs, we have been able to improve the analysis of this algorithm to show that it in fact achieves an approximation ratio of 2. The usefulness of factor-revealing LPs in the analysis of d-GREEDY AUGMENT suggests possible applicability of this technique to the study of other algorithms whose analysis involves multiple inequalities that interact in subtle ways.

Acknowledgements

The authors thank Jon Turner for many helpful discussions related to this research.

References

1. Kuby, M.: Programming models for facility dispersion: the p-dispersion and max-isum dispersion problems. Geographical Analysis **19** (1987) 315–329
2. Ravi, S.S., Rosencrantz, D.J., Tayi, G.K.: Heuristic and special case algorithms for dispersion problems. Operations Research **42** (1994) 299–310
3. Chandra, B., Halldorsson, M.M.: Approximation algorithms for dispersion problems. J. Algorithms **38** (2001) 438–465
4. Hassin, R., Rubinstein, S., Tamir, A.: Approximation algorithms for maximum dispersion. Operations Research Letters **21** (1997) 133–137
5. Jain, K., Mahdian, M., Markakis, E., Saberi, A., Vazirani, V.V.: Greedy facility location algorithms analyzed using dual fitting with factor-revealing LP. J. ACM **50** (2003) 795–824
6. Mehta, A., Saberi, A., Vazirani, U., Vazirani, V.: Adwords and generalized on-line matching. In: FOCS '05: Proceedings of the 46th Annual IEEE Symposium on Foundations of Computer Science, Washington, DC, USA, IEEE Computer Society (2005) 264–273
7. Birnbaum, B.E.: The remote-clique problem revisited: Undergraduate honors thesis. Technical Report WUCSE-2006-26, Washington University Department of Computer Science and Engineering, St. Louis, MO, USA (2006)

Tight Results on Minimum Entropy Set Cover

Jean Cardinal[1], Samuel Fiorini[2], and Gwenaël Joret[1],*

[1] Computer Science Department
Université Libre de Bruxelles CP 212
B-1050 Brussels, Belgium
{jcardin, gjoret}@ulb.ac.be
[2] Department of Mathematics
Université Libre de Bruxelles CP 216
B-1050 Brussels, Belgium
sfiorini@ulb.ac.be

Abstract. In the minimum entropy set cover problem, one is given a collection of k sets which collectively cover an n-element ground set. A feasible solution of the problem is a partition of the ground set into parts such that each part is included in some of the k given sets. The goal is to find a partition minimizing the (binary) entropy of the corresponding probability distribution, i.e., the one found by dividing each part size by n. Halperin and Karp have recently proved that the greedy algorithm always returns a solution whose cost is at most the optimum plus a constant. We improve their result by showing that the greedy algorithm approximates the minimum entropy set cover problem within an additive error of 1 nat = $\log_2 e$ bits $\simeq 1.4427$ bits. Moreover, inspired by recent work by Feige, Lovász and Tetali on the minimum sum set cover problem, we prove that no polynomial-time algorithm can achieve a better constant, unless P = NP. We also discuss some consequences for the related minimum entropy coloring problem.

1 Introduction

Let V be an n-element ground set and $\mathscr{S} = \{S_1, \ldots, S_k\}$ be a collection of subsets of V whose union is V. A *cover* is an assignment $f : V \to \mathscr{S}$ of each point of V to a set of $\mathscr{S}$ such that $v \in f(v)$ for all $v \in V$. For each $i = 1, \ldots, k$, we let $q_i = q_i(f)$ denote the fraction of points assigned by f to the i-th set of $\mathscr{S}$, i.e.,

$$q_i := \frac{|f^{-1}(S_i)|}{n}. \tag{1}$$

The *minimum entropy set cover problem* (MESC) asks to find a cover f minimizing the entropy of the distribution $(q_1, \ldots, q_k)$. Letting $\mathrm{ENT}(f)$ denote this latter quantity, we have

$$\mathrm{ENT}(f) := - \sum_{i=1}^{k} q_i \log q_i. \tag{2}$$

* Research Fellow of the Fonds National de la Recherche Scientifique (FNRS).

J. Diaz et al. (Eds.): APPROX and RANDOM 2006, LNCS 4110, pp. 61–69, 2006.
© Springer-Verlag Berlin Heidelberg 2006

Note that, throughout, all logarithms are to base 2. Note also that, for definiteness, we set $x \log x = 0$ when $x = 0$.

The minimum entropy set cover problem is a NP-hard variant of the classical minimum cardinality set cover problem. Its recent introduction by Halperin and Karp [8] was motivated by various applications in computational biology. The problem is closely related to the minimum entropy coloring problem, which itself originates from the problem of source coding with side information in information theory, see Alon and Orlitsky [1].

The well-known greedy algorithm readily applies to MESC. It iteratively assigns to some set of $\mathscr{S}$ all unassigned points in that set, until all points are assigned. In each iteration, the algorithm choses a set that contains a maximum number of unassigned points. Halperin and Karp [8] studied the performance of the greedy algorithm for MESC. They proved that the entropy of the cover returned by the algorithm is at most the optimum plus some constant[1]. Approximations within an additive error are considered because the entropy is a logarithmic measure. In the case of MESC, the optimum value always lies between 0 and $\log n$.

In this paper, we revisit the greedy algorithm and give a simple proof that it approximates MESC within 1 *nat*, that is, $\log e \simeq 1.4427$ bits. We then show that the problem is NP-hard to approximate to within $(1 - \varepsilon) \log e$ for all positive ε. At the end of the paper, we discuss some consequences for the minimum entropy coloring problem.

At first sight, it might seem surprising that MESC can be approximated so well whereas its father problem, the minimum cardinality set cover problem, is notoriously difficult to approximate, see Feige [3]. We conclude the introduction by offering an intuitive explanation to this phenomenon. A consequential difference between the two problems is the penalty incurred for using too many sets. A minimum entropy cover is allowed to use a lot more sets than a minimum cardinality cover, provided the parts of these extra sets are small.

The same phenomenon also appears when one compares the minimum cardinality set cover problem to the *minimum sum set cover problem* (MSSC), see Feige, Lovász and Tetali [5]. The approximability status of the latter problem is similar to that of MESC: the greedy algorithm approximates it within a factor of 4 and achieving a factor of $4 - \varepsilon$ is NP-hard, for all positive ε. Furthermore, the techniques used here for proving the corresponding results on MESC are comparable to the ones used in [5] for MSSC, especially for the inapproximability result.

2   Analysis of the Greedy Algorithm

We begin this section by exhibiting a family of instances on which the greedy algorithm perfoms poorly, namely, returns a solution whose cost exceeds the optimum by roughly $\log e$ bits. Below, we use the following bounds on the factorial. These bounds are implied by the more precise bounds given, e.g., in [6].

---

[1] They claim that the greedy algorithm gives a 3 bits approximation (which is correct). However, their proof is flawed (e.g., see their Lemma 6). A straightforward fix gives an approximation guarantee of $3 + 2 \log e \simeq 5.8854$ bits.

Lemma 1. *For any positive integer ℓ, we have*

$$\left(\frac{\ell}{e}\right)^{\ell} < \ell! < 2\sqrt{2\pi\ell}\left(\frac{\ell}{e}\right)^{\ell}.$$

Let ℓ be a positive integer. We let the points of V be the cells of a $\ell \times \ell!$ array and $\mathscr{S}$ be the union of two collections $\mathscr{S}_{col}$ and $\mathscr{S}_{line}$ each of which partitions V. The sets in $\mathscr{S}_{col}$ are the $\ell!$ columns of the array. For each $i = 1, \ldots, \ell$, collection $\mathscr{S}_{line}$ contains $\ell!/i$ sets of size i which partition the i-th line of the array. An illustration is given in Figure 1. (While in the figure each set of $\mathscr{S}_{line}$ consists of contiguous cells, we do not require this in general.) Each of the collections $\mathscr{S}_{col}$ and $\mathscr{S}_{line}$ directly yields a feasible solution for MESC, which we denote respectively by f_{col} and f_{line}. Clearly, f_{line} is one of the possible outcomes of the greedy algorithm (sets are produced from bottom to top on Figure 1).

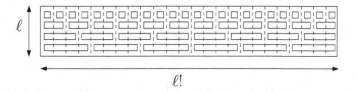

Fig. 1. The sets forming $\mathscr{S}_{line}$

The respective costs of f_{col} and f_{line} are as follows:

$$\text{ENT}(f_{col}) = -\sum_{j=1}^{\ell!} \frac{1}{\ell!} \log \frac{1}{\ell!} = \log \ell!,$$

$$\text{ENT}(f_{line}) = -\sum_{i=1}^{\ell} \frac{\ell!}{i} \frac{i}{\ell \cdot \ell!} \log \frac{i}{\ell \cdot \ell!} = \log \ell + \log \ell! - \frac{1}{\ell} \log \ell!.$$

By the second inequality of Lemma 1, we then have

$$\text{ENT}(f_{line}) \geq \log \ell + \log \ell! - \frac{1}{\ell} \log \left[2\sqrt{2\pi\ell}\left(\frac{\ell}{e}\right)^{\ell}\right] = \text{ENT}(f_{col}) + \log e - o(1).$$

This implies that the cost of f_{line} is at least the optimum plus $\log e - o(1)$. We now show that the previous instances are essentially the worst for the greedy algorithm. Because the two formulations of MESC given above are equivalent to each other, we can regard a cover f as a partition of the ground set. Accordingly, we refer to the sets $f^{-1}(S_i)$ as the *parts* of f.

Theorem 1. *Let f_{OPT} and f_G be a cover of minimum entropy and a cover returned by the greedy algorithm, respectively. Then we have $\text{ENT}(f_G) \leq \text{ENT}(f_{OPT}) + \log e$.*

Proof. For $i = 1, \ldots, k$, we let X_i denote the i-th part of f_{OPT} and $x_i = |X_i|$. For $v \in V$, we let a_v be the size of the part of f_G containing v. We claim that the following holds for all v and all i:

$$\prod_{v \in X_i} a_v \geq x_i!. \tag{3}$$

Let us consider the points of X_i in the order in which they were assigned to sets of $\mathscr{S}$ by the greedy algorithm, breaking ties arbitrarily. Consider the j-th element of X_i assigned, say v. In the iteration when v was assigned, the greedy algorithm could have picked set S_i. Because at that time at most $j - 1$ points of X_i were assigned, at least $x_i - j + 1$ points of S_i were unassigned, and we have $a_v \geq x_i - j + 1$. This implies the claim.

We now rewrite the entropy of f_G as follows:

$$\text{ENT}(f_G) = -\frac{1}{n} \sum_{v \in V} \log \frac{a_v}{n} = -\frac{1}{n} \sum_{i=1}^{k} \sum_{v \in X_i} \log \frac{a_v}{n} = -\frac{1}{n} \sum_{i=1}^{k} \log \prod_{v \in X_i} \frac{a_v}{n}.$$

By Inequality (3) and the first inequality of Lemma 1, we then have:

$$\text{ENT}(f_G) \leq -\frac{1}{n} \sum_{i=1}^{k} \log \frac{x_i!}{n^{x_i}} \leq -\frac{1}{n} \sum_{i=1}^{k} \log \frac{x_i^{x_i}}{n^{x_i} e^{x_i}} \leq \text{ENT}(f_{OPT}) + \log e.$$

$\square$

Finally, we mention that MESC has a natural weighted version in which each point $v \in V$ has some associated probability p_v. Again, we can associate to each cover f a probability distribution $(q_1, \ldots, q_k)$. This time, we let q_i denote the probability that a random point is assigned to S_i by f, that is,

$$q_i := \sum_{v \in f^{-1}(S_i)} p_v.$$

The goal is then to minimize (2), just as in the unweighted version. The greedy algorithm easily transposes to the weighted case, and so does our analysis. This is easily seen when the probabilities are rational. Indeed, let K be a positive integer such that Kp_v is integral for all points v. Now replicate each point in the ground set $Kp_v - 1$ times. Thus we obtain an unweighted instance which is equivalent to the original weighted instance, in the following sense. The optimum values of the two instances are equal (Lemma 2, given below, forbids replicated versions of a point to be assigned to different sets) and the behavior of the greedy algorithm on the new instance is identical to its behavior on the original instance. The case of real probabilities follows by a continuity argument.

3   Hardness of Approximation

Before turning to the main theorem of this section, we state a lemma which helps deriving good lower bounds on the optimum. Let $q = (q_i)$ and $r = (r_i)$ be two probability distributions over $\mathbb{N}^+$. If $\sum_{i=1}^{\ell} r_i \geq \sum_{i=1}^{\ell} q_i$ holds for all ℓ, we say that q is *dominated* by r. The lemma tells us that in such a case, the entropy of q is at least that of r, provided that q is non-increasing (see, e.g., [9] for a proof).

Lemma 2. *Let* $q = (q_i)$ *and* $r = (r_i)$ *be two probability distributions over* $\mathbb{N}^+$ *with finite support. Assume that* q *is non-increasing, that is,* $q_i \geq q_{i+1}$ *for* $i \geq 1$. *If* q *is dominated by* r, *then we have* $ENT(q) \geq ENT(r)$.

We now prove that no polynomial-time algorithm for MESC can achieve a better constant approximation guarantee than the greedy algorithm, unless P = NP. Halperin and Karp [8] gave a polynomial time approximation scheme (PTAS) for the problem. Our result does not contradict theirs since the PTAS they designed is multiplicative, i.e., returns a solution whose cost is most $(1 - \varepsilon)$ times the optimum.

Theorem 2. *For every* $\varepsilon > 0$, *it is NP-hard to approximate the minimum entropy set cover problem within an additive term of* $(1 - \varepsilon) \log e$. *This remains true on instances such that every point is in the same number of sets and every set has the same size.*

Proof. A *3SAT-6 formula* is a CNF formula in which every clause contains exactly three literals, every litteral appears in exactly three clauses, and a variable appears at most once in each clause. Such a formula is said to be δ-*satisfiable* if at most a δ-fraction of its clauses are satisfiable. It is known that distinguishing between a satisfiable 3SAT-6 formula and one which is δ-satisfiable is NP-hard for some δ with $0 < \delta < 1$, see Feige et al. [5]. In the latter reference, the authors slightly modified a reduction due to Feige [3] to design a polynomial-time reduction associating to any 3SAT-6 formula φ a corresponding set system $\mathbf{S}(\varphi) = (V, \mathscr{S})$. They used the new reduction to prove that the minimum sum set cover problem is NP-hard to approximate to within $2 - \varepsilon$ on uniform regular hypergraphs (see Theorem 12 in that paper). For any given constants $c > 0$ and $\lambda > 0$, it is possible to set the values of the parameters of the reduction in such a way that:

- the sets of $\mathscr{S}$ have all the same size n/t, where n denotes the number of points in V, and every point of V is contained in the same number of sets;
- if φ is satisfiable, then V can be covered by t disjoint sets of $\mathscr{S}$;
- if φ is δ-satisfiable, then every i sets chosen from $\mathscr{S}$ cover at most a $1 - (1 - 1/t)^i + \lambda$ fraction of the points of V, for $1 \leq i \leq ct$.

Suppose from now on that φ is a 3SAT-6 formula which is either satifiable or δ-satisfiable, and denote by f_{OPT} an optimal solution of MESC with input $\mathbf{S}(\varphi)$. For $1 \leq i \leq k$, let $q_i = q_i(f_{OPT})$ be defined as in (1). For $i > k$, we let $q_i = 0$. Letting q denote the sequence (q_i), we assume without loss of generality that q is non-increasing.

If φ is satisfiable, then it follows from Lemma 2 that the optimal solution consists in covering V with t disjoint sets. Hence, $ENT(f_{OPT}) = ENT(q) = \log t$ in this case. Assume now that φ is δ-satisfiable. Let $\alpha = \varepsilon/2$, $\lambda = \alpha^2/2 - \alpha^3/6$ and $c = -\ln \lambda$.

Claim 1. *The following lower bound on the optimum holds:*

$$ENT(q) \geq \log t + (1 - \varepsilon/2) \log e + o(1),$$

where $o(1)$ *tends to zero when* t *tends to infinity.*

Claim 1 implies that any algorithm approximating MESC within an additive term of $(1 - \varepsilon) \log e$ can be used to decide whether φ is satisfiable or δ-satisfiable. Indeed, as noted in [5], t may be assumed to be larger than any fixed constant. The theorem then follows.

In order to prove the claim, we define a sequence $r = (r_i)$ as follows (see Figure 2 for an illustration):

$$r_i = \begin{cases} 1/t & \text{for } 1 \le i \le \lceil \alpha t \rceil, \\ (1 - 1/t)^{i-1}/t & \text{for } \lceil \alpha t \rceil + 1 \le i \le \lfloor \tilde{c} t \rfloor, \\ 1 - \sum_{i=1}^{\lfloor \tilde{c} t \rfloor} r_i & \text{for } i = \lfloor \tilde{c} t \rfloor + 1, \\ 0 & \text{otherwise,} \end{cases}$$

where $\tilde{c}$ is a real such that

$$\frac{\lceil \alpha t \rceil}{t} + (1 - 1/t)^{\lceil \alpha t \rceil} - (1 - 1/t)^{\tilde{a}} = 1. \tag{4}$$

By our choice of parameters, we can assume $\lceil \alpha t \rceil + 1 \le \lfloor \tilde{c} t \rfloor$ by lowering ε if necessary. From the definition of $\tilde{c}$ we have

$$\sum_{i=1}^{\lfloor \tilde{c} t \rfloor} r_i = \frac{\lceil \alpha t \rceil}{t} + (1 - 1/t)^{\lceil \alpha t \rceil} - (1 - 1/t)^{\lfloor \tilde{a} \rfloor} \le 1.$$

Therefore, the sequence r is a probability distribution over $\mathbb{N}^+$.

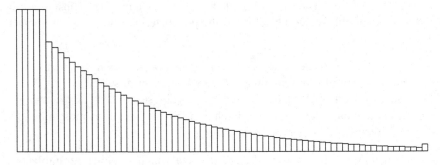

Fig. 2. The shape of distribution $r = (r_i)$ for $t = 20$ and $\varepsilon = 1/2$

By the properties of $\mathbf{S}(\varphi)$ we have

$$\sum_{i=1}^{\ell} q_i \le \ell/t \quad \text{and} \quad \sum_{i=1}^{\ell} q_i \le 1 - (1 - 1/t)^{\ell} + \lambda \tag{5}$$

for $1 \le \ell \le \lfloor ct \rfloor$, and it can be checked that $\tilde{c} \le c$ for t large enough.

Claim 2. *Sequence q is dominated by sequence r, that is, for all ℓ we have*

$$\sum_{i=1}^{\ell} q_i \le \sum_{i=1}^{\ell} r_i. \tag{6}$$

For $1 \leq \ell \leq \lceil \alpha t \rceil$, Inequality (6) readily follows from the definition of r and Equation (5). Notice that we have

$$1 - (1 - 1/t)^{\lceil \alpha t \rceil} + \lambda \leq 1 - (1 - \alpha + \alpha^2/2 - \alpha^3/6) + \lambda = \alpha \leq \lceil \alpha t \rceil / t \quad (7)$$

whenever t is large enough. Hence, for $\lceil \alpha t \rceil + 1 \leq \ell \leq \lfloor \tilde{c} t \rfloor$, from Equations (5) and (7) we derive

$$\sum_{i=1}^{\ell} q_i \leq 1 - (1 - 1/t)^{\ell} + \lambda = 1 - (1 - 1/t)^{\lceil \alpha t \rceil} + \lambda + \sum_{i=\lceil \alpha t \rceil + 1}^{\ell} (1 - 1/t)^{i-1}/t$$

$$\leq \lceil \alpha t \rceil / t + \sum_{i=\lceil \alpha t \rceil + 1}^{\ell} (1 - 1/t)^{i-1}/t = \sum_{i=1}^{\ell} r_i.$$

Finally, note that (6) is also true for $\ell > \lfloor \tilde{c} t \rfloor$, as the q_i's and r_i's both sum up to 1. It follows that q is dominated by r. In other words, Claim 2 holds true. By Lemma 2, we have $\mathrm{ENT}(q) \geq \mathrm{ENT}(r)$. In order to show Claim 1, it then suffices to prove the following claim.

Claim 3. *We have* $\mathrm{ENT}(r) \geq \log t + (1 - \varepsilon/2) \log e + o(1)$.

The entropy of r can be expressed as follows:

$$\mathrm{ENT}(r) = - \sum_{i=1}^{\lfloor \tilde{c} t \rfloor + 1} r_i \log r_i = - \sum_{i=1}^{\lfloor \tilde{c} t \rfloor} r_i \log r_i + o(1)$$

$$= \frac{\lceil \alpha t \rceil}{t} \log t - \sum_{i=\lceil \alpha t \rceil + 1}^{\lfloor \tilde{c} t \rfloor} \frac{(1 - 1/t)^{i-1}}{t} \log \frac{(1 - 1/t)^{i-1}}{t} + o(1)$$

$$= \alpha \log t + \frac{1}{t} \log \frac{t}{t-1} \sum_{i=\lceil \alpha t \rceil + 1}^{\lfloor \tilde{c} t \rfloor} (i-1)(1 - 1/t)^{i-1}$$

$$+ \frac{1}{t} \log t \sum_{i=\lceil \alpha t \rceil + 1}^{\lfloor \tilde{c} t \rfloor} (1 - 1/t)^{i-1} + o(1).$$

Let $\beta := \lim_{t \to \infty} \tilde{c}$. In the sum above, the second and third terms are asymptotically equal to respectively $\log e \cdot ((1 + \alpha)e^{-\alpha} - (1 + \beta)e^{-\beta})$ and $\log t \cdot (e^{-\alpha} - e^{-\beta})$ (proofs are omitted). It follows from Equation (4) that

$$\beta = -\ln(\alpha + e^{-\alpha} - 1).$$

In virtue of this equation and by what precedes, we can rewrite the entropy of r as

$$\mathrm{ENT}(r) = \alpha \log t + \log e \cdot ((1 + \alpha)e^{-\alpha} - (1 + \beta)e^{-\beta}) + \log t \cdot (e^{-\alpha} - e^{-\beta}) + o(1)$$

$$= (\alpha + e^{-\alpha} - e^{-\beta}) \log t + ((1 + \alpha)e^{-\alpha} - (1 + \beta)e^{-\beta}) \log e + o(1)$$

$$= \log t + ((1 + \alpha)e^{-\alpha} - (1 + \beta)e^{-\beta}) \log e + o(1).$$

We leave it to the reader to show that $\alpha e^{-\alpha} - \beta e^{-\beta}$ is nonnegative provided ε is sufficiently small. Claim 3 follows then by noticing

$$(1+\alpha)e^{-\alpha} - (1+\beta)e^{-\beta} = 1 - \alpha + \alpha e^{-\alpha} - \beta e^{-\beta} \geq 1 - \alpha = 1 - \varepsilon/2.$$

Hence, Claim 1 and the theorem follow.   □

4   Graph Colorings with Minimum Entropy

There are situations where the collection of sets $\mathscr{S} = \{S_1, \ldots, S_k\}$ input to the minimum entropy set cover problem is given implicitly. One possibility, which is the focus of this section, is to define $\mathscr{S}$ as the collection of all inclusion-wise maximal stable sets of some (simple, undirected) graph $G = (V, E)$. The corresponding variant of MESC is known as the *minimum entropy coloring problem* (MEC). It stems from information theory, having applications in zero-error coding with side information [1]. Notice that, by our choice of $\mathscr{S}$, every cover f can be regarded as a (proper) coloring of the graph G.

The results of Section 2 directly apply to MEC. The greedy algorithm, transposed to the setting of MEC, constructs a coloring of G by iteratively removing a maximum size stable set from G. Of course, its running time can no longer be guaranteed to be polynomial, unless P = NP. Theorem 1 implies the following result, which again holds in the weighted case.

Corollary 1. *Let f_{OPT} and f_G be a coloring of G with minimum entropy and a coloring returned by the greedy algorithm, respectively. Then we have $ENT(f_G) \leq ENT(f_{OPT}) + \log e$.*

The bound given in Corollary 1 is asymptotically tight because the bad MESC instances described in the beginning of Section 2 can be easily turned into MEC instances. Indeed, for a given ℓ, it suffices to consider the graph G obtained from the complete graph on V by removing every edge which is entirely included in some set of $\mathscr{S}_{col}$ or $\mathscr{S}_{line}$.

Clearly, the greedy algorithm runs in polynomial time when restricted to graphs in which a maximum weight stable set can be found in polynomial time. This includes perfect graphs [7] and claw-free graphs [10]. So MEC can be approximated within an additive term of $\log e$ on such graphs, in polynomial time. In contrast, for arbitrary graphs it is known that for any $\varepsilon > 0$ there is no polynomial-time approximation algorithm whose additive error is bounded by $(1 - \varepsilon) \log n$ unless ZPP=NP. This was proved by the authors in [2] using as a black-box an inapproximability result for the minimum cardinality coloring problem due to Feige [4].

References

[1] N. Alon and A. Orlitsky. Source coding and graph entropies. *IEEE Trans. Inform. Theory*, 42(5):1329–1339, September 1996.
[2] J. Cardinal, S. Fiorini, and G. Joret. Minimum entropy coloring. In *Proceedings of the 16th International Symposium on Algorithms and Computation (ISAAC 2005)*, volume 3827 of *Lecture Notes in Computer Science*, pages 819–828, Berlin, 2005. Springer.

[3] U. Feige. A threshold of ln n for approximating set cover. *J. ACM*, 45(4):634–652, 1998.

[4] U. Feige and J. Kilian. Zero knowledge and the chromatic number. *J. Comput. System Sci.*, 57(2):187–199, 1998. Complexity 96—The Eleventh Annual IEEE Conference on Computational Complexity (Philadelphia, PA).

[5] U. Feige, L. Lovász, and P. Tetali. Approximating min sum set cover. *Algorithmica*, 40(4): 219–234, 2004.

[6] W. Feller. *An introduction to probability theory and its applications. Vol. I*. Third edition. John Wiley & Sons Inc., New York, 1968.

[7] M. Grötschel, L. Lovász, and A. Schrijver. *Geometric algorithms and combinatorial optimization*, volume 2 of *Algorithms and Combinatorics*. Springer-Verlag, Berlin, second edition, 1993.

[8] E. Halperin and R. M. Karp. The minimum-entropy set cover problem. *Theoret. Comput. Sci.*, 348(2-3):240–250, 2005.

[9] G. H. Hardy, J. E. Littlewood, and G. Pólya. *Inequalities*. Cambridge Mathematical Library. Cambridge University Press, Cambridge, 1988. Reprint of the 1952 edition.

[10] D. Nakamura and A. Tamura. A revision of Minty's algorithm for finding a maximum weight stable set of a claw-free graph. *J. Oper. Res. Soc. Japan*, 44(2):194–204, 2001.

A Tight Lower Bound for
the Steiner Point Removal Problem on Trees

T.-H. Hubert Chan[1,*], Donglin Xia[2,**],
Goran Konjevod[2,**], and Andrea Richa[3,***]

[1] Computer Science Department, Carnegie Mellon University
hubert@cs.cmu.edu
[2] Department of Computer Science and Engineering, Arizona State University
{dxia, goran}@asu.edu
[3] Department of Computer Science and Engineering, Arizona State University
aricha@asu.edu

Abstract. Gupta (SODA'01) considered the Steiner Point Removal (SPR) problem on trees. Given an edge-weighted tree T and a subset S of vertices called *terminals* in the tree, find an edge-weighted tree T_S on the vertex set S such that the distortion of the distances between vertices in S is small. His algorithm guarantees that for any finite tree, the distortion incurred is at most 8. Moreover, a family of trees, where the leaves are the terminals, is presented such that the distortion incurred by any algorithm for SPR is at least $4(1 - o(1))$. In this paper, we close the gap and show that the upper bound 8 is essentially tight. In particular, for complete binary trees in which all edges have unit weight, we show that the distortion incurred by any algorithm for the SPR problem must be at least $8(1 - o(1))$.

1 Introduction

The Steiner Point Removal (SPR) problem was first considered by Gupta [1]. An instance of the problem is given by an edge-weighted tree $T = (V, E)$ and a subset $S \subseteq V$ of vertices called *terminals*. Informally, we would like to find an edge-weighted tree T_S on the terminal set S such that the new tree approximates all the distances between terminal pairs in the original tree. Formally, we say that a weighted tree T_S on the set S has *distortion* at most α if for all $u, v \in S$, the condition $d_T(u, v) \leq d_{T_S}(u, v) \leq \alpha \cdot d_T(u, v)$ holds, where $d_G(u, v)$ is the shortest path distance between two nodes u and v in the graph G. We say an instance has *distortion* at most α if such a tree T_S exists. The objective is to find the smallest constant $\alpha > 0$ such that every instance of the SPR Problem has distortion at most α.

In Gupta's original paper [1], it was shown that $\alpha \leq 8$, i.e., there exists a tree T_S with distortion at most 8. This shows that any submetric of a tree metric is

* Supported in part by the NSF CAREER award CCF-0448095, by an Alfred P. Sloan Fellowship, and by a fellowship from the Croucher Foundation.
** Supported in part by NSF grant CCR-0209138.
*** Supported in part by NSF CAREER grant CCR-9985284.

J. Diaz et al. (Eds.): APPROX and RANDOM 2006, LNCS 4110, pp. 70–81, 2006.
© Springer-Verlag Berlin Heidelberg 2006

"close" to a tree metric. Such a result leads to the first combinatorial proof of the fact that a graph of girth g embeds into a tree with distortion at least $\Omega(g)$, as opposed to the topological proof given by Rabinovich and Raz [2].

Moreover, such a result has potential applications in end system multicast [3, 4, 5, 6]. In a multicast routing protocol, a routing tree $T = (V, E)$ is defined on hosts S, which correspond to the terminals, and routers that connect the hosts and forward messages. The edges represent connections between hosts and routers, and their weights correspond to transmission costs. However, most routers are designed to handle only unicast, and hence a virtual routing tree T_S consisting of only the hosts is suggested for implementing the multicast protocol. Thus, it is important that the virtual tree T_S approximates the original costs well, which is ensured by the upper bound result.

The result has also been used subsequently for embedding k-outerplanar metrics into ℓ_1 by Chekuri et al. [7], embedding general metrics into distributions of tree metrics by Fakcharoenphol et al. [8], and solving the metric labeling problem via tree-rounding by Archer et al. [9].

A natural question to ask is whether the upper bound of 8 is tight. The original paper [1] only gives a lower bound of $4(1 - o(1))$ for some family of trees. In this paper, we close this gap and prove the following theorem showing that the upper bound of 8 is essentially tight.

Theorem 1. *For any $\epsilon > 0$, there exists an instance of the Steiner Point Removal Problem with distortion at least $8 - \epsilon$.*

We anticipate that the techniques presented in this paper may also be applicable to the several open problems in this area, in particular, to the open problems listed in Section 5.

1.1   Proof Strategy

Our lower bound examples will be complete binary trees with unit-weight edges, with the leaves being the terminals. We first show in Section 3 that as far as complete binary trees are concerned, the optimal distortion can always be achieved by a *minor* T_S of the original tree $T = (V, E)$, i.e., the tree T_S can be obtained by contracting edges of tree T of the following form: (1) an edge between two non-terminals; (2) an edge between a terminal x and a non-terminal node y, with the resulting merged node keeping the same name (and terminal status) as x. The weight assigned to each edge (x, y) in T_S will be $d_T(x, y)$, the distance between its two endpoints in the original tree T. Note that each node in V will eventually be contracted into a terminal in S. Thus the minor tree T_S can also be characterized by a mapping $f : V \to S$ that maps each vertex in V to the terminal in S to which it eventually contracts. We call such a mapping f a *minor mapping*.

In Section 4, we show that there exists a complete binary tree such that its minors must incur a large distortion, namely $8-o(1)$. Let us define some notation before giving the general idea on how one can get such a lower bound:

1. Denote by T_n the complete binary tree of height n, having 2^n leaves, with unit-weight edges, and denote by r_n the root of T_n.

2. *Expanding Parameter* $\rho_f(r)$: Suppose the tree T has its root r mapped under f to leaf l, i.e. $l = f(r)$. Suppose that w is a vertex furthest away from the root r in the subtree rooted at the child of r that is not an ancestor of l and $f(w) = l$. Set w to be r if no such vertex exists. The *expanding parameter* $\rho_f(r)$ at r with respect to f is defined to be the ratio $d_T(r,w)/d_T(r,l)$. See Figure 1(a).

3. For each complete binary tree T_n, let ρ_n be the maximum $\rho_f(r_n)$ for all the minor mappings f for T_n with distortion no more than α. Then define
$$\rho := \limsup_{n\to\infty} \rho_n.$$

First we show that $0 < \rho < 1$ (See Claims 4 and 4.). Thus there exists an arbitrarily small constant $\epsilon_1 > 0$ such that $0 < \rho - \epsilon_1 < \rho + \epsilon_1 < 1$. Then by the definition of ρ, there exists an arbitrarily large integer m such that $\rho - \epsilon_1 < \rho_m < \rho + \epsilon_1$. Now consider the complete binary tree T_m and the minor mapping f with distortion no more than α that achieves $\rho_f(r_m) = \rho_m$. As shown in Figure 1(a), let w be the lowest vertex that achieves the *expanding parameter* $\rho_f(r_m)$, vertices x and y be the children of vertex w, and $T(x)$ and $T(y)$ be subtrees rooted at x and y respectively.

The idea is to find leaves p and q in the subtrees $T(x)$ and $T(y)$ respectively such that the distortion exhibited by the pair (p, q) is large. First observe that the distance in T_m between any leaf in $T(x)$ and any leaf in $T(y)$ is $2m(1 - \rho_m) < 2m(1 - (\rho - \epsilon_1))$. Next, we want to argue that there is a leaf p in the subtree $T(x)$ such that the distance between p and $f(r_m)$ in the minor tree $f(T_m)$ is larger than $\frac{2m}{\rho + \epsilon_1}(1 - \epsilon_2)$ for any constant $\epsilon_2 > 0$ if m is large enough. Symmetrically, we can also find such a leaf q in the subtree $T(y)$, thereby the distance between p and q in the minor tree $f(T_m)$ is larger than $\frac{4m}{\rho + \epsilon_1}(1 - \epsilon_2)$. Therefore the distortion according the minor mapping f must be larger than $\frac{2}{(1-(\rho-\epsilon_1))(\rho+\epsilon_1)}(1 - \epsilon_2) \geq 8(1 - \epsilon_2)$. Since the distortion of f is no more than α, we get the lower bound $\alpha > 8 - o(1)$.

We still need to determine how to find such a leaf p in the subtree $T(x)$. We will use a recursive algorithm on the roots of the subtrees considered, starting with the subtree $T(x)$. First we limit p to be one of the leaves in $T(x)$, whose distances to $f(r_m)$ in T_m are all $2m$. Then, we limit p to be one of the leaves of $T(x)$ in the subtree of x that does not contain $f(x)$; the distances of those leaves to $f(x)$ in T_m are all $2m(1 - \rho_m) - 2 \gtrsim 2m(1 - (\rho + \epsilon_1))$. In general, as shown in Figure 1(b), we limit p to be one of the leaves of the subtree of $T(z)$ (initially $z = x$) that does not contain $f(z)$; we then let z be the root of the corresponding subtree, and recurse. Roughly speaking, the heights of these trees are no less than m, $m(1 - (\rho + \epsilon_1))$, $m(1 - (\rho + \epsilon_1))^2$, $m(1 - (\rho + \epsilon_1))^3$, $\cdots$, respectively, if m is large enough (See Lemma 1 for a formal proof). Thus the distance between p and $f(r_m)$ in the minor tree $f(T_m)$ must be larger than $\frac{2m}{\rho+\epsilon_1}(1 - \epsilon_2)$, where $\epsilon_2 > 0$ can be any constant and m is large enough. Therefore our algorithm finds such a leaf p, and it follows that $\alpha > 8 - o(1)$.

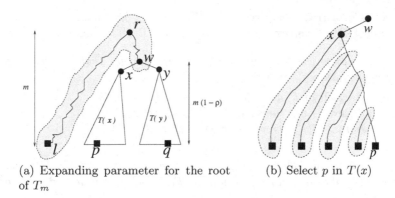

(a) Expanding parameter for the root of T_m

(b) Select p in $T(x)$

Fig. 1. The Minor Construction for Tree T_m (Shadow areas refer to components contracted to a terminal)

2 Notation

In this section, we will introduce and formalize some additional notation that will be used in Sections 3 and 4. Suppose T is a tree with edge set E and a positive distance associated with each edge. We denote the distance of the unique shortest path between two vertices u and v by $d_T(u, v)$. We use $\mathsf{L}(T)$ to denote the set of leaves, i.e. the degree-one vertices in T.

As defined in Section 1.1, we denote by T_n the complete binary tree of height n, having 2^n leaves with unit weight edges. We denote by r_n the root of T_n and the terms *child, parent, ancestor and descendant* are used with their usual meanings. From now on, we restrict the SPR Problem to such trees, with the leaves being the terminals.

Formally, we say f is a transformation from T to $\widehat{T}$, if $\widehat{T} = (\mathsf{L}(T), \widehat{E})$ is a tree on the vertex set $\mathsf{L}(T)$, and each edge $(u, v) \in \widehat{E}$ has weight $d_T(u, v)$. The distortion of such a transformation is

$$\mathsf{D}(f) := \max_{x \neq y \in \mathsf{L}(T)} \frac{d_{\widehat{T}}(x, y)}{d_T(x, y)}.$$

A transformation f from T to $\widehat{T}$ is *minor* if $\widehat{T}$ is a minor of T, i.e. $\widehat{T}$ can be obtained from T by edge contractions. Note that a minor transformation f for a tree T can be equivalently viewed as a mapping $f : \mathsf{V}(T) \rightarrow \mathsf{L}(T)$ that maps each vertex to the terminal to which it eventually contracts. We call such f a *minor mapping*.

3 Restricting to Minor Transformations

In this section, we show that in order to obtain a lower bound on the distortion of transformations for complete binary trees, it suffices to consider minor transformations.

The radius of a tree T is given by $R(T) = \min_{u \in V(T)} \max_{v \in V(T)} d_T(u, v)$. A *center point* of T is a vertex $u_0 \in V(T)$ such that $R(T) = \max_{v \in V(T)} d_T(u_0, v)$.

Theorem 2. *For any $n \geq 0$ and for any transformation f of T_n, there exists a minor transformation f' such that*

   (a) *the distortion of f' does not increase, $D(f') \leq D(f)$;*
   (b) *the radius does not increase, $R(f'(T_n)) \leq R(f(T_n))$;*
   (c) *the terminal $f'(r_n)$ is a center point of $f'(T_n)$.*

Proof: We argue by induction on n. The case $n = 0$ is trivial. For the case $n = 1$, there is only one transformation for T_1, which is minor and satisfies the requirements.

Assume the result holds true for any T_k, where $k < n$. Consider some transformation $f : T_n \to \widehat{T}_n$.

We denote by $[n]$ the set of integers $\{0, 1, \dots, n\}$.

For any $x \in L(T_n)$ and $i \in [n]$, denote by $T_i(x)$ the i-level complete binary subtree of T_n which contains x; denote the root of $T_i(x)$ by $r_i(x)$. For any $x \in L(T_n)$ and $i \in [n]$, denote by $S_i(x)$ the minimal subtree of $\widehat{T}_n$ that includes all the vertices in $L(T_i(x))$. Let k be the maximum integer such that for any $x \in L(T_n)$, $V(S_k(x)) \subseteq L(T_{n-1}(x))$. Since $k = 0$ satisfies the above conditions, such a k exists. Note that $k < n$; otherwise, $L(T_n(x)) \subseteq V(S_k(x)) \subseteq L(T_{n-1}(x))$, which is a contradiction.

From the maximality of k, there exists $u \in L(T_n)$ such that $V(S_{k+1}(u)) \nsubseteq L(T_{n-1}(u))$. Also, there exists $v \in L(T_{k+1}(u))$ such that $T_k(v) \neq T_k(u)$ and the u-v path in $\widehat{T}_n$ uses some vertex not in $L(T_{n-1}(u))$. Let vertex $w \notin L(T_{n-1}(u))$ be the first such vertex on the path from u to v, and $u' \in L(T_{n-1}(u))$ be the previous vertex of w on the path. Since $T_{n-1}(u') \neq T_{n-1}(w)$, it follows that (u', w) has weight $2n$.

Claim. Edge (u', w) is an edge of weight $2n$ that separates $S_k(u)$ and $S_k(v)$ in $\widehat{T}_n$.

Proof of Claim 3: By the definition of k, $V(S_k(u)) \subseteq L(T_{n-1}(u))$ and $V(S_k(v)) \subseteq L(T_{n-1}(v))$. Since $w \notin L(T_{n-1}(u))$, edge (u', w) separates $S_k(u)$ and v. Since $u' \in L(T_{n-1}(u))$ and $w \notin L(T_{n-1}(u))$, then exactly one of them is not in $L(T_{n-1}(v))$. Since $V(S_k(v)) \subseteq L(T_{n-1}(v))$, edge (u', w) separates $S_k(v)$ and u. Therefore edge (u', w) separates $S_k(u)$ and $S_k(v)$. $\square$

Thus in the tree $\widehat{T}_n$, there is a unique path connecting $S_k(u)$ and $S_k(v)$ with all its intermediate vertices not in $V(S_k(u)) \cup V(S_k(v))$. Let $u_0 \in V(S_k(u))$ and $v_0 \in V(S_k(v))$ be the two endpoints of the path. Then, vertex w is on the u_0-v_0 path and $d_{\widehat{T}_n}(u_0, w) \geq 2n$.

If $k + 1 < n$, then $v \in L(T_{k+1}(u)) \subseteq L(T_{n-1}(u))$, thereby $d_{\widehat{T}_n}(v_0, w) \geq 2n$; if $k + 1 = n$, we have the trivial bound $d_{\widehat{T}_n}(v_0, w) \geq 0$.

Consider vertices $u_1 \in V(S_k(u))$ and $v_1 \in V(S_k(v))$, which are furthest away from u_0 and v_0 respectively. Hence, we have $d_{\widehat{T}_n}(u_0, u_1) \geq R(S_k(u))$ and $d_{\widehat{T}_n}(v_0, v_1) \geq R(S_k(v))$. Without loss of generality, assume $R(S_k(u)) \leq R(S_k(v))$.

Observing that $d_{\widehat{T}_n}(u_1, v_1) = d_{\widehat{T}_n}(u_1, u_0) + d_{\widehat{T}_n}(u_0, w) + d_{\widehat{T}_n}(w, v_0) + d_{\widehat{T}_n}(v_0, v_1)$, we have

$$D(f) \geq \frac{d_{\widehat{T}_n}(u_1, v_1)}{d_{T_n}(u_1, v_1)} \geq \begin{cases} \frac{4n + 2R(S_k(u))}{2(k+1)} & \text{if } k+1 < n; \\ \frac{2n + 2R(S_k(u))}{2(k+1)} & \text{if } k+1 = n \end{cases} \tag{3.1}$$

Also,

$$R(f(T_n)) \geq 2n + R(S_k(u)) \tag{3.2}$$

Next, we construct a transformation g for the subtree $T_k(u)$. We obtain the transformed tree $\widehat{T}_k(u)$ from $S_k(u)$, the minimal subtree in $\widehat{T}_n$ containing $L(T_k(u))$, by contracting all the vertices $v \notin L(T_k(u))$ as follows:

1. Contract any edge neither of whose endpoints is in $L(T_k(u))$.
2. For each remaining vertex $x \notin L(T_k(u))$, contract one of the edges incident to x.
3. For each edge (x, y) in $\widehat{T}_k(u)$ set its weight as $d_{T_k(u)}(x, y)$, i.e. $d_{T_n}(x, y)$.

The following claim states the properties of the transformation g. Its proof is technical and will be deferred to the end of the section.

Claim. Suppose the transformation g from $T_k(u)$ to the tree $\widehat{T}_k(u) = (L(T_k(u)), \widehat{E})$ is as described above. Then, the distortion $D(g) \leq D(f)$ and the radius
$R(g(T_k(u))) \leq R(S_k(u))$.

By the induction hypothesis , there exists a minor transformation g' for $T_k(u)$ such that $D(g') \leq D(g)$, $R(g'(T_k(u))) \leq R(g(T_k(u)))$, and $r_k(u)$ is contracted into a center point of $g'(T_k(u))$. By Claim 3, we also have $D(g) \leq D(f)$ and $R(g(T_k(u))) \leq R(S_k(u))$. Hence, we have $D(g') \leq D(f)$ and $R(g'(T_k(u))) \leq R(S_k(u))$.

We next use the transformation g' to construct a minor transformation f' for T_n. Since all the k-level complete binary subtrees T_k of T_n are isomorphic to $T_k(u)$, the transformation g' also defines a minor transformation for each of these subtrees T_k. Then a minor transformation f' for T_n can be obtained by edge contractions as follows:

1. Remove internal nodes in each T_k via edge contraction using minor transformation g'.
2. Since the $(n - k - 1)$-level complete binary subtree rooted at r_n is the remaining component for contraction, we just contract the whole subtree into its adjacent vertex in $g'(T_k(u))$.

Therefore, r_n and $r_k(u)$ are contracted to the same leaf. Hence, r_n is contracted into a center point of $g'(T_k(u))$. In fact, the tree $f'(T_n)$ consists of components $g'(T_k)$ and additional edges connecting the center point of $g'(T_k(u))$ to the center points of the other components. Moreover if $k + 1 = n$, $f'(T_n)$ only has two components $g'(T_k)$, thereby its diameter is $2n + 2 \cdot R(g'(T_k(u)))$. And if $k + 1 < n$, $f'(T_n)$ has more than two components $g'(T_k)$, thereby its diameter is $4n + 2 \cdot R(g'(T_k(u)))$. Thus

$$\mathsf{D}(f') = \begin{cases} \max(\mathsf{D}(g'), \frac{4n+2\cdot R(g'(T_k(u)))}{2(k+1)}) & \text{if } k+1 < n; \\ \max(\mathsf{D}(g'), \frac{2n+2\cdot R(g'(T_k(u)))}{2(k+1)}) & \text{if } k+1 = n; \end{cases} \tag{3.3}$$

Thus, by Equation (3.1) and the relationship between the transformations g' and f, we have $\mathsf{D}(f') \leq \mathsf{D}(f)$, proving part (a) of the theorem. Moreover, by Equation (3.2), we obtain part(b)

$$R(f'(T_n)) = 2n + R(g'(T_k(u))) \leq R(f(T_n)), \tag{3.4}$$

and r_n is contracted into a center point of $g'(T_k(u))$, which can be verified to be a center point of $R(f'(T_n))$, hence proving part(c). $\square$

We next give the proof of Claim 3, as promised earlier.

Proof of Claim 3: We first observe that any maximal connected component C in the tree $S_k(u)$ that does not contain any vertex in $\mathsf{L}(T_k(u))$ will be contracted into a vertex of $\mathsf{L}(T_k(u))$.

We will use the following fact about distances between leaves.

Fact 3. *Any edge between two leaves in $\mathsf{L}(T_k(u))$ has weight at most $2k$; and any edge between a leaf in $\mathsf{L}(T_k(u))$ and one outside it has weight at least $2(k+1)$.*

1. To show $\mathsf{D}(g) \leq \mathsf{D}(f)$, we prove that $d_{\widehat{T}_k(u)}(x, y) \leq d_{\widehat{T}_n}(x, y)$ for any $x, y \in \mathsf{L}(T_k(u))$.
   Fix any $x, y \in \mathsf{L}(T_k(u))$. Let P be the x-y path in $\widehat{T}_k(u)$ and Q be the x-y path in $S_k(u)$.
   Since any maximal connected component C excluding vertices in $\mathsf{L}(T_k(u))$ in the tree $S_k(u)$ is contracted into one vertex of $\mathsf{L}(T_k(u))$, any maximal subpath Q' of Q excluding vertices in $\mathsf{L}(T_k(u))$ is contracted into some vertex c of $\mathsf{L}(T_k(u))$. By maximality of Q', there exists $a, b \in \mathsf{L}(T_k(u))$ on path Q such that a-Q'-b is a subpath of Q, which would become a subpath a-c-b in P. By Fact 3, the length of this subpath decreases.
   On the other hand, an edge in Q that joins two vertices in $\mathsf{L}(T_k(u))$ remains in P and its weight does not change.
   Hence, it follows that the length of P is at most that of Q.
   Therefore,

   $$d_{\widehat{T}_k(u)}(x, y) \leq d_{\widehat{T}_n}(x, y) \text{ for any } x, y \in \mathsf{L}(T_k(u)) \tag{3.5}$$

   Thus $\mathsf{D}(g) \leq \mathsf{D}(f)$.

2. Next we show that $R(g(T_k(u))) \leq R(S_k(u))$.

Let $u_0 \in V(S_k(u))$ be the center point of $S_k(u)$. By the minimality of $S_k(u)$, this radius must be realized by some vertex in $L(T_k(u))$.

$$R(S_k(u)) = \max_{x \in L_k(u)} (d_{\widehat{T}_n}(u_0, x)) \tag{3.6}$$

If $u_0 \in L(T_k(u)) = V(\widehat{T}_k(u))$, then by Equations (3.5) and (3.6),

$$R(\widehat{T}_k(u)) \leq \max_{x \in L(T_k(u))} d_{\widehat{T}_k(u)}(u_0, x) \leq \max_{x \in L(T_k(u))} (d_{\widehat{T}_n}(u_0, x)) = R(S_k(u)).$$

If $u_0 \notin L(T_k(u)) = V(\widehat{T}_k(u))$, then let $u_0' \in V(\widehat{T}_k(u))$ be the vertex into which u_0 is contracted. For any $x \in L(T_k(u)) = V(\widehat{T}_k(u))$, let P be the u_0'-x path in $\widehat{T}_k(u)$ and Q be the u_0-x path in $S_k(u)$.

Observe that the initial maximal subpath Q' of Q excluding vertices in $L(T_k(u))$ is contracted into u_0'. Let u_1 be the first vertex on Q in the direction from u_0 to x such that $u_1 \in L(T_k(u))$. Hence, the subpath Q'-u_1 becomes a subpath u_0'-u_1 in P, whose length decreases by Fact 3. By Equation (3.5), the length of the remaining subpath of P is at most that of the remaining subpath of Q. Hence, the length of P is at most that of Q. Therefore,

$$R(\widehat{T}_k(u)) \leq \max_{x \in V(\widehat{T}_k(u))} d_{\widehat{T}_k(u)}(u_0', x) \leq \max_{x \in V(S_k(u))} d_{\widehat{T}_n}(u_0, x) = R(S_k(u))$$

Thus, we also have $R(g(T_k(u))) \leq R(S_k(u))$ in this case. $\square$

4   A Lower Bound for Minor Transformations

In view of Theorem 2 in the previous section, we consider only minor transformations for complete binary trees.

Definition 4 (Optimal distortion for minor transformation). We define $\alpha \geq 1$ to be the smallest constant such that for any instance of the SPR Problem, there exists a *minor* transformation that achieves distortion at most α.

Observe that the algorithm given by Gupta [1] indeed produces a minor with distortion at most 8. Hence, the constant α is at most 8. We prove the following theorem, which implies that the constant $\alpha \geq 8$.

Theorem 5. *For any $\epsilon > 0$, the constant $\alpha \geq 8 - \epsilon$.*

Hence, combining Theorems 2 and 5, we obtain the result of Theorem 1, which states that:

For any $\epsilon > 0$, there exists an instance of the Steiner Point Removal Problem with distortion at least $8 - \epsilon$.

We first introduce some notation. Without causing ambiguity, we use $d(u,v)$ to denote the distance between nodes u and v in the original tree T, and $path(u,v)$ to denote the subset of vertices lying on the unique path between u and v in T. Let v be a vertex in T_n. We denote the subtree rooted at v by $T(v)$, which is identical to $T_{n-d(r_n,v)}$. For $u, v \in \mathsf{L}(T)$, we use $d_f(u,v)$ to denote the distance between them after the transformation f is applied to the tree.

Definition 6. Given a minor mapping $f : \mathsf{V}(T) \to \mathsf{L}(T)$, a vertex v is a *normal vertex* (with respect to f) if v is an ancestor of $f(v)$.

Consider a normal vertex v and suppose $u = f(v)$. Then, v is an ancestor of u and all the vertices along the path from v to u are mapped to u. Recall that $T(v)$ has two branches rooted at v. We wish to measure how far vertices down the branch *not* containing u are mapped to u under f.

Definition 7. For each normal vertex v, its *expanding parameter* with respect to some minor mapping f is defined to be
$$\rho_f(v) := \max\{\tfrac{d(v,w)}{d(v,f(v))} : w \in T(v),\ f(w) = f(v),$$
$$path(v, f(v)) \cap path(v,w) = \{v\}\}.$$

Since our lower bound is obtained from large trees, we consider how the expanding parameter behaves for large values of n.

Definition 8. For each $n \in \mathbb{N}$, let
$$\rho_n := \max\{\rho_f(r_n) \mid \text{minor mapping } f : T_n \to \mathsf{L}(T_n),\ \mathsf{D}(f) \le \alpha\}.$$
Define
$$\rho := \limsup_{n \to \infty} \rho_n. \tag{4.7}$$

Observe that since $\rho_n \in [0,1]$, it follows the limit supremum $\rho \in [0,1]$. We show in the next claim that ρ is strictly less than 1.

Claim. The limit supremum $\rho < 1$.

Proof: Assume on the contrary that $\rho = 1$. Then, by the definition of limit supremum ρ, there exists an integer n such that $\rho_n \ge 7/8$. Thus by the definition of ρ_n, there exists a minor mapping f on T_n with $\mathsf{D}(f) \le \alpha$ such that $\rho_f(r_n) \ge 7/8$.

Let w be a vertex that attains $\rho_f(r_n)$. Since every leaf of T_n is mapped into itself and $w \ne f(w)$, w is not a leaf. Then let p and q be two leaves from different branches of the subtree $T(w)$. Thus $d(p,q) = 2(1-\rho_f(r_n))n \le n/4$. On the other hand, $d_f(p,q) = d_f(p,f(w)) + d_f(f(w),q) \ge 4n$. Thus $\mathsf{D}(f) \ge \frac{d_f(p,q)}{d(p,q)} \ge \frac{4n}{n/4} \ge 16$, contradicting $\mathsf{D}(f) \le \alpha \le 8$. Thus $\rho < 1$. □

The following lemma shows the relationship between the expanding parameter ρ_n and the distorted distance d_f. Intuitively, if the expanding parameters for normal vertices of large heights are small, then there exists some vertex whose distorted distance to the image of the root is large.

Lemma 1. *Suppose $0 < \beta < 1$ and $N_0 \in \mathbb{N}$ such that for any integer $n > N_0$, the expanding parameter $\rho_n \leq \beta$. Then, for any real $0 < \epsilon < 1$, there exists integer $N > N_0$ such that for any integer $m \geq N$ and any minor mapping f on tree T_m with distortion $\mathsf{D}(f) \leq \alpha$, there exists a leaf p in T_m such that the distorted distance*

$$d_f(p, f(r_m)) \geq \frac{2m}{\beta}(1 - \epsilon).$$

Furthermore, if $\rho_f(r_m) > 0$, then $\mathsf{D}(f) \geq \frac{2(1-\epsilon)}{\beta(1-\rho_f(r_m))}$.

Proof: Given any real $\epsilon > 0$, fix a large enough integer k such that $(1-\beta)^k \leq \frac{\epsilon}{2}$. Let N be large enough such that $\frac{k}{N} \leq \frac{\epsilon}{2}$ and $(1-\beta)^k(N + \frac{1}{\beta}) - \frac{1}{\beta} > N_0$.

Let $m \geq N$ and let f be a minor mapping on T_m with $\mathsf{D}(f) \leq \alpha$. We define sequences of vertices $\{v_i\}_{i=0}^k$ and $\{w_i\}_{i=0}^{k-1}$ in T_m as follows. Let $v_0 = r_m$, and w_0 be the vertex that attains $\rho_f(v_0)$ under the minor mapping f with $\mathsf{D}(f) \leq \alpha$. For $1 \leq i \leq k$, let v_i be a child of vertex w_{i-1} such that $f(w_{i-1}) \notin T(v_i)$, and hence v_i is normal. Let w_i be the vertex that attains $\rho_f(v_i)$, for $1 \leq i < k$. Let h_i be the height of $T(v_i)$ for $0 \leq i \leq k$.

Claim. For $0 \leq i < k$, the height $h_i \geq (1 - \beta)^i(m + \frac{1}{\beta}) - \frac{1}{\beta} > N_0$.

Proof of Claim 4: The claim is trivial for $i = 0$. Assume that $h_{i-1} \geq (1 - \beta)^{i-1}(m+\frac{1}{\beta}) - \frac{1}{\beta} > N_0$, for some $0 < i < k$. Observe that $h_i + 1 + \rho_f(v_{i-1})h_{i-1} = h_{i-1}$ and $\rho_f(v_{i-1}) \leq \beta$, since $h_{i-1} > N_0$. Then $h_i = (1 - \rho_f(v_{i-1}))h_{i-1} - 1 \geq (1 - \beta)\{(1 - \beta)^{i-1}(m + \frac{1}{\beta}) - \frac{1}{\beta}\} - 1 = (1 - \beta)^i(m + \frac{1}{\beta}) - \frac{1}{\beta} > N_0$. $\square$

Thus, we set $p := f(v_k)$ and from Claim 4, we have

$$d_f(f(r_m), p) = 2\sum_{i=0}^{k-1} h_i \geq 2\sum_{i=0}^{k-1}\{(1 - \beta)^i(m + \frac{1}{\beta}) - \frac{1}{\beta}\}$$

$$= 2(m + \frac{1}{\beta})\frac{1 - (1 - \beta)^k}{\beta} - \frac{2k}{\beta} \geq \frac{2m}{\beta} \cdot \{1 - (1 - \beta)^k - \frac{k}{m}\} \quad (4.8)$$

$$\geq \frac{2m}{\beta}(1 - \epsilon),$$

where the last inequality follows from $(1 - \beta)^k \leq \frac{\epsilon}{2}$ and $\frac{k}{m} \leq \frac{k}{N} \leq \frac{\epsilon}{2}$.

Furthermore, if $\rho_f(r_m) > 0$, then $m \cdot \rho_f(r_m) > 0$. Thus w_0 is a proper descendant of r_m. Note that p is a leaf of $T(w_0)$ and $T(w_0)$ has two branches. Thus by symmetry, there exists another leaf q such that p and q are in the different branches of $T(w_0)$ and $d_f(q, f(r_m)) \geq \frac{2m}{\beta}(1-\epsilon)$. Observing that $f(w_0) = f(r_m)$, the distorted distance $d_f(p, q) = d_f(p, f(r_m)) + d_f(f(r_m), q) \geq \frac{4m}{\beta}(1 - \epsilon)$, and the original distance $d(p, q) = 2m(1 - \rho_f(r_m))$. Therefore, the distortion $\mathsf{D}(f) \geq \frac{d_f(p,q)}{d(p,q)} \geq \frac{2(1-\epsilon)}{\beta(1-\rho_f(r_m))}$. $\square$

Using Lemma 1, we can show that the limit supremum $\rho > 0$.

Claim. The limit supremum $\rho > 0$.

Proof of Lemma 4: On the contrary, suppose $\rho = 0$. Let $\beta = 1/32$. By the definition of limit supremum ρ, there exists N_0 such that for any $n > N_0$, $\rho_n < \beta$. Then by Lemma 1, for $\epsilon = 1/2$, there exists $m > N_0$ such that for any minor mapping f on T_m with $\mathsf{D}(f) \leq \alpha$, there exists a leaf p in T_m such that $d_f(p, f(r_m)) \geq \frac{2m}{\beta}(1 - \epsilon) = \frac{m}{\beta}$. Thus $\mathsf{D}(f) \geq \frac{d_f(p, f(r_m))}{d(p, f(r_m))} \geq \frac{m}{2m\beta} = \frac{1}{2\beta} = 16$, which contradicts $\mathsf{D}(f) \leq \alpha \leq 8$. Thus $\rho > 0$. □

Now, we are ready to prove the main theorem of this section.

Proof of Theorem 5: Let $\epsilon > 0$. Without loss of generality, we can assume $\epsilon < 1$. Suppose on the contrary, we have $\alpha < 8 - \epsilon$.

Since $0 < \rho < 1$, let $\epsilon_1 < \min\{\epsilon/48, \rho\}$ be a positive small constant such that $\rho + \epsilon_1 < 1$. By the definition of limit supremum ρ, there exists $N_0 > 0$ such that for all $n > N_0$, $\rho_n < \rho + \epsilon_1$. Then by Lemma 1, for $\epsilon_2 = \epsilon/24$ there exists N such that for any $m \geq N$ and any minor mapping f on tree T_m with distortion $\mathsf{D}(f) \leq \alpha$ and $\rho_f(r_m) > 0$ we have $\mathsf{D}(f) \geq \frac{2(1-\epsilon_2)}{(\rho+\epsilon_1)(1-\rho_f(r_m))}$.

By the definition of limit supremum ρ, there exists arbitrarily large m such that $\rho_m > \rho - \epsilon_1 > 0$. Hence, we can choose m such that $m > N$. By the definition of ρ_m, there exists a minor mapping f on tree T_m with distortion $\mathsf{D}(f) \leq \alpha$ and $\rho_f(r_m) = \rho_m > \rho - \epsilon_1 > 0$. Thus, the constant α is at least

$$
\begin{aligned}
\mathsf{D}(f) &\geq \frac{2(1-\epsilon_2)}{(\rho+\epsilon_1)(1-\rho_f(r_m))} \\
&\geq \frac{2(1-\epsilon_2)}{(\rho+\epsilon_1)(1-(\rho-\epsilon_1))} \geq \frac{8(1-\epsilon_2)}{(1+2\epsilon_1)^2} \text{ (The denominator is min when } \rho = \tfrac{1}{2}.) \\
&\geq \frac{8(1-\epsilon_2)}{(1+\epsilon_2)^2} \geq 8(1 - 3\epsilon_2) \quad \text{(Note: } 2\epsilon_1 \leq \epsilon_2; \text{ as } \epsilon_2 \geq 0, \tfrac{1-\epsilon_2}{(1+\epsilon_2)^2} \geq 1 - 3\epsilon_2) \\
&= 8 - \epsilon,
\end{aligned}
$$

obtaining the desired contradiction. Hence, for all $\epsilon > 0$, the constant $\alpha \geq 8 - \epsilon$. □

5   Open Problems

We conclude the paper by outlining some directions for future work.

1. Of course one final goal would be to consider the SPR problem on general graphs. Formally, there are two main questions to be addressed: (1) we would like to determine what is the smallest α (possibly depending on the size of input), such that given any edge weighted graph $G = (V, E)$ and a set of terminals $S \subset V$, there is a way to remove non-terminals by edge contractions to produce a minor $H = (S, E')$ where for any pair of terminals (u, v), $d_G(u, v) \leq d_H(u, v) \leq \alpha \cdot d_G(u, v)$; and (2) we would like to devise a constructive algorithm that outputs such a minor $H = (S, E')$ with distortion at most α. Since this task may prove to be quite hard to accomplish on general graphs, one could first consider other restricted classes of graphs — such as outerplanar graphs, planar graphs, series-parallel graphs, etc. — as an intermediate step. Note that no algorithm with proven nontrivial bounds on distortion for these classes of graphs is known.

2. Another interesting question is to be able to determine the *approximation bound on the optimal distortion* of a given algorithm for the SPR problem. For example, it would be interesting to determine, given any instance of the SPR problem on trees, how far from the optimal distortion for that instance can the distortion obtained by Gupta's algorithm [1] be (in that paper, Gupta only shows an absolute bound on the distortion of his algorithm; this paper confirms that for some instances of the problem, this is the best distortion possible).

3. We can also ask a similar question as that in Problem 1 in a probabilistic framework. What is the smallest α such that given any weighted graph $G = (V, E)$ and a set of terminals $T \subset V$, there exists a distribution $\mathcal{H}$ of minors $\{H = (T, E')\}$ such that $d_G(u, v) \leq E_{\mathcal{H}}[d_H(u, v)] \leq \alpha \cdot d_G(u, v)$? This task may be easier to accomplish than that in Problem 1, since some upper bounds on α under a probabilistic framework exist in the literature. For example, it follows from [7] that k-outerplanar graph can be embedded into a probability distribution over spanning trees with $O(c^k)$ distortion for some absolute constant c, implying that $\alpha = O(c^k)$ for k-outerplanar graphs; and a recent result by Elkin et. al. [10] shows that for general graphs, $\alpha = O(\log^2 n \log\log n)$, which is later improved to $O(\log^2 n)$ by Dhamdhere et. al. [11], shows that for general graphs, $\alpha = O(\log^2 n)$. Can we do any better?

References

1. Gupta, A.: Steiner points in tree metrics don't (really) help. SODA (2001) 220–227
2. Rabinovich, Y., Raz, R.: Lower bounds on the distortion of embedding finite metric spaces in graphs. Discrete Comput. Geom. **19**(1) (1998) 79–94
3. Chu, Y., Rao, S., Zhang, H.: A case for end system multicast. In: Proceedings of ACM Sigmetrics, Santa Clara, CA. (2000)
4. Xie, J., Talpade, R.R., Mcauley, A., Liu, M.: Amroute: ad hoc multicast routing protocol. Mob. Netw. Appl. **7**(6) (2002)
5. Chawathe, Y.: Scattercast: an adaptable broadcast distribution framework. Multimedia Syst. **9**(1) (2003)
6. Francis, P.: Yoid: Extending the internet multicast architecture. (2000)
7. Chekuri, C., Gupta, A., Newman, I., Rabinovich, Y., Sinclair, A.: Embedding k-outerplanar graphs into ℓ_1. In: Proceedings of the 14th Annual ACM-SIAM Symposium on Discrete Algorithms. (2003) 527–536
8. Fakcharoenphol, J., Rao, S., Talwar, K.: A tight bound on approximating arbitrary metrics by tree metrics. In: Proceedings of the thirty-fifth ACM symposium on Theory of computing, ACM Press (2003) 448–455
9. Archer, A., Fakcharoenphol, J., Harrelson, C., Krauthgamer, R., Talwar, K., Tardos, E.: Approximate classification via earthmover metrics. In: In 15th Annual ACM-SIAM Symposium on Discrete Algorithms. (2004) 1072–1080
10. Elkin, M., Emek, Y., Spielman, D.A., Teng, S.H.: Lower-stretch spanning trees. In: Proceedings of the 37th Annual ACM Symposium on Theory of Computing. (2005) 494–503
11. Dhamdhere, K., Gupta, A., Räcke, H.: (Improved embeddings of graph metrics into random trees)

Single-Source Stochastic Routing

Shuchi Chawla[1],[*] and Tim Roughgarden[2],[**]

[1] Microsoft Research, Silicon Valley Campus, Mountain View, CA 94043
shuchi@cs.wisc.edu
[2] Department of Computer Science, Stanford University, 462 Gates Building, 353
Serra Mall, Stanford, CA 94305
tim@cs.stanford.edu

Abstract. We introduce and study the following model for routing uncertain demands through a network. We are given a capacitated multicommodity flow network with a single source and multiple sinks, and demands that have known values but unknown sizes. We assume that the sizes of demands are governed by independent distributions, and that we know only the means of these distributions and an upper bound on the maximum-possible size. Demands are irrevocably routed one-by-one, and the size of a demand is unveiled only after it is routed.

A *routing policy* is a function that selects an unrouted demand and a path for it, as a function of the residual capacity in the network. Our objective is to maximize the expected value of the demands successfully routed by our routing policy. We distinguish between *safe* routing policies, which never violate capacity constraints, and *unsafe* policies, which can attempt to route a demand on any path with strictly positive residual capacity.

We design safe routing policies that obtain expected value close to that of an optimal unsafe policy in planar graphs. Unlike most previous work on similar stochastic optimization problems, our routing policies are fundamentally adaptive. Our policies iteratively solve a sequence of linear programs to guide the selection of both demands and routes.

1 Introduction

We introduce and study the following model for routing uncertain demands through a network. We are given a multicommodity flow network, defined by a directed graph $G = (V, E)$ with vertices V and edges E, a nonnegative capacity c_e on each edge $e \in E$, and a collection $(s_1, t_1), \ldots, (s_k, t_k)$ of source-sink pairs, also called *commodities*. Associated with each commodity i is a demand with a known nonnegative *value* v_i and an unknown size. Our goal is to choose routes for a subset of the demands to maximize the value of these demands without violating the edge capacities. In the special case of known demand sizes, this is the well known and difficult *unsplittable flow* problem.

[*] This work was performed while the author was visiting the Department of Computer Science, Stanford University and supported by DARPA grant W911NF-05-1-0224.
[**] Supported in part by ONR grant N00014-04-1-0725, DARPA grant W911NF-05-1-0224, and an NSF CAREER Award.

J. Diaz et al. (Eds.): APPROX and RANDOM 2006, LNCS 4110, pp. 82–94, 2006.

Inspired by recent work of Dean, Goemans, and Vondrak [4, 5] on stochastic versions of the Knapsack and Set Packing problems, we adopt the following model for unknown demand sizes. We assume that the size of the ith demand is governed by a distribution with known mean μ_i, and that the sizes of different demands are independent. We also assume that there is a known upper bound D_{max} on the maximum-possible size of a demand. No other information about the size distributions is available. We assume that commodities are routed one-by-one. When a commodity is selected, the size of its demand is unveiled only *after* it is routed. Decisions are irrevocable, and a previously routed demand cannot be removed from the network.

A *routing policy* is a function that selects an unrouted commodity (s_i, t_i) and an s_i-t_i path for it, as a function of the residual capacity in the network. While routing policies can be very complex, we will only be interested in routing policies defined by polynomial-time algorithms. A routing policy can be *adaptive*, in the sense that its decisions depend on the instantiated sizes of the previously routed commodities, or *non-adaptive*, in which case it simply specifies a fixed order in which the demands should be routed and fixed paths for routing them. There has been significant recent work proving upper and lower bounds on the *adaptivity gap*—the ratio between the objective function values of an optimal adaptive and non-adaptive policy, respectively—for various problems [6, 4, 5, 8]. We show in the full version of this paper [2] that the problems we consider have a large (polynomial) adaptivity gap, even in networks of parallel links. In contrast to previous work, which primarily studied non-adaptive policies for various problems, we focus on the design and analysis of near-optimal adaptive policies. Our objective is to maximize the expected value of the successfully routed commodities.

When demand sizes are stochastic, edge capacity constraints can be interpreted in several ways. The most stringent definition is to require that a routing policy respect every edge capacity with probability 1. We call a routing policy *safe* if it meets this definition and *unsafe* otherwise. When an unsafe routing policy routes a commodity in a way that violates some capacity constraints, we assume that no value is obtained for this unsuccessfully routed commodity, and that all violated edges drop out of the network.

Both safe and unsafe policies have their advantages. Unsafe policies are clearly more general than safe ones, and may obtain a much larger expected value. Safe policies guarantee successful transport for all admitted commodities; this property is clearly desirable, and could be essential in certain applications.

In this work, we seek the best of both worlds: we design *safe* routing policies, but bound their performance relative to an optimal *unsafe* routing policy. This goal is somewhat analogous to previous work [4, 5] that designed non-adaptive policies with expected value close to that of an optimal adaptive policy.

Pursuing this ambitious goal forces us to adopt an additional assumption. To motivate it, consider the following example. Fix a value $\alpha \in (0, 1]$, let $\epsilon > 0$ be much smaller than α, and let $\delta > 0$ be much smaller than ϵ. Consider a network with two vertices s, t and one directed edge (s, t) with unit capacity. Suppose there are a large number of commodities, each with source s, sink t, unit value, and with size equal to α with probability δ and to ϵ with probability $1 - \delta$. A safe routing policy must cease routing commodities after roughly $(1 - \alpha)/\epsilon$ commodities have been routed. On the other hand, an unsafe policy will typically route roughly $1/\epsilon$ commodities successfully, provided δ is sufficiently small. Thus safe policies might capture only a $1 - \alpha$ fraction of the expected value of an optimal unsafe policy, where α is the maximum-possible fraction of an edge that a demand can occupy. For this reason, we assume throughout this paper that the maximum-possible size D_{max} of a commodity is bounded above by an $\alpha < 1$ fraction of the minimum edge capacity c_{min}. Similar but weaker assumptions are often made in the classical single-sink unsplittable flow problem [7, 13, 14]. When this gap α is $O(1/\log n)$, even the general multicommodity stochastic routing problem can be approximated to within a constant factor using a straightforward randomized rounding algorithm. (See the full version [2] for details.) Our goal will be to design routing policies that have good (constant or logarithmic) approximation ratios for every fixed constant α less than 1.

Achieving this goal in general multicommodity networks would give, as a special case, a fundamental breakthrough for solving the disjoint paths problem with constant congestion in directed graphs. On the other hand, the single-source unsplittable flow problem (with known demands) admits constant-factor approximation algorithms [7, 13, 14]. These facts motivate our second crucial assumption: we assume that all commodities share a common source vertex s. We call the problem of designing a routing policy for such an instance the *Single-Source Stochastic Routing (SSSR)* problem.

Our Results. We first define a general algorithmic and analytical approach for designing near-optimal, safe, adaptive routing policies for SSSR instances. Our algorithm uses a linear program (LP), the optimal value of which is an upper bound on the expected value of an optimal (unsafe) routing policy, to guide the commodity and route selection at each stage. The algorithm re-solves this LP each time a new commodity is routed. Our analysis framework is based on tracking the successive expected changes in the optimal value of the LP, as our algorithm routes and instantiates demands.

As noted above, previous work on related stochastic optimization problems [6, 4, 5, 8] has concentrated primarily on the design and analysis of non-adaptive policies; few techniques for designing adaptive policies are currently known. We believe that our iterative LP rounding approach could form the basis of near-optimal adaptive policies for a range of stochastic optimization problems.

We apply this framework to obtain polynomial-time, safe routing policies with expected value close to that of an optimal unsafe policy for SSSR problems in planar graphs. (More generally, we only require that the supporting subgraph of a natural fractional flow relaxation is planar.) We achieve an approximation factor of $O((\log W)/(1 - \alpha))$, where $\alpha < 1$ is a constant satisfying $D_{max} \le \alpha c_{min}$, and W denotes the maximum ratio between the "expected per-unit value" v_i/μ_i of two different commodities. Recall from the above example that the dependence on $1/(1-\alpha)$ is necessary for this type of guarantee, even in single-link networks. We also obtain a superior approximation factor of $O(1/(1-\alpha))$ in the special case where all of the sinks lie on a common face. This special case includes all outerplanar networks and all single-source, single-sink planar networks.

Related Work. Starting with the work of Dantzig [3] in 1955, stochastic optimization problems have been studied extensively in Operations Research (see e.g. [1, 18]). Owing to the complexity of optimally solving[1] general stochastic problems, much of this work has focused on the special cases of stochastic linear programming and *k-stage recourse* problems. Several recent works by the theoretical CS community have studied the recourse model. Starting with [12, 15], constant-factor approximation algorithms have been developed for the 2-stage stochastic versions of problems such as Steiner tree, network design, facility location, and vertex cover (see e.g. [9, 10, 11, 17]). Some of this work has been extended to the k-stage versions of these problems [10, 16], albeit with approximation factors that depend linearly or even exponentially on k.

The work that is most closely related to ours is that of Dean, Goemans and Vondrak [4, 5, 8]. Dean et al. study stochastic versions of several packing and covering problems such as Knapsack, that are similar in flavor to our stochastic routing problem. For example, the Stochastic Knapsack problem is essentially SSSR in a single-link network, and SSSR in a general graph is similar to an instance of the Stochastic Multi-dimensional Knapsack problem, with a unique dimension corresponding to each edge of the graph.

However, our focus on routing applications leads to several key differences between their work and ours. First, in the SSSR problem, a routing policy must select both the next commodity to route, as well as *how* to route it. There is no analogue of this combinatorial route selection issue in the packing and covering problems studied in [4, 5], which primarily involve only binary decision variables. Second, capacity constraints are enforced differently in the work of Dean et al. than in the present paper. In [4, 5], unsafe policies are allowed, but such a policy must terminate as soon as a single constraint is violated. In the SSSR problem, an unsafe routing policy can continue to route the remaining commodities on edges that have not yet dropped out of the network. We believe that this less restrictive notion of an unsafe policy is more suitable for routing applications. Third, we design safe routing policies, whereas Dean et al. design policies that are unsafe in the above restricted sense. Thus while our guarantees are in some sense stronger than those in [4, 5], we prove such guarantees only under an additional assumption ($D_{max} \leq \alpha c_{min}$ for some $\alpha < 1$) that is not needed in the work of Dean et al. Finally, as noted earlier, Dean et al. focus on obtaining tight bounds on the adaptivity gap, whereas we seek adaptive solutions that achieve an approximation factor far smaller than the adaptivity gap.

---

[1] The optimal solution to a stochastic optimization problem such as SSSR can be a complex, exponential-size decision diagram. The number of possible solutions can be doubly-exponential in the number of stages.

2   The Stochastic Routing Model

We consider a directed network $G = (V, E)$ with edge capacities $c : E \to \Re^+$. We are given k commodities indexed by $i \in I$, each with a source-sink pair (s_i, t_i) and a value v_i. In Section 4, we will assume that all commodities share a common source s. The "size" or demand of a commodity i is given by the random variable D_i, drawn from an independent distribution with mean $\mu_i = \mathbf{E}[D_i]$. For every commodity i, let $w_i = v_i/\mu_i$ denote its "expected per-unit value". We assume that commodities are ordered such that $w_1 \geq w_2 \geq \cdots \geq w_k$.

Let $D_{\max}$ be the smallest value d such that $\mathbf{Pr}[D_i > d] = 0$ for all $i \in I$. We assume that D_{max} is known to the algorithm and that $D_{\max} < c_{\min}$, where $c_{\min} = \min_e c_e$ is the minimum edge capacity in the graph. Let $\alpha < 1$ denote the ratio between $D_{\max}$ and $c_{\min}$. As shown by the example in the Introduction, our approximation guarantees necessarily depend on the value of α.

Let $\mathcal{P}_i$ denote the s_i-t_i paths of G. A routing policy successively picks a commodity i and a path $P_i \in \mathcal{P}_i$ for routing it. After the algorithm picks a commodity and its corresponding path, the demand D_i for that commodity gets instantiated to some value d_i. If d_i is at most the minimum residual capacity of the edges of P_i, then the commodity is admitted and the algorithm obtains the value v_i. The algorithm continues until no more commodities can be admitted. The goal of the algorithm is to maximize the expectation of its total accrued value. As described previously, a routing policy is safe if every commodity picked by it gets admitted with probability one.

3   Approximation Algorithms Via Iterative Rounding

An LP Relaxation for the Optimal Routing Policy. We now give a general algorithmic and analytic approach for approximating stochastic routing problems; we apply these ideas to SSSR problems in planar graphs in the next section. We begin with a linear program giving an upper bound on the expected value of an optimal (unsafe) routing policy for a given stochastic routing instance:

$$LP(I, u): \quad \max \sum_{i \in I} w_i \sum_{e \in \delta^+(s_i)} f_e^{(i)} \text{ s.t.}$$
$$\sum_{i \in I} f_e^{(i)} \leq u_e \qquad\qquad \forall e \in E$$
$$\sum_{e \in \delta^+(s_i)} f_e^{(i)} \leq \mu_i \qquad\qquad \forall i \in I$$
$$\sum_{e \in \delta^-(v)} f_e^{(i)} = \sum_{e \in \delta^+(v)} f_e^{(i)} \quad \forall i \in I, v \in V \setminus \{s_i, t_i\}$$
$$f_e^{(i)} \geq 0 \qquad\qquad \forall i \in I, e \in E.$$

Recall that w_i denotes the ratio v_i/μ_i. Also, $\delta^+(v)$ and $\delta^-(v)$ denote the sets of edges directed out of and into the vertex v, respectively. Note that $LP(I, u)$ is simply a standard LP formulation of the maximum-value (w.r.t. "values" w) multicommodity flow subject to edge capacities u and per-commodity flow rate constraints μ.

Input: A stochastic routing instance G, c, I.
Output: A commodity $i \in I$ and a path $P \in \mathcal{P}_i$ at every step.

1. Initialize J to I and $\hat{c}_e = (1 - \alpha)c_e$ for every $e \in E$. Solve $LP(J, \hat{c})$, obtaining an optimal solution $\hat{f}$.
2. While $\hat{f}_e^{(i)} > 0$ for some commodity $i \in J$ and edge $e \in E$:
   (a) Pick $i \in J$ and $P \in \mathcal{P}_i$ such that $\hat{f}_e^{(i)} > 0$ for every $e \in P$, and route the commodity i on P.
   (b) Set $J := J \setminus \{i\}$.
   (c) Set $\hat{c}_e := \max\{0, \hat{c}_e - d_i\}$ for every edge $e \in P$, where d_i is the instantiated size of commodity i.
   (d) Re-solve $LP(J, \hat{c})$, obtaining a new optimal solution $\hat{f}$.

Fig. 1. High-level description of the algorithm IR

Proposition 1. *The expected value obtained by an optimal adaptive routing policy for a stochastic routing instance with commodities I and edge capacities c is at most $LP(I, (1 + \alpha)c)$, where $\alpha = D_{max}/c_{min}$.*

Proposition 1 is similar to a result by Dean, Goemans, and Vondrak [5] in the special case of a single-link network (Knapsack). Scaling, we also obtain the following corollary.

Corollary 1. *For every $\gamma \in (0, 1]$, the expected value obtained by an optimal routing policy for a stochastic routing instance with commodities I and edge capacities c is at most $\frac{1}{\gamma} \cdot LP(I, \gamma(1 + \alpha)c)$, where $\alpha = D_{max}/c_{min}$.*

An Iterative Rounding Algorithm. We next develop a safe, adaptive routing algorithm that iteratively uses linear programs of the form $LP(I, u)$ to guide both commodity and route selections. The high-level idea of the algorithm is to scale down the given edge capacities (to ensure safeness), and solve $LP(I, u)$. We then pick the fractionally routed commodity with largest ratio w_i, route it on one of its (fractional) flow paths, and repeat. This high-level algorithm is given in Figure 1.

Fact 1. *Algorithm IRis a safe routing policy.*

To obtain good approximation results, however, we need to choose the commodity i and the path $P \in \mathcal{P}_i$ in Step 2a carefully. One natural refinement of Algorithm IRis to always choose a commodity i in Step 2a with maximum-possible ratio w_i; we call this variant the GREEDY-IRalgorithm.

We next discuss the much more subtle issue of path selection. To motivate the next definition, suppose that in the first stage we pick a commodity i and an s_i-t_i flow path P. The size of commodity i might get instantiated to some value much larger than μ_i, which in turn could evict other commodities in the LP solution from the edges of P. Intuitively, our goal will be to pick a path to minimize the severity of this eviction. We make this idea precise with the following notion of r-*coverable* paths.

Definition 1. *Fix a stochastic routing instance. Let $\{\hat{f}_e^{(i)}\}_{i,e}$ be a feasible solution to $LP(I, u)$. Let $\{\hat{f}_P^{(i)}\}_{i,P \in S}$ be a flow decomposition of f, where $S \subseteq \cup_i P_i$ denotes the set of paths that carry a positive amount of flow.*

(a) Let $P^* \in S$ be a path with $f_{P^*}^{(i)} > 0$ and $S' \subseteq S$ a collection of flow paths for commodities other than i. Let $F^* \subseteq P^*$ denote the edges of P^* contained in some path of S'. The set S' r-covers P^* if there are $q \leq r$ paths $P_1, \ldots, P_q \in S'$ such that every edge of F^* lies in at least one path P_i.

(b) The path decomposition $\{\hat{f}_P^{(i)}\}$ r-covers the path $P^* \in S$ if for every subset $S' \subseteq S$ of flow paths for commodities other than i, S' r-covers P^*.

(c) An s_i-t_i path P^* with $\hat{f}_e^{(i)} > 0$ for every $e \in P^*$ is r-coverable if there exists a path decomposition with $\hat{f}_{P^*}^{(i)} > 0$ that r-covers P^*.

Intuitively, increasing the amount of flow on an r-coverable path only evicts flow from r other flow paths. For example, in a stochastic routing instance in a single-link network (i.e., Knapsack), every flow path is 1-coverable.

We next prove the central result of this section: if Algorithm GREEDY-IRcan be implemented to route commodities only on r-coverable paths, then its expected value is at least an $\Omega(1/r)$ fraction of the expected value of an optimal (unsafe) routing policy.

Lemma 1. *If Algorithm* GREEDY-IR*routes commodities only on r-coverable paths, then its expected value is at least a $(1 - \alpha)/(r + 1)(1 + \alpha)$ fraction of that of an optimal routing policy.*

Proof. Fix an execution of Algorithm GREEDY-IRon a stochastic routing instance. Let h denote the number of times that the main while loop executes. Relabel the commodities $I = \{1, \ldots, k\}$ so that the ith commodity is routed in iteration i. Set $I^0 = I$ and I^j equal to $\{j+1, \ldots, k\}$, the commodities remaining after the first $j \leq h$ iterations. Set $c^0 = (1 - \alpha)c$ and let c^j denote the residual capacities $\hat{c}$ after the first j commodities have been routed. By the stopping condition, $LP(I^h, c^h) = 0$.

Our key claim is that for every $j \in \{1, 2, \ldots, h\}$,

$$LP(I^{j-1}, c^{j-1}) - LP(I^j, c^j) \leq r \cdot w_j \cdot d_j + v_j, \tag{1}$$

where d_j is the instantiated size of commodity j. Conceptually, this claim asserts that each time we route a new commodity, the amount by which the value of $LP(I^j, c^j)$ decreases is not much more than the additional value that we accrue. Since the initial value $LP(I, c^0)$ is comparable to the expected value of an optimal routing policy (by Corollary 1), this ensures that, in expectation, Algorithm GREEDY-IRwill capture a significant (roughly $1/r$) fraction of the maximum-possible expected value.

To prove the claim, fix j and let P^* denote the path on which Algorithm GREEDY-IRroutes commodity j. By the definition of r-coverable, there is a flow decomposition $\{\hat{f}_P^{(i)}\}$ of an optimal solution $\hat{f}$ to $LP(I^{j-1}, c^{j-1})$ that r-covers P^*. Let S denote the paths that carry a positive amount of flow in this decomposition. We next massage this path decomposition into a feasible solution for $LP(I^j, c^j)$ in two steps. For an edge $e \in P^*$, let $\hat{f}_e^{(-j)}$ denote the flow

$\sum_{i \neq j} \hat{f}_e^{(i)}$ on edge e belonging to commodities other than j. We first decrease flow on paths of $\mathcal{S}$ for commodities other than j until the flow of every edge $e \in P^*$ has decreased by at least $\min\{\hat{f}_e^{(-j)}, d_j\}$. We then remove all flow paths corresponding to commodity j. Since $c_e^j = \max\{0, c_e^{j-1} - d_j\}$ for $e \in P^*$ and $c_e^j = c_e^{j-1}$ for $e \notin P^*$, these two steps define a flow g feasible for $LP(I^j, c^j)$.

We now elaborate on the first step. Initialize $g_P^{(i)}$ to $\hat{f}_P^{(i)}$ for all paths $P \in \mathcal{S}$. Let $F^* \subseteq P^*$ denote the edges of P^* from which flow still needs to be removed, in the sense that $\hat{f}_e^{(-j)} - g_e^{(-j)} < \min\{\hat{f}_e^{(-j)}, d_j\}$. While $F^* \neq \emptyset$, we decrease flow on paths of $\mathcal{S}$ as follows. Consider the paths P of $\mathcal{S}$ with $g_P^{(i)} > 0$, $i \neq j$, and $P \cap F^* \neq \emptyset$. Each edge of F^* lies in at least one such path. Since the original flow decomposition of $\hat{f}$ r-covers P^*, there are $q \leq r$ such paths $P_1, \ldots, P_q$ that collectively contain all of the edges of F^*. We decrease the corresponding value of $g_P^{(i)}$ for each of these paths at a uniform rate, until either $\hat{f}_e^{(-j)} - g_e^{(-j)} = \min\{\hat{f}_e^{(-j)}, d_j\}$ for some edge $e \in F^*$, or until $g_P^{(i)}$ is decreased to 0 for one of the paths $P_1, \ldots, P_q$. We denote by Δ_ℓ the amount by which the flow on $P_1, \ldots, P_q$ is decreased during the ℓth iteration of this procedure.

As long as $F^* \neq \emptyset$, we can perform the above operation to decrease flow. Every iteration strictly decreases the sum of $|F^*|$ and the number of paths of $\mathcal{S}$ that carry flow in g. The above procedure must therefore terminate with a final flow g. After deleting all of the flow paths corresponding to the commodity j, the flow g is feasible for $LP(I^j, c^j)$.

We complete the proof of the key claim by comparing the objective function values of $\hat{f}$ and g. First, we have

$$w_j \sum_{P \in \mathcal{P}_j} \hat{f}_P^{(j)} \leq w_j \cdot \mu_j = v_j. \tag{2}$$

Second, consider the flow decrease operations used to obtain the final flow g from $\hat{f}$. Every such operation decreases flow on at most r paths. Also, since every such operation decreases the amount of flow on every edge of F^*, the total flow decrease $\sum_{\ell \geq 1} \Delta_\ell$ over all such operations is at most d_j. Thus $\sum_{i \in I_j} \sum_{P \in \mathcal{P}_i} (\hat{f}_P^{(i)} - g_P^{(i)}) \leq r \cdot d_j$. By the definition of Algorithm GREEDY-IR, $w_j \geq w_i$ for every commodity $i \in I^j$ with $\hat{f}_e^{(i)} > 0$ for some $e \in E$. Hence

$$\sum_{i \in I^j} w_i \sum_{P \in \mathcal{P}_i} \hat{f}_P^{(i)} - \sum_{i \in I^j} w_i \sum_{P \in \mathcal{P}_i} g_P^{(i)} \leq r \cdot d_j \cdot w_j. \tag{3}$$

Since $\hat{f}$ and g are optimal and feasible solutions to $LP(I^{j-1}, c^{j-1})$ and $LP(I^j, c^j)$, respectively, adding the inequalities (2) and (3) proves the claim (1).

With the key claim in hand, we now complete the proof of the lemma. First, for a fixed execution of Algorithm GREEDY-IR, we can sum (1) over all $j \in \{1, 2, \ldots, h\}$ to obtain

$$\frac{1-\alpha}{1+\alpha} \cdot OPT \leq LP(I, (1-\alpha)c) \leq \sum_{i \in I^h} v_i \left(r \frac{d_i}{\mu_i} + 1 \right), \tag{4}$$

where the first inequality follows from Corollary 1 with $\gamma = (1-\alpha)/(1+\alpha)$, and in the second inequality we are using the equalities $w_i = v_i/\mu_i$ and $LP(I^h, c^h) = 0$.

Finally, consider a random execution of the algorithm GREEDY-IR. Label the commodities $1, 2, \ldots, k$ in an arbitrary way. Let X_i denote the indicator variable for the event that Algorithm GREEDY-IR attempts to route commodity i, and D_i the random variable equal to the size of commodity i. By the Principle of Deferred Decisions, the random variables X_i and D_i are independent for each i. Taking expectations in (4), we have

$$\frac{1-\alpha}{1+\alpha} \cdot OPT \leq \mathbf{E}\left[\sum_{i=1}^{k} X_i \cdot v_i \left(r \frac{D_i}{\mu_i} + 1\right)\right] = r \sum_{i=1}^{k} \frac{v_i}{\mu_i} \mathbf{E}[X_i \cdot D_i] + \sum_{i=1}^{k} v_i \mathbf{E}[X_i]$$

$$= r \sum_{i=1}^{k} \frac{v_i}{\mu_i} \mathbf{E}[X_i] \cdot \mathbf{E}[D_i] + \sum_{i=1}^{k} v_i \mathbf{E}[X_i] \tag{5}$$

$$= (r+1) \sum_{i=1}^{k} v_i \mathbf{E}[X_i], \tag{6}$$

where (5) follows from the independence of X_i and D_i. Since Algorithm GREEDY-IR is a safe routing policy (Fact 1), the sum on the right-hand side of (6) is precisely the expected value obtained by Algorithm GREEDY-IR. □

To usefully apply Lemma 1, there must be a commodity i that meets two orthogonal criteria: a large ratio w_i and a flow path that is r-coverable for small r. When the maximum variation w_1/w_k in expected per-unit values is small, the choice of commodity can be dictated by the second criterion alone. Precisely, we have the following variation on Lemma 1, which will be useful in Section 4.

Lemma 2. *If Algorithm IR routes commodities only on r-coverable paths, then its expected value is at least a $(1 - \alpha)/(rW + 1)(1 + \alpha)$ fraction of that of an optimal routing policy, where $W = w_1/w_k$.*

4   Iterative Rounding in Planar Graphs

We now consider the SSSR problem in planar graphs and show the existence of r-coverable paths in them. In particular, we show that there always exists a 2-coverable commodity in a planar flow and give an algorithm for finding it (Section 4.1). Unfortunately, this is not necessarily the commodity with the maximum per-unit value w_i. (See the full version [2] for a planar SSSR instance where the maximum per-unit value commodity is only $\Theta(\log k)$-coverable.) However, limiting our solution to a subset of commodities that have comparable w_i values, we obtain an $O(\log W)$ approximation for general planar graphs, where $W = w_1/w_k$ (Section 4.3).

We obtain a constant-factor approximation in the special case where all of the sinks lie on a common face in some embedding of the planar network. Here, we show that every commodity has a 2-coverable path (Section 4.2). Lemma 1 then implies that the GREEDY-IR algorithm achieves a constant-factor approximation for such instances.

4.1   Preliminaries

Let $G = (V, E)$ be a planar multicommodity flow network with a single source s, and f a feasible flow. Let $g : V \to \Re^2$ be a straight-line planar embedding of G. Such an embedding always exists [19].

A Non-crossing Path-Decomposition. Recall that $\{f_P^{(i)}\}_{P \in \mathcal{S}}$ denotes a path-decomposition of f with $\mathcal{S}$ being the set of flow-carrying paths. We are interested in path decompositions of planar flows that are *non-crossing*, as defined below.

Definition 2. *A path P crosses another path P' if there exists a bounded connected region X in $\Re^2$ with the following properties: P and P' each cross the boundary of X exactly twice and these crossings are interleaved. Precisely, if we scan the boundary of X in clockwise direction starting from the point where P enters it, we encounter P' exactly once before we see P again (Figure 2(a)). The set X is called a* witness *to this crossing of P and P'.*

Definition 3. *A set of paths is said to be* non-crossing *if every pair of paths is distinct and non-crossing.*

Given two crossing paths, we can "uncross" them (Figure 2(b)). We therefore get the following lemma (proof omitted for brevity).

Lemma 3. *Every single-source planar multicommodity flow f has a non-crossing path decomposition that can be found in polynomial time.*

Given a non-crossing path-decomposition $\{f_P\}_{P \in \mathcal{S}}$, we can pick a small cover for a path as follows. We order all the paths in anticlockwise order. (This is well defined because no two paths cross.) Then for any path, roughly speaking, the two paths immediately neighboring the path should cover all its intersections with other paths.

More formally, we define a linear order $\prec$ on paths as follows. We order all the edges incident on s in anticlockwise order, starting from an arbitrary edge. This divides the paths $P \in \mathcal{S}$ into groups $\mathcal{S}_e$ based on the first edge in each path. If the edge e_1 precedes edge e_2 in anticlockwise order, then for all $P_1 \in \mathcal{S}_{e_1}$ and $P_2 \in \mathcal{S}_{e_2}$, we have $P_1 \prec P_2$. We then refine the ordering in each group. For group $\mathcal{S}_e$ with $e = u \to v$, consider all edges outgoing from v, and order them in anticlockwise order starting from e. This subdivides the group $\mathcal{S}_e$ into subgroups $\mathcal{S}_{e'}$ based on the next edge e' in each path. As before, if the edge e_1' precedes edge e_2' in anticlockwise order, then for all $P_1 \in \mathcal{S}_{e_1'}$ and $P_2 \in \mathcal{S}_{e_2'}$, we have $P_1 \prec P_2$. We continue in this manner until we obtain a total order. We rename the paths according to this order so that $P_1 \prec \cdots \prec P_q$ with $q = |\mathcal{S}|$.

Undominated Commodities. Fix a non-crossing flow decomposition of a planar single-source multicommodity flow and a flow path P. Above, we suggested covering a path P using the two immediately neighboring paths. This is not sufficient to cover all of the intersections between P and other flow paths if, informally, the neighboring paths are "shorter" than P. To dodge this issue, we

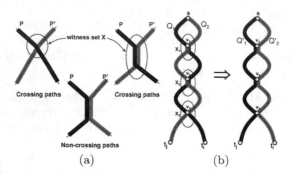

Fig. 2. (a) Crossing and non-crossing paths; (b) Converting a crossing path-decomposition to a non-crossing one

define a partial order on the commodities, roughly in order of the source-sink distance, and pick the commodity that is the "closest" to the source in this order.

For a commodity i, let E_i denote the set of edges from which t_i is reachable along flow-carrying edges. Let $\mathcal{A}_i$ denote the subset of $\Re^2$ enclosed by this set of edges (not including $g(t_i)$). We call this set the *region enclosed by i*.

Definition 4. *A commodity i dominates a commodity j if $g(t_i) \in \mathcal{A}_j$.*

It is easy to verify that the dominance relation defines a partial order on commodities.

Lemma 4. *If i dominates j, then $\mathcal{A}_i \subset \mathcal{A}_j$.*

Corollary 2. *The dominance relation is transitive and antisymmetric; hence, there exists an undominated commodity.*

4.2   Undominated Commodities Are 2-Coverable

We now show that for every planar single-source multicommodity flow, there is at least one 2-coverable flow path.

Lemma 5. *Let $\{f_P^{(i)}\}_{P \in \mathcal{S}}$ be a non-crossing path decomposition of the planar, single-source multicommodity flow f. Let i be an undominated commodity. Then every commodity i flow path in $\mathcal{S}$ is 2-covered by $\{f_P^{(i)}\}_{P \in \mathcal{S}}$.*

Proof. (Sketch) Let $P_1 \prec \cdots \prec P_q$ be a linear order on $\mathcal{S}$ defined as in the previous subsection. Consider a commodity i flow path $P = P_l \in \mathcal{S}$ and let $\mathcal{S}' \subseteq \mathcal{S}$. Let $x_1 = \operatorname{argmax}_{x<l}\{P_{x \bmod q} \in \mathcal{S}'\}$ and $x_2 = \operatorname{argmin}_{x>l}\{P_{x \bmod q} \in \mathcal{S}'\}$. Let $Q_1 = P_{x_1 \bmod q}$ and $Q_2 = P_{x_2 \bmod q}$. A reasonably straightforward argument then shows that $\{Q_1, Q_2\}$ covers P with respect to $\mathcal{S}'$. $\qquad\square$

Lemmas 2 and 5 easily imply a constant-factor approximation ratio for the GREEDY-IRalgorithm when all sinks like on a common face in some planar embedding. In particular, if we consider a planar embedding of the graph with all sinks on the outer face, then by definition, all the commodities are undominated.

Theorem 2. *In a planar instance of SSSR in which all sinks lie on a single face, algorithm GREEDY-IR achieves a $\left(3\frac{(1+\alpha)}{(1-\alpha)}\right)$-approximation.*

Of course, Theorem 2 includes the special cases of outerplanar networks and of single-source, single-sink planar instances of SSSR.

4.3   An $O(\log W)$-Approximation for General Planar Graphs

In the previous subsection we showed that there always exists a 2-coverable commodity in a planar flow. Unfortunately, we show in the full version that the commodity with the highest value of w_i may not be $o(\log k)$-coverable. However, as we show below, having at least one 2-coverable commodity in every planar graph instance is sufficient to obtain an $O(\log W)$-approximation, where $W = w_1/w_k$ is the ratio between the maximum and minimum per-unit values.

We can assume via scaling that the minimum per-unit value w_k is 1. We divide the commodities into $\log W$ groups: $I_x = \{i : w_i \in [2^x, 2^{x+1})\}$ for each $x \in \{0, \cdots, \log W\}$.

Algorithm PLANAR-IR proceeds as follows. We consider the optimal values $\mathcal{V}_x$ of $\log W$ linear programs $LP(I_x, (1-\alpha)c)$, one for each group I_x. These values give us an estimate of the total value that an optimal adaptive solution can derive from each group of commodities. Let x^* be the index of the group for which the maximum value $\mathcal{V}_x$ is achieved. We run the algorithm IR on the graph using only commodities in the group I_{x^*}. (In other words, we round the flow obtained by solving the $LP(I_{x^*}, (1-\alpha)c)$.) In step 2a of the algorithm, we pick any undominated commodity and route it along a flow path in a non-crossing path decomposition of the flow.

Theorem 3. *Algorithm PLANAR-IR is a $(5\frac{1+\alpha}{1-\alpha} \log W)$-approximation.*

Proof. Since we pick the best over $\log W$ groups of commodities, $\mathcal{V}_{x^*}$ is at least a $1/\log W$ fraction of the value of $LP(I, (1-\alpha)c)$. Now Lemma 5 implies that we always route a commodity along a 2-coverable path in step 2a of the algorithm PLANAR-IR. Furthermore, the per-unit value of the commodity routed in each step is at least half the per-unit value of any other commodity in the set I_{x^*}. Lemma 2 then implies that the expected value obtained by the PLANAR-IR algorithm is at least a $1/5$ fraction of $\mathcal{V}_{x^*}$, and is thus at least a $\frac{1-\alpha}{5(1+\alpha)}\frac{1}{\log W}$ fraction of the expected value obtained by an optimal routing policy for all of the demands.   □

References

1. John R. Birge and Francois Louveaux. *Introduction to stochastic programming.* Springer Series in Operations Research. Springer-Verlag, New York, 1997.
2. Shuchi Chawla and Tim Roughgarden.   Single-source stochastic routing. http://www.cs.cmu.edu/~shuchi/papers/stoch-routing.ps.
3. George B. Dantzig. Linear programming under uncertainty. *Management Science,* 1:197–206, 1955.

4. B. Dean, M. Goemans, and J. Vondrak. Adaptivity and approximation for stochastic packing problems. In *SODA '05*, pages 395–404.
5. B. Dean, M. Goemans, and J. Vondrak. The benefit of adaptivity: Approximating the stochastic knapsack problem. In *FOCS '04*, pages 208–217.
6. Brian Dean. *Approximation Algorithms for Stochastic Scheduling Problems*. PhD thesis, Massachusetts Institute of Technology, Massachusetts, 2005.
7. Yefim Dinitz, Naveen Garg, and Michel X. Goemans. On the single-source unsplittable flow problem. *Combinatorica*, 19(1):17–42, 1999.
8. M. Goemans and J. Vondrak. Stochastic covering and adaptivity. In *LATIN '06*, pages 532–543.
9. A. Gupta, M. Pal, R. Ravi, and A. Sinha. Boosted sampling: Approximation algorithms for stochastic optimization. In *STOC '04*, pages 417–426.
10. A. Gupta, M. Pal, R. Ravi, and A. Sinha. What about wednesday? approximation algorithms for multistage stochastic optimization. In *APPROX '05*.
11. A. Gupta, R. Ravi, and A. Sinha. An edge in time saves nine: Lp rounding approximation algorithms for stochastic network design. In *FOCS '04*, pages 218–227.
12. N. Immorlica, D. Karger, M. Minkoff, and V. Mirrokni. On the costs and benefits of procrastination: Approximation algorithms for stochastic combinatorial optimization problems. In *SODA '04*, pages 684–693.
13. Jon Kleinberg. Single-source unsplittable flow. In *FOCS '96*, pages 68–77.
14. S. G. Kolliopoulos and C. Stein. Approximation algorithms for single-source unsplittable flow. *SIAM Journal on Computing*, 31(3):919–946, 2001.
15. R. Ravi and A. Sinha. Hedging uncertainty: approximation algorithms for stochastic optimization problems. *Mathematical Programming*, 2005.
16. David Shmoys and Chaitanya Swamy. Sampling-based approximation algorithms for multi-stage stochastic optimization. In *FOCS '05*.
17. David Shmoys and Chaitanya Swamy. Stochastic optimization is (almost) as easy as deterministic optimization. In *FOCS '04*, pages 228–237.
18. Stochastic programming community homepage. http://stoprog.org/.
19. W. T. Tutte. How to draw a graph. *Proceedings of the London Mathematical Society*, 3(13):743–768, 1963.

An $O(\log n)$ Approximation Ratio for the Asymmetric Traveling Salesman *Path* Problem

Chandra Chekuri[1] and Martin Pál[2]

[1] Lucent Bell Labs, 600 Mountain Avenue, Murray Hill, NJ 07974, USA
[2] Google Inc., 1440 Broadway, New York, NY 10018, USA
chekuri@cs.uiuc.edu, mpal@google.com

Abstract. Given an arc-weighted directed graph $G = (V, A, \ell)$ and a pair of vertices s, t, we seek to find an *s-t walk* of minimum length that visits all the vertices in V. If ℓ satisfies the *asymmetric* triangle inequality, the problem is equivalent to that of finding an *s-t path* of minimum length that visits all the vertices. We refer to this problem as ATSPP. When $s = t$ this is the well known asymmetric traveling salesman tour problem (ATSP). Although an $O(\log n)$ approximation ratio has long been known for ATSP, the best known ratio for ATSPPis $O(\sqrt{n})$. In this paper we present a polynomial time algorithm for ATSPPthat has approximation ratio of $O(\log n)$. The algorithm generalizes to the problem of finding a minimum length path or cycle that is required to visit a subset of vertices in a given order.

1 Introduction

In the classical traveling salesman problem (TSP) we are given an undirected (directed) graph with edge (arc) lengths and we seek to find a Hamiltonian cycle of minimum length. It is one of the most extensively studied combinatorial optimization problems. TSP is not only NP-hard, it is also NP-hard to approximate to within any polynomial factor - both these facts follow easily from the NP-Completeness of the Hamiltonian cycle problem. We obtain a more tractable variant of the problem if we ask for a tour instead of a cycle; that is we allow a vertex to be visited multiple times. In the undirected graph setting this relaxation is equivalent to assuming that the edge lengths satisfy the triangle inequality and in directed graphs this is equivalent to assuming that the arc lengths satisfying the asymmetric triangle inequality. The relaxed problem is referred to as Metric-TSP in undirected graphs and ATSP in directed graphs. For Metric-TSP the best known approximation ratio is $3/2$ due to Christofides [6]. For ATSP an approximation ratio of $\log_2 n$ was obtained by Frieze, Galbiati and Maffioli [8]. This ratio has been slightly improved [3, 12] and the best ratio known currently is $0.842 \log_2 n$ [12].

In this paper we are concerned with the traveling salesman *path* problem. The input to the problem is a graph with edge (arc) lengths and two vertices s and t. We seek a path from s to t of minimum length that visits all the vertices. The path version is NP-hard and also hard to approximate to within any polynomial

J. Diaz et al. (Eds.): APPROX and RANDOM 2006, LNCS 4110, pp. 95–103, 2006.

factor via a reduction from the Hamiltonian path problem. We therefore consider the relaxed version where the objective is to find a walk instead of a path. We refer to undirected graph and directed graph versions as Metric-TSPP and ATSPPrespectively. For Metric-TSPP the best known approximation ratio is 5/3 due to Hoogeveen [11] (see [10] for a different proof). The ATSPPproblem does not seem to have been considered much in the literature and we are only aware of the recent work of Lam and Newman [14] who give an $O(\sqrt{n})$ approximation. Our main result is the following.

Theorem 1. *There is an $O(\log n)$ approximation algorithm for the* ATSPP*problem.*

We also consider a generalization of ATSPP. We are given a set of distinct vertices $\{v_1, v_2, \ldots, v_k\}$ and seek a minimum length path P (or cycle) that visits all vertices of the graph but visits $v_1, v_2, \ldots, v_k$ in that order. We can assume without loss of generality that the path P starts at v_1 and ends at v_k. In the undirected graph setting, this problem has been referred to as path-constrained TSP and is a special case of a more general problem called precedence-constrained TSP [4]. Bachrach *et al.* [2] gave a 3-approximation for the path-constrained TSP in metric spaces. Our approach for ATSPPgeneralizes to the asymmetric version of the path-constrained TSP.

Theorem 2. *There is an $O(\log n)$ approximation algorithm for the path-constrained* ATSPP*problem.*

ATSPP vs ATSP: It is easy to see that an α approximation for ATSPPimplies an α approximation for ATSP; we can reduce a given ATSPinstance to an ATSPPinstance on the same graph by choosing an arbitrary vertex v and and setting $s = t = v$. At first glance it might appear that ATSPPcan be reduced to ATSPby taking an instance of ATSPPand adding an arc (t, s) to the graph with an appropriate length. It is, however, not hard to convince oneself that such a reduction does not work. To better understand the difficulty in the directed setting and develop the main ingredient of our algorithm we give a brief overview of the algorithm of Frieze *et al.* [8] for ATSPand a variant proposed by Kleinberg and Williamson [13] (see [16] for a description and proof). Both algorithms work in an iterative fashion. We let OPT denote the value of an optimum solution to a given instance.

The algorithm in [8] finds a collection of directed cycles othat partition the vertex set (called a cycle-cover in some settings) such that the total length of the cycles is minimized. at most OPT. This can be achieved in polynomial time using a reduction to the minimum cost assignment problem. Note that the optimum solution to the given instance of ATSPis a single cycle that spans the vertices, and hence the length of the cycles computed is at most OPT. From each cycle an arbitrary vertex is chosen to be the cycles proxy and the problem is reduced to the graph induced on the proxy vertices. The number of proxies is no more than half the number of initial vertices since each cycle constains at least two vertices. A tour in the smaller graph can be extended to the original graph using the cycles. Further, it can be easily seen that there must be a tour of length OPT

in the new instance on the proxy vertices. Thus the algorithm incurs a cost of
OPT in each iteration and since the number of vertices is reduced by a factor
of 2 in each iteration, the total length of the final tour is upper bounded by
$\log_2 n \cdot$ OPT.

The algorithm in [13] works differently. It finds a single cycle in each iteration
such that the ratio of the length of the cycle to the number of vertices in the cycle is minimum. Such a cycle (also called a minimum mean cycle) can be found in
polynomial time [1]. An arbitrary vertex in the cycle is chosen as a proxy and the algorithm works in a reduced graph with the non-proxy vertices of the cycle removed.
The analysis is similar to that of the analysis of the greedy algorithm for covering
problems, in particular the set cover problem [7]. This results in an approximation
ratio of $2H_n$ where $H_n = 1 + 1/2 + \ldots + 1/n$ is the n-th harmonic number.

Both the algorithms described above crucially rely on the fact that cycles
allow the problem size to be reduced. Cycles can be used in a similar way for
ATSPPas well. However, in ATSPP, cycles cannot be relied on as the only
building blocks since the solution to the problem might not contain any cycle;
for example G can be a directed path. In addition to cycles, we also need to
consider paths. However there is no simple way to reduce the problem size using
paths. We therefore restrict ourselves to maintaining a single partial path from
s to t. A simple, and indeed the only natural way to augment a partial path P is
to replace one of the arcs (u, v) of P by a subpath P' from u to v that contains
some yet unvisited vertices. Our main technical contribution is the following: for
any partial path there *exists* an augmentation to a path that contains all vertices
such that the length of augmentation is at most 2OPT. We combine this with
the greedy approach similar to that in [13] to prove Theorem 1 and Theorem 2.

Related Work: TSP is a cornerstone problem for combinatorial optimization and
there is a vast amount of literature on many aspects including a large number
of variants. The books [15, 9] provide extensive pointers as well as details. Our
work is related to understanding the approximability of TSP and its variants.
In this context one of the major open problems is to resolve whether ATSPhas
a constant factor approximation. The natural LP relaxation for ATSPhas only
a lower bound of 2 on its integrality gap [5]. Resolving the integrality gap of
this formulation is also an important open problem. The path-constrained TSP
problem is a special case of the precedence-constrained TSP problem [4]: we are
given a partial order on the vertices and the goal is to seek a minimum length
cycle that visits vertices in an order that is consistent with the given partial
order. In [4] it is shown that this general problem is hard to approximate for
even special classes of metric spaces.

2   Preliminaries

Let G be an arc-weighted directed graph. For a path P in G let $V(P)$ and
$A(P)$ denote the vertices and arcs of P respectively. Let $\mathcal{P}(s, t)$ denote the set
of all $s \rightsquigarrow t$ paths in G. A path $P \in \mathcal{P}(s, t)$ is *non-trivial* if it contains internal

vertices, that is $|V(P)| > 2$. Let $\mathcal{C}(s,t)$ denote the set of cycles in G that do *not* contain either s or t. Let P be a non-trivial path in $\mathcal{P}(s,t)$. Then the *density* of P, denoted by $d(P)$, is the ratio of the total arc length of P to the number of internal vertices in P. In other words $d(P) = \sum_{a \in A(P)} \ell(a)/|V(P)-2|$. Similarly, the density of a cycle $C \in \mathcal{C}(s,t)$ is defined to be $d(C) = \sum_{a \in A(C)} \ell(a)/|V(C)|$.

Lemma 1. *Given a directed graph G and two vertices s,t, let λ^* be the density of a minimum density non-trivial path in $\mathcal{P}(s,t)$. There is a polynomial time algorithm that either finds a path $P \in \mathcal{P}(s,t)$ such that $d(P) = \lambda^*$ or finds a cycle $C \in \mathcal{C}(s,t)$ such that $d(C) < \lambda^*$.*

Proof. We give a polynomial time algorithm that takes a parameter $\lambda > 0$ in addition to G and s,t and outputs one of the following: (i) a non-trivial path $P \in \mathcal{P}(s,t)$ with $d(P) \leq \lambda$ (ii) a cycle $C \in \mathcal{C}(s,t)$ with $d(C) < \lambda$ (iii) a proof that no path in $\mathcal{P}(s,t)$ has a density at most λ. This can be combined with binary search to obtain the desired algorithm.

We remove arcs into s and out of t. This ensures that there are no cycles that contain s or t and does not affect the solution. Given λ we create a graph G_λ that differs from G only in the arc lengths. The arc lengths of G_λ, denoted by ℓ', are set as follows:

$$\begin{aligned} \ell'(s,u) &= \ell(s,u) - \lambda/2 & u &\in V - \{s,t\} \\ \ell'(u,t) &= \ell(u,t) - \lambda/2 & u &\in V - \{s,t\} \\ \ell'(u,v) &= \ell(u,v) - \lambda & u,v &\in V - \{s,t\} \end{aligned}$$

It is easy to verify that the density of a path $P \in \mathcal{P}(s,t)$ or a cycle $C \in \mathcal{C}(s,t)$ is at most λ iff its length in G_λ is non-positive. Thus we can use the Bellman-Ford algorithm [1] to compute a shortest path in G_λ between s and t. If the algorithm finds a negative length cycle we output it. Otherwise, if the shortest path length is non-positive then we obtain a path of density at most λ. If the shortest path is of positive length, we obtain a proof of the non-existence of a path in $\mathcal{P}(s,t)$ of density λ. □

We remark that the above proof only guarantees a weakly-polynomial time algorithm due the binary search for λ^*. A strongly polynomial time algorithm can be obtained by using a *parametric* shortest path algorithm. Our focus is on the approximation ratio and hence we do not go into the details of this well-understood area and refer the reader to [1, 17].

Given a directed path P and two vertices $u, v \in P$ we write $u \preceq_P v$ if u precedes v in P (we assume that u precedes itself). If $u \preceq_P v$ and $u \neq v$ we write $u \prec_P v$. If P is clear from the context we simply write $u \preceq v$ or $u \prec v$.

We call a path $P \in \mathcal{P}(s,t)$ *spanning* if $V(P) = V$, otherwise it is *partial*. Let P_1 and P_2 be two paths in $\mathcal{P}(s,t)$. We say that P_2 *dominates* P_1 iff $V(P_1) \subset V(P_2)$. We say that P_2 is an *extension* of P_1 if P_2 dominates P_1 and the vertices in $V(P_1)$ are visited in the same sequence in P_2 as they are in P_1. It is clear that if P_2 extends P_1 then we can obtain P_2 by replacing some arcs of P_1 by subpaths of P_2. Let $\ell(P_1, P_2)$ denote the *cost of extension* which is defined to be

$\sum_{a \in A(P_2) \setminus A(P_1)} \ell(a)$. Note that the cost of extension does not include the length of arcs in P_1.

3 Augmentation Lemma

Our main lemma is the following.

Lemma 2. *Let $G = (V, A, \ell)$ satisfy the asymmetric triangle inequality and let P_1, P_2 in $\mathcal{P}(s,t)$ such that P_2 dominates P_1. Then there is a path $P_3 \in \mathcal{P}(s,t)$ that dominates P_2, extends P_1, and satisfies $\ell(P_1, P_3) \leq 2\ell(P_2)$.*

We remark that the above lemma only guarantees the existence of P_3 but not a polynomial time algorithm to find it. Let us introduce some syntactic sugar before plunging into the proof. For a path P and two vertices $u \preceq_P v$, we use $P(u,v)$ to denote the subpath of P starting at u and ending at v. Specifically for the path P_1, we use the following notation: for a vertex $u \in P_1 \setminus \{t\}$, we denote by u^+ the successor of u on P_1.

Proof of Lemma 2. Consider the set $X \subseteq P_1$ of vertices u with the property that $u \prec_{P_2} u^+$. For each such vertex, we think of replacing the arc (u, u^+) of P_1 by the subpath $P_2(u, u^+)$. Naïvely, we could replace all arcs (u, u^+) by the corresponding subpaths of P_2. Unfortunately this might cause some arcs of P_2 to be used multiple times and thus incur a high cost of extension. To avoid this, we choose only some of the vertices in X to replace their corresponding arcs. We shall *mark* a subset of vertices $u \in X$ with their corresponding path segments $P_2(u, u^+)$ such that each vertex of P_2 occurs in some marked path segment at least once, while each arc of P_2 appears in at most *two* marked segments.

We construct a sequence $g_1, g_2, \ldots$ of marked vertices iteratively. To start, we let $g_1 = s$ be the first marked vertex. Given $g_1, \ldots, g_i$, we construct g_{i+1} as follows. Find the last vertex v on the subpath $P_1(g_i^+, t)$ such that $v \in P_2(s, g_i^+)$. Such a vertex v always exists, as g_i^+ belongs to both path segments. Note that, by the choice of v, $v^+ \notin P_2(s, g_i^+)$, which means that (unless $v = t$) $v \prec_{P_2} v^+$ and thus $v \in X$. If $v \neq t$, we let $g_{i+1} = v$ and continue to the next iteration. If $v = t$, we stop. Let g_l be the last vertex of the constructed sequence. To prove the lemma, it now suffices to prove the following two statements.

(P1) For every vertex $v \in P_2$, there is at least one marked segment $P_2(g_i, g_i^+)$ that contains v.

(P2) Every arc $a \in P_2$ belongs to at most two marked segments $P_2(g_i, g_i^+)$, with $i = 1, \ldots, l$.

These statements in turn follow from the following inequalities:

(I1) For $i = 1, \ldots, l - 1$, we have $g_i \prec_{P_1} g_{i+1}$.

(I2) For $i = 1, \ldots, l - 1$, we have $g_i \prec_{P_2} g_{i+1} \preceq_{P_2} g_i^+$.

(I3) For $i = 1, \ldots, l - 2$, we have $g_i^+ \preceq_{P_2} g_{i+2}$.

In particular, (I2) shows that any two consecutive path segments $P_2(g_i, g_i^+)$ and $P_2(g_{i+1}, g_{i+1}^+)$ overlap. Since the first segment contains s and the last segment contains t, the union of these segments must necessarily cover the whole path P_2. Hence (P1) holds. Inequalities (I2) and (I3) imply that two path segments $P_2(g_i, g_i^+)$ and $P_2(g_j, g_j^+)$ overlap only if $|i - j| \leq 1$, and thus each arc $a \in P_2$ can belong to at most two consecutive segments. This proves (P2).

We finish the proof by showing that (I1)–(I3) hold. (I1) holds by construction, as $g_{i+1} \in P_1(g_i^+, t)$. The second part of (I2), $g_{i+1} \preceq_{P_2} g_i^+$ is easily seen to hold as well, since g_{i+1} is defined to be the last vertex v along the path P_1 such that $v \preceq_{P_2} g_i^+$.

From (I1) we know that g_{i+2} occurs on the path P_1 later than g_{i+1}, thus it must be that $g_{i+2} \preceq_{P_2} g_i^+$ does not hold, and hence $g_i^+ \prec_{P_2} g_{i+2}$. This proves inequality (I3).

Finally, we prove the first part of inequality (I2), $g_i \prec_{P_2} g_{i+1}$. Since $g_1 = s$, this certainly holds for $i = 1$. For contradiction, suppose that $g_{i+1} \preceq_{P_2} g_i$ for some $i > 1$. Consider the iteration in which g_i got marked. Recall that by construction, g_i is the last vertex along the path P_1 that belongs to $P_2(s, g_{i-1}^+)$. But then, from $g_{i+1} \preceq_{P_2} g_i$ and $g_i \preceq_{P_2} g_{i-1}^+$ it follows that $g_{i+1} \preceq_{P_2} g_{i-1}^+$, and hence $g_{i+1} \in P_2(s, g_{i-1})$. This is a contradiction, because by (I1), g_{i+1} occurs on P_1 later than g_i. $\qquad\square$

We obtain the following useful corollary.

Corollary 1. *Let $P \in \mathcal{P}(s, t)$. Then there is a spanning path $P' \in \mathcal{P}(s, t)$ such P' extends P and $\ell(P, P') \leq 2\mathrm{OPT}$.*

Proof. In Lemma 2, we let $P_1 = P$ and we choose P_2 to be some fixed optimum spanning path. The path P_3 guaranteed by the lemma is the desired P'. $\qquad\square$

4   Algorithm for ATSPP

Our algorithm for ATSPP works in a greedy fashion, choosing a best ratio augmentation in every step similar in spirit to that in [13]. The approximation ratio follows from the same arguments as in the analysis of the greedy algorithm for set cover [7].

At any point in time, the algorithm maintains an s-t path P, where $P = (s = p_0, p_1, \ldots, p_k = t)$, and a list $\mathcal{C}$ of vertex disjoint cycles $C_1, \ldots, C_l$. The cycles are at all times disjoint from P and together with P partition the vertex set V. From each cycle C_i, we pick a vertex c_i as a proxy for that cycle. Initially, the path P consists of a single arc s-t, and every vertex $v \in V \setminus \{s, t\}$ is considered a separate (degenerate) cycle. (Thus initially, each vertex will be its own cycle's proxy.)

In each iteration, we seek to decrease the number of components by performing a *path or cycle augmentation*. In a path augmentation step, we pick a path π that starts at some vertex $p_i \neq t$ on the path P, visits one or more cycle proxy vertices, and ends at p_{i+1}, the successor of p_i on P. Let $R(\pi) = c_{i_1}, c_{i_2}, \ldots, c_{i_m}$

be the set of proxy vertices visited by π. Consider the union of the path π and the cycles $\{C_i\}_{c_i \in R(\pi)}$. In this graph, the in-degree of every vertex equals its out-degree, except for p_i and p_{i+1}. Thus, it is possible to construct an Eulerian walk from p_i to p_{i+1} that visits all arcs (and hence all vertices) of $\bigcup_{c_i \in R(\pi)} C_i$. Using triangle inequality and short-cutting, we convert the walk into a path π' that visits every vertex only once without increasing its cost. We then extend P by replacing the arc (p_i, p_{i+1}) by the path π'. Finally, we remove all cycles in $R(\pi)$ from $\mathcal{C}$.

The cycle augmentation step is very similar. We pick a non-degenerate cycle C on proxy vertices (that is, it contains two or more proxy vertices). We let $R(C)$ be the set of proxy vertices visited by C, and consider the graph $C \cup \bigcup_{c_i \in R(C)} C_i$. This graph is Eulerian: by following an Eulerian tour of it and short-cutting, we obtain a cycle C' visiting every vertex of $\bigcup_{c_i \in R(C)} C_i$. We add C' to the list $\mathcal{C}$ (we pick a proxy for C' arbitrarily). Again, we remove all cycles in $R(C)$ from the list $\mathcal{C}$.

In every iteration, we pick a path or a cycle augmentation step with minimum density. In the following, we use π to refer to either an augmenting path or augmenting cycle. For the purposes of this algorithm, we define the density of a path or cycle π to be $d(\pi) = \ell(\pi)/|R(\pi)|$ the ratio of the length of π to the number of proxy vertices covered by π. Note that although we consider only proxy vertices in the above definition of density, we can still use Lemma 1 to find, in polynomial time, an augmenting path of minimum density λ^*, or find an augmenting cycle with density no greater than λ^*.

Each augmenting path or cycle iteration reduces the size of the list $\mathcal{C}$, and hence it takes at most $|V| - 2$ iterations to exhaust it. At this point, all outstanding cycles must have been included in P, and hence P is a spanning path. We output P and stop.

4.1   Bounding the Cost

We now turn to bounding the cost of the resulting path. To do this, we observe the quantity $L = \ell(P) + \sum_{c \in \mathcal{C}} \ell(C)$. Initially, $L = \ell(s, t) \leq \text{OPT}$. Note that in every augmentation step, L increases by at most $\ell(\pi)$, where π is the current augmenting path or cycle. Hence, it is enough to bound the lengths of the augmenting paths and cycles.

Claim. In every iteration, if π is the augmenting path or cycle in that iteration,

$$\ell(\pi) \leq \frac{|R(\pi)|}{|\mathcal{C}|} \cdot 2\text{OPT}.$$

Proof. Let P^* be a minimum length s-t path that visits all proxy vertices of cycles in $\mathcal{C}$. One such path can be obtained by short-cutting the optimum ATSPPpath in G, hence $\ell(P^*) \leq \text{OPT}$. Lemma 2 states that the path P can be extended to a path P_3 such that $R(C) \subseteq P_3$ and the cost of the extension is at most $2\ell(P^*) \leq 2\text{OPT}$. The extension covers $|\mathcal{C}|$ proxy vertices, and hence has density at most $2\text{OPT}/|\mathcal{C}|$. The subpaths of this extension are also valid augmentation paths; and one of them must have density no greater than the density of the whole extension. Thus, there is an augmenting path with density $2\text{OPT}/|\mathcal{C}|$; the density of the best path or cycle can only be lower. $\square$

Lemma 3. *The overall cost of the path output by the algorithm is at most* $\max(4H_{n-2}, 1) \cdot \text{OPT}$.

Proof. At any given stage of the algorithm, let $k = |\mathcal{C}|$ be the number of components left. We claim that the cost of reducing k by one is at most $4\text{OPT}/k$. Assuming the claim and summing over $k = 1, \ldots, |V| - 2$ yields an upper bound of $4H_{n-2}\text{OPT}$ on the total cost of the augmentation steps. We also have to account for the arc (s, t) included in the initialization phase; note that if $n \geq 3$, this arc will be removed during the execution of the algorithm and hence does not contribute to the final cost. It is easy to verify that for $n = 2$, our algorithm finds an optimal solution.

To prove the claim, consider any fixed value of k and consider the augmentation step in which the value of $|\mathcal{C}|$ drops from some $k_1 \geq k$ to $k_2 < k$. The augmentation step was either a path step, or a cycle step. In a path step, $k_1 - k_2$ cycles are removed at cost $2\text{OPT}(k_1 - k_2)/k_1$, i.e. $2\text{OPT}/k_1 \leq 2\text{OPT}/k$ per cycle. In a cycle step, $k_1 - k_2 + 1$ cycles are removed and one cycle is added, at cost $2\text{OPT}(k_1 - k_2 + 1)/k_1$. The amortized cost per cycle is thus $2\frac{\text{OPT}}{k_1} \cdot \frac{k_1 - k_2 + 1}{k_1 - k_2}$. Since in a cycle step, $k_1 - k_2 \geq 1$, the amortized cost per cycle is at most $4\text{OPT}/k_1$. $\square$

We briefly discuss the running time of the algorithm. The number of augmenting iterations is, in the worst case, linear in n. In each iteration we need to find a parametric shortest path between every adjacent pair of vertices in the current partial path. Thus, in the worst case the algorithm requires $\Theta(n^2)$ parametric shortest path computations. Each parametric shortest path computation can be implemented in $O(nm + n^2 \log n)$ time in a graph with n vertices and m arcs [17]. One way to simplify the implementation is to use the transitive closure of the original graph: an arc (u, v) in the trantive closure has length equal to the shortest path from u to v in the original graph. A simple upper bound on the number of arcs in the closure is n^2. Thus a parametric shortest path computation takes $O(n^3)$ time. Putting together these bounds gives a total running time of $O(n^5)$ steps. The running time can be improved at the expense of a (slightly) worse approximation guarantee. In particular the density computation for the augmentation in each iteration can be approximate.

Path-constrained ATSPP. Our algorithm for ATSPPgeneralizes to the path-constrained version in a straight forward fashion. Recall that we are given a sequence of vertices $s = v_1, v_2, \ldots, v_k = t$ and seek a minimum length spanning path in $\mathcal{P}(s, t)$ that visits $v_1, v_2, \ldots, v_k$ in order. The only change from the algorithm for ATSPPis in the initialization step. Instead of starting with a path consisting of the arc (s, t) we start with a path P consisting of the arcs $(v_1, v_2), (v_2, v_3), \ldots, (v_{k-1}, v_k)$. Note that the length of this path is a lower bound on the length of an optimum path. The algorithm simply augments this path to a spanning path in exactly the same way as for ATSPP. The analysis is essentially the same as for ATSPP.

Acknowledgments. We thank Fumei Lam for an enlightening conversation, for sending us a copy of the manuscript [14] and for pointing out [2]. We thank Moses Charikar for pointing out [13]. Part of this work was done while the second author was at Lucent Bell Labs. Chandra Chekuri acknowledges support from an ONR basic research grant N00014-05-1-0256 to Lucent Bell Labs.

References

1. R. Ahuja, T. Magnanti and J. Orlin. Network Flows. Prentice Hall, 1993.
2. A. Bachrach, K. Chen, C. Harrelson, S. Rao and A. Shah. Lower Bounds for Maximum Parsimony with Gene Order Data. *RECOMB Comparative Genomics*, 1–10, 2005.
3. M. Bläser. A New Approximation Algorithm for the Asymmetric TSP with Triangle Inequality. *Proc. of ACM-SIAM SODA*, 638–645, 2002.
4. M. Charikar, R. Motwani, P. Raghavan and C. Silverstein. Constrained TSP and lower power computing. *Proc. of WADS*, 104–115, 1997.
5. M. Charikar, M. Goemans, and H. Karloff. On the Integrality Ratio for Asymmetric TSP. *Proc. of IEEE FOCS*, 101–107, 2004.
6. N. Christofides. Worst-case analysis of a new heuristic for the traveling salesman problem. Technical report, GSIA, CMU, 1976.
7. V. Chvatal. A greedy heuristic for the set-covering problem. *Math. of Oper. Res.*, Vol 4:233–235, 1979.
8. A. Frieze, G. Galbiati and M. Maffioli. On the worst-case performance of some algorithms for the asymmetric traveling salesman problem. *Networks* 12, 23–39, 1982.
9. G. Gutin and A. P. Punnen (Eds.). Traveling Salesman Problem and Its Variations. Springer, Berlin, 2002.
10. N. Guttmann-Beck, R. Hassin, S. Khuller and B. Raghavachari. Approximation Algorithms with Bounded Performance Guarantees for the Clustered Traveling Salesman Problem. *Algorithmica*, Vol 28 pp. 422–437, 2000. Preliminary version in *Proc. of FSTTCS*, 1998.
11. J. Hoogeveen. Analysis of Christofides' heuristic: Some paths are more difficult than cycles. *Operations Research Letters*, 10:291–295, 1991.
12. H. Kaplan, M. Lewenstein, N. Shafir and M. Sviridenko. Approximation Algorithms for Asymmetric TSP by Decomposing Directed Regular Multidigraphs. *Journal of ACM* vol. 52 (2005), pp. 602-626.
13. J. Kleinberg and D. Williamson. Unpublished note, 1998.
14. F. Lam and A. Newman. Traveling Salesman Path Problems. Manuscript, April 2005.
15. E. Lawler, J. K. Lenstra, A. H. G. Rinnooy Kan, and D. Shmoys (Eds.). The Traveling Salesman Problem: A Guided Tour of Combinatorial Optimization. John Wiley & Sons Ltd., 1985.
16. D. Williamson. Lecture Notes on Approximation Algorithms. IBM Research Report RC 21273, February 1999.
17. N. Young, R. Tarjan and J. Orlin. Faster parametric shortest path and minimum balance algorithms. *Networks*, 21(2): 205–221, 1991.

Online Algorithms to Minimize Resource Reallocations and Network Communication

Sashka Davis[1,*], Jeff Edmonds[2], and Russell Impagliazzo[1,*]

[1] Dept. of Computer Science, Univ. of California, San Diego
[2] Dept. of Computer Science, York University, Canada

Abstract. In this paper, we consider two new online optimization problems (each with several variants), present similar online algorithms for both, and show that one reduces to the other. Both problems involve a control trying to minimize the number of changes that need to be made in maintaining a state that satisfies each of many users' requirements. Our algorithms have the property that the control only needs to be informed of a change in a users needs when the current state no longer satisfies the user. This is particularly important when the application is one of trying to minimize communication between the users and the control.

The Resource Allocation Problem (RAP) is an abstraction of scheduling malleable and evolving jobs on multiprocessor machines. A scheduler has a fixed pool of resources of total size T. There are n users, and each user j has a resource requirement for $r_{j,t}$ resources. The scheduler must allocate resources $\ell_{j,t}$ to user j at time t such that each allocation satisfies the requirement ($r_{j,t} \leq \ell_{j,t}$) and the combined allocations do not exceed T ($\sum_j \ell_{j,t} \leq T$). The objective is to minimize the total number of changes to allocated resources (the number of pairs j,t where $\ell_{j,t} \neq \ell_{j,t+1}$).

We consider online algorithms for RAP whose resource pool is increased to sT and obtain an online algorithm which is $O(\log_s n)$-competitive. Further we show that the increased resource pool is crucial to the performance of the algorithm by proving that there is no online algorithm using T resources which is $f(n)$-competitive for any $f(n)$. Note that our upper bounds all have the property that the algorithms only know the list of users whose requirements are currently unsatisfied and never learn the precise requirements of users. We feel this is important for many applications, since users rarely report underutilized resources as readily as they do unmet requirements. On the other hand, our lower bounds apply to online algorithms that have complete knowledge about past requirements.

The Transmission-Minimizing Approximate Value problem is a generalization of one defined in [1], in which low-power sensors monitor real-time events in a distributed wireless network and report their results to a centralized cache. In order to minimize network traffic, the cache is allowed to maintain approximations to the values at the sensors, in the form of intervals containing the values, and to vary the lengths

---

* Research partially supported by NSF Award CCR-0515332, but views expressed are not endorsed by the NSF.

J. Diaz et al. (Eds.): APPROX and RANDOM 2006, LNCS 4110, pp. 104–115, 2006.

of intervals for the different sensors so that sensors with fluctuating values are measured less precisely than more stable ones. A constraint for the cache is that the sum of the lengths of the intervals must be within some precision parameter T. Similar models are described in [2,3]. We adapt the online randomized algorithm for the RAP problem to solve TMAV problem with similar competitive ratio: an algorithm can maintain sT precision and be $O(\log_s n)$-competitive in transmissions against an adversary maintaining precision T.

Further we show that solving TMAV is as hard as solving RAP, by reducing RAP to TMAV. This proves similar lower bounds for TMAV as we had for RAP, when s is near 1.

1 Introduction

Many applications have the following form: a central control is allocating resources among several users. Users have requirements and complain to the control when requirements are not met. The control may then reallocate resources to satisfy the complaints. However, such reallocation is expensive and should be minimized.

1.1 Resource Allocation Problems

For example, consider jobs competing for processor time on a parallel machine. In such time-shared multiprocessor machines, jobs with *rigid* requirements, which cannot run unless they obtain at least a fixed number of processors, can be a bottleneck, reducing system throughput and processor utilization. This overhead might be reduced by treating jobs as having *malleable, evolving* requirements rather than having a hard, fixed requirement for a specific number of processors. Malleable jobs are parallel tasks that can be performed with different numbers of processors, depending on how many they are allotted by the control. Evolving jobs have requirements that vary over time, and can request changes in their allotments when their needs change. Scheduling for such jobs is studied in [4,5,6,7].

We consider the problem of scheduling evolving parallel tasks that can request more processors from a scheduler of a multi-processor system with T identical processors while they are running. Preemption is possible, but expensive, since many parallel tasks have a high context switch cost (e.g. rendering applications). The goal of the scheduler is thus to minimize preemption while satisfying the processor requirements. We make this precise below. In the following problem, a *user* will represent a task/job and *resources* represent processors.

Definition 1. Resource Allocation Problem (RAP) *There are n users. The input specifies the amount of resources $r_{j,t}$ required by user j at time t. At each point in time, a scheduler A must allocate an amount $\ell_{j,t}^A$ to user j that is at least this required amount ($\ell_{j,t}^A \geq r_{j,t}$). The scheduler has only T resources to allocate and hence $\sum_j \ell_{j,t}^A \leq T$. The objective function is to minimize the number of times that the schedule changes the amount $\ell_{j,t}^A$ allocated to each user.*

We assume that at each point in time, the total resources requested $\sum_j r_{j,t}$ does not exceed the amount T available. Otherwise, all schedulers will fail. If these amounts are equal, then the scheduler has no choice. However, if it is less, then the scheduler must decide where to allocated the extra resources. If the scheduler knows the future then it will give the extra resources to users that will need more later.

An online scheduler lacks knowledge about future requirements of the users and so it must guess where to allocate its extra resources. Making it harder, in many situations, the online scheduler will not learn of changes to user's requirements until the user complains because it does not have enough. To formalize the above limitation on information available to a scheduler, we consider a restricted class of schedulers. A *restricted scheduler* only has access to input in the following way. At time t, it learns the set of currently unsatisfied users. It then repeatedly reallocates resources, paying a cost for each change. After each reallocation, it learns the set of unsatisfied users (some users given more might still not have enough resources, whereas others might become unsatisfied after part of their allocations were given to complaining users.) This repeats until all users are satisfied. (Note that the scheduler in this model may be charged repeatedly for reallocations involving the same processor in the same time step.) We denote the above **Restricted-scheduler Resource Allocation Problem** as **RRAP**. Note that unlike RAP, in RRAP the scheduler never learns the resource requirements exactly, only an upper bound for each.

In RAP, it is always possible to fulfill all the requirements. However, schedulers often have to deny some requests in order to preserve resources for others. Our techniques also apply to the following variant, where the scheduler can deny requests by paying a penalty.

Definition 2. Resource Allocation Problem with Penalties (RAPP)
There are n users and a pool of T resources. At time t some user $j \in [n]$ produces a request for resource $r_{j,t}$ with a **penalty** $p_{j,t} \geq 1$. *At each time, the scheduler allocates $\ell_{j,t}$ resources to user j, where the sum of the allocations is at most T. The scheduler's cost is the total number of changes made to allocations, plus the total penalties $p_{j,t}$ over all times when the scheduler fails to satisfy the t'th request.*

Note that, for simplicity, we define user requirements for this problem as instantaneous rather than continuing. However, we could model users with continuing requirements by having users issue requests whenever their current requirements are unmet (due to increase in their requirements or reallocation of their resources).

A restricted scheduler for the RAPP problem only learns of requests that are currently unsatisfied, but learns the required amount and penalty for such requests.

RAPP could model resource allocation for distributed grid computing systems, where the penalty represents the priority of the job requesting the resource. The higher the priority the higher the penalty if the algorithm chooses to not satisfy the request. It could also model malleable jobs, which would make requests for each possible number of processors from largest to smallest until a request is satisfied, with increasing penalties.

We will show that the RAP problem reduces to RAPP, by repeating requests until they must be satisfied. RAP could represent a situation where the jobs are rigid but evolving and the scheduler is not allowed to starve any job.

For our algorithms, we consider memoryless online algorithms that utilize sT resources, and compare them to all knowing all powerful adversaries with a total budget of T. The reason is that when given no extra resources, the online/off-line competitive ratio is infinite (See Sect. 5). Let $E(\mathcal{A}_s(\sigma))$ denote the expected cost of algorithm $\mathcal{A}$ with sT resource on input σ. Let $OPT_1(\sigma)$ denote the minimum cost of any solution with T resources on the same sequence σ. We call the algorithm $\mathcal{A}$ (s, c)-competitive if there exists a constant d, possibly depending on n and T, such that for all inputs σ, $E(\mathcal{A}_s(\sigma)) \leq c\ OPT_1(\sigma) + d$. Our goal is to find algorithms that are (s, c)-competitive for as small values of s and c as possible. Our results are $(s, O(\log_s n)$-competitive algorithms for RAP and RAPP problems.

1.2   Transmission-Minimizing Approximate Value Problem

Small power wireless sensor networks are used for variety of applications from monitoring seismological data to tracking wildlife. Communication in such networks is expensive, especially in terms of power usage. Consider such a network in which there is a set of sensors each capable of transmitting one value to a central cache. The cache needs to know the values read by each sensor, allowing for some imprecision. If the cache needs perfect precision then the cache must be updated each time a sensor's value changes. To minimize communication, the central station might relax the precision for the values read by the sensor. The cache would still need to be notified when the value changed by more than the allowable precision. Thus, some sensors (with fluctuating values) might be given a more relaxed standard of precision than sensors with more stable values.

The problem of setting the precision of approximated cached values is defined in [1] as follows. There are data sources $S_1, \ldots, S_n$. Each data source S_i hosts a value V_i. There is a cache C, which holds an interval approximation to the exact values $V_1, \ldots, V_n$. An interval $[L, H]$ is a valid approximation of a numeric value V if $V \in [L, H]$. If an interval becomes invalid, then the cache must be updated. We add a hard requirement that the sum of the lengths of intervals assigned to the data sources be at most a parameter T; in [1], this was a soft requirement. The goal is to maintain valid intervals of total size T in a way that minimizes the overall network traffic, assuming that each report of an invalid interval and each reassignment of intervals takes one message. We call our version of this problem **Transmission-Minimizing Approximate Value Problem (TMAV)**. [1] give an optimal solution to a related problem under the assumption that the sources' values arise from certain types of probability distributions. Similar problems were studied in [2, 3], who present heuristic approaches and experimentally evaluate their performance. However, we give the first analysis of algorithms when we allow the values read by the data sources to be adversarial.

Similarly to the RAP problem, we compare online algorithms whose sum of all intervals adds up to sT to an adversary bounded by the total precision T. Our algorithm is almost identical to that for RAP, and achieves a similar competitive ratio of $O(\log_s n)$. For scheduling, it would seem more compelling to take s as small as possible, but for TMAV, a larger value of s might still be reasonable, since a factor of s corresponds to a loss of only $\log s$ bits of precision. Again, our algorithm only uses very restricted access to the input: it only sees the current value at a sensor when it becomes invalid or when it changes the required precision at the sensor. This seems essential to any meaningful solution to the problem, since we are trying to minimize communication. Nevertheless, when we prove lower bounds, they apply to algorithms that have a complete history of all values up to the current time.

Our results and outline of paper: In Sect. 2 we present a randomized, memoryless, online algorithm for RRAP problem that is $(s, O(\log_s n)$-competitive for $s > 3$ and show how this algorithm can be used to solve RAP and RAPP. In Sect. 3 we show how the algorithm above is modified to give a similar result for the TMAV problem, for $s > 6$. In Sect. 4 we also show how to reduce RAP to TMAV and RAP to RAPP using competitive ratio preserving online reduction with respect to adaptive online adversaries. In Sect. 5 we prove lower bounds on the competitive ratio achieved by any online algorithm.

2   The Steal From the Rich Algorithm

Our algorithm, *Steal-From-the-Rich* (SFR) for the RRAP problem, is more or less as follows. When a user j requests more resources (but does not specify how much he needs), the scheduler chooses a random user k with probability proportional to the amount $\ell_{k,t}^A$ currently allocated to it. Then he moves resources from j to k so that neither changes by more than a factor of $r = \Theta(\sqrt{s})$.

The SFR algorithm is defined more precisely as follows. Initially, the resources are partitioned evenly, i.e. $\ell_{j,0}^{\mathrm{SFR}} = \frac{sT}{n}$. Let $r = \Theta(\sqrt{s})$ and $\mu > 0$ be parameters defined more precisely later. If at time t user j requests more resources, the algorithm repeats the following until all demands are satisfied; setting j to be the user whose demands are not satisfied:

- It selects another user k at random with probability proportional to its resource allocation, i.e. $\Pr[\, k \text{ is selected} \,] = \frac{\ell_{k,t-1}^{\mathrm{SFR}}}{sT - \ell_{j,t-1}^{\mathrm{SFR}}}$.
- $\delta \leftarrow \min\left\{ \ell_{k,t-1}^{\mathrm{SFR}} - \mu\frac{T}{n}, \frac{r-1}{r}\ell_{k,t-1}^{\mathrm{SFR}}, (r-1)\ell_{j,t-1}^{\mathrm{SFR}} \right\}$, $\ell_{k,t}^{\mathrm{SFR}} \leftarrow \ell_{k,t-1}^{\mathrm{SFR}} - \delta$,
  $\ell_{j,t}^{\mathrm{SFR}} \leftarrow \ell_{j,t-1}^{\mathrm{SFR}} + \delta$. The other allocations are left unchanged.

Note that the choice of δ is the maximum so that ℓ_k^{SFR} does not decrease below $\mu\frac{T}{n}$ nor by more than a factor of r and that ℓ_j^{SFR} does not increase by more than a factor of r. In addition SFR maintains that $\sum_j \ell_{j,t}^{\mathrm{SFR}} = sT$ and $\ell_{j,t}^{\mathrm{SFR}} \geq \mu\frac{T}{n}$ for all intervals, hence it is a valid schedule using sT resources.

Theorem 1. *For $s > 3$, the algorithm SFR for the Resource Allocation Problem is $(s, O(\log_s n))$-competitive against an adaptive online adversary.*

Proof. Let OPT be any adaptive online adversary strategy with a total of T resources. We prove the $O(\log_s n)$ competitive ratio using the potential function $\Phi: R^n \times R^n \to R^+$, where the first input describes the algorithm's configuration and the second the adversary's. More precisely,

$$\Phi_{j,t} = \frac{14}{\log r} \left| \log \left(\frac{\ell_{j,t}^{\mathrm{SFR}}}{r\ell_{j,t}^{\mathrm{OPT}} + \mu T/n} \right) \right| \quad \text{and} \quad \Phi_t = \sum_{j=1}^{j=n} \Phi_{j,t}.$$

The intuition is that this potential function is small when all allocations assigned by the SFR are proportional to those assigned by the adversary. In this case, SFR allocates more to each user and hence any cost incurred by the algorithm will also be incurred by the adversary. At any point when SFR has a cost and the adversary does not, SFR will grow an allocation that is short relative to the adversary's and will probably shrink an allocation that is long relative to the adversary's, thus reducing the potential.

Observe that $\Phi_{j,t} \le O(\log_s n)$. There are two cases in bounding $\Phi_{j,t}$.

1. If $\ell_{j,t}^{\mathrm{SFR}} \gg \ell_{j,t}^{\mathrm{OPT}}$ then

$$\Phi_{j,t} = \frac{14}{\log r} \log \left(\frac{\ell_{j,t}^{\mathrm{SFR}}}{r\ell_{j,t}^{\mathrm{OPT}} + \mu T/n} \right) \le \frac{14}{\log r} \log \left(\frac{sT}{\mu T/n} \right) \le O\left(\frac{\log n}{\log s} \right) = O(\log_s n).$$

2. If $\ell_{j,t}^{\mathrm{SFR}} \ll \ell_{j,t}^{\mathrm{OPT}}$, then

$$\Phi_{j,t} = \frac{14}{\log r} \log \left(\frac{r\ell_{j,t}^{\mathrm{OPT}} + \mu T/n}{\ell_{j,t}^{\mathrm{SFR}}} \right) \le \frac{14}{\log r} \log \left(\frac{rT + \mu T/n}{\mu T/n} \right) \le O\left(\frac{\log n}{\log s} \right) = O(\log_s n).$$

This implies that $\Phi_t \le O(n \log_s n)$, for all t.

Let SFR_t and OPT_t be the costs incurred by the algorithm and the adversary during the tth change in allocations and define $a_t = \mathrm{SFR}_t + (\Phi_t - \Phi_{t-1})$ to be the *amortized update cost* to the algorithm. We will show that for every t, $\mathrm{E}(a_t) \le O(\log_s n)OPT_t$, where the expectation is over the SFR's random choice for k conditioned on the configurations at time $t-1$. This establishes the claimed competitive ratio because:

$$\mathrm{E}(\mathrm{SFR}_s(\sigma)) = \mathrm{E}\left(\sum_t \mathrm{SFR}_t \right) = \mathrm{E}\left(\sum_t (a_t - \Phi_t + \Phi_{t-1}) \right)$$

$$= \sum_t \mathrm{E}(a_t) - \Phi_{end} + \Phi_0 \le \sum_t \mathrm{E}(a_t) + \Phi_0$$

$$\le \sum_t O\left(\frac{\log n}{\log s} \right) OPT_t + O(n \log n) = O(\log_s n)OPT(\sigma) + d,$$

for some $d \in O(n \log n)$ (Recall that for all j, t we bounded $\Phi_{j,t} \le O(\log_s n)$ and s is a constant, hence $\Phi_0 \le O(n \log_s n)$.

Our goal is to establish $\mathrm{E}(a_t) \le O(\log_s n)OPT_t$. We assume without loss of generality that the adversary reallocates resources before issuing an increased request for resources, since moving such a reallocation to before the request changes neither the adversary's nor the algorithm's cost. Thus, we can break the analysis into two cases where one type of two events described below happen.

Case 1: The adversary reallocates resources to users, and the algorithms does nothing (since no user's demands have changed).

Case 2: One iteration of SFR's main loop occurs. The algorithm moves resources from one user to another, the adversary does nothing.

Note that the claim, $\mathrm{E}(a_t) \leq O(\log_s n)OPT_t$, holds trivially when $\mathrm{SFR}_t = OPT_t = 0$, because $a_t = 0$.

Analysis Case 1: The adversary, anticipating that the user's needs will change, adjusts the allocations of some of the users, while the the algorithm not aware of them makes no changes to its configuration: $\mathrm{SFR}_t = 0$ and $OPT_t > 0$. For each such user j, Φ_j increases by at most its maximum value, which we saw is $O(\log_s n)$. Hence, $a_t \leq 0 + O(\log_s n)OPT_t$, giving the competitive ratio as stated.

Analysis Case 2: One iteration of SFR has occurred. $\mathrm{SFR}_t = 2$ because SFR has changed the allocations of two users j and k, and OPT does nothing, hence $OPT_t = 0$. Because user j is requesting more from SFR and not from OPT, we have that $\ell^{\mathrm{SFR}}_{j,t-1} \leq \ell^{\mathrm{OPT}}_{j,t-1}$. Having changed only the allocations of user j and the randomly chosen user k, $\Delta\Phi = \Phi_t - \Phi_{t-1} = \Delta\Phi_j + \Delta\Phi_k$. We bound the expectation of this change to be at most -2. This gives the required bound as $\mathrm{E}(a_t) = 2 + \mathrm{E}(\Phi_t - \Phi_{t-1}) \leq 0$. We consider two cases.

Event B: Let B be the unlikely and unfortunate event that user k which is randomly selected by SFR has a relatively small allocation, namely $\ell^{\mathrm{SFR}}_{k,t-1} < r^2\ell^{\mathrm{OPT}}_{k,t-1} + \mu r\frac{T}{n}$. This event is unlikely because on average the users are allocated $s = \Theta(r^2)$ times more under SFR and because the probability k is selected is proportionally to its allocation. More formally, let K denote the set of users k for which if selected event B occurs. We choose each such k with probability $\frac{\ell^{\mathrm{SFR}}_{k,t-1}}{sT - \ell^{\mathrm{SFR}}_{j,t-1}}$. Because j is requesting, we have that $\ell^{\mathrm{SFR}}_{j,t-1} \leq \ell^{\mathrm{OPT}}_{j,t-1} \leq T$. Because k causes event B, we have that $\ell^{\mathrm{SFR}}_{k,t-1} < r^2\ell^{\mathrm{OPT}}_{k,t-1} + \mu r\frac{T}{n}$. This gives

$$\Pr[B] = \sum_{k \in K} \Pr[\mathrm{SFR} \text{ chooses } k\ ] = \sum_{k \in K} \frac{\ell^{\mathrm{SFR}}_{k,t-1}}{sT - \ell^{\mathrm{SFR}}_{j,t-1}}$$

$$< \frac{\sum_{k \in K}(r^2\ell^{\mathrm{OPT}}_{k,t-1} + \mu r\frac{T}{n})}{sT - T} \leq \frac{r^2 \cdot T + \mu r \cdot T}{(s-1)T} = \frac{r^2 + \mu r}{s - 1} \leq \frac{3}{7}.$$

We set $r = \Theta(\sqrt{s})$ so that this is true. Note that for $r > 1$ and $\mu > 0$, we need $s > 3.34$ for this to be true. (We could decrease s to $3+\epsilon$ by setting $r = 1+\frac{\epsilon}{11}$, $\mu = \frac{\epsilon}{11}$, $\Pr[B] = \frac{1}{2} - \frac{\epsilon}{11}$, the multiplicative constant 14 in the formula for $\Phi_{t,j}$ to $\frac{11}{\epsilon}$, and the competitive ratio[1] to $O(\frac{11}{\epsilon \log r} \log(\frac{rn}{\mu})) = O(\frac{\log n}{\epsilon^2})$.)

Event $\overline{B}$: Suppose that the unlikely and unfortunate event B does not occur and the user selected has relatively big allocation, namely k has $\ell^{\mathrm{SFR}}_{k,t-1} \geq r^2\ell^{\mathrm{OPT}}_{k,t-1} + \mu r\frac{T}{n}$. Recall that $\ell^{\mathrm{SFR}}_{j,t-1}$ is increased and $\ell^{\mathrm{SFR}}_{k,t-1}$ is decreased by $\delta = \min\{\ell^{\mathrm{SFR}}_{k,t-1} - \mu\frac{T}{n}, \frac{r-1}{r}\ell^{\mathrm{SFR}}_{k,t-1},\ (r-1)\ell^{\mathrm{SFR}}_{j,t-1}\}$. The first of these possible values for δ does not occur when $\overline{B}$ happens, because $\ell^{\mathrm{SFR}}_{k,t-1} \geq \mu r\frac{T}{n}$ and hence $\ell^{\mathrm{SFR}}_{k,t-1} - \mu\frac{T}{n} \geq \frac{r-1}{r}\ell^{\mathrm{SFR}}_{k,t-1}$.

---

[1] If OPT was restricted so that it could never give more than αT to a single user, then we could decrease s to $2+\alpha+\epsilon$, because $\ell^{\mathrm{SFR}}_{j,t-1} \leq \ell^{\mathrm{OPT}}_{j,t-1} \leq \alpha T$ allows us to change the $s - 1$ to $s - \alpha$.

Now lets look at the change of the potential function for user j when we are in case 2. Here we have $\ell_{j,t-1}^{\mathrm{SFR}} < \ell_{j,t-1}^{\mathrm{OPT}}$, then $\ell_{j,t}^{\mathrm{SFR}} \le r\ell_{j,t-1}^{\mathrm{SFR}} \le r\ell_{j,t-1}^{\mathrm{OPT}} = r\ell_{j,t}^{\mathrm{OPT}} < r\ell_{j,t}^{\mathrm{OPT}} + \mu\frac{T}{n}$. Recall that $\Phi_j = \frac{14}{\log r}\left|\log\left(\frac{\ell_j^{\mathrm{SFR}}}{r\ell_j^{\mathrm{OPT}}+\mu T/n}\right)\right|$ and because of the absolute value operator, this function decreases as ℓ_j^{SFR} increases when ℓ_j^{SFR} is small and increases with it when it is large.

We now consider the effect of these changes on Φ_j and Φ_k in the two remaining cases: when j increases by a factor of r (event C) or when k decreases by a factor of r (event D).

- Let $C \cap \overline{B}$ be the event that $\delta = (r-1)\ell_{j,t-1}^{\mathrm{SFR}}$, and hence $\ell_{j,t-1}^{\mathrm{SFR}}$ increases by a factor of r. Then $\mathrm{E}(\Delta\Phi_j^{\mathrm{SFR}}|C \cap \overline{B}) = \frac{14}{\log r}\log\left(\frac{\ell_{j,t-1}^{\mathrm{SFR}}}{\ell_{j,t}^{\mathrm{SFR}}}\right) = -14$. Further because ℓ^{SFR} decreases then $\mathrm{E}(\Delta\Phi_k^{\mathrm{SFR}}|C\cap\overline{B}) = \frac{14}{\log r}\log\left(\frac{\ell_{k,t}^{\mathrm{SFR}}}{\ell_{k,t-1}^{\mathrm{SFR}}}\right) < 0$. Combined we have that $\mathrm{E}(\Delta\Phi^{\mathrm{SFR}}|C\cap\overline{B}) = \mathrm{E}(\Delta\Phi_k^{\mathrm{SFR}}|C\cap\overline{B})+\mathrm{E}(\Delta\Phi_j^{\mathrm{SFR}}|C\cap\overline{B}) < -14$.

- Let $D \cap \overline{B}$ be the event that $\delta = \frac{r-1}{r}\ell_{k,t-1}^{\mathrm{SFR}}$, and hence $\ell_{k,t-1}^{\mathrm{SFR}}$ decreases by a factor of r. Given event $\overline{B}$ happens, we have $\ell_{k,t}^{\mathrm{SFR}} \ge \frac{1}{r}\ell_{k,t-1}^{\mathrm{SFR}} \ge \frac{1}{r}\left(r^2\ell_{k,t-1}^{\mathrm{OPT}} + \mu r\frac{T}{n}\right) = r\ell_{k,t}^{\mathrm{OPT}} + \mu\frac{T}{n}$, then $\mathrm{E}(\Delta\Phi_k^{\mathrm{SFR}}|D \cap \overline{B}) = \frac{14}{\log r}\log\left(\frac{\ell_{k,t}^{\mathrm{SFR}}}{\ell_{j,t-1}^{\mathrm{SFR}}}\right) = -14$. As in the previous case since $\ell_{j,t-1}^{\mathrm{SFR}}$ increases we have $\mathrm{E}(\Delta\Phi_j^{\mathrm{SFR}}|D \cap \overline{B}) = \frac{14}{\log r}\log\left(\frac{\ell_{j,t-1}^{\mathrm{SFR}}}{\ell_{j,t}^{\mathrm{SFR}}}\right) < 0$. Combined we have $\mathrm{E}(\Delta\Phi^{\mathrm{SFR}}|D \cap \overline{B}) = \mathrm{E}(\Delta\Phi_k^{\mathrm{SFR}}|D \cap \overline{B}) + \mathrm{E}(\Delta\Phi_j^{\mathrm{SFR}}|D \cap \overline{B}) = -14$.

Now we bound the expectation of the change of the potential function when we are in Case 2, and event B has not occurred:

$$\mathrm{E}(\Delta\Phi^{\mathrm{SFR}}|\overline{B}) = \Pr[C \cap \overline{B}] \cdot \mathrm{E}(\Delta\Phi^{\mathrm{SFR}}|C \cap \overline{B}) + \Pr[D \cap \overline{B}] \cdot \mathrm{E}(\Delta\Phi^{\mathrm{SFR}}|D \cap \overline{B})$$
$$\le -14(\Pr[C \cap \overline{B}] + \Pr[D \cap \overline{B}]) \le -14\Pr[\overline{B}].$$

Event B: We now bound $\Delta\Phi$ when the unlikely and unfortunate event B does occur. Because we are in Case 2, the above argument that $\mathrm{E}(\Delta\Phi_j^{\mathrm{SFR}}|B) \le 0$ does not change. Since SFR does not change k by more than a factor of r and because OPT does not change the allocations at all, then $\mathrm{E}(\Delta\Phi_k^{\mathrm{SFR}}|B) = \frac{14}{\log r}\log(\frac{\ell_{k,t-1}^{\mathrm{SFR}}}{\ell_{k,t}^{\mathrm{SFR}}}) \le \frac{14}{\log r}\log r$.

$$\mathrm{E}(\Delta\Phi|B) \le \mathrm{E}(\Delta\Phi_j|B) + \mathrm{E}(\Delta\Phi_k|B) \le 14.$$

Conclude Case 2: We bound the change of the potential function as:

$$\mathrm{E}(\Delta\Phi) = \Pr[B]\cdot\mathrm{E}(\Delta\Phi \mid B)+(1-\Pr[B])\cdot\mathrm{E}(\Delta\Phi \mid \overline{B}) \le \frac{3}{7}\cdot(14)+\frac{4}{7}\cdot(-14) = -2.$$

We have established that $\mathrm{E}(a_t) = \mathrm{SFR}_t + \mathrm{E}(\Delta\Phi) \le 0 = O(\log_s n)OPT_t$. Hence $\mathrm{E}(a_t) \le O(\log_s n)OPT_t$, thus concluding the proof. $\square$

2.1   Using SFR to Solve RAPP

We can also use a variant of SFR to solve RAPP as follows. When at time t user j requests $r_{j,t}$ resources with penalty $p_{j,t}$ then we call the main loop for $\mathrm{SFR}(j)$ $\lceil p_{j,t} \rceil$ times or until $l_{j,t} \geq r_{j,t}$.

We argue that this is $O(\log_s n)$-competitive. First, note that the total penalty costs for SFR are at most half that of its communication costs, since we only suffer a penalty $p_{j,t}$ after calling the main loop $\mathrm{SFR}(j)$ at least $p_{j,t}$ times, and each time has communication cost 2. Thus, if we can bound the communication costs of SFR in terms of the total costs of OPT, the same bound, times 1.5, holds for the total costs for SFR.

We use the same potential function as for the analysis of SFR for RAPP. Note that the same bounds in the proof hold for changes in the adversary's allotments. Moreover, the same bound on the expected amortized cost of a loop of SFR holds when $\ell^{\mathrm{SFR}}_{j,t-1} \leq \ell^{\mathrm{OPT}}_{j,t-1}$. If $\ell^{\mathrm{SFR}}_{j,t-1} \geq r_{j,t}$, there is no complaint and the algorithm has no costs. In the final case, $r_{j,t} \geq \ell^{\mathrm{SFR}}_{j,t-1} \geq \ell^{\mathrm{OPT}}_{j,t-1}$, so the adversary fails to satisfy the request and pays penalty $p_{j,t}$. Since SFR performs at most $2p_{j,t}$ iterations of its main loop, and each iteration has communication cost 2, and changes the potential function by at most 28, the total amortized communication costs are at most 60 times the costs for OPT.

3   The Transmission Minimizing Approximate Value Problem

Recall that TMAV problem has n sensors reading values and a central cache must maintain an estimate of each value by knowing an interval $I_{j,t} = [a_{j,t}, b_{j,t}]$ containing this value. The constraint is that the sum of the intervals (or allowable errors) always be bounded by T, namely $\sum_j (b_{j,t} - a_{j,t}) \leq T$. Among other things, this assures that the cache knows the sum $\sum_j v_{j,t} \in \left[\sum_j a_{j,t}, \sum_j b_{j,t} \right]$ within an accuracy of T. In an online algorithm, the cache only learns the new value $v_{j,t}$ if it moves outside of its current interval. At such times, the cache sends a message to the node telling it its new interval. The objective is to minimize the number of messages sent between the cache and the nodes.

To solve the TMAV problem we will modify our algorithm for the RAP. The input to the TMAV problem specifies the value $v_{j,t}$ of node $j \in [n]$ at time $t \geq 0$. The algorithm for RAP on the other side computes allocations, based on resource requests, so we need to map: 1) the allocations computed to intervals and 2) resource pool sT to appropriate precision parameter.

The Steal-From-the-Rich algorithm for this problem mimics that for Resource Allocation Problem but with a total precision of $sT/2$. The algorithm partitions the precision $sT/2$ amongst the n nodes. When the cache learns a new value $v_{j,t}$, it sets $I_{j,t} = [v_{j,t} - \ell^{\mathrm{SFR}}_{j,t}, v_{j,t} + \ell^{\mathrm{SFR}}_{j,t}]$ to have width $2\ell^{\mathrm{SFR}}_{j,t}$ centered around this new value. We assume the algorithm learns a new value $v_{j,t}$, when the value moves out of the current interval or when the interval is changed.

When a node reports that its value is outside its current interval, then we mimic one iteration of SFR to increase its allocation $\ell_{j,t}^{\mathrm{SFR}}$, decrease one other nodes interval. We assign both nodes intervals of width their new allocations, centered around their current values. Note that since $\sum_i \ell_{j,t}^{\mathrm{SFR}} = \frac{sT}{2}$ then the total length of intervals is sT.

Theorem 2. *For $s > 6$, the algorithm SFR for TMAV problem is $(s, O(\log_s n))$-competitive against an adaptive online adversary.*

4   Reductions Between Online Problems

In general reductions between problems are used in at least two different ways. The first usage is: given an algorithm for problem A we can obtain an algorithm for problem B by combining an appropriate reduction function which maps instances of B to instances of A, then use a good algorithm for A. On the other hand, if we have a lower bound for problem B, we inherit the same lower bound for problem A from a reduction from B to A. Online reductions have been used to design algorithms in, for example, [8], which solves a fractional version of a Maximizing Switch Throughput problem and then reduces the more interesting discrete version to the fractional version. Unlike in complexity, online reductions have rarely been used to compare the likely hardness of online algorithms, probably because researchers have been successful at proving lower bounds directly. Since there is a large gap between our lower and upper bounds for these problems, however, we are interested in the relationship between them.

In the full version of the paper we formalize the notion of a general online reduction, which preserves competitive ratio especially against adaptive online adversaries. We use online reductions to relate the hardness of RAP and TMAV, RAP and RAPP, and later in Sect.5 Paging and RAP problems.

Theorem 3. *Let A be any (s, k)-competitive algorithm against adaptive online adversaries for the TMAV problem. Then there exists an (s, k)-competitive algorithm against adaptive online adversaries for the RAP problem.*

Theorem 4 (RAP $\leq_{\mathrm{AD\_ON}}$ RAPP). *There is an online adaptive reduction from RAP with resource factor s to RAPP with resource factor s.*

Proof. Let A be an algorithm for RAPP. We define an algorithm B for RAP as follows. At any time B will have the same allocation as A does. If at any time t, a user j becomes unsatisfied, B simulates a request for $r_{j,t}$ resources from user j to A, assigning it penalty 1, and changes allocations as A does. This continues while there is an unsatisfied user. B maintains the same budget as A, and has costs at most the costs of A.

Let Opt_B be an online adaptive adversary for RAP. When Opt_B reallocates resources, Opt_A reallocates resources accordingly. When Opt_B generates a request $r_{j,t}$ for resources, let Opt_A simulate B, generating the same sequence of requests to A. Note that, since Opt_B's allocation must satisfy all users' requirements, there is no penalty cost for Opt_A for this sequence of requests. Thus, Opt_A's total costs are the same as its reallocation costs, which are the same as Opt_B. Opt_A maintains the same budget as Opt_B. $\square$

5 Lower Bounds

Next we show that the factor of s extra resources is crucial to algorithm's performance. We consider the case when both the algorithm and the adversary have the same resources.

Theorem 5. *There is no online algorithm that is $(1, O(f(n)))$-competitive for the RAP problem or the TMAV, for any function f.*

Proof. (sketch) We give a strategy for the adversary. We consider RAP with $n = 2$ and $T = 1$. The adversary picks a random $r \in [0, 1]$ and places r resources with user 1 and $1-r$ resources with user 2. The request sequence is then generated as follows: To generate the t-th request, $s_t \in [0, 1]$ is chosen uniformly at random. If $s_t \le r$, a request for s_t resources is made by user 1. If $s_t > r$, a request for $1 - s_t$ resources is made by user 2. The adversary can always meet these requests without transferring resources, so the adversaries costs are constant for the entire sequence. We show that, for any online algorithm, the expected costs diverge as t goes to infinity, because to do better would require learning the real value r with a finite number of bits of information. □

In our upper bounds we compared the performance of an online algorithms using sT resources against adversary using T resources. We want to obtain lower bounds on the competitive ratio achievable by online algorithms for RAP using $(1 + \epsilon)$ resources against adversary using 1 resource. We use the following lower bound for Paging.

Theorem 6 ([9, 10]). *Any randomized online algorithm for (h, k)-paging problem has a competitive ratio of at least $\ln \frac{k}{k-h} - \ln \ln \frac{k}{k-h} + \frac{1}{2}$ against oblivious adversary.*

Lemma 1 ((h,k)-Paging $\le_{AD\_ON}$ RAP). *There is an online adaptive reduction from the (h,k)-Paging problem with algorithm using cache of size $k = \frac{1}{\epsilon}$, adversary using cache of size $h = k-1$, and total number of pages needed to be served $k + 1$ to RAP.*

The reduction is trivial. The $k+1$ pages are mapped each to one user of RAP instance. The RAP algorithm is given k resources while the RAP adversary will use h resources. Each request generated for page i at time t is mapped to $r_{i,t} = 1$ request to the RAP algorithm. If the interval assigned to user j by the RAP algorithm is at least 1 then the j-th page will be in the cache, else it will be left out.

Lemma 1, Theorem 6, and the the standard paging lower bound imply the following results.

Theorem 7. *Any online algorithm using $(1+\epsilon)$ resources for the RAP problem has competitive ratio $\Omega(\log(\frac{1}{\epsilon}))$ against an oblivious adversary using resource pool of size 1.*

Theorem 8. *Any online algorithm for the RAP problem using $(1+\epsilon)$ resources has competitive ratio $\Omega(\frac{1}{\epsilon})$ against an adaptive online adversary using resource pool of size 1.*

6   Future Work

We obtained $(s, O(\log_s n))$-competitive algorithms for the RRAP, RAP, RAPP, and TMAV problems and proved that the extra resource sT granted to the algorithm is vital. Our intuition is that the upper bound proved in Sect. 2 is tight, however we have not been able to prove a matching lower bound.

There are two new issues that we have raised. Although online reductions were used before our work ([8]) as a general technique for obtaining new online algorithms from algorithms for other problems. To our knowledge our work is the first that uses the notion of reduction between online problems to prove lower bounds on competitive ratio and relate hardness of one problem to that of another with respect to adaptive online adversaries using reductionist approach.

The second issue is that traditionally online algorithms have known the past history and were oblivious to the future. In this paper we study memoryless algorithms that not only were unaware of the past when making current decision, but also *did not know the current demands exactly*. Their knowledge of the current request is limited and in the process they only learn an upper bound approximation of the request. It will be interesting to know whether other online problems can have similar online solutions.

References

1. Olston, C., Loo, B.T., Widom, J.: Adaptive precision setting for cached approximate values. In: SIGMOD Conference. (2001) 355–366
2. Chandramouli, B., Yang, J., Vahdat, A.: Distributed network querying with bounded approximate caching. In: DASFAA. (2006) 374–388
3. Çetintemel, U., Keleher, P.J., Ahmad, Y.: Exploiting precision vs. efficiency tradeoffs in symmetric replication environments. In: PODC. (2002) 128
4. Kalé, L.V., Kumar, S., DeSouza, J.: A malleable-job system for timeshared parallel machines. In: CCGRID, IEEE Computer Society (2002)
5. Pruyne, J., Livny, M.: Parallel processing on dynamic resources with carmi. In: JSSPP. (1995) 259–278
6. Ioannidis, S., Rencuzogullari, U., Stets, R., Dwarkadas, S.: Craul: Compiler and run-time integration for adaptation under load. Scientific Programming **7** (1999)
7. Edmonds, J.: Scheduling in the dark. Theor. Comput. Sci. **235** (2000) 109–141
8. Azar, Y., Litichevskey, A.: Maximizing throughput in multi-queue switches. In: ESA. (2004) 53–64
9. Young, N.E.: On-line caching as cache size varies. In: SODA. (1991) 241–250
10. Young, N.E.: The k-server dual and loose competitiveness for paging. Algorithmica **11** (1994) 525–541
11. Borodin, A., El-Yaniv, R.: Online Computation and Competitive Analysis. Cambridge University Press (1998)

Weighted Sum Coloring in Batch Scheduling of Conflicting Jobs

Leah Epstein[1], Magnús M. Halldórsson[2], Asaf Levin[3], and Hadas Shachnai[4]

[1] Department of Mathematics, University of Haifa, 31905 Haifa, Israel
lea@math.haifa.ac.il
[2] Department of Computer Science, University of Iceland, IS-107 Reykjavik, Iceland
mmh@hi.is
[3] Department of Statistics, The Hebrew University, 91905 Jerusalem, Israel
levinas@mscc.huji.ac.il
[4] Department of Computer Science, The Technion, Haifa 32000, Israel
hadas@cs.technion.ac.il

Abstract. Motivated by applications in batch scheduling of interval jobs, processes in manufacturing systems and distributed computing, we study two related problems. Given is a set of jobs $\{J_1, \ldots, J_n\}$, where J_j has the processing time p_j, and an undirected intersection graph $G = (\{1, 2, \ldots, n\}, E)$; there is an edge $(i, j) \in E$ if the pair of jobs J_i and J_j cannot be processed in the same batch. At any period of time, we can process a *batch* of jobs that forms an independent set in G. The batch completes its processing when the last job in the batch completes its execution. The goal is to minimize the sum of job completion times. Our two problems differ in the definition of *completion time* of a job within a given batch. In the first variant, a job completes its execution when its batch is completed, whereas in the second variant, a job completes execution when its own processing is completed.

For the first variant, we show that an adaptation of the greedy set cover algorithm gives a 4-approximation for perfect graphs. For the second variant, we give new or improved approximations for a number of different classes of graphs. The algorithms are of widely different genres (LP, greedy, subgraph covering), yet they curiously share a common feature in their use of *randomized geometric partitioning*.

1   Introduction

Batching (see Chapter 8 in [3]) is usually defined as follows. A batch is a set of jobs that can be processed jointly. The completion time of all the jobs in a batch is the finishing time of the last job in the batch. In the p-batch set of problems, the length of a batch is defined as the maximum processing time of any job in the batch. The s-batch set of problems has a different definition for the length of a batch, namely, it is partitioned into a setup time and the sum of the processing times of the jobs in the batch. In this paper we study p-batch problems.

Consider a line communication network (e.g., an optical network), which consists of a set of points on the line, $V = \{1, \ldots, n\}$. Given is a set of requests

J. Diaz et al. (Eds.): APPROX and RANDOM 2006, LNCS 4110, pp. 116–127, 2006.
© Springer-Verlag Berlin Heidelberg 2006

$R_i = [a_i, b_i]$, where $a_i, b_i \in V$; each request R_i has a length p_i, meaning that during a period of p_i time units the communication links along the interval $[a_i, b_i]$ must be dedicated to the processing of R_i. In order to ease the allocation of resources, we need to divide time into disjoint periods; during such a period, each link is dedicated to at most a single request, so the processing of this request starts at the beginning of the period and ends at some point within the same period. When the processing of all jobs assigned to the same time period is completed, this time period ends, and we start serving the requests of the next time period. Note that the intersection graph of the corresponding requests is an interval graph, and at any time period it is possible to process an independent set in this graph. Motivated by the above scheduling problem and other batch scheduling problems arising in manufacturing systems and in distributed computing (see in [5]), we study two related problems.

In the MINIMUM SUM OF BATCH COMPLETION TIMES PROBLEM (**MSBCT**), we are given a set of jobs $\mathcal{J} = \{J_1, \ldots, J_n\}$, where J_j has the processing time p_j, and an undirected intersection graph $G = (\{1, 2, \ldots, n\}, E)$; there is an edge $(i, j) \in E$ if the pair of jobs J_i and J_j cannot be processed in the same batch. The goal is to minimize the sum of job completion times. In each time period we can process a *batch* of jobs that forms an independent set in G. A batch is completed when the last job in the batch finishes its processing, and all the jobs in a batch terminate once the batch is completed. Therefore, in **MSBCT**, the goal is to partition V into independent sets, and to sort these independent sets, so that the weighted sum of batch completion times is minimized; the weight of each batch (or, independent set) S is the number of vertices in S, and its processing time is equal to the maximum processing time of any job in S. Note that once we have a partition of V into independent sets, the optimal order of the sets can be found using Smith's rule [18] (see also [3]).

The MINIMUM SUM OF JOB COMPLETION TIMES PROBLEM (**MSJCT**) is similar to **MSBCT**, except that each job can terminate as soon as its processing is completed, i.e., a job need not wait until the end of processing of the entire batch. However, we still cannot start other jobs that conflict with this job until the end of the batch. Thus, at the beginning of each time period we start the processing of an independent set S in G, and until all the jobs in S are completed, we cannot start any other job. Upon completion of the last job in S, a new time period starts, and we can start processing another independent set in G.

Denote an (optimal) algorithm and its cost by (OPT) $\mathcal{A}$. Since the problem is scalable, we consider the absolute approximation ratio criterion. For a minimization (maximization) problem, the absolute approximation ratio of an algorithm $\mathcal{A}$ is the infimum (supremum) R such that for any input, $\mathcal{A} \leq$ R$\cdot OPT$ ($\mathcal{A} \geq$ R $\cdot OPT$). An algorithm $\mathcal{A}$ for a minimization (maximization) problem yields an R-approximation if its approximation ratio is at most (at least) R.

The SUM COLORING (SC) PROBLEM is the following. Given an input graph $G = (V, E)$, find a *proper coloring* of V, i.e., a function $f : V \to N$ satisfying $f(u) \neq f(v)$ whenever $(u, v) \in E$, such that $\sum_{v \in V} f(v)$ is minimized. Halldórsson et al. [11] considered a class of graphs, that we denote by $\mathcal{F}$, for which one can compute in polynomial time a maximum size ℓ-colorable induced subgraph, for any $\ell \geq 1$. This class includes interval graphs, comparability graphs and co-comparability graphs (see [7, 20]). They presented a (randomized) 1.796-approximation algorithm for SC when G belongs to $\mathcal{F}$, and a deterministic $(1.796 + \varepsilon)$-approximation algorithm for this problem on $\mathcal{F}$. This last result improves an earlier 2-approximation algorithm of Nicoloso et al. [16] for sum coloring of interval graphs. Bar-Noy et al. [1] presented a 4-approximation algorithm for SC of perfect graphs, a 2-approximation algorithm for SC of line graphs and a k-approximation algorithm for SC of $k + 1$-claw free graphs. They also showed a lower bound of $n^{1-\varepsilon}$ for SC of general graphs. We note that SC is the special case of both **MSBCT** and **MSJCT** where all the processing times are equal. Due to the hardness result of SC of general graphs that applies also to our problems, we consider **MSBCT** and **MSJCT** only on special families of graphs.

Feige et al. [6] extended the definition of SC to the following variant of set cover, called MINIMUM SUM SET COVER (MSSC). We are given a collection of subsets $S_1, S_2, \ldots, S_m$ of a ground set $S = \{1, 2, \ldots, n\}$. A feasible solution is an ordering π of a subset $\mathcal{S}'$ of $1, 2, \ldots, m$, such that $\bigcup_{\mathcal{X} \in \mathcal{S}'} \mathcal{X} = S$, and for each element j of the ground set we incur a cost i such that $j \in S_{\pi(i)}$ and $j \notin S_{\pi(k)}$ for all $k < i$. The goal is to find an ordering that minimizes the total cost. They extended the greedy algorithm of [1] to obtain a 4-approximation algorithm for MSSC. This greedy algorithm is equivalent to the well known greedy algorithm for the classical set cover problem [13, 14]. They further showed that this approximation ratio is best possible unless $P = NP$. A weighted generalization of this problem, motivated by database applications, was considered recently by Munagala et al. [15]. In this problem, each subset S_ℓ has a weight c_ℓ. For an element j of the ground set, let i be an index such that $j \in S_{\pi(i)}$ and $j \notin S_{\pi(k)}$ for all $k < i$. The cost incurred by j is $\sum_{\ell=1}^{i} c_\ell$. They showed that an application of the weighted greedy set cover algorithm (see [4]) gives a 4-approximation for this problem as well. To show this, they used linear programming. Note that the above results can be applied to **MSBCT** on perfect graphs. In a graph the sets S_i are given implicitly, however, in perfect graphs we can compute a maximum independent set in polynomial time.

Bar-Noy et al. [2] studied **MSJCT** with integer weights. They presented a 16-approximation for perfect graphs. Their method can be generalized and applied for arbitrary real non-negative weights.

In the MaxIS problem, we are given an undirected graph $G = (V, E)$, and the goal is to find a subset U of V where E does not contain an edge between a pair of vertices in U, such that the size of U is maximized. This problem is well-known to be NP-hard on general graphs (see problem [GT20] in [9]). However, there are graph classes for which it is solvable in polynomial time (e.g. perfect graphs and line graphs), and there are graph classes for which there are efficient approximation algorithms. Given a graph G, we denote by ρ the approximation ratio of an algorithm for MaxIS on the graph class that contains G.

Recall the following properties of perfect graphs (see e.g. [17]): a subgraph of a perfect graph is perfect, and for a perfect graph we can solve MaxIs in polynomial time; also, we can find a proper coloring that uses the minimum number of colors. Let $e \approx 2.71828$ denote the base of the natural logarithm.

Our Results: We describe below our main results for **MSBCT** and **MSJCT**. In Section 2 we define the extension of the greedy algorithm and its analysis (due to [6]) to obtain a $\frac{4}{\rho}$-approximation algorithm for **MSBCT**. Note that this result for the case $\rho = 1$ also follows from the results of [15], however their proof technique is different.

In Section 3 we study **MSJCT**: we first present a $2e \approx 5.43656$-approximation algorithm for the problem on interval graphs, and later obtain a better bound of $1.296e + 3/2 + \varepsilon \approx 5.022$ for any graph class that belongs to $\mathcal{F}$. The first two algorithms can be combined to obtain an improved bound of 4.912 for interval graphs. Then, we show a $\frac{4e}{\rho}$-approximation algorithm for any graph class that has a ρ-approximation algorithm for MaxIS. We also show that the classical Greedy algorithm (given in Section 2) provides alternative $\frac{4e}{\rho}$-approximation for **MSJCT**. Thus, both of these algorithms yield combinatorial $4e \approx 10.87313$-approximation for perfect graphs and line graphs, and a $(2ek + \epsilon)$-approximation for $(k + 1)$-claw free graphs.

In the full version of this paper (see in [5]), we present a general LP-based scheme for batch scheduling that yields a bound of 9.9 for perfect graphs, and $9.9 + o(1)$ for line graphs. Our results for **MSJCT** yield significant improvements over previous bounds, in particular, the bounds that we derive for interval graphs, line graphs and perfect graphs improve upon the uniform bound of 16 obtained for these graph classes in [2]. We summarize the known results for **MSJCT** in Table 1. New bounds given in this paper are shown in boldface, with the previous best known bound given in parenthesis. The reference to previous results is [2].

Table 1. Known results for batch scheduling of conflicting jobs under minsum criteria

	MSJCT
General graphs	$n/\log^2 n$ [1]
Perfect Graphs	**9.9** (16)
Family $\mathcal{F}$	**5.022**
Interval Graphs	**4.912** (16)
k-colorable Graphs	$1.544k + 1$ [2]
Bipartite Graphs	2.796 [2]
$k + 1$-claw free Graphs	**2ek** + ϵ
Line Graphs	**9.9 + o(1)** (16)

Techniques: While the algorithms that we develop for various graph classes are inherently different and tailored to exploit the intrinsic properties of each graph class, there is an interesting link among the approaches at the basis of these algorithms. Crucial to obtaining our approximation results is the usage of *randomized geometric partitioning*. Our main partitioning lemma (Lemma 1) uses randomized rounding to partition the jobs in the instance to weight classes. This randomized partitioning enables to improve the bound of 16 obtained in [2] for **MSJCT**, to $4e$. Our algorithm for the family $\mathcal{F}$ (see in Section 3.3) randomizes on ℓ, and then finds an ℓ-colorable subgraph, from which to construct batches. Finally, our LP-based scheme (omitted) combines randomized partitioning in two different ways, after solving a linear programming relaxation of the problem. We show that the resulting algorithms can be derandomized while preserving their approximation ratio within an additive of ε, for some small $\varepsilon > 0$. We believe that this powerful technique will find more applications in solving other scheduling and partitioning problems.

Due to space constraints, some of the proofs and implementation details are omitted. The detailed results appear in [5].

2   Approximating MSBCT

In this section, we present the greedy algorithm for **MSBCT** and show that it yields a $\frac{4}{\rho}$-approximation for the problem. Our proof follows a similar proof of [6] for the unweighted case.

The Greedy Algorithm
while G is non-empty do:
    for each $j = 1, 2, \dots, n$ do:
        Let G_j be the induced subgraph of G over the vertex set $\{j' : p_{j'} \leq p_j\}$.
        Find a ρ-approximate independent set S_j in G_j.
    Let $j = \mathrm{argmin}_{j'=1,2,\dots,n} \frac{p_{j'}}{|S_{j'}|}$.
    Schedule the jobs in S_j in the next p_j time units, and remove S_j from G

Theorem 1. *The greedy algorithm yields a $\frac{4}{\rho}$-approximation for* **MSBCT**.

Proof. The greedy algorithm clearly returns a feasible solution. To analyze the running time note the following facts. The ρ-approximation algorithm for maximum independent set runs in polynomial time (by definition). Therefore, each iteration of the greedy algorithm runs in polynomial time. Since in each iteration, the size of the vertex set decreases by at least one, the number of iterations is at most n. We get that the algorithm terminates in polynomial number of steps. Thus it remains to show the approximation ratio of the algorithm.

For $i = 1, 2, \dots$ let X_i denote the set of jobs that belongs to the independent set of the i-th iteration of the greedy algorithm. Let R_i be the set of jobs that are still not processed prior to the i-th iteration. Note that $X_i \subseteq R_i$. Denote by $P_i = \max_{j \in X_i} p_j$ the processing time of the batch X_i. Note that the cost of the greedy algorithm is $\sum_i P_i \cdot |R_i|$. For each $j \in X_i$ we define the *price of job* j to be $\mathbf{price}(j) = \frac{|R_i| \cdot P_i}{|X_i|}$. Then, clearly the cost of the greedy algorithm is $\sum_{j=1}^n \mathbf{price}(j)$.

Consider the following histogram corresponding to OPT. There are n columns, one for every job, where the jobs are ordered from left to right by the order in which they were processed by the optimal solution (we break ties arbitrarily). The height of a column is the time step at which the job was completed by the optimal solution (i.e., the time in which the batch that contains the job is completed). Hence, we get a histogram with nondecreasing heights. The total area beneath this histogram is exactly OPT.

Consider now a different diagram corresponding to the greedy solution. Again there are n columns, one for every job, and in analogy to the previous diagram, the jobs are ordered by the order in which they were processed by the greedy solution. But unlike the previous diagram, the height of a column is not the time step by which the job was completed, but rather its price. Hence the heights are not necessarily monotone. The total area of the histogram is exactly the total price, i.e., the cost of the greedy solution. We would like to show that the area of the second histogram is at most $\frac{4}{\rho}$ times that of the first. To do this we shrink the second histogram by a factor of $\frac{4}{\rho}$ as follows. We shrink the height of each column by a factor of $2/\rho$. Hence the column heights are $\mathbf{price}(j)\rho/2$. We also shrink the width of each column by a factor of two. Hence the total width of the second histogram is now $\frac{n}{2}$. We align the second histogram to the right. Namely, it now occupies the space that was previously allocated to columns $\frac{n}{2} + 1$ up to n (assume for simplicity of notation and without loss of generality that n is even). Now we claim that this shrunk version of the second histogram fits completely within the first histogram, implying that its total area is no more than that of the first histogram. This suffices in order to prove the approximation ratio of the greedy algorithm.

Consider an arbitrary point q in the original second histogram, let j be the job to which it corresponds, and let i denote the iteration of the greedy algorithm in which we chose to process job j (i.e., $j \in X_i$). Then the height of q is at most $\mathbf{price}(j)$, and the distance of q from the right hand side boundary is at most $|R_i|$. The shrinking of the second histogram maps q to a new point q'. We now show that q' must lie within the first histogram. The height of q' (which we denote by h) satisfies $h \leq \frac{|R_i| \cdot P_i \cdot \rho}{2|X_i|}$, and the distance of q' from the right hand side boundary (which we denote by r) satisfies $r \leq \frac{|R_i|}{2}$. For this point q' to lie within the first histogram, it suffices to show that by time step h, at least r jobs are still not completed by the optimal solution.

Consider now only the jobs in the set R_i. No independent set whatsoever can complete more than $\frac{|X_i|}{P_i \cdot \rho}$ jobs from R_i per time unit. Hence in h time units the optimal solution could complete processing at most $\frac{h|X_i|}{P_i \rho} \leq \frac{|R_i| \cdot P_i \cdot \rho}{2|X_i|} \cdot \frac{|X_i|}{P_i \cdot \rho} = \frac{|R_i|}{2}$ jobs from R_i, leaving at least $\frac{|R_i|}{2}$ jobs of R_i that the optimal solution still has not completed. Hence the point q' indeed lies within the first histogram. $\square$

Remark 1. If G is perfect or claw-free, then greedy is a 4-approximation algorithm for **MSBCT**.

3   Approximating MSJCT

The outline of the algorithms is as follows: In the preprocessing phase, all of our algorithms partition the jobs into classes according to their processing times and round up the processing time of each job to the maximum processing time of a job in its class. Then, each batch that we generate is devoted to a single class of jobs. For each class of graphs, we design a specialized algorithm to find an approximate partition of each class into batches. Then, all of our algorithms find an optimal sequence of the resulting batches, using Smith's rule. We first present our preprocessing step.

Preprocessing Algorithm
Pick uniformly a random number α in the range $[0, 1)$, i.e., $\alpha \sim \mathbf{U}[0, 1)$.
Partition the jobs into classes according to their processing times,
    i.e., let $\mathcal{J}_0 = \{\, j : p_j \le e^\alpha \,\}$, and $\mathcal{J}_i = \{\, j : e^{i-1+\alpha} < p_j \le e^{i+\alpha} \,\}$.
Denote by k the largest index of any non-empty class.
For all $i = 0, \ldots, k$ and for any job $j \in \mathcal{J}_i$, round up
    the processing time of job j; let $p'_j = e^{i+\alpha}$.

We now show that by partitioning the input into the classes $\mathcal{J}_i$, rounding up the processing times (as done in the Preprocessing Algorithm) and requiring that each batch contains jobs from a single class, we lose a factor of at most e.

Lemma 1. *Denote by OPT a fixed optimal solution, and for a fixed value of α denote by OPT_α an optimal solution to the instance in which we use the rounded up processing times p'_j and the solution must satisfy an additional requirement that is for each batch the jobs scheduled in the batch have a common class. Denote also by OPT and OPT_α the cost of OPT and OPT_α, respectively. Then, $E[OPT_\alpha] \le e \cdot OPT$, where the expectation is over the random choice of α.*

Proof. In the sequel, we prove that it is possible to replace every batch with a sequence of other batches, so that the new schedule satisfies the requirements of the lemma, and the total processing time processing time of each batch grows by an expected multiplicative factor of e.

Consider a fixed batch $\mathcal{B}$ of OPT with job set J, where the maximum processing time of a job in this batch is e^x (i.e., $x = \ln \max_{j \in J} p_j$). We replace $\mathcal{B}$ with a set of batches: the i-th batch in the set serves the jobs of $\mathcal{J}_i \cap J$ and has rounded-up processing time $e^{i+\alpha}$. Since $\mathcal{J}_i \cap J = \emptyset$ for $i > x - \alpha$, we add to the set of batches (that replaces $\mathcal{B}$) only the batches with integer index i such that $i \le \lceil x - \alpha \rceil$. Denote by i_{max} the maximum index of a batch in this set, i.e., $i_{max} = \lceil x - \alpha \rceil$. Then, the total processing time of the batches in this set of batches is at most $\sum_{i=0}^{i_{max}} e^{i+\alpha} \le e^{i_{max}+\alpha} \cdot \frac{1}{1-\frac{1}{e}} = \frac{e^{i_{max}+1+\alpha}}{e-1}$. Since the total processing time of this job is e^x, it suffices to show that $E[e^{i_{max}+\alpha-x}] = e - 1$. Since α is uniformly distributed in $[0, 1)$, so is the random variable $i_{max} + \alpha - x$. The claim regarding the length of the batch follows by noting that for $u \sim \mathbf{U}[0, 1)$, $E[e^u] = \int_{u=0}^{1} e^u du = e^1 - e^0 = e - 1$.

Next, we need to prove that the completion time of a job j within the set of batches which replaces its batch also grows by an expected multiplicative factor of e. However, the proof is identical to the above, since if a job has original length e^y, then its processing time within the batch used to be e^y. Let $i' = \lceil y - \alpha \rceil$, the time until the new batch of j is completed is now at most $\sum_{i=0}^{i'} e^{i+\alpha}$, whose expected value is $e^{y+1} = e \cdot e^y$.   □

We next fix a value of α and show how to design algorithms that approximate OPT_α within constant factors. Then, using Lemma 1, the approximation ratio of the combined algorithms will be e times these constants.

A result of Lemma 1 is that there exists a value of α whose effect on the rounding and classification is an increase of the total cost by a factor of at most e. In [5] we show that the preprocessing step can be derandomized, while maintaining the performance guarantee of Lemma 1. We comment that a similar derandomization method can be applied to the randomized algorithm of [11], for the SC problem on a graph that belongs to $\mathcal{F}$, to obtain a deterministic 1.796-approximation algorithm, improving the deterministic $1.796 + \varepsilon$ approximation algorithm of [11].

3.1 The Final Step of the Algorithms

To approximate OPT_α, we use a different algorithm for each class of graphs, so the choice of algorithm depends on the smallest class to which the graph G belongs. In the next sections we first present a 2-approximation for interval graphs, then we present an approximation algorithm for graphs that belong to $\mathcal{F}$, and finally we show that a greedy algorithm is a $\frac{4}{\rho}$-approximation algorithm. However, the final step of the algorithms is identical and is therefore presented now.

We assume that the algorithm has determined the partition of the jobs of each class into batches. Therefore, we have a set of batches, each is a set of jobs with a common (rounded-up) processing time, and we need to schedule the batches so as to minimize the sum of job completion times. We note that such an ordering can be found optimally using Smith's rule [18, 3]. Sort the batches according to a non-decreasing order of the ratio of the weight of the batch divided by the (common) processing time of the jobs in the batch, where a weight of a batch is the number of jobs assigned to this batch.

Since we find an optimal ordering of the batches, in our analysis we can consider some fixed ordering (that may depend on the structure of the optimal solution), and note that our algorithm will return a better solution than the one indicated by some other (sub-optimal) order of the same batches. It remains to show how to partition the jobs of each class into a set of batches.

3.2 Interval Graphs

Nicoloso et al. [16] designed a 2-approximation algorithm for SC of interval graphs. Their algorithm computes G_ℓ for all $\ell = 1, 2, \ldots, \chi(G)$, where G_ℓ is a maximum size ℓ-colorable induced subgraph of G. They considered the case when G_ℓ are computed from left to right (according to an interval representation of the graph, that can be found in polynomial time [10]) in a greedy fashion. Given the output of such a process, they showed that for all $\ell > 1$, G_ℓ contains $G_{\ell-1}$ and the difference graph $G_\ell - G_{\ell-1}$ is 2-colorable. Thus their algorithm simply colors G_1 using color 1, and the difference graph $G_\ell - G_{\ell-1}$ is colored using colors $2\ell - 2$ and $2\ell - 1$.

For the rounded-up instance resulting from the preprocessing step we apply the algorithm of [16] on each class $\mathcal{J}_i$ separately to find a partition of this job class into batches (and then apply the final step of the algorithm described in Section 3.1). Let $p(V) = \sum_{v \in V} p_v$ $(p'(V) = \sum_{v \in V} p'_v)$ denote the sum of (rounded) processing times of all jobs.

Theorem 2. *There is resulting algorithm yields a $2e$-approximation ratio for interval graphs. Moreover, the output solution has cost at most $2eOPT - 2p'(V) + p(V)$.*

3.3   Graphs That Belong to $\mathcal{F}$

Recall that, for a graph G that belongs to $\mathcal{F}$, for all ℓ, a maximum size induced subgraph of G that is ℓ-colorable can be found in polynomial time.

Let $\mathcal{A}$ denote an algorithm uses as a subroutine (defined later). We require that $\mathcal{A}$ is be a fully polynomial time dual approximation scheme for the following variant of the Knapsack problem: each item can be packed at most n times, and packing ℓ times of item i results a profit of $c_i(\ell)$ where c_i is a monotonically non-decreasing integer valued function of ℓ. We are given a budget B on the total size of all the packed items. I.e., $\mathcal{A}$ runs in polynomial time (where the polynomial is a function of n and $\frac{1}{\varepsilon}$) and returns a solution such that the total size of the returned solution is within $(1 + \varepsilon)$ multiplicative factor of the budget (i.e., at most $(1 + \varepsilon)B$), and the total profit of the packed items is at least the total profit of the optimal solution that uses only total size that is at most the given budget.

For each class $\mathcal{J}_i$, denote by G^i the subgraph of G induced by the vertices that correspond to the jobs of $\mathcal{J}_i$. The outline of our algorithm for this case is as follows. Partition the jobs into batches in iterations. In each iteration the algorithm decides to open a set of batches. The length of an iteration, defined to be the sum of completion times of the batches of this iteration, increases geometrically between the iterations. In each iteration, given the current bound on the length, i.e. the sum of completion times of the batches in the current iteration, pack a maximum number of jobs into the batches of the current iteration, using a total length of time which is within factor $1 + \varepsilon$ of the required length.

Algorithm Pack_Subgraphs

1. Apply the Preprocessing Algorithm.
2. Pick uniformly at random a number β from the range $[0, 1)$ (independently of α). I.e., $\beta \sim \mathbf{U}[0, 1)$. Set $t = 0$, and set q to a constant value to be fixed afterwards.
3. **While** there are jobs that still have not been assigned to batches **do:**
   (a)   **For** $i = 0, 1, \ldots, k$ **do:**
       $a_i = q^{i+\alpha}$.
       For $\ell = 1, 2, \ldots, |\mathcal{J}_i|$ **do:**
       Let $W_{i,\ell}$ be the vertex set of a maximum size ℓ-colorable subgraph of G^i, and denote by $c_i(\ell) = |W_{i,\ell}|$.
   (b)   Apply Algorithm $\mathcal{A}$ to approximate the Knapsack problem instance where the profit from packing ℓ copies of the i-th element is $c_i(\ell)$, and

the size of the i-th element is a_i. The budget on the total sizes of all elements that we place in the knapsack is $q^{t+\beta}$.

(c) **For $i = 0, 1, \ldots, k$ do:**
Assume that the approximate solution places b_i copies of the i-th element; then place exactly b_i batches for the class $\mathcal{J}_i$ in the current set of batches SB. For these batches, pick a maximum size b_i-colorable induced subgraph of G^i; assign to these batches all the jobs that participate in this induced subgraph.[1] Distribute the set of new assigned jobs among the b_i new batches of $\mathcal{J}_i$ so that each batch is an independent set. Add this set of batches to a pool of all batches. These batches will be assigned to times after all jobs have been assigned to batches.

(d)   Remove the jobs that have been assigned in the current iteration from G and increase t by 1.

4.   Apply the final step.

We now turn to analyze the performance of Pack_Subgraphs

Lemma 2. *For all $q > 1$, Algorithm Pack_Subgraphs always returns a feasible solution in polynomial time.*

Let f_r denote the completion time of the r-th job according to OPT_α. Then, clearly $OPT_\alpha = \sum_{r=1}^n f_r$. Note that f_r is monotonically non-decreasing sequence. Consider now the class of solutions, which similarly to OPT_α do not assign jobs of different classes (i.e., different rounded processing times) to a common batch. For each r denote by d_r the minimum time in any solution from the class above that is needed to complete at least r jobs. Then, clearly $f_r \geq d_r$, and therefore we establish the following lemma.

Lemma 3. $OPT_\alpha \geq \sum_{r=1}^n d_r.$

Consider the jobs sorted according to non-decreasing completion times in the solution returned by Algorithm Pack_Subgraphs. Consider now the r-th job according to this order. Assume that this job belongs to the set of batches created in an iteration in which $t = \tau$ (i.e., the value of t in the beginning of the iteration in which we allocate this job, is τ). We define the *modified completion time* of this job to be $\pi_r = (1 + \varepsilon) \cdot \left(\sum_{i'=1}^{\tau-1} q^{i'+\beta} + \frac{q^{\tau+\beta}}{2} \right)$.

Instead of analyzing the solution we achieved, which orders all batches optimally by applying the final step, we analyze a solution which assigns the batches of each iteration consecutively. Each time the algorithm defines a set of batches for a new iteration, it assigns them right after the batches of the previous iteration. This can be done using Smith's rule. This way, the algorithm selects the best order of these batches. Applying Smith's rule on the complete set of batches results in a solution that is not worse than this solution. Furthermore, for the analysis we use the fact that this order is not worse than the better one out of the following two orders: (*i*) an arbitrary order, and (*ii*) the batches are ordered in the exact opposite order.

---

[1] Note that assigned jobs are removed from G and thus from G^i, therefore all the chosen jobs were not assigned prior to this iteration.

Lemma 4. *The cost of the solution returned by Algorithm Pack_Subgraphs is at most $\sum_{r=1}^{n} \pi_r + \frac{\sum_{r=1}^{n} p'_r}{2}$.*

Note that $(1+\varepsilon) \cdot \left(\sum_{i'=1}^{\tau-1} q^{i'+\beta} + \frac{q^{\tau+\beta}}{2} \right) \leq (1+\varepsilon) \cdot q^{\tau+\beta} \cdot \left(\frac{1}{q-1} + \frac{1}{2} \right)$. We next present a bound on $\sum_{r=1}^{n} \pi_r$ in terms of $\sum_{r=1}^{n} d_r$.

Lemma 5. *Let q be chosen as the root of the equation $\ln x = \frac{x+1}{x}$. That is $q \sim 3.591$, then $\frac{E[\sum_{r=1}^{n} \pi_r]}{\sum_{r=1}^{n} d_r} \leq (1+\varepsilon) \cdot 1.796$.*

Theorem 3. *Pack_Subgraphs is a randomized $(1.296e + \frac{3}{2} + \varepsilon)$-approximation algorithm for **MSJCT** on graphs that belong to $\mathcal{F}$.*

In [5] we show how the algorithm can be derandomized. Hence, we get

Proposition 1. *There is a deterministic $(1.296 \cdot e + \frac{3}{2} + \varepsilon)$-approximation algorithm for **MSJCT** on graphs that belong to $\mathcal{F}$.*

Note that this also improves the bound of Theorem 2 for interval graphs, to 5.022. We next analyze an improved algorithm for interval graphs, that combines the two algorithms. We run both the algorithm for interval graphs of Section 3.2, and algorithm Pack_Subgraphs, and we pick the better solution.

Theorem 4. *The resulting algorithm is a 4.912-approximation algorithm for interval graphs.*

Proof. By Theorem 2, the cost of the solution is at most $2eOPT - 2p'(V) + p(V)$. Since we have already established that $E[p'(V)] = (e-1)p(V)$, we conclude that the expected cost of the returned solution $E[SOL]$ (whose cost is the minimum of the two outputs) satisfies $E[SOL] \leq 2eOPT - (2e - 3)p(V)$. By the proof of Proposition 1, we conclude that $E[SOL] \leq (1.796e + \varepsilon)OPT + \left(\frac{3-e}{2} \right) p(V)$. We multiply the first inequality by $\frac{3-e}{3e-3}$ and the second inequality by $\frac{4e-6}{3e-3}$. Then, we obtain $E[SOL] \leq (1.807e + \varepsilon)OPT \approx 4.912OPT$. $\qquad\square$

We note that since both algorithms can be derandomized without affecting the approximation ratio, so does this combined algorithm. It remains to show how to implement Algorithm $\mathcal{A}$ for our Knapsack problem. We give the details in the full version of the paper.

A $\frac{4}{\rho}$-approximation algorithm: We now analyze the greedy algorithm from Section 2, to approximate the SC instance defined as the set of jobs from $\mathcal{J}_i$, i.e., we first apply the preprocessing step and then each class is approximated separately. Then, we order the resulting batches according to our final step using Smith's rule.

Theorem 5. *There is a $\frac{4e}{\rho}$-approximation algorithm for problem **MSJCT**.*

Recall that MaxIS can be solved in polynomial time on line graphs, and within $k/2 + \epsilon$ factor on $k + 1$-claw free graphs for any $\epsilon > 0$ [12]. Thus, we have

Corollary 1. *MSJCT can be approximated within factor 4e on line graphs and $2ek + \epsilon$ for any $\epsilon > 0$ on $k + 1$-claw free graphs.*

We can also show that the greedy algorithm applied directly gives the same ratio. This is deferred to the full version.

References

1. A. Bar-Noy, M. Bellare, M. M. Halldórsson, H. Shachnai, and T. Tamir. On chromatic sums and distributed resource allocation. *Inf. Comput.*, 140(2):183–202, 1998.
2. A. Bar-Noy, M. M. Halldórsson, G. Kortsarz, R. Salman, and H. Shachnai. Sum multicoloring of graphs. *J. of Algorithms*, 37(2):422–450, 2000.
3. P. Brucker. *Scheduling Algorithms 4th ed.* Springer-Verlag, 2004.
4. V. Chvátal. A greedy heuristic for the set-covering problem. *Mathematics of Operations Research*, 4:233–235, 1979.
5. L. Epstein, M. M. Halldórsson, A. Levin and H. Shachnai. Weighted Sum Coloring in Batch Scheduling of Conflicting Jobs. full version. http://www.cs.technion.ac.il/~hadas/PUB/cosum.pdf.
6. U. Feige, L. Lovász, and P. Tetali. Approximating min sum set cover. *Algorithmica*, 40(4):219–234, 2004.
7. A. Frank. On chain and antichain families of a partially ordered set. *J. of Combinatorial Theory Series B*, 29:176–184, 1980.
8. R. Gandhi R., M. M Halldórsson, G. Kortsarz and H. Shachnai, Improved Bounds for Sum Multicoloring and Scheduling Dependent Jobs with Minsum Criteria. In *Proc. of WAOA*, 2004.
9. M. R. Garey and D. S. Johnson. *Computers and intractability*. W. H. Freeman and Company, New York, 1979.
10. M. C. Golumbic. *Algorithmic Graph Theory and Perfect Graphs*. Academic Press, 1980.
11. M. M. Halldórsson, G. Kortsarz, and H. Shachnai. Sum coloring interval and k-claw free graphs with application to scheduling dependent jobs. *Algorithmica*, 37(3):187–209, 2003.
12. C. A. J. Hurkens, and A. Schrijver. On the size of systems of sets every t of which have an SDR, with an application to the worst-case ratio of heuristics for packing problems. *SIAM J. Discrete Math.*, vol. 2, 1989, pp. 68–72.
13. D. S. Johnson. Approximation algorithms for combinatorial problems. *Journal of Computer and System Sciences*, 9:256–278, 1974.
14. L. Lovász. On the ratio of optimal integral and fractional covers. *Discrete Mathematics*, 13:383–390, 1975.
15. K. Munagala, S. Babu, R. Motwani, and J. Widom. The pipelined set cover problem. In *Proc. of ICDT* 2005.
16. S. Nicoloso, M. Sarrafzadeh, and X. Song. On the sum coloring problem on interval graphs. *Algorithmica*, 23(2):109–126, 1999.
17. A. Schrijver. *Combinatorial optimization polyhedra and efficiency*. Springer-Verlag, 2003.
18. W. E. Smith. Various optimizers for single-stage production. *Naval Research Logistics Quarterly*, 3:59–66, 1956.
19. G. J. Woeginger. When does a dynamic programming formulation guarantee the existence of a fully polynomial time approximation scheme (FPTAS)? *INFORMS J. on Computing*, 12(1):57–74, 2000.
20. M. Yannakakis and F. Gavril. The maximum k-colorable subgraph problem for chordal graphs. *Information Processing Letters*, 24(2):133–137, 1987.

Combinatorial Algorithms for Data Migration to Minimize Average Completion Time

Rajiv Gandhi[1,*] and Julián Mestre[2,**]

[1] Department of Computer Science, Rutgers University-Camden, Camden, NJ 08102
rajivg@camden.rutgers.edu
[2] Department of Computer Science, University of Maryland, College Park, MD 20742
jmestre@cs.umd.edu

Abstract. The *data migration* problem is to compute an efficient plan for moving data stored on devices in a network from one configuration to another. It is modeled by a transfer graph, where vertices represent the storage devices, and the edges represent the data transfers required between pairs of devices. Each vertex has a non-negative weight, and each edge has unit processing time. A vertex completes when all the edges incident on it complete; the constraint is that two edges incident on the same vertex cannot be processed simultaneously. The objective is to minimize the sum of weighted completion times of all vertices. Kim (*Journal of Algorithms, 55:42-57, 2005*) gave an LP-rounding 3-approximation algorithm. We give a more efficient primal-dual algorithm that achieves the same approximation guarantee, which can be extended to yield a 5.83-approximation for arbitrary processing times. We also study a variant of the open shop scheduling problem. This is a special case of the data migration problem in which the transfer graph is bipartite and the objective is to minimize the completion times of edges. We present a simple algorithm that achieves an approximation ratio of $\sqrt{2} \approx 1.414$, thus improving the 1.796-approximation given by Gandhi *et al.*(*ACM Transaction on Algorithms, 2(1):116-129*, 2006). We show that the analysis of our algorithm is almost tight.

1 Introduction

The *data migration* problem arises in large storage systems, such as *Storage Area Networks* [13], where a dedicated network of disks is used to store multimedia data. As the data access pattern changes over time, the load across the disks needs to be rebalanced so as to continue providing efficient service. This is done by computing a new data layout and then "migrating" data to convert the initial data layout to the target data layout. While migration is being performed, the storage system is running suboptimally, therefore it is important to compute a data migration schedule that converts the initial layout to the target layout quickly.

* Research partially supported by Rutgers University Research Council Grant.
** Research supported by NSF Awards CCR-0113192 and CCF-0430650, and the University of Maryland Dean's Dissertation Fellowship.

J. Diaz et al. (Eds.): APPROX and RANDOM 2006, LNCS 4110, pp. 128–139, 2006.

This problem can be modeled as a *transfer graph* [15], in which the vertices represent the storage disks and an edge between two vertices u and v corresponds to a data object that must be transferred from u to v, or vice-versa. Each edge has a length that represents the transfer time of a data object between the disks corresponding to the end points of the edge. An important constraint is that any disk can be involved in at most one transfer at any time. In this work, we assume that the edges have unit length, that is, all data transfers take the same amount of time.

Several variations of the data migration problem have been studied. These variations arise either due to different objective functions or due to additional constraints. One common objective function is to minimize the *makespan* of the migration schedule, i.e., the time by which all migrations complete. Coffman *et al.* [4] show that when the edges have equal (unit) lengths, the problem reduces to edge coloring of the transfer (multi)graph of the system. The best approximation algorithm known for minimum edge coloring [17] then yields an algorithm for data migration with unit edge length, whose makespan is $1.1\chi' + 0.8$, where χ' is the chromatic index of the graph. Approximation algorithms are also developed [9, 1, 13, 14] for generalizations of the makespan minimization problem in which there are storage constraints on disks and constraints on how the data can be transferred.

The data migration problem has also been studied with the objective of minimizing the sum of weighted completion time over all storage disks. Kim [15] proved that the problem is NP-hard when edge lengths are the same and showed that Graham's list scheduling algorithm [7], when guided by an optimal solution to a linear programming relaxation, gives an approximation ratio of 3. When edges have arbitrary lengths there are several constant factor approximation algorithms [5, 15, 20] with the best approximation guarantee being 5.03 [5].

A problem related to the data migration problem is *open shop scheduling*. In this problem, we have a set of jobs, $\mathcal{J}$, and a set of machines $M_1, \ldots, M_m$. Each job $J_j \in \mathcal{J}$ consists of a set of m_j operations. For $1 \leq i \leq m_j$, operation $o_{j,i}$ has processing time $p_{j,i}$ and must be processed on $M_{\phi(j,i)}$. Each machine can process a single operation at any time, and two operations that belong to the same job cannot be processed simultaneously. Each job J_j has a positive weight, w_j and the objective is to minimize the sum of weighted completion times of all jobs. This problem is a special case of the data migration problem [5]. Open shop scheduling problem has been studied in [3, 12, 19, 20].

There has also been interest in the study of data migration problem with the objective function being to minimize the average completion time over all data transfers. This corresponds to minimizing the average edge completion time in the transfer graph. For arbitrary edge lengths, several constant factor approximation algorithms [11, 15, 6] are known with the best approximation factor being 7.682 [6]. For the case of unit edges lengths, Bar-Noy *et al.* [2] showed that the problem is NP-hard and gave a simple 2-approximation algorithm. When restricted to bipartite graphs, the latter problem becomes a variant of open shop scheduling in which the operations have unit processing times and the objective is to minimize the sum of completion times of operations; for this problem Gandhi *et al.* [5] give a 1.796-approximate solution that uses an algorithm for sum coloring of interval graphs due to Halldórsson *et al.* [11].

Our Contribution: First we study the data migration problem with unit length edges and the objective of minimizing the average completion time over all storage disks. Kim [15] gave a 3-approximation algorithm that rounds the solution produced by a linear programming relaxation for the problem. This algorithm involves solving a linear program with an exponential number of constraints, though there are equivalent linear programs with a polynomial number of constraints (cf. [6]). Gandhi *et al.* [5] show that Kim's algorithm can not give an approximation guarantee better than 3. In this work, we present an efficient primal-dual algorithm that gives a 3-approximate solution; our scheme can be extended to yield a 5.83-approximation for arbitrary processing times.

The second problem we study is the data migration problem with the objective of minimizing the sum of completion times of edges. In other words, given a graph $G = (V, E)$ we want to partition the edge set E into matchings $M_1, M_2, \ldots$, so as to minimize $\sum_i i|M_i|$. Bar-Noy *et al.* [2] show that if M_i is maximal with respect to $G \setminus \cup_{j<i} M_j$ then we get a 2-approximate solution. We show that if, for all $b \geq 1$, the b-matching $\cup_{j \leq b} M_j$ is maximal in G then we get a $\sqrt{2}$-approximate schedule. We show that such schedules always exist in bipartite graphs and can be computed in polynomial time. Data migration in bipartite graphs is equivalent to a variant of open shop scheduling in which we want to minimize the sum of operation completion times. Marx [16] has shown that the problem is APX-hard. Gandhi *et al.* [5] show that using the sum-coloring algorithm of Halldórsson *et al.* [11] one can obtain a 1.796 approximation guarantee. We improve this ratio to $\sqrt{2} \approx 1.414$, though our guarantee does not extend to the objective of minimizing the sum of weighted edge completion times. We also show that the analysis is almost tight by giving an example on which the algorithm gives a 1.375-approximate solution.

2   Data Migration Problem

We are given a graph $G = (V, E)$. Let $E(u)$ denote the set of edges incident on a vertex u. The vertices and edges in G are jobs to be completed. Each vertex v has weight w_v. We assume that edges have unit length. The completion time of an edge is simply the time at which its processing is completed. The completion time, C_v, of v is the latest completion time of any edge in $E(v)$. The crucial constraint is that two edges incident on the same vertex cannot be processed at the same time. The objective is to minimize $\sum_{v \in V} w_v C_v$.

We first analyze the performance of the following natural and intuitive algorithm: Process the edges in any order scheduling them as early as possible without creating conflicts with the edges scheduled so far. While this algorithm gives a solution that is at most twice the cost of optimal for $\min \sum_e C_e$ [2], we show in Appendix A that for the objective of $\min \sum_v C_v$ it may produce a solution with cost $\Omega(\sqrt[3]{n})$ times the optimum.

2.1   A Linear Programming Relaxation

The linear programming relaxation for the data migration problem was given
by Kim [15]. Such relaxations have been proposed earlier by Wolsey [22] and
Queyranne [18] for single machine scheduling problems and by Schulz [21] and
Hall *et al.* [10] for parallel machines and flow shop problems. For the purpose
of clarity, we state only portions of the LP relaxation relevant for obtaining the
primal-dual algorithm.

For a vertex v, let C_v represent the completion time of v. Let $N(u)$ represent
the neighbors of vertex u.

$$\min \sum_{v \in V} w_v C_v$$

subject to

$$\sum_{v \in S_u} C_v \geq \frac{|S_u| (|S_u| + 1)}{2} \qquad \forall u \in V, S_u \subseteq N(u) \quad (1)$$

$$C_v \geq 0 \qquad \forall v \in V$$

The dual LP contains a variable y_{S_u} (for each set S_u) corresponding to each
constraint represented by (1). The dual LP is given below.

$$\max \sum_{\substack{u \in V \\ S_u \subseteq N(u)}} \frac{|S_u|(|S_u| + 1)}{2} y_{S_u}$$

subject to

$$\sum_{\substack{u \in V \\ S_u : v \in S_u}} y_{S_u} \leq w_v \qquad \forall v \in V \qquad (2)$$

$$y_{S_u} \geq 0 \qquad \forall u \in V, S_u \subseteq N(u)$$

2.2   Algorithm

The high level idea of the algorithm is as follows. There are two phases—*labeling*
and *scheduling*.

In the labeling phase, each vertex is initially unlabeled. This phase proceeds
in iterations; iteration i labels some neighbors of x_i. Vertex x_i is chosen to be
the one with maximum number of unlabeled neighbors. Let S_{x_i} be the unlabeled
neighbors of x_i. The value of the dual variable $y_{S_{x_i}}$ is incremented until the dual
constraint (2) is met with equality for some vertex $v \in S_{x_i}$. In other words, $y_{S_{x_i}}$
assumes the smallest value such that for some vertex $v \in S_{x_i}$ we have

$$\sum_{j < i \, : \, v \in S_{x_j}} y_{S_{x_j}} + y_{S_{x_i}} = w_v.$$

Let $T_{x_i} \subseteq S_{x_i}$ be the vertices for which the above equality holds. All vertices T_{x_i} are labeled $|S_{x_i}|$. The label of vertex u is denoted by $\ell(u)$.

In the scheduling phase, the edges in E are ordered so that edge (u, v) precedes (u', v') if:

(i) $\min\{\ell(u), \ell(v)\} < \min\{\ell(u'), \ell(v')\}$, or

(ii) $\min\{\ell(u), \ell(v)\} = \min\{\ell(u'), \ell(v')\} \wedge \max\{\ell(u), \ell(v)\} \leq \max\{\ell(u'), \ell(v')\}$.

The edges in E are then processed in order. When processing $(u, v) \in E$, the edge is scheduled at the earliest time such that no edge incident upon u or v is already scheduled at that time. The pseudo-code is given below.

```
PRIMAL-DUAL(G = (V, E))
1      // labeling phase
2      for each v ∈ V do
3          ℓ(v) ← nil            // v is unlabeled
4      i ← 0
5      while (there exists an unlabeled vertex) do
6          i ← i + 1
7          x_i ← vertex with the maximum number of unlabeled neighbors.
8          S_{x_i} ← unlabeled neighbors of x_i.
9          y_{S_{x_i}} ← min_{v ∈ S_{x_i}} {w_v}
10         T_{x_i} ← {v ∈ S_{x_i} | w_v = y_{S_{x_i}}}
11         for each v ∈ T_{x_i} do
12             ℓ(v) ← |S_{x_i}|    // v is now labeled
13         for each v ∈ S_{x_i} do
14             w_v ← w_v − y_{S_{x_i}}
15     // scheduling phase
16     sort edges (u, v) ∈ E in lex. order of ⟨min{ℓ(u), ℓ(v)}, max{ℓ(u), ℓ(v)}⟩
17     for each edge e = (u, v) ∈ E processed in order do
18         schedule e if no edge in E(u) ∪ E(v) is already scheduled.
```

2.3 Analysis

Let $\widetilde{C}_v$ be the completion time of vertex v in our algorithm. Recall that $E(v)$ is the set of edges incident on a vertex v and $N(v)$ denotes the neighbors of v.

Lemma 1. *For each $v \in V$, $\widetilde{C}_v \leq \ell(v) + |E(v)| - 1$.*

Proof. Let (w, v) be the last edge to finish among the edges in $E(v)$. Also, let $F(w, v) = \{y \in N(w) \mid \ell(y) \leq \ell(v)\}$. Observe that because of the order in which the edges are scheduled, $\widetilde{C}_v \leq |F(w, v)| + |E(v)| - 1$. Let i be the iteration of the algorithm in which the first vertex in $F(w, v)$ is labeled, and let y be that vertex. At the beginning of the ith iteration vertex w has at least $|F(w, v)|$ unlabeled neighbors. Because x_i is chosen (line 7 of the pseudocode) to be the vertex with the maximum number of unlabeled neighbors it must be the case that $|F(w,v)| \leq |S_{x_i}| = \ell(y)$, which by definition is at most $\ell(v)$. □

Theorem 1. *The data migration problem with edges having unit processing times has a 3-approximate primal-dual algorithm.*

Proof. Let $G = (V,E)$ be an instance of the data migration problem. Let $DFS(G)$ denote the cost of the dual feasible solution for instance G obtained by our algorithm. Let $OPT(G)$ denote the cost of an optimal solution for instance G. Clearly, $DFS(G) \leq OPT(G)$. Also, $OPT(G) \geq \sum_{v \in V} w_v |E(v)|$. Let $iter(v)$ be the iteration in which v gets labeled. The cost of our algorithm is given by

$$\sum_v w_v \widetilde{C}_v \leq \sum_v w_v \left(\ell(v) + |E(v)| \right) \qquad \text{(using Lemma 1)}$$

$$= \sum_v w_v \, \ell(v) + \sum_v w_v |E(v)|$$

$$= \sum_v \left(\sum_{i \,:\, v \in S_{x_i}} y_{S_{x_i}} \right) \ell(v) + OPT(G)$$

$$= \sum_v \left(\sum_{i \,:\, v \in S_{x_i}} y_{S_{x_i}} \right) |S_{x_{iter(v)}}| + OPT(G)$$

$$\leq \sum_v \sum_{i \,:\, v \in S_{x_i}} y_{S_{x_i}} |S_{x_i}| + OPT(G)$$

$$= \sum_i \sum_{v \in S_{x_i}} y_{S_{x_i}} |S_{x_i}| + OPT(G)$$

$$= \sum_i y_{S_{x_i}} |S_{x_i}|^2 + OPT(G)$$

$$= \sum_{\substack{u \in V \\ S_u \subseteq N(u)}} y_{S_u} |S_u|^2 + OPT(G)$$

$$\leq 2 \cdot DFS(G) + OPT(G)$$

$$\leq 3 \cdot OPT(G)$$

□

We finish this section by mentioning that the above labeling procedure coupled with the scheduling technique used by Gandhi *et al.* [5] yields a constant factor approximation for the case of arbitrary processing times.

Theorem 2. *The data migration problem with edges having arbitrary processing times has a 5.83-approximate primal-dual algorithm.*

Due to space limitation the proof of the above theorem is deferred to the journal version of this paper.

3   Minimizing Sum of Edge Completion Times

The problem of scheduling the edges of a graph to minimize the sum of their completion times can be cast as an edge coloring problem: Given $G = (V, E)$ we want to partition the edge set E into matchings $M_1, \ldots, M_k$ as to minimize $\sum_i i |M_i|$. Indeed, this problem is also known as minimum sum edge coloring.

Bar-Noy *et al.* [2] show that *any* minimal schedule is 2-approximate. In a minimal schedule every matching M_i is maximal with respect to $G \setminus \cup_{j<i} M_j$. The main result of this section is to identify a stronger minimality requirement that results in a better approximation guarantee.

Definition 1. *A schedule* $M_1, \ldots, M_k$ *of* G *is said to be* strongly minimal *if, for all* $1 \leq b \leq k$, *the* b-matching $\cup_{i \leq b} M_i$ *is maximal w.r.t.* G.

Theorem 3. *Any strongly minimal schedule is* $\sqrt{2}$-*approximate.*

Proof. The high level idea of the proof is to *assign* every edge to at least one of its endpoints. Each vertex is responsible for paying for the cost of the edges assigned to it. In order to pay for this cost each vertex charges a lower bound on the completion time of the edges assigned to it.

Let $(u, v) \in M_i$, we say endpoint u is *full* if u is matched in all $M_{j<i}$. We consider the endpoints of edges in M_1 to be full. Notice that every edge $(u, v) \in M_i$ must have at least one full endpoint, otherwise $\cup_{j<i} M_j + (u, v)$ would be a valid $(i - 1)$-matching, which contradicts the fact that the schedule is strongly minimal. If both endpoints of (u, v) are full then the edge is *half-assigned* to u and v. Otherwise the edge is *fully-assigned* to the one full endpoint.

Every vertex u is responsible for the cost of edges assigned to it. If an edge is half-assigned to u, then u pays for half of its completion time; if the edge is fully-assigned then u pays in full. Let s_1 and s_2 be the number of half-assigned and fully-assigned edges to u respectively. Notice that all edges assigned to u must belong to M_j for some $j \leq s_1 + s_2$. We think of u as paying $\frac{1}{2}$ of the completion time of *all* edges assigned to it, plus an additional $\frac{1}{2}$ for the fully-assigned edges, which in the worst case will be scheduled the latest,

$$u \text{ must pay} \leq \frac{1}{2} \sum_{i=1}^{s_1+s_2} i + \frac{1}{2} \sum_{i=s_1+1}^{s_1+s_2} i$$

Vertex u will pay this amount by charging the completion time (in the optimal solution) of the edges assigned to it. Fully-assigned will be charged a soon-to-be-determined ρ factor, and half-assigned edges will be charged $\frac{\rho}{2}$. This will be u's budget. How fast can the optimal solution possibly schedule these edges?

$$u\text{'s budget} \geq \frac{\rho}{2} \sum_{i=1}^{s_1+s_2} i + \frac{\rho}{2} \sum_{i=1}^{s_2} i$$

Notice that every edge is charged at most to an extent of ρ: fully-assigned edges are charged ρ once, from a single endpoint, and half-assigned edges are charged $\frac{\rho}{2}$ twice, once from each endpoint. Thus, strongly greedy schedules are ρ-approximate. The discrepancy between the upper and lower bound on the completion times of edges assigned to u is due to fully-assigned edges which are scheduled the latest in the upper bound, and the earliest in the lower bound. We need to determine the smallest ρ such that u's budget is enough to cover u's payment, namely

$$\frac{(s_1 + s_2)(s_1 + s_2 + 1)}{4} + \frac{(2s_1 + s_2 + 1)s_2}{4} \leq$$
$$\rho\frac{(s_1 + s_2)(s_1 + s_2 + 1)}{4} + \rho\frac{s_2(s_2 + 1)}{4}.$$

Or equivalently,

$$(s_1 + s_2)^2 + (2s_1 + s_2)s_2 \leq \rho(s_1 + s_2)^2 + \rho s_2^2 + (\rho - 1)(s_1 + 2s_2).$$

Let $\alpha = \frac{s_2}{s_1+s_2}$, since $\rho > 1$ the above follows provided

$$\frac{1 + 2\alpha - \alpha^2}{1 + \alpha^2} \leq \rho.$$

The left hand side is maximized for $\alpha = \sqrt{2} - 1$, which yields $\sqrt{2} \leq \rho$ □

While strongly minimal schedules are not guaranteed to exist for general graphs, we now show that in bipartite graphs they always exist and can be computed in polynomial time. The bipartite, is an interesting and nontrivial case: it is a variant of the open shop scheduling problem in which we want to minimize the sum of completion time of operations [5]. This problem is APX-hard [16]. The best known approximation guarantee for the problem is 1.796 [5].

Theorem 4. *The procedure* FIND STRONGLY MINIMAL *is a $\sqrt{2}$-approximation for minimizing the sum of completion times of unit length operations in open shop scheduling.*

FIND STRONGLY MINIMAL(G)

1   **for** $i \leftarrow \Delta$ **down to** 1 **do**
2        $M_i \leftarrow$ a matching incident to all vertices of G with degree i
3        $G \leftarrow G \setminus M_i$
4   **return** $M_1, M_2, \ldots, M_\Delta$

In each iteration, the procedure FIND STRONGLY MINIMAL computes a matching incident to the maximum degree vertices of G and removes the matching from G. This continues until all edges have been removed. The matchings found are then scheduled in reverse order. Because the degree of G decreases by one with each iteration, the algorithm finishes after Δ iterations, here Δ is the degree of the original graph.

Let us argue that the schedule found is strongly minimal. Let $e \in M_i$ and $b < i$, we want to show that e cannot be added to $\cup_{j \le b} M_j$ without violating the b-matching property. Let G' be the remaining graph when M_i was computed. One of the endpoint of e must have degree i in G', let u be that endpoint. After removing M_i the degree of u becomes $i - 1$, and thus u must be matched in M_{i-1}. In general u will be matched in all $M_{j<i}$. Therefore, the degree of u in $\cup_{j \le b} M_j$ is b, which in turn means the b-matching is maximal with respect to e.

In bipartite graphs a matching incident to all the maximum degree vertices always exists and can be computed in polynomial time (cf. [8]). Together with Theorem 3, this finishes the proof of Theorem 4.

3.1   An Almost Tight Example

While at first sight the analysis of the approximation factor of strongly minimal schedules may seem too pessimistic, it turns out it is almost tight. Consider the following bipartite graph with vertices $u_1, \ldots, u_n$ on one side and vertices $v_1, \ldots, v_n$ on the other side of the bipartition. There is an edge $(u_i, v_j) \in E$ if and only if $i \le j$.

It is not difficult to show that the optimal schedule uses matchings

$$M_k = \{(u_i, v_{i+k-1}) \,|\, i \le n - k + 1\}$$

and has cost $\sum_{i=1}^{n} i\,(n - i + 1) = \frac{1}{6}n^3 + 3n^2 + 2n$.

Now suppose we run FIND STRONGLY MINIMAL. Initially the maximum degree vertices are u_1 and v_n, and the algorithm finds the matching M_n consisting of $(u_1, v_{\frac{n}{2}})$ and $(u_{\frac{n}{2}+1}, v_n)$. After removing M_n the maximum degree vertices are u_1, u_2, v_{n-1}, and v_n. In general the algorithm may find, for $\frac{n}{2} < k \le n$,

$$M_k = \{(u_i, v_{i+k-\frac{n}{2}-1}) \,|\, i \le n - k + 1\} \cup \{(u_{j-k+\frac{n}{2}+1}, v_j) \,|\, j \ge k\}.$$

After these matchings are removed from the graph we are left with a complete bipartite graph on $u_1, \ldots u_{\frac{n}{2}}$ and $v_{\frac{n}{2}+1}, \ldots v_n$, thus $|M_k| = \frac{n}{2}$ for all $1 \le k \le \frac{n}{2}$. Therefore, the cost of this strongly minimal schedule is $\frac{11}{48}n^3 + \frac{5}{8}n^2 + \frac{1}{3}n$.

The ratio of the cost of the optimal and strongly minimal solutions approaches 1.375 as $n \to \infty$. Compare this to approximation guarantee of $\sqrt{2} \approx 1.414$ obtained in Theorem 3.

3.2   Integrality Gap

Let us now study the inherent limitations of the lower bounding technique used to prove Theorem 3. The lower bound used there can be generalized as follows: for any subset of edges S incident on a vertex, we know that *any* feasible schedule must spend at least $\frac{|S|(|S|+1)}{2}$ time on these edges. We can charge the cost of this set of edges a factor $y_S \ge 0$. If for every edge e the total charge $\left(\sum_{S \,:\, e \in S} y_S\right)$ on

e is at most 1, then $\sum_S \frac{|S|(|S|+1)}{2} y_S$ offers a lower bound on the cost an optimal schedule. The best such lower bound corresponds to the optimal solution of the following dual LP:

$$\max \sum_S \frac{|S|(|S|+1)}{2} y_S$$

subject to

$$\sum_{S:e\in S} y_S \leq 1 \qquad\qquad \forall e \in E \qquad\qquad (3)$$

$$y_S \geq 0 \qquad\qquad \forall S \subseteq E(u), u \in V$$

In hindsight, the proof of Theorem 3 can be viewed as a case of dual-fitting in which constraint (3) is violated a $\sqrt{2}$ factor. To determine how good a lower bound the dual offers, we derive the primal LP and study its integrality gap.

Theorem 5. *The integrality gap of the LP below is at least $\frac{4}{3}$ in general graphs and at least $\frac{10}{9}$ in bipartite graphs.*

$$\min \sum_{e\in E} C_e$$

subject to

$$\sum_{e\in S} C_e \geq \frac{|S|(|S|+1)}{2} \qquad\qquad \forall S \subseteq E(u), u \in V \qquad (4)$$

$$C_e \geq 0 \qquad\qquad\qquad \forall e \in E$$

Proof. For general graphs, consider a triangle. The optimal solution schedules one edge at the time, and incurs a cost of 6. The LP can schedule all edges at $C_e = 1.5$, with a cost of 4.5. Thus, the integrality gap for this graph is $\frac{4}{3}$.

For the bipartite case (our example is in fact a tree) consider a spider with three legs of length two. The graph is shown on the right along with the edge completion times of an optimal schedule (in black and to the left) and of the optimal LP solution (in gray and to the right). Optimum schedules three edges in M_1, two in M_2 and one in M_3, with a total cost of 10. On the 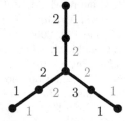 other hand, the LP solution manages to schedule all edges in two rounds, with a total cost of 9. Thus the integrality gap for bipartite graphs is at least $\frac{10}{9}$. $\square$

3.3   Limitations of Strongly Minimal Schedules

We conclude this section with a note on the limitations of strongly minimal schedules. One common generalization of our scheduling problem is to minimize the weighted sum of completion times. In this setting the proof of Theorem 3 does not go through as we make crucial use of the fact that the edges have uniform weight.

It would be natural to hope that the following slight modification of FIND STRONGLY MINIMAL would produce good schedules: Instead of finding any matching incident to the maximum degree vertices, we find one with minimum weight. Unfortunately, the following bipartite example shows that strongly minimal schedules are just not suited for the weighted case. Take a path of length four and replace each edge with a copy of $K_{t,t}$. The edges in the first and the last $K_{t,t}$ have weight 1, and the ones in the middle have weight 0. The optimal solution schedules the first and the last $K_{t,t}$ in the first t rounds and the remaining edges are scheduled in the next $2t$ rounds, with a total cost of $t(t + 1)$. On the other hand, a strongly minimal solution can schedule at most t edges with weight 1 per round, thus incurring a total cost of $t(2t+1)$. The ratio of the cost of the two solutions approaches 2 as $t \to \infty$.

Acknowledgements. We thank Yoo-Ah Kim for useful discussions.

References

1. E. Anderson, J. Hall, J. Hartline, M. Hobbes, A. Karlin, J. Saia, R. Swaminathan, and J. Wilkes. *An Experimental Study of Data Migration Algorithms.* Proc. of the Workshop on Algorithm Engineering, pages 145-158, 2001.
2. A. Bar-Noy, M. Bellare, M. M. Halldórsson, H. Shachnai, and T. Tamir. *On Chromatic Sums and Distributed Resource Allocation.* Information and Computation, 140:183-202, 1998.
3. S. Chakrabarti, C. A. Phillips, A. S. Schulz, D. B. Shmoys, C. Stein, and J. Wein. *Improved Scheduling Problems For Minsum Criteria.* Proc. of the 23rd International Colloquium on Automata, Languages, and Programming, LNCS 1099, 646-657, 1996.
4. E. G. Coffman, M. R. Garey, D. S. Johnson, and A. S. LaPaugh. *Scheduling File Transfers.* SIAM Journal on Computing, 14(3):744-780, 1985.
5. R. Gandhi, M. M. Halldórsson, G. Kortsarz, and H. Shachnai. *Improved Results for Data Migration and Openshop Scheduling.* ACM Transactions on Algorithms, 2(1):116-129, 2006.
6. R. Gandhi, M. M. Halldórsson, G. Kortsarz, and H. Shachnai. *Improved Bounds for Scheduling Conflicting Jobs with Minsum Criteria.* Proc. of the Second Workshop on Approximation and Online Algorithms, 68-82, 2004.
7. R. Graham. *Bounds for certain multiprocessing anomalies.* Bell System Technical Journal, 45:1563-1581, 1966.
8. H. Gabow and O. Kariv. *Algorithms for edge coloring bipartite graphs and multigraphs.* SIAM Journal of Computing, 11(1), February 1982.
9. J. Hall, J. Hartline, A. Karlin, J. Saia, and J. Wilkes. *On Algorithms for Efficient Data Migration.* Proc. of the 12th ACM-SIAM Symposium on Discrete Algorithms, 620-629, 2001.
10. L. Hall, A. S. Schulz, D. B. Shmoys, and J. Wein. *Scheduling to Minimize Average Completion Time: Off-line and On-line Approximation Algorithms.* Mathematics of Operations Research, 22:513-544, 1997.

11. M. M. Halldórsson, G. Kortsarz, and H. Shachnai. *Sum Coloring Interval Graphs and k-Claw Free Graphs with Applications for Scheduling Dependent Jobs.* Algorithmica, 37:187-209, 2003.
12. H. Hoogeveen, P. Schuurman, and G. Woeginger. *Non-approximability Results For Scheduling Problems with Minsum Criteria.* Proc. of the 6th International Conference on Integer Programming and Combinatorial Optimization, LNCS 1412, 353-366, 1998.
13. S. Khuller, Y. Kim, and Y. C. Wan. *Algorithms for Data Migration with Cloning.* In Proc. of the 22nd ACM Symposium on Principles of Database Systems, 27-36, 2003.
14. S. Khuller and A. Malekian. *Improved Algorithms for Data Migration.* To appear in APPROX 2006.
15. Y. Kim. *Data Migration to Minimize the Average Completion Time.* Journal of Algorithms,55:42-57, 2005.
16. D. Marx. *Complexity results for minimum sum edge coloring.* Manuscript, 2004.
17. T. Nishizeki and K. Kashiwagi. *On the 1.1 edge-coloring of multigraphs.* SIAM Journal on Discrete Mathematics, 3(3):391-410, 1990.
18. M. Queyranne. *Structure of a Simple Scheduling Polyhedron.* Mathematical Programming, 58:263-285, 1993.
19. M. Queyranne and M. Sviridenko. *A $(2+\epsilon)$-Approximation Algorithm for Generalized Preemptive Open Shop Problem with Minsum Objective.* Journal of Algorithms, 45:202-212, 2002.
20. M. Queyranne and M. Sviridenko. *Approximation Algorithms for Shop Scheduling Problems with Minsum Objective.* Journal of Scheduling, 5:287-305, 2002.
21. A. S. Schulz. *Scheduling to Minimize Total Weighted Completion Time: Performance Guarantees of LP-based Heuristics and Lower Bounds.* In Proc. of the 5th International Conference on Integer Programming and Combinatorial Optimization, LNCS 1084, 301-315, 1996.
22. L. Wolsey. *Mixed Integer Programming Formulations for Production Planning and Scheduling Problems.* Invited talk at the 12th International Symposium on Mathematical Programming, MIT, Cambridge, 1985.

A   Minimal Schedules and $\min \sum_v C_v$

Since *any* minimal schedule is 2-approximate with respect to the objectives $\min \sum_e C_e$ [2] and $\min \max_e C_e$, it may tempting to think that they also perform well for $\min \sum_v C_v$. Unfortunately that is not the case. Take a complete bipartite graph $K_{q,q}$ and attach to it $2q$ copies of $K_{1,\sqrt{q}}$ so that each node in $K_{q,q}$ is the center of one of the stars. The optimal solution first schedules the stars in parallel and then the edges in $K_{q,q}$, with a total cost of $\Theta(q^2)$. A minimal solution may schedule $K_{q,q}$ before the stars, incurring a cost of $\Theta(q^{2.5})$. Since the graph has $\Theta(q^{1.5})$ vertices, a minimal schedule can be a factor $\Omega(\sqrt[3]{n})$ away from the optimum.

LP Rounding and an Almost Harmonic Algorithm for Scheduling with Resource Dependent Processing Times*

Alexander Grigoriev[1], Maxim Sviridenko[2], and Marc Uetz[1]

[1] Maastricht University, Quantitative Economics, P.O. Box 616,
6200 MD Maastricht, The Netherlands
{a.grigoriev, m.uetz}@ke.unimaas.nl
[2] IBM T. J. Watson Research Center, P.O. Box 218, Yorktown Heights,
NY 10598, USA
sviri@us.ibm.com

Abstract. We consider a scheduling problem on unrelated parallel machines with the objective to minimize the makespan. In addition to its machine dependence, the processing time of any job is dependent on the usage of a scarce renewable resource, e.g. workers. A given amount of that resource can be distributed over the jobs in process at any time. The more of the resource is allocated to a job, the smaller is its processing time. This model generalizes the classical unrelated parallel machine scheduling problem by adding a time-resource tradeoff. It is also a natural variant of a generalized assignment problem studied by Shmoys and Tardos. On the basis of an integer linear programming formulation for (a relaxation of) the problem, we adopt a randomized LP rounding technique from Kumar et al. (FOCS 2005) in order to obtain a deterministic, integral LP solution that is close to optimum. We show how this rounding procedure can be used to derive a deterministic 3.75-approximation algorithm for the scheduling problem. This improves upon previous results, namely a deterministic 6.83-approximation, and a randomized 4-approximation. The improvement is due to the better LP rounding and a new scheduling algorithm that can be viewed as a restricted version of the harmonic algorithm for bin packing.

1 Introduction

Unrelated parallel machine scheduling to minimize the makespan, $R||C_{\max}$ in the three-field notation of Graham et al. [3], is one of the classical problems in combinatorial optimization. Given are n jobs that have to be scheduled on m parallel machines, and the processing time of job j if processed on machine i is p_{ij}. The goal is to minimize the latest job completion, the makespan $C_{\max}$. If the number of machines m is part of the input, the best approximation algorithm

---

* This work was done while the second author was visiting Maastricht University, supported by METEOR, the Maastricht Research School of Economics of Technology and Organizations.

J. Diaz et al. (Eds.): APPROX and RANDOM 2006, LNCS 4110, pp. 140–151, 2006.
© Springer-Verlag Berlin Heidelberg 2006

to date is a 2-approximation by Lenstra, Shmoys and Tardos [11]. Moreover, the problem cannot be approximated within a factor strictly smaller than 3/2, unless P=NP [11].

In this paper, we consider a generalization of the unrelated parallel machine scheduling problem $R||C_{\max}$ by adding a *time-resource* tradeoff. This generalization also involves a scarce renewable resource (e.g., workers) that can be used in order to speed up the processing times of the jobs. This generalization has recently been studied by Grigoriev et al. [4] and Kumar et al. [10]; it can be seen also as a variant of the unrelated machine scheduling problem with budget constraint that was considered by Shmoys and Tardos [13]. More precisely, a maximum number of k units of a resource is available at any time. It may be used to speed up the jobs, and the available amount of k units of that resource must not be exceeded at any time. In contrast to the linearity assumption of the relation of processing times and costs in [13], the only assumption we make in this paper is that the processing times p_{ijs}, which now depend also on the number s of allocated resources, are non-increasing in s for each job-machine pair. That is, we assume that $p_{ij0} \geq p_{ij1} \geq \cdots \geq p_{ijk}$ for all jobs j and all machines $i = 1, \ldots, m$.

As a matter of fact, machine scheduling problems with the additional feature of a *nonrenewable* resource constraint, such as a total budget constraint, have received quite some attention in the literature as *time-cost tradeoff* problems. To give a few references, see [1, 7, 8, 13, 14]. Surprisingly, time-resource tradeoff problems with a *renewable* resource constraint, such as personnel, seem to have received less attention, although they are not less appealing from a practical viewpoint. Apart from our previous paper [4], some (restricted) versions of the problem were considered in [5, 6, 9, 12, 15].

In this work we improve upon our previous deterministic 6.83-approximation from [4], and upon a randomized 4-approximation by Kumar, Marathe, Parthasarathy, and Srinivasan [10]. The main result of the paper is a deterministic 3.75-approximation algorithm for the unrelated machine scheduling problem with resource dependent processing times. Compared to our previous approach from [4], the improvement is obtained by using two new ingredients. The first ingredient is a more sophisticated LP rounding that can be seen as a derandomized version of a randomized rounding approach by Kumar et al. [10]. The second ingredient is a new scheduling algorithm that resembles (a restricted version of) the well-known harmonic algorithm for bin packing.

In fact, the new rounding procedure can be viewed as an extension of the Shmoys and Tardos rounding theorem for the generalized assignment problem [13]. In this extension we consider the generalized assignment problem on a bipartite *multigraph* instead of a simple bipartite graph. Note that the techniques from [13] do not seem to be extendible to the case of a multigraph.

Combining the greedy scheduling algorithm from our previous paper [4] with the new rounding theorem yields a deterministic 4-approximation algorithm for scheduling with resource dependent processing times; matching the corresponding result of Kumar et al. [10]. But using a more sophisticated linear programming relaxation that still fits the framework for the rounding procedure, combined with (a restricted version of) the harmonic algorithm for the classical bin packing problem, we design an even better 3.75-approximation algorithm. In particular, this considerably improves upon our previous 6.83-approximation [4].

2   Problem Definition

Let $V = \{1, \ldots, n\}$ be a set of jobs. Jobs must be processed non-preemptively on a set $M = \{1, \ldots, m\}$ of unrelated parallel machines. The objective is to find a schedule that minimizes the makespan $C_{\max}$, that is, the time of the last job completion. During its processing, a job j may be assigned an amount $s \in \{0, 1, \ldots, k\}$ of an additional resource, for instance additional workers, that may speed up its processing. If s resources are allocated to a job j, and the job is processed on machine i, the processing time of that job is p_{ijs}. The only assumption on the processing times, regarding their dependence on the amount of allocated resources, is monotonicity. That is, we assume that

$$p_{ij0} \geq p_{ij1} \geq \cdots \geq p_{ijk}$$

for every machine-job pair (i, j). Without loss of generality, we also assume that all processing times p_{ijs} are integral. Hence, we can restrict to feasible schedules where the jobs only start (and end) at integral points in time.

The allocation of resources to jobs is restricted as follows. At any time, no more than the available k units of the resource may be allocated to the set of jobs in process. Moreover, since we assume a discrete resource, the amount of resources assigned to any job must be integral, and we require it to be the same along its processing. In other words, if $\ell \leq k$ units of the resource are allocated to some job j, t_j and t'_j denote j's starting and completion time, respectively, only $k - \ell$ of the resources are available for other jobs between t_j and t'_j.

We finally introduce an additional piece of notation. Since we do not assume that the functions p_{ijs} are *strictly* decreasing in s, the only information that is effectively required is the *breakpoints* of p_{ijs}, that is, indices s where $p_{ijs} < p_{ij,s-1}$. Hence, define the 'relevant' indices for job j on machine i as

$$S_{ij} = \{0\} \cup \{s \mid s \leq k, \ p_{ijs} < p_{ij,s-1}\} \subseteq \{0, \ldots, k\}.$$

Considering this index sets obviously suffices, since in any solution, if s units of the resource are allocated to some job j, we may as well only use s' units, where $s' \leq s$ and $s' \in S_{ij}$, without violating feasibility.

3   LP-Based Approximations for Unrelated Parallel Machines

Integer programming relaxation. Let x_{ijs} denote binary variables, indicating that an amount of s resources is used for processing job j on machine i. Then consider the following integer linear program, referred to as (IP).

$$\sum_{i \in M} \sum_{s \in S_{ij}} x_{ijs} = 1 \ , \qquad\qquad \forall \, j \in V \, , \qquad (1)$$

$$\sum_{j \in V} \sum_{s \in S_{ij}} x_{ijs} \, p_{ijs} \leq C \ , \qquad\qquad \forall \, i \in M \, , \qquad (2)$$

$$\sum_{j \in V} \sum_{i \in M} \sum_{s \in S_{ij}} x_{ijs} \, s \, p_{ijs} \leq k \, C \ , \qquad\qquad (3)$$

$$x_{ijs} = 0 \ , \qquad\qquad \text{if } p_{ijs} > C, \qquad (4)$$

$$x_{ijs} \in \{0, 1\} \ , \qquad\qquad \forall \, i, j, s. \qquad (5)$$

Here, C represents the schedule makespan. Equalities (1) make sure that every job is assigned to one machine and uses a constant amount of resources during its processing. Inequalities (2) express the fact that the total processing on each machine is a lower bound on the makespan. Inequality (3) represents the aggregated resource constraint: In any feasible schedule, the left-hand side of (3) is the total resource consumption of the schedule. Because no more than k resources may be consumed at any time, the total resource consumption cannot exceed $k \, C$. Finally, constraints (4) make sure that we do not use machine-resource pairs such that the job processing time exceeds the schedule makespan. These constraints are obviously redundant for the integer program (IP), but they will play a role later when rounding a fractional solution for the linear relaxation of (IP). Summarizing the above observations, we have:

Lemma 1 ([4]). *If there is a feasible schedule with makespan C for the unrelated machine scheduling problem with resource dependent processing times, integer linear program (1)–(5) has a feasible solution (C, x).*

Linear programming relaxation. The integer linear program (IP) with the 0/1-constraints on x relaxed to

$$x_{ijs} \geq 0, \qquad j \in V, \ s \in S_{ij}, \ i \in M$$

also has a solution for value C if there is a feasible schedule for the original scheduling problem with makespan C. We note that it can be solved in polynomial time, because it has a polynomial number of variables and constraints. Since we assume integrality of data, we are actually only interested in integral values C. Moreover, an upper bound for C is given by $\sum_{j \in V} \min_{i \in M}\{p_{ijk}\}$. Therefore, by using binary search on possible values for C, we can find in polynomial time the smallest integral value C^{LP} such that the linear programming relaxation of (1)–(5) has a feasible solution x^{LP}. We therefore obtain the following.

Lemma 2 ([4]). *The smallest integral value value C^{LP} such that the linear programming relaxation of (1)–(5) has a feasible solution is a lower bound on on the makespan of any feasible schedule, and it can be computed in polynomial time.*

Notice that, as long as we insist on constraints (4), we can not just solve a single linear program minimizing C, since constraints (4) depend nonlinearly on C. Moreover, due to the fact that we only search for integral values C, the binary search on C does not entail any additional approximation error.

Rounding the LP solution. Given a feasible solution $(C^{\mathrm{LP}}, x^{\mathrm{LP}})$ for the linear programming relaxation of (1)–(5), the vector x^{LP} may clearly be fractional. We aim at rounding this fractional solution to an integer one without sacrificing too much in terms of violation of the constraints (2) or (3). We present a rounding procedure that is inspired by a recent paper by Kumar et al. [10]. In fact, it can be seen as a deterministic version of the randomized rounding algorithm of [10]. In the following lemma, we replace the total resource consumptions of jobs, $s\, p_{ijs}$, by arbitrary (nonnegative) coefficients c_{ijs}. This will come in handy in Section 4.

Lemma 3. *Let C^{LP} be the minimal integer for which the following linear program has a feasible solution*

$$\sum_{i \in M} \sum_{s \in S_{ij}} x_{ijs} = 1 , \qquad\qquad \forall\, j \in V, \qquad\qquad (6)$$

$$\sum_{j \in V} \sum_{s \in S_{ij}} x_{ijs}\, p_{ijs} \leq C^{\mathrm{LP}}, \qquad\qquad \forall\, i \in M , \qquad\qquad (7)$$

$$\sum_{j \in V} \sum_{i \in M} \sum_{s \in S_{ij}} x_{ijs} c_{ijs} \leq k C^{\mathrm{LP}} , \qquad\qquad\qquad (8)$$

$$x_{ijs} = 0 , \qquad\qquad\qquad\quad \textit{if } p_{ijs} > C, \qquad (9)$$

$$x_{ijs} \geq 0 , \qquad\qquad\qquad\quad \forall\, i, j, s , \qquad\qquad (10)$$

and let $(C^{\mathrm{LP}}, x^{\mathrm{LP}})$ be the corresponding feasible solution, then we can find a feasible solution $x^* = (x^*_{ijs})$ for the following integer linear program in polynomial time.

$$\sum_{i \in M} \sum_{s \in S_{ij}} x_{ijs} = 1 , \qquad\qquad \forall\, j \in V, \qquad\qquad (11)$$

$$\sum_{j \in V} \sum_{s \in S_{ij}} x_{ijs}\, p_{ijs} \leq C^{\mathrm{LP}} + p_{\max}, \qquad \forall\, i \in M , \qquad (12)$$

$$\sum_{j \in V} \sum_{i \in M} \sum_{s \in S_{ij}} x_{ijs} c_{ijs} \leq k C^{\mathrm{LP}} , \qquad\qquad\qquad (13)$$

$$x_{ijs} \in \{0, 1\} , \qquad\qquad\qquad\quad \forall\, i, j, s , \qquad\qquad (14)$$

where $p_{\max} = \max\{p_{ijs} \mid x^{\mathrm{LP}}_{ijs} > 0\}$ and $c_{ijs} \geq 0$ are arbitrary fixed coefficients.

One option to prove the lemma is to derandomize the corresponding randomized rounding algorithm of [10], using the method of conditional probabilities.

However, for reasons of self-containedness and accessibility, we prefer to present a direct proof here. Notice, however, that the basic elements of the proof are indeed the same as in [10].

Proof (of Lemma 3). The rounding algorithm works in stages. Let x denote the current fractional solution in a given stage. In the first stage, define $x = x^{\mathrm{LP}}$, and notice that x^{LP} fulfills (11)–(13). Subsequently, we alter the current solution x, while maintaining validity of constraints (11) and (13).

In each stage, we consider a bipartite multigraph $G(x) = (V \cup M, E)$, where the set E of edges is defined as follows. For every pair $i \in M$ and $j \in V$, E contains a set of parallel edges, namely one for each fractional value $0 < x_{ijs} < 1$, $s = 0, \ldots, k$. Therefore, we could have up to $k + 1$ parallel edges between every machine-job pair (i, j). Notice that the degree of any non-isolated vertex $v \in V$ is at least 2, due to constraint (11). We furthermore eliminate isolated vertices from graph $G(x)$.

We will encode each edge $e \in E$ by the triplet (i, j, s). For every vertex $w \in V \cup M$ let d_w denote its degree in $G(x)$. We define a variable ϵ_{ijs} for every edge $(i, j, s) \in E$ and a set of linear equations:

$$\sum_{(i,j,s) \in E_j} \epsilon_{ijs} = 0, \ j \in V, \tag{15}$$

$$\sum_{(i,j,s) \in E_i} p_{ijs} \epsilon_{ijs} = 0, \ i \in M \text{ and } d_i \geq 2. \tag{16}$$

Let c_1 and c_2 be the number of constraints in (15) and (16), respectively. Let $r \leq \min\{c_1 + c_2, |E|\}$ be the rank of that system. Now observe that $c_1 \leq |V| \leq |E|/2$, because of constraint (11). Moreover, $c_2 \leq |E|/2$ by definition. Thus we obtain that either $r \leq c_1 + c_2 \leq |E| - 1$ or $c_1 = c_2 = |E|/2$. In the latter case, constraints (11), the degree condition in (16), and the fact that there are no isolated vertices, imply that there are exactly $|E|$ vertices in $G(x)$. Hence, the degree of each vertex must equal 2 (and graph $G(x)$ is a collection of even cycles).

Consider the first case when $r \leq |E| - 1$. Since the system of linear equations (15)–(16) is underdetermined, by Gaussian elimination we can find a general solution of this system in the form $\epsilon_{ijs} = \sum_{t=1}^{|E|-r} \alpha_{tijs} \delta_t$ in polynomial time. Here, δ_t, $t = 1 \ldots, |E| - r$, are the real valued parameters representing the degrees of freedom of the linear system, and $\alpha_{tijs} \neq 0$ are the corresponding coefficients. Hence, by fixing $\delta_2 = \delta_3 = \cdots = \delta_{|E|-r} = 0$, we obtain a solution $\epsilon_{ijs} = \alpha_{1ijs} \delta_1$. For convenience of notation we just write $\epsilon_{ijs} = \alpha_{ijs} \delta$, and note that δ is an arbitrary parameter.

Next, we can define a new fractional solution of the original linear program by letting

$$\bar{x}_{ijs} = \begin{cases} x_{ijs} + \alpha_{ijs} \delta & \text{if } x_{ijs} \text{ is fractional,} \\ x_{ijs} & \text{otherwise.} \end{cases}$$

Due to constraints (15) and (16) we obtain that constraints (11) are satisfied for all $j \in V$, and constraints (12) are satisfied for all $i \in M$ except those vertices (machines) $i \in M$ that have $|E_i| = 1$. Finally, since $\sum_{j \in V} \sum_{i \in M} \sum_{s \in S_{ij}} \bar{x}_{ijs} c_{ijs}$ is a linear function of δ we obtain that constraint (13) is satisfied either for positive or for negative δ. Therefore, by choosing δ either maximal or minimal such that $0 \le \bar{x}_{ijs} \le 1$ and such that constraint (13) is still satisfied, we obtain a new solution with one more integral variable satisfying constraints (11) and (13).

Repeating the above procedure we either end up with an integral solution x, fulfilling constraints (11) and (13), together with an empty graph $G(x)$, or we end up with some fractional solution x such that the degree of each vertex in $G(x)$ is at most 2 (even exactly 2). This means that at most two fractional jobs are assigned to any machine, and each fractional job is assigned to at most two machines. If that happens, we continue with with a rounding procedure that is akin to the dependent rounding that was proposed by Gandhi et al. [2]. Let us therefore call the following rounding stages *late* stages, and the previous ones *early* stages.

In a late stage, the maximum vertex degree in $G(x)$ is 2. Moreover, since $G(x)$ is bipartite, we can partition $G(x)$ into two matchings M_1 and M_2. Thus we can define a new fractional solution

$$\bar{x}_{ijs} = \begin{cases} x_{ijs} & \text{for } x_{ijs} \text{ integral,} \\ x_{ijs} + \delta & \text{for } (i, j, s) \in M_1, \\ x_{ijs} - \delta & \text{for } (i, j, s) \in M_2, \end{cases}$$

for some δ. Again, since $\sum_{j \in V} \sum_{i \in M} \sum_{s \in S_{ij}} \bar{x}_{ijs} c_{ijs}$ is a linear function of δ we obtain that constraint (13) is satisfied either for positive or for negative δ. Therefore, by choosing δ either maximal or minimal such that $0 \le \bar{x}_{ijs} \le 1$ and such that constraint (13) is still satisfied, we obtain a new solution with at least one more integral variable, still satisfying constraint (13). Moreover, since the two edges incident to any vertex $v \in V$ must belong to different matchings M_1 and M_2, the assignment constraint (11) remains valid, too. Notice that the resulting graph $G(\bar{x})$ still has vertex degrees at most 2, since only edges are dropped due to the rounding. Hence, we can iterate the rounding until all variables are integral.

In the end of the rounding algorithm we obtain an integral solution that obviously satisfies constraints (11). Since on every step we were choosing a solution minimizing a linear function corresponding to constraint (13), we obtain that this constraint is satisfied too.

To show that constraints (12) are satisfied for each $i \in M$, we have to show that the left hand side of the original constraint (2) increases by at most $p_{\max} = \max\{p_{ijs} \mid x_{ijs}^{\mathrm{LP}} > 0\}$. We consider the rounding stage when (2) is violated for machine $i \in M$.

On the one hand, this might happen in an early stage when $d_i = 1$. In this case, however, since there is exactly one fractional edge incident to i, we could add at most $p_{\max}$ in any future rounding stages to the total load of machine i. Hence, constraint (12) is fulfilled by machine i.

On the other hand, the violation of the original constraint (2) might happen in a late stage, where all vertices in $G(x)$ have degree at most 2. When $d_i = 1$

we argue as before. So assume $d_i = 2$. Consider machine i together with its two incident edges (i, j, s) and (i, j', s'). Whenever $x_{ijs} + x_{ij's'} \geq 1$ before the rounding, we claim that the total load of machine i increases by at most $p_{\max}$ by any possible further rounding. This because the total remaining increase in the left hand side of (2) for machine i is at most $(1-x_{ijs})p_{ijs}+(1-x_{ij's'})p_{ij's'} \leq p_{\max}$. So assume that $x_{ijs} + x_{ij's'} < 1$. We claim that at most one of the jobs j and j' will finally be assigned to machine i. To see why, consider the stage where one of these variables was rounded to an integer. Recalling that edges (i, j, s) and (i, j', s') must belong to different matchings M_1 and M_2, we may assume that x_{ijs} is rounded up, and $x_{ij's'}$ is rounded down. Clearly, $x_{ijs} + x_{ij's'} < 1$ holds before that rounding stage. Assuming that x_{ijs} is rounded to 1, it must hold that $x_{ij's'} \geq 1 - x_{ijs}$, because otherwise $x_{ij's'}$ would become negative. In other words, $x_{ijs} + x_{ij's'} \geq 1$, a contradiction. Hence, the only way to round one of the variables x_{ijs} or $x_{ij's'}$ to an integer, is to round $x_{ij's'}$ down to 0. Therefore edge $(i, j's')$ disappears, and indeed, at most job j can be assigned to machine i. Clearly, the resulting increase in the left hand side of (2) for machine i is again at most $p_{\max}$. Hence, constraint (12) is fulfilled after the rounding. $\square$

4   Scheduling

We complete the paper by designing a new 3.75–approximation algorithm for the unrelated parallel machine scheduling problem with resource dependent processing times. Notice that this improves considerably upon the 6.83–approximation from [4], and also upon the 4–approximation from [10]. To achieve this result, we apply the same rounding as in Lemma 3 to another integer programming relaxation, and we use a scheduling algorithm that is inspired by the harmonic algorithm for bin packing.

Let $B_{ij} \subseteq S_{ij}$ be the set of breakpoints that lie in the interval $(k/2, k]$, i.e., $B_{ij} = \{s \in S_{ij} \mid k/2 < s \leq k\}$. If any two jobs are processed using s resources, where $s \in B_{ij}$, these two jobs cannot be processed in parallel. Then consider the following integer linear program.

$$\sum_{i \in M} \sum_{s \in S_{ij}} x_{ijs} = 1 \ , \qquad \forall\, j \in V , \tag{17}$$

$$\sum_{j \in V} \sum_{s \in S_{ij}} x_{ijs}\, p_{ijs} \leq C \ , \qquad \forall\, i \in M , \tag{18}$$

$$\sum_{j \in V} \sum_{i \in M} \left(1.5 \sum_{s \in S_{ij}} x_{ijs}\, \frac{s}{k}\, p_{ijs} + 0.25 \sum_{s \in B_{ij}} x_{ijs} p_{ijs} \right) \leq 1.75 C \ , \tag{19}$$

$$x_{ijs} = 0 \ , \qquad \text{if } p_{ijs} > C, \tag{20}$$

$$x_{ijs} \in \{0,1\} \ , \qquad \forall\, i, j, s. \tag{21}$$

Lemma 4. *If there is a feasible schedule with makespan C for the unrelated machine scheduling problem with resource dependent processing times, integer linear program (17)–(21) has a feasible solution $(C, \tilde{x})$.*

Proof. To prove the lemma we only have to verify validity of the new total resource constraint (19). For any feasible schedule, two jobs with resource consumption larger then $k/2$ cannot be processed in parallel, so

$$\sum_{j \in V} \sum_{i=1}^{m} \sum_{s \in B_{ij}} \tilde{x}_{ijs} p_{ijs} \leq C. \tag{22}$$

Combining (22) with valid inequality (3) we derive inequality (19).   □

As before, by binary search on C while using Lemma 4 instead of Lemma 1, we can find a lower bound C^{LP} on the makespan of an optimal solution for the unrelated machine scheduling problem.

Lemma 5. *Let C^{LP} be the lower bound on the makespan of an optimal solution, and let $(C^{\mathrm{LP}}, x^{\mathrm{LP}})$ be the corresponding feasible solution of the LP-relaxation of (17)–(21), then we can find a feasible solution $x^* = (x^*_{ijs})$ for the following integer linear program in polynomial time.*

$$\sum_{i \in M} \sum_{s \in S_{ij}} x_{ijs} = 1 \,, \qquad\qquad \forall \, j \in V, \tag{23}$$

$$\sum_{j \in V} \sum_{s \in S_{ij}} x_{ijs}\, p_{ijs} \leq C^{\mathrm{LP}} + p_{\max}, \qquad \forall \, i \in M \,, \tag{24}$$

$$\sum_{j \in V} \sum_{i \in M} \left(1.5 \sum_{s \in S_{ij}} x_{ijs} \frac{s}{k}\, p_{ijs} + 0.25 \sum_{s \in B_{ij}} x_{ijs} p_{ijs} \right) \leq 1.75 C^{\mathrm{LP}} \,, \tag{25}$$

$$x_{ijs} \in \{0,1\} \,, \qquad\qquad \forall \, i, j, s \,, \tag{26}$$

where $p_{\max} = \max\{p_{ijs} \mid x^{\mathrm{LP}}_{ijs} > 0\}$.

Proof. The proof follows from Lemma 3 with

$$c_{ijs} = \begin{cases} \left(\frac{1.5s}{1.75k} + \frac{0.25}{1.75} \right) p_{ijs} & \text{for all } s \in B_{ij} \,, \\ \frac{1.5s}{1.75k}\, p_{ijs} & \text{for all } s \in S_{ij} \setminus B_{ij} \,. \end{cases}$$

□

Now, we are ready to present a scheduling algorithm with performance guarantee 3.75. To that end, we first partition the set of jobs J into three groups J_1, J_2, and J_3 according to the amount of resources consumed. Define $J_1 = \{j \mid k/2 < s^* \leq k$ and $x_{ijs^*} = 1\}$, $J_2 = \{j \mid k/3 < s^* \leq k/2$ and $x_{ijs^*} = 1\}$, and $J_3 = \{j \mid s^* \leq k/3$ and $x_{ijs^*} = 1\}$.

Algorithm LP-GREEDY: Let the resource allocations and machine assignments be determined by the rounded LP solution as in Lemma 5. The algorithm schedules jobs group by group. In the first phase it schedules jobs from J_1 one after another (they cannot be processed in parallel since they consume too much resources). Let C_1 be the completion time of the last job from J_1. In the second phase the algorithm schedules jobs from J_2 starting at time C_1. The algorithm always tries to run two jobs from J_2 in parallel. Let C_2 be the first time when the algorithm fails to do so. This could happen either because J_2 is empty or all remaining jobs must be processed on the same machine, say M_1. In the last case the algorithm places all remaining jobs on M_1 without idle time between them. In the third phase the algorithm greedily schedules jobs from J_3, starting no earlier than time C_2. So if some job from J_3 can be started at the current time C_2, we start processing this job. When no jobs can start at the current time we increment the current time to the next job completion time and repeat until all jobs are scheduled. Let C_3 be the completion time of the last job from the set $J_2 \cup J_3$.

We now estimate the makespan C^{LPG} of the schedule. Consider the machine i with the job that finishes last in the schedule. Let B be the total time when machine i is busy and I be the total time when machine i is idle in the interval $[0, C^{\mathrm{LPG}}]$, i.e., $C^{\mathrm{LPG}} = B + I$. By constraint (24) in Lemma 5, we have $B \leq C^{\mathrm{LP}} + p_{max} \leq 2C^{\mathrm{LP}}$, where the last inequality follows from (20).

To bound the total idle time on machine i we consider two cases. We first consider the case where the job that finishes last belongs to J_1 or J_2. If the last job is from J_1, intervals $[C_1, C_2]$ and $[C_2, C_3]$ have length 0. If the last job is from J_2, there is no idle time on machine i in the interval $[C_2, C_3]$. Thus in both cases, there is no idle time on machine i in the interval $[C_2, C_3]$. Let I_1 be the total idle time on machine i during $[0, C_1]$ and I_2 be the total idle time during $[C_1, C_2]$. Then $I = I_1 + I_2$ is the total idle time on machine i. Since we process one job at a time from J_1 during the time interval $[0, C_1]$ and two jobs at a time from J_2 during $[C_1, C_2]$, the total resource consumption of the schedule during idle times on machine i is at least

$$I_1 \frac{k}{2} + I_2 \frac{2k}{3} \, .$$

Letting $R_I := I_1/2 + 2I_2/3$, the total resource consumption of the schedule during idle times on machine i is at least $R_I k$, and we therefore get

$$
\begin{aligned}
I \;=\; I_1 + I_2 \;=\; &\frac{3}{2} R_I + \frac{I_1}{4} \\
&\leq \sum_{j \in V} \sum_{i \in M} \left(1.5 \sum_{s \in S_{ij}} \tilde{x}_{ijs} \frac{s}{k} p_{ijs} + 0.25 \sum_{s \in B_{ij}} \tilde{x}_{ijs} p_{ijs} \right) \\
&\leq 1.75 C^{\mathrm{LP}} \, .
\end{aligned}
\tag{27}
$$

Here, the first inequality holds since $\sum_{j \in V} \sum_{i \in M} \sum_{s \in S_{ij}} \tilde{x}_{ijs} \frac{s}{k} p_{ijs}$ equals the total resource consumption of the schedule divided by k, and since $I_1 \leq C_1 = \sum_{j \in V} \sum_{i \in M} \sum_{s \in B_{ij}} \tilde{x}_{ijs} p_{ijs}$. The second inequality follows from (25).

Similarly, if the last job in the schedule belongs to J_3, let I_1 be the total idle time on machine i during $[0, C_1]$, I_2 be the total idle time on machine i during $[C_1, C_2]$ and I_3 be the total idle time on machine i during $[C_2, C_3]$. Then $I = I_1 + I_2 + I_3$ is the total idle time on machine i. Again, we process one job at a time from J_1 during the time interval $[0, C_1]$, and two jobs at a time from J_2 during $[C_1, C_2]$. Moreover, due to the resource constraint the last job –which is from J_3– could not be scheduled at idle times on machine i during $[C_2, C_3]$, so the total resource consumption of the schedule during idle times on machine i in $[C_2, C_3]$ is at least $2/3\,k$. Hence, the total resource consumption of the schedule during idle times on machine i is at least

$$ I_1 \frac{k}{2} + (I_2 + I_3) \frac{2k}{3} . $$

Again, letting $R_I := I_1/2 + 2(I_2 + I_3)/3$, the total resource consumption of the schedule during idle times on machine i is at least $R_I k$, and we get

$$ I = I_1 + (I_2 + I_3) = \frac{3}{2} R_I + \frac{I_1}{4} . $$

Exactly as before in (27) we conclude that $I \leq 1.75 C^{\mathrm{LP}}$. Therefore, in either of the two cases we have $C^{\mathrm{LPG}} = B + I \leq 2\,C^{\mathrm{LP}} + 1.75\,C^{\mathrm{LP}} = 3.75\,C^{\mathrm{LP}}$, and we have proved the following theorem.

Theorem 1. *Algorithm* LP-GREEDY *is a* 3.75*–approximation algorithm for unrelated parallel machine scheduling with resource dependent processing times.*

Acknowledgments

We thank the referees for some helpful remarks.

References

1. Z.-L. Chen, Simultaneous Job Scheduling and Resource Allocation on Parallel Machines, *Annals of Operations Research* **129** (2004), 135–153.
2. R. Gandhi, S. Khuller, S. Parthasarathy, and A. Srinivasan, Dependent Rounding in Bipartite Graphs, in *Proc. 43rd Annual IEEE Symposium on Foundations of Computer Science*, 2002, 323–332.
3. R. L. Graham, E. L. Lawler, J. K. Lenstra, and A. H. G. Rinnooy Kan, Optimization and approximation in deterministic sequencing and scheduling: A survey, *Annals of Discrete Mathematics* **5** (1979), 287–326.
4. A. Grigoriev, M. Sviridenko and M. Uetz, Unrelated Parallel Machine Scheduling with Resource Dependent Processing Times, Proceedings 11th Conference on Integer Programming and Combinatorial Optimization, M. Jünger and V. Kaibel (eds.), *Lecture Notes in Computer Science* 3509, Springer, 2005, 182–195.

5. A. Grigoriev and M. Uetz, Scheduling Jobs with Linear Speedup, Proceedings 3rd Workshop on Approximation and Online Algorithms, T. Erlebach and P. Persiano (eds.), *Lecture Notes in Computer Science* 3879, Springer, 2006, 203–215.

6. K. Jansen, Scheduling Malleable Parallel Tasks: An Asymptotic Fully Polynomial Time Approximation Scheme, *Algorithmica* **39** (2004), pp. 59-81.

7. K. Jansen and M. Mastrolilli, Approximation schemes for parallel machine scheduling problems with controllable processing times, *Computers and Operations Research* **31** (2004), 1565–1581.

8. J. E. Kelley and M. R. Walker, *Critical path planning and scheduling: An introduction*, Mauchly Associates, Ambler (PA), 1959.

9. H. Kellerer and V. A. Strusevich, Scheduling parallel dedicated machines under a single non-shared resource, *European Journal of Operational Research* **147** (2003), 345–364.

10. V. S. A. Kumar, M. V. Marathe, S. Parthasarathy, and A. Srinivasan, Approximation Algorithms for Scheduling on Multiple Machines, *Proc. 46th Annual IEEE Symposium on Foundations of Computer Science*, 2005, 254–263.

11. J. K. Lenstra, D. B. Shmoys and E. Tardos, Approximation algorithms for scheduling unrelated parallel machines, *Mathematical Programming*, Series A **46** (1990), 259–271.

12. G. Mounie, C. Rapine, and D. Trystram, Efficient Approximation Algorithms for Scheduling Malleable Tasks, *Proc. 11th Annual ACM Symposium on Parallel Algorithms and Architectures*, 1999, 23–32.

13. D. B. Shmoys and E. Tardos, An approximation algorithm for the generalized assignment problem, *Mathematical Programming*, Series A **62** (1993), 461–474.

14. M. Skutella, Approximation algorithms for the discrete time-cost tradeoff problem, *Mathematics of Operations Research* **23** (1998), pp. 909–929.

15. J. Turek, J. L. Wolf, and P. S. Yu, Approximate Algorithms for Scheduling Parallelizable Tasks, *Proc. 4th Annual ACM Symposium on Parallel Algorithms and Architectures*, 1992, 323–332.

Approximating Buy-at-Bulk and Shallow-Light k-Steiner Trees

M.T. Hajiaghayi[1,*], G. Kortsarz[2], and M.R. Salavatipour[3,**]

[1] Department of Computer Science, Carnegie Mellon University
hajiagha@cs.cmu.edu
[2] Department of Computer Science, Rutgers University-Camden
guyk@crab.rutgers.edu
[3] Department of Computing Science, University of Alberta
mreza@cs.ualberta.ca

Abstract. We study two related network design problems with two cost functions. In the buy-at-bulk k-Steiner tree problem we are given a graph $G(V, E)$ with a set of terminals $T \subseteq V$ including a particular vertex s called the root, and an integer $k \leq |T|$. There are two cost functions on the edges of G, a buy cost $b : E \longrightarrow \mathbb{R}^+$ and a distance cost $r : E \longrightarrow \mathbb{R}^+$. The goal is to find a subtree H of G rooted at s with at least k terminals so that the cost $\sum_{e \in H} b(e) + \sum_{t \in T-s} dist(t, s)$ is minimize, where $dist(t, s)$ is the distance from t to s in H with respect to the r cost. We present an $O(\log^4 n)$-approximation for the buy-at-bulk k-Steiner tree problem. The second and closely related one is bicriteria approximation algorithm for Shallow-light k-Steiner trees. In the shallow-light k-Steiner tree problem we are given a graph G with edge costs $b(e)$ and distance costs $r(e)$ over the edges, and an integer k. Our goal is to find a minimum cost (under b-cost) k-Steiner tree such that the diameter under r-cost is at most some given bound D. We develop an $(O(\log n), O(\log^3 n))$-approximation algorithm for a relaxed version of Shallow-light k-Steiner tree where the solution has at least $\frac{k}{8}$ terminals. Using this we obtain an $(O(\log^2 n), O(\log^4 n))$-approximation for the shallow-light k-Steiner tree and an $O(\log^4 n)$-approximation for the buy-at-bulk k-Steiner tree problem.

1   Introduction

We study network design problems on graphs with two cost functions on the edges. These are the *buy-at-bulk $k-Steiner$ tree* problem and the *shallow-light $k-Steiner$ tree* problem. In the buy at bulk $k-Steiner$ tree problem we are given an undirected graph $G(V, E)$ with a terminal set $T \subseteq V$, a specific vertex $s \in T$ called *the root*, and an integer $k \leq |V| = n$. We also have two (non-related) cost functions on the edges of G: buy cost $b : E \longrightarrow \mathbb{R}^+$ and distance cost (sometimes

* This research was supported in part by IPM under grant number CS1383-2-02.
** Supported by NSERC grant No. G121210990, and a faculty start-up grant from University of Alberta.

J. Diaz et al. (Eds.): APPROX and RANDOM 2006, LNCS 4110, pp. 152–163, 2006.

also called rent cost) $r : E \longrightarrow \mathbb{R}^+$. We use the term *non-uniform* to denote that b and r are not related. All variants studied here are non-uniform. Our goal is to find a Steiner tree H spanning at least k vertices of T including the root which minimizes the following:

$$\sum_{e \in H} b(e) + \sum_{t \in T-s} L(t, s), \tag{1}$$

where $L(t, s) = \sum_{e \in P(t,s)} r(e)$ with $P(t, s)$ being the unique path from t to s in H.

Buy-at-bulk network optimization problems with two cost functions, have been extensively studied, sometimes under different names such as cost-distance (where one function defines buy cost and another function defines length). These problems have practical importance (see e.g. [2, 8, 18, 17, 19, 20, 25, 27]).

The second problem we consider is a variant of the shallow-light network design problem. A graph $G(V, E)$ and a collection $T \subseteq V$ of terminals are given in addition to cost and length functions $b, r : E \longrightarrow \mathbb{R}^+$ and two bounds, a cost bound B and a length bound D. The cost of a spanning subtree $H(V, E')$ is $b(E') = \sum_{e \in E'} b(e)$. For a path P, $L(P) = \sum_{e \in P} r(e)$. The distance between u, v in H is $\text{dist}_H(u, v) = L(P_{u,v})$ so that $P_{u,v}$ is the unique path between u, v in H. The diameter of H is $\text{diam}(H) = \max_{u,v} \text{dist}_H(u, v)$. Throughout, whenever we talk about the cost of a path or the cost of a tree we mean the cost under b-cost and whenever we say length or diameter we mean under r-cost. Assuming a spanning subtree $H(V, E')$ with cost at most B and diameter at most D exists, the shallow-light spanning tree problem is to find H. The more general shallow-light k-Steiner tree problem requires to select for an input k a tree spanning k nodes that meets the diameter and cost bounds D and B, respectively. Even the shallow-light spanning tree ($k = n$) special case is NP-hard and also NP-hard to approximate within a factor better than $c \log n$ for some universal constant c [5]. Thus we focus on approximation algorithms. An (α, β) bi-criteria approximation algorithm for the shallow-light k-Steiner tree problem is an algorithm that delivers a tree H' with at least k terminals (vertices in T) whose diameter is at most $\alpha \cdot D$, and whose cost is at most β times the cost of a D-diameter minimum cost tree.

Our result for shallow-light $k-$Steiner trees has implication for the well known non-uniform multicommodity buy at bulk problem. For the most general case, the best known ratio for the non-uniform buy-at-bulk multicommodity problem is $exp(O(\sqrt{\log n \log \log n}))$ by Charikar and Karagiozova [8]. Recently, we [11] have improved this result to a polylogarithmic factor approximation using the results for shallow-light $k-$Steiner trees.

1.1   Related Work

In the buy-at-bulk multicommodity problem we are given p source-sink pairs, $\{s_i, t_i\}_{i=1}^p$. A subset E' of the edges is feasible if for every i, an s_i to t_i path exists in $G' = (V, E')$, namely, s_i, t_i belong to the same connected component in G'. The cost of E' is $\sum_{e \in E'} b(e) + \sum_i \text{dist}_{G'}(s_i, t_i)$ where the distance is with respect to r, and the goal is to find a minimum cost feasible E'. If we are also given an integer $k \leq p$ and must find a solution that connects k (out of p) s_i, t_i pairs then we have the buy-at-bulk k-multicommodity problem. It is easy to see that the buy-at-bulk Steiner (but not k-Steiner) tree problem is a special case of the buy-at-bulk multicommodity problem in which all the sinks are at a single vertex (namely the root). The buy-at-bulk k-Steiner tree problem is a special case of the buy-at-bulk k-multicommodity problem. However, it is shown in [21] that if the buy-at-bulk k-multicommodity problem admits a polylogarithmic ratio approximation then so does the dense-k-subgraph problem (see [14]). For a long time now (almost 10 years) the best known approximation for the dense k-subgraph problem is $O(n^{1/3-\varepsilon})$ for some positive $\varepsilon > 0$ [14], and it is widely believed that the dense k-subgraph problem admits no polylogarithmic ratio approximation. If indeed, the dense k-subgraph problem admits no polylogarithmic approximation, then the result in our paper shows that the case of single source (but many sinks) namely, buy-at-bulk k-Steiner tree is provably easier to approximate than the general case of arbitrary source-sink pairs.

In the uniform version of the buy-at-bulk multicommodity problem all the buy values along edges are equal. The best approximation known for the uniform case is $O(\log n)$ due to the results of Awerbuch and Azar [3], Bartal [6] and Fakcharoenphol et al. [13]. Kumar et al. [25] and Gupta et al. [19] present constant factor approximation for a the case the cost of buying each edge is equal to M times the cost of renting the edge (per unit length) for a fixed M. The single sink uniform version also admits constant-factor approximation algorithms [17, 20].

Meyerson et al. [27] study the buy-at-bulk Steiner tree or equivalently, the non-uniform single sink buy-at-bulk multicommodity problem for which they give a randomized $O(\log n)$-approximation that was derandomized by Chekuri, Khanna, and Naor [9] via an LP formulation. Note that none of these algorithms yield any polylogarithmic ratio approximation for the k-Steiner tree case.

On the lower bound side, Andrews [1] showed that unless NP $\subseteq$ ZPTIME $(n^{\text{polylog } n})$ the buy-at-bulk multicommodity problem has no $O(\log^{1/2-\varepsilon} n)$- approximation for any $\varepsilon > 0$. Under the same assumption, the uniform variant admits no $O(\log^{1/4-\varepsilon} n)$-approximation for any constant $\varepsilon > 0$. For the single sink case, Chuzhoy et al. [10] showed that the problem cannot be approximated better than $\Omega(\log \log n)$ unless NP $\subseteq$ DTIME$(n^{\log \log \log n})$.

The buy-at-bulk k-Steiner tree problem generalizes the classic Steiner tree, k-MST, and more generally k-Steiner tree problems when the rent cost is zero. See for example [16]. As we mentioned above, the buy-at-bulk Steiner tree problem first was studied by Meyerson et al. [27], but we are not aware of any result on buy-at-bulk k-MST or buy-at-bulk k-Steiner tree.

The shallow-light $k-$Steiner tree problem generalizes the Shallow-light Steiner problem [26] which is the special case of $k = |T|$. It generalizes the k-MST problem [29, 4, 7, 15, 16] which is the case $D = \infty$ and also the bounded diameter spanning tree problem [23] which is the zero costs case.

Even the $k = |T|$ special case is NP-hard and also NP-hard to approximate within a factor better than $c \log n$ for some universal constant c [5]. For $k = |T|$ an $(O(\log n), O(\log n))$-approximation is given in [26]. The constraint that only $k < n$ nodes have to be picked seems to make this problem harder to approximate than the usual shallow-light Steiner tree problem, namely, the $k = |T|$ case.

1.2   Our Results

The approximation for both problems use as a subroutine an approximation for a relaxed version of Shallow-light k-Steiner tree in which the algorithm finds a $\frac{k}{8}$-Steiner tree with diameter at most $O(D \log n)$ and b-cost at most $O(B \log^3 n)$.

Theorem 1. *Given an instance of the shallow-light k-Steiner tree problem with diameter bound D we can obtain a $\frac{k}{8}$-Steiner tree with diameter at most $O(\log n \cdot D)$ and cost at most $O(\log^3 n \cdot \text{OPT})$, where OPT is the cost of an optimum shallow-light k-Steiner tree with diameter bound D.*

Corollary 1. *We can obtain an $(O(\log^2 n), O(\log^4 n))$ bicriteria approximation for shallow-light k-Steiner tree.*

For general k, no approximation for the problem was known previous to our result. Theorem 1 is used to prove:

Theorem 2. *There is a polynomial time $O(\log^4 n)$-approximation for the buy-at-bulk k-Steiner tree problem.*

No approximation was known for general k prior to this paper.

It is worth mentioning that Theorem 1 is one of the main tools we use to obtain the first polylogarithmic approximation algorithm for the non-uniform multicommodity buy-at-bulk problem [11].

The technique used to approximate the shallow-light $k-$Steiner tree problem can be described as follows. Let a terminal be called a *true terminal* if it belongs to the optimum solution. Our procedure keeps discarding terminals from T by changing their status to "regular vertices". The crucial point is that we prove that even though "many" terminals are discarded, only "few" real terminals are deleted.

2   The Algorithms

2.1   Reducing the Buy-at-Bulk k-Steiner Tree Problem to a Shallow-Light k-Steiner Tree Problem

In this section, we show how to prove Theorem 2 and Corollary 1 using Theorem 1. A bicriteria network design problem [26] (A, B, S) is defined by identifying two objective functions, A and B, and specifying a membership requirement in a class of subgraphs S. Typically, there is a budget constraint on the first objective and we seek to minimize the second objective function. This way, the (diameter, cost, k-Steiner tree) problem is naturally defined as follows: we are given an undirected graph $G(V, E)$ with terminal set T, an integer $k \leq |T|$, diameter bound D, and two cost functions $b : E \longrightarrow \mathbb{R}^+$ and $r : E \longrightarrow \mathbb{R}^+$ on the edges. Our goal is to find a minimum b-cost (i.e. minimizing the cost under the b function) Steiner tree with k terminals in G such that the diameter of the tree under the r-cost is at most D. We can assume that a particular terminal $s \in T$, called the root belongs to the solution (we can simply guess this node s). Therefore, we are solving the rooted shallow-light k-Steiner tree. We may relax the condition of requiring at least k terminals being in the solution to at least σk terminals be in the solution for some constant $\sigma \leq 1$. We call this variation the relaxed shallow-light k-Steiner tree.

We say an algorithm is an (α, β)-approximation for an (A, B, S)-bicriteria problem if in the solution produced the first objective (A) value is within factor at most α of the budget and the second objective (B) value is at most β times the minimum for any solution that is within the budget on A. Marathe et al. [26] gave an $(O(\log n), O(\log n))$-approximation for the (diameter, cost, Spanning tree) problem. In Theorem 1 we show how to obtain an $(O(\log n), O(\log^3 n))$-approximation for the relaxed shallow-light k-Steiner tree where the solution has at least $\frac{k}{8}$ terminals.

First consider the buy-at-bulk k-Steiner tree problem. Note that by doing a binary search (or geometric-mean binary search):

Observation: We can assume we know the value of an optimum solution. Let OPT denote this value.

Lemma 1. *If there is an (α, β)-approximation for the relaxed shallow-light rooted k-Steiner tree problem such that the solution has at least $\frac{k}{8}$ terminals, then we have an $O((\alpha + \beta) \log k)$-approximation for (rooted) buy-at-bulk k-Steiner tree problem*

Proof. Consider the input graph $G(V, E)$ for the buy-at-bulk k-Steiner tree problem. By observation mentioned above we can assume we know OPT (the value of optimum solution). We mark every vertex with r-distance larger than OPT to s as "to be ignored". Clearly these vertices cannot be part of any optimal solution. Then, while $k > 0$ we do the following steps:

1. Run the (α, β)-approximation algorithm **A** for the relaxed shallow-light $\frac{k}{2}$-Steiner tree with diameter (under r-cost) bounded by $D = \frac{4\,\mathrm{OPT}}{k}$.
2. Mark all the terminals (except the root) of the solution of **A** as Steiner nodes.
3. Decrease k by the number of new terminals found in this stage.

Since the root belongs to all the (sub)trees found in each iteration of the while loop, at the end we will get a connected graph (tree) which spans k terminals. Now we upper bound the cost of the solution.

At some iteration let k' be the number of yet unspanned terminals. Consider an optimal solution H^* for buy-at-bulk k'-Steiner tree instance and iteratively delete every leaf of H^* with r-distance to s (the root) larger than $\frac{2\,\mathrm{OPT}}{k'}$. We delete at most $\frac{k'}{2}$ terminals. Otherwise, more than $k'/2$ terminals have rent distance at least $2\,\text{OPT}\,/k'$ to the root s and this is a contradiction as the total cost is more than OPT. So we are left with a tree rooted at s containing at least $\frac{k'}{2}$ terminals. An (α,β)-approximation for the relaxed shallow-light $\frac{k'}{2}$-Steiner tree finds a tree containing s with at least $\frac{k'}{16}$ new terminals with r-distance to S is at most $\beta \cdot \frac{2\,\text{OPT}}{k'}$ and cost bounded by $\alpha \cdot \text{OPT}$. Given the bound from the root s, this adds at most $k' \cdot \beta \cdot \frac{2OPT}{k'} = 2\beta \cdot \text{OPT}$ to the rent cost of the solution. The buy cost added is at most $\alpha \cdot \text{OPT}$. So we have covered a constant fraction of the remaining terminals at cost at most $\alpha \cdot \text{OPT}$ and the diameter increase is at most $2\beta \cdot \text{OPT}$. By a standard set-cover arguments (see [24]), after at most $O(\log k)$ iterations, we have a tree with k terminals whose total cost is at most $O((\alpha + \beta)OPT \log k)$. $\qquad\square$

Proof of Theorem 2: Follows from Lemma 1 and Theorem 1. $\qquad\square$

An argument similar to Lemma 1 (by iteratively using the algorithm for Theorem 1) proves Corollary 1.

2.2   Algorithm for Relaxed Shallow-Light k-Steiner Tree

In this subsection we prove Theorem 1. Our algorithm is inspired by the algorithms of [26] (for shallow-light Steiner tree) and [4] for the (standard) k-MST problem. Recall that the input consists of a graph $G(V, E)$ with two edge costs b and r, D is a bound on the diameter under the r cost, $T \subseteq V$ is the set of terminals including the root s, k is the number of terminals we wish to cover, and ε is an error parameter.

First we transform the input graph G into another graph, G^c, which we call the completion of G by doing the following. For every pair of vertices $u, v \in V$ we find a $(1 + \varepsilon)$-approximate minimum cost u, v-path under b-cost with length (under r-cost) at most $2D$. Let $p^*(u, v)$ denote this cost. For this, we use the FPTAS algorithm of Hassin [22] which runs in time $O(|E|(\frac{n^2}{\varepsilon} \log \frac{n}{\varepsilon}))$. We add a new edge between u and v with b-cost equal to the cost of $p^*(u, v)$ and r-cost equal to the length of $p^*(u, v)$. Later on, in any solution of G^c that uses this new edge, we can replace it with path $p^*(u, v)$ in G at no extra cost and without increasing the length (diameter). Therefore:

Lemma 2. *If we have a bicriteria solution of cost X and diameter Y in G^c then we can find (in polynomial time) a solution of cost at most X and diameter at most Y in G.*

By this lemma, and since $G \subseteq G^c$, it is enough to work with graph G^c. Note that we can delete every vertex which is not connected to s by a new edge (because the r-distance of it to s is larger than D). So all the vertices are at distance at most D from s and so are at distance at most $2D$ from each other in G. Thus we can assume G^c is a complete (multi)graph.

Before presenting the algorithm, we should note that the "rooted" and "unrooted" versions of this problem are reducible to each other at the cost of a constant factor loss in the approximation ratio. Clearly, if we can solve the rooted version we can also solve the un-rooted version by simply trying all the terminals as the root and choose the smallest solution. On the other hand, if we have an algorithm for the un-rooted version we can do the following. Assume that OPT is the cost of the optimum solution. Delete every node $v \in G^c$ for which the cost of edge sv is larger than $(1 + \varepsilon)$ OPT. Solve the un-rooted problem and if the solution does not contain the root v then add the root. This is done by arbitrarily adding an edge from v to some node in T. This will increase the cost by at most $(1 + \varepsilon)$ OPT and the diameter by at most $2D$. Hence, it is enough to present an approximation algorithm for "un-rooted" shallow-light k-Steiner tree.

We focus on graph G^c and give an algorithm which finds a shallow-light $\frac{k}{8}$-Steiner tree in it that has cost at most $O(\log^3 n \cdot \text{OPT})$ and diameter at most $O(D \log n)$. The algorithm runs in rounds and in every round we may have several iterations (of some loop). At every round we start with every terminal as a singleton connected component. Initially, every terminal is the center of its own component. In every iteration of a round we perform a *test*. Each test has one of two outcomes: "success" or "failure". If the test is a successes, we merge two connected components and go to the next iteration. A single failure in a round causes the entire round to be a failure (so we end that round). After a failed round some of the terminals are deleted, we exit the loop, and start the next round of algorithm with a new (smaller) set of terminals. As stated above we initialize again each terminal to be a component of size 1, ignoring any mergers that were done in the last failed round.

Our goal is to find a connected component (tree) containing at least $k/8$ terminals. We say that a round is *failure free* if it has no failures at all. The number of connected components is reduced by 1 by every test that ends with successes. Thus, a failure free round will eventually end with a connected component with at least $\frac{k}{8}$ terminals. Clearly, either we fail at every round, in which case the number of terminals turns eventually empty, or we will eventually have a failure free round. We later show that the first case above cannot happen.

Assume that in round i of the algorithm the number of terminals is t_i, where $t_1 = t = |T|$. At each iteration of the loop in each round i, we divide the connected components into $O(\log t_i)$ clusters, where cluster j contains the connected components whose number of terminals is between $t_i/2^{j+1}$ and $t_i/2^j$, for $j \geq 3$.

Definition 1. *In every iteration of round i (for every $i \geq 1$), a cluster is called light if the total number of terminals in the union of the connected components in that cluster is at most $\frac{t_i}{2 \log t_i}$. Otherwise, it is called heavy.*

Lemma 3. *In every iteration of round i (for every $i \geq 1$) there are at least $\frac{t_i}{2}$ terminals in heavy clusters.*

Proof. There are at most $\log t_i$ light clusters as there are at most $\log t_i$ clusters in total, and therefore they have a total of at most $\frac{t_i}{2}$ terminals. The rest of the terminals must belong to heavy clusters.   □

In any round i and any iteration of this round, we compute the light and heavy clusters. Assuming that there are at least $\frac{k}{2}$ terminals remaining in G^c, we show (in the Main Lemma) that there is a heavy cluster with at least two connected components. Then we pick such a heavy cluster arbitrarily, say cluster C_j. Assume that all the components of C_j have between p and $2p$ terminals where $p = t_i/2^{j+1}$. For every two components in C_j we consider the edge connecting their centers (recall that since we are in G^c this edge may be obtained from the approximate minimum cost path with length at most $2D$ between those vertices in G). Two connected components c_a and c_b in C_j are called *reachable* if the cost of the edge connecting their centers is at most $16 \log^2 t \cdot \text{OPT} \cdot p/k$. We test to see if there is a pair of reachable connected components in C_j. If there is such a pair of components, then we merge the components by adding the edge between their centers and then charge every node in the two components by $8 \log^2 t \cdot \text{OPT}/k$. Since there are at least $2p$ vertices in c_a and c_b combined, the total charge is enough to pay for the cost of connecting the two components. We make one of the centers of c_a or c_b (arbitrarily) to be the new center of the new (merged) component and proceed to the next iteration of this round.

Otherwise, if our test fails because there are no two reachable centers in C_j (i.e. the cost of every edge between the centers of components in C_j is larger than $16 \log^2 t \cdot \text{OPT} \cdot p/k$) then we delete all the centers of the connected components of C_j (which are all terminals). Assuming that C_j has x_j components, we set $t_{i+1} = t_i - x_j$, and then exit the loop and start round $i + 1$. Below is the formal description of the algorithm.

1. Set the counter i (for round) to 1 and let $t_1 = t = |T|$.
2. Every terminal is a connected component by itself and is the center of that component.
3. Repeat until there is a connected component with $\frac{k}{8}$ terminals:
   (a) Compute light and heavy clusters.
   (b) Throw away (ignore) every heavy cluster which has only one connected component and pick an arbitrary heavy cluster, say C_j, which has at least two components.
   (c) If there are two components c_a and c_b in C_j such that the cost of the edge connecting their centers is at most $16 \log^2 t \cdot \text{OPT}/k$ then we do the following merger:
   /* The test succeeded */
      i. Merge the components by adding that edge.
      ii. Charge every node in the two components by $8 \log^2 t \cdot \text{OPT}/k$.
      iii. Make one of the two centers the center of the new (merged) component and goto step (a).
   (d) Otherwise,   /* The test failed */
      i. Delete all the centers of components of C_j and reset the charges of all nodes to 0.
      ii. Set $t_{i+1} = t_i - x_j$ where x_j is the number of components of C_j.
      iii. Set $i = i + 1$, exit this loop and goto Step 2.

Lemma 4. *In any round $i \geq 1$, every component participates in at most $O(\log n)$ merger operations.*

Proof. Each time a component participates in a merger the number of terminals of the components it belongs to is multiplied by at least $\frac{3}{2}$. This follows as the size of the large component is at most $2p$ for some integer p and of the smaller one at least p. Therefore there are at most $O(\log n)$ (or more precisely $O(\log k)$) iterations involving that component. $\qquad \square$

Lemma 5. *In any round $i \geq 1$ of algorithm, for every component c_a that may be obtained from σ merge operations, the length (under r-cost) between the center of c_a and any other node in c_a is at most $2\sigma D$.*

Proof. The proof is by induction on σ and noting the fact that whenever we merge two components the length of the edge we add (between the centers) is at most $2D$. $\qquad \square$

Corollary 2. *In any round $i \geq 1$, every component has diameter at most $O(D \log n)$, always.*

Proof. Follows from Lemmas 4 and 5. $\qquad \square$

Lemma 6. *In any round $i \geq 1$, every terminal is charged at most $O(\log n)$ times and the total charge of every terminal is $O(\log^3 n \cdot \mathrm{OPT}/k)$.*

Proof. Recall that every time a terminals is charged, the number of terminals in its new (merged) cluster grows by at least a $3/2$ factor. Thus, each terminal participates in at most $O(\log n)$ mergers before we find a component with $\frac{k}{8}$ terminals or before the round fails (after which the charges are all reset to zero). Furthermore, each time a terminal is charged $8 \log^2 t \cdot \mathrm{OPT}/k$. So the total charge of every terminal at any given time is $O(\log^3 n \cdot \mathrm{OPT}/k)$ $\qquad \square$

By this lemma, if the algorithm terminates with a $\frac{k}{8}$-Steiner tree then the cost of the tree is at most $O(\log^3 n \cdot \mathrm{OPT})$. Also, by Corollary 2 the diameter is at most $O(D \log n)$. Thus we only need to argue that the algorithm does find a $\frac{k}{8}$-Steiner tree and for that we need to show that the algorithm terminates before the number of terminals goes down below $\frac{k}{8}$. Since at every failed round the number of terminals is reduced, after at most t rounds the number of terminals becomes zero unless the algorithm terminates earlier with a feasible solution. Hence, if we show that the the set of terminals can never be smaller than $\frac{k}{2}$ then it means that the algorithm terminates before we have fewer than $\frac{k}{2}$ terminals. We also need to prove that we can perform step 3(b) of algorithm (i.e. find a heavy cluster with at least two connected components). These are proved in our main lemma, below. For that, we use the following pairing lemma:

Lemma 7. [26] *Let T be an arbitrary tree and let $v_1, v_2, \ldots, v_{2q}$ be an even number of vertices in T. There exists a pairing of the v_i (into q pairs) so that the unique paths joining the respective pairs are edge-disjoint.*

In the following lemma, we claim some properties on terminals not previously discarded by some failed round. We fix some optimal tree OPT and use that tree for proving these claims. We use OPT to refer to both the optimal solution and its cost. As some of the terminals in OPT may have been deleted by the failed rounds, the original OPT as defined *over G* does not exist any longer (the removal of deleted terminals may have destroyed that tree). Nevertheless, we can still use this original OPT to prove properties on G^c.

Lemma 8 (Main Lemma). *At the beginning of any round $i \geq 1$, if the number of terminals is $t_i \geq \frac{k}{2}$ then:*

1. *There is at least one heavy cluster with at least two connected components (so we can perform step 3(b) of the algorithm).*
2. *The number of terminals in round $i+1$ (if there is such a round) is at least $k/2$.*

Proof. 1) By Lemma 3 there are at least $\frac{t_i}{2} \geq \frac{k}{4}$ terminals in heavy clusters. Throw away every cluster with only one connected component. These components have a total of at most $\frac{k}{8} + \frac{k}{16} + \ldots < \frac{k}{4}$ terminals. Therefore, there is at least one heavy cluster with at least two components.

2) We prove that the number of remaining terminals is always at least $\frac{k}{2}$. Let k_i be the number of terminals of OPT that are in G^c at the beginning of round i; so $k_1 = k$. Note that always $k_i \leq t_i$. Suppose at some iteration of round i and for some heavy cluster C_j chosen by the algorithm, no pair of centers are reachable to each other; so we have to delete all the centers of C_j from G^c. Assume that all the components of C_j have size between p_i and $2p_i$.

Proposition 1. *The number of centers of components of C_j that belong to OPT is at most $k/(8p_i \log^2 t)$.*

Proof. Otherwise, using the pairing lemma (Lemma 7), we can pair those centers in OPT such that the paths connecting the pairs in OPT are all edge-disjoint. By averaging, there is at least one path with cost at most $16p_i \log^2 t \cdot \text{OPT}/k$ contradicting our assumption (because if there was such a path we would have merged the two components). □

Therefore, by Proposition 1, the number of terminals of OPT in G^c goes down by a factor of at most $1 - 1/(8p_i \log^2 t)$. On the other hand, since C_j is a heavy cluster and we have at most $2p_i$ nodes in every component of C_j, there are at least $t_i/(2\log t_i)/(2p_i) = t_i/(4p_i \log t_i)$ components in C_j. This is also a lower bound on the number of centers (terminals) that are deleted in round i. Therefore, the number of terminals in G^c goes down by a factor of at least $1 - 1/(4p_i \log t_i)$. Hence:

$$k_i\left(1 - \frac{1}{8p_i \log^2 t}\right) \leq k_{i+1} \leq t_{i+1} \leq t_i\left(1 - \frac{1}{4p_i \log t_i}\right) \leq t_i\left(1 - \frac{1}{4p_i \log t}\right)$$

We now use the following two inequalities:

$$\text{If } x \leq 1/2, \text{ then } 1 - x \geq e^{-2x} \tag{2}$$

and
$$1 - x \le e^{-x} \tag{3}$$

Using Inequality (2) and since $1/(8p_i \log^2 t) < 1/2$, it follows that $1 - 1/(8p_i \log^2 t) \ge e^{-1/(4p_i \log^2 t)}$. On the other hand from Inequality (3): $(1 - \frac{1}{4p_i \log t}) \le e^{-1/(4p_i \log t)}$. Thus

$$k \cdot exp\left(-\sum_{\ell=1}^{i} \frac{1}{4p_\ell \log^2 t}\right) \le k \prod_{\ell=1}^{i}\left(1 - \frac{1}{8p_\ell \log^2 t}\right) \le k_{i+1} \le t_{i+1}$$

$$\le t \prod_{\ell=1}^{i}\left(1 - \frac{1}{4p_\ell \log t}\right) \le t \cdot exp\left(-\sum_{\ell=1}^{i} \frac{1}{4p_\ell \log t}\right).$$

Note that both sequences t_i and k_i are decreasing but at different rates and t_i is lower bounded by k_i.

Note that $\sum_{\ell=1}^{i} \frac{1}{4p_\ell} \le \log t \cdot \ln t$, because for this value $t_{i+1} \le t \cdot e^{-\ln t} = 1$. Plugging this upper bound on $\sum_{\ell=1}^{i} \frac{1}{4p_\ell}$ in the k_{i+1} lower bound we get that $k_{i+1} \ge k \ln t / \log t \ge k/2$. Therefore, k_j is always at least $k/2$ and so $t_j \ge k_j \ge k/2$. □

Acknowledgments

The first author would like to thank Kamal Jain and Kunal Talwar for some initial discussions on the buy-at-bulk k-Steiner tree problem.

References

1. M. Andrews, *Hardness of Buy-at-Bulk Network Design*, In Proceedings of FOCS 2004, 115-124.
2. M. Andrews and L. Zhang, *Approximation algorithms for access network design*, Algorithmica 34(2):197-215, 2002.
3. B. Awerbuch and Y. Azar, *Buy-at-bulk network design*, In Proceedings of FOCS 97, pp 542-547.
4. B. Awerbuch, Y. Azar, A. Blum, and S. Vempala, *New approximation guarantees for minimum-weight k-trees and prize-collecting salesmen*, SIAM Journal on Computing 28(1):254-262, 1999.
5. J. Bar-Ilan, G. Kortsarz, and D. Peleg, *Generalized submodular cover problems and applications*, Theoretical Computer Science 250:179-200, 2001.
6. Y. Bartal, *On approximating arbitrary matrices by tree metrics*, In Proceedings of STOC 1998, pp 161-168.
7. A. Blum, R. Ravi, and S. Vempala, *A constant-factor approximation algorithm for the k MST problem (extended abstract)*, In Proceedings of STOC 96, pp 442-448.
8. M. Charikar and A. Karagiozova, *On non-uniform multicommodity buy-at-bulk network design*, In Proceedings of STOC 2005, pp 176–182.
9. C. Chekuri, S. Khanna, and J. Naor, *A deterministic algorithm for the cost-distance problem*, In Proceedings of SODA 2001, 232-233.

10. J. Chuzhoy, A. Gupta,J. Naor, and A. Sinha, *On the approximability of some network design problems*, In Proceedings of SODA 2005, pp 943-951.

11. C. Chekuri, M. Hajiaghayi, G. Kortsarz, and M. Salavatipour, *Approximation Algorithms for Non-Uniform Buy-at-Bulk Network Design Problems*, submitted, 2006.

12. J. Cheriyan, F.S. Salman, R. Ravi, and S. Subramanian, *Buy-at-bulk network design: Approximating the single-sink edge installation problem*, SIAM Journal on Optimization, 11(3):595–610, 2000.

13. J. Fakcharoenphol, S. Rao, and K.Talwar, *A tight bound on approximating arbitrary metrics by tree metrics*, Journal of Computer and System Sciences 69(3):485-497, 2004.

14. U. Feige, G. Kortsarz, and D. Peleg, *The dense k-subgraph problem*, Algorithmica 29(3):410-421, 2001.

15. N. Garg, *A 3-Approximation for the minimum tree spanning k vertices*, In Proceedings FOCS 1996, pp 302-309.

16. N. Garg, *Saving an epsilon: a 2-approximation for the k-MST problem in graphs*, In Proceedings of STOC 2005, pp 396-402.

17. S. Guha, A. Meyerson, K Munagala, *A constant factor approximation for the single sink edge installation problems*, In Proceedings of STOC 2001, pp 383-388.

18. S. Guha and A. Meyerson and K. Munagala, *Hierarchical placement and network design problems*, In Proceedings of FOCS 2001, pp 603-612.

19. A. Gupta, A. Kumar, M. Pal, and T. Roughgarden, *Approximation Via Cost-Sharing: A Simple Approximation Algorithm for the Multicommodity Rent-or-Buy Problem*, In Proceedings of FOCS 2003, page 606-617.

20. A. Gupta, A. Kumar, and T. Roughgarden, *Simpler and better approximation algorithms for network design*, In Proceedings STOC 2003, pp 365-372.

21. M.T. Hajiaghayi and K. Jain, *The Prize-Collecting Generalized Steiner Tree Problem via a new approach of Primal-Dual Schema*, In Proceedings of SODA 2006, pp 631 - 640.

22. R. Hassin, *Approximation schemes for the restricted shortest path problem*, Mathematics of Operations Research 17(1):36-42, 1992.

23. R. Hassin and A. Levin, *Minimum Restricted Diameter Spanning trees*, In Proceedings of APPROX 2002, pp 175-184.

24. D.S. Johnson, *Approximation algorithms for combinatorial problems*, Journal of Computer and System Sciences 9:256-278, 1974.

25. A. Kumar, A. Gupta, and T. Roughgarden, *A Constant-Factor Approximation Algorithm for the Multicommodity Rent-or-Buy Problem*, In Proceedings of FOCS 2002, pages 333-342.

26. M. Marathe, R. Ravi, R. Sundaram, S.S. Ravi, D. Rosenkrantz, and H. Hunt, *Bicriteria network design problems*, J. Algorithms 28(1):141-171, 1998.

27. A. Meyerson, K. Munagala, and S. Plotkin, *Cost-Distance: Two Metric Network Design*, In Proceedings of FOCS 2000, pp 383–388.

28. A. Moss and Y. Rabani, *Approximation algorithms for constrained node weighted steiner tree problems*, In Proceedings of STOC 2001, pp 373-382.

29. R. Ravi, R. Sundaram, M.V. Marathe, D.J. Rosenkrantz, and S. Ravi, *Spanning trees short or small*, SIAM Journal on Discrete Mathematics 9(2):178-200, 1996.

Improved Algorithms for Data Migration

Samir Khuller[1,*], Yoo-Ah Kim[2], and Azarakhsh Malekian[1,*]

[1] Department of Computer Science, University of Maryland, College Park, MD 20742
{samir, malekian}@cs.umd.edu
[2] Department of Computer Science and Engineering, University of Connecticut,
Storrs, CT 06269
ykim@engr.uconn.edu

Abstract. Our work is motivated by the need to manage data on a collection of storage devices to handle dynamically changing demand. As demand for data changes, the system needs to automatically respond to changes in demand for different data items. The problem of computing a migration plan among the storage devices is called the *data migration problem*. This problem was shown to be NP-hard, and an approximation algorithm achieving an approximation factor of 9.5 was presented for the half-duplex communication model in [Khuller, Kim and Wan: Algorithms for Data Migration with Cloning, *SIAM J. on Computing*, Vol. 33(2):448–461 (2004)]. In this paper we develop an improved approximation algorithm that gives a bound of $6.5 + o(1)$ using various new ideas. In addition, we develop better algorithms using external disks and get an approximation factor of 4.5. We also consider the full duplex communication model and develop an improved bound of $4 + o(1)$ for this model, with no external disks.

1 Introduction

To handle high demand, especially for multimedia data, a common approach is to replicate data objects within the storage system. Typically, a large storage server consists of several disks connected using a dedicated network, called a *Storage Area Network*. Disks typically have constraints on storage as well as the number of clients that can access data from a single disk simultaneously. These systems are getting increasing attention since TV channels are moving to systems where TV programs will be available for users to watch with full video functionality (pause, fast forward, rewind etc.). Such programs will require large amounts of storage, in addition to bandwidth capacity to handle high demand.

Approximation algorithms have been developed [16, 17, 7, 11] to map known demand for data to a specific data layout pattern to maximize utilization, where the utilization is the total number of clients that can be assigned to a disk that contains the data they want. In the layout, we compute not only how many copies of each item we need, but also a layout pattern that specifies the precise subset of items on each disk. The problem is NP-hard, but there are polynomial-time

---

* Research supported by NSF Award CCF-0430650.

J. Diaz et al. (Eds.): APPROX and RANDOM 2006, LNCS 4110, pp. 164–175, 2006.
© Springer-Verlag Berlin Heidelberg 2006

approximation schemes [7, 16, 17, 11]. Given the relative demand for data, the algorithm computes an almost optimal layout. Note that this problem is slightly different from the *data placement problem* considered in [9, 15, 3] since all the disks are in the same location, it does not matter which disk a client is assigned to. Even in this special case, the problem is NP-hard [7].

Over time as the *demand for data changes*, the system needs to create *new* data layouts. The problem we are interested in is the problem of computing a data migration plan for the set of disks to convert an initial layout to a target layout. We assume that data objects have the same size (these could be data blocks, or files) and that it takes the same amount of time to migrate any data item from one disk to another disk. In this work we consider two models. In the first model (half-duplex) the crucial constraint is that each disk can participate in the transfer of only one item – either as a sender or as a receiver. In other words, the communication pattern in each round forms a matching. Our goal is to find a migration schedule to minimize the time taken to complete the migration (makespan). To handle high demand for popular objects, new copies will have to be dynamically created and stored on different disks. All previous work on this problem deals with the half-duplex model. We also consider the full-duplex model, where each disk can act as a sender and a receiver in each round for a single item. Previously we did not consider this natural extension of the half-duplex model since we did not completely understand how to utilize its power to prove interesting approximation guarantees.

The formal description of the *data migration problem* is as follows: data item i resides in a specified (source) subset S_i of disks, and needs to be moved to a (destination) subset D_i. In other words, each data item that initially belongs to a subset of disks, needs to be moved to another subset of disks. (We might need to create new copies of this data item and store it on an additional set of disks.) See Figure 1 for an example. If each disk had exactly one data item, and needs to copy this data item to every other disk, then it is exactly the problem of gossiping. The data migration problem in this form was first studied by Khuller, Kim and Wan [4], and it was shown to be NP-hard. In addition, a polynomial-time 9.5-approximation algorithm was developed for the half-duplex communication model.

A slightly different formulation was considered by Hall et al. [10] in which a particular transfer graph was specified. While they can solve the problem very well, this approach is limited in the sense that it does not allow (a) cloning (creation of several new copies) and (b) does not allow optimization over the space of transfer graphs. In [4] it was shown that a more general problem formulation is the one with source and destination subsets specified for each data item. However, the main focus in [10] is to do the transfers without violating space constraints. Another formulation has been considered recently where one can optimize over the space of possible target layouts [12]. The resulting problems are also NP-hard. However, no significant progress on developing approximation algorithms was made on this problem. A simple flow based heuristic was presented for the problem, and was demonstrated to be effective in finding good target layouts.

Job migration has also been considered in the scheduling context recently as well [2], where a fixed number of jobs can be migrated to reduce the makespan by as much as possible. There is a lot of work on data migration for minimizing completion time for a fixed transfer graph as well (see [6, 14] for references).

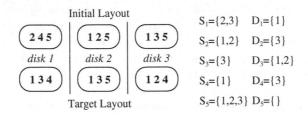

Fig. 1. An initial and target layout, and their corresponding S_i's and D_i's. For example, disk 1 initially has items $\{2, 4, 5\}$ and in the target layout has items $\{1, 3, 4\}$.

1.1 Communication Model

Different communication models can be considered based on how the disks are connected. In this paper we consider two models. The first model is the same model as in the work by Hall et al. [10, 1, 4, 13] where the disks may communicate on any matching; in other words, the underlying communication graph allows for communication between any pair of devices via a matching (a switched storage network with unbounded backplane bandwidth). Moreover, to model the limited switching capacity of the network connecting the disks, one could allow for choosing any matching of bounded size as the set of transfers that can be done in each round. We call this the *bounded-size matching model*. It was shown in [4] that an algorithm for the bounded matching model can be obtained by a simple simulation of the algorithm for the unbounded matching model with excellent performance guarantees.

In addition we consider the full duplex model where each disk may act as a sender and a receiver for an item in each round. Note that we do not require the communication pattern to be a matching any more. For example, we may have cycles, with disk 1 sending an item to disk 2, disk 2 to disk 3 and disk 3 to disk 1. In earlier work we did not discuss this model as we were unable to utilize the power of this model to prove non-trivial approximation guarantees. Note that this does not correspond directly to edge coloring anymore.

1.2 Our Results

Our approach is based on the approach initially developed in [4]. Using various new ideas lets us reduce the approximation factor to $6.5 + o(1)$. The main technical difficulty is simply that of "putting it all together" and making the analysis work.

In addition we show two more results. If we are allowed to use "external disks" (called bypass disks in [10]), we can improve the approximation guarantee further to $3 + \frac{1}{2}\max(3, \gamma)$. This can be achieved by using at most $\lceil \frac{\Delta}{\gamma} \rceil$ external disks, where Δ is the number of items that need to be migrated. We assume that each external disk can hold γ items. This gives an approximation factor of 4.5 by setting $\gamma = 3$.

Finally, we also consider the full-duplex model where each disk can be the source or destination of a transfer in each round. In this model we show that an approximation guarantee of $4 + o(1)$ can be achieved. Earlier, we did not focus on this model specifically as we were unable to utilize the power of this model in any non-trivial manner.

The algorithm developed in [4] has been implemented, and we performed an extensive set of experiments comparing its performance with the performance of other heuristics [8]. Even though the worst case approximation factor is 9.5, the algorithm performed very well in practice, giving approximation ratios within twice the optimal solution in most cases.

2   The Data Migration Algorithm

Our algorithms make use of known results on edge coloring of multigraphs. Given a graph G with max degree Δ_G and multiplicity μ the following results are known (see Bondy-Murty [5] for example). Let χ' be the edge chromatic number of G. Note that when G is bipartite, $\chi' = \Delta_G$ and such an edge coloring can be obtained in polynomial time [5].

Theorem 1. *(Vizing [20]) If G has no self-loops then $\chi' \leq \Delta_G + \mu$.*

Theorem 2. *(Shannon [18]) If G has no self-loops then $\chi' \leq \lfloor \frac{3}{2}\Delta_G \rfloor$.*

As in [4] let β_j be $|\{i | j \in D_i\}|$, i.e., the number of different sets D_i, to which a disk j belongs. We then define β as $\max_{j=1...N} \beta_j$. In other words, β is an upper bound on the number of items a disk may need. Note that β is a lower bound on the optimal number of rounds, since the disk j that attains the maximum, needs at least β rounds to receive all the items i such that $j \in D_i$, since it can receive at most one item in each round. Moreover, we may assume that $D_i \neq \emptyset$ and $D_i \cap S_i = \emptyset$. This is because we can define the destination set D_i as the set of disks that need item i and do not currently have it. Before performing data migrations, we first choose several representative sets from S_i and D_i.

2.1   Selecting Representative Sets

1. For an item i decide a primary source $s_i \in S_i$ so that $\alpha = \max_{j=1,...,N}(|\{i | j = s_i\}| + \beta_j)$ is minimized. In other words, α is the maximum number of items for which a disk may be a primary source (s_i) or destination. *Note that α is also a lower bound on the optimal number of rounds.* This step is the same as in [4].

2. Find $R_i (\subseteq D_i)$ for each item i.
   (a) We divide set D_i into $\lceil \frac{|D_i|}{q} \rceil$ subgroups of size at most q (q is a parameter that will be specified later.) That is, we create $\lfloor \frac{|D_i|}{q} \rfloor$ subgroups of size q and (if $|D_i|$ is not a multiple of q) one subgroup of size $|D_i| - q \cdot \lfloor \frac{|D_i|}{q} \rfloor$.
   (b) We find $R_i \subseteq D_i$ and assign subgroups to disks in R_i so that for each disk in R_i the total size of subgroups assigned to the disk is at most $\beta + q$. (We describe how to find R_i and the assignment, later in detail.) Let r_i be the disk in R_i to which the small subgroup (a subgroup with size strictly less than q) is assigned. Note that if $|D_i|$ is a multiple of q, there is no disk r_i. We define $\overline{R_i}$ to be $R_i \setminus r_i$.
3. For each item, we select $G'_i \subseteq D_i$ as follows.
   (a) Compute $G_i \subseteq D_i$ such that $|G_i| = \lfloor \frac{|D_i|}{\beta} \rfloor$ and they are mutually disjoint. This step is the same as in [4].
   (b) For each item i for which $G_i = \emptyset$ but $\overline{R_i} \neq \emptyset$, we select a disk g_i. Let $G'_i = G_i$ if G_i is not empty and $G'_i = \{g_i\}$ otherwise.

We now describe the details of Step 2 and Step 3.

Step 2: Select R_i for each item i. Let D_{ik} ($k = 1, \ldots, \lceil \frac{|D_i|}{q} \rceil$) be k-th subgroup of D_i. The size of D_{ik} is q for $k = 1, \ldots, \lfloor \frac{|D_i|}{q} \rfloor$ and D_{ik}, $k = \lfloor \frac{|D_i|}{q} \rfloor + 1$ consists of the remaining $|D_i| - q \cdot \lfloor |D_i|/q \rfloor$ disks (the last set is possibly empty). We make use of the following theorem by Shmoys and Tardos to choose R_i.

Theorem 3. *(Shmoys-Tardos [19]) We are given a collection of jobs $\mathcal{J}$, each of which is to be assigned to exactly one machine among the set $\mathcal{M}$; if job $j \in \mathcal{J}$ is assigned to machine $i \in \mathcal{M}$, then it requires p_{ij} units of processing time, and incurs a cost c_{ij}. Suppose that there exists a fractional solution (that is, a job can be assigned fractionally to machines) with makespan P and total cost C. Then in polynomial time we can find a schedule with makespan $P + \max p_{ij}$ and total cost C.*

We can think of each subgroup D_{ik} as a job and each disk as a machine. If disk j belongs to D_i, then we can assign job D_{ik} to disk j with zero cost. The processing time is the size of D_{ik}, which is at most q. If disk j does not belongs to D_i, then the cost to assign D_{ik} to j is ∞ (disk j cannot be in R_i).

Lemma 1. *There exists a fractional assignment such that the max load of each disk is at most β.*

Proof. We can assign $\frac{1}{|D_i|}$ fraction of subgroup D_{ik} to each disk $j \in D_i$. It is easy to check that every subgroup D_{ik} is completely assigned. The load on disk j is given by

$$\sum_{i:j \in D_i} \sum_k \frac{|D_{ik}|}{|D_i|} = \sum_{i:j \in D_i} \frac{1}{|D_i|} \sum_k |D_{ik}| = \sum_{i:j \in D_i} 1 \leq \beta$$

Lemma 2. *There is a way to choose R_i sets for each $i = 1 \dots \Delta$ and assign subgroups D_{ik} such that for each disk in R_i the total size of subgroups D_{ik} assigned to the disk is at most $\beta + q$.*

Proof. By Theorem 3, we can convert the fractional solution obtained in Lemma 1 to a solution such that each subgroup is completely assigned to one disk, and the maximum load on a disk is at most $\beta + q$ as maximum size of D_{ik} is q.

Let r_i be the disk in R_i that is assigned the small subgroup (a subgroup with size strictly less than q). Note that if $|D_i|$ is a multiple of q, there is no disk r_i. We define $\overline{R_i}$ to be $R_i \setminus r_i$. We will need the following fact later in the algorithm.

Fact. *For each disk j, at most $\beta/q + 1$ different large subgroups D_{ik} (of size exactly q) can be assigned to the disk j.*

Step 3: Select $G'_i \subseteq D_i$. We can find disjoint sets $G_i \subseteq D_i$ using the same algorithm as in [4]. To deal with the remaining items i for which $G_i = \emptyset$ but $\overline{R_i} \neq \emptyset$, we find a disk g_i. Note that if $|G_i| = 0$ then $|D_i| < \beta$, and therefore, $|\overline{R_i}| < \beta/q$. We define G'_i to be G_i if $G_i \neq \emptyset$ and $G'_i = g_i$ otherwise.

Lemma 3. *For each item i for which $G_i = \emptyset$ but $\overline{R_i} \neq \emptyset$, we can find g_i so that for a disk j, $\sum_{i:j=g_i} |\overline{R_i}| \leq 2\beta/q + 1$.*

Proof. We reduce the problem to the following scheduling problem. In this problem, each disk acts like a machine. For each item such that $|G_i| = 0$ we create a job of size $|\overline{R_i}|$. The cost of assigning job i to disk j is 1 iff $j \in \overline{R_i}$, otherwise it is infinite. Note that there is a fractional assignment such that the load to each disk is at most $\beta/q + 1$. (Assign each job fractionally ($\frac{1}{|\overline{R_i}|}$) to each machine (disk) in its $\overline{R_i}$ set. The load due to this job on the machine (disk) is 1. Since a disk is in at most $\beta/q + 1$ different $\bar{R}_i$ sets, its fractional load is at most $\beta/q + 1$.) By applying the Shmoys-Tardos [19] scheduling algorithm (see Theorem 3), we can find an assignment of jobs (items) to machines (disks) such that the total cost is at most the number of items and the load on each machine (disk) is at most $2\beta/q + 1$. (Note that the size of each job is at most β/q.) Let g_i denote the disk that item i is assigned to.

2.2 Performing Data Migrations

1. Send data item i from S_i to G'_i. For this step, we first send items from S_i to a subset of G'_i. We have to carefully choose which disk in S_i sends to a disk in G'_i (see Lemma 4). For sets with $|G'_i| > 1$, note that those G'_i sets are disjoint. Therefore, we can double the number of copies in every round (cloning) once each set receives at least one copy.
2. Send item i from G'_i to $\overline{R_i} \setminus G'_i$. We can create a transfer graph with maximum degree and multiplicity $O(\beta/q)$.
3. Send item i from s_i to r_i if r_i has not received item i. This step can be done in $3\alpha/2$ rounds.

4. We now create a transfer graph from R_i to $D_i \setminus R_i$. We find an edge coloring of the transfer graph and the number of colors used is an upper bound on the number of rounds required to ensure that each disk in D_i gets item i. In Lemma 5 we derive an upper bound on the number of required colors.

We describe the details of each step in data migration.

Step 1: Sending item i from S_i to G'_i. In the first step, we send data from S_i to G'_i. We claim that this can be done in $2OPT + O(\beta/q)$ rounds. We develop a lowerbound on the optimal solution by solving the following linear program **L(m)** for a given m.

$$\mathbf{L(m)}: \quad \sum_j \sum_{k=1}^m n_{ijk} x_{ijk} \geq |G'_i| \quad \text{for all } i \tag{1}$$

$$0 \leq x_{ijk} \leq 1 \tag{2}$$

where $n_{ijk} = \min(2^{m-k}, |G'_i|)$ if disk j belongs to S_i and $n_{ijk} = 0$ otherwise. Intuitively, x_{ijk} indicates that at time k, disk j send item i to some disk in G'_i. Let M be the minimum m such that **L(m)** has a feasible solution. Note that M is a lowerbound for the optimal solution.

Lemma 4. *We can perform migrations from S_i to G'_i in $2 \cdot M + O(\beta/q)$ rounds.*

Proof. Given a fractional solution x^* to $L(M)$, we can obtain an integral solution x^{**} such that for all i, $\sum_j \sum_k x^{**}_{ijk} \geq \lfloor \sum_j \sum_k x^*_{ijk} \rfloor$ (see [4] for details). For each item i, we arbitrarily select $\min(\sum_j \sum_k x^{**}_{ijk}, |G'_i|)$ disks from G'_i. Let H_i denote this subset. We create the following transfer graph from S_i to H_i: create an edge from a disk $j \in S_i$ to a disk H_i if $x^{**}_{ijk} = 1$. (Make sure every disk in H_i has an incoming edge from a disk in S_i.) Note the indegree of a disk in this transfer graph is $2 + \beta/q$ since a disk can belong to H_i for at most $2 + \beta/q$ different items i. (A disk can be g_i for at most $\beta/q + 1$ different items and also may belong to one G_i.) The outdegree is M and the multiplicity is $2\beta/q + 4$. Therefore, we can perform the migration from S_i to H_i in $M + O(\beta/q)$ rounds. For items with $|G'_i| = 1$, we are done for this step. For other items, since sets $G'_i(= G_i)$ are disjoint, we can double the number of copies in each round until the number of copies becomes $|G_i|$. After M rounds, the number of copies we can make for item i is at least

$$2^M |H_i| = 2^M \min(\sum_j \sum_k x^{**}_{ijk}, |G_i|)$$

$$\geq \min(2^{M-1} \cdot 2 \sum_j \sum_k x^{**}_{ijk}, |G_i|)$$

$$\geq \min(2^{M-1} \cdot (\sum_j \sum_k x^{**}_{ijk} + 1), |G_i|)$$

$$\geq \min(2^{M-1} \sum_j \sum_k x_{ijk}^*, |G_i|)$$

$$\geq \min(\sum_j \sum_k n_{ijk} x_{ijk}^*, |G_i|) \geq |G_i|.$$

The second inequality comes from the fact that $\sum_j \sum_k x_{ijk}^{**} \geq 1$. Therefore we can finish the first step in $2 \cdot M + O(\beta/q)$ rounds.

Step 2: Sending item i from G_i' to $\overline{R}_i$. We now focus on sending item i from the disks in G_i' to disks in $\overline{R}_i$. We construct a transfer graph to send data from G_i' to $\overline{R}_i$ sets so that each disk in $\overline{R}_i \setminus G_i'$ receives item i from one disk in G_i'. We create edges as follows: Add directed edges from disks in G_i to disks in $\overline{R}_i$ first. Recall that $|G_i| = \lfloor \frac{|D_i|}{\beta} \rfloor$ and $|\overline{R}_i| = \lfloor \frac{|D_i|}{q} \rfloor$. Since G_i sets are disjoint, there is a transfer graph where each disk in G_i has at most $\Theta(\beta/q)$ outgoing edges. For items with $G_i = \emptyset$, we create edges from g_i to all $\overline{R}_i$. The outdegree of the disks can be increased by at most $2\beta/q + 1$. The indegree of a disk in $\overline{R}_i$ is at most $\beta/q + 1$ and the multiplicity is $2\beta/q + 2$. Therefore, this step can be done in $O(\beta/q)$ rounds.

Step 3: Sending item i from s_i to r_i. We create a transfer graph where there is an edge from s_i to r_i if r_i has not received item i in the previous steps. The indegree of a disk j is at most β_j since a disk j is selected as r_i only if $j \in D_i$ and the outdegree of disk j is at most $\alpha - \beta_j$. Using Theorem 2, this step can be done in $3\alpha/2$ rounds.

Step 4: Sending item i from R_i to $D_i \setminus (R_i \bigcup G_i')$. We now create a transfer graph from R_i to $D_i \setminus (R_i \bigcup G_i')$ such that there is an edge from disk $a \in R_i$ to disk b if the subgroup that b belongs to is assigned to a in Lemma 2. We find an edge coloring of the transfer graph. The following lemma gives an upper bound on the number of rounds required to ensure that each disk in D_i gets item i.

Lemma 5. *The number of colors we need to color the transfer graph is at most $3\beta + q$.*

Proof. First, we compute the maximum indegree and outdegree of each node. The outdegree of a node is at most $\beta + q$ due to the way we choose R_i (See Lemma 2). The indegree of each node is at most β since in the transfer graph we send items only to the disks in their corresponding destination sets. Multiplicity of the graph is also at most β since we send data item i from disk j to disk k (or vice versa) only if both disk j and k belong to D_i. By Theorem 1, we see that the maximum number of colors needed is at most $3\beta + q$.

To wrap up, in the next theorem we show that the total number of rounds in this algorithm is bounded by $6.5+o(1)$ times the optimal solution.

Theorem 4. *The total number of rounds required for the data migration is at most $6.5 + o(1)$ times OPT.*

Proof. The total number of rounds we need is $2M + 3\alpha/2 + 3\beta + O(\beta/q) + q$. Since M, α, and β are the lowerbounds on the optimal solution, chooosing $q = \Theta(\sqrt{\beta})$ gives the desired result.

3   External Disks

Until now we assumed that we had N disks, and the source and destination sets were chosen from this set of disks and only essential transfers are performed. In other words, if an item i is sent to disk j, then it must be that $j \in D_i$ (disk j was in the destination set for item i), hence the total number of transfers done is the least possible. In several situations, we may have access to idle disks with available storage that we can make use of as temporary devices to enable a faster completion of the transfers we are trying to schedule. In addition, we exploit the fact that by performing a small number of non-essential transfers (this was also used in [13, 10]), we can further reduce the total number of rounds required. We show that indeed such techniques can considerably reduce the total number of rounds required for performing the transfers from S_i sets to D_i sets.

We assume that each external disk has enough space to pack γ items. If we are allowed to use $\lceil \frac{\Delta}{\gamma} \rceil$ external disks, the approximation ratio can be improved to $3 + \max(1.5, \frac{\gamma}{2})$. For example, choosing $\gamma = 3$ gives a bound of 4.5.

Define $\bar{\beta} = \sum_{i=1}^{\Delta} \frac{|D_i|}{N}$. We can see that $2\bar{\beta}$ is a lowerbound on the optimal number of rounds since in each round at most $\lfloor \frac{N}{2} \rfloor$ data items can be transferred. The high level description of the algorithm is as follows:

1. Assign γ items to each external disk. Send items to their assigned external disks.
2. For each item i, choose disjoint $\overline{G_i}$ sets of size $\lfloor \frac{D_i}{\bar{\beta}} \rfloor$.
3. Send item i to all disks in the $\overline{G_i}$ set.
4. Send item i from the $\overline{G_i}$ set to all the disks in D_i. We will also make use of the copy of item i on the external disk.

We now discuss the steps in detail.

First step can be done in at most $\max(\alpha, \gamma)$ rounds by sending the items from their primary sources to the external disks (for this step we will compute α as before, with the change that we can ignore the β_j term). The maximum degree of each disk is at most $\max(\alpha, \gamma)$. Since the graph is bipartite, transferring items to their assigned external disks can be finished in $\max(\alpha, \gamma)$ rounds.

We can easily choose disjoint set $\overline{G_i}$ as we are allowed to perform non-essential transfers (i.e., a disk j can belong to $\overline{G_i}$ even if j is not in D_i.) Hence we can use a simple greedy method to choose $\overline{G_i}$. Broadcasting items inside $\overline{G_i}$ can be done in $2M$ rounds as described in Section 2.

Next step is to send the item to all the remaining disks in the D_i sets. We make a transfer graph as follows: assign to each disk in $\overline{G_i}$ at most $\bar{\beta}$ disks in D_i

so that each disk in D_i is assigned to at most one disk in the $\overline{G_i}$ set. The number of unassigned disks from each D_i set is at most $\bar{\beta}$. Assign all of the remaining disks from D_i to the external disk containing that item. The outdegree of the internal disks is at most $\bar{\beta}$ since each disk belongs to at most one $\overline{G_i}$ set. The indegree of each internal disk is at most β since a disk will receive an item only if it is in its demand set. The multiplicity between two internal disks is at most 2. (Since each disk can belong to at most one $\overline{G_i}$ set.) So the total degree of each internal disk is at most $\beta + \bar{\beta}$. Each external disk has at most γ items and the number of remaining disks for each item is at most $\bar{\beta}$. So the outdegree of each external disk is at most $\gamma\bar{\beta} \leq \frac{\gamma}{2}OPT$.

So the maximum degree of each node in the whole graph is at most $\max(\beta + \bar{\beta}, \gamma\bar{\beta})$. and the maximum number of colors needed to color this graph is $\frac{1}{2}\max(3,\gamma)OPT + \max(2,\gamma)$. Adding up all these values the complete transfer can be done in $\alpha + 2m' + 3 + \frac{1}{2}\max(3,\gamma)OPT + \max(2,\gamma) \leq (3 + \frac{1}{2}\max(3,\gamma))OPT + 2\gamma + O(1)$.

4   Full Duplex Model

In this section we consider the full duplex communication model. In this model, we assume that each disk can send and receive at most one item in each round. In the half-duplex model, we assumed that at each round, a disk can either send or receive one item (but not both at the same time). In the full duplex model the communication pattern does not have to induce a matching since directed cycles are allowed (the direction indicates the data transfer direction).

We develop a $4 + o(1)$ approximation algorithm for this model. In this model, given a transfer graph G, we find an optimal migration schedule for G as follows: Construct a bipartite graph by putting one copy of each disk in each partition. We call the copy of vertex u in the first partition u_A, and in the other partition u_B. We add an edge from u_A to v_B in the bipartite graph if and only if there is a directed edge in the transfer graph from u to v. The bipartite graph can be colored optimally in polynomial time and the number of colors is equal to the maximum degree of the bipartite graph.

Note that β and M are still lower bounds on the optimal solution in the full-duplex model. The algorithm is the same as in Section 2 except the procedure to select primary sources s_i.

- For each item i, decide a primary source s_i so that $\alpha' = \max_{j=1...N}(\max(|\{j|j = s_i\}|, \beta_j))$ is minimized. Note that α' is also a lower bound for the optimal solution. We can find these primary sources as shown in Lemma 6 by adapting the method used in [4].

We show how to find the primary sources s_i.

Lemma 6. *By using network flow we can choose primary sources to minimize* $\max_{j=1...N}(\max(|\{j|j = s_i\}|, \beta_j))$

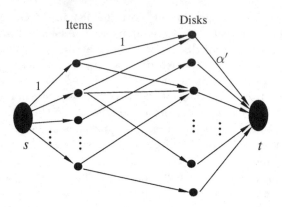

Fig. 2. Computing α'

Proof. Create two vertices s and t. (See Figure 4 for example.) Make two sets, one for the items and one for the disks. Add edges from s to each node corresponding to an item of unit capacity. Add a directed edge of infinite capacity between item j and disk i if $i \in S_j$. Add edges of capacity α' from each node in the set of disks to t. Find the minimum α' (initially $\alpha' = \beta$), so that we can find a feasible flow of value Δ. For each item j, choose the disk as its primary source s_j to which it sends one unit of flow.

Theorem 5. *There is a $4 + o(1)$ approximation algorithm for data migration in the full duplex model.*

Proof. Step 1 (from S_i to G'_i) and Step 2 (from G'_i to $\overline{R_i}$) still take $2M + O(\beta/q)$ rounds and $O(\beta/q)$ rounds, respectively. For Step 3, if we construct a bipartite graph, then the max degree is at most $\max(\alpha', \beta)$, which is the number of rounds required for this step. For Step 4, the maximum degree of the bipartite graph is $\beta + q$. Therefore, the total number of rounds we need is $2M + \max(\alpha', \beta) + \beta + O(\beta/q) + q$. By choosing $q = \Theta(\sqrt{\beta})$, we can obtain a $4 + o(1)$-approximation algorithm.

References

1. E. Anderson, J. Hall, J. Hartline, M. Hobbes, A. Karlin, J. Saia, R. Swaminathan and J. Wilkes. An Experimental Study of Data Migration Algorithms. *Workshop on Algorithm Engineering*, pages 145–158, London, UK, 2001. Springer-Verlag
2. G. Aggarwal, R. Motwani and A. Zhu. The load rebalancing problem. *Symp. on Parallel Algorithms and Architectures*, pages 258–265, (2003).
3. I. D. Baev and R. Rajaraman. Approximation algorithms for data placement in arbitrary networks. *Proc. of ACM-SIAM SODA*, pp. 661–670, 2001.
4. S. Khuller, Y.A. Kim and Y.C. Wan. Algorithms for Data Migration with Cloning, *Siam J. on Comput.*, Vol. 33, No. 2, pp. 448–461,Feb. 2004.

5. J. A. Bondy and U. S. R. Murty. Graph Theory with Applications. *American Elsevier*, New York, 1977.
6. R. Gandhi and J. Mestre. Combinatorial algorithms for Data Migration to minimize the average completion time. *APPROX* (2006) (to appear).
7. L. Golubchik, S. Khanna, S. Khuller, R. Thurimella and A. Zhu. Approximation Algorithms for Data Placement on Parallel Disks. *Proc. of ACM-SIAM SODA*, pages 661–670, Washington, D.C., USA, 2000. Society of Industrial and Applied Mathematics.
8. L. Golubchik, S. Khuller, Y. Kim, S. Shargorodskaya and Y. C. Wan. Data migration on parallel disks. *Proc. of European Symp. on Algorithms* (2004). LNCS 3221, pages 689–701. Springer. To appear in Special Issue of *Algorithmica* from ESA 2004.
9. S. Guha and K. Munagala. Improved algorithms for the data placement problem, 2002. *Proc. of ACM-SIAM SODA*, pages 106–107, San Fransisco, CA, USA, 2002. Society of Industrial and Applied Mathematics.
10. J. Hall, J. Hartline, A. Karlin, J. Saia and J. Wilkes. On Algorithms for Efficient Data Migration. *Proc. of ACM-SIAM SODA*, pp. 620–629, 2001.
11. S. Kashyap and S. Khuller. Algorithms for Non-Uniform Size Data Placement on Parallel Disks. *Conference on FST&TCS Conference*, LNCS 2914, pp. 265–276, 2003. Full version to appear in *Journal of Algorithms* (2006).
12. S. Kashyap, S. Khuller, Y. C. Wan and L. Golubchik. Fast reconfiguration of data placement in parallel disks. *2006 ALENEX Conference*, Jan 2006.
13. S. Khuller, Y. Kim and Y. C. Wan. On Generalized Gossiping and Broadcasting. *ESA Conference*. pages 373–384, Budapest, Hungary, 2003. Springer.
14. Y. Kim. Data Migration to minimize the average completion time. *Proc. of ACM-SIAM SODA*, pp. 97–98, 2003.
15. A. Meyerson, K. Munagala, and S. A. Plotkin. Web caching using access statistics. In *Symposium on Discrete Algorithms*, pages 354–363, 2001.
16. H. Shachnai and T. Tamir. On Two Class-constrained Versions of the Multiple Knapsack Problem. *Algorithmica*, 29:442–467, 2001.
17. H. Shachnai and T. Tamir. Polynomial Time Approximation Schemes for Class-constrained Packing Problems. *Workshop on Approximation Algorithms*, LNCS 1913, pp. 238–249, 2000.
18. C.E. Shannon. A Theorem on Colouring Lines of a Network. *J. Math. Phys.*, 28:148–151, 1949.
19. D.B. Shmoys and E. Tardos. An Aproximation Algorithm for the Generalized Assignment Problem. *Mathematical Programming*, A 62, pp. 461–474, 1993.
20. V. G. Vizing. On an Estimate of the Chromatic Class of a p-graph (Russian). *Diskret. Analiz.* 3:25–30, 1964.

Approximation Algorithms for Graph Homomorphism Problems

Michael Langberg[1,*], Yuval Rabani[2,**], and Chaitanya Swamy[3]

[1] Dept. of Computer Science, Caltech, Pasadena, CA 91125
mikel@cs.caltech.edu
[2] Computer Science Dept., Technion — Israel Institute of Technology,
Haifa 32000, Israel
rabani@cs.technion.ac.il
[3] Center for the Mathematics of Information, Caltech, Pasadena, CA 91125
cswamy@ist.caltech.edu

Abstract. We introduce the *maximum graph homomorphism* (MGH) problem: given a graph G, and a target graph H, find a mapping $\varphi : V_G \mapsto V_H$ that maximizes the number of edges of G that are mapped to edges of H. This problem encodes various fundamental NP-hardproblems including Maxcut and Max-k-cut. We also consider the *multiway uncut* problem. We are given a graph G and a set of terminals $T \subseteq V_G$. We want to partition V_G into $|T|$ parts, each containing exactly one terminal, so as to maximize the number of edges in E_G having both endpoints in the same part. Multiway uncut can be viewed as a special case of *prelabeled* MGH where one is also given a prelabeling $\varphi' : U \mapsto V_H$, $U \subseteq V_G$, and the output has to be an extension of φ'.

Both MGH and multiway uncut have a trivial 0.5-approximation algorithm. We present a 0.8535-approximation algorithm for multiway uncut based on a natural linear programming relaxation. This relaxation has an integrality gap of $\frac{6}{7} \simeq 0.8571$, showing that our guarantee is almost tight. For maximum graph homomorphism, we show that a $\left(\frac{1}{2} + \varepsilon_0\right)$-approximation algorithm, for any constant $\varepsilon_0 > 0$, implies an algorithm for distinguishing between certain average-case instances of the *subgraph isomorphism* problem that appear to be hard. Complementing this, we give a $\left(\frac{1}{2} + \Omega(\frac{1}{|H|\log|H|})\right)$-approximation algorithm.

1 Introduction

We introduce the *maximum graph homomorphism* (MGH) problem: given a graph $G = (V_G, E_G)$ and a target or "label" graph $H = (V_H, E_H)$, find a mapping $\varphi : V_G \mapsto V_H$ that maximizes the number of edges of G that are mapped to edges of H. This problem is trivially NP-hard; for example, deciding if G is k-colorable is equivalent to checking if the solution to MGH with graph G and the target graph H being a k-clique, has value $|E_G|$. Several fundamental

---

* Research supported in part by NSF grant CCF-0346991.
** Supported in part by ISF 52/03, BSF 2002282, and the Fund for the Promotion of Research at the Technion. Part of this work was done while visiting Caltech.

J. Diaz et al. (Eds.): APPROX and RANDOM 2006, LNCS 4110, pp. 176–187, 2006.

NP-hardoptimization problems can be encoded easily as special cases of MGH. For example, *Maxcut* is equivalent to MGH where the target graph H is a single edge; similarly *Max-k-cut* is the problem where H is a k-clique. This also shows that MGH is APX-hard even when H is fixed (i.e., not part of the input), that is, there is some absolute constant $\varepsilon_0 > 0$ such that it is NP-hardto approximate MGH better than a factor of $1 - \varepsilon_0$. The maximum graph homomorphism problem is an optimization version of the well-studied H-*coloring* problem [20], which is the problem of *deciding* whether there exists a mapping φ of value equal to $|E_G|$ (such a mapping is called a *homomorphism*).

We also consider a *prelabeled* version of the maximum graph homomorphism problem (prelabeled MGH), where the input also includes a partial mapping $\varphi' : U \mapsto V_H$ where $U \subseteq V_G$, and the output is restricted to extensions $\varphi : V_G \mapsto V_H$ of φ'. This problem, too, includes some natural NP-hardproblems as special cases. For example, consider the *multiway uncut* problem (the complement of *multiway cut*): given a graph G and a set of terminals $T \subseteq V_G$, partition V_G into $|T|$ parts, each containing exactly one element of T, so as to maximize the number of edges in E_G whose both endpoints lie in the same part. This is precisely prelabeled MGH where H consists of $|T|$ disconnected self-loops, and the prelabeling $\varphi' : T \mapsto V_H$ is a bijection.

Our Results. We present a 0.8535-approximation algorithm for the *multiway uncut* problem in Section 3. To the best of our knowledge, this is the first time anyone has considered this problem. From an exact optimization point of view, multiway uncut is equivalent to the complementary problem of multiway cut introduced by Dahlhaus et al. [9], and the APX-hardness reduction for multiway cut in [9] also shows that our problem is APX-hard. However, approximation results for multiway cut [9, 5, 23] do not directly yield guarantees for the maximization objective of multiway uncut. Our algorithm is based on a natural linear programming (LP) relaxation and rounding procedure that are motivated by the work of Calinescu, Karloff and Rabani [5] on multiway cut, and Kleinberg and Tardos [24] on the related uniform labeling problem.

In Section 4, we consider the *maximum graph homomorphism* (MGH) problem. MGH admits a simple 0.5-approximation algorithm: take any edge (i, j) of H, run the randomized/greedy algorithm for Maxcut on G to obtain a cut of value $\frac{1}{2}|E_G|$, and map the two sides of the cut to i and j. (The problem is trivial if H contains no edges, or self-loops.) This gives a solution of value at least $\frac{1}{2}|E_G|$. Our work focuses on the question of improving upon the ratio of 0.5.

We show that in general, any $\left(\frac{1}{2} + \varepsilon_0\right)$-approximation algorithm for a constant $\varepsilon_0 > 0$, would imply an algorithm for deciding certain average-case instances of the *subgraph isomorphism* problem that appear to be hard. This suggests an inherent difficulty in obtaining such an improvement. This result falls into the line of research, initiated by Feige [14], of using average-case complexity assumptions to derive hardness of approximation results. The basis of our reduction is the following key fact (that we prove): if H is a *triangle-free* graph, and G is a *random graph* drawn from the distribution $\mathcal{G}_{n,p}$ where $p = \Theta\left(\frac{\ln |V_H|}{n}\right)$, then with high probability, no mapping φ maps more than $\frac{|E_G|}{2}(1 + \epsilon)$ edges of G (the constant in $\Theta(.)$ depends on ϵ). So when G and H are drawn from a suitable distribution on triangle-free graphs, this establishes a factor 2 gap between the cases when G is a subgraph of H (so there is a mapping of value $|E_G|$), and when it is not. Thus, a $\left(\frac{1}{2} + \varepsilon_0\right)$-approximation algorithm would allow us to distinguish between these two cases.

Motivated by the known better bounds for some special cases of MGH (e.g., Maxcut [18]), we also study special families of label graphs H. We present a $\left(\frac{1}{2} + \Omega\left(\frac{1}{|V_H|\log|V_H|}\right)\right)$-approximation algorithm for MGH, by using an algorithm of Charikar and Wirth [6] for Maxcut that is based on rounding the semidefinite program for Maxcut used by Goemans and Williamson [18]. This gives an improvement over the approximation ratio of 0.5 for any fixed graph H. We obtain better improvements for some structured classes of graphs H. For the prelabeled problem, we show that an α-approximation algorithm for unlabeled MGH with label graph H yields an $\frac{\alpha}{1+\alpha}$-approximation algorithm for prelabeled MGH with graph H. Finally, we consider the problem on dense graphs G and obtain a PTAS for any fixed H, and a quasi-PTAS when H is part of the input.

Related Work. We are not aware of any previous work on the maximum graph homomorphism (or the prelabeled version) or the multiway uncut problems.

As mentioned earlier, the maximum graph homomorphism problem is an optimization version of the H-*coloring* problem, which is the problem of deciding if there exists a mapping $\varphi : V_G \mapsto V_H$ (called a homomorphism or H-coloring) that maps each edge of G to an edge of H. Homomorphisms, and the H-coloringproblem and its variants have been extensively studied from various perspectives; see, e.g., [21] and the references therein. Hell and Nešetřil [20] showed that H-coloringis in Pif H contains a self loop or is bipartite, and NP-completeotherwise. Dyer and Greenhill [12] established a similar dichotomy for the problem of counting the number of H-colorings, namely, that the problem is either in Por is #P-complete. Various variants of the H-coloringproblem and their counting versions have also been studied; see, e.g., [13, 11]. Cooper et al. [8] considered the problem of sampling a random H-coloring.

Minimization versions of the H-coloringproblem have been considered in [19, 7, 1]. Here there is a cost for assigning a label to a node of G and/or weights associated with the edges of H, and one seeks a mapping/homomorphism φ that minimizes the sum of the labeling costs and the weights of the images of the edges of G. (If the edge weights form a metric, then this is precisely the metric labeling problem [24].) Cohen et al. [7] consider the setting where the weight of assigning an edge $e \in E_G$ to an edge of H may even depend on e, and identify a class of cost functions for which the problem is in P. Aggarwal et al. [1] consider the problem with edge weights where H is a complete graph with self-loops at every node, and present various approximation and inapproximability results. Gutin et al. [19] consider the problem with only labeling costs, restricting φ to be a homomorphism, and classify the polynomial-time solvable and NP-hardcases.

A closely related problem is the *maximum common subgraph* problem: given two graphs G and H we want to find a subgraph of G with maximum number of edges that is isomorphic to a subgraph of H. MGH can be reduced to the maximum common subgraph problem by replacing each node of H by an independent set of size $|V_G|$, and each edge of H by the corresponding complete bipartite graph. Kann [22] presented a $B + 1$-approximation algorithm, where B is the maximum degree in G and H. Notice that the reduction outlined above does not preserve the degrees in the target graph H.

The complement of the multiway uncut problem, namely the *multiway cut* problem, was introduced by Dahlhaus et al. [9]. They showed that multiway cut is APX-hard, and gave a $\left(2 - \frac{2}{|T|}\right)$-approximation algorithm. Calinescu, Karloff and Rabani [5] proposed a new LP relaxation for the problem and used this to improve the factor to $\left(1.5 - \frac{1}{|T|}\right)$. The current best factor is 1.3438 due to Karger et al. [23]. Our LP-relaxation for multiway uncut is the same as the one in [5] (but with a maximization objective), and our algorithm uses a rounding procedure of Kleinberg and Tardos [24] for the uniform labeling problem (which is a generalization of the multiway cut problem).

Basing hardness of approximation results on average-case complexity is an evolving field of research which was initiated by the work of Feige [14]. Feige gave the first inapproximability results for various NP-hardoptimization problems assuming the complexity of refuting random-3CNF formulas. Subsequently, results of a similar nature (for other optimization problems, based on other hardness assumptions) were obtained by Alekhnovich [2] and Demaine et al. [10].

2   Definitions and Preliminaries

Maximum Graph Homomorphism. The input to the *maximum graph homomorphism* (MGH) problem consists of two graphs $G = (V_G, E_G)$ and $H = (V_H, E_H)$. The objective is to find a mapping $\varphi : V_G \mapsto V_H$ that maximizes the number of edges of G that are mapped to edges of H. More formally, we want to maximize $|\{(u, v) \in E_G : (\varphi(u), \varphi(v)) \in E_H\}|$. We will often refer to the mapping φ as a labeling, $\varphi(u)$ as the label of u, and H as the label graph or target graph. Let $OPT(G, H)$ denote the value of an optimal solution. Throughout, n will denote $|V_G|$ and k will denote $|V_H|$. We use variables u, v, w to denote vertices in V_G and i, j, ℓ to denote vertices in V_H.

We also consider a prelabeled version of maximum graph homomorphism where some of the nodes of G are already labeled, and we want to label the remaining vertices so as to maximize the objective function. More precisely, in the *prelabeled maximum graph homomorphism* problem, in addition to the graphs G and H, we are given a prelabeling $\varphi' : U \mapsto V_H$ where $U \subseteq V_G$, and the goal is to find an extension φ of φ' that maximizes $|\{(u, v) \in E_G : (\varphi(u), \varphi(v)) \in E_H\}|$. In general, the label graph H may also contain self-loops. However, note that if H has a self-loop, say at node i, then the unlabeled problem becomes trivial: we can simply map every vertex of G to label i to obtain $OPT(G, H) = |E_G|$. Thus, the problem with self-loops is only interesting in the prelabeled setting.

The Multiway Uncut Problem. In the *multiway uncut* problem, we are given a graph $G = (V, E)$ and a set of k terminals $T \subseteq V$. We want to find a partition of V into k subsets $V_1, \ldots, V_k$ such that each part V_i contains a distinct terminal, so as to maximize the number of *uncut* edges, that is, the quantity $\sum_{i=1}^{k} |\{(u,v) \in E : u, v \in V_i\}|$. Notice that the multiway uncut problem is a special case of the prelabeled MGH problem, where the label graph H consists of k disconnected self loops and the prelabeling is a bijection $\varphi' : T \mapsto V_H$.

3   The Multiway Uncut Problem

In this section, we consider the multiway uncut problem and present a 0.8535-approximation algorithm based on a natural linear programming (LP) relaxation. The integrality gap of this relaxation is at least $\frac{6}{7} \simeq 0.8571$, which shows that our guarantee is almost tight. Since multiway uncut is a special case of the prelabeled maximum graph homomorphism problem, we will use the terminology of MGH for consistency: we have k labels $i = 1, \ldots, k$, and the prelabeling φ' is given by $\varphi'(t_i) = i$ for the i-th terminal $t_i \in T$. Note that we may assume that there are no edges between two labeled vertices since such edges contribute 0 to the value of any solution. We consider the following LP relaxation. We use u to index the vertices of $G = (V, E)$, and i to index the labels.

$$\max \quad \sum_{(u,v) \in E} \sum_{i} c_{uv}^{i} \qquad \text{(MU-LP)}$$

$$\text{s.t.} \quad \sum_{i} x_u^i = 1 \qquad \text{for all } u,$$

$$x_t^{\varphi'(t)} = 1 \qquad \text{for all } t \in T,$$

$$c_{uv}^i = \min(x_u^i, x_v^i) \qquad \text{for all } (u,v) \in E \qquad (1)$$

$$x_u^i, c_{uv}^i \geq 0 \qquad \text{for all } u, v, i.$$

Here x_u^i indicates if vertex u is assigned label i, and c_{uv}^i indicates if both endpoints of edge (u,v) are assigned label i. The first constraint states that every node must be assigned a label, and the second enforces that this labeling is an extension of φ' (i.e., the label of a terminal does not change). The term $\sum_i c_{uv}^i$ measures the *similarity* along edge (u,v). Although (1) is not written as a linear constraint, it is easy to see that one can encode (1) using linear constraints.

One can show that the LP relaxation (MU-LP) is identical to the relaxation introduced by Calinescu et al. [5] for the multiway cut problem, i.e., any solution of value *Val* to (MU-LP) is a solution of value $|E| - $ *Val* to the relaxation in [5]. For the multiway cut problem, Calinescu et al. showed that the integrality gap of the relaxation is at most $1.5 - \frac{1}{k}$, which was improved to 1.3438 [23], whereas Freund and Karloff [16] showed that the integrality gap is at least $\frac{8}{7+1/(k-1)}$.

Our result shows that the integrality gap of (MU-LP) (which is now less than 1) is at most 0.8535, that is, there is always an integer solution of value at least 0.8535 times the optimum of (MU-LP). This also holds in the weighted setting (non-negative edge weights) where the goal is to maximize the weight of the uncut edges. The integrality-gap example in [16] also yields an integrality gap of

$\frac{6k^2-10k+4}{7k^2-13k+6} \to \frac{6}{7} \simeq 0.8571$, as $k \to \infty$, for (MU-LP) (with weighted edges). Thus, our guarantee is very close to the best possible using this LP relaxation.

A similar LP relaxation was used by Kleinberg and Tardos [24] for the uniform labeling problem. We will use a randomized rounding procedure from [24], but we will need a more refined analysis of this procedure than that in [24]. The algorithm is simple: we return the better of the following two labelings.

1. The first labeling picks an arbitrary label i, and sets $\varphi(u) = i$ for every vertex $u \notin T$. We call this the "trivial labeling".
2. The second labeling is obtained via the randomized rounding procedure of Kleinberg and Tardos, which we describe below for completeness. They also show how to derandomize the rounding, so we could use this and obtain a deterministic algorithm with the same performance guarantee. We consider the randomized version for ease of exposition and analysis.

   Let $\{x, c\}$ be an optimal solution to (MU-LP). The rounding proceeds in several rounds. Initially all vertices in $V \setminus T$ are unassigned. In each round, we independently pick a label $i \in \{1, \ldots, k\}$ uniformly at random, and a threshold ρ uniformly in $[0, 1]$. For each unassigned vertex $u \in V$, we assign u the label i (i.e., set $\varphi(u) = i$) if $x_u^i \geq \rho$. We repeat this until all the vertices in $V \setminus T$ are assigned. We call this the "LP labeling".

Analysis. Let $C_{uv} = \sum_i c_{uv}^i$. We analyze the algorithm by considering a "hybrid labeling", where we choose the LP-labeling with probability λ and the trivial labeling with probability $1 - \lambda$, for some $\lambda \in [0, 1]$. We will compare the expected contribution of an edge (u, v) in the hybrid labeling against the LP-value C_{uv}. Let $E_0 = \{(u, v) \in E : u, v \notin T\}$ and $E_1 = \{(u, v) \in E : u \text{ or } v \in T\}$. Note that $E = E_0 \cup E_1$ since there are no edges with both endpoints in T. The trivial labeling obtains a value of 1 for every edge in E_0. We now analyze the LP-labeling. For an edge (u, v), let X_{uv} denote the random variable that is 1 if u and v are assigned the same label in the LP-labeling, and 0 otherwise. We will use "$u \mapsto i$" and "$u \mapsto *$" as a shorthand to denote that "u is assigned label i", and "u is assigned some label" respectively. Let X_u^i be a random variable that is 1 if $u \mapsto i$ in the LP-labeling, and 0 otherwise.

Fact 3.1. *Suppose u is unassigned before a round. Then,* $\Pr[u \mapsto i \text{ in the round}] = \frac{1}{k} \cdot x_u^i$. *Therefore* $\Pr[u \mapsto * \text{ in the round}] = \frac{1}{k} \cdot \sum_i x_u^i = \frac{1}{k}$.

Claim 3.2. $\Pr[X_u^i = 1] = x_u^i$. *Thus, for an edge* $(u, v) \in E_1$, $\mathrm{E}[X_{uv}] = C_{uv}$.

Proof. $\Pr[X_u^i = 1] = \sum_{r=1}^{\infty}(1 - \Pr[u \mapsto * \text{ before round } r]) \cdot \Pr[u \mapsto i \text{ in round } r] = \sum_{r=1}^{\infty}(1 - \frac{1}{k})^{r-1} \cdot \frac{x_u^i}{k} = x_u^i$. For an edge $(u, v) \in E_1$, where $v \in T$ has label i, we have $\mathrm{E}[X_{uv}] = \Pr[X_u^i] = x_u^i = c_{uv}^i = C_{uv}$. $\square$

Lemma 3.3. *For an edge* $(u, v) \in E_0$, *we have* $\mathrm{E}[X_{uv}] \geq \frac{C_{uv}}{2 - C_{uv}}$.

Proof. We can lower bound $\mathrm{E}[X_{uv}]$ by the probability that both u and v are assigned a label in the *same* round. Observe that if both u and v are unassigned before a given round, then (a) the probability that u *and* v are both assigned in the round is $\frac{1}{k}\sum_i \min(x_u^i, x_v^i) = \frac{1}{k}\cdot C_{uv}$, and (b) the probability that u *or* v is assigned in the round is $\frac{1}{k}\sum_{i=1}^k \max(x_u^i, x_v^i) = \frac{1}{k}\cdot(2-C_{uv})$, since $\sum_i (\min(x_u^i, x_v^i) + \max(x_u^i, x_v^i)) = 2$. Thus, $\Pr[u$ and v are assigned in the same round$]$ is exactly

$$\sum_{r=1}^{\infty} (1 - \Pr[u \mapsto * \text{ or } v \mapsto * \text{ before round } r]) \cdot \Pr[u \mapsto * \text{ and } v \mapsto * \text{ in round } r]$$

$$= \sum_{r=1}^{\infty} \left(1 - \frac{2 - C_{uv}}{k}\right)^{r-1} \cdot \frac{C_{uv}}{k} = \frac{C_{uv}}{2 - C_{uv}}. \qquad \square$$

Fact 3.1 and Claim 3.2 were proved in [24], but for edges in E_0 their analysis proves the weaker bound $\mathrm{E}[X_{uv}] \geq 1 - \|x_u - x_v\|_1 = 2C_{uv} - 1$ which only yields a $\frac{2}{3}$-approximation guarantee for the overall algorithm.

Theorem 3.4. *The solution returned has value at least* $\left(\frac{1}{2} + \frac{\sqrt{2}}{4}\right) \cdot \left(\sum_{(u,v)\in E} C_{uv}\right)$. *Thus the approximation ratio of the above algorithm is at least* $\frac{1}{2} + \frac{\sqrt{2}}{4} \simeq 0.8535$.

Proof. We prove the stated bound for the expected value of the random hybrid labeling; the theorem then follows. For an edge $(u, v) \in E_0$, we get an expected value of (at least) $\frac{C_{uv}}{2-C_{uv}}$ in the LP-labeling by Lemma 3.3, and 1 in the trivial labeling. So the expected contribution of this edge in the hybrid labeling is at least $C_{uv} \cdot \left(\frac{\lambda}{2-C_{uv}} + \frac{1-\lambda}{C_{uv}}\right) \geq \left(\frac{1}{2} + \sqrt{\lambda(1-\lambda)}\right)C_{uv}$. The last inequality follows since $\min_{C\in[0,1]}\left(\frac{\lambda}{2-C} + \frac{1-\lambda}{C}\right) \geq \frac{1}{2} + \sqrt{\lambda(1-\lambda)}$ by simple calculus. For an edge $(u, v) \in E_1$, using Claim 3.2, the (expected) contribution in the hybrid labeling is at least λC_{uv}. Therefore the expected total value of the hybrid labeling is at least $\min\left(\lambda, \frac{1}{2} + \sqrt{\lambda(1-\lambda)}\right) \cdot \left(\sum_{(u,v)\in E} C_{uv}\right)$. Taking $\lambda = \frac{1}{2} + \frac{\sqrt{2}}{4} = \frac{1}{2} + \sqrt{\lambda(1-\lambda)} \simeq 0.8535$ maximizes this expression and yields a solution of value at least $0.8535 \cdot \left(\sum_{(u,v)\in E} C_{uv}\right)$. As mentioned earlier, the rounding procedure can be derandomized to yield a deterministic algorithm with the same guarantee. $\square$

Extensions. We can also handle the weighted case where we have non-negative weights on the edges and we want to maximize the weight of the uncut edges. The algorithm remains unchanged and the analysis requires only notational changes. One can also consider the problem where we have non-negative *profits* $\{p_u^i\}$ for assigning label i to node u, and we want to maximize the sum of the profits and the weight of the uncut edges. This problem is the complement of the uniform labeling problem considered in [24]. We can reduce this to the no-profit setting by adding an edge (u, i) with weight p_u^i for every node $u \in V$ and label i.

4   The Maximum Graph Homomorphism Problem

We now consider the maximum graph homomorphism (MGH) problem (with an arbitrary label graph H). Recall that we are given graphs G and H, and the goal is to find a mapping $\varphi : V_G \mapsto V_H$ that maximizes the number of edges of G

mapped to edges of H. In Section 4.1, we give a $\left(\frac{1}{2} + \Omega(\frac{1}{k \log k})\right)$-approximation algorithm (where $k = |V_H|$) for this problem. In Section 4.2, we present some evidence suggesting that obtaining a $\left(\frac{1}{2} + \Omega(1)\right)$-approximation algorithm may be inherently difficult. We argue that such an approximation algorithm would yield an algorithm for distinguishing between certain average-case instances of the subgraph isomorphism problem. In Section 4.3, we consider some extensions and refinements. We show that any approximation guarantee for the unlabeled problem yields a corresponding guarantee for prelabeled MGH. We also obtain a quasi-PTAS for the problem on dense graphs G (i.e., $|E_G| = \Omega(|V_G|^2)$).

4.1 A $\left(\frac{1}{2} + \Omega(\frac{1}{k \log k})\right)$-Approximation Algorithm

We now present the $\left(\frac{1}{2} + \Omega(\frac{1}{k \log k})\right)$-approximation algorithm. Recall that $k = |V_H|$. We assume that H contains at least one edge and has no self-loops (otherwise the problem is trivial). We start with some simple observations. Observe that any cut $(U, V_G \setminus U)$ of G yields a labeling φ of value equal to the size of the cut, since we can consider any edge $(i, j) \in E_H$ and map all the nodes in U to i, and all the nodes in $V_G \setminus U$ to j. Thus, since one can easily obtain a cut of value at least $\frac{|E_G|}{2}$ (e.g., by using the greedy, or randomized, algorithm where we assign each vertex greedily, or independently and uniformly at random, to one of the two parts), there is a trivial 0.5-approximation algorithm for the maximum graph homomorphism problem. Conversely, for bipartite graphs H, one can show that $MaxCut(G) = OPT(G, H)$.

Fact 4.1. *Any cut of G yields a mapping φ of value equal to the size of the cut. Thus, $OPT(G, H) \geq MaxCut(G) \geq \frac{|E_G|}{2}$.*

Claim 4.2. *If H is bipartite, the MGH problem on graphs G and H is equivalent to the Maxcut problem on G, that is, $MaxCut(G) = OPT(G, H)$.*

We improve upon this factor of 0.5 for any fixed graph H, by using a result of Charikar and Wirth [6]. They used the semidefinite program for Maxcut in [18], along with the RPR^2 rounding technique of [15] to obtain the following theorem.

Theorem 4.3 (Charikar and Wirth). *Let G be a graph with non-negative edge weights, having a cut of weight $|E_G|(\frac{1}{2} + \delta)$, where $\delta > 0$. One can obtain a cut of G with weight $|E_G|(\frac{1}{2} + \frac{c\delta}{\log(1/\delta)})$ in polynomial time, where c is a constant.*

Notice that the algorithm mentioned in the above theorem always returns a cut of value at least $\frac{|E_G|}{2}$. Our algorithm for MGH simply uses the algorithm mentioned in Theorem 4.3 to obtain a cut of G; this induces a labeling of the same value and the algorithm returns this labeling. The idea behind the algorithm is that if $OPT(G, H)$ is small compared to $|E_G|$, then $\frac{|E_G|}{2}$ would be strictly larger than $\frac{OPT(G,H)}{2}$. Otherwise, we will show that there exists a *bipartite subgraph H'* of H that captures more than half the edges of G, which in turn implies that G has a cut of value strictly larger than $\frac{|E_G|}{2}$. Thus, using Theorem 4.3 we obtain a cut of G, and hence a labeling, of value strictly larger than $\frac{|E_G|}{2} \geq \frac{OPT(G,H)}{2}$.

Theorem 4.4. *There is a $\left(\frac{1}{2} + \frac{c}{k \log k}\right)$-approximation algorithm for MGH, where $c > 0$ is a constant independent of k.*

Proof. Let G and H be the input graphs. If $OPT(G, H) \leq |E_G|\left(1 - \frac{1}{2k}\right)$, then our algorithm returns a solution of value at least $\frac{|E_G|}{2} \geq \frac{OPT(G,H)}{2(1-1/2k)} \geq \frac{OPT(G,H)}{2}\left(1 + \frac{1}{2k}\right)$. So suppose that $OPT(G, H) \geq |E_G|\left(1 - \frac{1}{2k}\right)$. Consider an optimal mapping φ^*. For each edge (i, j) in H, let $m_{ij} = \left|\{(u, v) \in E_G : \{\varphi^*(u), \varphi^*(v)\} = \{i, j\}\}\right|$. Thus, $OPT(G, H) = \sum_{(i,j) \in E_H} m_{ij}$. We claim that there is a bipartite subgraph H' of H such that $OPT(G, H') \geq \sum_{(i,j) \in H'} m_{ij} \geq \frac{|E_G|}{2}\left(1 + \frac{1}{4k}\right)$. Consider the cut $(U_H, V_H \setminus U_H)$ where U_H is a random subset of vertices of H of size $k/2$. The probability that an edge is cut by such a partition is $\frac{k^2}{4} / \binom{k}{2} = \frac{1}{2}\left(1 + \frac{1}{k-1}\right)$. Therefore, the expected weight of the cut edges is $\left(\sum_{(i,j) \in E_H} m_{ij}\right) \cdot \frac{1}{2}\left(1 + \frac{1}{k-1}\right) = \frac{OPT(G,H)}{2}\left(1 + \frac{1}{k-1}\right) \geq \frac{|E_G|}{2}\left(1 + \frac{1}{4k}\right)$. Thus, there exists such a partition of at least this value, and we can take H' to be the associated bipartite subgraph of H. Now by Claim 4.2, G must have a cut of value at least $\frac{|E_G|}{2}\left(1 + \frac{1}{4k}\right)$. So applying Theorem 4.3, our algorithm finds a cut, and hence a labeling, of value at least $|E_G|\left(\frac{1}{2} + \frac{c}{k \log k}\right)$. The theorem follows since $OPT(G, H) \leq |E_G|$. $\square$

4.2   Connection to the Subgraph Isomorphism Problem

Given two graphs G and H, the *subgraph isomorphism* problem is the problem of deciding whether G is a subgraph of H. The subgraph isomorphism problem is a well-known NP-completeproblem. We show that a $\left(\frac{1}{2} + \varepsilon_0\right)$-approximation algorithm for MGH, where $\varepsilon_0 > 0$ is an absolute constant, implies an algorithm for distinguishing between certain average-case instances of the subgraph isomorphism problem (this is defined precisely below). This hints at an inherent difficulty in obtaining an approximation ratio better than 0.5 for MGH.

The main technical result of this section (Lemma 4.5) is as follows. For any $\epsilon > 0$, if H is a *triangle-free graph*, and G is a *random graph* drawn from the distribution $\mathcal{G}_{n,p}$, for a suitable $p = p(\epsilon) \in [0, 1]$ and large enough n, then $OPT(G, H) \leq \frac{|E_G|}{2}(1 + \epsilon)$ with high probability. If however G is a subgraph of H, then $OPT(G, H) = |E_G|$. The gap between these two cases motivates the definition of a refutation problem for certain average-case instances of the subgraph isomorphism problem, which allows us to encode the difficulty of obtaining a better than 0.5-approximation algorithm for MGH. Let $\triangle_n$ be the set of all triangle-free graphs on n vertices. For $p \in [0, 1]$, let $\triangle_{n,p}$ be the distribution over $G \in \triangle_n$ obtained by choosing a random graph $G \in \mathcal{G}_{n,p}$, and then considering the edges of G in a random order and deleting any edge that is part of a triangle.

Refutation problem (with parameter $c > 0$). Find a polynomial time algorithm $\mathcal{A}$ such that given a pair of random graphs $G \in \triangle_{n,p_G}, H \in \triangle_{n,p_H}$, where $p_G = \frac{c \ln n}{n}$, $p_H \gg p_G$, (a) $\mathcal{A}$ returns "yes" if H contains G as a subgraph, and (b) $\mathcal{A}$ returns "no" on most instances, more precisely $\Pr_{G,H}[\mathcal{A}(G, H) = \text{"no"}] \geq \frac{1}{2}$.

Intuitively, the *refutation algorithm* $\mathcal{A}$ refutes most tuples (G, H) as being "no" instances of the subgraph isomorphism problem, but always announces "yes" when G is a subgraph of H. As mentioned earlier, with very high probability G will not be a subgraph of H, thus conditions (a) and (b) do not conflict. We will show that a $\left(\frac{1}{2} + \varepsilon_0\right)$-approximation algorithm for MGH yields such a refutation algorithm; thus the non-existence of such an algorithm implies that MGH cannot be approximated to a factor better than 0.5.

We mention a few remarks. First, one could also define the refutation problem in terms of an approximation version of subgraph isomorphism by requiring (a'): $\mathcal{A}$ always return "yes" if G contains a subgraph of size $|E_G|(1 - \epsilon)$ that is isomorphic to a subgraph of H. Such a modification was also considered by Feige (see Hypothesis 2 in [14]). An algorithm satisfying (a'), (b) refutes average-case instances of the *maximum common subgraph problem* [22], and is also a refutation algorithm for the exact-version of the problem. Thus, the non-existence of an algorithm satisfying (a'), (b) is a weaker hardness assumption (implying a $\left(\frac{1}{2} + \varepsilon_0\right)$ inapproximability for MGH). Moreover, this version of the refutation problem might be more robust than the exact-version. Second, we take $p_H \gg p_G$ to avoid the case where $p_H \simeq p_G$. In this setting, the problem is closely related to the graph isomorphism problem on random graphs, which is known to be solvable on average in polynomial time; see [4], and §6 of the survey [17] and its references.

Lemma 4.5. *For any $\epsilon \in (0,1)$, there exist constants $n_0(\epsilon), c_0(\epsilon)$, such that if $G = (V_G, E_G)$ is a random graph in $\mathcal{G}_{n,p}$, where $n \geq n_0(\epsilon)$, $p = \frac{c \ln k}{n}$, $c \geq c_0(\epsilon)$, and $H = (V_H, E_H)$ is a simple triangle-free graph with k vertices, then*
(i) $OPT(G, H) < \frac{cn \ln k}{4}(1 + \epsilon/2)$ with probability at least $1 - e^{-n \ln k}$, and
(ii) $OPT(G, H) < \frac{|E_G|}{2}(1 + \epsilon)$ with probability at least $1 - 2e^{-n \ln k}$.

Proof. Set $n_0(\epsilon) = \frac{8}{\epsilon}$, $c_0(\epsilon) = \frac{2048}{7\epsilon^2}$. Let $m = p\binom{n}{2}$ be the expected number of edges in G. Fix a mapping $\varphi : V_G \mapsto V_H$. We will show that with very high probability, mapping φ has value at most $\frac{m}{2}(1 + \epsilon/2)$. Applying the union bound over all mappings then yields that $OPT(G, H) < \frac{m}{2}(1 + \epsilon/2)$ with high probability, proving part (i). Since $|E_G|$ is strongly concentrated around its expectation, this will also prove part (ii).

Given the mapping φ, consider the following graph H': H' also has $n = |V_G|$ vertices, and we include an edge (u, v) in H' iff $(\varphi(u), \varphi(v))$ is an edge in H. It is easy to see that H' is also triangle-free: a triangle (v_1, v_2, v_3) in H' implies that H has edges $(\varphi(v_1), \varphi(v_2))$, $(\varphi(v_2), \varphi(v_3))$, and $(\varphi(v_3), \varphi(v_1))$, and therefore contains a triangle. Since H' is triangle-free, by Turán's Theorem [25] it has at most $\frac{n^2}{4}$ edges. Let $X(\varphi)$ denote the (random) value of the mapping φ for G. Observe that $X(\varphi)$ is simply the number of edges of H' that are also edges of G. For every pair u, v, (u, v) is in E_G with probability p, so we have $E[X(\varphi)] = p \cdot |E_{H'}| \leq p \cdot \frac{n^2}{4} = \frac{cn \ln k}{4}$. Since $X(\varphi)$ is the sum of independent indicator random variables, using Chernoff bounds, we get $\Pr[X(\varphi) \geq \frac{cn \ln k}{4}(1 + \epsilon/2)] \leq e^{-(\epsilon^2 cn \ln k)/48} \leq e^{-2n \ln k}$. The number of mappings φ is k^n. So by the union bound, $\Pr[OPT(G, H) \geq \frac{cn \ln k}{4}(1 + \epsilon/2)] = \Pr[\exists \varphi, X(\varphi) \geq \frac{cn \ln k}{4}(1 + \epsilon/2)] \leq e^{-n \ln k}$.

The expected number of edges in G is $p\binom{n}{2} = \frac{cn \ln k}{2}(1 - \frac{1}{n}) \geq \frac{cn \ln k}{2}(1 - \frac{\epsilon}{8})$. Again using Chernoff bounds, we get that $\Pr[|E_G| \leq \frac{cn \ln k}{2}(1 - \epsilon/4)] \leq$

$e^{-(7\epsilon^2 cn \ln k)/2048} \le e^{-n \ln k}$. So using part (i), with probability at least $1 - 2e^{-n \ln k}$ it is the case that $OPT(G, H) < \frac{cn \ln k}{4}(1 + \epsilon/2) < \frac{|E_G|}{2}(1 + \epsilon)$. $\qquad\square$

Theorem 4.6. *For any $\varepsilon_0 > 0$, a $\left(\frac{1}{2} + \varepsilon_0\right)$-approximation algorithm $\mathcal{A}$ for MGH yields an algorithm for the refutation problem with parameter $c \ge c_0(\varepsilon_0) = \frac{2048}{7\varepsilon_0^2}$.*

Proof Sketch. Let G and H be the two input graphs. Let n be sufficiently large. If we are in case (a), then $OPT(G, H) = |E_G|$, so running $\mathcal{A}$ on (G, H) will produce a solution of value at least $|E_G|(\frac{1}{2} + \varepsilon_0)$. Otherwise, we can use Lemma 4.5 to show that that $OPT(G, H) < |E_G|(\frac{1}{2} + \varepsilon_0)$ with high probability; thus, one can use $\mathcal{A}$ to distinguish between the two cases. Let G by obtained by deleting edges from $G' \in \mathcal{G}_{n,p}$. Lemma 4.5 shows that $OPT(G, H) \le OPT(G', H) < \frac{cn \ln n}{4}(1 + \varepsilon_0/2)$ and $|E'_G| \ge \frac{cn \ln n}{2}(1 - \varepsilon_0/4)$, with high probability. Although we delete edges from G', with high probability, the number of triangles in G' is a negligible fraction of $|E_{G'}|$. So we obtain that $|E_G| \ge \frac{cn \ln n}{2}(1 - \varepsilon_0/2)$ and therefore we have $OPT(G, H) < |E_G|(\frac{1}{2} + \varepsilon_0)$. $\qquad\square$

4.3   Extensions and Refinements

Prelabeled MGH. Recall that in prelabeled MGH, we are given a prelabeling $\varphi' : U \mapsto V_H$, $U \subseteq V_G$ and the output has to be an extension of φ'. We can show that for any label graph H, an α-approximation algorithm for MGH on instances (G, H) (α could depend on H) gives an $\frac{\alpha}{1+\alpha}$-approximation algorithm for prelabeled MGH on instances (G, H).

Dense Graphs G. We obtain much better results when G is dense, i.e., when $|E_G| = \Omega(n^2)$ ($n = |V_G|$). One can adapt the techniques of Arora, Karger and Karpinski [3] to obtain a solution φ of value $OPT(G, H) - \epsilon n^2$ in time $O\left((nk)^{\log k/\epsilon^2}\right)$ (although MGH does not directly fall into the problem-class detailed in [3]). Since $OPT(G, H) \ge \frac{|E_G|}{2} = \Omega(n^2)$, we can obtain a quasi-PTAS by setting ϵ suitably. This also yields a PTAS for any fixed graph H.

Special graphs H. When H if bipartite, by Claim 4.2 it follows that one can obtain a 0.878-approximation algorithm for MGH using the Maxcut algorithm of Goemans and Williamson [18]. One can also obtain an approximation ratio better than 0.5 if H has a dense subgraph. Let $\rho_H = \max_U \rho(U)$, where $\rho(U) = \left(2|\{(u, v) \in E_H : u, v \in U\}|\right)/|U|^2$. Let $U^* \subseteq V_H$ be such that $\rho(U^*) = \rho_H$. The randomized algorithm that maps each node of G to a node of U^* chosen uniformly at random, returns a solution of expected value $\rho(U^*)|E_G|$ and is thus a ρ_H-approximation algorithm. This immediately implies an approximation ratio of at least $2/3$ if H contains a triangle.

References

[1] G. Aggarwal, T. Feder, R. Motwani, and A. Zhu. Channel assignment in wireless networks and classification of minimum graph homomorphism. In *ECCC: TR06-040*, 2006.

[2] M. Alekhnovich. More on average case vs approximation complexity. In *Proceedings, 44th FOCS*, pages 298–307, 2003.

[3] S. Arora, D. Karger, and M. Karpinski. Polynomial time approximation schemes for dense instances of NP-hard problems. *J. Comput. Syst. Sci.*, 58:193–210, 1999.

[4] L. Babai, P. Erdös, and S. Selkow. Random graph isomorphism. *SICOMP*, 9:628–635, 1980.

[5] G. Calinescu, H. Karloff, and Y. Rabani. An improved approximation algorithm for multiway cut. *Journal of Computer and System Sciences*, 60:564–574, 2000.

[6] M. Charikar and A. Wirth. Maximizing quadratic programs: Extending Grothendieck's inequality. In *Proceedings, 45th FOCS*, pages 54–60, 2004.

[7] D. Cohen, M. Cooper, P. Jeavons, and A. Krokhin. A maximal tractable class of soft constraints. *Journal of Artificial Intelligence Research*, 22:1–22, 2004.

[8] C. Cooper, M. Dyer, and A. Frieze. On Markov chains for randomly H-coloring a graph. *Journal of Algorithms*, 39:117–134, 2001.

[9] E. Dahlhaus, D. Johnson, C. Papadimitriou, P. Seymour, and M. Yannakakis. The complexity of multiterminal cuts. *SICOMP*, 23:864–894, 1994.

[10] E. D. Demaine, U. Feige, M. T. Hajiaghayi, and M. Salavatipour. Combination can be hard: Approximability of the unique coverage problem. In *Proceedings, 17th SODA*, pages 162–171, 2006.

[11] J. Díaz, M. J. Serna, and D. M. Thilikos. The complexity of restrictive H-coloring. In *Proceedings, 28th International Workshop (WG 2002)*, pages 126–137, 2002.

[12] M. E. Dyer and C. S. Greenhill. The complexity of counting graph homomorphisms. *Random Structures and. Algorithms*, 25:346–352, 2004.

[13] T. Feder and P. Hell. List homomorphisms to reflexive graphs. *Journal of Combinatorial Theory, Series B*, 72:236–250, 1998.

[14] U. Feige. Relations between average case complexity and approximation complexity. In *Proceedings, 34th STOC*, pages 534–543, 2002.

[15] U. Feige and M. Langberg. The RPR^2 rounding technique for semidefinite programs. In *Proceedings, 28th ICALP*, pages 213–224, 2001.

[16] A. Freund and H. Karloff. A lower bound of $8/\left(7 + \frac{1}{k-1}\right)$ on the integrality ratio of the Calinescu-Karloff-Rabani relaxation for multiway cut. *Information Processing Letters*, 75:43–50, 2000.

[17] A. Frieze and Colin McDiarmid. Algorithmic theory of random graphs. *Random Structures and Algorithms*, 10:5–42, 1997.

[18] M.X. Goemans and D.P. Williamson. Improved approximation algorithms for maximum cut and satisfiability problems using semidefinite programming. *Journal of the ACM*, 42:1115–1145, 1995.

[19] G. Gutin, A. Rafiey, A. Yeo, and M. Tso. Level of repair analysis and minimum cost homomorphisms of graphs. *Discrete Applied Mathematics*, 154:881–889, 2006.

[20] P. Hell and J. Nešetřil. On the complexity of H-coloring. *Journal of Combinatorial Theory, Series B*, 48:92 – 110, 1990.

[21] P. Hell and J. Nešetřil. *Graphs and Homomorphisms*. Oxford Univ. Press, 2004.

[22] V. Kann. On the approximability of the maximum common subgraph problem. In *Proceedings, 9th STACS*, pages 377–388, 1992.

[23] D. Karger, P. Klein, C. Stein, M. Thorup, and N. Young. Rounding algorithms for a geometric embedding of minimum multiway cut. *M. of OR*, 29:436–461, 2004.

[24] J. Kleinberg and É. Tardos. Approximation algorithms for classification problems with pairwise relationships: metric labeling and Markov random fields. *Journal of the ACM*, 49:616–639, 2002.

[25] P. Turán. On an extremal problem in graph theory. *Mat. Fiz. Lapok*, 48:436–452, 1941.

Improved Approximation Algorithm for the One-Warehouse Multi-Retailer Problem

Retsef Levi and Maxim Sviridenko

IBM T. J. Watson Research Center, P.O. Box 218, Yorktown Heights, NY 10598

1 Introduction

In this paper, we will consider a well-studied inventory model, called the *one-warehouse multi-retailer problem* (OWMR) and its special case the *joint replenishment problem* (JRP). As the name suggests, in this model there is one warehouse that orders a particular commodity from a supplier, in order to serve demand at N distinct retailers. We consider a discrete finite planning horizon of T periods, and are given the demand d_{it} required for each retailer $i = 1, \ldots, N$ in each time period $t = 1, \ldots, T$. There are two types of costs incurred: *ordering costs* (to model that there are fixed costs incurred each time the warehouse replenishes its supply on hand from the supplier, as well as the analogous cost for each retailer to be stocked from the warehouse) and *holding costs* (to model the fact that maintaining inventory, at both the warehouse and the retail store, incurs a cost). The aim of the model is to provide an optimization framework to balance the fact that ordering too frequently is inefficient for ordering costs, whereas ordering too rarely incurs excessive holding costs.

The details of this model are as follows. At the beginning of each period s, each retailer i can place an order for any number of units from the warehouse, to replenish its on-hand inventory. The order is assumed to arrive instantaneously (this is without loss of generality), and can be used to satisfy demand in period s, or in subsequent periods. Any such order placed by retailer i incurs a *fixed* ordering cost K^i, that is independent of the size of the order and of the time period in which the order is placed. However, all orders placed by the different retailers in each period s must be satisfied only from the on-hand inventory at the warehouse in that period. So in turn, at the beginning of each period r the warehouse can place an order for any number of units from a supplier. This order is again assumed to arrive instantaneously, and can be used to satisfy retailers orders in period r, or in subsequent periods. Any such order of the warehouse in period r incurs a *fixed* ordering cost K^0_r, which also is independent of the size of the order. All demands must be satisfied on time, i.e., any unit that is used by retailer i to satisfy its demand in period t, d_{it}, must be ordered by the warehouse from the supplier in some period r, and then by retailer i from the warehouse in some period s, where $r \leq s \leq t$. The goal is to find an ordering policy that satisfies all demands on time with minimum total ordering and holding costs. Throughout the paper, we will use $\lceil r, s \rfloor$ $(r \leq s)$ to denote a pair of warehouse and retailer orders in periods r and s, respectively. We note that while the

J. Diaz et al. (Eds.): APPROX and RANDOM 2006, LNCS 4110, pp. 188–199, 2006.

warehouse ordering cost K_r^0 is time-dependent, the retailer ordering cost K^i is stationary over time.

The standard models for holding cost make two natural linearity assumptions: (1) that the cost is proportional to the number of units of the commodity held, and (2) that there is a per-unit cost h_t^i associated with holding a unit of item i from period t to $t+1$. We use the more general holding cost structure that has been introduced by Levi, Roundy and Shmoys for the JRP problem and has been extended later to the OWMR problem [4, 5, 6]. While still maintaining (1), relaxes (2) in a way that preserves the most useful properties of an optimal solution (as well as of an optimal solution to the natural LP relaxation), but captures much more general phenomena, such as the notion of perishable goods (where the holding cost becomes infinite, when the good is held too long). For each demand point (i, t), we introduce a holding cost parameter h_{rs}^{it} associated with ordering one unit of the demand at retailer i for period t according to the pair $\lceil r, s \rfloor$. The parameters h_{rs}^{it} are assumed to satisfy certain natural monotonicity properties:

- **Property 1:** *Non-negativity* ($h_{rs}^{it} \geq 0$).
- **Property 2:** *Monotonicity with respect to* r. For each demand point (i, t) and fixed retailer order in period s ($s \leq t$), we assume that h_{rs}^{it} is non-increasing in $r \in [1, s]$.
- **Property 3:** *Monotonicity with respect to* s. Here we assume that each of the retailers has exactly one of the following properties. For each fixed demand point (i, t) and warehouse order in period r ($r \leq t$), h_{rs}^{it} is *either* non-increasing in $s \in [r, t]$ (for *all* demand points (i, t), *or* it is non-decreasing in $s \in [r, t]$ (for *all* demand points (i, t)). It is straightforward to see that in an optimal policy, the warehouse does not hold inventory of J-retailers. (The joint replenishment problem is the special case with only J-retailers.)
- **Property 4:** *Dominance of r-Monotonicity*. We also assume that if $r \leq r'$ and $s \leq s'$, then $h_{rs}^{it} \geq h_{r's'}^{it}$, for each demand point (i, t), and regardless of whether i is a J-retailer of a W-retailer.
- **Property 5:** *Monge Property*. For each demand point (i, t) with $i \in I_W$ and any four periods $r_2 < r_1 \leq s_2 < s_1 \leq t$, the inequality, $h_{r_2, s_1}^{it} + h_{r_1, s_2}^{it} \geq h_{r_2, s_2}^{it} + h_{r_1, s_1}^{it}$ is satisfied.

Previous work. Arkin, Joneja and Roundy [1] have shown that OWMR is NP-hard even for the special case of the JRP, where the warehouse serves only as a cross-docking point (i.e., no inventory is ever held at the warehouse). Federgrun and Tzur [3] have proposed an interesting heuristic based on dynamic programming. However, for the theoretical analysis of the worst-case performance of their algorithm, they have assumed that the cost parameters and the demands are bounded by uniform constants. Chan, Muriel, Shen, Shimchi-Levi and Teo [2] have considered a variant of OWMR, in which the ordering costs are piecewise-linear functions, and the holding cost is linear and additive. They considered the class of *zero-inventory ordering* (ZIO) policies, in which the warehouse and retailers order if and only if their current on hand inventory is 0. They established the effectiveness of these policies, showing that the cost of the optimal ZIO policy is at most $\frac{4}{3}$ times the cost of the optimal policy. In [2] and in a subsequent paper by Shen, Simchi-Levi and Teo [7], they have proposed an LP-based algorithm for approximating the best ZIO policy. However, the performance guarantee of their algorithm is $O(\log(N + T))$. For the problem we consider in this paper, it is well known that ZIO policies are optimal.

In several recent results, Levi, Roundy and Shmoys have provided constant approximation algorithms for a broad class of deterministic inventory models, including the one-warehouse and multi-retailer problem and the joint replenishment problem. They have provided a general primal-dual algorithmic framework that solves the single-item lot-sizing problem, and provides a 2-approximation for the JRP and assembly problem (which is yet another basic inventory model) [4, 5]. However, the more complex cost structure of holding inventory in two different "levels" appears so far to be an impediment in extending the primal-dual approach to the OWMR problem. They have also provided an LP-based rounding 2.39-approximation algorithm for the OWMR problem [6].

Our techniques and results. We use the same LP relaxation as Levi, Roundy and Shmoys [6]. Like their algorithms, we round the fractional solution of the LP in two phases. In the first phase we determine the warehouse orders; based on that, we determine the retailer orders in the second phase, and this is done separately for each retailer. Our algorithms are based on new dependent randomized rounding techniques that better exploit the special "line" structure of the inventory model that is induced by the notion of time. This enables us to bound the *average* holding costs incurred by the algorithm. This is in contrast to Levi, Roundy and Shmoys [6] who only bound the *worst* (most expensive) holding costs incurred, using the dual of the corresponding LP relaxation. Our techniques lead to conceptually simpler rounding algorithms with elegant worst-case analysis that is entirely based on the primal solution.

We show that the solution produced by the randomized algorithms has expected cost, that is guaranteed to be at most 1.8 times the cost of an optimal solution to the OWMR problem. We then show how to derandomize these algorithms and this yields a deterministic 1.8-approximation algorithm for the OWMR problem. When specialized to the JRP problem our LP is identical to the one used in [4, 5]. Thus, our approach can be applied to the JRP problem and improve on the primal-dual 2-approximation of Levi, Roundy and Shmoys [4, 5].

2   A Linear Program

The integer programming formulation and the corresponding LP relaxation are based on the well- known fact that there exists an optimal solution to the OWMR problem in which each demand d_{it} is satisfied from a unique pair of orders $\lceil r, s \rfloor$, where again $r \leq s \leq t$ (see [10] for details). That is, the warehouse orders the entire demand d_{it} in some period $r \leq t$, and keeps it in inventory over the time interval $[r, s)$ $(r \leq s \leq t)$. Then in period s, the entire demand d_{it} is ordered from the warehouse by retailer i and is kept in inventory (at the retailer's premises) until time t. We define $H_{rs}^{it} := h_{rs}^{it} d_{it}$ to be the total cost of providing the demand d_{it} from the pair of orders $\lceil r, s \rfloor$. This gives rise to the following LP formulation:

$$\min \quad \sum_{r=1}^{T} y_r^0 K_r^0 + \sum_{i=1}^{N} \sum_{s=1}^{T} y_s^i K^i + \sum_{i=1}^{N} \sum_{t=1}^{T} \sum_{r,s:r \leq s \leq t} x_{rs}^{it} H_{rs}^{it} \tag{1}$$

$$\sum_{r,s:r \leq s \leq t} x_{rs}^{it} = 1, \quad \forall i, t, \tag{2}$$

$$\sum_{r:r \leq s} x_{rs}^{it} \leq y_s^i, \quad \forall i, t, s \leq t, \tag{3}$$

$$\sum_{s:r \leq s \leq t} x_{rs}^{it} \leq y_r^0, \quad \forall i, t, r \leq t, \tag{4}$$

$$x_{rs}^{it}, \ y_r^i \geq 0, \quad \forall i, r, s, t, r \leq s \leq t. \tag{5}$$

The variable x_{rs}^{it} (for $r \leq s \leq t$) indicates whether demand point (i, t) (i.e., demand d_{it}) was provided from the pair of orders in periods r (warehouse order) and s (retailer i order). The variable y_s^i (for each $i = 1, .., N$) indicates whether retailer i placed an order in period s. Finally, the variable y_r^0 indicates whether the warehouse placed an order in period r. Constraint (2) ensures that each positive demand point (i, t) is satisfied from some pair of warehouse-retailer orders in periods $\lceil r, s \rfloor$, no later than period t. Constraint (3) ensures that no demand d_{it} can be satisfied by a retailer order in period $s \leq t$ (and some warehouse order in period $r \leq s$), unless retailer i indeed has placed an order in period s. Lastly, constraint (4) ensures that no demand point d_{it} can be satisfied by a warehouse order in period r (and some retailer order $r \leq s \leq t$), unless the warehouse has placed an order in period r. It is straightforward to see that the corresponding integer program provides a correct formulation to the OWMR problem. Hence, the LP-relaxation provides a lower bound on the cost of any feasible solution to the OWMR problem. For the rest of this paper we let $(\hat{x}, \hat{y})$ and opt_{LP} be the optimal solution and the value of (P), respectively.

We note that for each J-retailer i, it suffices to consider only the variables x_{rs}^{it} with $r = s$ (the warehouse does not hold inventory of J-retailers). Hence, for the all J-retailers, we can adapt accordingly the constraints (2), (3) and (4). In particular, the modified constraints (3) and (4), imply that, in an optimal solution, $y_s^i \leq y_s^0$.

The Monge Property. Recall the Monge property of the holding cost, i.e., property 5 of h. We say that a feasible solution (x, y) to (P) satisfies the Monge property, if $x_{rs}^{it} > 0$ ($r \leq s \leq t$) implies that $x_{\tilde{r}, \tilde{s}}^{it} = 0$ for any $\lceil \tilde{r}, \tilde{s} \rfloor$ such that $\tilde{r} < r$ *and* $\tilde{s} > s$. Without loss of generality, we will assume that $(\hat{x}, \hat{y})$ (the optimal solution of (P)) satisfies the Monge property. We note that because of the Monge property on the holding cost, any feasible solution to (P) can be converted in polynomial time to one that satisfies the Monge property and has no greater cost.

3   The Random Shifts Algorithms

In this section, we will show how to round the optimal solution of (P), denoted again by $(\hat{x}, \hat{y})$, to a feasible solution to the OWMR problem with cost at most 1.8 times the optimal cost. We shall first describe two different randomized rounding procedures that we call *random shift algorithm with retailer two-sided push* and *random shift algorithm with retailer one-sided push*. Our rounding procedures run in two phases. In the first phase, we determine the warehouse orders, using a simple mechanism that we call *random shift*. In the second phase, we use the output of the first rounding phase to determine the orders of each retailer. This phase is done separately for each retailer. We shall show that the expected cost of each one of the algorithms is guaranteed to be at most twice the cost of an optimal policy for the OWMR problem. Moreover, the expected cost of the cheapest among these algorithms is guaranteed to be at most 1.8 times the optimal cost of the OWMR problem. Due to the lack of space we omit all the proofs (for details see the full version of the paper).

Random shifts. The warehouse orders are placed using a simple randomized procedure that is based on the respective values of the y_r^0 variables in the optimal solution of (P), i.e., on the values $\hat{y}_1^0, \ldots, \hat{y}_T^0$. For the description of the random shift procedure, consider the interval $(0, \sum_{r=1}^{T} \hat{y}_r^0]$. Each period $m = 1, \ldots, T$ is then associated with the respective interval $(\sum_{r=1}^{m-1} \hat{y}_r^0, \sum_{r=1}^{m} \hat{y}_r^0]$ that is of length $\hat{y}_m^0$. The input for this procedure is a *step parameter* $c \in (0, 1]$. Given c, choose a *shift parameter* α_0 uniformly at random from $(0, c]$. Let W be the upper ceiling of the total accumulated weight of fractional warehouse orders in the optimal LP solution $(\hat{x}, \hat{y})$ scaled by $\frac{1}{c}$. That is, $W = \lceil \frac{1}{c} \sum_{r=1}^{T} \hat{y}_r^0 \rceil$. It is clear that the interval $(0, \sum_{r=1}^{T} \hat{y}_r^0]$ is contained in the interval $[0, cW]$. Within the interval $[0, cW]$ focus on the sequence of points $0, c, \ldots, c(W - 1)$. The shift parameter α_0 induces a sequence of what we call *warehouse shift points*. Specifically, the set of warehouse shift points is defined as $\{\alpha_0 + cw : w = 0, \ldots, W - 1\}$. This set is constructed through a shift of length α_0 to the right of the points $0, c, \ldots, c(W - 1)$. Thus, there are W shift points that are all located within the interval $[0, cW]$. Observe that the sequence of warehouse shift points is a-priori random and is realized with the shift parameter α_0.

The warehouse shift points determine the periods in which warehouse orders are placed. For each period $m = 1, \ldots, T$, we place a warehouse order in that period if there is at least one shift point within the interval $(\sum_{r=1}^{m-1} \hat{y}_r^0, \sum_{r=1}^{m} \hat{y}_r^0]$ that is associated with it. That is, we place a warehouse order in period m, if for some $0 \leq w \leq W - 1$ there exists a warehouse shift point $\alpha_0 + cw$ that falls within the interval $(\sum_{r=1}^{m-1} \hat{y}_r^0, \sum_{r=1}^{m} \hat{y}_r^0]$.

Lemma 1. *Consider the random shift procedure described above with input length parameter $c \in (0, 1]$. Then, for each period $m = 1, \ldots, T$, the probability to place a warehouse order in period m is at most $\frac{1}{c} \hat{y}_m^0$. Thus, the total expected warehouse ordering cost of the random shift procedure, denoted by K_0 is at most $\frac{1}{c}$ times the total warehouse ordering costs in the optimal LP solution. That is, $K_0 \leq \frac{1}{c} \sum_{r=1}^{T} \hat{y}_r^0 K_r^0$.*

Let $\mathcal{T}_W := \{r_1 < r_2 < ... < r_M\}$ be the set of periods of the warehouse orders as determined in the first phase of the algorithm. Note that once we decide upon the warehouse orders, then the OWMR problem decomposes into N single-location, single-item lot-sizing problems. These problems can be solved optimally using dynamic programming (see [9] for details) to achieve the minimum overall retailer ordering cost and holding cost under the assumption that warehouse orders are placed at $r_1 < r_2 < ... < r_M$. The collection of the solutions to these single-location problems provides a solution to the OWMR problem. However, as part of the worst-case analysis, we next describe the second phase of the randomized rounding procedures, in which we consider each retailer i separately, and determine its orders. More specifically, we shall describe two different randomized algorithms and analyze their worst-case expected performance. The corresponding algorithms might not yield the optimal solution with respect to the warehouse orders placed in phase one. Nevertheless, we shall show that, regardless of the instance of the problem, the cheapest among the two algorithms will produce a solution with expected cost at most 1.8 times the cost of an optimal solution for the OWMR problem.

3.1 Random Shift Algorithm with Retailer Two-Sided Push Algorithm

Throughout the rest of the paper, we shall refer to the random shift algorithm with two-sided retailer push algorithm as *Algorithm 1*. As we have already mentioned, Algorithm 1 has two phases. The first phase is the random shift procedure described above with step parameter $c = 1$. Consider again $\mathcal{T}_W := \{r_1 < r_2 < ... < r_M\}$, the set of warehouse orders placed in the first phase of the algorithm.

Next we consider each retailer i separately $(i = 1, ..., N)$, and determine its orders using what we call *two-sided push* procedure. First, we shall construct a sequence of (random) *retailer-i shift points* in a way similar to how warehouse shift points are constructed. Let W_i be the upper ceiling of the accumulated fractional retailer orders in the LP solution. That is $W_i = \lceil \sum_{s=1}^{T} \hat{y}_s^i \rceil$. Similar to the random warehouse shift procedure above, choose a retailer shift parameter α_i uniformly at random from $(0, 1]$ and construct a sequence of W_i retailer-i shift points $\{\alpha_i + cw : w = 0, ..., W_i - 1\}$. In contrast to the warehouse shift points, the retailer-i shift points are used to determine only *tentative retailer-i orders*. The reason is that placing retailer orders depends also on the output of the first phase, in which warehouse orders are determined. Thus, the way tentative retailer-i are determined is similar to how warehouse orders are determined. For each period $m = 1, ..., T$, we say that there is a tentative retailer order placed in period m, if there is a retailer-i shift point within the interval $(\sum_{s=1}^{m-1} \hat{y}_s^i, \sum_{s=1}^{m} \hat{y}_s^i]$. The tentative orders are used to determine the *permanent retailer orders*. The way this is done depends on whether retailer i is a J-retailer or a W-retailer. Suppose that there is a tentative retailer i order placed in some period m, then one of the following two cases applies:

Case I: Retailer i is a J-retailer. Recall that, without loss of generality, for J-retailers we restrict attention only to policies in which warehouse and retailer orders are placed in the same periods. That is, permanent retailer orders in the second phase must be placed in periods $s \in \mathcal{T}_W$, where again $\mathcal{T}_W$ is the set of periods of all warehouse orders placed in the first phase of the algorithm. Since we place retailer orders only in periods $s \in \mathcal{T}_W$, if $m \notin \mathcal{T}_W$ we wish to *push* this tentative retailer order. In particular, for each tentative retailer order, we place up to two permanent retailer orders: one order is placed in the latest period with warehouse order in $\mathcal{T}_W$ prior to period m, if such order exists (i.e., 'pushed' earlier in time); a second order is placed in the earliest period with warehouse order in $\mathcal{T}_W$ after period m (i.e., 'pushed' later in time), if such order exists. In other words, we place permanent retailer-i orders in $\max\{r \in \mathcal{T}_W : r \leq m\}$ and $\min\{r \in \mathcal{T}_W : r \geq m\}$.

Case II: Retailer i is a W-retailer. In this case we can place a permanent retailer order in each period m for which there is a tentative order. However, we also place a second permanent retailer order in the earliest period with warehouse in $\mathcal{T}_W$ (strictly) after m, if such warehouse order exists. That is, we place one permanent order at m and possibly a second permanent order in $\min\{r \in \mathcal{T}_W : r > m\}$.

The reason that we push tentative retailer orders both earlier and later in time will be made clear in the following discussion. Intuitively, we place additional retailer orders to guarantee that the holding costs incurred are not too high (see also Lemmas 3 and 4 below).

Let $\mathcal{T}_i$ be the set of permanent retailer orders placed by Algorithm 1. We claim that the sets $\mathcal{T}_W$ and $\mathcal{T}_i$ (for $i = 1, \ldots, T$) induce a feasible solution, in which, each demand point (i, t), is satisfied by the cheapest pair of orders $\lceil r, s \rfloor$, such that $r \in \mathcal{T}_W$ and $s \in \mathcal{T}_i$. This is established is Lemma 3 below.

From Lemma 1 above it follows that the total expected warehouse ordering cost of Algorithm 1 is bounded by $\sum_{r=1}^{T} \hat{y}_r^0 K_r^0$. Next we bound the total expected retailer ordering costs, which is denoted by $\mathcal{K}_I$.

Lemma 2. *The total expected retailer ordering costs of Algorithm 1 is at most twice the total retailer ordering costs in $(\hat{x}, \hat{y})$, the optimal solution of the LP. That is, $\mathcal{K}_I \leq 2 \sum_{i=1}^{N} \sum_{s=1}^{T} \hat{y}_s^i K^i$.*

Finally, we wish to bound the total expected holding costs incurred by Algorithm 1, that is denoted by $\mathcal{H}$. Each demand point (i, t) is considered separately (for $i = 1, \ldots, N$ and $t = 1, \ldots, T$), and its expected holding costs is bounded by the holding cost that this demand point incurs in the optimal LP solution $(\hat{x}, \hat{y})$. In particular, focus on some demand point (i, t), and let $\hat{H}^{it} = \hat{H}$ be the random holding costs that Algorithm 1 incurs in satisfying this demand point. (Since the following discussion is focused on a fixed demand point, we simplify notation and omit the subscript of it whenever possible.) We wish to bound $E[\hat{H}]$, the expectation of $\hat{H}$.

Service points. Consider demand point (i, t), and let $\mathcal{S}^{it} = \mathcal{S}$ be set of all pairs of warehouse and retailer orders, which fractionally serve (i, t) in the optimal LP solution $(\hat{x}, \hat{y})$. Specifically, let $\mathcal{S} = \{\lceil r_m, s_m \rfloor : \hat{x}_{r_m, s_m}^{it} > 0\}$. Without loss of generality assume that $\mathcal{S} = \{\lceil r_m, s_m \rfloor : m = 1, \ldots, L\}$, where $H^{it}_{r_1,s_1} \leq H^{it}_{r_2,s_2} \leq \ldots, \leq H^{it}_{r_L,s_L}$. That is, the order pairs $\lceil r_1, s_1 \rfloor, \ldots, \lceil r_L, s_L \rfloor$ are sorted in an increasing order according to the holding costs they incur. However, since the solution $(\hat{x}, \hat{y})$ is assumed to have the Monge Property, we conclude that $\lceil r_m, s_m \rfloor \geq \lceil r_{m'}, s_{m'} \rfloor$, i.e., $r_m \geq r_{m'}$ and $s_m \geq s_{m'}$, for each $1 \leq m < m' \leq L$. Moreover, if i is a J-retailer, we have $s_m = r_m$, for each $m = 1, \ldots, L$. To simplify notation, we use H_m to denote $H^{it}_{r_m,s_m}$, for each $m = 1, \ldots, L$, assuming $H_1 \leq H_2 \leq \ldots \leq H_L$. Thus, the holding cost incurred by (i,t) in the optimal LP solution $(\hat{x}, \hat{y})$ can be expressed as $\sum_{m=1}^{L} \hat{x}^{it}_{r_m,s_m} H_m$.

Next we show that the holding costs incurred by demand point (i,t) under Algorithm 1 is, with probability 1, at most H_L, that is, $\hat{H} \leq H_L$.

Lemma 3. *For each demand point (i,t), the holding cost it incurs under Algorithm 1 is guaranteed to be at most H_L. That is, $\hat{H} \leq H_L$ with probability 1.*

Lemma 3 above implies that under Algorithm 1, demand point (i,t) is served by a pair of orders $\lceil r', s' \rfloor$, such that $r_L \leq r'$ and $s_L \leq s'$. Thus, we can express $E[\hat{H}]$ as

$$\sum_{\lceil r,s \rfloor: \, r_L \leq r, \, s_L \leq s} H^{it}_{rs} Pr(\hat{H} = H^{it}_{rs}), \tag{6}$$

where $Pr(\hat{H} = H^{it}_{rs})$ denotes the corresponding probability that under Algorithm 1 demand point (i,t) is served by pair of orders $\lceil r, s \rfloor$.

Given (6) above, it is straightforward to derive an upper bound on the expected holding costs incurred by demand point (i,t) under Algorithm 1. Let $H_0 = 0$ and observe that

$$E[\hat{H}] = \sum_{\lceil r,s \rfloor: \, r_L \leq r, \, s_L \leq s} H^{it}_{rs} Pr(\hat{H} = H^{it}_{rs}) \tag{7}$$

$$\leq H_1 Pr(H_0 \leq \hat{H} \leq H_1) + \sum_{m=2}^{L} H_m Pr(H_{m-1} < \hat{H} \leq H_m)$$

$$= H_1 Pr(\hat{H} \leq H_1) + \sum_{m=2}^{L} H_m [Pr(\hat{H} \leq H_m) - Pr(\hat{H} \leq H_{m-1})]$$

$$= H_L + \sum_{m=1}^{L-1} Pr(\hat{H} \leq H_m)[H_m - H_{m+1}].$$

Recall that $Pr(\cdot)$ refers to the corresponding probability induced by Algorithm 1. The inequality in (7) follows from the fact that, for each $m = 1, \ldots, L$, we weight the probability $Pr(H_{m-1} < \hat{H} \leq H_m)$ by H_m, which is the highest holding costs within this range. The first equality follows from the fact that $Pr(\hat{H} < 0) = 0$ and the identity $Pr(H_{m-1} < \hat{H} \leq H_m) = Pr(\hat{H} \leq H_m) - Pr(\hat{H} \leq H_{m-1})$. The last equality follows from Lemma 3 in which we show that $Pr(\hat{H} \leq H_L) = 1$. Moreover, observe that the term $\sum_{m=1}^{L-1} Pr(\hat{H} \leq H_m)[H_m - H_{m+1}]$ above is non-positive, since $H_m - H_{m+1} \leq 0$. This implies that if we consider (7) above, but, for each $m = 1, \ldots, L - 1$, replace $Pr(\hat{H} \leq H_m)$ with a (non-negative) lower bound on that probability, then the upper bound developed in (7) is still maintained.

Next we shall establish lower bounds on the corresponding probabilities $Pr(\hat{H} \leq H_m)$.

Lemma 4. *For each $m = 1, \ldots, L - 1$, the probability that the holding costs incurred by (i, t) under Algorithm 1 are lower than H_m is at least $(\sum_{u=1}^{m} \hat{x}_{r_u,s_u}^{it})^2$. That is, $Pr(\hat{H} \leq H_m) \geq (\sum_{u=1}^{m} \hat{x}_{r_u,s_u}^{it})^2$.*

Lemma 4 and (7) above imply that

$$
E[\hat{H}] \leq H_1(\hat{x}_{r_1,s_1}^{it})^2 + \sum_{m=2}^{L} H_m \left[(\sum_{u=1}^{m} \hat{x}_{r_u,s_u}^{it})^2 - (\sum_{u=1}^{m-1} \hat{x}_{r_u,s_u}^{it})^2 \right] \tag{8}
$$

$$
= \sum_{m=1}^{L} H_m \left[(\sum_{u=1}^{m} \hat{x}_{r_u,s_u}^{it})^2 - (\sum_{u=1}^{m-1} \hat{x}_{r_u,s_u}^{it})^2 \right].
$$

The inequality follows because, for each $m = 1, \ldots, L - 1$, we replace $Pr(\hat{H} \leq H_m)$ by the lower bound established in Lemma 4 above. Moreover, in Lemma 3 we have already observed that $Pr(\hat{H} \leq H_L) = 1$ and Constraint (2) implies that $(\sum_{u=1}^{L} \hat{x}_{r_u,s_u}^{it})^2 = 1$.

To conclude the analysis, we next introduce the *density holding cost function* $\bar{H}^{it}(\alpha) = \bar{H}(\alpha)$. This function is defined for each demand point and based on the optimal LP solution $(\hat{x}, \hat{y})$. For a given value of $\alpha \in (0, 1]$, let $m(\alpha)$ be the index such that $\alpha \in (\sum_{u=1}^{m(\alpha)-1} \hat{x}_{r_u,s_u}^{it}, \sum_{u=1}^{m(\alpha)} \hat{x}_{r_u,s_u}^{it}]$. Then we define $\bar{H}(\alpha) = H_{m(\alpha)}$. The function $\bar{H}(\alpha)$ is a step function with steps at the points $0, \hat{x}_{r_1,s_1}^{it}, \ldots, \sum_{u=1}^{L-1} \hat{x}_{r_u,s_u}^{it}$ and step heights $H_1, \ldots, H_L$, respectively. Moreover, the integral of $\bar{H}(\alpha)$ over $(0, 1]$ is equal to the holding costs incurred by (i, t) in the LP optimal solution $(\hat{x}, \hat{y})$. That is, $\int_0^1 \bar{H}(\alpha)d\alpha = \sum_{u=1}^{L} \hat{x}_{r_u,s_u}^{it} H_u$. We note that Shmoys, Tardos and Aardal [8] have used a similar function to $\bar{H}(\alpha)$ in their seminal paper that provides the first constant approximation algorithm for the classical metric facility location problem. Next we shall describe another application of this function.

Specifically, Inequality (8) and the properties of the function $\bar{H}(\alpha)$ imply

$$
E[\hat{H}] \leq \sum_{m=1}^{L} H_m \left[(\sum_{u=1}^{m} \hat{x}_{r_u,s_u}^{it})^2 - (\sum_{u=1}^{m-1} \hat{x}_{r_u,s_u}^{it})^2 \right] \tag{9}
$$

$$
= 2 \int_0^1 \alpha \bar{H}(\alpha)d\alpha \leq 2 \int_0^1 \bar{H}(\alpha)d\alpha \leq 2 \sum_{u=1}^{L} \hat{x}_{r_u,s_u}^{it} H_u.
$$

The second equality follows from the properties of $\bar{H}$, being a step function. The second inequality follows from the fact that we integrate over $[0, 1]$. This implies the following lemma.

Lemma 5. *Let $\mathcal{H}$ denotes the total expected holding costs incurred by Algorithm 1. Then these costs are at most twice the total holding costs incurred in the optimal LP solution $(\hat{x}, \hat{y})$. That is, $\mathcal{H} \leq 2 \sum_{t=1}^{T} \sum_{r,s:r \leq s \leq t} \hat{x}_{rs}^{it} H_{rs}^{it}$.*

3.2 Random Shift Algorithm with One-Sided Shift

Next we describe the second algorithm that we refer to as *Algorithm 2*. In this algorithm, we shall place the warehouse orders more frequently, incurring more warehouse ordering costs, in order to save some of the retailer ordering costs and holding costs incurred. In the first phase of Algorithm 2, we determine the warehouse orders by applying the random shift procedure described above with some step parameter $c \in (0, 0.5]$. Let $\mathcal{T}_W$ be again the set of periods of the warehouse orders placed by the algorithm. From Lemma 1 above, we conclude that the total expected warehouse ordering costs of Algorithm 2 is at most $\frac{1}{c}$ times the warehouse ordering costs in the LP solution. That is, $\mathcal{K}_0 \leq \frac{1}{c} \sum_{r=1}^{T} \hat{y}_r^0 K_r^0$.

In the second phase of the algorithm we determine the retailer orders, and this is again done separately for each retailer. First we generate retailer-i shift points in a way similar to what described above for Algorithm 1 but with a different length parameter. Let W_i be the upper ceiling of the accumulated fractional retailer orders in the LP solution scaled by $\frac{1}{1-c}$. That is $W_i = \lceil \frac{1}{1-c} \sum_{s=1}^{T} \hat{y}_s^i \rceil$. Similar to Algorithm 1 above, choose a retailer shift parameter α_i uniformly at random from $(0, 1 - c]$ and construct the sequence of retailer-i shift points, $\{\alpha_i + (1 - c)w : w = 0, \ldots, W_i - 1\}$. We again use the retailer shift points to determine tentative retailer orders. For each period $m = 1, \ldots, T$, we say that there is a tentative retailer order placed in period m, if there is a retailer shift point that falls within the interval $(\sum_{s=1}^{m-1} \hat{y}_s^i, \sum_{s=1}^{m} \hat{y}_s^i]$. The permanent retailer orders are again placed according to whether retailer i is a J-retailer or a W-retailer. Suppose that there is a tentative retailer i order placed in some period m, then one of the following two cases applies:

1. Retailer i is a J-retailer. In this case we simply push the tentative order to the earliest period in $\mathcal{T}_W$ later than in time, if such order exists. That is, we place the permanent order in $\min\{r \in \mathcal{T}_W : r \geq m\}$.
2. Retailer i is a W-retailer. In this case we can simply place a permanent retailer order in period m.

The next Lemma bounds the total expected retailer ordering costs incurred by Algorithm 2. The proof is similar to that of Lemma 2.

Lemma 6. *Let $\mathcal{K}_I$ be the overall expected retailer ordering costs incurred by Algorithm 2. Then these costs are at most $\frac{1}{1-c}$ times the total retailer ordering costs incurred in the LP optimal solution $(\hat{x}, \hat{y})$. That is, $\mathcal{K}_I \leq \frac{1}{1-c} \sum_{i=1}^{N} \sum_{s=1}^{T} \hat{y}_s^i K^i$*

For each $i = 1, \ldots, N$, let $\mathcal{T}_i$ be again the set of periods of permanent retailer-i orders placed by the algorithm. It is readily verified that together with $\mathcal{T}_W$ this induces a feasible solution to the OWMR, in which each demand point is served from the cheapest possible pair of orders (see Lemma 7 below).

Finally, we again bound the total expected holding costs incurred by Algorithm 2. Similar to the analysis of Algorithm 1, we shall consider each demand point separately, and bound the expected holding costs it incurs under Algorithm 2.

Let $\hat{H}$ be again the holding cost demand point (i, t) incurs in the solution obtained by Algorithm 2. First we show that the holding cost incurred by demand point (i, t) under Algorithm 2 is, with probability 1, at most H_L, that is, $\hat{H} \leq H_L$.

Lemma 7. *For each demand point (i, t), the holding cost it incurs under Algorithm 2 is at most H_L. That is, $\hat{H} \leq H_L$ with probability 1.*

Lemma 7 above implies that Inequality (7) is valid (but with the respective probabilities defined with respect to Algorithm 2). Similar to the analysis of algorithm 1, the next step will be to develop lower bounds on $Pr(\hat{H} \leq H_m)$, which is now defined with respect to Algorithm 2.

Lemma 8. *Let $Pr(\hat{H} \leq H_m)$ denote the probability that demand point (i, t) is satisfied from a pair of orders $\lceil r, s \rfloor$ with cost at most H_m. Then*

$$Pr(\hat{H} \leq H_m) \geq \max\{0, \frac{\sum_{u=1}^{m} \hat{x}_{r_u, s_u}^{it} - c}{1 - c}\}.$$

Lemma 8 and Inequality (7) imply that

$$E[\hat{H}] \leq \frac{1}{1-c} \sum_{u=m(c)}^{L} H_u \hat{x}_{r_u, s_u}^{it} \leq \frac{1}{1-c} \sum_{u=1}^{L} \hat{x}_{r_u, s_u}^{it} H_u.$$

We have established the following lemma.

Lemma 9. *Let $\mathcal{H}$ denote the overall expected holding costs incurred by Algorithm 2 with a step parameter $c \in (0, 0.5]$. Then $\mathcal{H}$ is at most $\frac{1}{1-c}$ times the holding costs incurred by the optimal LP solution $(\hat{x}, \hat{y})$. That is,*
$\mathcal{H} \leq \frac{1}{1-c} \sum_{i=1}^{N} \sum_{t=1}^{T} \sum_{r,s:r \leq s \leq t} \hat{x}_{rs}^{it} H_{rs}^{it}.$

Lemmas 6 and 9 imply the following theorem.

Theorem 1. *The overall expected costs incurred by Algorithm 2 with a step parameter $c \in (0, 0.5]$ is at most*

$$\frac{1}{c} \sum_{r=1}^{T} \hat{y}_r^0 K_r^0 + \frac{1}{1-c} \sum_{i=1}^{N} \sum_{s=1}^{T} \hat{y}_s^i K^i + \frac{1}{1-c} \sum_{i=1}^{N} \sum_{t=1}^{T} \sum_{r,s:r \leq s \leq t} \hat{x}_{rs}^{it} H_{rs}^{it}.$$

It is readily verified that for $c = 0.5$ Algorithm 2 is a randomized 2-approximation for the OWMR problem.

An Improved Approximation Algorithm. Next we use both Algorithm 1 and Algorithm 2 together and show that taking the algorithm with the minimum expected cost among them yields an improved worst-case guarantee of 1.8. First, choose the step parameter of Algorithm 2 to be $c = 1/3$. Using the fact that $\min\{a, b\} \le \lambda a + (1 - \lambda)b$, for each $0 \le \lambda \le 1$, we apply Lemmas 1, 2, 5 and Theorem 1 (with $c = 1/3$) and take $\lambda = 3/5$ to conclude that the best out of two solutions has expected value at most 1.8 the optimal expected cost.

Theorem 2. *There exists a randomized 1.8-approximation algorithm for the OWMR problem and its special case the JRP problem.*

We note that the algorithm can be derandomized (see the full version for details).

References

1. E. Arkin, D. Joneja, and R. Roundy. Computational complexity of uncapacitated multi-echelon production planning problems. *Operations Research Letters*, 8:61–66, 1989.
2. A. Chan, A. Muriel, Z.-J. Shen, D. Simchi-Levi, and C.-P. Teo. Effectiveness of zero inventory ordering policies for an one-warehouse multi-retailer problem with piecewise linear cost structures. *Management Science*, 48:1446–1460, 2000.
3. A. Federgrun and M. Tzur. Time-partitioning heuristics: Application to one warehouse, multi-item, multi-retailer lot-sizing problems. *Naval Research Logistics*, 46:463–486, 1999.
4. R. Levi, R. O. Roundy, and D. B. Shmoys. Primal-dual algorithms for deterministic inventory problems. Technical Report TR1042, ORIE Department, Cornell University, 2004. Submitted.
5. R. Levi, R. O. Roundy, and D. B. Shmoys. Primal-dual algorithms for deterministic inventory problems. In *Proceedings of the 36th Annual ACM Symposium on Theory of Computing*, pages 353–362, 2004.
6. R. Levi, D. B. Shmoys, and R. O. Roundy. A constant approximation algorithm for the one-warehouse multi-retailer problem. In *Proceedings of the 15th Annual SIAM-ACM Symposium on Discrete Algorithms*, pages 365–374, 2005.
7. Z. J. Shen, D. Simchi-Levi, and C. P. Teo. Approximation algorithms for the single-warehouse multi-retailer problem with piecewise linear cost structures. url: citeseer.nj.nec.com/439759.html.
8. D. B. Shmoys, E. Tardos, , and K. I. Aardal. Approximation algorithms for facility location problems. In *Proceedings of the 29th Annual ACM Symposium on Theory of Computing*, pages 265–274, 1997.
9. H. M. Wagner and T. M. Whitin. Dynamic version of the economic lot sizing model. *Management Science*, 5:89–96, 1958.
10. P. H. Zipkin. *Foundations of inventory management.* The McGraw-Hill Companies, Inc, 2000.

Hardness of Preemptive Finite Capacity Dial-a-Ride

Inge Li Gørtz*

Technical University of Denmark
ilg@imm.dtu.dk

Abstract. In the *Finite Capacity Dial-a-Ride problem* the input is a metric space, a set of objects $\{d_i\}$, each specifying a source s_i and a destination t_i, and an integer k—the capacity of the vehicle used for making the deliveries. The goal is to compute a shortest tour for the vehicle in which all objects can be delivered from their sources to their destinations while ensuring that the vehicle carries at most k objects at any point in time. In the *preemptive* version an object may be dropped at intermediate locations and picked up later and delivered. Let N be the number of nodes in the input graph. Charikar and Raghavachari [FOCS '98] gave a $\min\{O(\log N), O(k)\}$-approximation algorithm for the preemptive version of the problem. In this paper we show that the preemptive Finite Capacity Dial-a-Ride problem has no $\min\{O(\log^{1/4-\varepsilon} N), k^{1-\varepsilon}\}$-approximation algorithm for any $\varepsilon > 0$ unless all problems in NP can be solved by randomized algorithms with expected running time $O(n^{\text{polylog} n})$.

1 Introduction

Vehicle routing and delivery problems have been widely studied in Computer Science and Operations Research. These problems occur in many practical settings such as transportation of goods or passengers and robotics (see Christofedes [5] and Golden and Assad [10]). Many of these problems are NP-hard and there has been a great deal of research in finding and analyzing heuristics to solve these problems. One such problem is the *Finite Capacity Dial-a-Ride problem*—or *Dial-a-Ride* for short—which is defined as follows. The input is a metric space, a set of objects, where each object d_i specifies a source s_i and a destination t_i, and an integer k—the capacity of the vehicle used for making the deliveries. The goal is to compute a shortest tour for the vehicle in which all objects can be delivered to their destinations (from their sources) while ensuring that the vehicle carries at most k objects at any point in time. There are two variants of the problem: the *non-preemptive* case, in which an object once loaded on the vehicle stays on it until delivered to its destination, and the *preemptive* case in which an object may be dropped at intermediate locations and then picked

---

* This work was performed while the author was a Ph.D. student at the IT University of Copenhagen.

J. Diaz et al. (Eds.): APPROX and RANDOM 2006, LNCS 4110, pp. 200–211, 2006.

up later by the vehicle and delivered. The Dial-a-Ride problem generalizes the Traveling Salesman problem (TSP) even for $k = 1$ and is thus NP-hard.

Let N denote the number of nodes in the input graph, i.e., the number of points that are either sources or destinations. In this paper we show that the preemptive Dial-a-Ride problem has no $\min\{O(\log^{1/4-\varepsilon} N), k^{1-\varepsilon}\}$-approximation algorithm for any $\varepsilon > 0$ unless $\mathsf{NP} \subseteq \mathsf{ZPTIME}(n^{\mathrm{polylog} n})^1$. To our knowledge, the TSP lower bound—which is a small constant—was the best known so far.

The Dial-a-Ride problem has several practical applications such as transportation of elderly and/or disabled persons and courier services. In practice, multi-vehicle systems, where there are more than one vehicle, are more common. Since single-vehicle Dial-a-Ride is a special case of the multi-vehicle Dial-a-Ride problem, the hardness results in this paper holds for these problems a well.

Previous and Related Results. Guan [12] proved that the preemptive case is NP-hard for trees when $k \geq 2$. Frederickson and Guan [8] showed that the unit-capacity non-preemptive case is NP-hard on trees. For this case Frederickson *et al.* [9] gave an 1.8-approximation algorithm on general graphs. The first non-trivial approximation algorithms for the Dial-a-Ride problem for general k were given by Charikar and Raghavachari [4]. For the preemptive case they gave a 2-approximation algorithm for trees. Using the results on probabilistic approximation of metric spaces by tree metrics [7] this gives an $O(\log N)$-approximation for arbitrary metrics. For the non-preemptive case they gave an $O(\sqrt{k})$-approximation algorithm for special instances on height-balanced trees. As above this implies an $O(\sqrt{k} \log N)$-approximation for arbitrary metrics. For points on a line they note that they have a 2-approximation. They also show that the ratio of the cost of the optimal non-preemptive solution to the cost of the optimal preemptive solution can be as large as $\Omega(k^{2/3})$. As noted by Charikar and Raghavachari an $O(k)$-approximation algorithm can be obtained by taking the $O(1)$-approximation algorithm for the unit-capacity case. We note that there is a simple $\frac{3N}{k}$-approximation algorithm (due to [14] for $k = N$).

Several papers have presented exact exponential time algorithms and heuristic algorithms for the Dial-a-Ride problem. For a description of many of these approaches see [6]. A related problem is the k-delivery TSP where all objects are identical and can be delivered to any of the destination points. Charikar *et al.* [3] gave a 5-approximation algorithm for both the preemptive and the non-preemptive problem. Haimovich and Rinnooy Kan [13] gave a 3-approximation for the problem when all objects initially are located at one central depot.

Our Results and Techniques. Our results rely on the hardness results for the two network design problems *Buy-at-Bulk* and *SumFiber-ChooseRoute(SFCR)* (defined in the next section). Andrews [1] and Andrews and Zhang [2] showed that there is no $O(\log^{1/4-\varepsilon} N)$-approximation algorithm for uniform Buy-at-Bulk and SFCR, respectively, for any $\varepsilon > 0$ unless $\mathsf{NP} \subseteq \mathsf{ZPTIME}(n^{\mathrm{polylog} n})$. The

---

[1] $\mathsf{ZPTIME}(n^{\mathrm{polylog} n})$ is the class of problems solvable by a randomized algorithm that always returns the right answer and has expected running time $O(n^{\mathrm{polylog} n})$, where n is the size of the input.

result for SFCR uses a network constructed from an interactive 2-prover system for MAX3SAT. They show that if the MAX3SAT formula ϕ is satisfiable then the optimal solution to the SFCR instance has small cost, and if ϕ is unsatisfiable then it has high cost. More precisely, the cost if ϕ is unsatisfiable is a factor of γ more than if ϕ is satisfiable for $\gamma = O(\log^{1/4-\varepsilon} N)$. Hence if there were an α-approximation for SFCR with $\alpha < \gamma$, then we would be able to determine if ϕ was satisfiable. Using almost the same construction we show that Buy-at-Bulk with cost function $h(x) = \lceil \frac{x}{k} \rceil$ has no $O(\log^{1/4-\varepsilon} N)$-approximation for any $\varepsilon > 0$ unless $\mathsf{NP} \subseteq \mathsf{ZPTIME}(n^{\mathrm{polylog}n})$, when k is between $\log^{11/(8\varepsilon)-9/2} n =$ $\Omega(\log^{1/4+(7\varepsilon)/11} N)$ and $O(2^{\log^2 n}/\log n)$. Here n is the size of ϕ. By changing some of the parameters in the construction we are able to show that the problem is not approximable within a factor of $k^{1-\varepsilon}$ for any $\varepsilon > 0$ when $k < \log^{1/4} N$.

We then show the same hardness results for the preemptive Dial-a-Ride problem by showing a relation between this problem and the Buy-at-Bulk problem with cost function $h(x)$ in the network constructed from the 2-prover system. This is the main technical contribution of this paper. Due to lack of space many proofs are omitted. They can be found in the full version of the paper [11].

2   Definitions

Uniform Buy-at-Bulk. Given an undirected network $\mathcal{N}$, with lengths l_e on the edges and a set $\{(s_i, t_i)\}$ of source-destination pairs. Each pair (s_i, t_i) has an associated demand δ_i. There is a cost function f on the edges, which is a function of the amount of demand x_e using edge e. Function f is subadditive[2], and $f(0) = 0$. The goal is to route all demands δ_i from their source s_i to their destination t_i minimizing the total cost. The demands are unsplittable, i.e., demand δ_i must follow a single path from s_i to t_i. The total cost of the solution is $\sum_e f(x_e) l_e$.

SumFiber-ChooseRoute (SFCR). Here we are given $\mathcal{N}$, l_e, $\{(s_i, t_i)\}$, and δ_i as in Buy-at-Bulk. Each demand requires bandwidth equivalent to one wavelength. Each fiber can carry k wavelengths, and the cost of deploying x fibers on edge e is $x \cdot l_e$. The problem is to specify a path from s_i to t_i for all demands δ_i, and a wavelength for the demand λ_i, minimizing the total cost. Let $f_e(\lambda)$ be the number of demands assigned to wavelength λ that are routed through edge e. Then $\max_\lambda f_e(\lambda)$ is the number of fibers needed on edge e. Thus the total cost of the solution is $\sum_e l_e \max_\lambda f_e(\lambda)$.

Interactive Proof Systems. A Raz-verifier is an *interactive two-prover system*. An interactive two-prover system for MAX3SAT(5) consists of a polynomial time *verifier* with access to a source of randomness and two computationally unbounded *provers*. The verifier sends a polynomial size query to each prover and receives a polynomial size answer. The provers try to convince the verifier that the formula is satisfiable. The provers cannot communicate with each other

---

[2] $f(x + y) \le f(x) + f(y)$.

and are restricted to see only the queries addressed to them. Based on the random bits and the answers to the queries the verifier decides whether or not to accept the input. The verifier accepts with probability 1 if ϕ is satisfiable. If ϕ is unsatisfiable then regardless of how the provers answer the verifier accepts with a very low probability, η, called the *error probability*.

Proof System Parameters. Let R be the random bits, Q_i the random query sent to prover i, and A_i the answer returned by prover i. We will use lowercase letters to denote specific values of these strings. Each random string r uniquely identifies a pair of queries q_0 and q_1. Each query may have many different answers. We say $a \in q$ if a is an answer to query q. We assume that the verifier appends the name of the prover to the query and the provers append the query name to its answer string. This way, an interaction is uniquely identified by the triple (r, a_0, a_1). If the verifier accepts the answers a_0 and a_1 from the provers we say that (r, a_0, a_1) is an *accepting interaction*. Note that two different random strings might result in the same prover-0 query (or prover-1 query), but in that case they will result in different prover-1 (prover-0) queries. Let $m(Q_i)$ denote the number of distinct possible values of Q_i. By padding random bits, we can assume, $m(Q_0) \le m(Q_1) < 2m(Q_0)$. We can ensure that the Raz verifier has the following properties (here $|x|$ denotes the number of bits in the string x): $|R| = O(\log^2 n)$, $|Q_i| = O(\log^2 n)$, $|A_i| = O(\log^2 n)$, and $\eta = 2^{-\Omega(\log n)}$. For each i and for any $q \in \{0,1\}^{|Q_i|}$: $Pr[Q_i = q] \in \{0, 1/m(Q_i)\}$.

3   Relation Between Buy-at-Bulk and Dial-a-Ride

The following lemma shows a relation between Buy-at-Bulk and Dial-a-Ride.

Lemma 1. *Let* OPT_B *be the value of an optimal solution to a Buy-at-Bulk instance* B *with source destination pairs* S *in graph* G *and cost function* $h(x) = \lceil \frac{x}{k} \rceil$, *and let* OPT_D *be the value an optimal solution to the Dial-a-Ride instance* D *with the same source-destination pairs* S *in* G. *Then* $OPT_B \le OPT_D$.

Proof. We will abuse notation and let OPT_i stand for both the value of the optimal solution and the solution itself. We can turn OPT_D into a solution to instance B as follows: Route a demand δ_i from its source s_i to its destination t_i by the same edges as object δ_i passes in OPT_D. It is straightforward to verify that this is a valid solution and that the cost is no larger than OPT_D. ☐

Since the optimal solution to B might be disconnected, there is in general no way to turn OPT_B into a solution to D at a cost bounded in terms of OPT_B. However, on the network used to construct the hardness result for Buy-at-Bulk we will show that in the case were the MAX3SAT instance ϕ is satisfiable it is possible to turn the solution to B into a solution to D at cost at most $7 \cdot OPT_B$.

4   The Network

In this section we describe the network that is used to show hardness of SFCR in [2]. The network is constructed randomly from an interactive proof system

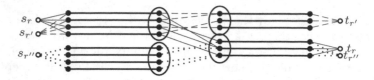

Fig. 1. The basic network $\mathcal{N}_0$. For each of the three random strings r, r' and r'', four canonical paths corresponding to four accepting interactions, are shown (r solid, r' dashed, and r'' dotted). The long thick edges are the answer edges.

for MAX3SAT. The idea is for each demand to define a set of *canonical paths* on which the demand can be carried. These canonical paths correspond to accepting interactions and are short paths directly connecting the source and destination.

We first construct a basic network $\mathcal{N}_0$, which is used as the base case in the random construction. Given an instance ϕ, first construct the interactive two-prover system. This is then turned into an instance of SFCR as follows. For each possible answer a there is an *answer edge* (also denoted by a). For each random string r there is a source node s_r, a destination node t_r, and a demand d_r of one to be routed from s_r to t_r. For each accepting interaction (r, a_0, a_1) there is a canonical path p. This path starts at node s_r, passes through a_0 and a_1 and ends at t_r. To make this possible we place edges between s_r and a_0, between a_0 and a_1, and between a_1 and t_r. The edge between a_0 and a_1 is referred to as a *center edge*, and the edge between s_r and a_0, and between a_1 and t_r as a *demand edge*. For each query q the answer edges $a \in q$ are grouped together (see Fig. 1). Answer edges have length $h > 1$ and the other edges have length 1.

Before defining the final network, we define a random network $\mathcal{N}_1$ in terms of $\mathcal{N}_0$ and two parameters X and Z. The network essentially replicates $\mathcal{N}_0$ in the vertical direction XZ times. Each answer edge a_0 (resp. a_1) of $\mathcal{N}_0$ has XZ copies, denoted by $a_{0,x,z}$ ($a_{1,x,z}$) where $0 \le x < X$ and $0 \le z < Z$. For each random string r, create X demands $d_{r,x}$ and X source and destination nodes, $s_{r,x}$ and $t_{r,x}$, where $0 \le x < X$. Each of the X demands $d_{r,x}$ routes one unit of flow from $s_{r,x}$ to $t_{r,x}$. For each accepting interaction (r, a_0, a_1), the demand $d_{r,x}$ has a canonical path that starts at $s_{r,x}$, passes through $a_{0,x',z'}$ and $a_{1,x'',z''}$ and ends at $t_{r,x}$. The answer edges $a_{0,x',z'}$ and $a_{1,x'',z''}$ are chosen randomly. More precisely, x' and x'' are chosen uniformly at random from the range $\{0, 1, \ldots, X-1\}$ and z' and z'' are chosen uniformly at random from the range $\{0, 1, \ldots, Z-1\}$. To make the canonical paths feasible, $\mathcal{N}_1$ has center edges connecting $a_{0,x',z'}$ and $a_{1,x'',z''}$, and edges connecting $s_{r,x}$ to $a_{0,x',z'}$, and $a_{1,x'',z''}$ to $t_{r,x}$.

The final network $\mathcal{N}_2$ is essentially a concatenation of $\mathcal{N}_1$ in the horizontal direction Y times for some parameter Y, where each level is constructed randomly and independently. Each answer edge is indexed by $a_{0,x,z,y}$ (resp. $a_{1,x,z,y}$) where $y \in \{0, 1, \ldots, Y-1\}$. As in $\mathcal{N}_1$, X demands $d_{r,x}$, $0 \le x < X$, are created for each random string r. For each accepting interaction (r, a_0, a_1), the demand $d_{r,x}$ has a canonical path starting at $s_{r,x}$ followed by answer edges $a_{0,x,z,0}$ and $a_{1,x,z,0}$ chosen uniformly at random at level $y = 0$. At each subsequent level y, the answer edges are chosen uniformly at random until the path ends at $t_{r,x}$.

The center edges and demand edges are defined by the canonical paths. Each canonical path also requires an edge between each consecutive pair of levels.

5   Hardness of Buy-at-Bulk with Cost Function $\lceil \frac{x}{k} \rceil$

In this section we use the network $\mathcal{N}_2$ to show hardness of Buy-at-Bulk with cost function $\lceil \frac{x}{k} \rceil$. The results are obtained by changing some of the parameters in the network compared to paper by Andrews and Zhang [2], but otherwise the proofs in this section are similar to the ones in the [2]. We use the following parameters to show hardness with dependence on N.

- $\ell = \log^\alpha n$ for some constant α.
- $Z = \frac{2^{|r|}}{k \min\{m(Q_0), m(Q_1)\}}$
- $X = (2^{6+|r|+|a_0|+|a_1|}YZ)^{2l+1} = 2^{O(\log^{\alpha+2} n)}$
- $k = \log^{\frac{\alpha}{4}+4} n$

- $\sigma = \log^{\frac{\alpha}{4}} n$
- $Y = \sqrt{\ell} = \log^{\frac{\alpha}{2}} n$
- $h = \frac{2^{|r|}}{(m(Q_0)+m(Q_1))Z}$
- $\eta = \frac{1}{\sigma^2 \log n}$

The only parameter changed compared to [2] is h. To show hardness with dependence on k we allow k to be smaller than $\log^{\alpha/4+4} n$. To make the proofs go through we change Z and h as follows. Let $c > 1$ be a constant such that $k = \log^{\alpha/4+4} n/c$ and set

- $Z = \frac{2^{|r|}}{ck \min\{m(Q_0), m(Q_1)\}} = \frac{2^{|r|}}{\log^{\frac{\alpha}{4}+4} n \cdot \min\{m(Q_0), m(Q_1)\}}$ • $h = \frac{2^{|r|}}{c(m(Q_0)+m(Q_1))Z}$

The next two lemmas hold for both definitions of Z and h. An answer edge is said to be *bought* if any demand is routed through it.

Lemma 2. *If ϕ is satisfiable, then the Buy-at-Bulk instance has a solution of total cost at most $2^{|r|}(2Y + 1)X + 2(m(Q_0) + m(Q_1))hXYZ$.*

Proof. Since ϕ is satisfiable there are two provers that always cause the verifier to accept. We route the demand on answer edge a if and only if for these two provers a is the answer to query q. For each string r there must be some accepting interaction (r, a_0, a_1) for which both a_0 and a_1 have been bought. Each of the demands $d_{r,x}$, for $0 \le x < X$, has one canonical path that corresponds to (r, a_0, a_1). The demand $d_{r,x}$ is routed along this path. There are $2Y + 1$ length one edges on this path and thus the total number of edges of length one needed is at most $2^{|r|}(2Y + 1)X$. It is possible to show that the expected cost of an answer edges is two. The details are omitted due to lack of space. The expected total cost of the answer edges is therefore $2XZY(m(Q_0) + m(Q_1))h$. The total solution has expected cost $2^{|r|}(2Y + 1)X + 2(m(Q_0) + m(Q_1))hXYZ$, and the cost of the optimal solution must therefore have cost no higher than that.   □

The second lemma gives a lower bound on the cost of the solution when ϕ is unsatisfiable. The proof is omitted due to lack of space.

Lemma 3. *With probability $\frac{2}{3} - o(1)$, if the instance ϕ of 3SAT is unsatisfiable then the cost of any solution to our instance of Buy-at-Bulk is at least*

$$\min\{\frac{\sigma h}{10}(m(Q_0) + m(Q_1))XYZ, \frac{Y^2}{4k}((X2^{|r|})(1 - \frac{77}{375} - o(1)) - X)\}.$$

Combining Lemma 2 and 3 we get the following hardness result for Buy-at-Bulk with cost function $h(x) = \lceil \frac{x}{k} \rceil$. The proof is omitted due to lack of space.

Corollary 1. *For any $\varepsilon > 0$, there is no $\min\{O(\log^{\frac{1}{4}-\varepsilon} N), k^{1-\varepsilon}\}$-approximation algorithm for Buy-at-Bulk with cost function $h(x) = \lceil \frac{x}{k} \rceil$ unless all problems in* NP *can be solved by a randomized algorithm with expected running time $O(n^{\text{polylog } n})$.*

6   Routing in the Network

Let B be the instance of Buy-at-Bulk constructed in Section 5, and let D be an instance of preemptive Dial-a-Ride with the same source-destination pairs in the same network. Let SOL_B denote the solution used to give the bound on the cost of the optimal solution in Lemma 2, and let OPT_D be the optimal solution to D. In this section we show how to construct a solution to D of cost at most $7 \cdot \mathsf{SOL}_B$ when ϕ is satisfiable.

Let $\mathcal{N}_2^f$ be the network induced by the edges bought in SOL_B. Recall that in SOL_B all demands are routed on canonical paths. For each demand d, let p_d be the canonical path which d is routed on in SOL_B. We say that edge $e \in \mathcal{N}_2^f$ is *used* by an object d if e is on the path p_d. Let u_e be all the objects using edge e.

6.1   The Tour When $\mathcal{N}_2^f$ Is Connected

We will first explain how to construct the tour when $\mathcal{N}_2^f$ is connected. We will say that the tour is using an edge in the forward direction if it uses it in the same direction as the demands routed on this edge and backwards otherwise. Assume that any edge in $\mathcal{N}_2^f$ is used by at most k objects (we show later how to get rid of this assumption). We will ensure that the tour has the following properties:

(i) The tour only uses edges from $\mathcal{N}_2^f$.
(ii) An object d will only be in the vehicle when the vehicle is on an edge $e \in p_d$.
(iii) When the vehicle goes forward on an edge it is either empty or carries all objects using that edge.

The algorithm to construct the tour has two kinds of phases—a *delivery phase* and a *pickup phase*—which are intermixed. In a delivery phase we are in the process of delivering a certain object. In a pickup phase the vehicle is on its way to pick up the next object to be delivered. The vehicle is always empty in a pickup phase. The algorithm calls the following two procedures.

Deliver(d,s): Follow p_d. For each edge on p_d there are two cases:
    1. All objects from u_e are present at u: Pick up all the objects and traverse e. At node v drop off all objects not going in the same direction as d.
    2. One or more objects from u_e are not present at u: Drop off d at node u, and go to pick up these objects as follows. Let d' be such an object. Follow $p_{d'}$ backwards from e until encountering d'. Pick up d' and deliver d' at *node* u (not $s_{d'}$) by recursively calling Deliver(d',u).

Route(d): First deliver d by calling Deliver(d,s_d) (this is the *delivery phase for object d*). Then follow the route constructed during this call to Deliver backwards until d_d is reached (this is a *pickup phase*). Whenever encountering an undelivered object d' on the way, pick it up and deliver it to its destination by recursively calling Route(d').

Algorithm. The algorithm starts at a node $s_{r,x}$ for some r and x, pick up $d_{r,x}$ and call Route($d_{r,x}$). Below we will show that when the vehicle returns to $d_{r,x}$ all objects are delivered.

Analysis of the Algorithm. It is easy to verify that the tour made by the algorithm satisfies property (i), (ii), and (iii). We will denote the route constructed during the delivery phase for object d by r_d.

Lemma 4. *For any object d, the route r_d, has the following properties:*

(iv) r_d *only goes backwards on an edge e to fetch "missing" objects. If d' is such an object then $e \in p_{d'}$.*

(v) *If r_d goes backwards on edge e it returns to the right endpoint of e through e.*

(vi) *When route r_d traverses an edge e in the forward direction the vehicle contains all objects using e.*

Proof. Property (iv) and (vi) follows immediately from the description of the algorithm. It remains to prove property (v). All canonical paths go through all levels of the network in increasing order. Therefore an object missing at the left endpoint of some edge at level i can be fetched at a level smaller than i or at i if the edge is not the first edge on level i. It is thus possible to fetch all objects missing at a certain node, since there are no cyclic dependencies.   □

Lemma 4 gives us the following two corollaries.

Corollary 2. *For any object d, the route r_d traverses each edge in $\mathcal{N}_2^f$ at most once in each direction.*

Corollary 3. *For any two objects d_1 and d_2 the routes r_{d_1} and r_{d_2} are disjoint.*

Lemma 5. *All objects are delivered to their destination.*

Proof. By contradiction. Recall, we assumed $\mathcal{N}_2^f$ is connected. Assume some subset of objects S are not delivered. Consider an object $d \in S$. If d is at a node $u \neq s_d$ then it was left at u during the delivery phase of some object d'. But then it would have been picked up and delivered to its destination when the vehicle traversed $r_{d'}$ backwards. Thus d must still be at its source s_d. Since d is still at s_d the path p_d does not share any edges with any path $p_{d'}$ where d' is a delivered object. To see this assume d shared an edge e with a delivered object d'. Due to property (ii) the vehicle crossed e containing d', since d' is delivered. Due to property (vi) of Lemma 4 d must have been in the vehicle when it crossed e, and thus d would no longer be at s_d. Since SOL_B are using canonical paths for each object, the graph $\mathcal{N}_2^f$ has the property that if two canonical paths p_d and $p_{d''}$ meet at some vertex then they must share an edge adjacent to that vertex. Therefore p_d cannot share any vertices with any path $p_{d'}$ where d' is a delivered object. This is true for all objects $d \in S$, contradicting that $\mathcal{N}_2^f$ is connected. $\square$

Lemma 6. *When $\mathcal{N}_2^f$ is connected the tour has length at most $4 \cdot \mathrm{SOL}_B$.*

Proof. Let $l(r_d)$ denote the length of the route r_d. The total length of the parts of the tour constructed during delivery phases is $\sum_{d \in D} l(r_d)$.

Now consider the parts of the tour constructed during a pickup phase. Here we are going backwards on the route r_d for some object d. During this pickup phase we stop each time we meet an object d' and deliver it by calling Route(d'). Due to Corollary 3 the part of the tour constructed during the call to Route(d') is disjoint from r_d, since it only contains edges on $r_{d'}$. The route r_d is thus traversed at most once during the pickup phases. Thus the total length of the parts of the tour constructed during delivery phases is at most $\sum_{d \in D} l(r_d)$.

Adding together the total length of the tours constructed during the delivery phases and the pickup phases, we get that the total length of the tour is at most $2 \cdot \sum_{d \in D} l(r_d)$. Using Corollary 2 and Corollary 3 we get that the tour uses each edge in $\mathcal{N}_2^f$ at most 4 times, and thus the cost of the tour is at most $4 \cdot \mathrm{SOL}_B$. $\square$

Edges used by more than k Objects. We assumed that any edge in $\mathcal{N}_2^f$ is used by at most k objects. We can get rid of this assumption by a minor modification of the algorithm. Let S_e be the set of objects using edge e. Then the solution SOL_B paid $\lceil \frac{S_e}{k} \rceil \cdot l_e$ for this edge. As before, when we want to traverse e we go backwards and pick up all objects in S_e. We then go forward and back on e carrying as many objects from S_e as possible each time until all objects from S_e are on the right endpoint of e. The number of times we traverse e is $\lceil \frac{S_e}{k} \rceil$, and thus Lemma 6 still holds.

6.2 $\mathcal{N}_2^c$ Connected and $\mathcal{N}_2^f$ Disconnected

Let $\mathcal{N}_2^c$ be the graph induced by the canonical paths ($\mathcal{N}_2$ can contain answer edges that are not part of any canonical path). If $\mathcal{N}_2^c$ is connected but $\mathcal{N}_2^f$ is disconnected we can add edges from $\mathcal{N}_2^c$ to $\mathcal{N}_2^f$ to connect it. We can do this by adding edges of total length equal to the number of connected components minus one times the length of a canonical path in $\mathcal{N}_2^c$.

First we note that since $\mathcal{N}_2^c$ consists of the union of canonical paths, then for any component C in $\mathcal{N}_2^f$ there must be another component C' in $\mathcal{N}_2^f$ such that some object d routed in C has a canonical path p that intersect with a canonical path p' for an object d' routed in C'. We connect C and C' by adding the following edges: All edges on p from s_d to the intersecting edge e (including e), and all edges on p' from e to $t_{d'}$. We call these added edges a *connecting path from C' to C*. Since $\mathcal{N}_2^c$ is connected we can make $\mathcal{N}_2^f$ connected by adding $c - 1$ connecting paths, where c is the number of connected components in $\mathcal{N}_2^f$. We add these connecting paths in such a way that all components can be reached from one component—called the *start component*—using a path that when going from component C to a component C' uses a connecting path from C to C' (not from C' to C). Since the length of a connecting path is the same as the length of a canonical path the total length is $c - 1$ times the length of a canonical path. Since each connected component consists of at least one canonical path the total length of the connecting paths is at most the same as the sum of all edges in $\mathcal{N}_2^f$, i.e., SOL_B.

Constructing the Tour. Start in the start component C_s in $\mathcal{N}_2^f$ and deliver the objects in this component as described in the previous section. Whenever the vehicle gets to a node d_d which is the starting point of a connecting path from this component to another component C, it follows this connecting path to C and delivers the objects in C the same way. When all objects in a component are delivered the vehicle returns to the starting point in this component and from there to the previous component C' if such a component exists. It then carries on delivering the objects in C'.

Lemma 7. *When $\mathcal{N}_2^c$ is connected the tour has length at most $6 \cdot \mathsf{SOL}_B$.*

Proof. If $\mathcal{N}_2^f$ is connected it follows from Lemma 6. If $\mathcal{N}_2^f$ is disconnected we use the approach described above. To deliver the objects in a single component we use no more time than in the previous section. By Lemma 6 the contribution from these parts of the tour is at most $4 \cdot \mathsf{SOL}_B$ in total. To get to the next component and back again we use a connecting path and the sum of the edges used to get to and from connected components is thus at most $2 \cdot \mathsf{SOL}_B$.   □

6.3 $\mathcal{N}_2^c$ Disconnected

If $\mathcal{N}_2^c$ is disconnected we connect it by adding edges of length one between a source node in one component and a source node in another component . We call these edges *component* edges. We add the minimum number of component edges, i.e., $l - 1$ where l is the number of connected components. This can be seen as constructing a tree on the components.

Since we add the component edges between disjoint components in $\mathcal{N}_2^c$, which are also disjoint components in $\mathcal{N}_2$, we do not introduce any new cycles in $\mathcal{N}_2$. Therefore the component edges cannot decrease the cost of the optimal solution to the Buy-at-Bulk instance or to the Dial-a-Ride instance: Let C_1 and C_2 be two components connected by a component edge e. If some object d with source s_d in C_1 is using e, then it has to use it again to get back to C_1, since $s_d \in C_1$ and the only connection between C_1 and C_2 is e.

Constructing the Tour. The vehicle first delivers the objects in a component C in $\mathcal{N}_2^c$ as described in the previous section. When it gets to the source node in the component that has a component edge to a source node in another component C', it goes to C' and delivers the objects in C' the same way. When all objects in a component are delivered it returns to the starting point of this component and follows the component edge back to the previous component C if such a component exists. It then carries on delivering the objects in component C.

Lemma 8. *The optimal solution to D has cost at most* $7 \cdot \text{OPT}_B$.

Proof. The cost of delivering the objects in the original components of $\mathcal{N}_2$ is at most $6 \cdot \text{SOL}_B$ due to Lemma 7. The total length of the new edges is $l-1$ which is less than $1/2 \cdot \text{SOL}_B$, since each connected component has a canonical path of at least three. The new edges are used twice: once in each direction. $\qquad\square$

7 Hardness of Preemptive Dial-a-Ride

From Lemma 8 and Lemma 2 we get,

Lemma 9. *If ϕ is satisfiable, then the Dial-a-Ride instance has a solution of total cost* $7 \cdot 2^{|r|}(2Y+1)X + 2(m(Q_0)+m(Q_1))hXYZ$.

We can now use Lemma 1, Lemma 3, and Lemma 9 to show hardness of the Dial-a-Ride problem.

Lemma 10. *Let $\gamma = \log^{\frac{\alpha}{4}-5} n$. If there exists a γ-approximation algorithm for the Finite Capacity Dial-a-Ride problem, then there exists a randomized $O(n^{\text{polylog } n})$ time algorithm for 3SAT.*

Proof. For any 3SAT instance ϕ we construct the network $\mathcal{N}_2$ from the two-prover system and then apply a γ-approximation algorithm A for Dial-a-Ride.

If the 3SAT instance ϕ is satisfiable then by Lemma 9 and our choice of h there is a solution to our instance of Dial-a-Ride of cost at most $7 \cdot 2^{|r|}(2Y+1)X + 2(m(Q_0)+m(Q_1))hXYZ = 7 \cdot 2^{|r|}(4Y+1)X$. Hence, the γ-approximation algorithm returns a solution of cost at most $\gamma \cdot 7 \cdot 2^{|r|}(4Y+1)X$, and we declare ϕ satisfiable. If ϕ is unsatisfiable then by Lemma 1, Lemma 3 and our choice of h, with probability $2/3 - o(1)$, any solution have cost at least the minimum of $\Omega(\sigma 2^{|r|}XY)$ and $\Omega(\frac{\ell}{k}X2^{|r|})$. Both these expressions are strictly larger than $\gamma \cdot 7 \cdot 2^{|r|}(4Y+1)X$.

The construction of the network takes time $O(n^{\text{polylog } n})$ since $\mathcal{N}_2$ has size $O(n^{\text{polylog } n})$. Hence we have described a randomized $O(n^{\text{polylog } n})$ time algorithm for 3SAT that has one-sided error probability at most $1/3 + o(1)$. It is possible to convert this into a randomized algorithm that never makes an error and has expected running time $O(n^{\text{polylog } n})$. $\qquad\square$

In the Dial-a-Ride instance N is the number of sources and destinations. We have $2^{|r|}X$ sources and $2^{|r|}X$ destinations, and thus $N = 2 \cdot 2^{|r|}X = 2^{O(\log^{\alpha+2} n)}$. For any constant $\varepsilon > 0$, if we set $\alpha = \frac{11}{2\varepsilon} - 2$ then $\gamma = \Omega(\log^{1/4-\varepsilon} N)$. This gives us the following corollary.

Corollary 4. *There is no $O(\log^{\frac{1}{4}-\varepsilon} N)$-approximation algorithm to the preemptive Finite Capacity Dial-a-Ride problem on general graphs for any constant $\varepsilon > 0$ unless* $\mathsf{NP} \subseteq \mathsf{ZPTIME}(n^{\text{polylog} n})$.

In the above construction we had $k = \log^{\alpha/4+4} n$. The proofs hold for larger k too, but since Z should be a positive integer we require $k \leq 2^{|r|} / \min(m(Q_0), m(Q_1))$. To get a hardness result for small k we chang the variables Z and h as described in Section 5. Using Lemma 1 and Lemma 3, we get

Lemma 11. *Let $k < \log^{\frac{1}{4}} N$. Then there is no $k^{1-\varepsilon}$-approximation algorithm to the preemptive Finite Capacity Dial-a-Ride problem on general graphs for any constant $\varepsilon > 0$ unless* NP $\subseteq$ ZPTIME$(n^{\text{polylog}n})$.

The proof is omitted due to lack of space. To summarize we have shown,

Theorem 1. *There is no $\min\{O(\log^{\frac{1}{4}-\varepsilon} N), k^{1-\varepsilon}\}$-approximation algorithm to the preemptive Finite Capacity Dial-a-Ride problem on general graphs for any constant $\varepsilon > 0$ unless* NP $\subseteq$ ZPTIME$(n^{\text{polylog}n})$.

Acknowledgments. The author wants to thank Moses Charikar and Matthew Andrews for many helpful and useful discussions.

References

1. M. Andrews. Hardness of buy-at-bulk network design. In *45th Annual IEEE Symposium on Foundations of Computer Science*, pages 115–124, October 2004.
2. M. Andrews and L. Zhang. Bounds on fiber minimization in optical networks with fixed fiber capacity. In *IEEE INFOCOM*, 2005.
3. M. Charikar, S. Khuller, and B. Raghavachari. Algorithms for capacitated vehicle routing. *SICOMP: SIAM Journal on Computing*, 31(3):665–682, 2002.
4. M. Charikar and B. Raghavachari. The finite capacity dial-a-ride problem. In *IEEE Symposium on Foundations of Computer Science*, pages 458–467, 1998.
5. N. Christofedes. Vehicle routing. In E. L. Lawler, J. K. Lenstra, A. H. G. Rinnooy Kan, and D. B. Shmoys, editors, *The Traveling Salesman Problem*, pages 431–448. John Wiley & Sons, 1985.
6. G. Desaulniers, J. Desrosiers, A. Erdmann, M. M. Solomon, and F. Soumis. VRP with pickup and delivery. In P. Toth and D. Vigo, editors, *The vehicle routing problem*, pages 225–242. Society for Industrial and Applied Mathematics, 2001.
7. J. Fakcharoenphol, S. Rao, and K. Talwar. A tight bound on approximating arbitrary metrics by tree metrics. *J. Comput. System Sci.*, 69(3):385–497, 2004.
8. G. N. Frederickson and D. J. Guan. Non-preemptive ensemble motion planning on a tree. *Journal of Algorithms*, 15(1):29–60, 1993.
9. G. N. Frederickson, M. S. Hecht, and C. E. Kim. Approximation algorithms for some routing problems. *SIAM Journal on Computing*, 7(2):178–193, 1978, May.
10. B. L. Golden and A. A. Assad. *Vehicle Routing: Methods and Studies*. Studies in Management Science and Systems, 16. Elsevier, 1991.
11. I. L. Gørtz. Hardness of preemptive finite capacity dial-a-ride. IMADA Preprints 2006 No. 4, University of Southern Denmark, 2006.
12. D. J. Guan. Routing a vehicle of capacity greater than one. *Discrete Applied Mathematics*, 81(1-3), 1998.
13. M. Haimovich and A. H. G. Rinnooy Kan. Bounds and heuristics for capacitated routing problems. *Mathematics of Operations Research*, 10(4):527–542, 1985.
14. H. N. Psaraftis. An exact algorithm for the single vehicle many-to-many dial-a-ride problem with time windows. *Transportation Science*, 17(3):351–357, 1983.

Minimum Vehicle Routing with a Common Deadline

Viswanath Nagarajan* and R. Ravi**

Tepper School of Business, Carnegie Mellon University, Pittsburgh PA 15213
{viswa, ravi}@cmu.edu

Abstract. In this paper, we study the following vehicle routing problem: given n vertices in a metric space, a specified root vertex r (the depot), and a length bound D, find a minimum cardinality set of r-paths that covers all vertices, such that each path has length at most D. This problem is $\mathcal{NP}$-complete, even when the underlying metric is induced by a weighted star. We present a 4-approximation for this problem on tree metrics. On general metrics, we obtain an $O(\log D)$ approximation algorithm, and also an $(O(\log \frac{1}{\epsilon}), 1 + \epsilon)$ bicriteria approximation. All these algorithms have running times that are almost linear in the input size. On instances that have an optimal solution with one r-path, we show how to obtain in polynomial time, a solution using at most 14 r-paths.

We also consider a linear relaxation for this problem that can be solved approximately using techniques of Carr & Vempala [7]. We obtain upper bounds on the integrality gap of this relaxation both in tree metrics and in general.

1 Introduction

A common version of vehicle routing problems involves locations that demand service, and a single depot that has to send vehicles to satisfy these demands. It may be important to service *all* demands before a deadline, so several vehicles may need to be deployed. In this context, meeting demands are hard constraints which must be satisfied, while the objective is to minimize the number of vehicles used.

Vehicle routing problems are extensively studied in the Operations Research literature [10, 11, 13, 15, 16]. Most of these papers focus on developing heuristic solutions or solving the problems optimally. The methods used in these papers include branch and bound, cutting plane algorithms, local search, and genetic algorithms. There has been considerably less work on these problems in the approximation algorithms literature, perhaps due to the inapproximability of natural formulations of these problem. The version that we study is more tractable from the point of view of obtaining approximation guarantees.

Approximation guarantees for the problem we consider have been studied in Li et al. [9], and Bazgan et al. [5]. Li et al. [9] suggested a tour-splitting heuristic

---

* Supported by NSF ITR grant CCR-0122581. (The ALADDIN project)
** Supported in part by NSF grants CCF-0430751 and ITR grant CCR-0122581. (The ALADDIN project)

which has a performance guarantee which depends on some *values* in the input instance, but does not yield any worst case approximation bounds. Bazgan et al. [5] suggested algorithms achieving provable worst case bounds under a *differential approximation* measure. However, bounds in this measure do not imply any bounds in the standard approximation measure. In this paper, we study standard approximation algorithms for the common deadline vehicle routing problem.

There has been some interesting recent work [3, 6] on approximating the related *orienteering* problem. In this problem, there is a single vehicle, and the goal is to find a bounded length path from the depot, that maximizes the number of vertices covered. Improving on work by Blum et al. [6], Bansal et al. [3] presented a 3-approximation algorithm for orienteering on general metrics. Bansal et al. [3] also considered extensions where vertices can only be covered in individual time windows, giving poly-logarithmic approximations for this case. Improved algorithms for special classes of metrics were obtained in [2, 4].

Problem definition: We model locations as points in a metric space (V, d), with $|V| = n$. Here d is a function $d : V \times V \to \mathbb{Z}^+$ that is symmetric and satisfies the triangle inequality. We assume throughout that all distances are integral. The input to the *minimum vehicle routing problem* consists of a metric space (V, d), one designated *root* vertex $r \in V$, and a length bound D. The root corresponds to the depot, and the length bound D represents the common deadline of the demand locations. The objective is to find a minimum cardinality set of paths originating from r, that covers all vertices in V. In addition, all these paths are required to have length at most D. Paths originating from r are called r-paths. We also refer to the minimum vehicle routing problem as rooted vehicle routing.

A related problem is the *unrooted vehicle routing* problem. In this problem, there is no designated root, and vehicles can start at *any* vertex. The goal here is to find a minimum cardinality set of paths that cover all the vertices. The paths are again required to have length at most the length bound D. This problem has been studied recently in Arkin et al. [1] as minimum path cover, and they gave a 3-approximation algorithm for it.

Our results: For minimum vehicle routing on a tree metric, we obtain a 4-approximation algorithm in Section 2. We note that the problem is $\mathcal{NP}$-complete even in this special case. For minimum vehicle routing on general metrics, we obtain an $O(\log D)$-approximation algorithm, and a bi-criteria result in Section 3. We also consider an integer programming formulation of this problem. We show that the integrality gap of its linear relaxation is upper bounded by a constant in the case of tree metrics (Section 2.1), and by $O(\min\{\log n, \log D\})$ in case of general metrics (Section 3.1). Determining a tight bound on the integrality gap of this relaxation is still open. We consider the following promise problem in Section 4: *given* an instance of minimum vehicle routing, where the optimal solution uses just a single vehicle, *find* a solution using a small number of vehicles. We show how the minimum excess path problem [6] can be used to obtain (in polynomial time) a solution to this promise problem having at most 14 vehicles.

Due to lack of space, we omit some proofs in this version of the paper. The interested reader may refer to [14] for the proofs missing here.

2   Minimum Vehicle Routing on a Tree

In this section, we consider the special case of minimum vehicle routing, when the metric space $T = (V, d)$ is induced by a tree. Even in the special case of a star, the problem remains $\mathcal{NP}$-complete (reduction from 3-partition). Here we present a 4-approximation for minimum vehicle routing on trees.

We assume without loss of generality, that the tree is binary, and rooted at r. This can be ensured by splitting high degree vertices, and adding edges of zero length. Suppose the input consists of tree $T = (V, d)$, root $r \in V$, and length bound D. Algorithm minTVR for minimum vehicle routing on trees is as follows.

1. Initialize $T' = T$.
2. While $(T' \neq \{r\})$ do
   (a) Pick a deepest vertex $v \in T'$ s.t. the subtree T'_v below v can *not* be covered by just one r-path, of length at most D. If no such v exists, add an r-path covering T', and exit loop.
   (b) Let w_1 and w_2 be the two children of v. For $i = 1, 2$, set W_i to be the minimum length r-path traversing subtree T'_{w_i}.
   (c) Add r-paths W_1 and W_2.
   (d) $T' = T' \setminus T'_v$.

Note that it is easy to find the minimum length r-path covering all the vertices of a tree - the longest r to leaf path is traversed once, and all other edges are traversed 2 times. See Figure 1b for the structure of an r-path on a tree. Thus the condition in step 2a can be checked efficiently.

Theorem 1. *Algorithm minTVR obtains a 4-approximation to the minimum vehicle routing problem on trees.*

Proof: It is not hard to see that algorithm minTVR can be implemented in a single depth-first search of the tree; so the time complexity is linear in the input size. Suppose we are given an instance of minimum vehicle routing on a tree $T = (V, d)$, with root $r \in V$.

A *heavy cluster* is a set of vertices $C \subseteq V$ such that the induced subgraph $T[C]$ is connected, and the vertices in C can not all be covered by a single r-path of length at most D. Note that the subtrees T'_v seen in step 2a of the algorithm are heavy clusters. Suppose, in its entire execution, the algorithm finds k heavy clusters $C_1, \cdots C_k$ (these vertex sets will be disjoint). Then algorithm minTVR uses at most $2k+1$ r-paths to cover all the vertices. From the definition of vertex v (in step 2a), each r-path W_i added in step 2c (corresponding to the children of v), has length at most D. So the algorithm indeed produces a feasible solution. The following lemma shows that the optimal solution requires at least $\frac{k+1}{2}$ vehicles, and thus proves Theorem 1.

Lemma 1. *If there are k disjoint heavy clusters $C_1, \cdots C_k \subseteq V$ in the tree T, the minimum number of r-paths of length at most D required to cover $\bigcup_{i=1}^{k} C_i$ is more than $\lfloor \frac{k+1}{2} \rfloor$.*

Proof: The proof of this lemma is by induction on k. For $k = 1$, the lemma is trivially true. Suppose $k > 1$, and assume that the lemma holds for all values up to $k - 1$. Suppose the minimum number of r-paths required to cover all these clusters, $OPT \leq \lfloor \frac{k+1}{2} \rfloor$. Note that OPT can not be smaller than $\lfloor (k+1)/2 \rfloor$: taking any $k - 1$ of these k clusters, we get a contradiction to the induction hypothesis with $k-1$ clusters! Similarly, k can not be even because in that case, $\lfloor \frac{k+1}{2} \rfloor = \lfloor \frac{(k-1)+1}{2} \rfloor$. So we may assume that k is odd, and $OPT = (k+1)/2$.

Every cluster C_i forms a connected subgraph of T. It will be convenient to think of the lengths associated with C_i in the following parts - the path from r to the highest vertex in C_i, and the internal part of C_i (see Figure 1a).

Now consider the bipartite graph $H = (\Gamma, \mathcal{C}, E)$ where $\Gamma = \{t_1, \cdots, t_{(k+1)/2}\}$ is the set of r-paths in the optimal cover (note that $|\Gamma| = OPT = (k+1)/2$), and $\mathcal{C} = \{C_1, \cdots, C_k\}$ is the set of the k heavy clusters. There is an edge $(t_j, C_i) \in E$ iff path t_j visits some vertex of cluster C_i. A set of edges M is said to be a *1-2-matching* from $\mathcal{C}$ to Γ, if the number of edges of M incident on a vertex of $\mathcal{C}$ is exactly 1, and the number of edges of M incident on a vertex of Γ is at most 2. In other words, it is a perfect matching of $\mathcal{C}$ in the graph H' obtained from H by duplicating all the vertices in Γ and the edges in E.

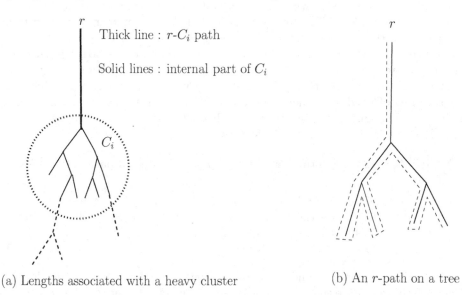

Thick line : r-C_i path

Solid lines : internal part of C_i

(a) Lengths associated with a heavy cluster

(b) An r-path on a tree

Fig. 1. Structures of a heavy cluster and an r-path

We claim that H must have a 1-2-matching from $\mathcal{C}$ to Γ. Suppose not - then by Hall's Theorem, we get a set $S \subseteq \mathcal{C}$ such that S has fewer than $|S|/2$ neighbors in Γ. Note that $S \neq \mathcal{C}$, as $\mathcal{C}$ has $OPT > \frac{|\mathcal{C}|}{2}$ neighbors. This implies that the clusters in S are visited completely by fewer than $|S|/2$ r-paths, which contradicts the induction hypothesis with clusters S ($|S| < k$). Let $\pi : \mathcal{C} \to \Gamma$ be a 1-2-matching in H. Since there are $(k+1)/2$ vertices in Γ, and only k vertices in $\mathcal{C}$, there is one vertex in Γ which is matched to only one cluster. Let this vertex be $t_{(k+1)/2}$.

Let $l_1, l_2, \cdots, l_{(k+1)/2}$ denote the lengths of the paths in Γ. Clearly each $l_i \leq D$. Assign a *capacity* to each edge $e \in T$, equal to $n_e(t_{(k+1)/2}) + 2\sum_{j=1}^{(k-1)/2} n_e(t_j)$, where $n_e(t_j)$ is the number of times e is traversed in path t_j. Note that the total weighted capacity over all edges is exactly $2\sum_{j=1}^{(k-1)/2} l_j + l_{(k+1)/2} \leq kD$. As observed before, every r-path on T has a unique path from r to some leaf which is traversed only once, and all other edges on the r-path are traversed 2 times each (see Figure 1b). Let P denote this r to leaf path in the r-path $t_{(k+1)/2}$. Note that the capacity of every edge $e \in T \setminus P$ is at least twice the *number* of paths of Γ containing e.

We will now *charge* each edge an amount at most its capacity, and show that the total charge over all edges is larger than kD, which would be a contradiction. For cluster C_i, charge an amount equal to the path from r to C_i: each edge on this path is charged one unit against the capacity on that edge attributed to r-path $\pi(C_i)$. So far no edge has a charge more than its capacity - as edges of r-path $t_{(k+1)/2}$ are charged against only once, and other r-paths were doubled. Now we will show that we can further charge an additional amount corresponding to a path on the internal part of each cluster $C_1, \cdots, C_k$.

Consider an edge $e \notin P$ which is on the internal part of some cluster C_i. Let m denote the number of clusters (C_i not included) that appear *below* e in tree T. If $m = 0$, this edge has never been charged so far, and thus has at least 2 units of residual capacity. If $0 < m \leq k - 1$, by induction on the set of clusters below e, there are at least $(m+2)/2$ r-paths using e. *i.e.* e has a capacity of at least $m + 2$. But we have charged e exactly m times so far. So, again we have at least 2 units of residual capacity. For an edge $e \in P$ on the internal part of C_i, a similar argument shows that there is at least 1 unit of residual capacity. The total charge can now be written as follows:

$$\sum_{i=1}^{k} \left[d(r, C_i) + 2 \cdot d((\text{internal part of } C_i) \setminus P) + d((\text{internal part of } C_i) \cap P) \right]$$

The i-th term above corresponds to an r-path covering C_i : where the edges charged just 1 are all on the path P. Since each C_i is a heavy cluster, this is more than D. So the total charge is more than kD, the total capacity! Thus $OPT > (k+1)/2$, and the lemma is proved. ∎

We note that the lower bound in Lemma 1 does not hold in general metrics. In fact, even if we require the distance between the heavy clusters $C_1, \cdots, C_k$ to be 'large', there are instances in which $\cup_{i=1}^k C_i$ can be covered using $\frac{k}{\log D}$ r-paths.

2.1   An LP Relaxation

We consider the following integer programming formulation for the minimum vehicle routing problem, which is valid even for general metrics. For every r-path T, having length at most D, there is a binary variable x_T. The constraints require that every vertex be covered by at least one such path. The LP relaxation is obtained by dropping the integrality on the variables and is as follows.

$$\min \sum_T x_T$$
$$s.t.$$
$$(\mathcal{LP}) \sum_{T:v \in T} x_T \geq 1 \; \forall v \in V \setminus r$$
$$x_T \geq 0 \qquad \forall T : r\text{-path of length at most } D$$

Although this LP has an exponential number of variables, it can be approximately solved in polynomial time using the framework of Carr & Vempala [7]. The dual separation problem is orienteering, for which there is a 3-approximation algorithm [3]. This implies that we can solve $\mathcal{LP}$ within a factor of 3 in polynomial time, via the ellipsoid method. In this section, we show that the integrality gap of $\mathcal{LP}$ on tree metrics is at most a constant.

We may assume, without loss of generality that the tree is binary. Recall the definition of a heavy cluster from Theorem 1. Then, similar to Lemma 1, we have the following lemma.

Lemma 2. *If $C_1, \cdots, C_k$ are k disjoint heavy clusters in the tree, the optimal value of $\mathcal{LP}$ is at least $\frac{k}{32}$.*

Proof: The dual of $\mathcal{LP}$ is the following.

$$\max \sum_{v \in V \setminus r} p_v$$
$$s.t.$$
$$\sum_{v \in T} p_v \leq 1 \quad \forall T : r\text{-path of length at most } D$$
$$p_v \geq 0 \qquad \forall v \in V \setminus r$$

We will construct an appropriate dual solution which has value at least $\frac{k}{32}$. Then the lemma would follow by weak duality. The proof also uses the following claim, which we state without a proof.

Claim 1. *For any edge weighted tree H with root s and a weight function w, it is possible to distribute a total profit of 1 among the leaves of H such that the profit contained in any rooted subtree F of H is at most $\frac{w(F)}{w(H)}$.*

First, we preprocess the set of clusters. The internal part of a cluster C_i is the subtree that C_i induces (see Figure 1a). The internal part of C_i is divided into two parts: the edges that lie on the r-path to some other cluster constitute the *through part* of C_i (length denoted by t_i); all other edges constitute the *local part* of C_i (length denoted by l_i). A leaf cluster is one that has no cluster below it in the tree. Note that leaf clusters have zero through length. It is clear that the number of clusters with a branching in their through part is at most the number of leaf clusters. Thus the number of clusters with *no* branching in their through part is $m \geq k/2$. In the rest of the proof we restrict our attention to only these m clusters. For a cluster C_i with no branching in its through part, one r-path that covers it is as follows: take the path from r to C_i, the through part, and twice the local part. This r-path has length $d(r, C_i) + t_i + 2l_i$ which is more than D, as C_i is a heavy cluster. We divide these m clusters into two sets : $\mathcal{A}$ consisting of clusters with $l_i \geq t_i/2$, and $\mathcal{B}$ consisting of clusters with $l_i < t_i/2$. We consider the following two cases.

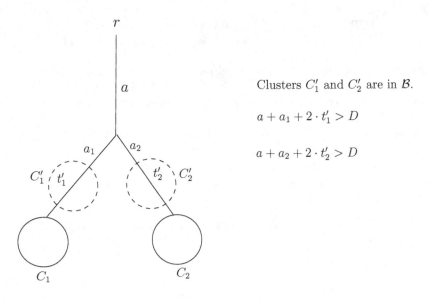

Clusters C'_1 and C'_2 are in $\mathcal{B}$.

$$a + a_1 + 2 \cdot t'_1 > D$$

$$a + a_2 + 2 \cdot t'_2 > D$$

Fig. 2. r-path Π in case 2

Case 1: $|\mathcal{A}| \geq m/4$. In this case, we only consider clusters in $\mathcal{A}$. The dual solution is as follows: for each such cluster $C_i \in \mathcal{A}$, we shrink the through part to a root and distribute a total profit of $1/4$ among the vertices in its local part using Claim 1. Let Π be any r-path with profit more than 1 w.r.t. this dual solution. Let α_i denote the fraction of the local part of C_i in Π. From Claim 1, we have $\frac{1}{4}\sum \alpha_i > 1$. Let C_m denote the cluster of minimum local length visited by Π. Then we have $len(\Pi) \geq d(r, C_m) + \sum \alpha_i \cdot l_i > d(r, C_m) + 4l_m \geq d(r, C_m) + 2l_m + d_m > D$. So the profit in any r-path of length at most D is at most 1.

Case 2: $|\mathcal{B}| \geq 3m/4$. Note that the number of clusters appearing immediately below a branching in the tree is at most twice the number of leaf-clusters. But the number of leaf clusters is at most $|\mathcal{A}| \leq m/4$. Thus, ignoring clusters of $\mathcal{B}$ appearing just after a branching, leaves us with at least $m/4$ clusters. The dual solution here assigns a profit of $1/4$ to each remaining cluster, in the same manner as in case 1. Let Π be any r-path with profit more than 1 w.r.t. this dual solution. Suppose clusters $C_1, C_2 \in \mathcal{B}$ appear as leaves in Π. Since we ignore clusters of $\mathcal{B}$ just after any branching, there are clusters $C'_1, C'_2 \in \mathcal{B}$ above C_1 and C_2 (but before any branching) as in Figure 2. We have $D < d(r, C'_j) + t'_j + 2l'_j = a + a_j + t'_j + 2l'_j \leq a + a_j + 2t'_j$ for $j = 1, 2$. So $len(\Pi) \geq a + a_1 + t'_1 + a_2 + t'_2 \geq a + 2\min\{a_1 + t'_1, a_2 + t'_2\} > D$. Thus we may assume that Π has at most one cluster of $\mathcal{B}$ that is a leaf in it. Ignoring the profit from this cluster, we are left with a profit of at least $3/4$ from clusters whose through parts are contained in Π. By an argument similar to case 1, we can show that in this case also $len(\Pi) > D$.

In both cases above, we have a feasible dual solution of value $\frac{m}{16} \geq \frac{k}{32}$. ∎

Algorithm $minTVR$ (Section 2) finds k^* disjoint heavy clusters such that there is an integral solution of value at most $2k^*$. Using Lemma 2 on these k^* clusters, we obtain that the integrality gap of $\mathcal{LP}$ is $O(1)$.

3 Minimum Vehicle Routing on General Metrics

In this section, we present an approximation algorithm achieving a guarantee of $O(\log D)$ for minimum vehicle routing on general metrics. Using orienteering as a subproblem in a greedy algorithm, one can obtain an $O(\log n)$ approximation algorithm for minimum vehicle routing.[1] However, due to the large running time of the orienteering algorithm, this approach yields an algorithm with a running time of $O(n^{12})$. The algorithm that we present here is simpler and has a running time of $O(n^2 \cdot \log n \cdot \log D)$. This algorithm uses an algorithm for unrooted vehicle routing, which is obtained from Arkin et al. [1].

The basic idea of the algorithm for rooted vehicle routing is that, if an r-path visits some points a "large" distance from the root, it resembles an unrooted path (with smaller length) over just those vertices. More concretely, we divide the vertices of the graph into $\lg D$ parts, roughly according to their distance from the root, and solve an unrooted vehicle routing in each part (with appropriate path length). We state (without proof) the following theorem.

Theorem 2. *There is an $O(\log D)$-approximation algorithm for the minimum vehicle routing problem on general metrics, that runs in $O(n^2 \cdot \lg n \cdot \lg D)$ time.*

As a consequence of this Theorem, we also obtain a bi-criteria approximation algorithm for minimum vehicle routing. In particular, if we are allowed to violate the deadline D by a small factor ϵ, we can cover all the vertices using $O(\log \frac{1}{\epsilon}) \cdot OPT$ r-paths. Here OPT is the minimum number of r-paths of length at most D, required to cover all vertices. This result can be compared to the tour-splitting heuristic discussed in Li et al. [9]. The best guarantee one can obtain using tour-splitting is an $(O(\frac{1}{\epsilon}), 1 + \epsilon)$ bi-criteria approximation.

Corollary 3. *For every $0 < \epsilon < 1$, there is an $(O(\log \frac{1}{\epsilon}), 1 + \epsilon)$ bi-criteria approximation algorithm for minimum vehicle routing.*

We note that the above bicriteria approximation is also obtained independently by Khuller et al. [12].

3.1 Integrality Gap of the Linear Relaxation

We consider the LP relaxation $\mathcal{LP}$ for minimum vehicle routing introduced in Section 2.1, and show that its integrality gap is at most $O(\log D)$ in general

---

[1] Even though we use a constant factor approximation for orienteering, the greedy framework only gives an $O(\log n)$ guarantee.

metrics. We first show that a similar linear program for unrooted vehicle routing has a constant integrality gap. Then in a manner similar to the algorithm of Section 3, we obtain the result for rooted vehicle routing. We state (without proofs) the following theorems.

Theorem 4. *The LP relaxation for unrooted vehicle routing on general metrics has an integrality gap of at most 49/3.*

Corollary 5. *The LP relaxation for rooted vehicle routing on general metrics has an integrality gap of at most $O(\log \frac{D}{D-d_{max}+1})$, where d_{max} is the maximum distance of any vertex from the root.*

Note also that a trivial randomized rounding (as in set cover) shows that the integrality gap of $\mathcal{LP}$ is at most $\log n$. So $\mathcal{LP}$ has an integrality gap at most $O(\min\{\log D, \log n\})$. It will be interesting to know if the integrality gap of $\mathcal{LP}$ is bounded above by a constant, irrespective of the metric.

4   The $OPT = 1$ Promise Problem

In this section, we look at the problem of *finding* a small set of r-paths covering all the vertices, given a promise that there exists a single r-path that covers all vertices. Note that even testing whether all vertices can be covered by one path is NP-complete. So unless P=NP, it is not possible to find in polynomial time, a path that covers V, even if we know that there exists one. Here we present an algorithm that finds a cover using at most 14 r-paths.

This algorithm is based on guessing the structure of the optimal path, and approximating it. First we need a definition from Blum et al. [6]. For an s-t path P, we define the *excess* of path P to be $\epsilon(P) = d(P) - d(s,t)$, where $d(P)$ denotes the length of path P, and $d(s,t)$ is the shortest path distance between s and t. Given vertices s and t, and a target k, the minimum excess path problem is to find an s-t path of minimum possible excess that contains *at least* k vertices. Blum et al. [6] gave a $2 + \delta$ approximation algorithm for minimum excess path, for any fixed $\delta > 0$.[2]

We divide the vertex set into roughly $\lg D$ *blocks* as follows:

$$V_j = \begin{cases} \{v : D - 1 < d(r,v) \le D\} & j = 0 \\ \{v : D - 2^j < d(r,v) \le D - 2^{j-1}\} & 1 \le j \le \lfloor \lg D \rfloor \\ \{v : 0 < d(r,v) \le D - 2^{\lfloor \lg D \rfloor}\} & j = \lfloor \lg D \rfloor + 1 \end{cases}$$

Note that if an r-path visits a vertex in V_{j+2} after a vertex in V_j, its length would be more than $D - 2^j + d(V_{j+2}, V_j) > D - 2^j + 2^j = D$. So the optimal path visits vertices of V_j strictly after all vertices of V_{j+2}. Thus we can split the optimal path O into two paths O_1 and O_2 such that O_1 visits the vertices

---

[2] We note that a 2-approximation to minimum excess path would imply a 2 approximation to the $OPT = 1$ promise problem considered here. However, no such algorithm is currently known.

in even numbered blocks, and O_2 visits vertices in odd numbered blocks. So O_1 (O_2) is obtained by restricting O to vertices from even (odd) numbered blocks, and short-cutting over the other vertices. O_1 is monotone across even blocks, and O_2 is monotone across odd blocks. This suggests that we can approximate these paths using a dynamic program.

4.1 Approximating Path O_1

We know that O_1 is monotone over the even blocks $V_0, V_2, \cdots$. Let l_j denote the length of the part of O_1 in V_j (which is contiguous), and d_j the shortest path distance between the first and last vertices of V_j in O_1. Let $\epsilon_j = l_j - d_j$ denote the excess in block V_j. Also, let Δ denote the total length of edges going from one block (say V_j) to the next (V_{j-2}) in path O_1. Then the length of O_1 is $\sum_{j:even}(d_j + \epsilon_j) + \Delta \leq D$. We will denote $\sum_{j:even} \epsilon_j$ by ϵ.

In this Section, we show a weaker approximation guarantee of covering O_1 with 16 r-paths. A block V_j with $\epsilon_j > \frac{\epsilon}{6}$ is called *heavy*. Clearly there are at most 5 heavy blocks, and these can be guessed (there are at most $\log^5 D$ possibilities). Consider the non-heavy blocks: each has excess $\epsilon_j \leq \epsilon/6$. So the set of non-heavy blocks can be partitioned (in a greedy fashion) into 6 groups such that each group is contiguous and has a total excess of at most $\frac{\epsilon}{3}$. Since these groups are contiguous, they can also be guessed (there are at most $\log^5 D$ possibilities). We now describe two algorithms: one that covers heavy blocks, and the other for the groups of non-heavy blocks defined above.

Algorithm TS: Each heavy block is covered separately using a tour splitting algorithm TS, which works as follows. Suppose V_j is the heavy block to be covered.

1. Compute an approximate minimum Hamilton path H on V_j.
2. Split H into 3 pieces $\sigma_1, \sigma_2, \sigma_3$, such that each has length at most $\frac{d(H)}{3}$.
3. Output the r-paths $\{(r \cdot \sigma_i) : 1 \leq i \leq 3\}$. Here, $r \cdot \sigma_i$ is the path obtained by concatenating an edge from r to path σ_i.

To verify that each of the paths output has length at most D, note that the minimum Hamilton path on V_j has length at most 2^j. This is because every vertex in V_j is distant at least $D - 2^j$ from r, and O_1 (of length $\leq D$) when restricted to $r \cup V_j$ is a Hamilton path along with an edge from r. Christofides' heuristic [8] achieves an approximation guarantee of $3/2$ for the Hamilton path problem. Using this, $d(H) \leq \frac{3}{2} \cdot 2^j$, and $d(\sigma_i) \leq 2^{j-1}$ ($i = 1, 2, 3$). Thus $d(r \cdot \sigma_i) \leq D - 2^{j-1} + 2^{j-1} = D$.

Algorithm EXS: Every group in the partition of non-heavy blocks is covered separately, using a dynamic program EXS. Let $W_1, \cdots, W_q$ denote the blocks in the group to be covered, in increasing order of distance from r. Let ϵ'_j be the excess of block W_j ($1 \leq j \leq q$). For every block W_j and a pair of vertices $u_j, v_j \in W_j$, we use the minimum excess algorithm to compute a u_j-v_j path in W_j, covering at least $|W_j|$ vertices (*i.e.* all of W_j). Let $A[u_j, v_j, j]$ denote the length of this path. The following dynamic program finds the best way to piece such paths together to obtain a single monotone path covering $\bigcup_{j=1}^{q} W_j$.

$$PATH[v_1, 1] = \min\{d(r, u_1) + A[u_1, v_1, 1] : u_1 \in W_1\}, \qquad v_1 \in W_1$$

$$PATH[v_j, j] = \min \left\{ \begin{array}{c} PATH[v_{j-1}, j-1] + d(v_{j-1}, u_j) + A[u_j, v_j, j] : \\ v_{j-1} \in W_{j-1}, u_j \in W_j \end{array} \right\},$$
$$v_j \in W_j \quad \& \quad 2 \leq j \leq q$$

Finally we output the path corresponding to $\min\{PATH[*, q]\}$.

Suppose $u_j^*, v_j^* \in W_j$ denote the entry and exit points of O_1, for each block W_j, $1 \leq j \leq q$. This dynamic program will consider a path corresponding to these points. The length of such a path would be at most $d(r, u_1^*) + d(u_1^*, v_1^*) + \sum_{j=2}^{q}(d(v_{j-1}^*, u_j^*) + d(u_j^*, v_j^*)) + 3\sum_{j=1}^{q} \epsilon_j'$. This follows since there is a u_j^*-v_j^* path (namely $O_1 \cap W_j$) covering W_j of excess ϵ_j', and the minimum excess algorithm is a 3-approximation. But from the way a group is constructed, $\sum_{j=1}^{q} \epsilon_j' \leq \frac{\epsilon}{3}$. So the length of this path is at most the length of O_1, which is at most D.

To summarize, the overall algorithm for approximating O_1 is as follows.

1. Guess the heavy blocks $b_1, \cdots, b_5$ (repetitions allowed).
2. Guess the partition of non-heavy blocks into groups $G_1, \cdots, G_6$.
3. For each block b_l ($l = 1, \cdots, 5$), use algorithm TS to cover it.
4. For each group G_t ($1 \leq t \leq 6$), use algorithm EXS to cover it.
5. Over all the guesses, return the solution of minimum size which is feasible.

We now argue the performance guarantee of this algorithm. Suppose there are $0 \leq h \leq 5$ heavy blocks. Then the total excess of the remaining blocks is at most $\epsilon - \frac{h}{6}\epsilon$. So the number of groups of non-heavy blocks will be at most $6 - h$. Now, the total number of paths used by this algorithm would be at most $6 - h + 3h \leq 16$. With some more work, we can approximate O_1 using at most 7 r-paths (see [14]). Since O_2 can also be approximated by the same algorithm, we obtain the following.

Theorem 6. *Given an instance of minimum vehicle routing on general metrics, having an optimal solution that uses just one vehicle, there is a polynomial time algorithm that obtains a solution using at most 14 vehicles.*

Acknowledgements

We thank Shuchi Chawla for helpful discussions.

References

1. Esther M. Arkin, Refael Hassin, and Asaf Levin. Approximations for Minimum and Min-max Vehicle Routing Problems. *Journal of Algorithms*, 2005.
2. Esther M. Arkin, Joseph S. B. Mitchell, and Giri Narasimhan. Resource-constrained Geometric Network Optimization. *SCG '98: Proceedings of the Fourteenth Annual Symposium on Computational Geometry*, pages 307–316, 1998.

3. Nikhil Bansal, Avrim Blum, Shuchi Chawla, and Adam Meyerson. Approximation Algorithms for Deadline-TSP and Vehicle Routing with Time Windows. *Proceedings of the Thirty-sixth Annual ACM Symposium on Theory of Computing*, pages 166–174, 2004.
4. Reuven Bar-Yehuda, Guy Even, and Shimon (Moni) Shahar. On Approximating a Geometric Prize-Collecting Traveling Salesman Problem with Time Windows. *Proc. of ESA*, pages 55–66, 2003.
5. C. Bazgan, R. Hassin, and J. Monnot. Approximation Algorithms for Some Vehicle Routing Problems. *Discrete Applied Mathematics*, 146:27–42, 2005.
6. Avrim Blum, Shuchi Chawla, David R. Karger, Terran Lane, Adam Meyerson, and Maria Minkoff. Approximation Algorithms for Orienteering and Discounted-Reward TSP. *Proceedings of the 44th Annual IEEE Symposium on Foundations of Computer Science*, pages 46–55, 2003.
7. B. Carr and S. Vempala. Randomized meta-rounding. *32nd ACM Symposium on the Theory of Computing*, pages 58–62, 2000.
8. N. Christofides. Worst-case analysis of a new heuristic for the travelling salesman problem. *GSIA, CMU-Report 388*, 1977.
9. C.L.Li, D. Simchi-Levi, and M. Desrochers. On the distance constrained vehicle routing problem. *Operations Research*, 40:790–799, 1992.
10. M. Desrochers, J.Desrosiers, and M. Solomon. A New Optimization Algorithm for the Vehicle Routing Problem with Time Windows. *Operation Research*, 40:342–354, 1992.
11. M. Kantor and M. Rosenwein. The Orienteering Problem with Time Windows. *Journal of the Operational Research Society*, 43:629–635, 1992.
12. Samir Khuller, Azarakhsh Malekian, and Julian Mestre. To Fill or not to Fill: The Gas Station Problem. *Manuscript*, 2006.
13. A. Kohen, A. R. Kan, and H. Trienekens. Vehicle Routing with Time Windows. *Operations Research*, 36:266–273, 1987.
14. Viswanath Nagarajan and R. Ravi. Minimum Vehicle Routing with a Common Deadline. *https://server1.tepper.cmu.edu/gsiadoc/WP/2006-E53.pdf*, 2006.
15. M. Savelsbergh. Local Search for Routing Problems with Time Windows. *Annals of Operations Research*, 4:285–305, 1985.
16. K. C. Tan, L. H. Lee, K. Q. Zhu, and K. Ou. Heuristic Methods for Vehicle Routing Problems with Time Windows. *Artificial Intelligence in Engineering*, pages 281–295, 2001.

Stochastic Combinatorial Optimization with Controllable Risk Aversion Level*

(Extended Abstract)

Anthony Man–Cho So[1], Jiawei Zhang[2], and Yinyu Ye[3]

[1] Department of Computer Science, Stanford University, Stanford, CA 94305, USA
manchoso@cs.stanford.edu
[2] Department of Information, Operations, and Management Sciences, Stern School of Business, New York University, New York, NY 10012, USA
jzhang@stern.nyu.edu
[3] Department of Management Science and Engineering and, by courtesy, Electrical Engineering, Stanford University, Stanford, CA 94305, USA
yinyu-ye@stanford.edu

Abstract. Due to their wide applicability and versatile modeling power, stochastic programming problems have received a lot of attention in many communities. In particular, there has been substantial recent interest in 2–stage stochastic combinatorial optimization problems. Two objectives have been considered in recent work: one sought to minimize the expected cost, and the other sought to minimize the worst–case cost. These two objectives represent two extremes in handling risk — the first trusts the average, and the second is obsessed with the worst case. In this paper, we interpolate between these two extremes by introducing an one–parameter family of functionals. These functionals arise naturally from a change of the underlying probability measure and incorporate an intuitive notion of risk. Although such a family has been used in the mathematical finance [11] and stochastic programming [13] literature before, its use in the context of approximation algorithms seems new. We show that under standard assumptions, our risk–adjusted objective can be efficiently treated by the Sample Average Approximation (SAA) method [9]. In particular, our result generalizes a recent sampling theorem by Charikar et al. [2], and it shows that it is possible to incorporate some degree of robustness even when the underlying probability distribution can only be accessed in a black–box fashion. We also show that when combined with known techniques (e.g. [4, 14]), our result yields new approximation algorithms for many 2–stage stochastic combinatorial optimization problems under the risk–adjusted setting.

1 Introduction

A fundamental challenge that faces all decision–makers is the need to cope with an uncertain environment while trying to achieve some predetermined objectives.

---

* This research is supported in part by The Boeing Company.

J. Diaz et al. (Eds.): APPROX and RANDOM 2006, LNCS 4110, pp. 224–235, 2006.

One certainly does not need to go far to encounter such situations — for example, an office clerk trying to get to work as fast as possible while avoiding possibly congested roads; a customer in the supermarket trying to checkout while avoiding lines that may take a long time, and so on. From a decision–maker's perspective, it is then natural to ask whether one can determine the optimal decision given one's assessment of the uncertain environment. This is a motivating question in the field of stochastic optimization. To keep our discussion focused, we shall consider the class of 2–stage stochastic programs with recourse [1, 3], particularly in the context of combinatorial optimization problems. Roughly speaking, in the 2–stage recourse model, one commits to some initial (i.e. first stage) action x based on one's knowledge of the underlying probability distribution. The actions in the second stage cannot be determined in advance, since they depend on the actions of the first stage as well as the uncertain parameters of the problem. However, once those parameters are realized (according to the distribution), a recourse (i.e. second stage) action r can be taken so that, together with the first stage actions, all the requirements of the problem are satisfied. Naturally, one would seek for the action (x, r) that minimizes the "total cost". However, since the outcome is random, such an objective can have many possible interpretations. In this paper, we shall consider the problem of risk minimization. Specifically, let X be the set of permissible actions, and let $(\Omega, \mathscr{B}, \mathbb{P})$ be the underlying probability space. In accordance with the convention in the literature, we shall assume that the probability distribution is specified via one of the following models:

(a) **Scenario Model:** The set of scenarios $\mathcal{S}$ and their associated probabilities are explicitly given. Hence, under this model, an algorithm is allowed to take time polynomial in $|\mathcal{S}|$.

(b) **Black–Box Model:** The distribution of the scenarios is given as a black box. An algorithm can use this black box to draw independent samples from the distribution of scenarios.

We are interested in solving problems of the form:

$$\min_{x \in X} \{g(x) \equiv c(x) + \Phi(q(x, \omega))\} \tag{1}$$

where $c : X \to \mathbb{R}_+$ is a (deterministic) cost function, $q : X \times \Omega \to \mathbb{R}_+$ is another cost function that depends both on the decision $x \in X$ and some uncertain parameter $\omega \in \Omega$, and $\Phi : L^2(\Omega, \mathscr{B}, \mathbb{P}) \to \mathbb{R}$ is some *risk measure*. We shall refer to Problem (1) as a *risk–adjusted 2–stage stochastic program with recourse*. Two typical examples of Φ are the *expectation* operator and the *max* operator. The former gives rise to a risk–neutral objective, while the latter gives rise to an extremely risk–averse objective. Both of these risk measures have been studied in recent works on approximation algorithms for stochastic combinatorial optimization problems (see, e.g., [2, 4, 5, 6, 7, 8, 12, 14]). For the case where Φ is the expectation operator, it turns out that under the black–box model, one can obtain a near–optimal solution to Problem (1) with high probability by the so–called Sample Average Approximation (SAA) method [9]. Roughly speaking, the

SAA method works as follows. Let $\omega^1, \ldots, \omega^N$ be N i.i.d. samples drawn from the underlying distribution, and consider the sampled problem:

$$\min_{x \in X} \frac{1}{N} \sum_{i=1}^{N} \left(c(x) + q(x, \omega^i) \right) \tag{2}$$

Under some mild assumptions, it has been shown [9] that the optimal value of (2) is a good approximation to that of (1) with high probability, and that the number of samples N can be bounded. Unfortunately, the bound on N depends on the maximum variance V (over all $x \in X$) of the random variables $q(x, \omega)$, which need not be polynomially bounded. However, in a recent breakthrough, Shmoys and Swamy [14] have been able to circumvent this problem for a large class of 2–stage stochastic linear programs. Specifically, by bounding the relative factor by which the second stage actions are more expensive than the first stage actions by a parameter λ (called the *inflation factor*), they are able to show that an adaptation of the ellipsoid method will yield an $(1 + \epsilon)$–approximation with the number of samples (i.e. black–box accesses) bounded by a polynomial of the input size, λ and $1/\epsilon$. Subsequently, Charikar et al. [2] have established a similar but more general result using the SAA method. We should mention, however, that both of these results assume that the objective function is linear. Thus, in general, they do not apply to Problem (1).

On another front, motivated by robustness concerns, Dhamdhere et al. [4] have recently considered the case where Φ is the max operator and developed approximation algorithms for various 2–stage stochastic combinatorial optimization problems with recourse under that setting. Their framework works under the scenario model. In fact, since only the worst case matters, it is not even necessary to specify any probabilities in their framework. However, such a model can be too pessimistic. Also, as the worst–case scenario may occur only with an exponentially small probability, it seems unlikely that sampling techniques would apply to such problems.

From the above discussion, a natural question arises whether we can incorporate a certain degree of robustness (possibly with some other risk measures Φ) in the problem while still being able to solve it in polynomial time under the black–box model. If so, can we also develop approximation algorithms for some well–studied combinatorial optimization problems under the new robust setting?

Our Contribution. In this paper, we answer both of the above questions in the affirmative. Using techniques from the mathematical finance literature [11], we provide a unified framework for treating the aforementioned risk–adjusted stochastic optimization problems. Specifically, we use an one–parameter family of functionals $\{\varphi_\alpha\}_{0 \leq \alpha < 1}$ to capture the degree of risk aversion, and we consider the problem $\min_{x \in X} \{c(x) + \varphi_\alpha(q(x, \omega))\}$. As we shall see, such a family arises naturally from a change of the underlying probability measure $\mathbb{P}$ and possesses many nice properties. In particular, it includes $\Phi = \mathbb{E}$ as a special case and $\Phi = \max$ as a limiting case. Thus, our framework provides a generalization of those in previous works. Moreover, our framework works under the most general black–box model, and we show that as long as one does not insist on considering the worst–case scenario, one can use sampling techniques to obtain near–optimal solutions to the problems discussed above efficiently. Our sampling theorem and its analysis can be viewed as a generalization of those by Charikar et al. [2]. Consequently, our result extends the class of problems that can be efficiently treated by the SAA method. Finally, by combining with techniques developed in earlier works [4, 6, 12, 14], we obtain new approximation algorithms for a large class of 2–stage stochastic combinatorial optimization problems under the new robust setting.

2 Motivation: Risk Aversion as Change of Probability Measure

We begin with the setup and some notations. Let $(\Omega, \mathscr{B}, \mathbb{P})$ be a probability space, and let $L^2(\Omega, \mathscr{B}, \mathbb{P})$ be the Hilbert space of square–integrable random variables with inner product $\langle \cdot, \cdot \rangle$ given by $\langle U, V \rangle = \int_\Omega UV \, d\mathbb{P}$. We shall assume that the second stage cost function q satisfies the following: (i) $q(x, \cdot)$ is measurable w.r.t. $\mathscr{B}$ for each $x \in X$, (ii) q is continuous w.r.t. x, and (iii) $\mathbb{E}[q(x, \omega)] < \infty$ for each $x \in X$. To motivate our approach, let us investigate how the following problems capture risk:

$$\min_{x \in X} \{c(x) + \mathbb{E}[q(x, \omega)]\} \tag{3}$$

$$\min_{x \in X} \left\{c(x) + \sup_{\omega \in \Omega} q(x, \omega)\right\} \tag{4}$$

Problem (3) is a standard stochastic optimization problem, in which a first stage decision $x^* \in X$ is sought so that the sum of the first stage cost $c(x^*)$ and the expected second stage cost $\mathbb{E}[q(x^*, \omega)]$ is minimized. In particular, we do not consider any single scenario as particularly important, and hence we simply weigh them by their respective probabilities. On the other hand, Problem (4) is a pessimist's version of the problem, in which one considers the worst–case second stage cost over all scenarios. Thus, for each $x \in X$, we consider the scenario ω_x that gives the maximum second stage cost as most important, and we put a weight of 1 on ω_x and 0 on all $\omega \neq \omega_x$, regardless of what their respective probabilities are. These observations suggest the following approach for capturing risk. For each $x \in X$, let $f_x : \Omega \to \mathbb{R}_+$ be a measurable weighing function such that:

$$\int_\Omega f_x(\omega) \, d\mathbb{P}(\omega) = 1$$

Now, consider the problem:

$$\min_{x \in X} \{c(x) + \mathbb{E}[f_x(\omega)q(x, \omega)]\} \tag{5}$$

Observe that Problem (5) captures both Problems (3) and (4) as special cases. Indeed, if we set $f_x \equiv 1$, then we recover Problem (3). On the other hand, suppose that Ω is finite, with $\mathbb{P}(\omega) > 0$ for all $\omega \in \Omega$. Consider a fixed $x \in X$, and let $\omega' = \arg\max_{\omega \in \Omega} q(x, \omega)$. Then, by setting $f_x(\omega') = \frac{1}{\mathbb{P}(\omega')}$ and $f_x(\omega) = 0$ for all $\omega \neq \omega'$, we recover Problem (4).

From the above discussion, we see that one way of addressing risk is by changing the underlying probability measure $\mathbb{P}$ using a weighing function. Indeed, the new probability measure is given by:

$$\mathbb{Q}_x(\omega) \equiv f_x(\omega)\mathbb{P}(\omega) \tag{6}$$

and we may write $\mathbb{E}_{\mathbb{P}}[f_x(\omega)q(x, \omega)] = \mathbb{E}_{\mathbb{Q}_x}[q(x, \omega)]$. Alternatively, we can specify the probability measure $\mathbb{Q}_x$ directly without using weighing functions. As long as the new measure $\mathbb{Q}_x$ is absolutely continuous w.r.t. $\mathbb{P}$ for each $x \in X$ (i.e. $\mathbb{P}(\omega) = 0$ implies that $\mathbb{Q}_x(\omega) = 0$), there will be a corresponding weighing function f_x given precisely by (6). Thus, in this context, we see that f_x is simply the Radon–Nikodym derivative of $\mathbb{Q}_x$ w.r.t. $\mathbb{P}$.

Note that in the above formulation, we are allowed to choose a different weighing function f_x for each $x \in X$. Clearly, there are many possible choices for f_x. However, our goal is to choose the f_x's so that Problem (5) is computationally tractable. Towards that end, let us consider the following strategy. Let $\alpha \in [0, 1)$ be a given parameter (the *risk–aversion level*), and define:

$$\mathcal{Q} = \left\{ f \in L^2(\Omega, \mathcal{B}, \mathbb{P}) : 0 \leq f(\omega) \leq \frac{1}{1 - \alpha} \text{ for all } \omega \in \Omega, \langle f, 1 \rangle = 1 \right\} \tag{7}$$

For each $x \in X$, we take f_x to be the optimal solution to the following optimization problem:

$$f_x = \arg\max_{f \in \mathcal{Q}} \mathbb{E}_{\mathbb{P}}[f(\omega)q(x, \omega)] \tag{8}$$

Note that such an f_x always exists (i.e. the maximum is always attained), since the functional $f \mapsto \langle f, q(x, \cdot) \rangle$ is continuous, and the set $\mathcal{Q}$ is compact (in the weak*–topology) by the Banach–Alaoglu theorem (cf. p. 120 of [10]). Intuitively, the function f_x boosts the weights of those scenarios ω that have high second stage costs $q(x, \omega)$ by a factor of at most $(1 - \alpha)^{-1}$, and zeroes out the weights of those scenarios that have low second stage costs. Note also that when $\alpha = 0$, we have $f_x \equiv 1$; and as $\alpha \nearrow 1$, f_x tends to a delta function at the scenario ω that has the highest cost $q(x, \omega)$. Thus, the definition of f_x in (8) captures the intuitive notion of risk as discussed earlier. We then define φ_α by $\varphi_\alpha(q(x, \omega)) \equiv \mathbb{E}_{\mathbb{P}}[f_x(\omega)q(x, \omega)]$, where f_x is given by (8).

At this point, it may seem that we need to perform the non–trivial task of computing f_x for many $x \in X$. However, it turns out that this can be circumvented by the following representation theorem of Rockafellar and Uryasev [11]. Such a theorem forms the basis for our sampling approach.

Fact 1. *(Rockafellar and Uryasev [11]) Let $\alpha \in (0, 1)$, and for $x \in X$ and $\beta \in \mathbb{R}$, define:*

$$F_\alpha(x, \beta) = \beta + \frac{1}{1 - \alpha} \mathbb{E}_{\mathbb{P}}\left[(q(x, \omega) - \beta)^+ \right]$$

Then, $F_\alpha(x, \cdot)$ is finite and convex, with $\varphi_\alpha(q(x, \omega)) = \min_\beta F_\alpha(x, \beta)$. In particular, if q is convex w.r.t. x, then φ_α is convex w.r.t. x as well. Indeed, F_α is jointly convex in (x, β).

The power of the above representation theorem lies in the fact that it reduces the risk–adjusted stochastic optimization problem:

$$\min_{x \in X} \{c(x) + \varphi_\alpha(q(x, \omega))\} \tag{9}$$

to the well–studied problem of minimizing the expectation of a certain random function. Thus, it seems plausible that the machineries developed for solving the latter can be applied to Problem (9) as well. Moreover, when c, q are convex w.r.t. x and X is convex, Problem (9) is a convex optimization problem and hence can be solved (up to any prescribed accuracy) in polynomial time. In Section 3, we will show how the SAA method can be applied to obtain a near–optimal solution to (9).

3   Sampling Theorem for Risk–Adjusted Stochastic Optimization Problems

In this section, we show that for any fixed $\alpha \in [0, 1)$, it suffices to have only a polynomial number of samples in order for the Sample Average Approximation (SAA) method [9] to yield a near–optimal solution to Problem (9). Our result and analysis generalize those in [2]. To begin, let X be a finite set, and let us assume that the functions $c : X \to \mathbb{R}$ and $q : X \times \Omega \to \mathbb{R}$ satisfy the following properties:

(a) **(Non–Negativity)** The functions c and q are non–negative for every first stage action $x \in X$ and every scenario $\omega \in \Omega$.

(b) **(Empty First Stage)** There exists a first stage action $\phi \in X$ such that $c(\phi) = 0$ and $q(x, \omega) \le q(\phi, \omega)$ for every $x \in X$ and $\omega \in \Omega$.

(c) **(Bounded Inflation Factor)** There exists an $\lambda \ge 1$ such that $q(\phi, \omega) - q(x, \omega) \le \lambda c(x)$ for every $x \in X$ and $\omega \in \Omega$.

We remark that the assumptions above are the same as those in [2] and capture those considered in recent work (see, e.g., [6, 8, 12, 14]). Now, let $g_\alpha(x) = c(x) + \varphi_\alpha(q(x, \omega))$. By Fact 1, we have $\min_{x \in X} g_\alpha(x) = \min_{(x,\beta) \in X \times \mathbb{R}} g'_\alpha(x, \beta)$, where $g'_\alpha(x, \beta) \equiv c(x) + \mathbb{E}_\mathbb{P}[q'(x, \beta, \omega)]$, and

$$q'(x, \beta, \omega) \equiv \beta + \frac{1}{1 - \alpha}(q(x, \omega) - \beta)^+$$

Let $(x^*, \beta^*) \in X \times [0, \infty)$ be an exact minimizer of g'_α, and set $Z^* = g'_\alpha(x^*, \beta^*)$. It is easy to show that $\beta^* \in [0, Z^*]$. Furthermore, we have the following observation:

Lemma 1. *Let $\alpha \in [0,1)$, and let c, q and q' be as above.*

(a) *Let $\kappa \geq 1$ be fixed. For every $x \in X$, $\omega \in \Omega$ and $\beta \in [0, \kappa Z^*]$, we have:*

$$q'(x, \beta, \omega) \leq q'(\phi, \beta, \omega) \leq \max \left\{ q'(\phi, 0, \omega), q'(\phi, \kappa Z^*, \omega) \right\}$$

(b) *For every $x \in X$, $\omega \in \Omega$ and $\beta \in [0, \infty)$, we have:*

$$q'(\phi, \beta, \omega) - q'(x, \beta, \omega) \leq \frac{\lambda c(x)}{1 - \alpha}$$

Before we proceed, let us first make a definition and state the version of the Chernoff bound that we will be using.

Definition 1. *We say that $x^* \in X$ is an exact (resp. γ–approximate) minimizer of a function f if we have $f(x^*) \leq f(x)$ (resp. $f(x^*) \leq \gamma f(x)$) for all $x \in X$.*

Lemma 2. (Chernoff Bound) *Let $V_1, \ldots, V_n$ be independent random variables with $V_i \in [0,1]$ for $i = 1, \ldots, n$. Set $V = \sum_{i=1}^n V_i$. Then, for any $\epsilon > 0$, we have $P\left(\left| V - \mathbb{E}\left[V\right] \right| > \epsilon n \right) \leq 2e^{-\epsilon^2 n}$.*

Here is our main sampling theorem.

Theorem 1. *Let $g'_\alpha(x, \beta) = c(x) + \mathbb{E}_\mathbb{P}\left[q'(x, \beta, \omega) \right]$, where c and q' satisfy the assumptions above, and $\alpha \in [0, 1)$ is the risk–aversion level. Let $\epsilon \in (0, 1/3)$ and $\delta \in (0, 1/2)$ be given. Set:*

$$\lambda_\alpha = \frac{\lambda}{1 - \alpha}; \quad \eta = \max \left\{ 1, \frac{\alpha}{1 - \alpha} \right\}$$

and define:

$$\hat{g}^N_\alpha(x, \beta) = c(x) + \beta + \frac{1}{N(1 - \alpha)} \sum_{i=1}^N \left(q(x, \omega^i) - \beta \right)^+$$

to be the SAA of g'_α, where $\omega^1, \ldots, \omega^N$ are N i.i.d. samples from the underlying distribution, and

$$N = \Theta \left(\frac{\lambda_\alpha^2}{\epsilon^4 (1 - \alpha)^2} \log \left(\frac{\eta}{\epsilon} \cdot |X| \cdot \frac{1}{\delta} \right) \right)$$

Let $\kappa \geq 1$ be fixed, and suppose that $(\bar{x}, \bar{\beta}) \in X \times [0, \kappa Z^*]$ is an exact minimizer of $\hat{g}^N_\alpha$ over the domain $X \times [0, \kappa Z^*]$. Then, with probability at least $1 - 2\delta$, the solution $(\bar{x}, \bar{\beta})$ is an $(1 + \Theta(\epsilon \kappa))$–approximate minimizer of g'_α.

Remarks:

(a) Note that $(\bar{x}, \bar{\beta})$ needs not be a global minimizer of $\hat{g}^N_\alpha$ over $X \times [0, \infty)$, since such a global minimizer may have $\beta > \kappa Z^*$. In particular, the optimal solutions to the problems:

$$\min_{(x, \beta) \in X \times [0, \infty)} \hat{g}^N_\alpha(x, \beta) \tag{10}$$

and

$$\min_{(x, \beta) \in [0, \kappa Z^*]} \hat{g}^N_\alpha(x, \beta) \tag{11}$$

could be different. From a practitioner's point of view, it may be easier to solve (10) than (11), because in many applications, it is difficult to estimate Z^* without actually solving the problem. However, it can be shown (see Theorem 2) that by repeating the sampling sufficiently many times, we can obtain a sample average approximation $\hat{g}_\alpha^N$ whose exact minimizers $(\bar{x}^*, \bar{\beta}^*)$ over $X \times [0, \infty)$ satisfy $\bar{\beta}^* \leq (1+\epsilon)Z^*$ with high probability. Thus, we can still apply the theorem even though we are solving Problem (10).

(b) Note that this theorem does not follow from a direct application of Theorem 3 of [2] for two reasons. First, the domain of our optimization problem is $X \times [0, \kappa Z^*]$, which is compact but not finite. However, this can be circumvented by using a suitably chosen grid on $[0, \kappa Z^*]$. A second, and perhaps more serious, problem is that there may not exist an $\beta_0 \in [0, \kappa Z^*]$ such that $q'(x, \beta, \omega) \leq q'(\phi, \beta_0, \omega)$ for *all* $x \in X$ and $\omega \in \Omega$. Such an assumption is crucial in the analysis in [2]. On the other hand, we have the weaker statement of Lemma 1(a), and that turns out to be sufficient for establishing our theorem.

Proof. Let (x^*, β^*) be an exact minimizer of g'_α. Then, we have $Z^* = g'_\alpha(x^*, \beta^*)$. Our proof consists of the following three steps.

Step 1: Isolate the high–cost scenarios and bound their total probability mass. We divide the scenarios into two classes. We say that a scenario ω is *high* if $q(\phi, \omega)$ exceeds some threshold M; otherwise, we say that ω is *low*. Let $p = \mathbb{P}(\omega : \omega \text{ is high})$, and define:

$$\hat{l}_\alpha^N(x, \beta) = \frac{1}{N} \sum_{i:\, \omega^i \text{ low}} q'(x, \beta, \omega^i); \quad \hat{h}_\alpha^N(x, \beta) = \frac{1}{N} \sum_{i:\, \omega^i \text{ high}} q'(x, \beta, \omega^i)$$

Then, it is clear that $\hat{g}_\alpha^N(x, \beta) = c(x) + \hat{l}_\alpha^N(x, \beta) + \hat{h}_\alpha^N(x, \beta)$. Similarly, we define:

$$l'_\alpha(x, \beta) = \mathbb{E}_\mathbb{P}\left[q'(x, \beta, \omega) \cdot \mathbf{1}_{\{\omega \text{ is low}\}}\right] = (1-p) \cdot \mathbb{E}_\mathbb{P}\left[q'(x, \beta, \omega) \,|\, \omega \text{ is low}\right]$$

$$h'_\alpha(x, \beta) = \mathbb{E}_\mathbb{P}\left[q'(x, \beta, \omega) \cdot \mathbf{1}_{\{\omega \text{ is high}\}}\right] = p \cdot \mathbb{E}_\mathbb{P}\left[q'(x, \beta, \omega) \,|\, \omega \text{ is high}\right]$$

whence $g'_\alpha(x, \beta) = c(x) + l'_\alpha(x, \beta) + h'_\alpha(x, \beta)$. Now, using the arguments of [2], one can show that $p \leq \frac{\epsilon}{\lambda_\alpha(1-\epsilon)}$. In particular, by the Chernoff bound (cf. Lemma 2), we have the following lemma:

Lemma 3. *Let N_h be the number of high scenarios in the samples $w^1, \ldots, w^N$. Then, with probability at least $1 - \delta$, we have $N_h/N \leq 2\epsilon/\lambda_\alpha$.*

Step 2: Establish the quality of the scenario partition. We claim that each of the following events occurs with probability at least $1 - \delta$:

$$A_1 = \left\{ |l'_\alpha(x, \beta) - \hat{l}_\alpha^N(x, \beta)| \leq 2\epsilon\kappa Z^* \text{ for every } (x, \beta) \in X \times [0, \kappa Z^*] \right\}$$

$$A_2 = \left\{ \hat{h}_\alpha^N(\phi, \beta) - \hat{h}_\alpha^N(x, \beta) \leq 2\epsilon c(x) \text{ for every } (x, \beta) \in X \times [0, \infty) \right\}$$

$$A_3 = \left\{ h'_\alpha(\phi, \beta) - h'_\alpha(x, \beta) \leq 2\epsilon c(x) \text{ for every } (x, \beta) \in X \times [0, \infty) \right\}$$

A crucial observation needed in the proof and in the sequel is the following:

Lemma 4. *For each $x \in X$ and $\omega \in \Omega$, the function $q'(x, \cdot, \omega)$ is η–Lipschitz (i.e. $|q'(x, \beta_1, \omega) - q'(x, \beta_2, \omega)| \leq \eta|\beta_1 - \beta_2|$), where $\eta = \max\left\{1, \frac{\alpha}{1-\alpha}\right\}$.*

Due to space limitations, we defer the proofs to the full version of the paper.

Step 3: Establish the approximation guarantee.

With probability at least $1 - 2\delta$, we may assume that all of the above events occur. Then, for any $(x, \beta) \in X \times [0, \kappa Z^*]$, we have:

$$l'_\alpha(x, \beta) \leq \hat{l}^N_\alpha(x, \beta) + 2\epsilon\kappa Z^* \qquad \text{(Event } A_1)$$
$$h'_\alpha(x, \beta) \leq h'_\alpha(\phi, \beta) \qquad \text{(Lemma 1(a))}$$
$$0 \leq \hat{h}^N_\alpha(x, \beta) + 2\epsilon c(x) - \hat{h}^N_\alpha(\phi, \beta) \qquad \text{(Event } A_2)$$

Upon summing the above inequalities, we obtain:

$$g'_\alpha(x, \beta) - \hat{g}^N_\alpha(x, \beta) \leq 2\epsilon\kappa Z^* + 2\epsilon c(x) + h'_\alpha(\phi, \beta) - \hat{h}^N_\alpha(\phi, \beta) \qquad (12)$$

By a similar maneuver, we can also obtain:

$$\hat{g}^N_\alpha(x, \beta) - g'_\alpha(x, \beta) \leq 2\epsilon\kappa Z^* + 2\epsilon c(x) + \hat{h}^N_\alpha(\phi, \beta) - h'_\alpha(\phi, \beta) \qquad (13)$$

Now, let $(\bar{x}, \bar{\beta}) \in X \times [0, \kappa Z^*]$ be an exact minimizer of $\hat{g}^N_\alpha$ over $[0, \kappa Z^*]$. Upon instantiating (x, β) by $(\bar{x}, \bar{\beta})$ in (12) and by (x^*, β^*) in (13) and summing, we have:

$$g'_\alpha(\bar{x}, \bar{\beta}) - g'_\alpha(x^*, \beta^*) + \hat{g}^N_\alpha(x^*, \beta^*) - \hat{g}^N_\alpha(\bar{x}, \bar{\beta})$$
$$\leq 4\epsilon\kappa Z^* + 2\epsilon c(\bar{x}) + 2\epsilon c(x^*) + h'_\alpha(\phi, \bar{\beta}) - h'_\alpha(\phi, \beta^*) + \hat{h}^N_\alpha(\phi, \beta^*) - \hat{h}^N_\alpha(\phi, \bar{\beta})$$

Using Lemma 4 and the fact that $p \leq \frac{\epsilon}{\lambda_\alpha(1-\epsilon)}$, we bound:

$$\left|h'_\alpha(\phi, \bar{\beta}) - h'_\alpha(\phi, \beta^*)\right| \leq p \cdot \eta|\bar{\beta} - \beta^*| \leq \frac{\epsilon\eta\kappa Z^*}{\lambda_\alpha(1-\epsilon)} \leq 2\epsilon\kappa Z^*$$

where the last inequality follows from the facts that $\alpha \in [0, 1)$, $\epsilon \in (0, 1/2)$ and $\lambda \geq 1$. Similarly, together with Lemma 3, we have:

$$\left|\hat{h}^N_\alpha(\phi, \beta^*) - \hat{h}^N_\alpha(\phi, \bar{\beta})\right| \leq \frac{N_h}{N} \cdot \eta|\beta^* - \bar{\beta}| \leq \frac{2\epsilon\eta\kappa Z^*}{\lambda_\alpha} \leq 2\epsilon\kappa Z^*$$

Since we have $\hat{g}^N_\alpha(\bar{x}, \bar{\beta}) \leq \hat{g}^N_\alpha(x^*, \beta^*)$, we conclude that:

$$(1 - 2\epsilon)g'_\alpha(\bar{x}, \bar{\beta}) \leq g'_\alpha(\bar{x}, \bar{\beta}) - 2\epsilon c(\bar{x})$$
$$\leq g'_\alpha(x^*, \beta^*) + 2\epsilon c(x^*) + 4\epsilon\kappa Z^* + 4\epsilon\kappa Z^*$$
$$\leq (1 + 10\epsilon\kappa)Z^*$$

It follows that $g'_\alpha(\bar{x}, \bar{\beta}) \leq (1 + \Theta(\epsilon\kappa))Z^*$ as desired.   □

The next theorem shows that by repeating the sampling sufficiently many times, we can obtain an SAA $\hat{g}^N_\alpha$ whose exact minimizers $(\bar{x}^*, \bar{\beta}^*)$ over $X \times [0, \infty)$ satisfy $\bar{\beta}^* \leq (1+\epsilon)Z^*$ with high probability. Due to space limitations, we defer its proof to the full version of the paper.

Theorem 2. *Let $\alpha \in [0, 1)$, $\epsilon \in (0, 1/3]$ and $\delta \in (0, 1/3)$ be given, and let*

$$k = \Theta\left(\left(1 + \frac{1}{\epsilon}\right)\log\frac{1}{\delta}\right); \quad N = \Theta\left(\frac{\lambda_\alpha^2}{\epsilon^4(1-\alpha)^2}\log\left(\frac{\eta}{\epsilon}\cdot|X|\cdot\frac{1}{\delta}\cdot k\right)\right)$$

Consider a collection $\hat{g}_\alpha^{1,N}, \ldots, \hat{g}_\alpha^{k,N}$ of independent SAAs of g'_α, where each $\hat{g}_\alpha^{i,N}$ uses N i.i.d. samples of the scenarios. For $i = 1, 2, \ldots, k$, let $(\bar{x}^i, \bar{\beta}^i)$ be an exact minimizer of $\hat{g}_\alpha^{i,N}$ over $X \times [0, \infty)$. Set $v = \arg\min_i \hat{g}_\alpha^{i,N}(\bar{x}^i, \bar{\beta}^i)$. Then, with probability at least $1 - 3\delta$, the solution $(\bar{x}^v, \bar{\beta}^v)$ satisfies $\bar{\beta}^v \leq (1 + \epsilon)Z^*$ and is an $(1 + \Theta(\epsilon))$–minimizer of g'_α.

Note that in Theorems 1 and 2, we assume that the problem of minimizing $\hat{g}_\alpha^N$ can be solved exactly. In many cases of interest, however, we can only get an approximate minimizer of $\hat{g}_\alpha^N$. The following theorem shows that we can still guarantee a near–optimal solution in this case. Again, its proof can be found in the full version of the paper.

Theorem 3. *Let $\alpha \in [0, 1)$, $\epsilon \in (0, 1/3]$ and $\delta \in (0, 1/5)$ be given. Let $k = \Theta((1 + \epsilon^{-1})\log\delta^{-1})$, $k' = \Theta((1 + \epsilon^{-1})\log(k\delta^{-1}))$, and set:*

$$N = \Theta\left(\frac{\lambda_\alpha^2}{\epsilon^4(1-\alpha)^2}\log\left(\frac{\eta}{\epsilon}\cdot|X|\cdot\frac{1}{\delta}\cdot kk'\right)\right)$$

Consider a collection $\left\{\hat{g}_\alpha^{(i,j),N}\right\}_{i=1,j=1}^{i=k,j=k'}$ of independent SAAs of g'_α, where each $\hat{g}_\alpha^{(i,j),N}$ uses N i.i.d. samples of the scenarios. Then, with probability at least $1 - 5\delta$, one can find a pair of indices (u, v) such that any γ–approximate minimizer of $\hat{g}_\alpha^{(u,v),N}$ is an $(1 + \Theta(\epsilon))\gamma$–minimizer of g'_α.

As we shall see, Theorems 2 and 3 will allow us to obtain efficient approximation algorithms for a large class of risk–adjusted stochastic combinatorial optimization problems under the black–box model. Thus, we are able to generalize the recent results of [2, 4, 14].

4 Applications

In this section, we consider two stochastic combinatorial optimization problems that are special cases of Problem (9) and develop approximation algorithms for them. Due to space limitations, we refer those readers who are interested in more applications to the full version of the paper. Our techniques rely heavily on the following easily–checked properties of φ_α: for random variables $Z_1, Z_2 \in L^2(\Omega, \mathcal{B}, \mathbb{P})$ and any $\alpha \in [0, 1)$, we have (i) (*Translation Invariance*) $\varphi_\alpha(c + Z_1) = c + \varphi_\alpha(Z_1)$ for any constant c; (ii) (*Positive Homogeneity*) $\varphi_\alpha(cZ_1) = c\varphi_\alpha(Z_1)$ for any constant $c > 0$; (iii) (*Monotonicity*) if $Z_1 \leq Z_2$ a.e., then $\varphi_\alpha(Z_1) \leq \varphi_\alpha(Z_2)$. We shall assume that the cost functions satisfy the properties in Section 3. In view of Theorems 2 and 3, we shall also assume that, for each of the problems under consideration, there is only a polynomial number of scenarios.

4.1   Covering Problems

The 2–stage stochastic set cover problem is defined as follows. We are given a universe U of elements $e_1, \ldots, e_n$ and a collection of subsets of U, say $S_1, \ldots, S_m$. There is a probability distribution over scenarios, and each scenario specifies a subset $A \subset U$ of elements to be covered by the sets $S_1, \ldots, S_m$. Each set S_i has an a priori weight w_i^I and an a posteriori weight w_i^{II}. In the first stage, one selects some of these sets, incurring a cost of w_S^I for choosing set S. Then, a scenario $A \subset U$ is drawn according to the underlying distribution, and additional sets may then be selected, thus incurring their a posteriori costs. Following [14], we formulate the 2–stage problem as follows:

$$\text{minimize} \quad \sum_S w_S^I x_S + \varphi_\alpha(q(x, A)) \quad \text{subject to } x_S \in \{0, 1\} \; \forall S$$

where $q(x, A) = \min\{\sum_S w_S^{II} r_{A,S} : r_A \in \mathcal{F}(x, A)\}$, and

$$\mathcal{F}(x, A) = \left\{ r_A : \sum_{S:e \in S} r_{A,S} \geq 1 - \sum_{S:e \in S} x_S \; \forall e \in A; \; r_{A,S} \in \{0, 1\} \; \forall S \right\}$$

By relaxing the binary constraints, we obtain a convex program that can be solved (up to any prescribed accuracy) in polynomial time. Now, using the properties of φ_α and the arguments in [14], one can show the following: if there exists a deterministic algorithm that finds an integer solution whose cost, for each scenario, is at most ρ times the cost of the solution of the relaxed problem, then one can obtain an 2ρ–approximation algorithm for the risk–adjusted stochastic covering problem. In particular, for the risk–adjusted stochastic vertex cover problem, we obtain an 4–approximation algorithm.

4.2   Facility Location Problem

In the 2–stage stochastic facility location problem, we are given a set of facilities F and a set of clients D. Each scenario $A \in \{1, 2, \ldots, N\}$ specifies a subset $D_A \subseteq D$ of clients to be served. The connection cost between client j and facility i is c_{ij}, and we assume that the c_{ij}'s satisfy the triangle inequality. Facility i has a first–stage opening cost of f_i^0 and a recourse cost of f_i^A in scenario A. The goal is to open a subset of facilities in F and assign each client to an open facility. In the full version of the paper, we show how to adapt an algorithm by Shmoys et al. [15] to obtain an 8–approximation algorithm for the risk–adjusted version of this problem.

5   Conclusion and Future Work

In this paper, we have motivated the use of a risk measure to capture robustness in stochastic combinatorial optimization problems. By generalizing the sampling theorem in [2], we have shown that the risk–adjusted objective can be efficiently treated by the SAA method. Furthermore, we have exhibited approximation algorithms for various stochastic combinatorial optimization problems under the risk–adjusted setting. Our work opens up several interesting directions for future research. For instance, it would be interesting to develop approximation algorithms for other stochastic combinatorial optimization problems under our risk–adjusted setting. Also, there are other risk measures that can be used to capture robustness (see [13]). Can theorems similar to those established in this paper be proven for those risk measures?

References

1. E. M. L. Beale, *On Minimizing a Convex Function Subject to Linear Inequalities*, J. Royal Stat. Soc., Ser. B (Methodological) 17(2):173–184, 1955.
2. M. Charikar, C. Chekuri, M. Pál, *Sampling Bounds for Stochastic Optimization*, Proc. 9th RANDOM, pp. 257–269, 2005.
3. G. B. Dantzig, *Linear Programming under Uncertainty*, Manag. Sci. 1(3/4):197–206, 1955.
4. K. Dhamdhere, V. Goyal, R. Ravi, M. Singh, *How to Pay, Come What May: Approximation Algorithms for Demand–Robust Covering Problems*, Proc. 46th FOCS, pp. 367–378, 2005.
5. S. Dye, L. Stougie, A. Tomasgard, *The Stochastic Single Resource Service–Provision Problem*, Naval Research Logistics 50(8):869–887, 2003.
6. A. Gupta, M. Pál, R. Ravi, A. Sinha, *Boosted Sampling: Approximation Algorithms for Stochastic Optimization*, Proc. 36th STOC, pp. 417–426, 2004.
7. A. Gupta, R. Ravi, A. Sinha, *An Edge in Time Saves Nine: LP Rounding Approximation Algorithms for Stochastic Network Design*, Proc. 45th FOCS, pp. 218–227, 2004.
8. N. Immorlica, D. Karger, M. Minkoff, V. Mirrokni, *On the Costs and Benefits of Procrastination: Approximation Algorithms for Stochastic Combinatorial Optimization Problems*, Proc. 15th SODA, pp. 691–700, 2004.
9. A. J. Kleywegt, A. Shapiro, T. Homem–De–Mello, *The Sample Average Approximation Method for Stochastic Discrete Optimization*, SIAM J. Opt. 12(2):479–502, 2001.
10. P. D. Lax, *Functional Analysis*, Wiley–Interscience, 2002.
11. R. T. Rockafellar, S. Uryasev, *Conditional Value–at–Risk for General Loss Distributions*, J. Banking and Finance 26:1443–1471, 2002.
12. R. Ravi, A. Sinha, *Hedging Uncertainty: Approximation Algorithms for Stochastic Optimization Problems*, Proc. 10th IPCO, pp. 101–115, 2004.
13. A. Ruszczyński, A. Shapiro, *Optimization of Risk Measures*, in *Probabilistic and Randomized Methods for Design under Uncertainty* (G. Calafiore and F. Dabbene eds.), Springer–Verlag, 2005.
14. D. B. Shmoys, C. Swamy, *Stochastic Optimization is (Almost) as Easy as Deterministic Optimization*, Proc. 45th FOCS, pp. 228–237, 2004.
15. D. B. Shmoys, É. Tardos and K. I. Aardal, *Approximation Algorithms for Facility Location Problems*, Proc. 29th STOC, pp. 265–274, 1997.

Approximating Minimum Power Covers of Intersecting Families and Directed Connectivity Problems

Zeev Nutov

The Open University of Israel, 108 Ravutski Str., Raanana 43107, Israel
nutov@openu.ac.il

Abstract. Given a (directed) graph with costs on the edges, the power of a node is the maximum cost of an edge leaving it, and the power of the graph is the sum of the powers of its nodes. Motivated by applications for wireless networks, we consider fundamental directed connectivity network design problems under the power minimization criteria: the k-outconnected and the k-connected spanning subgraph problems. For $k = 1$ these problems are at least as hard as the Set-Cover problem and thus have an $\Omega(\ln|V|)$ approximation threshold, while for arbitrary k a polylogarithmic approximation algorithm is unlikely. We give an $O(\ln|V|)$-approximation algorithm for any constant k. In fact, our results are based on a much more general $O(\ln|V|)$-approximation algorithm for the problem of finding a min-power edge-cover of an intersecting set-family; a set-family $\mathcal{F}$ on a groundset V is intersecting if $X \cap Y, X \cup Y \in \mathcal{F}$ for any intersecting $X, Y \in \mathcal{F}$, and an edge set I covers $\mathcal{F}$ if for every $X \in \mathcal{F}$ there is an edge in I entering X.

1 Introduction and Preliminaries

1.1 The Problem, Motivation, and Previous Work

Wireless networks are an important subject of study due to their extensive applications. A large research effort focused on performing network tasks while minimizing the power consumption of the radio transmitters of the network. In wired networks, one wants to find a subgraph of the minimum cost instead of the minimum power. This is the main difference between the optimization problems for wired versus wireless networks. In wireless networks, a range (power) assignment to radio transmitters determines the resulting communication network. We consider finding a power assignment to the nodes of a network such that the resulting communication network satisfies prescribed connectivity properties, and such that the total power is minimized. For motivation and applications to wireless networks (which is the same as of their min-cost variant for wired networks), see, e.g., [1, 2, 10, 16].

Let $G = (V, E)$ be a (possibly undirected) graph with cost c_e on the edges. For $v \in V$, the *power* $p(v) = p_c(v)$ *of* v in G (w.r.t. c) is the maximum cost of an edge leaving v in G (or zero, if no such edge exists). The power $p(G) = \sum_{v \in V} p(v)$

J. Diaz et al. (Eds.): APPROX and RANDOM 2006, LNCS 4110, pp. 236–247, 2006.

of G is the sum of powers of its nodes. Note that $p(G)$ differs from the ordinary cost $c(G) = \sum_{e \in E} c(e)$ of G even for unit costs; for unit costs, if G is undirected then $c(G) = |E|$ and $p(G) = |V|$. For example, if E is a perfect matching on V then $p(G) = 2c(G)$. If G is a clique then $p(G)$ is roughly $c(G)/\sqrt{|E|/2}$. For directed graphs, the ratio between the power and the cost can be equal to the maximum outdegree of a node in G, e.g., for stars with unit costs. The following statement shows that these are the extremal cases for general edge costs.

Proposition 1 ([10]). $c(G)/\sqrt{|E|/2} \le p(G) \le 2c(G)$ *for any undirected graph* $G = (V, E)$, *and if* G *is a forest then* $c(G) \le p(G) \le 2c(G)$. *For any directed graph* G *holds:* $c(G)/d_{\max}(G) \le p(G) \le c(G)$, *where* $d_{\max}(G)$ *is the maximum outdegree of a node in* G.

A simple connectivity requirement is when there should be a path from a specified node r to any other node. In this case, the min-cost variant is just the Min-Cost Directed Tree problem which is solvable in polynomial time while the Min-Power Directed Tree problem is at least as hard as the Set-Cover problem; combined with the result of [22] this implies an $\Omega(\ln n)$-approximation threshold for this problem (namely, it cannot be approximated within $C \ln n$ for some universal constant $C < 1$, unless P=NP).

An important network property is fault-tolerance. A graph is k-*outconnected from* r if it has k internally disjoint rv-paths for any $v \in V$. When the paths are required only to be edge-disjoint, the graph is k-*edge outconnected from* r. A graph is k-*connected* (resp., k-*edge-connected*) if it is k-outconnected (resp., k-edge-outconnected) from every node. We consider the following generalization of the problems from [2] (where the case $k = 1$ was considerd):

Min-Power k-Outconnected Subgraph (MPk-OS):
Instance: A graph $\mathcal{G} = (V, \mathcal{E})$ with costs on the edges, $r \in V$, and an integer k.
Objective: Find a min-power k-outconnected from r spanning subgraph G of $\mathcal{G}$.

Min-Power k-Connected Subgraph (MPk-CS):
Instance: A graph $\mathcal{G} = (V, \mathcal{E})$ with costs on the edges and an integer k.
Objective: Find a min-power k-connected spanning subgraph G of $\mathcal{G}$.

When G is required to be k-edge-outconnected or k-edge-connected, we get the *Min-Power k-Edge Outconnected Subgraph* (MPk-EOS) and the *Min-Power k-Edge Connected Subgraph* (MPk-ECS) problems, respectively (for undirected graphs they are equivalent).

Min-cost versions of these problems were studied extensively for both directed and undirected graphs, see, e.g., [4, 7, 8, 6, 14, 3, 23, 11, 17, 18], and surveys in [5, 13, 19]. For directed graphs the min-cost versions of MPk-OS and MPk-EOS are polynomially solvable, see [4] and [7], respectively; more efficient algorithms are given in [8, 6]. The min-cost k-edge connected subgraph problem admits a 2-approximation algorithm for both directed and undirected graphs [14]. For the min-cost k-(node-)connected subgraph problem the best known approximation ratios are: $O(\ln^2 k \cdot \min\{\frac{n}{n-k}, \frac{\sqrt{k}}{\ln k}\})$ for both directed and undirected graphs [17], and $O(\ln k)$ for undirected graphs with $n \ge 2k^2$ [3].

For undirected graphs, the best known approximation ratio for MPk-ECS is $O(\min\{k, \sqrt{n}\})$ [10], and for MPk-CS is $O(\ln n) + \alpha$ [16], where α is the best known approximation ratio for the min-cost case. Directed min-power connectivity problems are usually much harder to approximate, and the methods used in [10, 16] do not seem to work for the directed case. For example, for $k = 1$ undirected MPk-CS/MPk-ECS admits an easy 2-approximation algorithm by just taking the min-cost spanning tree (the 2-approximation follows from Proposition 1), while its directed variant is "Set-Cover hard".

The problems MPk-OS and MPk-CS that we study are closely related to the undirected Node Weighted Steiner Forest problem considered by Klein and Ravi [15]; one difference is that in our problems the "weight" of a node v is not fixed but depends on the chosen edges leaving v. The Klein-Ravi algorithm [15] uses the set-cover greedy approach [12]. At each step a "spider" (a subtree having at most one node of degree more than 2) is chosen that maximizes the ratio of the number of terminal pairs the spider connects over its weight. They prove that greedily adding spiders yields a $2H(n)$-approximation algorithm ($H(n)$ denotes the nth Harmonic number). The approximation ratio was improved by Guha and Khuller [9] to $1.5H(n)$ using slight generalizations of spiders. For MPk-CS with $k = 1$, [2] defined spider as a rooted subtree having at most one node of outdegree more than 2.

1.2   Our Result and Its Comparison to Previous Work

Henceforth we consider mainly directed graphs, so, unless stated otherwise, "graph" means "directed graph". Suppose that $\mathcal{G}$ has a subgraph $G_0 = (V, E_0)$ of power zero which is k_0-outconnected from r, and the goal is to augment G_0 by a min-power edge-set $F \subseteq \mathcal{E} - E_0$ so that the resulting graph $G = G_0 + F$ is k-outconnected from r. Formally:

Min-Power (k_0, k)-Outconnectivity Augmentation (MP(k_0, k)-OA):
Instance: A graph $G_0 = (V, E_0)$ which is k_0-outconnected from r, an edge set $\mathcal{I}$ on V with costs $\{c_e : e \in I\}$, and an integer $k > k_0$.
Objective: Find min-power $F \subseteq \mathcal{I}$ so that $G = G_0 + F$ is k-outconnected from r.

In a similar way we define the augmentation versions of MPk-EOS, MPk-CS and MPk-ECS, respectively:

Min-Power (k_0, k)-Edge-Outconnectivity Augmentation (MP(k_0, k)-EOA)
Min-Power (k_0, k)-Connectivity Augmentation (MP(k_0, k)-CA);
Min-Power (k_0, k)-Edge-Connectivity Augmentation (MP(k_0, k)-ECA)

In [2], approximation algorithms are given for $k_0 = 0$ and $k = 1$: a $2H(n)$-approximation for the Min-Power Directed Tree problem and a $(2H(n) + 1)$-approximation for the Min-Power Strongly Connected Subgraph problem. As was mentioned, each one of these problems generalizes the Set-Cover problem (c.f., [2]), and thus the results in [2] are essentially tight up to a constant factor. For arbitrary k_0, k we prove:

Theorem 1. *There exist approximation algorithms with approximation ratios:*

(i) $2(k - k_0)H(n) = O(k \ln n)$ *for directed* $MP(k_0, k)$-OA *and* $MP(k_0, k)$-EOA;
(ii) $(2(k - k_0)H(n) + k) = O(k \ln n)$ *for directed* $MP(k_0, k)$-ECA;
(iii) $k(2(k - k_0)H(n) + k) = O(k^2 \ln n)$ *for directed* $MP(k_0, k)$-CA.

The approximation ratios in Theorem 1 are $O(\ln n)$ for any fixed k, which is tight (up to a constant factor) if k is "small" (usually, $k \leq 3$ in practical networks), but may seem weak if k is large. However, it might be that a much better approximation algorithm does not exists: in [20] it is proved that for $k = \Theta(n)$ MPk-EOS cannot be approximated within $O(2^{\log^{1-\varepsilon} n})$ for any fixed $\varepsilon > 0$, unless $NP \subseteq DTIME(n^{\text{polylog}(n)})$. The same hardness result is valid for the "reverse" problem of MPk-EOS when there should be k edge-disjoint vr-paths for every $v \in V$; however, unlike MPk-EOS, this problem admits a k-approximation algorithm [21], and, in particular, is in P for $k = 1$. In contrast, for *undirected* MPk-OS [21] gives an $O(\ln n)$-approximation algorithm for any k.

In fact, Theorem 1 is just a summary of (some) applications of a much more general approximation algorithm for finding a min-power edge-cover of an intersecting family. A family $\mathcal{F}$ of subsets of a groundset V is an *intersecting family* if $X \cap Y, X \cup Y \in \mathcal{F}$ for any intersecting $X, Y \in \mathcal{F}$. An edge set I covers $\mathcal{F}$ if for every $X \in \mathcal{F}$ there is an edge in I entering X, that is, there is $uv \in I$ with $u \in V - X$ and $v \in X$. We give an $O(\ln n)$-approximation algorithm for the problem of finding a min-power cover of an intersecting family $\mathcal{F}$, but its polynomial implementation (in case $\mathcal{F}$ is not given explicitly) requires that certain queries related to $\mathcal{F}$ can be answered in polynomial time. Given an edge set I on V, the *residual family* $\mathcal{F}_I$ of $\mathcal{F}$ (w.r.t. I) consists of all members of $\mathcal{F}$ that are uncovered by edges of I. It is well known that if $\mathcal{F}$ is intersecting so is $\mathcal{F}_I$ for any I. A set $C \in \mathcal{F}$ is an $\mathcal{F}$-*core*, or simply a *core* if $\mathcal{F}$ is understood, if C does not contain two disjoint members of $\mathcal{F}$. Clearly, the maximal $\mathcal{F}$-cores are pairwise disjoint if $\mathcal{F}$ is intersecting. Given a maximal core C let $\mathcal{F}(C) = \{X \in \mathcal{F} : X \subseteq C\}$. For any edge set I on V, make the following two assumptions:

Assumption 1:
The maximal $\mathcal{F}_I$-cores can be computed in polynomial time.
Assumption 2:
For any maximal $\mathcal{F}_I$-core C, a min-cost $\mathcal{F}_I(C)$-cover can be computed in polynomial time.

Theorem 2. *The problem of finding a min-power edge-cover of an intersecting family on on n elements admits a $2H(n)$-approximation algorithm under Assumptions 1 and 2.*

A set function f defined on subsets of a groundset V is *intersecting supermodular* if $f(X) + f(Y) \leq f(X \cap Y) + f(X \cup Y)$ for any intersecting $X, Y \subset V$. An edge set I covers f if in the graph (V, I) the indegree of every $X \subset V$ is at least $f(X)$. A $\{0, 1\}$-valued set function is intersecting supermodular if, and only if, its support is an intersecting family. A natural question is whether Theorem 2 extends to intersecting supermodular set functions. As MPk-EOS is a particular case of the problem of finding a min-power cover of an intersecting supermodular set function, such an extension is unlikely due to the hardness result of [20].

To prove Theorem 2 we combine "set-cover approximation techniques" of [15] used in [2] for $k = 1$ that are based on "density" considerations (c.f., [12]) with the techniques used for min-cost connectivity problems. However, unlike [15, 2], we cannot use specific graph properties. To prove that we can find an edge set of appropriate density, we use the method of "uncrossing" sets (c.f., [23]). We define an analogue of spiders which we call "star-covers": unlike [15, 2] a star cover is not necesarilly a tree. Showing that any inclusion minimal $\mathcal{F}$-cover can be decomposed into such star-covers is harder than showing a decomposition of a tree into spiders; we do not know whether such a decomposition exists for set families related to the undirected Node Weighted Steiner Network problem – a generalization of the Node Weighted Steiner Forest problem.

Theorems 2 and 1 are proved in Sections 2 and 3, respectively.

1.3   Notation

Let $G = (V, E)$ be a directed graph. For disjoint $X, Y \subseteq V$ let $\delta_G(X, Y) = \delta_E(X, Y)$ be the set of edges from X to Y in E. For brevity, $\delta_E(X) = \delta_E(X, V - X)$ is the set of edges in E leaving X, $d_E(X) = |\delta_E(X)|$, $\delta_E^+(X) = \delta_E(V - X, X)$ the set of edges in E entering X, and $d_E^+(X) = |\delta_E^+(V - X)|$ is the indegree of X. Thus given edge costs $\{c(e) : e \in E\}$, the power of a node v in G (with respect to c) is $p(v) = \max_{e \in \delta_E(v)} c(e)$, and the power of G is $p(G) = p_E(V) = \sum_{v \in V} p(v)$. Throughout the paper, let $\mathcal{G} = (V, \mathcal{E})$ denote the input graph with nonnegative costs on the edges. Let $n = |V|$ and $m = |\mathcal{E}|$. Given $\mathcal{G}$, our goal is to find a minimum power spanning subgraph $G = (V, E)$ of $\mathcal{G}$ that satisfies some prescribed property. We assume that a feasible solution exists; otherwise our algorithms can be easily modified to return an error message. Let opt denote the optimal solution value of an instance at hand.

2   Proof of Theorem 2

We use a result about the performance of an *Approximate Greedy Algorithm* for a certain type of covering problems, defined as follows:

Covering Problem

Instance: A groundset $\mathcal{I}$ and functions ν, p on $2^{\mathcal{I}}$ given by an evaluation oracle.
Objective: Find $I \subseteq \mathcal{I}$ with $\nu(I) = \nu(\mathcal{I})$ and with $p(I)$ minimized.

We call ν the *deficiency function* (it is assumed to be decreasing and measures how far I from being a feasible solution) and p the *payment function* (assumed to be increasing). In our case p is just the power function. Let $\rho > 1$ and let opt be the optimal solution value for the Covering Problem. The ρ-*Approximate Greedy Algorithm* starts with $I = \emptyset$ and iteratively adds subsets of $\mathcal{I} - I$ to I one after the other using the following rule. As long as $\nu(I) > \nu(\mathcal{I})$ it adds to I a set $F \subseteq \mathcal{I} - I$ so that

$$\sigma_I(F) = \frac{\nu(I) - \nu(I + F)}{p(F)} \geq \frac{\nu(I) - \nu(\mathcal{I})}{\rho \cdot \text{opt}}. \tag{1}$$

The following known statement is proved using the same methods as in [12] where the Set-Cover problem was considered.

Theorem 3. *For any covering problem so that the payment function p satisfies*

$$p(I_1 \cup I_2) \leq p(I_1) + p(I_2) \quad \forall I_1, I_2 \subseteq \mathcal{I} \tag{2}$$

and ν is monotone decreasing, the ρ-Approximate Greedy Algorithm computes a solution I with $p(I) \leq \rho H(\nu(\emptyset) - \nu(\mathcal{I})) \cdot$ opt, where $H(n)$ denotes the nth Harmonic number.

In the rest of this section we prove the following Lemma:

Lemma 1. *Let $\nu(I)$ be the number of minimal cores in $\mathcal{F}_I$. Then an edge set F satisfying (1) with $\rho = 2$ can be found in polynomial time under Assumptions 1 and 2.*

For simplicity of exposition, let us revise our notation and use $\mathcal{F}$ instead of $\mathcal{F}_I$, and let $\nu = \nu(\emptyset)$. We assume that $\mathcal{I}$ is a feasible solution, thus $\nu(\mathcal{I}) = 0$. Then we need to show that under Assumptions 1 and 2 one can find in polynomial time an edge set F so that:

$$\sigma(F) = \frac{\nu - \nu(F)}{p(F)} \geq \frac{\nu}{2 \cdot \text{opt}}. \tag{3}$$

Before presenting a formal proof of Lemma 1, we give a sketch. Let $\mathcal{C}$ be the set of maximal $\mathcal{F}$-cores. For $C \in \mathcal{C}$ let $E(C) = \{uv \in E : u, v \in C\}$ be the edges in E with both endpoints in C, and let $\mathcal{F}(C) = \{X \in \mathcal{F} : X \subseteq C\}$. Let E be a minimal $\mathcal{F}$-cover. We prove that (Corollary 1 and Lemma 3):
(i) $d_E(v) \leq 1$ for any $v \in C$ and $d_E^+(C) = 1$.
(ii) $E(C)$ plus the unique edge e_C in E that enters C cover $\mathcal{F}(C)$.
An edge set F is a *star-cover with root s* (an analogue of [15, 2] spiders) if for every $e \in \delta_F(s)$ there exists $C \in \mathcal{C}$ with $\delta_F(C) = \{e\}$ such that $e + F(C)$ is a minimal $\mathcal{F}(C)$-cover (Definition 1). We prove that adding a star cover F decreases the number of cores by at least $\Delta(F)$, where $\Delta(F) = d_F(s) - 1$ if $d_F(s) \geq 2$ and $\Delta(F) = 1$ if $d_F(s) = 1$ (Lemma 4). By (ii), the set E' of edges in E which head lies in some core is decomposed into star-covers $F_1, \ldots, F_t$, and adding all these star-covers decreases ν by at least $\sum_{i=1}^t \Delta(F_i) \geq \nu/2$. As $p(E') \leq$ opt, we use an averaging argument as in [15, 2] to conclude that there exists a star-cover F for which (3) holds (Lemma 5). By (i), the power of a star-cover equals the power of s plus the *cost* of its edges that are not incident to s (Corollary 2). This, together with Assumptions 1 and 2 enables us to find in polynomial time a star-cover F that maximizes $\Delta(F)/p(F)$ (Lemma 6).

A formal proof of Lemma 1 follows. We need to establish some properties of minimal $\mathcal{F}$-covers of an intersecting family $\mathcal{F}$. Let E be a minimal $\mathcal{F}$-cover. By the minimality of E, for every $e \in E$ there exists $W_e \in \mathcal{F}$ such that $\delta_E^+(W_e) = \{e\}$; we call such W_e a *witness set for e*; note that e might have several distinct witness sets.

Lemma 2. *Let $\mathcal{F}$ be an intersecting family and let E be a minimal $\mathcal{F}$-cover. Let W_e, W_f be intersecting witness sets of two distinct edges $e, f \in E$. Then $W_e \cap W_f$ is a witness for one of e, f and $W_e \cup W_f$ is a witness for the other.*

Proof. Note that there is an edge in E entering $W_e \cap W_f$ and there is an edge in E entering $W_e \cup W_f$; this is since $W_e, W_f \in \mathcal{F}$ implies that $W_e \cap W_f, W_e \cup W_f$ belong to $\mathcal{F}$ and thus each of them is covered by some edge in E. However, if for arbitrary sets X, Y an edge covers one of $X \cap Y, X \cup Y$ then it also covers one of X, Y, and if some edge covers both $X \cap Y$ and $X \cup Y$ then it must cover both X and Y. Thus no edge in $E - \{e, f\}$ can cover $W_e \cap W_f$ or $W_e \cup W_f$, so one of e, f covers $W_e \cap W_f$, and thus the other must cover $W_e \cup W_f$. ∎

Corollary 1. *Let X be a minimal core of an intersecting family $\mathcal{F}$ and let E be a minimal $\mathcal{F}$-cover. Then $d_E^+(X) = 1$.*

Proof. Clearly $d_E^+(X) \geq 1$, since E is an $\mathcal{F}$-cover and $X \in \mathcal{F}$. Assume that there are distinct $e, f \in \delta_E^+(X)$, and let W_e, W_f be their witness sets. Then $X \subseteq W_e \cap W_f$ (in particular, W_e, W_f intersect), and thus $e, f \in \delta_E^+(W_e \cap W_f)$. This contradicts Lemma 2. ∎

Lemma 3. *Let C be a maximal core of an intersecting family $\mathcal{F}$ and let E be a minimal $\mathcal{F}$-cover. Let $E(C)$ be the set of edges in E with both endpoints in C, let X be the minimal core of $\mathcal{F}_{E(C)}$ contained in C (possibly $X = C$), and let e_C be the unique edge in E that enters X. Then $E(C) + e_C$ covers $\mathcal{F}(C) = \{X \in \mathcal{F} : X \subseteq C\}$, and $d_{E(C)}(v) \leq 1$ for every $v \in C$; thus $p(E(C)) = c(E(C))$, namely, the power of $E(C)$ equals its cost.*

Proof. Let X_1 be the minimal $\mathcal{F}$-core contained in C. By Corollary 1 there is a unique edge in E entering X_1, say e_1. If e_1 covers C, then $E(C) = \emptyset$, and it is easy to see that the statement holds. Otherwise, let X_2 be the minimal $\mathcal{F}_{e_1}$-core contained in C and let e_2 be the unique edge in E entering X_2, and so on, until C is covered by some edge e_q. In such a way we obtain sequences $e_1, \ldots, e_{q-1}$ of edges in $E(C)$ together with an additional edge e_q that enters C, and $X_1 \subset X_2 \cdots \subset X_q \subseteq C$ of sets in $\mathcal{F}(C)$ so that X_{i+1} is the core of $\mathcal{F}_{E_i}$ where $E_i = \{e_1, \ldots, e_i\}$ and e_i is the unique edge in E entering X_i. The statement follows, since we must have $E_{q-1} = E(C)$, $X = X_q$, and $e_q = e_C$. In particular, $E(C) + e_C = E_q$ covers $\mathcal{F}(C)$ and no two edges in $E(C)$ share a tail. ∎

Definition 1. *An edge set F is a star-cover (with root s) if for every $e \in \delta_F(s)$ there exists $C \in \mathcal{C}$ with $\delta_F(C) = \{e\}$ such that $e + F(C)$ is a minimal $\mathcal{F}(C)$-cover.*

As the family $\mathcal{F}(C)$ is intersecting for any maximal $\mathcal{F}$-core C, by Lemma 3 we get:

Corollary 2. *Let F be a star-cover with root s. Then $\delta_F(v) \leq 1$ for any $v \neq s$, and thus $p(F) = p_F(s) + c(F - \delta_F(s))$.*

Lemma 4. *For a star-cover F with root s let $\Delta(F) = d(s) - 1$ if $d_F(s) \geq 2$ and $\Delta(F) = 1$ if $d_F(s) = 1$. Then $\nu - \nu(F) \geq \Delta(F)$.*

Proof. Let $\mathcal{F}'$ be the residual family of the sets that are uncovered by F. The minimal $\mathcal{F}$-cores not covered by F are also minimal $\mathcal{F}'$-cores, while any other minimal $\mathcal{F}'$-core X' must contain at least one $\mathcal{F}$-core covered by F. We claim that $s \in X'$ must hold for any such X', and thus: if $d(s) = 1$ no such X' exists, and if $d(s) \geq 2$ there is at most one such X', since the minimal $\mathcal{F}'$-cores are disjoint. To see that $s \in X'$, let X be a minimal $\mathcal{F}$-core contained in X' and let C be the maximal $\mathcal{F}$-core containing X. Let $Y = X \cap C$. Then $Y \in \mathcal{F}(C)$, thus there is an edge $uv \in F$ entering Y. Since uv does not cover X', we must have $u \in X' - C$. But then uv covers C, implying $u = s$.

Lemma 5. *There exists a star-cover F for which (3) holds.*

Proof. Let E be an inclusion minimal optimal $\mathcal{F}$-cover. For every maximal core C of $\mathcal{F}$ let E_C and e_C be as in Lemma 3. Let E' be the union taken over all maximal cores $C \in \mathcal{C}$ of the edge sets $F_C = E_C + e_C$. Then E' is decomposed into node disjoint star-covers $F_1, \ldots, F_t$. Now the statement follows by a simple averaging argument. Let $p_i = p(F_i)$ and let $\Delta_i = \Delta(F_i)$. We have $\sum_{i=1}^{t} p_i = p(E') \leq p(E) = \mathsf{opt}$ and $\sum_{i=1}^{t} \Delta_i \geq \nu/2$. Thus:

$$\frac{\sum_{i=1}^{t} \Delta_i}{\sum_{i=1}^{t} p_i} \geq \frac{\nu/2}{p(E')} \ .$$

From number theory we know that there must be index i so that $\Delta_i/p_i \geq \nu/(2p(E'))$. Let $F = F_i$. Then $\nu - \nu(F) \geq \Delta_i$, by Lemma 4. Consequently

$$\sigma(F) = \frac{\nu - \nu(F)}{p(F)} \geq \frac{\nu}{2 \cdot p(E')} \geq \frac{\nu}{2 \cdot p(E)} = \frac{\nu}{2 \cdot \mathsf{opt}} \ .$$

Lemma 6. *A star-cover F that maximizes $\Delta(F)/p(F)$ can be found in polynomial time under Assumptions 1 and 2.*

Proof. We first compute the maximal cores; this can be done in polynomial time by Assumption 1. Second, for every node v that belongs to a maximal core C we define the weight $w(v)$ of v to be the minimum cost of an $\mathcal{F}_e(C)$-cover, where $e = uv$ is an arbitrary edge that has head v and enters C. This can be done in polynomial time by Assumption 2. Let us say that a star F is proper if every its edge enters some maximal $\mathcal{F}$-core. Given a proper star F with root s, let $w(F) = p(s) + w(L_F)$ where L_F is the set of leaves of F. We now see that our goal is to compute a proper star F that maximizes $\max\{|L_F| - 1, 1\}/w(F)$. We may assume that we know the root s and its power $p = p_F(s)$ in F; there are $O(n^2)$ distinct choices. Delete all the edges, except that for every core $C \in \mathcal{C}$ among the edges sv with $v \in C$ and $p(sv) \leq p(s)$, if any, choose one with $w(v)$ minimal. This defines an auxiliary star T. Let $v_1, \ldots, v_q$ be the leaves of T sorted by increasing weight, so $w(v_1) \geq w(v_2) \geq \ldots \geq w(v_q)$. Let $W_i = \sum_{j=1}^{i} w_i$, and let $\sigma_1 = 1/(p + W_1)$ and $\sigma_i = (i - 1)/(p + W_i)$, $i = 1, \ldots, q$. We find the index j for which σ_j is maximum, which will determine the required star-cover.

3 Proof of Theorem 1

3.1 Part (i)

We give a $2H(n)$-approximation algorithm for $MP(\ell, \ell+1)$-OA (resp., $MP(\ell, \ell+1)$-EOA), that is for the problems of finding a min-power augmenting edge set that increases the outconnectivity (resp., edge-outconnectivity) from r by 1. We then apply this algorithm sequentially for $\ell = k_0, \ldots, k-1$, to produce edge sets $F_{k_0}, \ldots, F_{k-1}$ so that $G_0 + (F_{k_0} + \cdots + F_\ell)$ is $(\ell+1)$-outconnected (resp., $(\ell+1)$-edge-outconnected) from r, and $p(F_\ell) \leq 2H(n) \cdot$ opt. Consequently, $F = F_{k_0} + \cdots + F_{k-1}$ is k-outconnected from r, and

$$p(F) \leq \sum_{\ell=k_0}^{k-1} p(F_\ell) \leq \sum_{\ell=k_0}^{k-1} 2H(n) \cdot \text{opt} = 2(k-k_0)H(n) \cdot \text{opt}.$$

A graph $G = (V, E)$ is ℓ-*edge outconnected from* r *to* T if there are ℓ edge-disjoint rt-paths for every $t \in T$. Using Theorem 2, we give a $2H(n)$-approximation algorithm for the following augmentation problem, that generalizes both $MP(k_0, k_0 + 1)$-OA and $MP(k_0, k_0 + 1)$-EOA.

Instance: A graph $G_0 = (V, E_0)$ which is k_0-outconnected from r to T and an edge set $\mathcal{I}$ on V with costs $\{c_e : e \in \mathcal{I}\}$ so that every edge in $\mathcal{I}$ has its head in T.

Objective: Find a min-power edge-set $I \subseteq \mathcal{I}$ so that $G = G_0 + I$ is (k_0+1)-edge-outconnected from r to T.

$MP(k_0, k_0+1)$-EOA is a special case of this problem when $T = V$. For $MP(k_0, k_0+1)$-OA apply the following approximation ratio preserving reduction. Given an instance $G_0 = (V, E_0), k_0, r, \mathcal{I}$ of $MP(k_0, k_0+1)$-OA obtain an instance $G_0' = (V', E_0'), T', k_0, r, \mathcal{I}', c'$ of the above problem as follows. Replace every node $v \in V$ by the two nodes v_t, v_h connected by the edge $v_t v_h$ of cost zero, and replace every edge $uv \in E_0 \cup \mathcal{I}$ by the edge $u_h v_t$ having the same cost as uv (which is zero if $uv \in E_0$). Let $r' = r_h$, $T' = \{v_t : v \in V\}$, and

$$E_0' = \{u_h v_t : uv \in E_0\} + \{v_t v_h : v \in V\}, \quad \mathcal{I}' = \{u_h v_t : uv \in \mathcal{I}\}.$$

This establishes a bijective correspondence between edges in $\mathcal{I}$ and the edges in $\mathcal{I}'$. It is not hard to verify (see [6] for details) that $G_0' = (V', E_0')$ is k_0-edge-connected from r' to T'. Furthermore, if $I' \subseteq \mathcal{I}'$ corresponds to $I \subseteq \mathcal{I}$ then:
(i) I is a feasible solution if, and only if, I' is a feasible solution.
(ii) $d_I(v) = d_{I'}(v_h)$ and $d_{I'}(v_t) = 0$ for every $v \in V$; thus $p(I) = p(I')$.

We now show that above problem can be reduced to the min-power intersecting family cover problem, so that Assumptions 1 and 2 are valid. We say that $X \subseteq V - s$ is *tight* in G_0 if $X \cap T \neq \emptyset$ and $d^+(X) = k_0$. From Menger's Theorem we have:

Fact 4. *Let* $G_0 = (V, E_0)$ *be* k_0-*edge-outconnected from* r *to* T. *Then* $G = G_0 + I$ *is* $(k_0 + 1)$-*edge-outconnected from* r *to* T *if, and only if,* I *covers all the tight sets.*

We now see that the augmentation problem is equivalent to the problem of finding a cover of the family of tight sets. However, since only edges with head in T can be added, this is equivalent to covering the family:

$$\mathcal{F} = \{X \cap T : X \text{ is tight in } G_0\} . \tag{4}$$

It is well known (c.f. [6]) that:

Fact 5. *The family $\mathcal{F}$ defined in (4) is intersecting.*

It remains to show that Assumptions 1 and 2 are valid for $\mathcal{F}$ defined by (4). For Assumption 1 we need to show that given a graph, the maximal $\mathcal{F}$-cores can be found in polynomial time (if some edges were added at previous steps, we consider the graph after these edges were added). We first show how to find the minimal $\mathcal{F}$-cores. Then, for Assumption 1, we will show that finding maximal $\mathcal{F}$-cores can be done using n max-flow computations; for Assumption 2 we will show that finding a min-cost $\mathcal{F}(C)$-cover for a given maximal core C can be done using one min-cost $(k_0 + 1)$-flow computation.

The minimal cores can be found using $|T|$ max-flow computations as follows. For every $t \in T$, compute a maximum rt-flow. If its value is k_0, then in the corresponding residual network the set of nodes $\{v \in T : t \text{ is reachable from } v\}$ is the minimal core containing t; otherwise, no minimal core containing t exists. After the minimal $\mathcal{F}$-cores are found, to find the maximal cores, for every minimal core X do the following. Add an edge from r to every minimal core distinct from X. Then choose $t \in X$ and compute a maximum rt-flow; in the corresponding residual network the set of nodes $T \cap \{v \in T : v \text{ is reachable from } r\}$ is the maximal core containing X. Now we show how to find a min-cost $\mathcal{F}(C)$-cover for a maximal core C that contains a minimal core X. The construction is similar to the previous one: construct a network $H = G_0 + \mathcal{I}$, assigning zero costs to edges in E_0. Then add an edge from r to every minimal core distinct from X, and compute a min-cost $(k_0 + 1)$-flow f from r to some $t \in X$. The edge set $\{e \in \mathcal{I} : f(e) = 1\}$ is the desired $\mathcal{F}(C)$-cover.

3.2   Part (ii)

Let us say that a graph is k-*inconnected to r* (resp., k-*edge-inconnected to r*) if its reverse graph is k-outconnected from r (resp., k-edge-outconnected from r). The algorithm for $\mathsf{MP}(k_0, k)$-ECA is as follows. Let r be an arbitrary node of G.

1. Using the algorithm as in part (i) of Theorem 1 compute an edge set F' so that $G_0 + F'$ is k-edge-outconnected from r.
2. Compute a min-cost augmenting edge set F'' so that $G_0 + F''$ is k-edge-inconnected to r.

Let $F = F' + F''$. Note that $G = G_0 + F$ is both k-edge-outconnected from r and k-edge-inconnected to r. This implies that G is k-edge connected, so F is a feasible solution. To bound its power, let OPT be an optimal solution. Since $G_0 + OPT$ is k-edge-outconnected from r we have $p(F') \leq 2H(k - k_0)p(OPT)$.

A graph G is *minimally k-inconnected to r* if it is k-inconnected to r but $G - e$ is not k-inconnected to r for any edge e of G. In a similar way a *minimally k-edge-inconnected to r* graph is defined. To bound the power of F'' we need the following known statement, c.f., [4, 7, 5].

Fact 6. *Let G be minimally k-inconnected or minimally k-edge-inconnected to r. Then every node of G distinct from r has outdegree exactly k.*

Let $OPT'' \subseteq OPT$ be minimal so that $G_0 + OPT''$ is k-edge-inconnected to s. By Fact 6, the outdegree of every node w.r.t. OPT'' is at most k. Thus $c(OPT'') \le k \cdot p(OPT'')$, by Proposition 1. Using this, the cost optimality of F'', and Proposition 1 we get:

$$p(F'') \le c(F'') \le c(OPT'') \le k \cdot p(OPT'') \le k \cdot \text{opt}.$$

Consequently,

$$p(F) = p(F' + F'') \le p(F') + p(F'') \le 2H(k - k_0) \cdot \text{opt} + k \cdot \text{opt} = (2H(k - k_0) + k) \cdot \text{opt}.$$

The proof of part (ii) of Theorem 1 is complete.

We note that Theorem 2 can be extended to so called "crossing families". Two intersecting sets $X, Y \subset V$ *cross* if $X \cap Y, X - Y, Y - X$ are all nonempty and if $X \cup Y \ne V$. A set family $\mathcal{F}$ on V is a *crossing family* if $X \cap Y, X \cup Y \in \mathcal{F}$ for any crossing $X, Y \in \mathcal{F}$. Any crossing family can be naturally represented by two intersecting families as follows. Let $r \in V$ be arbitrary. For a set family $\mathcal{F}$ on V define $\mathcal{F}_r^+ = \{X \in \mathcal{F} : r \notin X\}$ and $\mathcal{F}_r^- = \{V - X : X \in \mathcal{F} - \mathcal{F}_r^+\}$. Summarizing, we get the following statement:

Corollary 3. *The problem of finding a min-power edge-cover of a crossing family $\mathcal{F}$ on V admits a $(2H(n) + 1)$-approximation algorithm, if for some $r \in V$ Assumptions 1 and 2 are valid for $\mathcal{F}_r^+$ and if the min-cost reverse cover of $\mathcal{F}_r^-$ can be computed in polynomial time.*

3.3   Part (iii)

The algorithm is as follows. Let $S \subset V$ be a subset of nodes of size k (so $|S| = k$). For every $r \in S$ do:

1. Using the algorithm as in part (i) of Theorem 1 compute an edge set F_r' so that $G_0 + F_r'$ is k-outconnected from r.
2. Compute a min-cost augmenting edge set F_r'' so that $G_0 + F_r''$ is k-edge-inconnected to r.

The fact that $F = \cup\{F_r' + F_r'' : r \in S\}$ is a feasible solution is known, c.f., [17] (this fact is independent from the cost/power of the edge sets computed). For every $r \in S$ we have $p(F_r') \le 2(k - k_0)H(n) \cdot \text{opt}$ and $p(F_r'') \le k \cdot \text{opt}$, by the same argument as in the proof of part (ii). Consequently,

$$p(F) \le |S|(2(k - k_0)H(n) + k) \cdot \text{opt} = k(2(k - k_0)H(n) + k) \cdot \text{opt}.$$

We note that by combining part (i) of Theorem 1 with the method used in [18] one can obtain an algorithm with approximation ratio $(2H(n) + k) \cdot (k - k_0) \cdot O(\frac{n}{n-k} \ln k)$ which for large values of k might be better.

References

1. Althaus E., Calinescu G., Mandoiu I., Prasad S, Tchervenski N., Zelikovsky A.: Power efficient range assignment in ad-hoc wireless networks. WCNC Proc. (2003) 1889–1894
2. Calinescu G., Kapoor S., Olshevsky A., Zelikovsky A.: Network lifetime and power assignment in ad hoc wireless networks. ESA Proc. (2003) 114-126
3. Cheriyan J., Vempala S., Vetta A.: An Approximation Algorithm for the Minimum-Cost k-Vertex Connected Subgraph. SIAM J. on Computing $32(4)$ (2003) 1050–1055
4. Edmonds J.: Matroid intersection. Annals of Discrete Math. 4 (1979) 185–204
5. Frank A.: Connectivity and network flows. In Handbook of Combinatorics, eds. R. Graham, M. Grötschel, and L. Lovász, Elsvier Science (1995) 111–177.
6. Frank A.: Increasing the rooted-connectivity of a digraph by one. Mathematical Programming $84(3)$ (1999) 565–576
7. Frank A., Tardos É.: An application of submodular flows. Linear Algebra and its Applications $114/115$ (1989) 329–348
8. Gabow H. N.: A representation for crossing set families with application to submodular flow problems. SODA Proc. (1993) 202–211
9. Guha S., Khuller S.: Improved methods for approximating node weighted Steiner trees and connected dominating sets. Inf. Comput. $150(1)$ (1999) 57–74
10. Hajiaghayi M. T., Kortsarz G., Mirokni V. S., Nutov Z.: Power optimization for connectivity problems IPCO Proc. (2005) 349–361
11. Jain K.: A Factor 2 Approximation Algorithm for the Generalized Steiner Network Problem. Combinatorica $21(1)$ (2001) 39–60
12. Johnson D. S.: Approximation Algorithms for Combinatorial Problems. Journal of Computing and System Sciences $9(3)$ (1974) 256–278
13. Khuller S.: Approximation algorithms for for finding highly connected subgraphs. Chapter 6 in: Approximation Algorithms for NP-hard problems, D. S. Hochbaum Ed., PWS, (1995) 236–265
14. Khuller S., Vishkin U.: Biconnectivity approximations and graph carvings. Journal of the Association for Computing Machinery $41(2)$ (1994) 214–235
15. Klein C., Ravi R: A nearly best-possible approximation algorithm for node-weighted steiner trees. Journal of Algorithms $19(1)$ (1995) 104–115
16. Kortsarz G, Mirokni V. S., Nutov Z., Tsanko E.: Approximation algorithms for minimum power fault tolerant network design. Manuscript (2006)
17. Kortsarz G., Nutov Z.: Approximating node-connectivity problems via set covers. Algorithmica 37 (2003) 75–92
18. Kortsarz G., Nutov Z.: Approximating k-node connected subgraphs via critical graphs. SIAM J. on Computing $35(1)$ (2005) 247–257
19. Kortsarz G., Nutov Z.: Approximating minimum cost connectivity problems. To appear in: Approximation Algorithms and Metaheuristics, T. F. Gonzalez ed.
20. Lando Y., Nutov Z.: On hardness of minimum power connectivity problems. Manuscript (2006)
21. Nutov Z.: Approximating minimum power connectivity problems. Manuscript (2006)
22. Raz R., Safra S.: A sub-constant error-probability low-degree test and a sub-constant error-probability PCP characterization of NP, STOC Proc. (1997) 475-484
23. Williamson D. P., Goemans M. X., Mihail M., Vazirani V. V.: A primal-dual approximation algorithm for generalized Steiner network problems. Combinatorica 15 (1995) 435-454

Better Approximations for the Minimum Common Integer Partition Problem

David P. Woodruff*

[1] MIT
dpwood@mit.edu
[2] Tsinghua University

Abstract. In the k-Minimum Common Integer Partition Problem, abbreviated k-MCIP, we are given k multisets $X_1, \ldots, X_k$ of positive integers, and the goal is to find an integer multiset T of minimal size for which for each i, we can partition each of the integers in X_i so that the disjoint union (multiset union) of their partitions equals T. This problem has many applications to computational molecular biology, including ortholog assignment and fingerprint assembly.

We prove better approximation ratios for k-MCIP by looking at what we call the *redundancy* of $X_1, \ldots, X_k$, which is a quantity capturing the frequency of integers across the different X_i. Namely, we show $.614k$-approximability, improving upon the previous best known $(k - 1/3)$-approximability for this problem. A key feature of our algorithm is that it can be implemented in almost *linear time*.

Keywords: minimum common integer partition problem, approximation algorithms, computational biology.

1 Introduction

In a recent work [2] a new combinatorial optimization problem called the *Minimum Common Integer Partition* problem was introduced. This problem is one of the many recent combinatorial problems with applications to computational molecular biology, including ortholog assignment [1, 3, 4, 5] and DNA fingerprint assembly [10]. The problem also poses interesting new algorithmic challenges.

Formally, the problem is as follows. Consider two multisets $X = \{x_1, \ldots, x_m\}$ and T of positive integers. If there is a partition of T into multisets T_i such that for each i the sum of integers in T_i equals x_i, then T is called an *integer partition* of X. We say that T is a *common integer partition* of multisets $X_1, \ldots, X_k$ if it is an integer partition of each X_i. The *k-Minimum Common Integer Partition Problem*, abbreviated k-MCIP$(X_1, \ldots, X_k)$, is to find a common integer partition T of minimum cardinality.

As an example, for a pair of multisets $X_1 = \{2, 2, 3\}$ and $X_2 = \{1, 1, 5\}$, the integer partition $T = \{1, 1, 2, 3\}$ is a minimum common integer partition of the

---

* The author would like to thank Andrew Yao and Tsinghua University for hospitality and support while performing this researh.

J. Diaz et al. (Eds.): APPROX and RANDOM 2006, LNCS 4110, pp. 248–259, 2006.

X_i. Indeed, to see that T is an integer partition of X_1, partition T into multisets $T_1 = \{1,1\}, T_2 = \{2\}$, and $T_3 = \{3\}$. Then the sum of integers in T_1 is 2, the sum in T_2 is 2, and the sum in T_3 is 3. To see that T is an integer partition of X_2, partition it into $T_1 = \{1\}$, $T_2 = \{1\}$, and $T_3 = \{2,3\}$. To see that T has minimal size, observe that any integer partition of either X_1 or X_2 must have size at least 3. Further, the only integer partition of X_1 of size 3 is X_1 itself. However, X_1 is not an integer partition of X_2. Thus every common partition has size at least 4, which is the size of T.

A common partition T exists if and only if the integers in each X_i have the same sum. As this property is easy to verify, we will assume it holds for the rest of the paper. Let $m = \sum_{i=1}^{k} |X_i|$. We will think that k is much smaller than m, as is the case in practice. Nevertheless, in our asymptotic notation we will write the dependence on both m and k.

In [2], it is shown that k-MCIP is NP-hard[1], and in fact APX-hard [9] for every $k \geq 2$. To show the former, the authors present a Cook-reduction from Set-Partition, while for the latter they present an L-reduction from Maximum-3-Dimensional-Matching with a bounded number of occurrences, which is known to be APX-hard [7]. The authors also give a 5/4-approximation algorithm when $k = 2$ and a $\frac{3k(k-1)}{3k-2}$-approximation algorithm for general k. Note that $\frac{3k(k-1)}{3k-2} \approx k - 1/3$. The former is based on an approximation algorithm for Set-Packing with small sets, and the latter is described below. Although their algorithm for $k = 2$ is polynomial-time, its running time[2] is $\Omega(m^9)$, which is likely to make it impractical. Indeed, as mentioned in the applications below, it is likely that $m \approx 2^{12}$, for which this running time is much too large to be of use. Their algorithm for general k is much more efficient, running in time $O(mk)$.

We note that an $O(m \log k)$-time k-approximation for k-MCIP is straightforward, though in [2] the authors only provide an $O(mk)$-time k-approximation. To see this, first suppose $k = 2$ and the multisets are X, Y. Repeatedly choose an element $x \in X$ and $y \in X$, and add $\min(x, y)$ to the common partition. Remove x from X and y from Y if $x = y$. Otherwise remove $\min(x, y)$ from the multiset it occurs in and replace $\max(x, y)$ with $\max(x, y) - \min(x, y)$ in the other multiset. This procedure produces at most $m - 1$ numbers in the common partition. Since the optimal solution has size at least $\max(|X|, |Y|) \geq m/2$, the algorithm provides a 2-approximation. It runs in $O(m)$ time. To solve k-MCIP, divide the input multisets into $\lfloor k/2 \rfloor$ pairs (plus one multiset if k is odd), run the above algorithm on each pair, and repeat the process on pairs of output multisets. The running time is now $O(m \log k)$ and the output size is again at most m, so we get an $m/(m/k) = k$ approximation. We refer to this algorithm as k-Greedy.

In fact, it is not hard to achieve ratio $\frac{5k}{8}$ for even k and $(\frac{5k}{8} + \frac{3}{8})$ for odd k. This was also missed in [2], and already improves the previous best known ratio for every $k \geq 3$. To see this, for simplicity suppose that k is even. Partition the k multisets into $k/2$ pairs (X_{2i-1}, X_{2i}). Run the algorithm of [2] for 2-MCIP on

---

[1] Lan Liu and the author have shown that k-MCIP is NP-hard in the strong sense.

[2] We assume the unit-cost RAM model on words of size $O(\log m)$ and that arithmetic operations on words can be done in constant time.

each pair. Let the output partition of the algorithm on inputs $X_{2i-1}X_{2i}$ be Y_i. Finally, output k-Greedy$(Y_1, \ldots, Y_{k/2})$.

Let opt_i denote the size of the minimum common partition of X_{2i-1} and X_{2i}. Then $|Y_i| \leq 5opt_i/4$. Moreover, if opt denotes the size of the minimum common partition of all of the X_i, then $opt_i \leq opt$ for all i. As the common partition output by k-Greedy never has size larger than its total input size, we get that

$$|\mathsf{k} - \mathsf{Greedy}(Y_1, \ldots, Y_{k/2})| \leq \sum_{i=1}^{k/2} |Y_i| \leq \sum_{i=1}^{k/2} \frac{5opt_i}{4} \leq \frac{5}{4} \sum_{i=1}^{k/2} opt = \frac{5k}{8} \cdot opt,$$

and the ratio of $5k/8$ follows.

The main problem with the above algorithm is that it invokes the algorithm for $k = 2$ given in [2], and thus its running time is also $\Omega(m^9)$. Thus the algorithm is likely to be very impractical.

In this paper we give a new approximation algorithm for k-MCIP which runs in almost *linear time*. More precisely, we have a randomized $O(m \log k)$ and a deterministic $O(m\text{poly}(k))$-time algorithm. Both running times are $O(m)$ for constant k. Moreover their ratios are bounded above by $.614k(1 + o(1))$. Since $.614 < 5/8$, we not only reduce the running time to linear, we even improve the approximation ratio of the natural (though inefficient) algorithm sketched above. Although the algorithm in [2] for general k is also efficient, it was only shown to achieve ratio $k - 1/3$. We improve the analysis of [2], and show their algorithm actually provides a $(k - 1/2)$-approximation. We also provide an instance to their algorithm for which this is best-possible, which turns out to be a bit non-trivial. Finally, for the special case when the multisets X_i are disjoint, we improve the analysis of our algorithm to show a ratio of $(k + 1)/2$.

Applications: Suppose we are given a collection of k genomes, one for each of k different species. We look at the following special case: each genome consists of the same number of copies of a single gene, but the copies are clustered into different substrings in the different genomes. Thus, we may view each genome i as a sequence of integer substring sizes $x_1^i, \ldots, x_r^i$, with the property that for all pairs of genomes i, j, $\sum_\ell x_\ell^i = \sum_{\ell'} x_{\ell'}^j$. The goal in this application is to partition the substrings into the same collection of strings, minimizing the number of strings in the common partition. This provides a measure of similarity between the different genomes, and has been proposed in practice. This is exactly the Minimum Common Integer Partition problem. For more detail, see [2, 3, 4, 5].

Actually, the main motivating example for $k > 2$ is *DNA fingerprint assembly*, as described in great detail on page 3 of [2]. This is a problem that has arisen in the ongoing *Oligonucleotide Fingerprinting Ribosomal Genes (OFRG)* project [10]. The goal of this project is to identify different microbial organisms using fingerprints obtained in the lab. Here k is a parameter determined by a specific measuring device, while m refers to a quantity known as the number of probe subsets of a fingerprint. We refer the reader to [2, 10] for the details, but we merely state that from [6] we have learned that a typical setting of MCIP parameters likely to occur in practice is $k = 28$ and $m = 2^{12}$.

2   Overview of the Algorithms

To illustrate our techniques, we first recall the algorithm CommonElements given in [2] which invokes the subroutine 2-Greedy described in the introduction. For a formal treatment of 2-Greedy, see [2] where it is shown to terminate with output partition size less than m (so the ratio is $m/(m/2) = 2$) in $O(m)$ time.

The algorithm CommonElements first adds the integers common to all of the X_i to a common partition, and then repeatedly invokes 2-Greedy. Let $X_1, \ldots, X_k$ be an instance of k-MCIP.

CommonElements($X_1, \ldots, X_k$):

1. $T \leftarrow \emptyset$.
2. While there is an x occurring in all of the X_i, choose such an x, add x to T, and remove one copy of x from each X_i.
3. Let $X'_1, \ldots, X'_k$ denote the resulting multisets.
4. $T' \leftarrow$ 2-Greedy(X'_1, X'_2).
5. For $i = 3, \ldots, k$,
   (a) $T' \leftarrow$ 2-Greedy(T', X'_i).
6. Output $T \cup T'$.

In [2], it is shown that this algorithm is a $(k - 1/3)$-approximation. We will later show that it is in fact a $(k - 1/2)$-approximation. However, let us first define our new algorithm to see how it contrasts with this one.

The structure of our algorithm for k-MCIP is as follows. Let $[k] = \{1, 2, \ldots, k\}$.

HighFrequency($X_1, \ldots, X_k$):

1. $T \leftarrow \emptyset$.
2. Choose a set-partition π of $[k]$ into pairs of integers, with one unpaired integer r if k is odd.
3. For each pair $(i, j) \in \pi$,
   (a) Compute $C_{i,j} \leftarrow$ CommonElements(X_i, X_j).
4. If there is only a single pair $(1, 2)$, output $C_{1,2}$, else
   - k even: output HighFrequency($\{C_{i,j} \mid (i,j) \in \pi\}$).
   - k odd: output HighFrequency($\{X_r\} \cup \{C_{i,j} \mid (i,j) \in \pi\}$).

We have not yet specified how to choose the partition π in step 2 of HighFrequency. We will try to choose π so that the output in step 4 has minimal size. For constant k, this is easy to do by an exhaustive enumeration of partitions. For larger k, we show a random π is a good choice, and in fact this choice can be efficiently derandomized. For now the choice is not essential, as we merely wish to compare the structure of HighFrequency with that of CommonElements.

At a high level, the main differences between HighFrequency and CommonElements are the following. In CommonElements, the multisets $X_1, \ldots, X_k$ (or more precisely, $X'_1, \ldots, X'_k$) are traversed sequentially, invoking 2-Greedy on each new

X_i, together with the current common partition of $X_1, \ldots, X_{i-1}$. In our algorithm, we traverse $X_1, \ldots, X_k$ in parallel, and we recurse. Moreover, the traversal order is not fixed, but rather determined by π. Also, instead of invoking 2-Greedy on each instance of 2-MCIP we encounter, we invoke CommonElements, which has a better approximation ratio and still can be implemented in linear time.

To get a feeling for the algorithms, consider the following example. Suppose $k = 4$ and the input multisets are $X_1 = \{2, 3\}$, $X_2 = \{1, 4\}$, $X_3 = \{2, 3\}$, and $X_4 = \{2, 3\}$. When we run CommonElements, step 2 has no effect since although items 2 and 3 occur many times, they do not occur in X_2. In step 4 we may assume that $T' = \{1, 1, 3\}$ (we are constructing a worst-case execution of 2-Greedy). Then after the first iteration of step 5a, we have $T' = \{1, 1, 1, 2\}$, and after the last iteration we obtain $T' = \{1, 1, 1, 1, 1\}$ (again, in the worst-case).

However, let $\pi = \{1, 2\}, \{3, 4\}$. Then $C_{1,2} = \{1, 1, 3\}$ or $C_{1,2} = \{1, 2, 2\}$, but $C_{3,4} = \{2, 3\}$, so that the output of step 4 is $\{3, 1, 1\}$ or $\{2, 2, 1\}$, which are of minimal size. Thus, our algorithm HighFrequency is able to exploit the high frequency of integers $2, 3$ in the input, even though CommonElements is not. This is the reason we've named our algorithm HighFrequency.

One of the main technical aspects of this paper is how to handle the case when there are not many integers occurring in multiple input mutisets X_i. In this case we show that even the optimal solution must be large, as intuitively if many integers have low frequency, then most of the integers in the X_i will have to be split into at least two new integers in any common partition. We show this by developing a framework for capturing the frequency of integers across the different input mutisets.

In the next section we prove a key lemma for *lower-bounding* the size of the optimal common partition, and in section 4 we use this lemma to analyze the performance of HighFrequency. We believe our lower bound can lead to future results. For example, in the next section we use this characterization to improve the analysis of the main algorithm of [2].

3   A Key Lemma and Two Quickies

Consider an instance S of k-MCIP consisting of k multisets of integers $S = \{X_1, \ldots, X_k\}$. We will define a certain quantity of S, called its *redundancy*, which captures the distribution of the number of occurrences, across the different X_i, of integers occurring in S.

At first glance it may seem that our definition is needlessly complicated. After presenting it, we explain the need for this complication.

Recall that the X_i are multisets, but may also be viewed as ordered lists. Thus, we may refer to the element in the jth position of X_i for $1 \le j \le |X_i|$.

Consider elements T of $[|X_1| + 1] \times [|X_2| + 1] \times \cdots \times [|X_k| + 1]$. T translates naturally into a multiset $\tilde{T}$ as follows: if its ith coordinate j does not equal $|X_i| + 1$, add the integer in the jth position of X_i to $\tilde{T}$. We say that T is *lonely* if the multiset $\tilde{T}$ has the form $\{t, t, \ldots, t\}$. In this case we use the notation $int(T)$ to denote the integer t. We say a set $\mathcal{C}$ of lonely elements of $[|X_1| + 1] \times [|X_2| +$

$1] \times \cdots \times [|X_k| + 1]$ is *consistent* if there are no two distinct elements $T, T' \in \mathcal{C}$ and an i for which $T_i = T'_i \neq |X_i| + 1$. That is, no two elements of $\mathcal{C}$ can agree on any coordinate i, unless they both have the value $|X_i| + 1$ on that coordinate.

We define the *weighted-size* of a set $\mathcal{C}$ of lonely elements T_j to be $\sum_{j=1}^{|\mathcal{C}|} |\tilde{T}_j|$.

Definition 1. *The r-redundancy of S, denoted $\mathbf{Red(r, S)}$, is the maximum, over all consistent sets $\mathcal{C}$ of at most r lonely elements, of the weighted-size of $\mathcal{C}$.*

We note that a simpler alternative, though incorrect, definition is the following: define the degree of a variable x as $deg(x, S) = |\{i \mid x \in X_i\}|$. Then define the redundancy $Red(r, S)$ to be $\max_{x_1,\ldots,x_r}$ distinct $\sum_{i=1}^{r} deg(x_i, S)$.

Although simpler, this definition fails to capture the following example: $X_1 = \{1, 1\}$, $X_2 = \{1, 1\}$. Here, $Red(2, S) = 4$. Indeed, consider $\mathcal{C} = \{(1, 1), (2, 2)\}$. Then the elements $(1, 1), (2, 2)$ are both lonely since their corresponding multisets have the form $\{1, 1\}$. Moreover, they are consistent. Finally, the weighted size of $\mathcal{C}$ is 4. However, the alternative definition would put $Red(2, s) = deg(1, S) = 2$. One could instead remove the word "distinct" from the definition, but this also does not solve the problem, since then for $X_1 = \{1, 3, 4\}$ and $X_2 = \{1, 2, 5\}$ it would return $Red(3, s) = 6$ since $x_1 = x_2 = x_3 = 1$, but our definition gives $Red(3, s) = 4$ with say $T_1 = (1, 1), T_2 = (2, 4)$, and $T_3 = (3, 4)$.

Define $opt(S)$ to be the size of a minimum common partition of S. When S is clear from the context, we will often just write opt. Recall that $m = \sum_{i=1}^{k} |X_i|$. The following lemma lower bounds opt in terms of the redundancy of S.

Lemma 1. $opt \geq (2m - Red(opt, S))/k$.

Proof. Let T be a minimum common integer partition of $X_1, \ldots, X_k$. Define the bipartite graph with right partition T and left partition S (here S is the multiset union[3] of the X_i). Each $x \in S$ is incident exactly to those elements $t_i \in T$ which partition x. So, for instance, the sum over all neighbors of x is equal to x.

Then $Red(opt, S)$ is an upper bound on the number of degree-1 vertices in the left part. To see this, we construct a consistent set $\mathcal{C}$ of opt lonely elements whose weighted-size is exactly the number of degree-1 vertices in the left part. For each vertex v on the right, let $\tilde{S}(v)$ denote v's neighbors on the left with degree 1. As each such v is incident to exactly 1 element in each X_i, we can naturally associate $\tilde{S}(v)$ with an element $S(v)$ of $[|X_1| + 1] \times [|X_2| + 1] \times \cdots \times [|X_k| + 1]$, where $S(v)_j = |X_j| + 1$ iff v partitions a vertex in S_j with degree more than 1. Then $S(v)$ is lonely since each integer in $\tilde{S}(v)$ equals the integer corresponding to v. The set $\{S(v) \mid v$ on the right $\}$ is consistent since if $w = S(v)_j = S(v')_j$ for $v \neq v'$ and $j \leq |X_j|$, then w would have degree more than 1. Finally, $\{S(v) \mid v$ on the right $\}$ has exactly opt elements. Thus, its weighted size is at most $Red(opt, S)$. Since every degree-1 vertex on the left is counted exactly once in the weighted-size of $\{S(v) \mid v$ on the right $\}$, there are at most $Red(opt, S)$ such vertices.

---

[3] The multiset union of two multisets is defined by the following rule: if x occurs f_1 times in the first multiset and f_2 times in the second, then x occurs $f_1 + f_2$ times in the multiset union.

Resuming the proof of the lemma, there are at least $m - Red(opt, S)$ remaining vertices in the left part, and each has degree at least 2. Thus, there are at least $Red(opt, S) + 2(m - Red(opt, S)) = 2m - Red(opt, S)$ edges in the graph. On the other hand, every vertex on the right has degree exactly k. Thus, $2m - Red(opt, S) \le k|T| = k \cdot opt$, and the lemma follows by dividing by k.

Corollary 1. *If for all $j \ne j'$, X_j and $X_{j'}$ are disjoint, then k-MCIP is $(k + 1)/2$-approximable in $O(m \log k)$ time.*

Proof. In this case $Red(r, S) \le r$ for any r, and the bound above gives $opt \ge 2m/(k + 1)$. The claim follows by running k-Greedy whose output size is $\le m$.

We now look at the approximation ratio of CommonElements. In [2], it is shown the ratio is $3k(k-1)/(3k-2) \le k - 1/3$. On the other hand, $3k(k-1)/(3k-2) \ge k - 1/3 - \epsilon$ for any constant $\epsilon > 0$ and large enough k. We show,

Corollary 2. CommonElements *outputs a $(k - 1/2)$-approximation.*

Proof. Recall the notation of section 2. Suppose CommonElements adds ℓ integers to T in step 2. It follows that T' is of size at most $m - \ell k$. Thus, $|T \cup T'| \le \ell + (m - \ell k) = m - \ell(k - 1)$. On the other hand, there are at most ℓ elements with corresponding multisets of size k in any consistent set $\mathcal{C}$ of lonely elements. It follows that the weighted-size of $\mathcal{C}$, and thus $Red(opt, S)$, can be at most $\ell k + (opt - \ell)(k - 1)$. Applying Lemma 1, $k \cdot opt \ge 2m - (\ell k + (opt - \ell)(k - 1))$, which, after rearranging, shows $opt \ge (2m - \ell)/(2k - 1)$. Using that $k \ge 2$ and $\ell \ge 0$, the corollary follows from the following bound on the approximation ratio,

$$(2k - 1)\frac{m - \ell(k - 1)}{2m - \ell} \le (2k - 1)\frac{m - \ell/2}{2m - \ell} = (2k - 1)\frac{1}{2} = k - \frac{1}{2}.$$

Claim. The approximation ratio of CommonElements is at least $k - 1/2 - o(1)$.

Proof. Let r be a large positive integer, and consider $X_1 = X_2 = \cdots = X_{k-1} = \{1, 1, 3, 3, 5, 5, 7, 7, \ldots, 2r + 1, 2r + 1\}$, and $X_k = \{2, 6, 10, 14, \ldots, 4r + 2\}$. Then $\sum_{x \in X_i} x = \sum_{x \in X_j} x = 2(r + 1)^2$ for all $i \ne j$. Thus, $S = \{X_1, \ldots, X_k\}$ is an instance of k-MCIP. The optimal solution is X_1, which has size $opt = 2r + 2$.

The output of CommonElements on S is just the output of steps 3-6 on S (e.g., T') since no integer occurs in all of the X_i, and thus step 2 does not modify S. In 2-Greedy it is not specified how to choose the two integers x, y, and our strategy is to present a sequence of choices for which T' is of size at least $(2k-1)r - O(k^2)$. It will follow that the approximation ratio is at least

$$\frac{(2k - 1)r - O(k^2)}{2r + 2} = \frac{(2k - 1)(r + 1) - (2k - 1) - O(k^2)}{2r + 2} = (k - 1/2) - \frac{O(k^2)}{2r + 2},$$

which can be made arbitrarily close to $k - 1/2$ by increasing r.

We show by induction, after i invocations of 2-Greedy, $0 \le i \le k - 2$ (recall that there are $k - 1$ invocations in total - we handle the last one separately), the common partition of $X_1, \ldots, X_{i+1}$ generated by CommonElements has the form:

$$\{1, 1, 3, 3, \ldots, 2(r - i) + 1, 2(r - i) + 1\} \cup 1^{(2i)} \cup 2^{(s)}, \tag{1}$$

where $a^{(b)}$ indicates b copies of a, and where $s = 2\sum_{j=0}^{i-1}(r-j)$.

Base Case: When $i = 0$, we have not yet invoked 2-Greedy, and so the multiset in expression 1 should be equal to X_1. Since $2i = 0$ and $s = 0$ in this case, this holds by definition of X_1.

Inductive Step: Suppose expression 1 is the common partition after $i \geq 1$ invocations, and consider the $(i+1)$st invocation, in which the common partition after i invocations is invoked together with $X_{i+2} = \{1, 1, 3, 3, \ldots, 2r+1, 2r+1\}$. We claim 2-Greedy may first repeatedly subtract 1s and 2s from X_{i+2} until the two multisets both have the form $\{1, 1, 3, 3, \ldots, 2r+1-2i, 2r+1-2i\}$. To see this, since each integer in X_{i+2} is odd, and there are $2i$ integers in X_{i+2} larger than $2r+1-2i$, 2-Greedy may subtract $2i$ different 1s so that X_{i+2} has the form

$$\{1, 1, \ldots, 2r+1-2i, 2r+1-2i, 2r+2-2i, 2r+2-2i, 2r+4-2i, 2r+4-2i, \ldots, 2r, 2r\}.$$

Next, observe that the sum of the last $2i$ terms of X_{i+2}, $2\sum_{j=1}^{i}(2r+2j-2i)$, is equal to s. Thus, 2-Greedy may subtract s different 2s so that the two multisets become $\{1, 1, 3, 3, \ldots, 2r+1-2i, 2r+1-2i\}$, as claimed, and the current partition is $1^{(2i)} \cup 2^{(s)}$.

Next 2-Greedy may choose pairs, $(1,3), (1,3), (3,5), (3,5), \ldots, (2r-1-2i, 2r+1-2i), (2r-1-2i, 2r+1-2i)$, where the first element in each pair is from the common partition after i invocations, and the second is from X_{i+2}. The first element in each pair is added to the new partition. The multisets now have the form $\{2r+1-2i, 2r+1-2i\}$ and $1^{(2)} \cup 2^{(2(r-i))}$. Finally, 2-Greedy may subtract 1 from the two different $2r+1-2i$, and then repeatedly subtract 2. Thus the common partition after $i+1$ invocations has the form

$$1^{(2i)} \cup 2^{(s)} \cup \{1, 1, 3, 3, \ldots, 2r-1-2i, 2r-1-2i\} \cup 1^{(2)} \cup 2^{(2(r-i))},$$

which is easily seen to satisfy the inductive hypothesis.

Last Invocation: By the inductive argument, the common partition of $X_1, \ldots, X_{k-1}$ has the form of expression 1 with $i = k-2$, namely, the form $\{1, 1, 3, 3, \ldots, 2r-2k+5, 2r-2k+5\} \cup 1^{(2k-4)} \cup 2^{(s)}$, where $s = 2\sum_{j=0}^{k-3}(r-j) = (2k-4)r - O(k^2)$. Recall $X_k = \{2, 6, 10, 14, \ldots, 4r+2\}$. First 2-Greedy may repeatedly subtracts 1s and 2s from the two mutisets, so that they become

$$\{1, 1, 3, 3, \ldots, 2r-2k+5, 2r-2k+5\} \text{ and } \{2, 6, \ldots, 4r-4k+10\},$$

and the common partition has the form $1^{(2k-4)} \cup 2^{(s)}$. Note that since every integer in X_k is even, this can be accomplished by first subtracting s different 2s from the largest integers of X_k, followed by $2k-4$ different 1s. Now 2-Greedy may choose pairs $(1,2), (3,6), (5,10), \ldots, (2r-2k+5, 4r-4k+10)$, so that the multisets both have the form $\{1, 3, \ldots, 2r-2k+5\}$.

Then it chooses pairs $(1,3), (3,5), \ldots, (2r-2k+3, 2r-2k+5)$, so that the multisets have the form $\{2r-2k+5\}$ and $1 \cup 2^{(r-k+2)}$. Finally, 2-Greedy may add $1 \cup 2^{(r-k+2)}$ to the common partition, so the output of Common-Elements is

$$1^{(2k-3)} \cup 2^{(r-k+2+s)} \cup \{1, 3, 5, \ldots, 2r-2k+5\} \cup \{1, 3, 5, \ldots, 2r-2k+3\}.$$

Since $2k-3+r-k+2+s \geq (2k-3)r - O(k^2)$, and there are $2r - O(k)$ elements in the last two sets, the output partition has size $(2k-1)r - O(k^2)$, as needed.

4   Analysis of HighFrequency

In this section we prove our main theorem, Theorem 1. We use the probabilistic method to show that there are good set-partitions π that HighFrequency can choose in step 2. We quantify how well HighFrequency performs in terms of the average size f of a multiset from an optimal consistent set of lonely elements. On the other hand, we also use Lemma 1 to lower bound the size of the minimum common partition in terms of f. We then choose f so that the ratio between this upper and lower bound is maximized, which is a worst-case ratio.

In the following, we will use $O(), \Omega(), o(), \omega()$ to denote functions of k which are independent of m, e.g., $o(1)$ is a function which tends to 0 as $k \to \infty$.

Theorem 1. HighFrequency *outputs a* $.614k(1 + o(1))$-*approximation.*

Proof. First observe that for two multisets X_i, X_j containing $c(X_i, X_j)$ elements in common [4], the output size of CommonElements(X_i, X_j) is at most

$$((|X_i| + |X_j| - 2c(X_i, X_j)) - 1) + c(X_i, X_j) < |X_i| + |X_j| - c(X_i, X_j).$$

In particular, its output size is always less than its input size.

Suppose in the ith invocation of HighFrequency, the algorithm is called with multisets $Y_1, \ldots, Y_r$. Then HighFrequency will partition these multisets into pairs (with one extra Y_j if r is odd) and invoke CommonElements on each pair. For any call to CommonElements in the ith invocation of HighFrequency, say CommonElements(Y_a, Y_b), $|$CommonElements$(Y_a, Y_b)| < |Y_a| + |Y_b| - c(Y_a, Y_b)$. Let c_i be the sum of $c(Y_a, Y_b)$ over all pairs (Y_a, Y_b) in the ith invocation.

Let m_i denote the output size of the ith invocation of HighFrequency, so for example, m_1 is the *input size* to the first recursive call (or the output size of HighFrequency if there are no recursive calls). Define $m_0 = m$. Let x be such that $2^x = o(\sqrt{k})$. Since $2^x \leq k$ for large enough k, there are at least x invocations of HighFrequency (and in fact, there may be many more, though we will only need to consider the first x). Then for $1 \leq i \leq x$, $m_i < m_{i-1} - c_i$. Summing these inequalities up for all i and canceling common terms, $m_x < m - \sum_{i=1}^{x} c_i$. Since $|$HighFrequency$(X_1, \ldots, X_k)| \leq m_x$, we have $|$HighFrequency$(X_1, \ldots, X_k)| < m - \sum_{i=1}^{x} c_i$. It follows that,

$$\mathbf{E}_{\pi_1, \ldots, \pi_x}[|\text{HighFrequency}(X_1, \ldots, X_k)|] < m - \mathbf{E}_{\pi_1, \ldots, \pi_x}\left[\sum_{i=1}^{x} c_i\right],$$

where π_i is a uniformly random set-partition chosen in the ith invocation of HighFrequency, and thus each of the integer pairs (a, b) in each invocation is (by itself) a uniformly random pair of integers. Indeed, by symmetry the first chosen pair and also all other pairs have the same probability distribution, namely they

---

[4] By elements in common, we mean we can find $c(X_i, X_j)$ *disjoint* pairs of elements, each pair containing one element from X_i and one element from X_j, such that the elements within each pair are equal as integers. So if $X_i = \{1, 1, 3, 4\}$ and $X_j = \{1, 1, 2, 5\}$, then $c(X_i, X_j) = 2$, even though 1 is the only integer value in common.

are uniformly drawn at random from all possible pairs. Thus, $\mathbf{E}[c(Y_a, Y_b)] = \mathbf{E}[c(Y_{a'}, Y_{b'})]$ for every two integer pairs $(a, b), (a', b')$ determined by π_i.

We may bound $\mathbf{E}[c(Y_a, Y_b)]$ as follows. Consider the largest (in terms of weighted size) consistent set of opt lonely elements of $[\|X_1\| + 1] \times \cdots \times [\|X_k\| + 1]$. Suppose the sizes of their corresponding multisets are $f_1, \ldots, f_{opt}$, and let $f = \lfloor \sum_j f_j / opt \rfloor$. Observe that f is a positive integer since each $f_j \geq 1$. Now from Lemma 1 we know that $opt \geq (2m - Red(opt, S))/k$. But $Red(opt, S) = \sum_{i=1}^{opt} f_i \leq (f+1)opt$, and after rearranging, we have $opt \geq 2m/(k + f + 1)$.

Suppose $f < k/5$. Then, since $|\mathsf{HighFrequency}(X_1, \ldots, X_k)| < m$, we have

$$\frac{|\mathsf{HighFrequency}(X_1, \ldots, X_k)|}{opt} < \frac{m}{\frac{2m}{k+f+1}} = \frac{k+f+1}{2} < \frac{6k}{10}(1+o(1)) = .6k(1+o(1)),$$

and the theorem is proven in this case.

Let us now handle the case when $f \geq k/5$. Consider two input multisets Y_a, Y_b in the ith invocation of $\mathsf{HighFrequency}$. Each is formed by successively applying $\mathsf{CommonElements}$ on at most 2^{i-1} different input multisets X_i. Suppose a lonely element S with $int(S) = y$ intersects each of the (at most) 2^{i-1} input multisets corresponding to Y_a. Then y will occur in Y_a. This also holds for Y_b. Thus, if y occurs in the (at most) 2^i different input multisets corresponding to Y_a and Y_b, y will be common to Y_a and Y_b. By our choice of $\pi_1, \ldots, \pi_i$, the set of these (at most) 2^i input multisets is uniformly random amongst all such sets. Thus,

$$\mathbf{E}[c(Y_a, Y_b)] \geq \sum_j \frac{\binom{f_j}{2^i}}{\binom{k}{2^i}} = \sum_j \frac{f_j(f_j - 1) \cdots (f_j - (2^i - 1))}{k(k - 1) \cdots (k - (2^i - 1))}.$$

We claim the above expession is minimized when all of the f_j are at least as large as $f = \lfloor \sum_j f_j / opt \rfloor$. To see this, suppose, w.l.o.g., that $f_1 \geq f_2 \geq \cdots \geq f_{opt}$. If this were not the case, then $f_1 \geq f + 1$ and $f_{opt} \leq f - 1$. Suppose we decrease f_1 by 1 and increase f_{opt} by 1. Then the average is the same and we still have $f_1 \geq f$. On the other hand, the expression changes by

$$\frac{2^i}{k \cdots (k - (2^i - 1))} \left((f_1 - 1) \cdots (f_1 - (2^i - 1)) - f_{opt} \cdots (f_{opt} - (2^i - 1) + 1) \right).$$

Now, $f_1 > f \geq k/5 > 2^i$ for large enough k (since $i \leq x$) and $f_1 - j > f_{opt} - j + 1$ for all j, so the above expression is non-negative. This substitution of variables did not cause the value of the sum to increase, so the sum is minimized when all the f_j are at least f. Moreover, since $f > 2^i$,

$$\mathbf{E}[c(Y_a, Y_b)] \geq \sum_j \frac{f(f-1) \cdots (f - (2^i - 1))}{k(k-1) \cdots (k - (2^i - 1))} \geq \sum_j \left(\frac{f - 2^i}{k - 2^i} \right)^{2^i} \quad \text{since } k \geq f > 2^i$$

$$\geq \sum_j c^{2^i} \left(\frac{1 - \frac{5 \cdot 2^i}{k}}{1 - \frac{2^i}{k}} \right)^{2^i} \quad \text{where } c = f/k, \text{ and } f \geq k/5.$$

To analyze this, observe that $\Theta(\frac{2^i}{k}) = 1/\omega(2^i)$ since $i \leq x$ and $2^x = o(\sqrt{k})$. We use the following inequality, which follows from Proposition B.3 of [8].

$$\left(1 - \frac{1}{\omega(2^i)}\right)^{2^i} / \left(1 - \frac{1}{\omega(2^i)}\right) \geq e^{-2^i/\omega(2^i)} \geq \left(1 - \frac{1}{\omega(2^i)}\right)^{2^i}.$$

Plugging these inequalities into our bound above, we have that, $\mathbf{E}[c(Y_a, Y_b)] \geq \sum_j c^{2^i} (1 - o(1)) = opt \cdot c^{2^i} (1 - o(1))$. In the ith invocation there are at least $\lfloor k/2^i \rfloor$ pairs. By linearity of expectation, $\mathbf{E}_{\pi_1,\ldots,\pi_x}[c_i] \geq \lfloor \frac{k}{2^i} \rfloor \mathbf{E}[c(Y_a, Y_b)]$, and so $\mathbf{E}_{\pi_1,\ldots,\pi_x}[c_i] \geq \lfloor \frac{k}{2^i} \rfloor \cdot opt \cdot c^{2^i} (1 - o(1))$. Thus,

$$\mathbf{E}_{\pi_1,\ldots,\pi_x}[|\mathsf{HighFrequency}(X_1,\ldots,X_k)|] < m - opt \sum_{i=1}^{x} \left\lfloor \frac{k}{2^i} \right\rfloor c^{2^i} (1 - o(1)).$$

Since HighFrequency chooses the optimal $\pi_1, \ldots, \pi_x$, it follows that

$$|\mathsf{HighFrequency}(X_1,\ldots,X_k)| < m - opt \sum_{i=1}^{x} \left\lfloor \frac{k}{2^i} \right\rfloor c^{2^i} (1 - o(1)).$$

The approximation ratio R of HighFrequency is $|\mathsf{HighFrequency}(X_1,\ldots,X_k)|/opt$. Dividing the expression above by opt gives $R < \frac{m}{opt} - \sum_{i=1}^{x} \lfloor \frac{k}{2^i} \rfloor c^{2^i} (1 - o(1))$. Now since $c \leq 1$ and $x = o(k)$, we can drop the floors,

$$R < \frac{m}{opt} - \sum_{i=1}^{x} \left(\frac{k}{2^i} - 1\right) c^{2^i} (1 - o(1)) < \frac{m}{opt} + o(k) - \sum_{i=1}^{x} \frac{k}{2^i} \cdot c^{2^i}.$$

Recall that we have shown $\frac{k+f+1}{2} \geq \frac{m}{opt}$. Using this and $f = ck$, we have

$$R < \frac{k}{2} + o(k) + k \max_c \left(\frac{c}{2} - \sum_{i=1}^{x} \frac{c^{2^i}}{2^i}\right).$$

We upper bound R by $\frac{k}{2} + o(k) + k \max_c \left(\frac{c}{2} - \frac{c^2}{2} - \frac{c^4}{4} - \frac{c^8}{8}\right)$, as looking at higher terms turns out to only negligibly reduce the approximation ratio further. Set $p(c) = \frac{c}{2} - \frac{c^2}{2} - \frac{c^4}{4} - \frac{c^8}{8}$. Then $p'(c) = \frac{1}{2} - c - c^3 - c^7$. We solve $p'(c^*) = 0$. By continuity, it is easy to see that there is exactly one positive real solution c^*. A MATLAB routine shows that this value c^* satisfies $.4222 < c^* < .4223$. Moreover, $p''(c)$ is non-positive for any c, and thus c^* is a local maximum. Again by computation, $p(c^*) < .11391$. At the extremes $p(1/5) \leq 1/10$ and $p(1) < 0$, and thus c^* is a global maximum. It follows that $R < \frac{k}{2} + o(k) + .114k = .614k(1 + o(1))$, and the proof is complete.

Remark 1. We claim that our analysis cannot show $R < k/2$. Indeed, one can construct S for which $|\mathsf{HighRedundancy}(S)| = m - (k - 1)$. Then, using Lemma 1, the best lower bound we can obtain for opt is $2m/k$. Thus, $R > k/2 - o(1)$.

Theorem 2. k-*MCIP is* $.614k(1+o(1))$-*approximable in* $O(m \log k)$ *probabilistic time and* $O(m\text{poly}(k))$ *deterministic time. Here,* $o(1) \to 0$ *as* $k \to \infty$.

Proof. It remains to establish the running time. The proof of Theorem 1 actually shows that only 3 invocations of HighFrequency are necessary to achieve the bound $.61391k(1 + o(1))$. So if we choose π_1, π_2, and π_3 judiciously, we may choose the π_i, $i \geq 4$, arbitrarily. By a Markov bound, the probability over the choices of π_1, π_2, and π_3, that the approximation ratio is less than $.614k(1+o(1))$ is $\Omega(1)$. To evaluate HighFrequency and all recursive calls on a given set of set-partitions π_i takes $O(m \log k)$ time since (1) there are $O(\log k)$ recursive calls, (2) CommonElements can be implemented in time proportional to its input size, and (3) the sum of input sizes across all calls to CommonElements in a given invocation of HighFrequency is at most m. By a Chernoff bound, we can output a $.614k(1+o(1))$-approximation in $O(m \log k)$ time with probability at least $99/100$ by running HighFrequency on $O(1)$ different triples (pi_1, π_2, π_3) and outputting the smallest partition found. The choice of (π_1, π_2, π_3) can be derandomized in $m\text{poly}(k)$ time with the method of conditional expectations. We omit the details.

Conclusions. We have given an $O(m \log k)$-time algorithm for k-MCIP with approximation ratio $.614k$, improving the previous bound of $k - 1/3$. The best lower bound is $\Omega(1)$. We believe it may be possible to slightly improve our aproximation ratio, but that significant progress will require a new approach.

Acknowledgment. We thank the referees and Lan Liu for helpful comments.

References

[1] X. Chen. *The minimum common partition revisited,* manuscript, 2005.

[2] X. Chen, L. Liu, Z. Liu, and T. Jiang. *On the minimum common integer partition problem,* CIAC 2006.

[3] X. Chen, J. Zheng, Z. Fu, P. Nan, Y. Zhong, S. Lonardi, and T. Jiang. *Computing the assignment of orthologous genes via genome rearrangement,* APBC, 2005.

[4] X. Chen, J. Zheng, Z. Fu, P. Nan, Y. Zhong, S. Lonardi, and T. Jiang. Assignment of orthologous genes via genome rearrangement. *IEEE/ACM Transactions on Computational Biology and Bioinformatics (TCBB)* 2-4, pp. 302-315, 2005

[5] Z. Fu, X. Chen, V. Vacic, P. Nan, Y. Zhong, and T. Jiang. *A parsimony approach to genome-wide ortholog assignment,* RECOMB, 2006.

[6] Tao Jiang. *Personal Communication.*

[7] V. Kann. *Maximum bounded 3-dimensional matching is MAX SNP-complete.* Information Processing Letters (IPL) 37: 27-35, 1991.

[8] R. Motwani and P. Raghavan. *Randomized Algorithms,* Cambridge University Press, 1995.

[9] C. H. Papadimitriou and M. Yannakakis. *Optimization, approximation, and complexity classes.* J. Computer and System Sciences (JCSS) 43: 425-440, 1991.

[10] L. Valinsky A. Schupham, G. D. Vedova, Z. Liu, A. Figueroa, K. Jampachaisri, B. Yin, E. Bent, R. Mancini-Jones, J. Press, T. Jiang, and J. Borneman. *Oligonucleotide fingerprinting of ribosomal RNA genes (OFRG),* pp. 569-585. In *Molecular Microbial Ecology Manual* (2nd ed). Kluwer Academic Publishers, Dordrecht, The Netherlands, 2004.

On Pseudorandom Generators with Linear Stretch in NC^0 ⋆

Benny Applebaum, Yuval Ishai, and Eyal Kushilevitz

Computer Science Department, Technion, Haifa 32000, Israel
{abenny, yuvali, eyal}@technion.ac.il

Abstract. We consider the question of constructing cryptographic pseudorandom generators (PRGs) in NC^0, namely ones in which each bit of the output depends on just a constant number of input bits. Previous constructions of such PRGs were limited to stretching a seed of n bits to $n + o(n)$ bits. This leaves open the existence of a PRG with a linear (let alone superlinear) stretch in NC^0. In this work we study this question and obtain the following main results:

1. We show that the existence of a linear-stretch PRG in NC^0 implies nontrivial hardness of approximation results *without relying on PCP machinery*. In particular, that Max 3SAT is hard to approximate to within some constant.
2. We construct a linear-stretch PRG in NC^0 under a specific intractability assumption related to the hardness of decoding "sparsely generated" linear codes. Such an assumption was previously conjectured by Alekhnovich [1].

We note that Alekhnovich directly obtains hardness of approximation results from the latter assumption. Thus, we do not prove hardness of approximation under new *concrete* assumptions. However, our first result is motivated by the hope to prove hardness of approximation under more general or standard cryptographic assumptions, and the second result is independently motivated by cryptographic applications.

1 Introduction

A cryptographic pseudorandom generator (PRG) [8, 24] is a deterministic function that stretches a short random seed into a longer string which cannot be distinguished from random by any polynomial-time observer. In this work, we study the existence of PRGs that are both (1) extremely parallel and (2) stretch their seed by a significant amount.

Considering the first goal alone, it was recently shown in [3] that the ultimate level of parallelism can be achieved under most standard cryptographic assumptions. Specifically, any PRG in NC^1 (the existence of which follows, for example, from the intractability of factoring, discrete logarithm, or lattice problems) can be efficiently "compiled" into a PRG in NC^0, namely one in which each output bit depends on just a constant number of input bits. However, the PRGs produced by this compiler can only stretch their seed by a sublinear amount: from n bits to $n + O(n^\epsilon)$ bits for some constant $\epsilon < 1$. Thus, these PRGs do not meet our second goal.

Considering the second goal alone, even a PRG that stretches its seed by just one bit can be used to construct a PRG that stretches its seed by any polynomial number of

---

⋆ Research supported by grant 36/03 from the Israel Science Foundation.

J. Diaz et al. (Eds.): APPROX and RANDOM 2006, LNCS 4110, pp. 260–271, 2006.

bits. However, all known constructions of this type are inherently sequential. Thus, we cannot use known techniques for turning an NC^0 PRG with a sublinear stretch into one with a linear, let alone superlinear, stretch.

The above state of affairs leaves open the existence of a *linear-stretch* PRG (LPRG) in NC^0; namely, one that stretches a seed of n bits into $n + \Omega(n)$ output bits.[1] (In fact, there was no previous evidence for the existence of LPRGs even in the higher complexity class AC^0.) This question is the main focus of our work. The question has a very natural motivation from a cryptographic point of view. Indeed, most cryptographic applications of PRGs either require a linear stretch (for example Naor's bit commitment scheme [19]), or alternatively depend on a larger stretch for efficiency (this is the case for the standard construction of a stream cipher or stateful symmetric encryption from a PRG, see [14]). Thus, the existence of an LPRG in NC^0 would imply better parallel implementations of other cryptographic primitives.

1.1 Our Contribution

LPRG in NC^0 implies hardness of approximation. We give a very different, and somewhat unexpected, motivation for the above question. We observe that the existence of an LPRG in NC^0 *directly* implies non-trivial and useful hardness of approximation results. Specifically, we show (via a very simple argument) that an LPRG in NC^0 implies that Max 3SAT cannot be efficiently approximated to within some multiplicative constant. This continues a recent line of work, initiated by Feige [12] and followed by Alekhnovich [1], that provides simpler alternatives to the traditional PCP-based approach by relying on stronger assumptions. Unlike these previous works, which rely on very specific assumptions, our assumption is of a more general flavor and may serve to further motivate the study of cryptography in NC^0. On the down side, the conclusions we get are weaker and in particular are implied by the PCP theorem. In contrast, some inapproximability results from [12, 1] could not be obtained using PCP machinery. It is instructive to note that by applying our general argument to the sublinear-stretch PRGs in NC^0 from [3] we only get "uninteresting" inapproximability results that follow from standard padding arguments (assuming P$\neq$NP). Furthermore, we do not know how to obtain stronger inapproximability results based on a superlinear-stretch PRG in NC^0. Thus, our main question of constructing LPRGs in NC^0 captures precisely what is needed for this application.

Constructing an LPRG in NC^0. We present a construction of an LPRG in NC^0 under a specific intractability assumption related to the hardness of decoding "sparsely generated" linear codes. Such an assumption was previously conjectured by Alekhnovich in [1]. The starting point of our construction is a modified version of a PRG from [1] that has a large output locality (that is, each output bit depends on many input bits) but has a simple structure. The main technical tool we employ in order to reduce its locality is a randomness extractor in NC^0 that can use a "sufficiently short" seed for sources with a "sufficiently high" entropy. We construct the latter by combining the known

---

[1] Note that an NC^0 LPRG can be composed with itself a constant number of times to yield an NC^0 PRG with arbitrary constant stretch.

construction of randomness extractors from ϵ-biased generators [18, 6] with previous constructions of ϵ-biased generator in NC^0 [17]. Our LPRG can be implemented with locality 4; this LPRG is essentially optimal, as it is known that no PRG with locality 4 can have a *superlinear* stretch [17]. However, the existence of superlinear-stretch PRG with a higher (but constant) locality remains open.

By combining the two main results described above, one gets non-trivial inapprox-imability results under the intractability assumption from [1]. These (and stronger) re-sults were *directly* obtained in [1] from the same assumption *without* constructing an LPRG in NC^0. Our hope is that future work will yield constructions of LPRGs in NC^0 under different, perhaps more standard, assumptions, and that the implications to hard-ness of approximation will be strengthened.

LPRG in NC^0 and Expanders. Finally, we observe that any LPRG in NC^0 contains a copy of a graph with some non-trivial expansion property. This connection implies that a (deterministic) construction of an LPRG in NC^0 must use some non-trivial com-binatorial objects. (In particular, one cannot hope for "simple" transformations, such as those given in [3], to yield LPRGs in NC^0.) The connection with expanders also allows to rule out the existence of *exponentially*-strong PRGs with *superlinear* stretch in NC^0.

1.2   Related Work

The existence of PRGs in NC^0 has been recently studied in [10, 17, 3]. Cryan and Mil-tersen [10] observe that there is no PRG in NC_2^0 (i.e., where each output bit depends on at most two input bits), and prove that there is no PRG in NC_3^0 achieving a superlinear stretch; namely, one that stretches n bits to $n + \omega(n)$ bits. Mossel et al. [17] extend this impossibility to NC_4^0. Viola [23] shows that an LPRG in AC^0 cannot be obtained from a OWF via non-adaptive black-box constructions. This result can be extended to rule out such a construction even if we start with a PRG whose stretch is sublinear.

On the positive side, Mossel et al. [17] constructed (non-cryptographic) ε-biased generators with linear stretch and exponentially small bias in NC_5^0. Later, in [3] it was shown that, under standard cryptographic assumptions, there are pseudorandom gener-ators in NC_4^0. However, these PRGs have only *sublinear-stretch*.

The first application of average-case complexity to inapproximability was suggested by Feige [12], who derived new inapproximability results under the assumption that re-futing 3SAT is hard on average on some natural distribution. Alekhnovich [1] continued this line of research. He considered the problem of determining the maximal number of satisfiable equations in a linear system chosen at random, and made several conjectures regarding the average case hardness of this problem. He showed that these conjectures imply Feige's assumption as well as several new inapproximability results. While the works of Feige and Alekhnovich derived *new* inapproximability results (that were not known to hold under the assumption that $P \neq NP$), they did not rely on the relation with a standard cryptographic assumption or primitive, but rather used specific average case hardness assumptions tailored to their inapproximability applications. A relation between the security of a cryptographic primitive and approximation was implicitly used in [17], where an approximation algorithm for Max 2LIN was used to derive an upper bound on the stretch of a PRG whose locality is 4.

2 Preliminaries

Probability notation. We use U_n to denote a random variable uniformly distributed over $\{0,1\}^n$. If X is a probability distribution, or a random variable, we write $x \leftarrow X$ to indicate that x is a sample taken from X. The *min-entropy* of a random variable X is defined as $\mathrm{H}_\infty(X) \stackrel{\text{def}}{=} \min_x \log(\frac{1}{\Pr[X=x]})$. The *statistical distance* between discrete probability distributions Y and Y', denoted $\|Y - Y'\|$, is defined as the maximum, over all functions A, of the *distinguishing advantage* $|\Pr[A(Y) = 1] - \Pr[A(Y') = 1]|$.

A function $\varepsilon(\cdot)$ is said to be *negligible* if $\varepsilon(n) < n^{-c}$ for any constant $c > 0$ and sufficiently large n. We will sometimes use $\mathrm{neg}(\cdot)$ to denote an unspecified negligible function. For two distribution ensembles $\{X_n\}_{n\in\mathbb{N}}$ and $\{Y_n\}_{n\in\mathbb{N}}$, we write $X_n \equiv Y_n$ if X_n and Y_n are identically distributed, and $X_n \stackrel{s}{\approx} Y_n$ if the two ensembles are *statistically indistinguishable*; namely, $\|X_n - Y_n\|$ is negligible in n. A weaker notion of closeness between distributions is that of *computational* indistinguishability: We write $X_n \stackrel{c}{\approx} Y_n$ if for every (non-uniform) polynomial-size circuit family $\{A_n\}$, the distinguishing advantage $|\Pr[A_n(X_n) = 1] - \Pr[A_n(Y_n) = 1]|$ is negligible. By definition, $X_n \equiv Y_n$ implies that $X_n \stackrel{s}{\approx} Y_n$ which in turn implies that $X_n \stackrel{c}{\approx} Y_n$. A distribution ensemble $\{X_n\}_{n\in\mathbb{N}}$ is said to be *pseudorandom* if $X_n \stackrel{c}{\approx} U_n$.

We will use the following definition of a pseudorandom generator.

Definition 1. (Pseudorandom generator) *A pseudorandom generator (PRG) is a deterministic function* $G : \{0,1\}^* \to \{0,1\}^*$ *satisfying the following two conditions:*

- *Expansion: There exists a* stretch function $s : \mathbb{N} \to \mathbb{N}$ *such that* $s(n) > n$ *for all* $n \in \mathbb{N}$ *and* $|G(x)| = s(|x|)$ *for all* $x \in \{0,1\}^*$.
- *Pseudorandomness: The ensembles* $\{G(U_n)\}_{n\in\mathbb{N}}$ *and* $\{U_{s(n)}\}_{n\in\mathbb{N}}$ *are computationally indistinguishable.*

When $s(n) = n + \Omega(n)$ *we say that* G *is a* linear-stretch *pseudorandom generator (LPRG). By default, we require* G *to be polynomial time computable.*

It will sometimes be convenient to define a PRG by an infinite family of functions $\{G_n : \{0,1\}^{m(n)} \to \{0,1\}^{s(n)}\}_{n\in\mathbb{N}}$. Such a family can be transformed into a single function that satisfies Definition 1 via padding. We will also rely on ε-*biased generators*, defined similarly to PRGs except that the pseudorandomness holds only against linear functions. Namely, for a bias function $\varepsilon : \mathbb{N} \to (0,1)$ we say that $G : \{0,1\}^n \to \{0,1\}^{s(n)}$ is an ε-biased generator if for every non-constant linear function $L : \mathrm{GF}_2^n \to \mathrm{GF}_2$ and all sufficiently large n's it holds that $|\Pr[L(G(U_n)) = 1] - \frac{1}{2}| < \varepsilon(n)$.

Locality. We say that $f : \{0,1\}^n \to \{0,1\}^s$ is c-*local* if each of its output bits depends on at most c input bits, and that $f : \{0,1\}^* \to \{0,1\}^*$ is c-local if for every n the restriction of f to n-bit inputs is c-local. The uniform versions of these classes contain functions that can be computed in polynomial time.

3 LPRG in NC^0 Implies Hardness of Approximation

In the following we show that if there exists an LPRG in NC^0 then there is no polynomial-time approximation scheme (PTAS) for Max 3SAT; that is, Max 3SAT cannot be efficiently approximated within some multiplicative constant $r > 1$. Recall that in the Max 3SAT problem we are given a 3CNF boolean formula with s clauses over n variables, and our goal is to find an assignment that satisfies the largest possible number of clauses. The Max ℓ-CSP problem is a generalization of Max 3SAT in which instead of s clauses we get s boolean constraints $C = \{C_1, \ldots, C_s\}$ of arity ℓ. Again, our goal is to find an assignment that satisfies the largest possible number of constraints. (Recall that a constraint C of arity ℓ over n variables is a pair $(f : \{0,1\}^k \to \{0,1\}, (i_1, \ldots, i_k))$. A constraint C is satisfied by an assignment $(\sigma_1, \ldots, \sigma_n)$ if $f(\sigma_{i_1}, \ldots, \sigma_{i_k}) = 1$.)

The following standard lemma shows that in order to prove that Max 3SAT is hard to approximate, it suffices to prove that Max ℓ-CSP is hard to approximate. This follows by applying Cook's reduction to transform every constraint into a 3CNF.

Lemma 1. *Assume that, for some constants $\ell \in \mathbb{N}$ and $\varepsilon > 0$, there is no polynomial time $(1 + \varepsilon)$-approximation algorithm for Max ℓ-CSP. Then there is an $\varepsilon' > 0$ such that there is no polynomial time $(1 + \varepsilon')$-approximation algorithm for Max 3SAT.*

A simple and useful corollary of the PCP Theorem [5, 4] is the inapproximability of Max 3SAT.

Theorem 1. *Assume that* $P \neq NP$. *Then, there is an $\varepsilon > 0$ such that there is no $(1 + \varepsilon)$-approximatation algorithm for Max 3SAT.*

We now prove a similar result under the (stronger) assumption that there exists an LPRG in NC^0 without relying on the PCP Theorem.

Theorem 2. *Assume that there exists an LPRG in NC^0. Then, there is an $\varepsilon > 0$ such that there is no $(1 + \varepsilon)$-approximation algorithm for Max 3SAT.*

Proof. Let $s(n) = cn$ for some constant $c > 1$, and let $s = s(n)$. Let $G : \{0,1\}^n \to \{0,1\}^{s(n)}$ be an LPRG which is computable in NC^0_ℓ. Let $0 < \varepsilon < 1/2$ be a constant that satisfies $H_2(\varepsilon) < 1/2 - 1/(2c)$, where $H_2(\cdot)$ is the binary entropy function. Assume towards a contradiction that there exists a PTAS for Max 3SAT. Then, by Lemma 1, there exists a PTAS for Max ℓ-CSP. Hence, there exists a polynomial-time algorithm A_ε that distinguishes satisfiable instances of ℓ-CSP from instances of ℓ-CSP for which any assignment fails to satisfy a fraction ε of the constraints. We show that, given A_ε, we can "break" the LPRG G; that is, we can construct an efficient (non-uniform) adversary that distinguishes between $G(U_n)$ and U_s. Our adversary B_n will translate a string $y \in \{0,1\}^s$ into an ℓ-CSP instance ϕ_y with s constraints such that,

1. If $y \leftarrow G(U_n)$ then ϕ_y is always satisfiable.
2. If $y \leftarrow U_s$ then, with probability $1 - \mathrm{neg}(n)$, no assignment satisfies more than $(1 - \varepsilon)s$ constraints of ϕ_y.

Then, B_n will run A_ε on ϕ_y and will output $A_\varepsilon(\phi_y)$. The distinguishing advantage of B is $1 - \mathrm{neg}(n)$ in contradiction to the pseudorandomness of G.

It is left to show how to translate $y \in \{0,1\}^s$ into an ℓ-CSP instance ϕ_y. We use n boolean variables $x_1, \ldots, x_n$ that represent the bits of an hypothetical pre-image of y under G. For every $1 \leq i \leq s$ we add a constraint $G_i(x) = y_i$ where G_i is the function that computes the i-th output bit of G. Since G_i is an ℓ-local function the arity of the constraint is at most ℓ.

Suppose first that $y \leftarrow G(U_n)$. Then, there exists a string $\sigma \in \{0,1\}^n$ such that $G(\sigma) = y$ and hence ϕ_y is satisfiable. We move on to the case in which $y \leftarrow U_s$. Here, we rely on the fact that such a random y is very likely to be far from every element in the range of G. More formally, we define a set $\mathrm{BAD}_n \subseteq \{0,1\}^s$ such that $y \in \mathrm{BAD}_n$ if ϕ_y is $(1 - \varepsilon)$-satisfiable; that is, if there exists an assignment $\sigma \in \{0,1\}^n$ that satisfies a fraction $(1 - \varepsilon)$ of the constraints of ϕ_y. In this case, the Hamming distance between y and $\mathrm{Im}(G)$ is at most εs. Therefore, the size of BAD_n is bounded by

$$|\mathrm{Im}(G)| \cdot \binom{s}{\varepsilon s} \leq 2^n 2^{\mathrm{H}_2(\varepsilon)s} = 2^{n(1+c\mathrm{H}_2(\varepsilon))} \leq 2^{n(1+c(\frac{1}{2}-\frac{1}{2c}))}.$$

Hence,

$$\Pr_{y \leftarrow U_s}[\phi_y \text{ is } (1 - \varepsilon) \text{ satisfiable}] = \mathrm{BAD}_n \cdot 2^{-s} \leq 2^{n(-c+1+c(\frac{1}{2}-\frac{1}{2c}))} = 2^{(1-c)\frac{n}{2}},$$

which completes the proof. □

Remark 1. Theorem 2 can tolerate some relaxations to the notion of LPRG. In particular, since the advantage of B_n is exponentially close to 1, we can consider an LPRG that satisfies a weaker notion of pseudorandomness in which the distinguisher's advantage is bounded by $1 - 1/p(n)$ for some polynomial $p(n)$.

Papadimitriou and Yannakakis showed in [20] that if Max 3SAT does not have a PTAS (i.e., it cannot be approximated up to an arbitrary constant), then several other problems do not have PTAS as well (e.g., Max Cut, Max 2SAT, Vertex Cover). In fact, [20] defined the class Max SNP, and showed that Max 3SAT is complete for this class in the sense that any problem in Max SNP does not have a PTAS unless Max 3SAT has a PTAS. Hence, we get the following corollary (again, without the PCP machinery):

Corollary 1. *Assume that there exists LPRG in* NC^0. *Then, all Max SNP problems do not have a PTAS.*

4   A Construction of LPRG in NC^0

For ease of presentation, we describe our construction in a non-uniform way. We will later discuss a uniform variant of the construction.

4.1   The Assumption

Let $m = m(n)$ be an output length parameter where $m(n) > n$, let $\ell = \ell(n)$ be a locality parameter (typically a constant), and let $0 < \mu < 1$ be a noise parameter. Let $\mathcal{M}_{m,n,\ell}$ be the set of all $m \times n$ matrices over GF$_2$ in which each row contains exactly ℓ ones. For a matrix $M \in \mathcal{M}_{m,n,\ell}$ we denote by $D_\mu(M)$ the distribution of the random vector

$$Mx + e,$$

where $x \leftarrow U_n$ and $e \in \{0,1\}^m$ is a random error vector in which each entry is chosen to be 1 with probability μ (independently of other entries), and arithmetic is over GF$_2$.

The following assumption is a close variant of a conjecture suggested by Alekhnovich in [1, Conjecture 1]. [2]

Assumption 3. *For any $m(n) = O(n)$, and any constant $0 < \mu < 1$, there exists a positive integer ℓ, and an infinite family of matrices $\{M_n\}_{n \in \mathbb{N}}$, $M_n \in \mathcal{M}_{m(n),n,\ell}$, such that*

$$D_\mu(M_n) \overset{c}{\approx} D_{\mu+1/m(n)}(M_n)$$

(Note that since we consider non-uniform distinguishers, we can assume that M_n is public and is available to the distinguisher.)

Alekhnovich [1] shows that if the distribution $D_\mu(M_n)$ satisfies the above assumption then it is pseudorandom. (In fact, the original claim proved in [1, Thm. 3.1] deals with slightly different distributions. However, the proof can be adapted to our setting.)

Lemma 2. *For any polynomial $m(n)$ and constant $0 < \mu < 1$, and any infinite family, $\{M_n\}_{n \in \mathbb{N}}$, of $m(n) \times n$ matrices over GF_2, if $D_\mu(M_n) \overset{c}{\approx} D_{\mu+1/m(n)}(M_n)$, then $D_\mu(M_n) \overset{c}{\approx} U_{m(n)}$.*

Proof sketch. The proof follows by combining the following easy claims:

1. $D_{\mu+1/m(n)}(M_n) \equiv D_\mu(M_n) + r_n$ where $r_n \in \{0,1\}^{m(n)}$ is a random vector in which each entry is chosen to be 1 with probability $c/m(n)$ (independently of other entries) for some constant $c > 1$.

2. Let $r_n^{t(n)}$ be the distribution resulting from summing $t(n)$ independent samples from r_n. Then, for some polynomial $t(n)$ it holds that $r_n^{t(n)} \overset{s}{\approx} U_{m(n)}$.

3. Let $\{A_n\}$ be a polynomial-time samplable distribution ensemble over $\mathrm{GF}_2^{m(n)}$. For a polynomial $t(n)$, let $A_n^{t(n)}$ be the sum (over GF_2) of $t(n)$ independent samples from A_n. Suppose that $D_n \overset{c}{\approx} D_n + A_n$ for some distribution ensemble $\{D_n\}$. Then, for every polynomial $t(n)$ we have $D_n \overset{c}{\approx} D_n + A_n^{t(n)}$.

By the first claim and the Lemma's hypothesis, we have $D_\mu(M_n) \overset{c}{\approx} D_\mu(M_n) + r_n$. Hence, for some polynomial $t(n)$,

$$D_\mu(M_n) \overset{c}{\approx} D_n + r_n^{t(n)} \overset{s}{\approx} D_n + U_{m(n)} \equiv U_{m(n)},$$

where the first transition is due to the third claim and the second transition is due to the second claim.   □

By combining Assumption 3 and Lemma 2, we get the following proposition:

Proposition 1. *Suppose that Assumption 3 holds. Then, for any $m(n) = O(n)$, and any constant $0 < \mu < 1$, there exists a constant $\ell \in \mathbb{N}$, and an infinite family of matrices $\{M_n\}_{n \in \mathbb{N}}$ where $M_n \in \mathcal{M}_{m(n),n,\ell}$ such that $D_\mu(M_n) \overset{c}{\approx} U_{m(n)}$.*

---

[2] Our assumption is essentially the same as Alekhnovich's. The main difference between the two assumptions is that the noise vector e in [1] is a random vector of weight $\lceil \mu m \rceil$, as opposed to our noise vector whose entries are chosen to be 1 independently with probability μ. It can be shown that our assumption is implied by Alekhnovich's assumption (since our iid noise vectors can be viewed as a convex combination of noise vectors of fixed weight).

Remark 2. If the restriction on the density of the matrices M_n is dropped, the above proposition can be based on the conjectured (average case) hardness of decoding a random linear code (cf., [7, 15]). In fact, under the latter assumption we have that D_μ $(M_n) \stackrel{c}{\approx} U_{m(n)}$ for *most* choices of M_n's.

4.2 The Construction

From here on, we let $\mu = 2^{-t}$ for some $t \in \mathbb{N}$. Then we can sample each bit of the error vector e by taking the product of t independent random bits. In this case, we can define an NC^0 function whose output distribution is pseudorandom. Namely,

$$f_n(x, \hat{e}) = M_n x + E(\hat{e})$$

where

$$x \in \{0,1\}^n, \qquad \hat{e} \in \{0,1\}^{t \cdot m(n)}, \qquad E(\hat{e}) = \left(\prod_{j=1}^{t} \hat{e}_{t \cdot (i-1)+j} \right)_{i=1}^{m(n)}. \qquad (1)$$

Since $f_n(U_n, U_{t \cdot m(n)}) \equiv D_\mu(M_n)$, the distribution $f_n(U_n, U_{t \cdot m(n)})$ is pseudorandom under Assumption 3 (when the parameters are chosen appropriately). Moreover, the locality of f_n is $\ell + t = O(1)$. However, f_n is not a pseudorandom generator as it uses $n + t \cdot m(n)$ input bits while it outputs only $m(n)$ bits. To overcome this obstacle, we note that most of the entropy of $\hat{e}$ was not "used". Hence, we can apply an *extractor* to regain the lost entropy. Of course, in order to get a PRG in NC^0 the extractor should also be computed in NC^0. Moreover, to get a linear stretch we should extract all the $t \cdot m(n)$ random bits from $\hat{e}$ by investing less than n additional random bits. In the following, we show that such extractors can be implemented by using *ε-biased generators*.

First, we show that the distribution of $\hat{e}$ given $E(\hat{e})$ contains (with high probability) a lot of entropy. In the following we let $m = m(n)$.

Lemma 3. *Let $\hat{e} \leftarrow U_{t \cdot m}$ and $E(\hat{e})$ be defined as in Eq. 1. Denote by $[\hat{e}|E(\hat{e})]$ the distribution of $\hat{e}$ given the outcome of $E(\hat{e})$. Then, except with probability $e^{-(2^{-t}m)/3}$, it holds that*

$$\mathrm{H}_\infty([\hat{e}|E(\hat{e})]) \geq m(1 - 2^{-t+1}) \log(2^t - 1) \geq t \cdot m(1 - \delta(t)), \qquad (2)$$

where $\delta(t) = 2^{-\Omega(t)}$.

Proof. We view $E(\hat{e})$ as a sequence of m independent Bernoulli trials, each with a probability 2^{-t} of success. Recall that $\hat{e}$ is composed of m blocks of length t, and that the i-th bit of $E(\hat{e})$ equals the product of the bits in the i-th block of $\hat{e}$. Hence, whenever $E(\hat{e})_i = 1$ all the bits of the i-th block of $\hat{e}$ equal to 1, and when $E(\hat{e})_i = 0$ the i-th block of $\hat{e}$ is uniformly distributed over $\{0,1\}^t \setminus \{1^t\}$. Consider the case in which at most $2 \cdot 2^{-t} m$ components of $E(\hat{e})$ are ones. By a Chernoff bound, the probability of this event is at least $1 - e^{-(2^{-t}m)/3}$. In this case, $\hat{e}$ is uniformly distributed over a set of size at least $(2^t - 1)^{m(1 - 2^{-t+1})}$. Hence, $\mathrm{H}_\infty([\hat{e}|E(\hat{e})]) \geq m(1 - 2^{-t+1}) \log(2^t - 1) \geq tm(1 - \delta(t))$, for $\delta(t) = 2^{-\Omega(t)}$. $\qquad \square$

ε-biased generators can be used to extract random bits from distributions that contain sufficient randomness. Extractors based on ε-biased generators were previously used in [6, 11]. Formally,

Lemma 4 ([18, 2, 16]). *Let $g : \{0,1\}^s \to \{0,1\}^n$ be an ε-biased generator, and let X_n be a random variable taking values in $\{0,1\}^n$ whose min-entropy is at least p. Then,*

$$\|(g(U_s) + X_n) - U_n\| \le \varepsilon \cdot 2^{(n-p-1)/2} \ .$$

It can be shown that for some fixed exponentially small bias $\varepsilon(n) = 2^{-\Omega(n)}$ and every constant c there exists an ε-biased generator in NC^0 that stretches n bits into cn bits. (The locality of this generator depends on c). Hence, whenever p exceed some linear threshold we can extract n bits from X_n in NC^0 by investing only n/c random bits for any arbitrary c. (Details are deferred to the full version.) However, in our case p is very close to n and so we can rely on a weaker ε-biased generator with an arbitrary linear stretch c and bias $\varepsilon = 2^{-n/poly(c)}$. Recently, Mossel et al. [17] constructed such an ε-biased generator in NC_5^0.

Lemma 5 ([17], Thm. 14). *For every constant c, there exists an ε-biased generator $g : \{0,1\}^n \to \{0,1\}^{cn}$ in NC_5^0 whose bias is at most $2^{-bn/c^4}$ (where b is some universal constant that does not depend on c).*

We remark that the above construction can be implemented in *uniform* NC^0 by using the results of [9, Theorem 7.1]. [3]
    We can now describe our LPRG.

Construction 4. *Let t and ℓ be positive integers, and $c, k > 1$ be real numbers that will be used as stretch factors. Let $m = kn$ and let $\{M_n \in \mathcal{M}_{n,m,\ell}\}$ be an infinite family of matrices. Let $g : \{0,1\}^n \to \{0,1\}^{cn}$ be the ε-biased generator promised by Lemma 5. We define the function*

$$G_n(x, \hat{e}, r) = (M_n x + E(\hat{e}), g(r) + \hat{e}),$$

where $x \in \{0,1\}^n, \hat{e} \in \{0,1\}^{t \cdot m}, r \in \{0,1\}^{t \cdot m/c}, E(\hat{e}) = \left(\prod_{j=1}^t \hat{e}_{t \cdot (i-1)+j} \right)_{i=1}^m$.

Observe that G_n is an NC^0 function. We show that if the parameters are chosen properly then G_n is an LPRG.

Lemma 6. *Under Assumption 3, there exist constants $t, \ell \in \mathbb{N}$, constants $c, k > 1$, and a family of matrices $\{M_n \in \mathcal{M}_{n,m,\ell}\}$ such that the function G_n defined in Construction 4 is an LPRG.*

Proof. Set $k > 1$ to be some arbitrary constant and let $m = kn$. Let $c = 2t/(1 - 1/k)$ and choose t to be a constant satisfying:

$$\Delta \stackrel{\text{def}}{=} \frac{bt}{c^5} - \delta(t) > 0, \tag{3}$$

---

[3] Theorem 7.1 of [9] gives an explicit family of asymmetric constant-degree bipartite expanders, which can replace the probabilistic construction given in [17, Lemma 12]. We note that the locality of the resulting generator depends on c. See full version for details.

where $\delta(\cdot)$ is the negligible function from Eq. 2 and b is the bias constant of Lemma 5. There exists a (large) constant t satisfying the above since $\delta(t) = 2^{-\Omega(t)}$ while $bt/c^5 = \Theta(1/t^4)$. Let $\ell \in \mathbb{N}$ be a constant and $\{M_n \in \mathcal{M}_{n,m,\ell}\}$ be an infinite family of matrices satisfying Assumption 3.

First, we show that G_n has linear stretch. The input length of G_n is $n + tm + tm/c = n(tk + k/2 + 1/2)$. The output length is $m(t + 1) = n(tk + k)$. Hence, since $k > 1$, the constant $tk + k/2 + 1/2$ is smaller than the constant $tk + k$, and so the function G_n has a linear stretch.

Let $x, \hat{e}$ and r be uniformly distributed over $\{0,1\}^n$, $\{0,1\}^{t\cdot m}$ and $\{0,1\}^{t\cdot m/c}$ respectively. We prove that the distribution $G_{M_n}(x, \hat{e}, r)$ is pseudorandom. By Lemmas 3, 4 and 5 it holds that

$$\|(E(\hat{e}), \hat{e} + g(r)) - (E(\hat{e}), U_{t\cdot m})\| \leq e^{-(2^{-t}m)/3} + 2^{-b(tm/c)/c^4}2^{(tm-(t-\delta(t))m-1)/2}$$
$$\leq e^{-(2^{-t}m)/3} + 2^{(\delta(t)-bt/c^5)m}$$
$$\leq e^{-(2^{-t}m)/3} + 2^{-\Delta m} = \text{neg}(m) = \text{neg}(n),$$

where the last inequality is due to Eq. 3. Therefore, by Proposition 1, we get that

$$(M_n x + E(\hat{e}), g(r) + \hat{e}) \overset{s}{\approx} (M_n x + E(\hat{e}), U_{t\cdot m}) \equiv (D_{2^{-t}}(M_n), U_{t\cdot m}) \overset{c}{\approx} (U_m, U_{t\cdot m}) \ .$$

$\square$

By the above Lemma we get a construction of LPRG in NC⁰ from Assumption 3. In fact, in [3] it is shown that such an LPRG can be transformed into an LPRG whose locality is 4. Hence, we have:

Theorem 5. *Under Assumption 3, there exists an LPRG in* NC$_4^0$.

Mossel et al. [17] showed that a PRG in NC$_4^0$ cannot achieve a superlinear stretch. Hence, Theorem 5 is essentially optimal with respect to stretch.

Remarks on Theorem 5.

1. (Uniformity) Our construction uses a family of matrices $\{M_n\}$ satisfying Assumption 3 as a non-uniform advice. We can eliminate this advice and construct an LPRG in *uniform* NC$_4^0$ by slightly modifying Assumption 3. In particular, we follow Alekhnovich (cf. [1, Remark 1]) and conjecture that any family $\{M_n\}$ of good expanders satisfy Assumption 3. Hence, our construction can be implemented by using an explicit family of asymmetric constant-degree bipartite expanders such as the one given in [9, Theorem 7.1].

2. (The stretch of the construction) Our techniques do not yield a *superlinear* stretch PRG in NC⁰. To see this, consider a variant of Assumption 3 in which we allow $m(n)$ to be superlinear and let $\mu(n)$ be subconstant. (These modifications are necessary to obtain a superlinear PRG.) In this case, the noise distribution cannot be sampled in NC⁰ (since $\mu(n)$ is subconstant). This problem can be bypassed by extending Assumption 3 to alternative noise models in which the noise is not iid. However, it is not clear how such a modification affects the hardness assumption.

5 The Necessity of Expansion

As pointed out inthe previous section, our construction of LPRG makes use of expander graphs. This is also the case in several constructions of "hard functions" with low locality (e.g., [13, 17, 1]). We now show that this is not coincidental at least in the case of PRGs. Namely, we show that the structure of any LPRG in NC^0 contains a copy of a graph with some expansion property. (In fact, this holds even in the case of ε-biased generators.) Then, we use known lower bounds for expander graphs to rule out the possibility of exponentially strong PRG with superlinear stretch in NC^0.

Let $g : \{0,1\}^n \to \{0,1\}^s$ be a PRG. We claim that every set S of output bits whose size is $O(\log n)$ touches at least $|S|$ input bits. Otherwise, there exists a small set S of output bits and a string $y \in \{0,1\}^{|S|}$ such that $\Pr[g_S(U_n) = y] = 0$ (where $g_S(\cdot)$ is the restriction of g to the output bits of S). Hence, an efficient adversary can distinguish between $g_S(U_n)$ and $U_{|S|}$ with advantage $2^{-O(logn)} = 1/\text{poly}(n)$, in contradiction to the pseudorandomness of g. More generally, if g is ε-strong (i.e., cannot be broken by any efficient adversary with probability ε), then every set of $t \leq \log(1/\varepsilon)$ output bits touches at least t input bits. This claim extends to the case of ε-biased generators by using the Vazirani XOR Lemma [22].

In graph theoretic terms, we have a bipartite graph $G = ((\text{In} = [n], \text{Out} = [s]), E)$ that enjoys some output expansion property. This property is trivial when the output degree of G is high (as in standard constructions of PRGs) or when s is not much larger than n (as in the NC^0 constructions of [3]). However, when the locality is constant and the stretch is linear, G is a sparse bipartite graph having n input vertices, $s = n + \Omega(n)$ output vertices, and a constant output degree. In the standard cryptographic setting, when $\varepsilon(n)$ is negligible, we get expansion for sets of size $O(\log(n))$. That is, G expands (output) sets of size smaller than $\omega(\log n)$. When $\varepsilon < 2^{-\Omega(n)}$ (as in the ε-biased construction of [17]), we get expansion for sets of size at most $\Omega(n)$.

Radhakrishnan and Ta-Shma [21] obtained some lower bounds for similar graphs. In particular, by using [21, Thm. 1.5] it can be shown that if $g : \{0,1\}^n \to \{0,1\}^s$ is an NC^0_ℓ function that enjoys the above expansion property for sets of size $\leq t$, then $\ell \geq \Omega(\log(s/t)/\log(n/t))$. We therefore conclude that there is no $2^{-\Omega(n)}$-strong PRG (resp. $2^{-\Omega(n)}$-biased generator) with superlinear stretch in NC^0.

Acknowledgments. We thank Eli Ben-Sasson and Amir Shpilka for helpful discussions.

References

1. M. Alekhnovich. More on average case vs approximation complexity. In *Proc. 44th FOCS*, pages 298–307, 2003.
2. N. Alon and Y. Roichman. Random cayley graphs and expanders. *Random Struct. Algorithms*, 5(2):271–285, 1994.
3. B. Applebaum, Y. Ishai, and E. Kushilevitz. Cryptography in NC^0. *SIAM J. Comput.* To appear. Preliminary version in FOCS 04.
4. S. Arora, C. Lund, R. Motwani, M. Sudan, and M. Szegedy. Proof verification and hardness of approximation problems. *J. of the ACM*, 45(3):501–555, 1998.

5. S. Arora and S. Safra. Probabilistic checking of proofs: A new characterization of np. *J. of the ACM*, 45(1):70–122, 1998.
6. E. Ben-Sasson, M. Sudan, S. Vadhan, and A. Wigderson. Randomness-efficient low-degree tests and short pcps via epsilon-biased sets. In *Proc. 35th STOC*, pages 612–621, 2003.
7. A. Blum, M. Furst, M. Kearns, and R. J. Lipton. Cryptographic primitives based on hard learning problems. In *Advances in Cryptology: Proc. of CRYPTO '93*, volume 773 of *LNCS*, pages 278–291, 1994.
8. M. Blum and S. Micali. How to generate cryptographically strong sequences of pseudo-random bits. *SIAM J. Comput.*, 13:850–864, 1984.
9. M. Capalbo, O. Reingold, S. Vadhan, and A. Wigderson. Randomness conductors and constant-degree lossless expanders. In *Proc. 34th STOC*, pages 659–668, 2002.
10. M. Cryan and P. B. Miltersen. On pseudorandom generators in NC^0. In *Proc. 26th MFCS*, 2001.
11. Y. Dodis and A. Smith. Correcting errors without leaking partial information. In *Proc. 37th STOC*, pages 654–663, 2005.
12. U. Feige. Relations between average case complexity and approximation complexity. In *Proc. of 34th STOC*, pages 534–543, 2002.
13. O. Goldreich. Candidate one-way functions based on expander graphs. *ECCC*, 7(090), 2000.
14. O. Goldreich. *Foundations of Cryptography: Basic Tools*. Cambridge University Press, 2001.
15. O. Goldreich, H. Krawczyk, and M. Luby. On the existence of pseudorandom generators. *SIAM J. Comput.*, 22(6):1163–1175, 1993.
16. O. Goldreich and A. Wigderson. Tiny families of functions with random properties: A quality-size trade-off for hashing. *Random Struct. Algorithms*, 11(4):315–343, 1997.
17. E. Mossel, A. Shpilka, and L. Trevisan. On ϵ-biased generators in NC^0. In *Proc. 44th FOCS*, pages 136–145, 2003.
18. J. Naor and M. Naor. Small-bias probability spaces: Efficient constructions and applications. *SIAM J. Comput.*, 22(4):838–856, 1993.
19. M. Naor. Bit commitment using pseudorandomness. *J. of Cryptology*, 4:151–158, 1991.
20. C. Papadimitriou and M. Yannakakis. Optimization, approximation, and complexity classes. *J. of Computer and Systems Sciences*, 43:425–440, 1991.
21. J. Radhakrishnan and A. Ta-Shma. Tight bounds for depth-two superconcentrators. *SIAM J. Discrete Math.*, 13(1):2–24, 2000.
22. U. Vazirani. *Randomness, Adversaries and Computation*. Ph.d. thesis, UC Berkeley, 1986.
23. E. Viola. On constructing parallel pseudorandom generators from one-way functions. In *Proc. 20th CCC*, pages 183– 197, 2005.
24. A. C. Yao. Theory and application of trapdoor functions. In *Proc. 23rd FOCS*, pages 80–91, 1982.

A Fast Random Sampling Algorithm for Sparsifying Matrices

Sanjeev Arora*, Elad Hazan*, and Satyen Kale*

Computer Science Department, Princeton University
35 Olden Street, Princeton, NJ 08540
{arora, ehazan, satyen}@cs.princeton.edu

Abstract. We describe a simple random-sampling based procedure for producing sparse matrix approximations. Our procedure and analysis are extremely simple: the analysis uses nothing more than the Chernoff-Hoeffding bounds. Despite the simplicity, the approximation is comparable and sometimes better than previous work.

Our algorithm computes the sparse matrix approximation in a single pass over the data. Further, most of the entries in the output matrix are quantized, and can be succinctly represented by a bit vector, thus leading to much savings in space.

1 Introduction

Eigenvector computations are ubiquitous in numerous algorithmic tasks: a few applications include clustering in high dimensional data, principal component analysis, spectral graph partitioning, semidefinite programming, and Google's PageRank algorithm. Because of the central importance of eigenvector computations, this problem has been very well studied by numerical analysts.

In practical applications one frequently needs to compute eigenvectors of matrices arising from massive data sets such as web corpora, images, or video. Any superlinear computation quickly becomes infeasible as the matrix size becomes large. If approximate eigenvectors are allowed, then one can use the power method and the Lanczos method [TB97] which are very efficient in practice. These two methods spend the bulk of their processing time in computing matrix-vector products. Computing a matrix-vector product takes time proportional to the number of non-zero entries in the matrix (i.e. the *sparsity* of the matrix), and suggests that the eigenvector computation could be sped up by sparsifying the matrix first. This involves computing a different matrix that has fewer non-zero entries than the original, yet remains close to it by some metric. Section 2 makes these notions precise and describes how such a sparse matrix approximation can be used as a proxy for the original in the eigenvector computation.

Frieze, Kannan and Vempala [FKV04] considered the problem of efficiently computing low-rank approximations to matrices, and gave an algorithm to perform the task via random sampling of the columns of the input matrix with

---

* Supported by Sanjeev Arora's NSF grants MSPA-MCS 0528414, CCF 0514993, ITR 0205594.

J. Diaz et al. (Eds.): APPROX and RANDOM 2006, LNCS 4110, pp. 272–279, 2006.

carefully chosen probabilities. Since many columns are discarded in the random sampling process, the algorithm can be interpreted as computing a sparse representation of the input matrix (albeit only for the specific application of computing low-rank approximations). This work was later refined and extended by Drineas *et al* [DFK+04] and Drineas and Kannan [DK03]. Recently, Deshpande and Vempala [DV06] and Drineas *et al* [DMM06] gave algorithms for fast computation of low-rank approximations of matrices with multiplicative rather than additive error.

Achlioptas and McSherry [AM01] gave an algorithm that sparsifies the input matrix via random sampling of the entries rather than the columns. They applied their sparsification algorithm to the problem of computing low-rank approximations to matrices: the idea was to simply use the sparsified matrix in the orthogonal or Lanczos iteration algorithms for computing the approximations. They gave precise error estimates for the low-rank approximation in terms of the sparsification quality. Their best algorithm has better performance than [DFK+04] and [DK03] in minimizing the ℓ_2 norm of the difference matrix, and comparable performance in the Frobenius norm. In addition, they require only one pass over the input matrix instead of the two passes needed for previous work. Furthermore, [AM01] also describe a *quantization* algorithm: this algorithm transforms all non-zero entries of the input matrix into entries with the same magnitude, and so the output matrix can be succinctly represented by a bit vector corresponding to the sign of the entries.

The purpose of this note is to give a new and simple sparsification algorithm that has comparable performance to the algorithm of Achlioptas and McSherry, and is better in situations when the allowed approximation error is small. This is because the dependence on the approximation error ϵ is $\frac{1}{\epsilon}$ for our algorithm vs. $\frac{1}{\epsilon^2}$ for [AM01], so our algorithm scales better when the error that can be tolerated goes down. Another advantage is that it runs in a single pass over the input matrix and produces quantized entries directly (without the need for an extra quantization step like [AM01]). The analysis is particularly simple: all that is needed are the well-known Chernoff-Hoeffding bounds. This algorithm arose in applications in fast semidefinite programming [AHK05], where our algorithm gives better performance than that of [AM01]. In this paper, we abstract out the algorithm and refine the details that were hidden in the specific applications of [AHK05].

2 Preliminaries

Given an input symmetric matrix A, an ϵ-approximation for A is a symmetric matrix $\tilde{A}$ such that $\|A - \tilde{A}\|_2 \leq \epsilon$. Here, $\|A\|_2 := \max_{\|x\|_2=1} \|Ax\|_2$ is the ℓ_2 norm of A. For symmetric matrices A, $\|A\|_2$ is the magnitude of the largest eigenvalue in absolute value. We assume without loss of generality that the input matrix A is a symmetric, $n \times n$ real matrix. This is because given an arbitrary $n \times m$ real matrix B, we can instead consider the symmetric $(n+m) \times (n+m)$ matrix

$$A = \begin{pmatrix} 0 & B \\ B^\top & 0 \end{pmatrix}$$

which has the same ℓ_2 norm as B, and whose ϵ-approximation gives an ϵ-approximation for B in the obvious way.

Now, we will make precise what it means to compute an approximate eigenvector. For a matrix A, let v be a unit eigenvector corresponding to the largest eigenvalue, so that the largest eigenvalue of A is $v^\top A v$. A unit vector u will be called an ϵ-approximate largest eigenvector of A if $u^\top A u \geq v^\top A v - \epsilon$. Let $\tilde{A}$ be an ϵ-approximation of A, and let u be an arbitrary unit vector. Then we have

$$|u^\top (A - \tilde{A}) u| \leq \|A - \tilde{A}\|_2 = \epsilon.$$

Let u be the unit eigenvector corresponding to the largest eigenvalue of $\tilde{A}$. Then

$$v^\top A v \leq v^\top \tilde{A} v + \epsilon \leq u^\top \tilde{A} u + \epsilon.$$

which implies that u is an ϵ-approximate largest eigenvector of A.

3   Algorithm and Comparison of Results

The procedure SPARSIFY in Figure 1 computes a sparse, $O(\epsilon)$-approximation to an input matrix A.

Procedure SPARSIFY(A, ϵ)
for each $i \leq j \in [n]$ **do**
if $|A_{ij}| > \frac{\epsilon}{\sqrt{n}}$ **then**
$\quad \tilde{A}_{ji} = \tilde{A}_{ij} = A_{ij}$
else

$$\tilde{A}_{ji} = \tilde{A}_{ij} = \begin{cases} \text{sgn}(A_{ij}) \cdot \frac{\epsilon}{\sqrt{n}} & \text{with probability } p_{ij} = \frac{\sqrt{n}|A_{ij}|}{\epsilon} \\ \\ 0 & \text{with probability } 1 - p_{ij} \end{cases}$$

return $\tilde{A}$

Fig. 1. Procedure SPARSIFY

Theorem 1 below gives the performance guarantees for the procedure SPARSIFY. We defer the proof of the theorem to Section 4.

Theorem 1. *Let $A \in \mathbb{R}^{n \times n}$ be a matrix with N non-zero entries and let $S = \sum_{ij} |A_{ij}|$. Let $\epsilon > 0$ be a given error parameter. Then the procedure SPARSIFY runs in $O(N)$ time (a single pass over the input matrix) and produces a matrix $\tilde{A}$ such that:*

1. *With probability at least $1 - \exp(-\Omega(\frac{\sqrt{n}S}{\epsilon}))$, $\tilde{A}$ has $O(\frac{\sqrt{n}S}{\epsilon})$ non-zero entries, and*
2. *With probability at least $1 - \exp(-\Omega(n))$, we have $\|A - \tilde{A}\|_2 \leq O(\epsilon)$.*

Now, we give a comparison of our results with previous work. Achlioptas and McSherry [AM01] have a very detailed comparison of the use of the sparsification algorithm with the algorithms of [FKV04], [DFK+04], and [DK03] for the task of computing low-rank matrix approximations, so we refer the interested reader to [AM01] for this specific application. In this section, we only compare the algorithm of [AM01] to ours for the task of sparsifying an input matrix.

The strongest result[1] of [AM01] is a random sampling algorithm, that, in one pass over the input matrix A, computes a matrix $\tilde{A}$ such that with probability at least $1-1/n$, we have $\|A-\tilde{A}\|_2 \le \epsilon$, and which retains an expected $\tilde{O}(\frac{n}{\epsilon^2}\sum_{ij} A_{ij}^2 + n)$ non-zero entries.

In comparison, our algorithm computes, in one pass over the input matrix A, a matrix $\tilde{A}$ such that with probability at least $1 - \exp(-\Omega(n))$, we have $\|A - \tilde{A}\|_2 \le \epsilon$, and which retains an expected $\tilde{O}(\frac{\sqrt{n}}{\epsilon}\sum_{ij} |A_{ij}|)$ non-zero entries.

Thus, our algorithm has exponentially lower failure probability and better dependence on the error parameter ϵ (linear, rather than quadratic), and on the input matrix order n.

Our algorithm also has the advantage that barring a few large entries, the sampled entries are all quantized: since their magnitude is always $\frac{\epsilon}{\sqrt{n}}$, they can be represented very succinctly by just their sign. This can result in considerable savings in the space needed to store the sampled matrix. [AM01] also have an algorithm which can quantize a matrix, but it does not lead to any sparsification by itself. It can be applied to the sparsified matrix rather than the original one to obtain some amount of quantization. However, the error bound of the quantization process depends on the largest entry in the sparsified matrix, which curtails the benefits of quantization.

To elucidate how the choice of the error parameter affects the performance of the two algorithms, we consider two cases. In the first case, the error ϵ is of the order of the ℓ_2 norm of A, viz. $\epsilon = \delta\|A\|_2$. This situation arises in applications such as solving semidefinite programs efficiently [AHK05]. In the second case, the error ϵ is of the order of the Frobenius norm of A, viz. $\epsilon = \delta\|A\|_F = \delta\sqrt{\sum_{ij} A_{ij}^2}$. This situation arises in applications such as computing low-rank approximations to matrices [FKV04], [DFK+04], [DK03], and [AM01].

The first error is typically much smaller than the second, so our algorithm can be expected to perform better in the first case, and Achlioptas and McSherry's in the second. Figure 2 gives examples of the level of sparsification achieved by the algorithms in various cases.

In summary, the our algorithm is better in some situations than that of [AM01] and worse in others. Exactly which algorithm to use in a given situation depends on the input parameters. A general guideline is that our algorithm is preferable when one needs a high degree of accuracy.

---

[1] It has been suggested to the authors that an algorithm with similar parameters to ours can be derived using the techniques of [AM01]. However, for the purpose of comparison, we only consider the algorithms described explicitly in their paper.

Algorithm	Matrix	$\epsilon = \delta\|A\|_2$	$\epsilon = \delta\|A\|_F$
This paper	$I + \frac{1}{n}J$	$O\left(\frac{n^{1.5}}{\delta}\right)$	$O\left(\frac{n}{\delta}\right)$
[AM01]	$I + \frac{1}{n}J$	no sparsification	$O\left(\frac{n}{\delta^2}\right)$
This paper	H	no sparsification	$O\left(\frac{n^{1.5}}{\delta}\right)$
[AM01]	H	no sparsification	$O\left(\frac{n}{\delta^2}\right)$
This paper	C	$O\left(\frac{n^{1.5}}{\delta}\right)$	$O\left(\frac{n}{\delta}\right)$
[AM01]	C	no sparsification	$O\left(\frac{n}{\delta^2}\right)$

Fig. 2. Comparison of algorithms for sparsification under different error tolerances. The matrices are: (i) $I + \frac{1}{n}J$, where I is the identity matrix, and J is the all 1's matrix, (ii) the matrix H which is the Hadamard matrix of order n (assuming it exists), and (iii) C, the combinatorial Laplacian of a d-regular graph on n nodes.

4   Analysis: Proof of Theorem 1

We prove the first part of Theorem 1 in the following Lemma:

Lemma 1. *With probability at least $1 - \exp(-\Omega(\frac{\sqrt{n}S}{\epsilon}))$, the matrix $\tilde{A}$ contains at most $O(\frac{\sqrt{n}S}{\epsilon})$ non-zero entries.*

Proof. Since $\sum_{ij} |A_{ij}| = S$, the number of entries with magnitude larger than $\frac{\epsilon}{\sqrt{n}}$ is at most $\frac{\sqrt{n}S}{\epsilon}$. So without loss of generality, we may assume that all the entries have magnitude smaller than $\frac{\epsilon}{\sqrt{n}}$.

The Chernoff bound [MR95] asserts that if $X_1, X_2, \ldots, X_n$ are indicator random variables and $X = \sum_i X_i$ with $\mathbb{E}[X] = \mu$, then

$$\mathbf{Pr}[X > (1+\delta)\mu] < \left[\frac{e^\delta}{(1+\delta)^{1+\delta}}\right]^\mu$$

In our case, we set up indicator random variables X_{ij} for $i \leq j$ which are 0 or 1 depending on whether $\tilde{A}_{ij} = 0$ or not. Let $X = \sum_{i \leq j} X_{ij}$. Then $2X$ is an upper bound on the number of non-zero entries of $\tilde{A}$. We have

$$\mathbb{E}[X] = \sum_{i \leq j} p_{ij} = \sum_{i \leq j} \frac{\sqrt{n}|A_{ij}|}{\epsilon} \leq \frac{\sqrt{n}S}{\epsilon}.$$

The claim follows by using the Chernoff bound with $\delta = e - 1$.   ∎

Now, we will proceed to prove the second part of Theorem 1. For this, define $M = A - \tilde{A}$. We will show that with high probability, for all unit vectors x, we have $|x^\top M x| \leq O(\epsilon)$, which implies $\|A - \tilde{A}\|_2 \leq O(\epsilon)$.

Notice that for all coordinates i, j such that $|A_{ij}| \geq \frac{\epsilon}{\sqrt{n}}$, we have $M_{ij} = 0$. For the rest of the coordinates, since $\mathbb{E}[\tilde{A}_{ij}] = \mathrm{sgn}(A_{ij}) \cdot \frac{\epsilon}{\sqrt{n}} \times \frac{\sqrt{n}|A_{ij}|}{\epsilon} = A_{ij}$, we conclude that $\mathbb{E}[M_{ij}] = 0$. We will now consider a $\frac{\epsilon_0}{\sqrt{n}}$-grid on the unit sphere (ϵ_0 is set to some constant, say $\frac{1}{2}$),

$$T = \left\{ x : x \in \frac{\epsilon_0}{\sqrt{n}}\mathbb{Z}^n, \, \|x\|_2 \leq 1 \right\}.$$

Feige and Ofek [FO05] give a bound on the size of T and show that it suffices to consider only vectors in T (we reprove this in Appendix A):

Lemma 2. *The size of $|T|$ is at most $\exp(cn)$ for $c = (\frac{1}{\epsilon_0} + 2)$. If for every $x, y \in T$ we have $|x^\top M y| \leq \epsilon$, then for every unit vector x, we have $|x^\top M x| \leq \frac{\epsilon}{(1-\epsilon_0)^2}$.*

Let $x, y \in T$. Since $\mathbb{E}[M_{ij}] = 0$, we conclude that $\mathbb{E}[x^\top M y] = 0$. We now a prove strong concentration bound:

Lemma 3. *With probability at least $1 - \exp(-\Omega(n))$, for every $x, y \in T$ it holds that $|x^\top M y| \leq c\epsilon$.*

Proof. We use the following bound from Hoeffding's original paper [Hoe63]: let $X_1, ..., X_n$ be independent random variables, such that X_i takes values in the range $[a_i, b_i]$. Let $X = \sum_i X_i$, and $\mathbb{E}[X] = \mu$. Then for any $t > 0$

$$\Pr[|X - \mu| \geq t] \leq 2\exp\left(-\frac{2t^2}{\sum_i (b_i - a_i)^2}\right).$$

Consider the random variables $Z_{ij} = M_{ij}x_iy_j$, then $x^\top M y = \sum_{ij} M_{ij}x_iy_j = \sum_{ij} Z_{ij}$. Since $\tilde{A}_{ij}$ is either $\text{sgn}(A_{ij}) \cdot \frac{\epsilon}{\sqrt{n}}$ or 0, the squared range of M_{ij} is $\frac{\epsilon^2}{n}$. Thus, the sum of squared ranges for the variables $\{Z_{ij}, i \leq j\}$ at most $\sum_{i \leq j} \frac{\epsilon^2}{n}x_i^2y_j^2 \leq \frac{\epsilon^2}{n}\sum_i x_i^2 \sum_j y_j^2 \leq \frac{\epsilon^2}{n}$, and similarly the sum of squared ranges for the variables $\{Z_{ij}, i > j\}$ is bounded by $\frac{\epsilon^2}{n}$. Since $\mathbb{E}[Z_{ij}] = 0$, by the Hoeffding bound we have:

$$\Pr\left[\left|\sum_{i \leq j} Z_{ij}\right| \geq c\epsilon\right] \leq 2\exp\left(-\frac{2c^2\epsilon^2}{\frac{\epsilon^2}{n}}\right) = 2\exp(-2c^2 n).$$

A similar bound holds for $\Pr[|\sum_{i>j} Z_{ij}| \geq c\epsilon]$. Since $|x^\top M y| = |\sum_{i \leq j} Z_{ij} + \sum_{i>j} Z_{ij}|$, by the union bound we have

$$\Pr[|x^\top M y| \geq 2c\epsilon] \leq 4\exp(-2c^2 n).$$

Since there are $\exp(2cn)$ pairs of vectors $x, y \in T$, the union bound implies that with probability at least $1 - \exp(-\Omega(n))$, for all vectors $x, y \in T$, we have $|x^\top M y| \leq c\epsilon$. ∎

5   Conclusions

In this paper, we presented a fast and simple random sampling algorithm to sparsify matrices, with comparable performance guarantees to previous work. The analysis of the algorithm is also fairly easy, relying only on the well-known Chernoff-Hoeffding bounds. The algorithm has better dependence on the error parameter than previous work, which makes it preferable when low error is desired.

However, its dependence on the input matrix size may be worse than previous algorithms in situations where all entries are roughly the same magnitude. This suggests that in practice, a hybrid algorithm combining ours with that of Achlioptas and McSherry may be able to strike a better balance between the dependence on the error parameter and input size.

References

[AHK05]   Sanjeev Arora, Elad Hazan, and Satyen Kale. Fast algorithms for approximate semidefinite programming using the multiplicative weights update method. In *46th FOCS*, pages 339–348, 2005.

[AM01]   Dimitris Achlioptas and Frank McSherry. Fast computation of low rank matrix approximations. In *32nd STOC*, pages 611–618, 2001.

[DFK+04]   Petros Drineas, Alan M. Frieze, Ravi Kannan, Santosh Vempala, and V. Vinay. Clustering large graphs via the singular value decomposition. *Machine Learning*, 56(1-3):9–33, 2004.

[DK03]   Petros Drineas and Ravi Kannan. Pass efficient algorithms for approximating large matrices. In *SODA*, pages 223–232, 2003.

[DMM06]   P. Drineas, M. Mahoney, and S. Muthukrishnan. Column-based relative-error. In *RANDOM*, 2006.

[DV06]   Amit Deshpande and Santosh Vempala. Adaptive sampling and fast low-rank matrix approximation. In *RANDOM*, 2006.

[FKV04]   Alan M. Frieze, Ravi Kannan, and Santosh Vempala. Fast monte-carlo algorithms for finding low-rank approximations. *J. ACM*, 51(6):1025–1041, 2004.

[FO05]   U. Feige and E. Ofek. Spectral techniques applied to sparse random graphs. *Random Structures and Algorithms*, 27(2):251–275, September 2005.

[Hoe63]   W. Hoeffding. Probability inequalities for sums of bounded random variables. *Journal of the American Statistical Association*, 58(301):13–30, 1963.

[MR95]   R. Motwani and P. Raghavan. *Randomized Algorithms*. Cambridge Univ. Press, 1995.

[TB97]   Lloyd N. Trefethen and David Bau. *Numerical Linear Algebra*. SIAM, 1997.

A   Discretization

In this section, we prove Lemma 2. We restate it here for convenience:

Lemma 4. *Let* $T = \left\{ x : x \in \frac{\epsilon_0}{\sqrt{n}} \mathbb{Z}^n, \|x\|_2 \leq 1 \right\}$. *The size of T is at most* $\exp(cn)$ *for* $c = (\frac{1}{\epsilon_0} + 2)$. *If for every $x, y \in T$ we have $|x^\top M y| \leq \epsilon$ then for every unit vector x, we have $|x^\top M x| \leq \frac{\epsilon}{(1-\epsilon_0)^2}$.*

Proof. Map every point in $x \in T$ in a one-to-one correspondence with a n-dimensional hypercube of side length $\frac{\epsilon_0}{\sqrt{n}}$ on the grid:

$$x \mapsto C_x = \left\{ x + u : \ u \geq \mathbf{0}, \ \|u\|_\infty \leq \frac{\epsilon_0}{\sqrt{n}} \right\}.$$

The maximum length of any vector in C_x is bounded by $\|x\| + \epsilon_0 \leq 1 + \epsilon_0$, and thus the union of these cubes is contained in the n-dimensional ball B of radius $(1 + \epsilon_0)$. We conclude:

$$|T| \times \left(\frac{\epsilon_0}{\sqrt{n}} \right)^n = \sum_{x \in T} \mathbf{Vol}(C_x) \leq \mathbf{Vol}(B) = \frac{\pi^{n/2}}{\Gamma(n/2 + 1)} (1 + \epsilon_0)^n.$$

And so:

$$|T| \leq \frac{\pi^{n/2}}{\Gamma(n/2 + 1)} \left(\frac{(1 + \epsilon_0)\sqrt{n}}{\epsilon_0} \right)^n \leq \exp\left(\left(\frac{1}{\epsilon_0} + 2 \right) n \right).$$

Next, given any unit vector, x, let $y = (1 - \epsilon_0)x$. By "rounding down" the coordinates of y to the nearest multiple of $\frac{\epsilon_0}{\sqrt{n}}$, we get a grid point z such that $y \in C_z$. Thus, the maximum length of any vertex of C_z is bounded by $\|y\| + \epsilon_0 = 1$, so all vertices of C_z are grid points in T. Express y as a convex combination of the vertices v_i of C_z; viz. $y = \sum_i \alpha_i v_i$ with $\alpha_i \geq 0$ and $\sum_i \alpha_i = 1$. Then we have

$$|y^\top M y| = |(\sum_i \alpha_i v_i)^\top M (\sum_i \alpha_i v_i)| \leq \sum_{i,j} \alpha_i \alpha_j |v_i^\top M v_j| \leq \sum_{i,j} \alpha_i \alpha_j \epsilon = \epsilon.$$

The second inequality above follows because we assumed that for all $x, y \in T$, $|x^\top M y| \leq \epsilon$. Finally, since $y = (1 - \epsilon_0)x$, we have

$$|x^\top M x| = \frac{|y^\top M y|}{(1 - \epsilon_0)^2} \leq \frac{\epsilon}{(1 - \epsilon_0)^2}.$$

∎

The Effect of Boundary Conditions on Mixing Rates of Markov Chains

Nayantara Bhatnagar[1,*], Sam Greenberg[2,*], and Dana Randall[1,*]

[1] College of Computing, Georgia Institute of Technology, Atlanta, GA 30332-0280
[2] School of Mathematics, Georgia Institute of Technology, Atlanta, GA 30332-0160

Abstract. Many natural Markov chains undergo a phase transition as a temperature parameter is varied; a chain can be rapidly mixing at high temperature and slowly mixing at low temperature. Moreover, it is believed that even at low temperature, the rate of convergence is strongly dependent on the environment in which the underlying system is placed. It is believed that the boundary conditions of a spin configuration can determine whether a local Markov chain mixes quickly or slowly, but this has only been verified previously for models defined on trees. We demonstrate that the mixing time of Broder's Markov chain for sampling perfect and near-perfect matchings does have such a dependence on the environment when the underlying graph is the square-octagon lattice. We show the same effect occurs for a related chain on the space of Ising and "near-Ising" configurations on the two-dimensional Cartesian lattice.

1 Introduction

Boundary conditions play a crucial role in statistical physics for determining the uniqueness of Gibbs states, or the limiting distributions of families of configurations on the infinite lattice. Consider the Ising model on the $n \times n$ Cartesian lattice, a fundamental physical model for ferromagnetism. Each configuration σ in the state space $S = \{+,-\}^{n^2}$ consists of an assignment of a $+$ or $-$ spin to each of the vertices, and the *Gibbs distribution* assigns weight

$$\pi(\sigma) = \lambda^{-D(\sigma)}/Z,$$

where $D(\sigma) = |\{(i,j) \in E \mid \sigma(i) \neq \sigma(j)\}|$ and Z is the normalizing constant or *partition function*. In the classical description of the Ising model, $\lambda = e^{2\beta}$, where $\beta > 0$ is inverse temperature.

To characterize when there is a phase transition in a physical model, physicists study whether there is a unique limiting distribution as $n \to \infty$. The vertices on the boundary of an $n \times n$ grid are hard-wired to be $+$ in one case and $-$ in another. The Gibbs measure on the interior is defined as the limiting distribution conditioned on the boundary. It is well known that there is a critical value λ_c such that, for $\lambda < \lambda_c$, the limiting distribution is unique, yet for $\lambda > \lambda_c$,

* Supported in part by NSF grants CCR-0515105 and DMS-0505505.

J. Diaz et al. (Eds.): APPROX and RANDOM 2006, LNCS 4110, pp. 280–291, 2006.

correlations between the spins of vertices inside a finite region and the spins on the boundary of that region persist over long distances and there are multiple limiting distributions (see, e.g., [3]). A related effect has been observed in the context of mixing times of local chains on finite regions. The *mixing time* of a chain, i.e., the number of steps required so that probabilities of reaching each configuration is close to the stationary distribution, undergoes a similar phase change. When λ is sufficiently small, local dynamics are efficient, while when λ is large, local chains require exponential time to converge to equilibrium [16]. This is because at low enough temperature the Gibbs distribution strongly favors configurations that are predominantly one spin; it takes exponential time to move between mostly $+$ and mostly $-$ states using local chains [8, 9].

A natural question that integrates these two perspectives is: *Can one type of boundary condition cause a Markov chain mix slowly, while another causes the same chain to mix rapidly?* Martinelli, Sinclair, and Weitz [10,11] answered this question in the affirmative in the context of spin systems on trees. The question remains unresolved when configurations are defined on lattices, although the same effect is believed to occur. Martinelli [8] showed that mixing times of Glauber (local) dynamics on Ising configurations of the 2-dimensional lattice can vary by an exponential factor, though the mixing time for both of his boundary conditions were shown to be exponential – in reality the boundary that leads to faster mixing is believed to converge in polynomial time.

Models and Results: The first problem we consider is sampling matchings on finite regions of the square-octagon lattice. This is the lattice formed by tightly packing octagons so that the uncovered space forms smaller squares (see Figure 1). For certain finite regions R of this lattice, there is an ergodic Markov chain on the set of perfect matchings; it starts at any matching and repeatedly does the following: choose a square or octagonal face uniformly, and if the matching alternates edges around this face, then "rotate" to the other matching. Propp [12] used coupling-from-the-past [13] on certain regions to generate so-called "diabolo tilings of fortresses," and conjectured that the chain mixes slowly. The only proof of slow mixing for this model requires "activities" on the edges that weigh matchings according to the number of edges that bound squares on the lattice [4]. It has been conjectured that there is a region such that one boundary condition will cause this local Markov chain to mix quickly while another will mix slowly; however, like the Ising model at sufficiently low temperature, it remains a challenge to show fast mixing for such a contour model, even though there is a boundary for which the chain is believed to mix rapidly.

In this paper, we consider instead the Broder-chain on the set of perfect and near-perfect matchings on the square-octagon lattice. For a finite, simply-connected region R on this lattice, let the *boundary* of R be the set of vertices that have neighbors both inside and outside R on the infinite lattice. The boundary condition is defined by specifying, for each vertex on the boundary, whether it is to be included in the matchings or not. We hardwire a boundary condition and start with a perfect matching on the remaining region. The Broder-chain successively chooses an edge and this edge is added, deleted, or exchanged with another edge (if exactly one endpoint was matched). The chain converges to the uniform distribution on perfect and near-perfect matchings. We show that, for a family of regions, there are two types of boundary conditions, one that causes the Broder-chain to mix slowly and another that causes the chain to mix rapidly. Remarkably, these two boundary conditions differ by the deletion of only four vertices. This is the first proof of slow mixing for matchings on the square-octagon lattice without activities on the edges.

The second model we consider is the Ising model on $\mathbb{Z}^2$. It is strongly believed that Glauber dynamics are very sensitive to boundary conditions, even at low temperature, and that they will be fast for the all plus boundary and slow for two sides fixed to plus and two to minus. It is useful to view the Ising model in terms of contours. Given an Ising configuration, take the union of all edges that separate a $+$ spin from a $-$ spin. For the all-plus boundary condition, these edges form an even degree subgraph that can be thought of as sets of closed contours. Glauber dynamics perform local changes to these contours and, at low temperature, short contours are thermodynamically favorable. Fernandez, Ferrari, and Garcia [2] proposed a chain that, in one step, moves between two configurations that differ by a single contour. They show that at sufficiently low temperature, this chain converges quickly to stationarity, but unfortunately there is no efficient way to perform a step of the chain.

Instead, we consider a Broder-type chain on the set of Ising and "near-Ising" configurations on finite regions of the Cartesian lattice. For the all-plus boundary condition, for example, near-Ising configurations allow exactly one contour to be open while the others must be closed. A step of the chain removes or adds an edge, with transition probabilities chosen so that we converge to the Gibbs distribution on Ising and near-Ising configurations. We show that, at sufficiently low temperature, there are two boundary conditions so that the chain mixes slowly with one and quickly with the other. The fast mixing results are of independent interest because they demonstrate how to extend the result by Fernandez et al. to define a rapidly mixing chain at low temperature that can be efficiently implemented. Moreover, this gives a much more efficient algorithm for sampling Ising configurations at low temperature than the only other rigorous method known previously [14].

Techniques: A key fact underlying our results is that both the matching and Ising models considered here can be reformulated as contour models. Contours for the Ising model are unrestricted, while the contours arising from matchings on the square-octagon lattice are required to turn left or right at every step. The contours arising from these matchings behave similarly to Ising contours at low temperature where long contours are penalized.

In the first (fast) cases, boundaries are defined so that initially all vertices, including the boundary, have even degree in the contour representation. During the simulation, configurations must have zero or one open path. We show that the weight of perfect and near-perfect configurations are polynomially related. Following the canonical path technique introduced by Jerrum and Sinclair [5], this suffices to show polynomial mixing for both models.

In the second (slow) cases, our proofs are based on a *Peierls argument* from statistical physics (see, e.g., [3]). We introduce four designated vertices on the boundary that initially have odd degree in the contour representation, and thus are the endpoints of two paths. At any point during the simulatation of the Broder-chains, at most two vertices in the contour representation have changed parity. Once one of the initial paths is disconnected, the two pieces tend to shrink. We prove, via sensitive injections, that it will take exponential time for these to reconnect — this is sufficient to show slow mixing. While this is an artifact of the extended state space, these models do correctly capture the effects of the two boundaries for models where the Gibbs measure favors short contours. On the square-octagon lattice these bounds are especially sensitive because we cannot modify the temperature to establish the required inequalities.

2   Preliminaries

Let $\mathcal{M}$ be an ergodic (i.e., irreducible and aperiodic), reversible Markov chain with finite state space S, transition probability matrix P, and stationary distribution π. Let $P^t(x, y)$ be the t-step transition probability from x to y and let $||\cdot, \cdot||$ denote the total variation distance.

Definition 1. *For $\varepsilon > 0$, the* mixing time $\tau = \min\{t : ||P^{t'}, \pi|| \leq 1/4, \forall t' \geq t\}$.

A Markov chain is *rapidly mixing* if the mixing time is bounded above by a polynomial in n, the size of each configuration in the state space. If the mixing time is exponential in n, the chain is *slowly mixing*. Jerrum and Sinclair defined the conductance of a chain and showed that it bounds mixing time [5].

Definition 2. *If a Markov chain has stationary distribution π, we define the* conductance Φ *as*

$$\Phi = \min_{S:\pi(S)\leq 1/2} \frac{\sum_{x\in S, y\notin S} \pi(x)P(x, y)}{\pi(S)}.$$

Theorem 1. *An ergodic, reversible chain with conductance Φ is rapidly mixing if and only if $\Phi > 1/p(n)$ for some polynomial $p(\cdot)$.*

Jerrum and Sinclair use this theorem to analyze a natural Markov chain on matchings due to Broder [1]. Here we consider the chain on the state space consisting of perfect and near-perfect matchings. At each step, **the Broder-chain $\mathcal{M}_B$** does the following:

Choose an edge e uniformly at random.
◊ If the endpoints of e are unmatched, add e to the matching.
◊ If e is in the matching and the matching is perfect, remove e.
◊ If exactly one endpoint of e is matched, remove the matched edge and add e.
◊ Otherwise, do nothing.

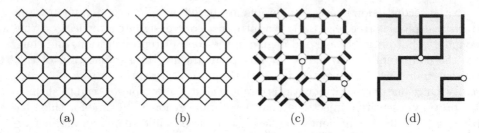

Fig. 1. For $n = 5$, (a) **L**, (b) **L′**, (c) a near-perfect matching, (d) that matching's contraction

This chain converges to the uniform distribution on matchings and near-perfect matchings. Jerrum and Sinclair [5, 15] find the following characterization for when the Broder-chain can be used to efficiently sample perfect matchings.

Theorem 2. *Let $\mathbf{S}_\mathcal{P}$ be the set of perfect matchings in $\mathbf{S}$ and $\mathbf{S}_\mathcal{N}$ be the set of near-perfect matchings. If $|\mathbf{S}_\mathcal{N}| \leq p(n)|\mathbf{S}_\mathcal{P}|$ for some polynomial $p(\cdot)$, then Φ is at least inverse-polynomial.*

Theorem 2 was proven using the *canonical path* technique. The key idea in this proof is to define paths between every pair of states $(I, F) \in S \times S$ in the transition graph of the chain. If not too many paths go through any specific transition then there cannot be a bottleneck in the transition graph. The following summary of their method will be useful in Section 4.

Theorem 3. *Suppose there exists a function η such that, for a transition $T = (G, G')$ along the canonical path from I to F,*
   1. given T and $\eta(T, I, F)$, we can reconstruct both I and F
   2. $\mu(I)\mu(F) \leq \mu(G)\mu(\eta(T, I, F))P(G, G')$.
Then $\Phi = \Omega(n^{-c})$ for some constant c.

This theorem is the key ingredient that establishes our fast mixing results. For our slow mixing results, we show that the conductance is exponentially small. For this, it suffices to identify a bad cut $(S, \bar{S})$ in the state space.

3   Perfect Matchings in the Square-Octagon Lattice

Let **L** to be the square-octagon lattice with n squares on a side, for n odd (see Figure 1a). Define **L′** to be the same lattice, but with one vertex missing from each of the corner squares (Figure 1b). Regions **L** and **L′** capture the two boundary conditions we study.

We let **S** (resp. **S′**) be the set of perfect and near-perfect matchings on **L** (resp. **L′**) and let $\mathcal{M}_B$ be the Broder-chain on these state spaces, as described in Section 2. Our first main result is the following:

Theorem 4. *There exist constants $c_1, c_2 > 1$ such that the mixing time of $\mathcal{M}_B$ on $\mathbf{S}$ is $O(n^{c_1})$ while the mixing time of $\mathcal{M}_B$ on $\mathbf{S′}$ is $\Omega(e^{c_2 n})$.*

3.1   Contraction to Contours

Before presenting the proof of Theorem 4, it will be convenient to define a bijection between perfect matchings and a related contour representation. We "contract" the lattice regions $\mathbf{L}$ and $\mathbf{L}'$ by replacing each of the squares with vertices so that only the edges bounded by octagons on each side survive. The result is isomorphic to a subregion of the Cartesian lattice (see Figure 1c and 1d). We use bold script ($\mathbf{L}$, $\mathbf{S}$, $\mathcal{M}_B$, etc.) when referring to the square-octagon graphs, and normal script (L, S, $\mathcal{M}_B$, etc.) for the contracted case. To that end, define L and L' to be the integer lattice with n vertices on each side.

Consider the effect of this contraction on a perfect matching of $\mathbf{L}$. The contraction is an even degree subgraph where all vertices of degree 2 are incident to edges that bound two sides of a unit square; if the endpoints were collinear, the corresponding square in $\mathbf{L}$ would have two unmatchable vertices. For $\mathbf{L}'$, we get an even degree subgraph with this turning property, only now there are four vertices of odd degree in the corners, as they correspond squares in $\mathbf{L}'$ with only three vertices present. Finally, near-perfect matchings of $\mathbf{L}$ and $\mathbf{L}'$ contract as above, except two vertices might have the opposite parity (corresponding to the square(s) in the lattice containing the two unmatched vertices).

We define new ground-states S and S' with this in mind. We call a subgraph of L a *turning graph* if all vertices have even degree and vertices of degree 2 "turn corners." We call a subgraph of L' a turning graph if this applies to all but the corner vertices. We call a subgraph of L or L' a *near-turning graph* if all vertices of degree 2 turn, and exactly two vertices have parity different from what was prescribed for turning graphs. Let $S_\mathcal{T}$ and $S_\mathcal{N}$ (resp. $S'_\mathcal{T}$ and $S'_\mathcal{N}$) be the turning and near-turning graphs of L (resp. L'). Let S and S' be the union of the turning and near-turning graphs in each case.

Notice that the contraction map is not one-to-one. Let $G \in S$ and let $v \in L$. If the degree of v in G is 2, 3, or 4, there is a unique way to expand v to a square face of $\mathbf{G} \in \mathbf{S}$, but if $d(v) = 0$ or 1, then there are two ways to recover the matched edge(s) that were deleted from the corresponding square, illustrated in Figure 1. For $G \in S \cap S'$, the number of matchings which contract to G is to within a small polynomial factor $q(n)$ of $2^{|V(\mathbf{L})|-|V(G)|}$.

Our first goal is to show that the Broder-chain is fast on $\mathbf{L}$, and for this we need to show a polynomial relationship between the numbers of near-perfect and perfect matchings. For simplicity, we instead consider their contracted versions in L and introduce a weight μ on turning and near-turning contours as follows: for $G \in S \cup S'$ let $\mu(G) := 2^{-|V(G)|}$. After a normalization, $\mu(G)$ is within $q(n)$ of the number of matchings which contract to G. It then follows that showing a polynomial relationship between $\sum_{G \in S_\mathcal{T}} \mu(G)$ and $\sum_{G \in S_\mathcal{N}} \mu(G)$ is sufficient to establish fast mixing of the Broder-chain.

The following lemma will be crucial in our proof of slow mixing. It shows that we can encode near-turning contours as a function of the number of vertices it hits, rather than its total length.

Lemma 1. *For $G \in S \cup S'$, let $\mathcal{N}_a(G)$ be the set of near-turning components A, edge-disjoint from G, such that $|V(G \cup A)| = |V(G)| + a$. Then $|\mathcal{N}_a(G)| \leq 4n^4 2^a$.*

Proof. Using a standard "Euler-like" decomposition, any A can be described as a single turning path between the odd-degree vertices. If we are given the coordinates of the first odd vertex and the initial direction of the path, we can encode the rest in a binary string, with 0 representing a left-turn and 1 representing a right-turn. This is a natural encoding of the turning-path, but requires $|E(A)|$ bits, possibly more than a. It will be necessary to define an encoding which focuses on vertices instead of edges.

Fortunately, not all turns need to be encoded; whenever the path touches the graph G, either because it turns back on itself or because it touches $G\backslash A$, the next turn is forced. We therefore only encode those turns when our path hits a previously-empty vertices, creating a bitstream of length $a - 1$. This encoding is not completely unique. After the last recorded turn, the path might proceed for any number of forced-turns; our encoding would fail to represent how many. However, knowing the length of the turning-path determines this uniquely, and we can upper bound the length by $2n^2$ edges. Hence each binary string corresponds to at most a polynomial number of turning-paths.

The total size of $\mathcal{N}_a(G)$ is therefore upper bounded by the number of possible lengths times the number of starting vertices and directions, times the number of binary strings of length $a - 1$. This gives us $|\mathcal{N}_a(G)| \leq 2n^2 \cdot n^2 \cdot 4 \cdot 2^{a-1}$. $\square$

3.2 Fast Mixing of $\mathcal{M}_B$ on S

For the rapid mixing result of Theorem 4, it is sufficient to show a polynomial relationship between the size of $\mathbf{S}_\mathcal{P}$ and the size of $\mathbf{S}_\mathcal{N}$. Focusing on the contracted representations, this is equivalent to the following.

Lemma 2. *For some polynomial $p(\cdot)$, $\mu(S_\mathcal{N}) \leq p(n) \cdot \mu(S_\mathcal{T})$.*

Proof. We define a function $f : S_\mathcal{N} \to S_\mathcal{T}$. For $G' \in S_\mathcal{T}$, define the pre-image of G' to be $f^{-1}(G') = \{G \in S_\mathcal{N} : f(G) = G'\}$. We define f in such a way that, although $f^{-1}(G')$ contains many graphs, their total weight is within a polynomial factor of the weight of G'.

Let $A(G)$ be the component of G containing the two odd vertices and let $f(G) = G\backslash A(G)$. partitioning according to the size of $A(G)$, for $G' \in \mathrm{Img}(f)$,

$$\mu(f^{-1}(G')) = \sum_{a=1}^{n^2} \sum_{A\in\mathcal{N}_a(G')} \mu(G'\cup A) = \sum_{a=1}^{n^2} |\mathcal{N}_a(G')|\cdot\mu(G')\cdot 2^{-a} \leq 16n^6\mu(G').$$

where the last inequality is due to Lemma 1. Then

$$\mu(S_\mathcal{N}) = \sum_{G'\in\mathrm{Img}(f)} \mu(f^{-1}(G')) \leq 16n^6 \sum_{G'\in\mathrm{Img}(f)} \mu(G') \leq 16n^6\mu(S_\mathcal{T}).   \square$$

Hence, by Theorem 2, the Broder-chain is fast on **L**.

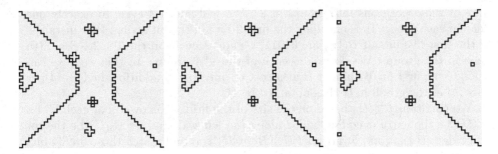

Fig. 2. Stages of f: (a) W, (b) removing $B(W)$ and shifting, (c) adding squares along the wall to obtain U

3.3 Slow Mixing of $\mathcal{M}_B$ on $\mathbf{S'}$

We turn now to the behavior of the Broder-chain on L'. Define a *bridge* to be a turning path connecting two corner vertices of L'. Perfect matchings of $\mathbf{L'}$ contract to turning subgraphs of L' with two distinct bridges, while near-perfect matchings of $\mathbf{L'}$ might map to a graph with only one bridge.

Our strategy will be to show that there is a bad cut in the state space. Let σ, τ be two configurations that each have only one bridge, and suppose that these bridges connect different pairs of vertices on the corners of the boundary. To move from σ to τ it is necessary to pass through a configuration with two bridges. We will show that the set of configurations with two bridges is exponentially smaller than configurations with any one bridge, and so this will establish slow mixing of the Broder-chain. We must define a very sensitive map from configurations with two bridges to those with one. This is accomplished by the following lemma.

Lemma 3. *Let $\mathcal{W}$ be the set of graphs in S' with two bridges and $\mathcal{U}$ be the graphs with only one. There is a constant $c > 1$ such that $\mu(\mathcal{W}) \leq c^{-n}\mu(\mathcal{U})$.*

Proof. We define a function $f : \mathcal{W} \to \mathcal{U}$ in such a way that $f^{-1}(U)$ is exponentially smaller than the weight of U. Informally, first remove the larger of the two bridges in W; then shift all components between that bridge and the wall by 1 unit (away from the wall). (See Figure 2.) This allows us to use cells adjacent to the wall to encode the initial part of the bridge, which is crucial to the result.

More precisely, for $W \in \mathcal{W}$, let $B(W)$ be the maximal bridge (with respect to vertices) in W. Let $\mathcal{W}_T$ be the set of turning graphs of S' and let $\mathcal{W}_N$ be the near-turning graphs. If $W \in \mathcal{W}_T$, we can remove $B(W)$ leaving a graph in $\mathcal{U}$. If $W \in \mathcal{W}_N$, we remove both $B(W)$ and the near-turning component $A(W)$. Suppose $B(W)$ connects the upper-left and lower-left corners of L'. Shift all of the components between $B(W)$ and the left wall one square to the right. This allows us to add edges along this left wall. It would be convenient to be able tho add a cycle to any face along the wall; this is not always possible, as the original components might obstruct the addition, even after shifting. However, it can be seen that at least $n/2$ of the faces *do* allow such an addition after shifting. We use these $n/2$ positions to encode the initial segment of the deleted bridge. Break

$n/2$ of these positions into groups of 2^5. We will add a 4-cycle to exactly one face in each group, thus encoding the first 5 bits of the bit string. (For instance, if the first five bits of $B(W)$ are 01011, we add a cycle on the $1 + 2 + 8 = 11$th face in the group.) We do this in each of the $\frac{n}{2 \cdot 32}$ groups. In this way, we can encode the first $5n/64$ bits of B at a cost of only $4n/64$ additional edges (4 per face). Let the graph in $\mathcal{U}$ thus obtained be U.

We partition $f^{-1}(U)$ based on the size of the bridge removed. For each U, let $E(U)$ be the extra encoding added along the left wall and let $\mathcal{B}_b(U)$ be the set of bridges B that add b vertices ($|U \cup B| = |U| + b$) and match those $5n/64$ bits encoded in $E(U)$. Then, partitioning according to b,

$$\mu(f^{-1}(U) \cap \mathcal{W}_\mathcal{T}) = \sum_{b=1}^{n^2} \sum_{B \in \mathcal{B}_b(U)} \mu(U \backslash E(U) \cup B)$$

$$\leq \sum_{b=1}^{n^2} |\mathcal{B}_b(U)| \cdot \mu(U) \cdot 2^{\frac{4n}{64} - b} \leq 32n^6 2^{\frac{-n}{64}} \cdot \mu(U).$$

For $f^{-1}(U) \cap \mathcal{W}_\mathcal{N}$, we must further partition according to size.

$$\mu(f^{-1}(U) \cap \mathcal{W}_\mathcal{N}) = \sum_{b=1}^{n^2} \sum_{B \in \mathcal{B}_b(U)} \sum_{a=1}^{n^2} \sum_{A \in \mathcal{N}_a(U \cup B)} \mu(U \backslash E(U) \cup A \cup B)$$

$$\leq \sum_{b=1}^{n^2} \sum_{B \in \mathcal{B}_b(U)} \sum_{a=1}^{n^2} |\mathcal{N}_a(U \cup B)| \mu(U) 2^{\frac{4n}{64} - b - a}$$

$$\leq \sum_{b=1}^{n^2} \sum_{a=1}^{n^2} |\mathcal{B}_N(U)| 16n^4 2^a \cdot \mu(U) 2^{\frac{4n}{64} - b - a} \leq 64n^{12} 2^{\frac{-n}{64}} \mu(U),$$

where the last two inequalities come from Lemma 1. Then, for some $c > 1$,

$$\mu(\mathcal{W}) = \sum_{U \in \mathrm{Img}(f) \cap \mathcal{W}_\mathcal{T}} \mu(f^{-1}(U)) + \sum_{U \in \mathrm{Img}(f) \cap \mathcal{W}_\mathcal{N}} \mu(f^{-1}(U)) < c^n \sum_{U \in \mathrm{Img}(f)} \mu(U) \leq c^{-n} \mu(\mathcal{U}). \quad \square$$

This establishes an exponentially small cut in S' because the set of turning graphs with a bridge from upper-left to upper-right has the same weight as the set with bridges from lower-left to lower-right, by symmetry. This upper bounds conductance, and verifies the slow mixing result in Theorem 4.

4   Ising Model

There is an analogous dichotomy for the mixing time of a Broder-type Markov chain defined for the Ising model. In the Ising model, each face of an $n \times n$ region on the Cartesian lattice is assigned one of two spins, $+$ or $-$. (Traditionally, the spins are assigned to the vertices; we use the equivalent model on faces for

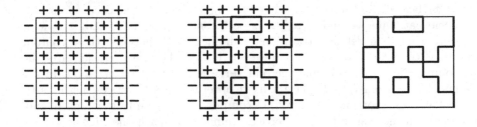

Fig. 3. Paths arising from the plus-minus boundary

reasons that will soon become clear.) Given a fixed assignment of spins to the faces just outside the boundary, our goal is to sample from the set of possible configurations with that boundary according to the Gibbs measure. We will consider two boundaries, the all-plus boundary β, and the plus-minus boundary β' where the we fix $+$ along the horizontal sides and $-$ along the vertical sides.

Given any Ising configuration with a prescribed boundary, we can uniquely reconstruct the spins on the interior from the set of edges that separate faces with unequal spins. We will concentrate on this contour representation of Ising configurations. The Ising model is defined so that the Gibbs measure of a configuration mapping to a subgraph G is proportional to the weight $\mu(G) = \lambda^{-|E(G)|}$. Notice that an Ising configuration with boundary β maps to a graph that can be decomposed into edge-disjoint contours, while an Ising configuration with boundary β' maps to a graph that can be decomposed into a set of contours and two paths connecting the four corners, as in Figure 3. These contours are no longer forced to turn, but the setting is otherwise reminiscent of Section 3.

Following Section 3, we first enlarge our state-space. Let Λ be the set of contours and near-contours of L, where every vertex has even degree except possibly two vertices. (We can think of near-contours representing "near-Ising" configurations, although this does not have a natural interpretation in the spin representation.) Let Λ' be the set of contours and near-contours of L', where the contours contain only vertices of even degree except in the four corners, and near-contours have this parity everywhere except at two vertices.

We now define a Markov chain $\mathcal{M}_I$ on Λ. Given G in Λ (or Λ'), choose an edge e uniformly at random in L. If $e \in G$, let $G' = G \backslash e$. If $e \notin G$, let $G' = G \cup e$. Then, if G' is in Λ (or Λ'), $\mathcal{M}_I$ sends G to G' with probability $\min(\frac{\mu(G')}{\mu(G)}, 1)$ and does nothing otherwise. Our second main theorem establishes that the mixing time of $\mathcal{M}_I$ is also very sensitive to the boundary.

Theorem 5. *For any $\lambda > 3$, there exists constants $c_1, c_2 > 1$ such that the mixing-time of $\mathcal{M}_I$ on Λ is $O(n^{c_1})$, but the mixing-time of $\mathcal{M}_I$ on Λ' is $\Omega(e^{c_2 n})$.*

4.1   Fast Mixing of $\mathcal{M}_I$ on Λ

To prove the fast mixing part of Theorem 5, we again bound conductance. This proof is almost identical to the arguments underlying the fast mixing of the Broder-chain on the square-octagon lattice.

Let Λ_T be the set of contours of Λ and let Λ_N be the set of near-contours. Our canonical paths map I and F to the closest contour graphs (if they are near-contours) and then define a traditional canonical path between these contour graphs. If $G \in \Lambda_T$, then $\overline{G} = G$. However, if $G \in \Lambda_N$, we again let $A(G)$ be the component of G containing the odd vertices and let $\overline{G} = G \backslash A(G)$. In the canonical path from I to $\overline{I}$, place a fixed ordering on all vertices, perhaps by vertical and then horizontal location. Decompose $A(I)$ as an Eulerian path, and, starting from the first of the odd vertices, remove this path one edge at a time. The path from $\overline{F}$ to F is defined to be the inverse of the path from F to $\overline{F}$.

For any contour G, the set of near-contours mapped to G in this manner has small weight. To show this, we use the trivial bound that the number of self-avoiding walks of length ℓ starting at a particular vertex is bounded by c^ℓ, where $c < 3$, see [7]. Observe that the set of H s.t. $\overline{H} = G$ has weight

$$\mu(\{H \in \Lambda_N : \overline{H} = G\}) = \sum_{l=1}^{n^2} \sum_{\substack{A \\ |A|=l}} \mu(G \cup A) < \sum_{l=1}^{n^2} n^2 3^l \mu(G) \lambda^{-l} < n^4 \mu(G)$$

We define a canonical path between contours I and $F \in \Lambda_T$ by "unwinding" the cycles and paths in their symmetric difference. Let $C = I \oplus F$. Order the components $\{C_i\}$ of C by $\{c_i\}$, where c_i is the earliest vertex in C_i. Each C_i is a contour, so it can be written as cycle starting from c_i. The path from I to F unwind each of the components in turn, starting at c_i and complementing each edge in turn. For a transition $T = (G, G')$, define the graph $\eta(T, I, F) := I \oplus F \oplus (G \cup G')$. Given $\eta(T, I, F)$ we will be able to reconstruct I and F given T. To reconstruct I and F from $\eta(T, I, F)$ and G, let $C = \eta(T, I, F) \oplus (G \cup G')$, and divide the edges of G and $\eta(T, I, F)$ according to the components of C. Note that every edge in both I and F is in both G and $\eta(T, I, F)$. Any edge in only one of I or F is in only one of G or $\eta(T, I, F)$. This implies that the weight $\mu(I)\mu(F) \leq \mu(\eta(T, I, F))\mu(G)$. Using Theorem 3, this is sufficient to prove that the conductance is at least $p(n)$, for some polynomial $p(\cdot)$, and thus is implies that the chain is rapidly mixing when we have the all-plus boundary.

4.2   Slow Mixing of $\mathcal{M}_I$ on Λ'

We proceed as in Section 3.3 by finding a cut set $\mathcal{W}$ that has exponentially smaller weight than the parts of the state space it separates. This implies that the conductance is exponentially small and that the chain is slowly mixing.

As before, define a *bridge* to be a path connecting corners of L'. (Recall they need no longer turn.) We define $\mathcal{W}$ to be the set of graphs in Λ' with two bridges and $\mathcal{U}$ to be the set of graphs with only one. For $W \in \mathcal{W}$, let $B(W)$ be the maximal (in terms of vertices) bridge in W. If W is an near-contour, let $A(W)$ be the component with internal odd vertices.

We define the function $f : \mathcal{W} \to \mathcal{U}$ such that, for $W \in \mathcal{W}$, f removes $B(W)$ (and $A(W)$ if one exists). By the choice of λ and the bound on the number of self-avoiding walks, there exists $c > 1$ such that, for any $U \in Img(f)$,

$$\mu(f^{-1}(U)) = \sum_A \sum_B \mu(U \cup A \cup B) = \sum_{a=1}^{n^2} \sum_{b=2n}^{n^2} \sum_{\substack{A,B \\ |A|=a \\ |B|=b}} \mu(U)\mu(A)\mu(B)$$

$$< \sum_{a=1}^{n^2} \sum_{b=2n}^{n^2} 3^a 3^b \mu(U)\lambda^{-a}\lambda^{-b} < c^n \mu(U).$$

Then, $\mu(\mathcal{W}) = \sum_{U \in \mathrm{Img}(f)} \mu(f^{-1}(U)) < c^{-n} \sum \mu(U) = c^{-n}\mu(\mathcal{U})$ which, by Theorem 1, shows slow mixing.

References

1. A.Z. Broder. How hard is it to marry at random? (On the approximation of the permanent). *Proc. 8th ACM Symposium on Theory of Computing*, 50–58, 1986.
2. R. Fernandez, P.A. Ferrari, and N.L. Garcia. Loss network representation of Ising contours. *Annals of Probability* **29**: 902–937, 2001.
3. H.O. Georgii. *Gibbs measures and phase transitions.* de Gruyter Studies in Mathematics **9**, Walter de Gruyter & Co., Berlin, 1988.
4. S. Greenberg and D. Randall. Slow mixing of Glauber dynamics on perfect matchings of the square-octagon lattice. Preprint, 2006.
5. M.R. Jerrum and A.J. Sinclair. Approximate counting, uniform generation and rapidly mixing Markov chains. *Information and Computation* **82**: 93–133, 1989.
6. M.R. Jerrum and A.J. Sinclair. Polynomial-time approximation algorithms for the Ising model. *SIAM Journal on Computing* **22**: 1087–1116, 1993.
7. N. Madras and G. Slade. *The Self-Avoiding Walk.* Boston, MA, Birkhuser, 1993.
8. F. Martinelli. Lectures on Glauber dynamics for discrete spin models. *Lectures on Probability Theory and Statistics (Saint-Flour, 1997)*, Lecture notes in Mathematics **1717**: 93-191, Springer, Berlin, 1998.
9. F. Martinelli, E. Olivieri and R. Schonmann. For 2-D lattice spin systems weak mixing implies strong mixing. *Comm. Mathematical Physics* **165**: 33–47, 1994.
10. F. Martinelli, A. Sinclair, D. Weitz, The Ising model on trees: Boundary conditions and mixing time. *Comm. Mathematical Physics* **250**: 301–334, 2004.
11. F. Martinelli, A. Sinclair, D. Weitz, Fast mixing for independent sets, colorings and other models on trees. *Proc. 15th ACM/SIAM Symposium on Discrete Algorithms*, 449–458, 2004.
12. J. Propp. Diabolo Tilings of Fortress. Talk given at the MIT Combinatorics Seminar, 1998 (http://www.math.wisc.edu/ propp/diabolo.ps.gz).
13. J. Propp and D.B. Wilson. Exact Sampling with Coupled Markov Chains and Applications to Statistical Mechanics. *Random Structures and Algorithms* **9**: 223–252, 1996.
14. D. Randall and D.B. Wilson. Sampling Spin Configurations of an Ising System. *Proc. 10th ACM/SIAM Symposium on Discrete Algorithms*, S959–960, 1999.
15. A.J. Sinclair and M.R. Jerrum. Approximate counting, uniform generation and rapidly mixing Markov chains. *Information and Computation* **82**: 93–133, 1989.
16. L. Thomas. Bound on the mass gap for the finite volume stochastic Ising models at low temperature. *Comm. Mathematical Physics* **126**: 1–11, 1989.

Adaptive Sampling and
Fast Low-Rank Matrix Approximation

Amit Deshpande and Santosh Vempala

Mathematics Department and CSAIL, MIT
amitd@mit.edu, vempala@mit.edu

Abstract. We prove that any real matrix A contains a subset of at most $4k/\epsilon + 2k \log(k + 1)$ rows whose span "contains" a matrix of rank at most k with error only $(1 + \epsilon)$ times the error of the best rank-k approximation of A. We complement it with an almost matching lower bound by constructing matrices where the span of any $k/2\epsilon$ rows does not "contain" a relative $(1 + \epsilon)$-approximation of rank k. Our existence result leads to an algorithm that finds such rank-k approximation in time

$$O\left(M\left(\frac{k}{\epsilon} + k^2 \log k\right) + (m + n)\left(\frac{k^2}{\epsilon^2} + \frac{k^3 \log k}{\epsilon} + k^4 \log^2 k\right)\right),$$

i.e., essentially $O(Mk/\epsilon)$, where M is the number of nonzero entries of A. The algorithm maintains sparsity, and in the streaming model [12, 14, 15], it can be implemented using only $2(k+1)(\log(k+1)+1)$ passes over the input matrix and $O\left(\min\{m, n\}(\frac{k}{\epsilon} + k^2 \log k)\right)$ additional space. Previous algorithms for low-rank approximation use only one or two passes but obtain an additive approximation.

1 Introduction

Given an $m \times n$ matrix A of reals and an integer k, the problem of finding a matrix B of rank at most k that minimizes $\|A - B\|_F^2 = \sum_{i,j}(A_{ij} - B_{ij})^2$ has received much attention in the past decade. The classical optimal solution to this problem is the matrix A_k consisting of the first k terms in the Singular Value Decomposition (SVD) of A:

$$A = \sum_{i=1}^{n} \sigma_i u_i v_i^T$$

where $\sigma_1 \geq \sigma_2 \geq \ldots \geq \sigma_n \geq 0$ are the singular values and $\{u_i\}_1^n, \{v_i\}_1^n$ are orthonormal sets of vectors called left and right singular vectors, respectively. Computing the SVD and hence the best low-rank approximation takes $O(\min\{mn^2, m^2n\})$ time.

Recent work on this problem has focussed on reducing the complexity while allowing an approximation to A_k. Frieze et al. [13] introduced the following sampling approach where rows of A are picked with probabilities proportional to their squared lengths.

J. Diaz et al. (Eds.): APPROX and RANDOM 2006, LNCS 4110, pp. 292–303, 2006.

Theorem 1 ([13]). *Let S be an i.i.d. sample of s rows of an $m \times n$ matrix A, from the following distribution: row i is picked with probability*

$$P_i \geq c\frac{\|A^{(i)}\|^2}{\|A\|_F^2}.$$

Then there is a matrix $\tilde{A}_k$ whose rows lie in span(S) *such that*

$$\mathsf{E}\left[\|A - \tilde{A}_k\|_F^2\right] \leq \|A - A_k\|_F^2 + \frac{k}{cs}\|A\|_F^2.$$

Setting $s = k/c\epsilon$ in the theorem, we get

$$\mathsf{E}\left[\|A - \tilde{A}_k\|_F^2\right] \leq \|A - A_k\|_F^2 + \epsilon\|A\|_F^2.$$

The theorem suggests a randomized algorithm (analyzed in [13], [7] and later in [9]) that makes two passes through the matrix A and finds such an approximation using $O(\min\{m, n\}k^2/\epsilon^4)$ additional time. So overall, it takes $O(M + \min\{m, n\}k^2/\epsilon^4)$ time, where M is the number of non-zero entries of A. A different sampling approach that uses only one pass and has comparable guarantees (in particular, additive error) was given in [2], and further improved in [1].

The additive error $\epsilon\|A\|_F^2$ could be arbitrarily large compared to the true error, $\|A - A_k\|_F^2$. Is it possible to get a $(1+\epsilon)$-relative approximation efficiently, i.e., in linear or sublinear time? Related to this, is there a small witness, i.e., is there a $(1 + \epsilon)$-approximation of rank k whose rows lie in the span of a small subset of the rows of A? Addressing these questions, it was shown in [11] that any matrix A contains a subset S of $O(k^2/\epsilon)$ rows such that there is a matrix $\tilde{A}_k$ of rank at most k whose rows lie in span(S) and

$$\|A - \tilde{A}_k\|_F^2 \leq (1 + \epsilon)\|A - A_k\|_F^2.$$

This existence result was applied to derive an approximation algorithm for a projective clustering [3, 16] problem: find j linear subspaces, each of dimension at most k, that minimize the sum of squared distances of each point to its nearest subspace. However, the question of efficiently finding such a $(1 + \epsilon)$-relative approximation to A_k was left open.

In recent independent work, Drineas et al. [6, 10] have shown that, *using the SVD*, one can find a subset of $O(k \log k/\epsilon)$ rows whose span "contains" such a relative approximation. They also provide practical motivation for this problem.

1.1   Our Results

Our first result is the following improved existence theorem.

Theorem 2. *Any $m \times n$ matrix A contains a subset S of $4k/\epsilon + 2k \log(k + 1)$ rows such that there is a matrix $\tilde{A}_k$ of rank at most k whose rows lie in* span(S) *and*

$$\|A - \tilde{A}_k\|_F^2 \leq (1 + \epsilon)\|A - A_k\|_F^2.$$

Based on this, we give an efficient algorithm in Section 3.2 that exploits any sparsity of the input matrix. For a matrix with M nonzero entries, a rank-k approximation is computed in

$$O\left(M\left(\frac{k}{\epsilon} + k^2 \log k\right) + (m+n)\left(\frac{k^2}{\epsilon^2} + \frac{k^3 \log k}{\epsilon} + k^4 \log^2 k\right)\right)$$

time using $O(\min\{m,n\}(\frac{k}{\epsilon} + k^2 \log k))$ space (Theorem 5). In the streaming model, the algorithm requires $2(k+1)(\log(k+1) + 1)$ passes over the input matrix. The running time is $O\left(M(k/\epsilon + k^2 \log k)\right)$ for M sufficiently larger than m, n; when k is a constant it is $O(M/\epsilon + 1/\epsilon^2)$. We note that while some of the analysis is new, most of the algorithmic ideas were proposed in [11].

We complement the existence result with the following lower bound (Prop. 4): there exist matrices for which the span of any subset of $k/2\epsilon$ rows does not contain a $(1+\epsilon)$-relative approximation.

Finally, the improved existence bound also leads to better PTAS for the projective clustering problem. The complexity becomes $d(n/\epsilon)^{O(jk^2/\epsilon + jk^2 \log k)}$ reducing the dependence on k in the exponent from k^3 and resolving an open question of [11].

Notation. Henceforth, we will use $\pi_V(A)$ to denote the matrix obtained by projecting each row of A onto a linear subspace V. If V is spanned by a subset S of rows, we denote the projection of A onto V by $\pi_{\text{span}(S)}(A)$. We use $\pi_{\text{span}(S),k}(A)$ for the best rank-k approximation to A whose rows lie in $\text{span}(S)$. Thus, the approximation $\tilde{A}_k$ in Theorem 2 is $\tilde{A}_k = \pi_{\text{span}(S),k}(A)$ for a suitable S.

2 Sampling Techniques

We now describe the two sampling techniques that will be used.

2.1 Adaptive Sampling

One way to generalize the sampling procedure of Frieze et al. [13] is to do the sampling in multiple rounds, and in an adaptive fashion. The rows in each new round get picked with probabilities proportional to their squared distance from the span of the rows that we have already picked in the previous rounds.

Here is the t-round adaptive sampling algorithm, introduced in [11].

1. Start with a linear subspace V. Let $E_0 = A - \pi_V(A)$, and $S = \emptyset$.
2. For $j = 1$ to t, do:
   (a) Pick a sample S_j of s_j rows of A independently from the following distribution: row i is picked with probability $P_i^{(j-1)} \geq c\frac{\|E_{j-1}^{(i)}\|^2}{\|E_{j-1}\|_F^2}$.
   (b) $S = S \cup S_j$.
   (c) $E_j = A - \pi_{\text{span}(V \cup S)}(A)$.

The next theorem, from [11], is a generalization of Theorem 1.

Theorem 3 ([11]). *After one round of the adaptive sampling procedure described above,*

$$\mathsf{E}_{S_1}\left[\|A - \pi_{\mathrm{span}(V \cup S_1),k}(A)\|_F^2\right] \leq \|A - A_k\|_F^2 + \frac{k}{cs_1}\|E_0\|_F^2.$$

We can now prove the following corollary of Theorem 3, for t-round adaptive sampling, using induction on the number of rounds.

Corollary 1. *After t rounds of the adaptive sampling procedure described above,*

$$\mathsf{E}_{S_1,\ldots,S_t}\left[\|A - \pi_{\mathrm{span}(V \cup S),k}(A)\|_F^2\right]$$
$$\leq \left(1 + \frac{k}{cs_t} + \frac{k^2}{c^2 s_t s_{t-1}} + \cdots + \frac{k^{t-1}}{c^{t-1} s_t s_{t-1} \cdots s_2}\right)\|A - A_k\|_F^2$$
$$+ \frac{k^t}{c^t s_t s_{t-1} \cdots s_1}\|E_0\|_F^2.$$

Proof. We prove the theorem by induction on t. The case $t = 1$ is precisely Theorem 3. For the inductive step, using Theorem 3 with $\mathrm{span}(V \cup S_1 \cup \cdots \cup S_{t-1})$ as our initial subspace, we have

$$\mathsf{E}_{S_t}\left[\|A - \pi_{\mathrm{span}(V \cup S),k}(A)\|_F^2\right] \leq \|A - A_k\|_F^2 + \frac{k}{cs_t}\|E_{t-1}\|_F^2.$$

Combining this inequality with the fact that

$$\|E_{t-1}\|_F^2 = \|A - \pi_{\mathrm{span}(V \cup S_1 \cup \cdots \cup S_{t-1})}(A)\|_F^2 \leq \|A - \pi_{\mathrm{span}(V \cup S_1 \cup \cdots \cup S_{t-1}),k}(A)\|_F^2,$$

we get

$$\mathsf{E}_{S_t}\left[\|A - \pi_{\mathrm{span}(S'),k}(A)\|_F^2\right] \leq \|A - A_k\|_F^2 + \frac{k}{cs_t}\|A - \pi_{\mathrm{span}(V \cup S_1 \cup \cdots \cup S_{t-1}),k}(A)\|_F^2.$$

Finally, taking the expectation over $S_1, \ldots, S_{t-1}$:

$$\mathsf{E}_{S_1,\ldots,S_t}\left[\|A - \pi_{\mathrm{span}(V \cup S),k}(A)\|_F^2\right]$$
$$\leq \|A - A_k\|_F^2 + \frac{k}{cs_t}\mathsf{E}_{S_1,\ldots,S_{t-1}}\left[\|A - \pi_{\mathrm{span}(V \cup S_1 \cup \cdots \cup S_{t-1}),k}(A)\|_F^2\right]$$

and the result follows from the induction hypothesis for $t - 1$.

From Corollary 1, it is clear that if we can get a good initial subspace V such that $dim(V) = k$ and the error given by V is within some multiplicative factor of $\|A - A_k\|_F^2$, then we can hope to prove something about relative rank-k approximation. This motivates a different generalization of the sampling method of [13].

2.2   Volume Sampling

Another way to generalize the sampling scheme of Frieze et al. [13] is by sampling subsets of rows instead of individual rows. Let S be a subset of k rows of A, and $\Delta(S)$ be the simplex formed by these rows and the origin. Volume sampling corresponds to the following distribution: we pick subset S with probability equal to

$$P_S = \frac{\text{vol}(\Delta(S))^2}{\sum_{T:|T|=k} \text{vol}(\Delta(T))^2}.$$

Remark: Volume sampling can also be thought of as squared length sampling in the exterior product space. Consider a matrix A' with rows $A'_S = A^{(i_1)} \wedge A^{(i_2)} \wedge \ldots \wedge A^{(i_k)} \in \bigwedge^k \mathbb{R}$, indexed by all k-subsets $S = \{i_1, i_2, \ldots, i_k\} \subseteq [m]$. It is easy to see that the topmost singular value of A' is $\sigma_1 \sigma_2 \ldots \sigma_k$ with $v_1 \wedge v_2 \wedge \ldots \wedge v_k$ as its corresponding right singular vector. Moreover, determinant (i.e., normalized volume) defines a norm on the wedge product of k vectors, and therefore, rank-k approximation of A by volume sampling k-subsets of rows can be thought of as rank-1 approximation of A' by squared length sampling of its rows.

Volume sampling technique was introduced in [11] to prove the following theorem.

Theorem 4 ([11]). *Let S be a random subset of k rows of a given matrix A chosen with probability P_S defined as above. Then.*

$$\mathsf{E}_S\left[\|A - \pi_{\text{span}(S)}(A)\|_F^2\right] \leq (k+1)\|A - A_k\|_F^2.$$

The next lemma was used crucially in the analysis of volume sampling.

Lemma 1 ([11]).

$$\sum_{S,|S|=k} \text{vol}(\Delta(S))^2 = \frac{1}{(k!)^2} \sum_{1 \leq t_1 < t_2 < \ldots < t_k \leq n} \sigma_{t_1}^2 \sigma_{t_2}^2 \ldots \sigma_{t_k}^2,$$

where $\sigma_1, \sigma_2, \ldots, \sigma_r > 0 = \sigma_{r+1} = \ldots = \sigma_n$ are the singular values of A.

2.3   Approximate Volume Sampling Via Adaptive Sampling

Here we give an algorithm for approximate volume sampling. In brief, we run a k-round adaptive sampling procedure, picking one row in each round.

1. $S = \emptyset$, $E_0 = A$.
2. For $j = 1$ to k, do:

   (a) Pick row i with probability proportional to $P_i^{(j-1)} \geq c \frac{\|E_{j-1}^{(i)}\|^2}{\|E_{j-1}\|_F^2}$.
   (b) Add this new row to subset S.
   (c) $E_j = A - \pi_{\text{span}(S)}(A)$.

Next we show that the above procedure gives an approximate implementation of volume sampling.

Proposition 1. *Suppose the k-round adaptive procedure mentioned above picks a subset S with probability $\tilde{P}_S$. Then,*

$$\tilde{P}_S \leq k!\, P_S$$

Proof. Let $S = \{A^{i_1}, A^{i_2}, \ldots, A^{i_k}\}$ be a subset of k rows, and let $\tau \in \Pi_k$, the set of all permutations of $\{i_1, i_2, \ldots, i_k\}$. By $H_{\tau,t}$ we denote the linear subspace span$(A^{\tau(i_1)}, A^{\tau(i_2)}, \ldots, A^{\tau(i_t)})$, and by $d(A^i, H_{\tau,t})$ we denote the orthogonal distance of A^i from this subspace. Our adaptive procedure picks a subset S with probability equal to

$$
\begin{aligned}
\tilde{P}_S &= \sum_{\tau \in \Pi_k} \frac{\|A^{\tau(i_1)}\|^2}{\|A\|_F^2} \frac{d(A^{\tau(i_2)}, H_{\tau,1})^2}{\sum_{i=1}^m d(A^i, H_{\tau,1})^2} \cdots \frac{d(A^{\tau(i_k)}, H_{\tau,k-1})^2}{\sum_{i=1}^m d(A^i, H_{\tau,k-1})^2} \\
&\leq \frac{\sum_{\tau \in \Pi_k} \|A^{\tau(i_1)}\|^2\, d(A^{\tau(i_2)}, H_{\tau,1})^2 \cdots d(A^{\tau(i_k)}, H_{\tau,k-1})^2}{\|A\|_F^2\, \|A - A_1\|_F^2 \cdots \|A - A_{k-1}\|_F^2} \\
&= \frac{\sum_{\tau \in \Pi_k} (k!)^2 \mathrm{vol}(\Delta(S))^2}{\|A\|_F^2\, \|A - A_1\|_F^2 \cdots \|A - A_{k-1}\|_F^2} \\
&= \frac{(k!)^3\, \mathrm{vol}(\Delta(S))^2}{\sum_{i=1}^m \sigma_i^2 \sum_{i=2}^m \sigma_i^2 \cdots \sum_{i=k}^m \sigma_i^2} \\
&\leq \frac{(k!)^3\, \mathrm{vol}(\Delta(S))^2}{\sum_{1 \leq i_1 < i_2 < \ldots < i_k \leq m} \sigma_{i_1}^2 \sigma_{i_2}^2 \cdots \sigma_{i_k}^2} \\
&= \frac{k!\, \mathrm{vol}(\Delta(S))^2}{\sum_{T:|T|=k} \mathrm{vol}(\Delta(T))^2} \qquad \text{(using Lemma 1)} \\
&= k!\, P_S
\end{aligned}
$$

Now we will show why it suffices to have just the approximate implementation of volume sampling. If we sample subsets S with probabilities $\tilde{P}_S$ instead of P_S, we get an analog of Theorem 4 with a weaker multiplicative approximation.

Proposition 2. *If we sample a subset S of k rows using the k-round adaptive sampling procedure mentioned above, then*

$$\mathsf{E}_S\left[\|A - \pi_S(A)\|_F^2\right] \leq (k+1)!\, \|A - A_k\|_F^2.$$

Proof. Since we are picking a subset S with probability $\tilde{P}_S$ the expected error is

$$
\begin{aligned}
\mathsf{E}_S\left[\|A - \pi_{\mathrm{span}(S)}(A)\|_F^2\right] &= \sum_{S:|S|=k} \tilde{P}_S \|A - \pi_{\mathrm{span}(S)}(A)\|_F^2 \\
&\leq k! \sum_{S:|S|=k} P_S \|A - \pi_{\mathrm{span}(S)}(A)\|_F^2 \\
&\leq k!\, (k+1)\|A - A_k\|_F^2 \qquad \text{(using Theorem 4)} \\
&= (k+1)!\, \|A - A_k\|_F^2
\end{aligned}
$$

3   Low-Rank Approximation with Multiplicative Error

In this section, we combine adaptive sampling and volume sampling to prove the existence of a small witness and then to derive an efficient algorithm.

3.1   Existence

We now prove Theorem 2.

Proof. From Theorem 4, we know that there exists a subset S_0 of k rows of A such that

$$\|A - \pi_{\mathrm{span}(S_0)}(A)\|_F^2 \leq (k+1)\|A - A_k\|_F^2.$$

Let $V = \mathrm{span}(S_0)$, $t = \log(k+1)$, $c = 1$ in Corollary 1, we know that there exist subsets $S_1, \ldots, S_t$ of rows with sizes $s_1 = \ldots = s_{t-1} = 2k$ and $s_t = 4k/\epsilon$, respectively, such that

$$
\begin{aligned}
\|A - \pi_{\mathrm{span}(V \cup S_1 \cup \ldots \cup S_t), k}(A)\|_F^2 &\leq \left(1 + \frac{\epsilon}{4} + \frac{\epsilon}{8} + \ldots\right)\|A - A_k\|_F^2 + \frac{\epsilon}{2^{t+1}}\|E_0\|_F^2 \\
&\leq \left(1 + \frac{\epsilon}{2}\right)\|A - A_k\|_F^2 + \frac{\epsilon}{2^{t+1}}\|A - \pi_V(A)\|_F^2 \\
&\leq \left(1 + \frac{\epsilon}{2}\right)\|A - A_k\|_F^2 + \frac{\epsilon}{2^{t+1}}(k+1)\|A - A_k\|_F^2 \\
&= \left(1 + \frac{\epsilon}{2}\right)\|A - A_k\|_F^2 + \frac{\epsilon}{2}\|A - A_k\|_F^2 \\
&= (1 + \epsilon)\|A - A_k\|_F^2.
\end{aligned}
$$

Therefore, for $S = S_0 \cup S_1 \cup \ldots \cup S_t$ we have

$$|S| \leq \sum_{j=0}^{t}|S_j| = k + 2k(\log(k+1) - 1) + \frac{4k}{\epsilon} \leq \frac{4k}{\epsilon} + 2k\log(k+1)$$

and

$$\|A - \pi_{\mathrm{span}(S'), k}(A)\|_F^2 \leq (1 + \epsilon)\|A - A_k\|_F^2.$$

3.2   Efficient Algorithm

In this section we describe an algorithm that given a matrix $A \in \mathbb{R}^{m \times n}$, finds another matrix $\tilde{A}_k$ of rank at most k such that $\|A - \tilde{A}_k\|_F^2 \leq (1 + \epsilon)\|A - A_k\|_F^2$. The algorithm has two phases. In the first phase, we pick a subset of k rows using the approximate volume sampling procedure described in Subsection 2.3. In the second phase, we use the span of these k rows as our initial subspace and perform $(k+1)\log(k+1)$ rounds of adaptive sampling. The rows chosen are all from the original matrix A.

Linear Time Low-Rank Matrix Approximation

Input: $A \in \mathbb{R}^{m \times n}$, integer $k \leq m$, error parameter $\epsilon > 0$.
Output: $\tilde{A}_k \in \mathbb{R}^{m \times n}$ of rank at most k.

1. Pick a subset S_0 of k rows of A using the approximate volume sampling procedure described in Subsection 2.3. Compute an orthonormal basis $\mathcal{B}_0$ of span(S_0).
2. Initialize $V = \mathrm{span}(S_0)$. Fix parameters as $t = (k+1)\log(k+1)$, $s_1 = s_2 = \ldots = s_{t-1} = 2k$, and $s_t = 16k/\epsilon$.
3. Pick subsets of rows $S_1, S_2, \ldots, S_t$, using t-round adaptive sampling procedure described in Subsection 2.1. After round j, extend the previous orthonormal basis $\mathcal{B}_{j-1}$ to an orthonormal basis $\mathcal{B}_j$ of span($S_0 \cup S_1 \cup \ldots \cup S_j$).
4. $S = \bigcup_{j=0}^{t} S_j$, and we have an orthonormal basis $\mathcal{B}_t$ of span(S).
5. Compute $h_1, h_2, \ldots, h_k$, the top k right singular vectors of $\pi_{\mathrm{span}(S)}(A)$.
6. Output matrix $\tilde{A}_k = \pi_{\mathrm{span}(h_1,\ldots,h_k)}(A)$, written in the standard basis.

Here are some details about the implementations of these steps.

In Step 1, we use the k-round adaptive procedure for approximate volume sampling. In the j-th round of this procedure, we sample a row and compute its component v_j orthogonal to the span of the rows picked in rounds $1, 2, \ldots, j-1$. The residual squared lengths of the rows are computed using $\|E_j^{(i)}\|^2 = \|E_{j-1}^{(i)}\|^2 - A^{(i)} \cdot v_j$, and $\|E_j\|_F^2 = \|E_{j-1}\|_F^2 - \|Av_j\|^2$. In the end, we have an orthonormal basis $\mathcal{B}_0 = \{v_1/\|v_1\|, \ldots, v_k/\|v_k\|\}$.

In Step 3, there are $(k+1)\log(k+1)$ rounds of adaptive sampling. In the j-th round, we extend the orthonormal basis from $\mathcal{B}_{j-1}$ to $\mathcal{B}_j$ by Gram-Schmidt orthonormalization. We compute the residual squared lengths of the rows $\|E_j^{(i)}\|^2$, as well as the total, $\|E_j\|_F^2$, by subtracting the contribution $\pi_{\mathrm{span}(\mathcal{B}_j \setminus \mathcal{B}_{j-1})}(A)$ from the values that they had during the previous round.

Each round in Steps 1 and 3 can be implemented using 2 passes over the matrix: one pass to figure out the sampling distribution, and an another one to sample a row (or a subset of rows) according to this distribution. So Steps 1 and 3 require $2(k+1)\log(k+1) + 2k$ passes.

Finally, in Step 5, we compute $\pi_{\mathrm{span}(S)}(A)$ in terms of basis $\mathcal{B}_t$ using one pass (now we have an $m \times O(k/\epsilon + k^2 \log k)$ matrix), and we compute its top k right singular vectors using SVD. In Step 6, we rewrite them in the standard basis and project matrix A onto their span, which requires one additional pass.

So the total number of passes is $2(k+1)(\log(k+1) + 1)$.

Theorem 5. *With probability at least $3/4$, the algorithm outputs a matrix $\tilde{A}_k$ such that*
$$\|A - \tilde{A}_k\|_F^2 \leq (1 + \epsilon)\|A - A_k\|_F^2.$$
Moreover, the algorithm takes
$$O\left(M\left(\frac{k}{\epsilon} + k^2 \log k\right) + (m+n)\left(\frac{k^2}{\epsilon^2} + \frac{k^3 \log k}{\epsilon} + k^4 \log^2 k\right)\right)$$
time and $O\left(\min\{m, n\}(\frac{k}{\epsilon} + k^2 \log k)\right)$ space.

Proof. We begin with a proof of correctness. After the first phase of approximate volume sampling, using Proposition 2, we have

$$\mathsf{E}_{S_0}\left[\|A - \pi_{\text{span}(S_0)}(A)\|_F^2\right] \le (k+1)! \, \|A - A_k\|_F^2.$$

Now using $V = \text{span}(S_0)$, $c = 1$, $t = (k+1)\log(k+1)$, $s_t = 16k/\epsilon$, $s_{t-1} = \ldots = s_1 = 2k$ in Theorem 1 we get that

$$\mathsf{E}_{S_1,\ldots,S_t}\left[\|A - \pi_{\text{span}(S),k}(A)\|_F^2\right]$$
$$\le \left(1 + \frac{\epsilon}{16} + \frac{\epsilon}{32} + \ldots\right)\|A - A_k\|_F^2 + \frac{\epsilon}{2^{t+3}}\|A - \pi_{\text{span}(S_0)}(A)\|_F^2$$
$$\le \left(1 + \frac{\epsilon}{8}\right)\|A - A_k\|_F^2 + \frac{\epsilon}{8 \cdot 2^t}\|A - \pi_{\text{span}(S_0)}(A)\|_F^2.$$

Now taking expectation over S_0 we have

$$\mathsf{E}_{S_0,\ldots,S_t}\left[\|A - \pi_{\text{span}(S),k}(A)\|_F^2\right]$$
$$\le \left(1 + \frac{\epsilon}{8}\right)\|A - A_k\|_F^2 + \frac{\epsilon}{8 \cdot 2^t}\,\mathsf{E}_{S_0}\|A - \pi_{\text{span}(S_0)}(A)\|_F^2$$
$$\le \left(1 + \frac{\epsilon}{8}\right)\|A - A_k\|_F^2 + \frac{\epsilon}{8 \cdot 2^t}\,(k+1)!\,\|A - A_k\|_F^2$$
$$\le \left(1 + \frac{\epsilon}{8}\right)\|A - A_k\|_F^2 + \frac{\epsilon}{8 \cdot 2^t}\,(k+1)^{(k+1)}\,\|A - A_k\|_F^2$$
$$\le \left(1 + \frac{\epsilon}{8}\right)\|A - A_k\|_F^2 + \frac{\epsilon}{8}\,\|A - A_k\|_F^2$$
$$= \left(1 + \frac{\epsilon}{4}\right)\|A - A_k\|_F^2.$$

This means

$$\mathsf{E}_{S_0,\ldots,S_t}\left[\|A - \pi_{\text{span}(S),k}(A)\|_F^2 - \|A - A_k\|_F^2\right] \le \frac{\epsilon}{4}\|A - A_k\|_F^2.$$

Therefore, using Markov's inequality, with probability at least $3/4$ the algorithm gives a matrix $\tilde{A}_k = \pi_{\text{span}(S),k}(A)$ satisfying

$$\|A - \tilde{A}_k\|_F^2 \le (1 + \epsilon)\|A - A_k\|_F^2.$$

Now let us analyze its complexity.

Step 1 has k rounds of adaptive sampling. In each round, the matrix-vector multiplication requires $O(M)$ time and storing vector v_j requires $O(n)$ space. So overall, Step 1 takes $O(Mk + nk)$ time, $O(nk)$ space.

Step 3 has $2(k+1)\log(k+1)$ rounds of adaptive sampling. The j-th round (except for the last round), involves Gram-Schmidt orthonormalization of $2k$ vectors in $\mathbb{R}^n$ against an orthonormal basis of size at most $(2j+1)k$, which takes time $O(njk^2)$. Computing $\pi_{\text{span}(B_j \setminus B_{j-1})}(A)$ for updating the values $\|E_j^{(i)}\|^2$ and $\|E_j\|_F^2$ takes time $O(Mk)$. Thus, the total time for the j-th round is $O(Mk + njk^2)$. In the last round, we pick $O(k/\epsilon)$ rows. The Gram-Schmidt orthonormalization of these $O(k/\epsilon)$ vectors against an orthonormal basis of $O(k^2 \log k)$

vectors takes $O(nk^3 \log k/\epsilon)$ time; storing this basis requires $O(nk/\epsilon + nk^2 \log k)$ space. So overall, Step 3 takes $O\left(Mk^2 \log k + n(k^3 \log k/\epsilon + k^4 \log^2 k)\right)$ time and $O(nk/\epsilon + nk^2 \log k)$ space (to store the basis $\mathcal{B}_t$).

In Step 5, projecting A onto span(S) takes $O\left(M(k/\epsilon + k^2 \log k)\right)$ time. Now we have $\pi_{\mathrm{span}(S)}(A)$ in terms of our basis $\mathcal{B}_t$ (which is a $m \times O(k^2 \log k + k/\epsilon)$ matrix) and computation of its top k right singular vectors takes time $O\left(m(k/\epsilon + k^2 \log k)^2\right)$.

In Step 6, rewriting $h_1, h_2, \ldots, h_k$ in terms of the standard basis takes time $O\left(n(k^3 \log k + k^2/\epsilon)\right)$. And finally, projecting the matrix A onto span($h_1, \ldots, h_k$) takes time $O(Mk)$.

Putting it all together, the algorithm takes

$$
O\left(M\left(\frac{k}{\epsilon} + k^2 \log k\right) + (m+n)\left(\frac{k^2}{\epsilon^2} + \frac{k^3 \log k}{\epsilon} + k^4 \log^2 k\right)\right)
$$

time and $O\left(\min\{m,n\}(k/\epsilon + k^2 \log k)\right)$ space (since we can do the same with columns instead of rows), and $O(k \log k)$ passes over the data.

This algorithm can be made to work with high probability, by running independent copies of the algorithm in each pass and taking the best answer found at the end. The overhead to get a probability of success of $1 - \delta$ is $O(\sqrt{\log(1/\delta)})$.

4   Lower-Bound for Relative Low-Rank Matrix Approximation

Here we show a lower bound of $\Omega(k/\epsilon)$ for rank-k approximation using a subset of rows.

Proposition 3. *Given $\epsilon > 0$ and n large enough so that $n\epsilon \geq 2$, there exists an $n \times (n+1)$ matrix A such that for any subset S of its rows with $|S| \leq 1/2\epsilon$,*

$$
\|A - \pi_{\mathrm{span}(S),1}(A)\|_F^2 \geq (1+\epsilon)\|A - A_1\|_F^2
$$

Proof. Let $e_1, e_2, \ldots, e_{n+1}$ be the standard basis for $\mathbb{R}^{n+1}$, considered as rows. Consider the $n \times (n+1)$ matrix A, whose i-th row is given by $A^{(i)} = e_1 + \epsilon\, e_{i+1}$, for $i = 1, 2, \ldots, n$. The best rank-1 approximation for this is A_1, whose i-th row is given by $A_1^{(i)} = e_1 + \sum_{i=1}^{n} \frac{1}{n} e_{i+1}$. Therefore,

$$
\|A - A_1\|_F^2 = \sum_{i=1}^{n} \|A^{(i)} - A_1^{(i)}\|^2 = n\left(\frac{(n-1)^2\epsilon^2}{n^2} + (n-1)\frac{\epsilon^2}{n^2}\right) = (n-1)\epsilon^2.
$$

Now let S be any subset of the rows with $|S| = s$. It is easy to see that the best rank-1 approximation for A in the span of S is given by $\pi_{\mathrm{span}(S),1}(A)$, whose i-th row is given by $\pi_{\mathrm{span}(S),1}(A)^{(i)} = e_1 + \frac{\epsilon}{s} \sum_{i \in S} e_{i+1}$, for all i (because it has to be a symmetric linear combination of them). Hence,

$$\|A - \pi_{\text{span}(S),1}(A)\|_F^2 = \sum_{i \in S} \|A^{(i)} - \pi_{\text{span}(S),1}(A)^{(i)}\|^2 + \sum_{i \notin S} \|A^{(i)} - \pi_{\text{span}(S),1}(A)^{(i)}\|^2$$

$$= s \left(\frac{(s-1)^2 \epsilon^2}{s^2} + (s-1)\frac{\epsilon^2}{s^2} \right) + (n-s)\left(s\frac{\epsilon^2}{s^2} + \epsilon^2 \right)$$

$$= \frac{(s-1)^2 \epsilon^2}{s} + \frac{(s-1)\epsilon^2}{s} + \frac{n\epsilon^2}{s} + n\epsilon^2 - \epsilon^2 - s\epsilon^2$$

$$= \frac{n\epsilon^2}{s} + n\epsilon^2 - 2\epsilon^2.$$

Now if $s \le \frac{1}{2\epsilon}$ then $\|A - \pi_{\text{span}(S),1}(A)\|_F^2 = (1+2\epsilon)n\epsilon^2 - 2\epsilon^2 \ge (1+\epsilon)n\epsilon^2 \ge (1+\epsilon)\|A - A_1\|_F^2$, for n chosen large enough so that $n\epsilon \ge 2$.

Now we will try to extend this lower bound for relative rank-k approximation.

Proposition 4. *Given $\epsilon > 0$, k, and n large enough so that $n\epsilon \ge 2k$, there exists a $kn \times k(n+1)$ matrix B such that for any subset S of its rows with $|S| \le k/2\epsilon$,*

$$\|B - \pi_{\text{span}(S),k}(A)\|_F^2 \ge (1+\epsilon)\|B - B_k\|_F^2.$$

Proof. Consider B to be a $kn \times k(n+1)$ block-diagonal matrix with k blocks, where each of the blocks is equal to A defined as in Proposition 3 above. It is easy to see that

$$\|B - B_k\|_F^2 = k\|A - A_1\|_F^2.$$

Now pick any subset S of rows with $|S| \le \frac{k}{2\epsilon}$. Let S_i be the subset of rows taken from the i-th block, and let $|S_i| = \frac{k}{2\epsilon_i}$. We know that $\sum_{i=1}^k |S_i| = \sum_{i=1}^k \frac{k}{2\epsilon_i} \le \frac{k}{2\epsilon}$, and hence $n\epsilon_i \ge n\epsilon \ge 2$.

Therefore,

$$\|B - \pi_{\text{span}(S),k}(B)\|_F^2 = \sum_{i=1}^k \|A - \pi_{\text{span}(S_i),1}(A)\|_F^2$$

$$\ge \sum_{i=1}^k (1 + \frac{\epsilon_i}{k})\|A - A_1\|_F^2 \qquad \text{(using Proposition 3)}$$

$$= (k + \frac{\sum_{i=1}^k \epsilon_i}{k})\|A - A_1\|_F^2$$

$$\ge (k + \frac{k}{\sum_{i=1}^k 1/\epsilon_i})\|A - A_1\|_F^2 \quad \text{(by A.M.-H.M. inequality)}$$

$$\ge (k + k\epsilon)\|A - A_1\|_F^2$$

$$= k(1+\epsilon)\|A - A_1\|_F^2$$

$$= (1+\epsilon)\|B - B_k\|_F^2.$$

5 Discussion

Our algorithm implements approximate volume sampling using $2k$ passes over the matrix. Can we do it using fewer passes? Can *exact* volume sampling be implemented efficiently?

It would also be nice to close the gap between the upper bound $O(k/\epsilon + k \log k)$ and the lower bound $\Omega(k/\epsilon)$ on the number of rows whose span "contains" a $(1 + \epsilon)$-approximation of rank at most k.

Acknowledgements. We would like to thank Sariel Har-Peled, Prahladh Harsha, Ravi Kannan, Frank McSherry, Luis Rademacher and Grant Wang.

References

1. S. Arora, E. Hazan, S. Kale, "A Fast Random Sampling Algorithm for Sparsifying Matrices." to appear in the Proceedings of RANDOM, 2006.
2. D. Achlioptas, F. McSherry, "Fast Computation of Low Rank Approximations." Proceedings of the 33rd Annual Symposium on Theory of Computing, 2001.
3. C. Aggarwal, C. Procopiuc, J. Wolf, P. Yu, J. Park. "Fast Algorithms for Projected Clustering." Proceedings of SIGMOD, 1999.
4. Z. Bar-Yosseff. "Sampling Lower Bounds via Information Theory." Proceedings of the 35th Annual Symposium on Theory of Computing, 2003.
5. W.F. de la Vega, M. Karpinski, C. Kenyon, Y. Rabani. "Approximation schemes for clustering problems." Proceedings of the 35th Annual ACM Symposium on Theory of Computing, 2003.
6. P. Drineas, personal communication, 2006.
7. P. Drineas, A. Frieze, R. Kannan, S. Vempala, V. Vinay. "Clustering in large graphs and matrices." Proceedings of the 10th SODA, 1999.
8. P. Drineas, R. Kannan. "Pass Efficient Algorithm for approximating large matrices." Proceedings of 14th SODA, 2003.
9. P. Drineas, R. Kannan, M. Mahoney. "Fast Monte Carlo Algorithms for Matrices II: Computing a Low-Rank Approximation to a Matrix." Yale University Technical Report, YALEU/DCS/TR-1270, 2004.
10. P. Drineas, M. Mahoney, S. Muthukrishnan. "Polynomial time algorithm for column-row based relative error low-rank matrix approximation." DIMACS Technical Report 2006-04, 2006.
11. A. Deshpande, L. Rademacher, S. Vempala, G. Wang. "Matrix Approximation and Projective Clustering via Volume Sampling." Proceedings of the 17th ACM-SIAM Symposium on Discrete Algorithms (SODA), 2006.
12. J. Feigenbaum, S. Kannan, A. McGregor, S. Suri, J. Zhang. "On Graph Problems in a Semi-Streaming Model." Proceedings of the 31st ICALP, 2004.
13. A. Frieze, R. Kannan, S. Vempala. "Fast Monte-Carlo algorithms for finding low-rank approximations." Journal of the ACM, 51(6):1025-1041, 2004.
14. S. Guha, N. Koudas, K. Shim. "Data-streams and histograms." Proceedings of 33rd ACM Symposium on Theory of Computing, 2001.
15. M. Henzinger, P. Raghavan, S. Rajagopalan. "Computing on Data Streams." Technical Note 1998-011, Digital Systems Research Center, Palo Alto, CA, May 1998.
16. J. Matoušek. "On approximate geometric k-clustering." Discrete and Computational Geometry, pg 61-84, 2000.

Robust Local Testability of Tensor Products of LDPC Codes[*]

Irit Dinur[1], Madhu Sudan[2], and Avi Wigderson[3]

[1] Hebrew University, Jerusalem, Israel
dinuri@cs.huji.ac.il
[2] Massachusetts Institute of Technology, Cambridge, MA
madhu@mit.edu
[3] Institute for Advanced Study, Princeton, NJ
avi@ias.edu

Abstract. Given two binary linear codes R and C, their tensor product $R \otimes C$ consists of all matrices with rows in R and columns in C. We analyze the "robustness" of the following test for this code (suggested by Ben-Sasson and Sudan [6]): Pick a random row (or column) and check if the received word is in R (or C). Robustness of the test implies that if a matrix M is far from $R \otimes C$, then a significant fraction of the rows (or columns) of M are far from codewords of R (or C).

We show that this test *is* robust, provided one of the codes is what we refer to as *smooth*. We show that expander codes and locally-testable codes are smooth. This complements recent examples of P. Valiant [13] and Coppersmith and Rudra [9] of codes whose tensor product is not robustly testable.

1 Introduction

A binary linear code is a linear subspace $C \subseteq \{0,1\}^n$. A code is *locally testable* if given a word $x \in \{0,1\}^n$ one can verify whether $x \in C$ by reading only few (randomly chosen) bits from x. More precisely such a code has a *tester*, which is a randomized algorithm with oracle access to the received word x. The tester reads at most q symbols from x and based on this "local view" decides if $x \in C$ or not. It should accept codewords with probability one, and reject words that are "far" (in Hamming distance) from the code with "noticeable" probability.

Locally testable codes (LTCs) are related to probabilistically checkable proofs (PCPs). LTCs were first explicitly studied by Goldreich and Sudan [12], who describe them as the "combinatorial core of PCPs". They constructed LTCs relying on some of the PCP machinery [11, 2, 1]. Since locally testable codes are simpler than PCPs, it seems natural to seek alternative constructions for them, possibly departing from the PCP framework.

---

[*] Most of the research was done while the authors were visiting Microsoft Research Theory group. Additionally, Irit Dinur's work was supported in part by ISF grant 984/04, Madhu Sudan's work was supported in part by NSF Award CCR-0514915, and Avi Wigderson's work was supported in part by NSF Award CCR-0324906.

J. Diaz et al. (Eds.): APPROX and RANDOM 2006, LNCS 4110, pp. 304–315, 2006.

One of the most interesting challenges in constructing LTCs, is to come up with an LTC that has constant relative distance and highest possible (maybe linear?) rate. Several steps in this direction were made in recent years, see [12, 8, 3, 4, 6, 7, 10].

All known efficient constructions of LTCs rely on some form of "composition" of two (or more) codes. In this paper we focus on composition by tensor product, which is an elementary way to compose two codes. Given two binary codes $R \subseteq \{0, 1\}^m$ and $C \subseteq \{0, 1\}^n$, their tensor product is the code $R \otimes C$ consisting of all binary $n \times m$ matrices whose rows belong to R and whose columns belong to C.

Ben-Sasson and Sudan [6] suggested using the tensor operation for constructing LTCs. They introduce the notion of *robust LTCs*: An LTC is called robust if whenever the received word is far from the code, then with noticeable probability the local view of the tester is *far* from an accepting local view. It is very easy to compose testers for robust LTCs: If it so happens that restriction of the code to the local view of the tester is itself an LTC, then instead of reading the entire local view, a tester for the smaller LTC can be invoked thereby saving on the query complexity of the tester.

Ben-Sasson and Sudan [6] showed that a code obtained by tensoring three or more codes (i.e. a code of the form $C_1 \otimes C_2 \otimes C_3$) is robustly testable, and used this result to construct LTCs. For the tensor product of two codes R and C, they considered the following natural test, and asked whether it is robust:

Test for $R \otimes C$: Pick a random row (or column), accept iff it belongs to R (or C).

Rather than providing a general definition of robustness (which can be found in Section 2.2), let us spell out the meaning of robustness for this particular test. Let x be an $n \times m$ matrix. Let $\delta^{\text{row}}(x)$ denote the expected distance of a random row of x from R, and let $\delta^{\text{col}}(x)$ denote the expected distance of a random column of x from C. Let $\delta_{R \otimes C}(x)$ denote the distance of x from the tensor product code $R \otimes C$. The robustness of the test is the largest value of α that satisfies

$$\frac{\delta^{\text{row}}(x) + \delta^{\text{col}}(x)}{2} \geq \alpha \cdot \delta_{R \otimes C}(x)$$

for every x. We say that the test is *robust* if its robustness is bounded away from 0.

Paul Valiant [13] showed a surprising example of two linear codes R and C for which the test above is not robust, by exhibiting a word x that is far from $R \otimes C$ but such that the rows of x are very close to being in R (i.e. $\delta^{\text{row}}(x)$ is small) and the columns of x are very close to being in C (i.e. $\delta^{\text{col}}(x)$ is small). An additional example of [9] gives a code whose tensor product with itself is not robust, and a similar result is shown for some non-linear code.

Results. Despite these examples, in this paper we show that the test above is robust for two important classes of Low Density Parity Check (LDPC) codes: Expander codes, and LTCs (see Proposition 1). We note that these are almost disjoint classes, as [5] prove that random expander LDPC codes are *not* locally testable.

We do this by introducing *smooth* codes which are a class of low density parity check codes. The smoothness property captures how badly the code is affected if some of the parity checks are removed from it.

We first show that if either R or C are smooth, then $R \otimes C$ has the following property. Any given word x that has small $\delta^{row}(x)$ and small $\delta^{col}(x)$, must have a large sub-matrix that completely agrees with some word in $R \otimes C$ (so x is close to $R \otimes C$). This implies that $R \otimes C$ is robust. We then argue that both LTCs and expander codes are smooth.

2 Notation, Definitions, and Results

All codes we consider will be binary linear codes. A binary linear code is a linear subspace $C \subseteq \{0,1\}^n$, whose dimension is denoted by $\dim(C)$. Every member of C is called a codeword.

We define the *distance* between two words $x, y \in \{0,1\}^n$ to be $\delta(x,y) = \Pr_i[x_i \neq y_i]$. We also define the weight of a string to be $\mathrm{wt}(x) = \delta(x, \mathbf{0})$. The distance of a code is denoted $\delta(C)$, and defined to be the minimal value of $\delta(x,y)$ for two distinct codewords $x, y \in C$. Clearly the distance of a linear code is equal to weight of the minimal-weight non-zero codeword.

Let $I_n = \{0,1\}^n$ denote the trivial code. For $x \in I_n$ and $C \subseteq I_n$, let $\delta_C(x) = \min_{\{y \in C\}}\{\delta(x,y)\}$ denote the distance of x from the code C.

2.1 Tensor Products of Codes

For $x \in I_m$ and $y \in I_n$ we let $x \otimes y$ denote the tensor product of x and y (i.e., the $n \times m$ matrix xy^T).

Let $R \subseteq I_m$ and $C \subseteq I_n$ be linear codes. We define the tensor product code $R \otimes C$ to be the linear subspace spanned by words $r \otimes c \in \{0,1\}^{n \times m}$ for $r \in R$ and $c \in C$. The following facts are immediate:

- The code $R \otimes C$ consists of all $n \times m$ matrices whose rows belong to R and whose columns belong to C.
- $\dim(R \otimes C) = \dim(R) \cdot \dim(C)$
- $\delta(R \otimes C) = \delta(R) \cdot \delta(C)$.

Fix $R \subseteq I_m$ and $C \subseteq I_n$ of distance δ_R and δ_C respectively for the rest of the manuscript.

Let $M \in I_m \otimes I_n$ and let $\delta(M) = \delta_{R \otimes C}(M)$. Let $\delta^{row}(M) = \delta_{R \otimes I_n}(M)$ denote its distance from the space of matrices whose rows are codewords of R. This is the expected distance of a random row in x from R. Similarly let $\delta^{col}(M) = \delta_{I_m \otimes C}(M)$.

2.2 Robust Locally Testable Codes

Locally testable codes, as described in the introduction, are codes for which one can test whether a given word x is in the code by reading only few (randomly chosen) symbols from x. We discuss here only *non adaptive* and *bi-regular* testers.

Non adaptive means that which queries are read is determined before any query is made, and bi-regular means that every test queries the same number of bits, and every bit is queried by the same number of tests. It would be interesting to extend our result for locally testable codes without these restrictions.

Definition 1 ((Non adaptive, bi-regular) Locally Testable Code). *We say that a code $C \subseteq I_n$ is $(d, \delta, \epsilon, \rho)$-locally-testable if $\delta(C) \geq \delta$ and there is a randomized algorithm (called a tester) T, which selects d indices from $[n]$, and for any given word $x \in I_n$, T reads the bits of x in these locations, satisfying:*

- *If $x \in C$ then $\Pr[T^x \text{ accepts}] = 1$.*
- *If $\delta_C(x) \geq \rho$ then $\Pr[T^x \text{ rejects}] > \epsilon$.*

Moreover, the probability that a given index is chosen to be read by T is the same for all indices in $[n]$.

A somewhat stronger notion of LTCs is that of robust-LTCs. Such a code has a stronger soundness requirement: Whenever $x \notin C$ the local view of the tester is *far* (in expectation) from an accepting view. For a formal definition let us introduce a little notation. The tester algorithm T has two inputs: the random string r, and the word x that is being tested. The tester reads the string r and computes a predicate T_r and a d-tuple of indices $i_1, \ldots, i_d$ in which it queries the word x. It accepts iff $T_r(x[i_1], \ldots, x[i_d]) = 1$. Let $acc(T_r) = \left\{ w \in \{0,1\}^d \;\middle|\; T_r(w) = 1 \right\}$ be the set of local-views on which the tester accepts. Define the robustness of T on x to be

$$\rho^T(x) = \mathbb{E}_r[\delta((x[i_1], \ldots, x[i_d]), acc(T_r))],$$

which is the expected distance of the local view from an accepting one. The robustness of T is the minimal ratio between the robustness of T on x, and the distance of x from the code:

$$\rho^T = \min_{x \notin C} \frac{\rho^T(x)}{\delta_C(x)}.$$

Definition 2 (Robust Code). *We say that a code $C \subseteq I_n$ is α-robust if there is a tester T that accepts every word in C with probability 1, such that $\rho^T \geq \alpha$.*

2.3 Low Density Parity Check (LDPC) Codes

A bipartite graph $([n], [m], E)$ is a parity check graph for a code $C \subseteq I_n$ if the following holds (let $\Gamma(j)$ denote the neighbors of j in the graph):

$$x \in C \qquad \Longleftrightarrow \qquad \forall j \in [m] \quad \sum_{i \in \Gamma(j)} x_i = 0 \bmod 2$$

In other words, every right-hand-side vertex $j \in [m]$ corresponds to a parity constraint, and a word is in the code if and only if it satisfies all of the constraints.

A code is referred to as an LDPC code if it has a "low-density" parity check graph, e.g. a graph with constant[1] average degree.

We first remark that LTCs are low density parity check codes, since a parity check graph can be constructed from the tester algorithm. Moreover, since our LTCs are bi-regular, so is their parity check graph.

Proposition 1. *Every* $(d, \delta, \epsilon, \rho)$*-LTC* C *with* $\rho < \delta$ *has a parity check graph with right degree* d *and such that for every word* x*, if* $\delta_C(x) \geq \rho$ *then it violates at least* ϵ *fraction of the parity checks.*

Proof. Let T be a tester for C. The predicates computed by T are parity checks (perhaps redundant) of C, since the code is linear. The construction of a parity graph (L, R, E) from T is immediate, with the nodes of R corresponding to the enumeration of the random strings of T.

Another important class of LDPC codes is that of expander codes.

Definition 3 (((c, d)-regular (γ, δ)-expander). *Let* $c, d \in \mathbb{N}$ *and let* $\gamma, \delta \in (0, 1)$*. Define a* (c, d)*-regular* (γ, δ)*-expander to be a bipartite graph* (L, R, E) *with vertex sets* L, R *such that all vertices in* L *have degree* c*, and all vertices in* R *have degree* d*; and the additional property that every set of vertices* $L' \subset L$*, such that* $|L'| \leq \delta |L|$*, has at least* $(1 - \gamma)c |L'|$ *neighbors.*

We say that a code C is an (c, d, γ, δ)-expander code if it has a parity check graph that is a (c, d)-regular (γ, δ)-expander.

The following is an important (and straightforward) property of expander codes,

Proposition 2. *If* C *is a* (c, d, γ, δ)*-expander code and* $\gamma < \frac{1}{2}$*, then* $\delta(C) \geq \delta$*.*

Proof. We prove that every non-zero word in C must have weight more than δn. Indeed let (L, R, E) be a parity check graph of C that is a (c, d)-regular (γ, δ)-expander. The proposition follows by examining the unique neighbor structure of the graph. Let $x \in C$ be a non-zero codeword, and let $L' \subseteq L$ be the set of indices in which x is 1. If $|L'| \leq \delta n$ then L' has at least $(1 - \gamma)c |L'| > \frac{c}{2} |L'|$ neighbors in R. At least one of these sees *only one* element of L', so the parity of its neighbors is one, violating the corresponding constraint and contradicting $x \in C$.

2.4   Results

Let R, C be codes. We study the robustness of the following test (described also in the introduction) for a given word $M \in I_m \otimes I_n$.

Test T for $R \otimes C$:

1. Select $b \in \{0, 1\}$ at random.
2. If $b = 0$ select $i \in [n]$ at random, and accept iff the i-th row of M is in R.
3. If $b = 1$ select $j \in [m]$ at random, and accept iff the j-th column of M is in C.

---

[1] Implicit throughout this manuscript is the notion that we are working with infinite families of codes/graphs, where the parameters such as the degree or the distance do not change with the length of the code/graph etc.

Obviously, T accepts every word of $R \otimes C$ with probability 1. We are interested in studying the robustness of T which we sometimes refer to as ρ instead of ρ^T.

Recall our notation $\delta(M) = \delta_{R \otimes C}(M)$ and our definition of $\delta^{\mathrm{row}}(M) = \delta_{R \otimes I_n}(M)$ and $\delta^{\mathrm{col}}(M) = \delta_{I_m \otimes C}(M)$. In other words $\delta^{\mathrm{row}}(M)$ equals the average distance of a row of M from R, and similarly $\delta^{\mathrm{col}}(M)$ equals the average distance of a column of M from C. The following proposition is immediate:

Proposition 3. *The robustness of T on input M is* $\rho(M) = \frac{\delta^{\mathrm{row}}(M) + \delta^{\mathrm{col}}(M)}{2}$.

$\square$

In order to establish robustness for T, say $\rho^T \geq \alpha > 0$, we must be able to prove for all M that $\frac{(\delta^{\mathrm{row}}(M) + \delta^{\mathrm{col}}(M))/2}{\delta(M)} \geq \alpha$.

As already mentioned in the introduction, for general codes R and C this is false. Paul Valiant [13] described a pair of codes R and C and a word M that is very far from $R \otimes C$, yet both $\delta^{\mathrm{row}}(M)$ and $\delta^{\mathrm{col}}(M)$ are very small.

Nevertheless, we observe that if C (or R) is somewhat "nice", then such a bound can be proven.

Theorem 1 (Tensoring Expander-codes). *Let $R \subset I_m$ be a code of distance at least $\delta_R > 0$. Let $C \subset I_n$ be a (c, d, γ, δ)-expander code for some $c, d \in \mathbb{N}, \delta > 0$, and $0 < \gamma < 1/6$. Then*

$$\rho^T \geq \frac{(\frac{1}{3} - 2\gamma)\delta\delta_R}{4d}.$$

Theorem 2 (Tensoring LTCs). *Let $R \subset I_m$ and $C \subset I_n$ be codes of relative distance at least δ_R, δ_C respectively. Furthermore, let C be a $(d, \delta_C, \epsilon, \rho)$-LTC, with $\rho \leq \frac{\delta_C}{16}$. Then,*

$$\rho^T \geq \min\left\{ \frac{\epsilon\delta_R}{2d^2}, \frac{\delta_R\delta_C}{16} \right\}.$$

3 Smooth Codes

We prove the two theorems by a common technique, where we show that the tensor product has nice testing properties if the underlying codes are nice in a certain sense that we refer to as "smooth". To motivate this notion, consider a code $C \subseteq I_n$ given by a (possibly redundant[2]) parity check graph $B = (L, R, E)$, where every vertex of R has degree d.

We consider how badly the code is affected if we remove some constraints $R_0 \subseteq R$. Let $C(R_0)$ denote the resulting code. $C(R_0)$ clearly contains C, but may now contain codewords of lesser weight. For instance we may remove all the neighbors of some vertex $u \in L$ (for the vertex u of minimum degree, this only requires us to remove a $d/|L|$ fraction of the right vertices), and now u is unconstrained, leading to a code of distance one. However if we delete the uth coordinate of $C(R_0)$ one may hope that the resulting code still has large

---

[2] A parity check graph is redundant if removing a node from the right still results in a parity check graph for the same code.

distance. More generally, we may hope that the negative effect of deleting some subset R_0 of the constraints may be recovered by dropping some subset L_0 of the coordinate vertices. If a code exhibits such a property, we call it smooth, defined quantitatively below.

For a set $S \subset [n]$ we always denote $\overline{S} = [n] - S$. For a code $C \subseteq I_n$ and $L_0 \subseteq L = [n]$ let $C|_{L_0}$ be the projection of the codewords of C to the coordinates of $\overline{L_0}$. (Such a code is called a punctured code. For reasons that will be evident later, it is nicer to highlight the set of coordinates that are being deleted.)

For a code C defined by a bipartite graph $B = (L, R, E)$, let $C(R_0)$ denote the "supercode" given by the parity check graph $B' = (L = [n], R - R_0, E' = E \cap (L \times (R - R_0)))$.

Definition 4 (Smooth Code). *A code $C \subseteq I_n$ is $(d, \alpha, \beta, \delta)$-smooth if it has a parity check graph $B = (L, R, E)$ where all the right vertices R have degree d, the left vertices have degree $c = d|R|/|L|$, and for every set $R_0 \subseteq R$ such that $|R_0| \leq \alpha|R|$, there exists a set $L_0 \subseteq L$, $|L_0| \leq \beta|L|$ such that the code $C(R_0)|_{L_0}$ has distance at least δ.*

We next turn to prove that the test T described in the previous section is robust when one of the codes being tensored is smooth. More specifically we prove that for any word M, if $\rho(M) = (\delta^{\mathrm{row}}(M) + \delta^{\mathrm{col}}(M))/2$ is small then $\delta(M)$ is proportionally small.

Lemma 1 (Main Lemma). *Let $R \subseteq I_m$ and $C \subseteq I_n$ be codes of distance δ_R and δ_C. Let C be $(d, \alpha, \frac{\delta_C}{2}, \frac{\delta_C}{2})$-smooth, and let $M \in I_m \otimes I_n$. If $\rho(M) \leq \min\left\{\alpha\frac{\delta_R}{2d^2}, \frac{\delta_R\delta_C}{8}\right\}$ then $\delta(M) \leq 8\rho(M)$.*

Proof. For row $i \in [n]$, let $r_i \in R$ denote the codeword of R closest to the ith row of M. For column $j \in [m]$, let $c^{(j)} \in C$ denote the codeword of C closest to the jth column of M. Let M_R denote the $n \times m$ matrix whose ith row is r_i, and let M_C denote the matrix whose jth column is $c^{(j)}$. Let $E = M_R - M_C$.

In what follows the matrices M_R, M_C and (especially) E will be the central objects of attention. We refer to E as the error matrix. Note that $\delta(M, M_R) = \delta^{\mathrm{row}}(M)$ and $\delta(M, M_C) = \delta^{\mathrm{col}}(M)$ and so

$$\mathrm{wt}(E) = \delta(M_R, M_C) \leq \delta(M, M_R) + \delta(M, M_C) = \delta^{\mathrm{row}}(M) + \delta^{\mathrm{col}}(M) = 2\rho(M). \tag{1}$$

Our proof strategy is to show that the error matrix E is actually very structured. We do this in two steps. First we show (Proposition 4) that its columns satisfy most constraints of the column code. Then we show (Proposition 5) that E contains a large submatrix which is all zeroes. Finally using this structure of E we show (Proposition 6) that M is close to some codeword of $R \otimes C$. Proposition 4 is the crux of our analysis (while Proposition 5 follows more or less in a straightforward way from the definition of smoothness, and Proposition 6 is a standard property of tensor product codes).

Proposition 4. *Let $\{i_1, \ldots, i_d\}$ be a constraint of C (i.e., every codeword of $y \in C$ satisfies $y_{i_1} + \ldots + y_{i_d} = 0$). Let e_i denote the ith row of E. Suppose $\mathrm{wt}(e_{i_j}) < \delta_R/d$ for every $j \in [d]$. Then $e_{i_1} + \cdots + e_{i_d} = \mathbf{0}$.*

Proof. Let c_i denote the i-th row of the matrix M_C. (Recall that these rows are not necessarily codewords of any nice code - it is only the columns of M_C that are codewords of C). For every column j, we have $(c_{i_1})_j + \cdots + (c_{i_d})_j = 0$ (since the columns of M_C are codewords of C). Thus we conclude that $c_{i_1} + \cdots + c_{i_d} = \mathbf{0}$ as a vector.

Now consider $r_{i_1} + \cdots + r_{i_d}$ (recall that r_i is the i-th row of M_R). Since each one of the r_i's is a codeword of R, we have $r_{i_1} + \cdots + r_{i_d} \in R$. But this implies

$$e_{i_1} + \cdots + e_{i_d} = (r_{i_1} - c_{i_1}) + \cdots + (r_{i_d} - c_{i_d}) = (r_{i_1} + \cdots + r_{i_d}) - (c_{i_1} + \cdots + c_{i_d})$$

$$= (r_{i_1} + \cdots + r_{i_d}) - \mathbf{0} \in R$$

Now we use the fact that the e_is have small weight. This implies that $\mathrm{wt}(e_{i_1} + \cdots + e_{i_d}) \leq \sum_j \mathrm{wt}(e_{i_j}) < \delta_R$. But R is an error-correcting code of minimum distance δ_R so the only word of weight less than δ_R in it is the zero codeword, yielding $e_{i_1} + \cdots + e_{i_d} = \mathbf{0}$.

Combined with the smoothness of C, the above proposition gives us sufficient structure to show that E has a large clean submatrix. We argue this below.

Proposition 5. *There exist subsets* $U \subseteq [m]$ *and* $V \subseteq [n]$ *with* $|U|/m < \delta_R/2$ *and* $|V|/n < \delta_C/2$ *such that* $E(i,j) \neq 0$ *implies* $i \in V$ *or* $j \in U$.

Proof. First, we consider the rows of E that have weight above δ_R/d. Let

$$V_1 = \{i \in [n] \mid \mathrm{wt}(e_i) \geq \delta_R/d\} .$$

We use $\delta^{\mathrm{row}}(M) \leq 2\rho(M) \leq \frac{\alpha\delta_R}{d^2}$ and Markov's inequality to deduce $|V_1|/n \leq \frac{2\rho(M)}{\delta_R/d} \leq \frac{\alpha}{d}$.

Next, we consider every constraint of C that involves an index in V_1. Recall that the code C is $(d, \alpha, \frac{\delta_C}{2}, \frac{\delta_C}{2})$-smooth, and let $B = ([n], [\ell], F)$ be the corresponding parity check graph of C (with right degree d and left degree $c = \frac{d\ell}{n}$). Viewing V_1 as a subset of the left vertices of B, let $W \subseteq [\ell]$ be the set of neighbors of V_1 in B. First notice that $|W| \leq c|V_1| \leq c \cdot \alpha n/d = \alpha\ell$. Next, observe that constraints in $[\ell] - W$ touch only indices outside V_1, i.e., indices j with $w(e_j) < \delta_R/d$. By Proposition 4, such constraints are satisfied by the rows of E. It is clear that if an equality holds for row-vectors, it also holds for each column separately. Thus, *every column* of the error matrix E, denoted $e^{(j)}$, is contained in the code $C(W)$.

Now we use the smoothness of C to define the sets V and U. Since $|W| \leq \alpha\ell$, there must be a set $V \subseteq [n]$ of cardinality at most $\frac{\delta_C}{2}n$ such that the code $C(W)|_V$ has distance at least $\frac{\delta_C}{2}n$. Let U be the set of indices corresponding to columns of E that have $\frac{\delta_C}{2}n$ or more non-zero elements in the rows outside V. This means that for every j, $e^{(j)}$ is either all zero on $\overline{V}$ or has at least $\frac{\delta_C}{2}n$ non-zero values on $\overline{V}$. If also $j \notin U$ then $e^{(j)}$ must be zero outside V. We conclude that if we throw away from the matrix E all the rows corresponding to V and all the columns corresponding to U, we are left with the zero matrix.

The fraction of rows thrown away is at most $\frac{|V|}{n} \leq \delta_C/2$. The fraction of columns thrown away is at most $\frac{\delta^{col}(M)}{\delta_C/2} \leq \frac{4\rho(M)}{\delta_C} \leq \delta_R/2$, where we used Markov's inequality and $\delta^{col}(M) \leq 2\rho(M) \leq \frac{\delta_C \delta_R}{4}$.

We now use a standard property of tensor products to claim M_R (and M_C and M) is close to a codeword of $R \times C$. Recall that $M \in \{0,1\}^{n \times m}$ and that $\delta(M_C, M_R) \leq 2\rho(M)$.

Proposition 6. *Assume there exist sets $U \subseteq [m]$ and $V \subseteq [n]$, $|U|/m \leq \delta_R/2$ and $|V|/n \leq \delta_C/2$ such that $M_R(i,j) \neq M_C(i,j)$ implies $j \in U$ or $i \in V$. Then $\delta(M) \leq 8\rho(M)$.*

Proof. This is a standard proposition. First we note that there exists a matrix $N \in R \otimes C$ that agrees with M_R and M_C on $\overline{V} \times \overline{U}$ (See [6, Proposition 3][3]). Recall also that $\delta(M, M_R) = \delta^{row}(M) \leq 2\rho(M)$. So it suffices to show $\delta(M_R, N) \leq 6\rho(M)$. We do so in two steps. First we show that $\delta(M_R, N) \leq 2\rho(M_R)$. We then show that $\rho(M_R) \leq 3\rho(M)$ concluding the proof.

For the first part we start by noting that M_R and N agree on every row in $\overline{V}$. This is the case since both rows are codewords of R which may disagree only on entries from the columns of U, but the number of such columns is less that $\delta_R m/2$. Next we claim that for every column $j \in [m]$ the closest codeword of C to the $M_R(\cdot, j)$, the jth column of M_R, is $N(\cdot, j)$, the jth column of N. This is true since $M_R(i,j) \neq N(i,j)$ implies $i \in V$ and so the number of such i is less than $\delta_C n/2$. Thus for every j, we have $N(\cdot, j)$ is the (unique) decoding of the jth column of M_R. Averaging over j, we get that $\delta^{col}(M_R) = \delta(M_R, N)$. In turn this yields $\rho(M_R) \geq \delta^{col}(M_R)/2 = \delta(M_R, N)/2$. This yields the first of the two desired inequalities.

Now to bound $\rho(M_R)$, note that for any pair of matrices M_1 and M_2 we have $\rho(M_1) \leq \rho(M_2) + \delta(M_1, M_2)$. Indeed it is the case that $\delta^{row}(M_1) \leq \delta^{row}(M_2) + \delta(M_1, M_2)$ and $\delta^{col}(M_1) \leq \delta^{col}(M_2) + \delta(M_1, M_2)$. To see the former, for instance, note that if the ith row of M_2 is within ρ_i of some codeword of R, then the ith row of M_1 is within $\rho_i + \delta(M_1(i, \cdot), M_2(i, \cdot))$ of the same codeword of R. Averaging over i yields $\delta^{row}(M_1) \leq \delta^{row}(M_2) + \delta(M_1, M_2)$. A similar argument yields $\delta^{col}(M_1) \leq \delta^{col}(M_2) + \delta(M_1, M_2)$, when combined the two yield $\rho(M_1) \leq \rho(M_2) + \delta(M_1, M_2)$. Applying this inequality to $M_1 = M_R$ and $M_2 = M$ we get $\rho(M_R) \leq \rho(M) + \delta(M_R, M) \leq 3\rho(M)$. This yields the second inequality and thus the proof of the proposition as well as Lemma 1.

In what follows we will show that expander codes, as well as LTCs are smooth.

4   Expander Codes Are Smooth

Lemma 2. *Every (c, d, γ, δ)-expander code C is $(d, \alpha, \beta, \delta)$-smooth, provided $\gamma < \frac{1}{6}$, $\alpha < (\frac{1}{3} - 2\gamma)\delta d$ and $\beta = \frac{\alpha}{(\frac{1}{3} - 2\gamma)d}$.*

---

[3] Erase from the matrix M_R entries in rows V or columns U. Observe that decoding from erasures first each row and then each column, must result in the same matrix as decoding first each column and then each row (due to the distances of the codes).

Proof. Let $B = (L, R, E)$ be the (c, d) regular (γ, δ)-expanding parity check graph of the code C. Let $R_0 \subseteq R$ of size $|R_0| \leq \alpha \cdot |R|$ be given. We will construct sets L', R' satisfying $L' \subseteq L$, $|L'| \leq \beta|L|$ and $R_0 \subseteq R' \subseteq R$ such that every subset of $L - L'$ of size at most δn expands sufficiently in the induced subgraph on $(L - L') \cup (R - R')$. This will suffice to prove that $C(R_0)|_{L'} \subseteq C(R')|_{L'}$ has distance at least δn.

We construct the sets L' and R' iteratively. Initially we set $L' = \emptyset$ and $R' = R_0$. We then iterate as follows: While there exists a vertex $u \in L - L'$ such that u has more than $\frac{1}{3}c$ neighbors in R', we add u' to L' and add all the neighbors of u' to R'. We prove below that this process stops in $t \leq \beta n$ steps, and that the induced graph on $(L - L') \cup (R - R')$ is a (good) expander.

We claim that this process must stop after at most βn steps. To see this, we count the number of unique neighbors of the set L' in the graph B. Initially this number is at most $|R_0|$. At each iteration this number goes up by at most $\frac{2}{3}c$. Assume we have completed some $t \leq \delta n$ iterations (and recall $\beta n < \delta n$). We have $|L'| = t$. Denote $\Gamma_{\text{unique}}(L')$ the set of vertices in R that have exactly one neighbor in L'. So $|\Gamma_{\text{unique}}(L')| \leq |R_0| + \frac{2}{3}ct$. Observe that $|\Gamma_{\text{unique}}(L')| \geq (1 - 2\gamma)c|L'|$, otherwise L' couldn't have $(1 - \gamma)c|L'|$ distinct neighbors (here we use $t \leq \delta n$). Putting these inequalities together we have

$$(1 - 2\gamma - \frac{2}{3})ct \leq |\Gamma_{\text{unique}}(L')| - \frac{2}{3}ct \leq |R_0|$$

and so $t \leq \frac{1}{(\frac{1}{3}-2\gamma)c}|R_0| \leq \frac{\alpha}{(\frac{1}{3}-2\gamma)c}|R| = \frac{\alpha}{(\frac{1}{3}-2\gamma)d}|L| = \beta n$.

Now we claim that the induced subgraph on $(L - L') \cup (R - R')$ is an expander. For this part consider any set $S \subseteq L - L'$ with $|S| \leq \delta n$. Let T be the neighborhood of S in the graph B. Then $|T| \geq (1 - \gamma)c|S|$. Now each vertex of S may have upto $\frac{1}{3}c$ neighbors in R'. Even allowing for these neighborhoods to be disjoint, we get $|T \cap (R - R')| \geq (1 - \gamma)c|S| - \frac{1}{3}c|S| = (\frac{2}{3} - \gamma)c|S|$. Since $\frac{2}{3} - \gamma > \frac{1}{2}$, we have that the induced subgraph on $(L - L') \cup (R - R')$ has the property that every set of size at most δn expands by more than a factor of $c/2$, thus implying that $C(R')|_{L'}$ is a code of minimum distance at least δn (see Proposition 2). This concludes the proof.

Proof (Theorem 1). Note that C is a code of distance at least δ (by Proposition 2). By Lemma 2 it follows that C is $(d, \alpha, \beta, \delta)$-smooth for any $\alpha \leq (\frac{1}{3} - 2\gamma)d\delta$ and $\beta = \frac{\alpha}{(\frac{1}{3}-2\gamma)d}$. Set $\alpha = (\frac{1}{3} - 2\gamma)d\delta/2$, and so $\beta = \frac{\alpha}{(\frac{1}{3}-2\gamma)d} = \delta/2$. The code is certainly $(d, \alpha, \frac{\delta}{2}, \frac{\delta}{2})$-smooth.

Fix any $M \notin R \otimes C$, and let us lower bound $\frac{\rho(M)}{\delta(M)}$. Set $\rho_0 = \min\left\{\alpha\frac{\delta_R}{2d^2}, \frac{\delta_R\delta}{8}\right\}$. If $\rho(M) \geq \rho_0$ then surely $\frac{\rho(M)}{\delta(M)} \geq \rho_0$. Otherwise, we note that the conditions necessary for the application of Lemma 1 are satisfied, and we get $\delta(M) \leq 8\rho(M)$. All in all, we have proven that

$$\rho^T = \min_{M \notin R \otimes C} \frac{\rho(M)}{\delta(M)} \geq \min\left\{\rho_0, \frac{1}{8}\right\} = \rho_0 = \frac{(\frac{1}{3} - 2\gamma)\delta\delta_R}{4d}$$

where the last equality follows by plugging the value for α into ρ_0 and assuming $d \geq 2$.

5   LTCs Are Smooth

Lemma 3. *Every* $(d, \delta, \epsilon, \rho)$-*LTC code* C *is* $(d, \epsilon, \delta', \delta')$-*smooth, provided* $\rho \leq \delta'/4$ *and* $\delta' \leq \delta/4$.

Proof. Let $B = (L, R, E)$ be a parity check graph for C whose right-hand-side corresponds to the tests of a tester for C (Proposition 1). Fix $R_0 \subseteq R$ of size $|R_0| \leq \epsilon \cdot |R|$ and consider the code $C(R_0)$. If all the non-zero words in $C(R_0)$ have weight at least δ' then setting $L_0 = \emptyset$ satisfies the definition of smoothness and so we have nothing to prove. So we assume $C(R_0)$ has some non-zero words of weight at most δ'. Let $\{c_1, \ldots, c_m\}$ be the set of all codewords of $C(R_0)$ whose weight is at most $2\delta'$. Let S_i be the set of coordinates where c_i is non-zero, and let $L_0 = \cup_i S_i$.

If $|L_0| \leq \delta'n$, we claim that $C(R_0)|_{L_0}$ has distance at least $\delta'n$ as needed. This is true since every codeword of $C(R_0)$ of weight less than $2\delta'n$ is non-zero only on some subset of L_0 and so projects to the zero codeword in $C(R_0)|_{L_0}$. On the other hand, codewords of weight greater than $2\delta'n$ in $C(R_0)$ project to words of weight at least $\delta'n$ when we delete the $\delta'n$ coordinates corresponding to L_0. Thus $C(R_0)|_{L_0}$ is a code of weight at least $\delta'n$. Thus it remains to show below that $|L_0| \leq \delta'n$.

Assume for contradiction that $|L_0| > \delta'n$. We show first that $C(R_0)$ must have a codeword of weight between $\frac{\delta'}{4}n$ and $2\delta'n$. We then show that this violates the local testability of C.

For the first part, note that if one of the c_i's has weight between $\frac{\delta'}{2}n$ and $2\delta'n$, then we are already done. So we may assume each c_i has weight less than $\frac{\delta'}{2}n$. Now pick a subset $\{c_1, \ldots, c_j\}$ of the low weight codewords so that $\frac{\delta'}{2}n \leq |\cup_{i=1}^{j} S_i| \leq \delta'n$. This is obviously possible since the cardinality of this union starts at 0, as j varies from 0 to m, ends at $|L_0| > \delta'n$ and goes up by at most $\frac{\delta'}{2}n$ in each step. For this setting of j, consider words of the form $\sum_{i=1}^{j} x_i c_i$ where $x_i \in \{0, 1\}$. For every choice of x_i's we get a codeword of $C(R_0)$ of weight at most $|\cup_{i=1}^{j} S_i| \leq \delta'n$. The expected weight of such a word, when $x_i \in \{0, 1\}$ are chosen uniformly and independently is $\frac{1}{2}|\cup_{i=1}^{j} S_i| \geq \frac{\delta'}{4}n$. Thus the maximum weight codeword in this set has weight between $\frac{\delta'}{4}n$ and $\delta'n$, as desired.

Now let $c_1 \in C(R_0)$ be a codeword of weight between $\frac{\delta'}{4}n$ and $2\delta'n$. Since $2\delta' < \delta/2$ we have that c_1 is a word at distance more than $\delta'n \geq \rho n$ from C but is rejected only by the tests in R_0 which form at most ϵ fraction of all parity checks in B, contradicting the assumption that C is a $(d, \delta, \epsilon, \rho)$-LTC. ∎

Theorem 2 follows from Lemma 3 analogous to the way Theorem 1 followed from Lemma 2.

Proof (Theorem 2). The code C is a $(d, \delta_C, \epsilon, \rho)$-LTC, with $\rho \leq \delta_C/16$. By Lemma 3, it must be $(d, \epsilon, \frac{\delta_C}{4}, \frac{\delta_C}{4})$-smooth. Fix any $M \notin R \otimes C$, and let us lower bound $\frac{\rho(M)}{\delta(M)}$.

Set $\rho_0 = \min\{\frac{\epsilon\delta_R}{2d^2}, \frac{\delta_R\delta_C}{16}\}$. If $\rho(M) \geq \rho_0$ then surely $\frac{\rho(M)}{\delta(M)} \geq \rho_0$. Otherwise, we apply Lemma 1 and deduce that $\rho(M) < \rho_0$ implies that $\delta(M) \leq \frac{6}{\max\{\delta_R, \delta_C/2\}}\rho(M)$.

All in all, we have proven that

$$\rho^T = \min_{M \notin R \otimes C} \frac{\rho(M)}{\delta(M)} \geq \min\left\{\rho_0, \frac{1}{8}\right\} = \min\left\{\frac{\epsilon\delta_R}{2d^2}, \frac{\delta_R\delta}{16}\right\}.$$

References

1. Sanjeev Arora, Carsten Lund, Rajeev Motwani, Madhu Sudan, and Mario Szegedy. Proof verification and the hardness of approximation problems. *Journal of the ACM*, 45(3):501–555, May 1998.

2. Sanjeev Arora and Shmuel Safra. Probabilistic checking of proofs: A new characterization of NP. *Journal of the ACM*, 45(1):70–122, January 1998.

3. Eli Ben-Sasson, Oded Goldreich, Prahladh Harsha, Madhu Sudan, and Salil Vadhan. Robust PCPs of proximity, shorter PCPs and applications to coding. In *Proceedings of the 36th Annual ACM Symposium on Theory of Computing*, page (to appear), 2004.

4. Eli Ben-Sasson, Oded Goldriech, Prahladh Harsha, Madhu Sudan, and Salil Vadhan. Short PCPs verifiable in polylogarithmic time. In *Proceedings of the Twelfth Annual IEEE Conference on Computational Complexity*, pages 120–134, June 12–15 2005.

5. E. Ben-Sasson and P. Harsha and S. Raskhodnikova, Some 3CNF properties are hard to test. In SIAM Journal on Computing, 35(1):1-21.

6. E. Ben-Sasson and M. Sudan. Robust locally testable codes and products of codes. In *Proc. RANDOM: International Workshop on Randomization and Approximation Techniques in Computer Science*, pages 286–297, 2004.

7. Eli Ben-Sasson and Madhu Sudan. Short PCPs with poly-log rate and query complexity. In *Proceedings of the 37th Annual ACM Symposium on Theory of Computing*, pages 266–275, 2005.

8. Eli Ben-Sasson, Madhu Sudan, Salil Vadhan, and Avi Wigderson. Randomness efficient low-degree tests and short PCPs via ε-biased sets. In *Proceedings of the 35th Annual ACM Symposium on Theory of Computing*, pages 612–621, 2003.

9. D. Coppersmith and A. Rudra. On the robust testability of product of codes. ECCC TR05-104, 2005.

10. Irit Dinur. The PCP theorem by gap amplification. In *Proceedings of the 38th Annual ACM Symposium on Theory of Computing*, pages 241–250, 2006.

11. Uriel Feige, Shafi Goldwasser, Laszlo Lovasz, Shmuel Safra, and Mario Szegedy. Interactive proofs and the hardness of approximating cliques. *Journal of the ACM*, 43(2):268–292, 1996.

12. O. Goldreich and M. Sudan. Locally testable codes and PCPs of almost-linear length. In *Proc. 43rd IEEE Symp. on Foundations of Computer Science*, pages 13–22, 2002.

13. P. Valiant. The tensor product of two codes is not necessarily robustly testable. In *APPROX-RANDOM*, pages 472–481, 2005.

Subspace Sampling and Relative-Error Matrix Approximation: Column-Based Methods

Petros Drineas[1], Michael W. Mahoney[2,*], and S. Muthukrishnan[3]

[1] Department of Computer Science, RPI
[2] Yahoo Research Labs
[3] Department of Computer Science, Rutgers University

Abstract. Given an $m \times n$ matrix A and an integer k less than the rank of A, the "best" rank k approximation to A that minimizes the error with respect to the Frobenius norm is A_k, which is obtained by projecting A on the top k left singular vectors of A. While A_k is routinely used in data analysis, it is difficult to interpret and understand it in terms of the *original data*, namely the columns and rows of A. For example, these columns and rows often come from some application domain, whereas the singular vectors are linear combinations of (up to all) the columns or rows of A. We address the problem of obtaining low-rank approximations that are directly interpretable in terms of the *original* columns or rows of A. Our main results are two polynomial time randomized algorithms that take as input a matrix A and return as output a matrix C, consisting of a "small" (i.e., a low-degree polynomial in k, $1/\epsilon$, and $\log(1/\delta)$) number of actual columns of A such that

$$\left\| A - CC^+A \right\|_F \leq (1 + \epsilon) \left\| A - A_k \right\|_F$$

with probability at least $1 - \delta$. Our algorithms are simple, and they take time of the order of the time needed to compute the top k right singular vectors of A. In addition, they sample the columns of A via the method of "subspace sampling," so-named since the sampling probabilities depend on the lengths of the rows of the top singular vectors and since they ensure that we capture entirely a certain subspace of interest.

1 Introduction

1.1 Motivation and Overview

In many applications, the data are represented by a real $m \times n$ matrix A. Such a matrix may arise if the data consist of m objects, each of which is described by n features. Examples of objects include documents, genomes, stocks, hyperspectral images, and web groups, while examples of the corresponding features are terms, environmental conditions, temporal resolution, frequency resolution, and individual users. In each of these application areas, practitioners spend vast

---

* Part of this work was done while at the Department of Mathematics, Yale University.

J. Diaz et al. (Eds.): APPROX and RANDOM 2006, LNCS 4110, pp. 316–326, 2006.

amounts of time analyzing the data in order to understand, interpret, and ultimately use this data. Often the central task in this analysis is to develop a compressed representation of A that may be easier to analyze and interpret.

The most common compressed representation of A used by data analysts is that obtained by truncating the SVD at some number $k \ll \min\{m, n\}$ terms, in large part because this provides the "best" rank-k approximation to A when measured with respect to any unitarily invariant matrix norm. However, there is a fundamental difficulty with this representation: the new "dimensions" (the so-called eigencolumns and eigenrows) of A_k are linear combinations of (up to all) the original dimensions. As such, they are notoriously difficult to interpret in terms of the underlying data and processes generating that data. For example, the vector $[(1/2)$ age - $(1/\sqrt{2})$ height $+ (1/2)$ income$]$, being one of the significant uncorrelated "factors" from a dataset of people's features is not particularly informative. From an analyst's point of view, it would be highly preferable to have a low-rank approximation that is nearly as good as that provided by the SVD but that is expressed in terms of a small number of *actual columns* and/or *actual rows* of a matrix, rather than linear combinations of those columns and rows. For example, consider recent data analysis work in DNA microarray and DNA Single Nucleotide Polymorphism (SNP) analysis [15, 16, 18], where linear combinations of genes or loci in the human genome have no clear biological interpretation.

In this paper, we focus on choosing columns of a matrix A in order to approximate very precisely a data matrix A as the product CX, where C consists of a few columns of A and where X is a matrix that expresses every column of A in terms of the basis provided by the columns of C.

1.2   Review of Linear Algebra

Let $[n]$ denote the set $\{1, 2, \ldots, n\}$. For any matrix $A \in \mathbb{R}^{m \times n}$, let $A_{(i)}, i \in [m]$ denote the i-th row of A as a row vector, and let $A^{(j)}, j \in [n]$ denote the j-th column of A as a column vector. The Singular Value Decomposition (SVD) of A will be denoted by $A = U \Sigma V^T$, where $U \in \mathbb{R}^{m \times \rho}$, $\Sigma \in \mathbb{R}^{\rho \times \rho}$, $V \in \mathbb{R}^{n \times \rho}$, and where ρ is the rank of A. The "best" rank-k approximation to A (with respect to, e.g., the Frobenius norm, $||A||_F = \sqrt{\sum_{i,j} A_{ij}^2}$) will be denoted by $A_k = U_k \Sigma_k V_k^T$, where $U_k \in \mathbb{R}^{m \times k}$ is the first k columns of U, etc. The SVD and hence the best rank-k approximation of a general matrix A can be computed in $O(\min\{n^2 m, nm^2\})$ time, and optimal rank-k approximations to it can be computed more rapidly with, e.g., Lanczos methods. We will use $SVD(A_k)$ to denote the time to compute A_k. For more details on linear algebra, see [1, 12, 14, 17], and for more details on notation and our sampling matrix formalism, see [5, 9].

1.3   Problem Definition

We start with the following definition.

Definition 1. *Let A be an $m \times n$ matrix, and let C be an $m \times c$ matrix whose columns consist of a small number c of columns of the matrix A. Then the $m \times n$*

matrix A' is a column-based low-rank matrix approximation *to A, or a* CX matrix approximation, *if it may be explicitly written as $A' = CX$ for some $c \times n$ matrix X.*

We prefer not to provide too precise a characterization of what we mean by a "small" number of columns, but one should think of $c \ll n$. Also, the low-rank matrix approximation provided by truncating the SVD at some value of $k < \rho = \text{rank}(A)$ will not in general satisfy the conditions of the definition. Finally, given a set of columns C, the approximation $A' = P_C A = CC^+ A$ clearly satisfies the requirements of Definition 1. Indeed, this is the "best" such approximation to A, in the sense that $\|A - C(C^+A)\|_F = \min_{X \in \mathbb{R}^{c \times n}} \|A - CX\|_F$.

The quality of a CX matrix approximation depends on the choice of C as well as on the matrix X. We consider the following problem.

Problem 1 *(Column-based low-rank matrix approximation problem.)*
Given a matrix $A \in \mathbb{R}^{m \times n}$ and an integer $k \ll \min\{m, n\}$, choose a sufficient number of columns of A such that

$$\left\|A - CC^+A\right\|_F \leq (1 + \epsilon) \left\|A - A_k\right\|_F. \tag{1}$$

Here, C is a matrix consisting of the chosen columns of A, CC^+A is the projection of A on the subspace spanned by the chosen columns, and A_k is the best rank k approximation to A. The number of columns of C should be a function of k, $1/\epsilon$, and – in the case of randomized algorithms – a failure probability δ, and the running time of the algorithm should be a low-degree polynomial in m and n.

Note that is not obvious whether there exist, and if so whether one can efficiently find, a small (depending on k, $1/\epsilon$, and $1/\delta$, but independent of m and n) number of columns that provide such relative-error guarantees.

1.4   "Subspace Sampling" and Our Main Result

Our main result is the following theorem, which asserts the existence of two related algorithms to solve Problem 1.

Theorem 1. *There exists randomized algorithms that solve Problem 1.*

- *In one algorithm, exactly $c = O(k^2 \log(1/\delta)/\epsilon^2)$ columns of A are chosen to construct C.*
- *In the other algorithm, $c = O(k \log k \log(1/\delta)/\epsilon^2)$ columns in expectation are chosen to construct C.*

Both algorithms satisfy (1) with probability at least $1 - \delta$, both run in time $O(SVD(A_k))$, and both use the method of "subspace sampling" to sample columns to form C.

The algorithms of Theorem 1 for constructing a matrix C consisting of a few columns of A are simple:

1. Construct sampling probabilities $\{p_i\}_{i=1}^n$ satisfying the "subspace sampling" Condition (2) below.
2. Use these probabilities to randomly sample columns from A and construct a matrix C using one of two sampling procedures.
3. Repeat these two steps $O(\log(1/\delta))$ times, and return the set of columns C such that $\|A - CC^+A\|_F$ is smallest over all $O(\log(1/\delta))$ trials.

The first sampling procedure, which we call the EXACTLY(c) sampling algorithm, picks *exactly* c columns of A to be included in C in c i.i.d. trials, where in each trial the i-th column of A is picked with probability p_i. Notice that some columns of A may be included in the sample more than once. The second sampling procedure, which we call the EXPECTED(c) sampling algorithm, picks *in expectation* at most c columns of A to create C, by including the i-th column of A in C with probability $\min\{1, cp_i\}$. No column of A is included in the sample more than once.

The key technical insight that leads to the relative-error guarantees is that the columns are selected by a novel sampling procedure that we call "subspace sampling." Rather than sample columns from A with a probability distribution that depends on the Euclidean norms of the columns of A (which gives provable additive-error bounds [5, 6, 7]), in "subspace sampling" we randomly sample columns of A with a probability distribution that depends on the Euclidean norms of the rows of the top k right singular vectors of A. This allows us to capture entirely a certain subspace of interest. The "subspace sampling" probabilities $p_i, i \in [n]$ will satisfy

$$p_i \geq \frac{\beta \left|(V_k)_{(i)}\right|^2}{k} \qquad \forall i \in [n], \tag{2}$$

for some $\beta \in (0, 1]$. Note that $\sum_{j=1}^n \left|(V_k)_{(j)}\right|^2 = k$ and that $\sum_{i \in [n]} p_i = 1$. To construct sampling probabilities satisfying Condition (2), it is sufficient to spend $O(SVD(A_k))$ time to compute (exactly or approximately, in which case $\beta = 1$ or $\beta < 1$, respectively) the top k right singular vectors of A.

1.5   Related Work

The seminal work of Frieze, Kannan and Vempala [10, 11] can be viewed, in our parlance, as sampling columns from a matrix A to form a matrix C such that $\|A - CX\|_F \leq \|A - A_k\|_F + \epsilon \|A\|_F$. The matrix C has $poly(k, 1/\epsilon, 1/\delta)$ columns and is constructed after making only two passes over A using $O(m+n)$ work space. Under similar resource constraints, a series of papers have followed [10, 11] in the past seven years [4, 6, 20], improving the dependency of c on $k, 1/\epsilon$, and $1/\delta$, and analyzing the spectral as well as the Frobenius norm, yielding bounds of the form

$$\|A - CX\|_\xi \leq \|A - A_k\|_\xi + \epsilon \|A\|_F \tag{3}$$

for $\xi = 2, F$, and thus providing additive-error guarantees for column-based low-rank matrix approximations.

Most relevant for our relative-error column-based low-rank matrix approximation of Problem 1 is the recent work of Rademacher, Vempala and Wang [19] and Deshpande, Rademacher, Vempala and Wang [2]. Using two different methods (in one case iterative sampling in a backwards manner and an induction on k argument [19] and in the other case an argument which relies on estimating the volume of the simplex formed by each of the k-sized subsets of the columns [2]), they reported the *existence* of a set of $O(k^2/\epsilon^2)$ columns that provide relative-error CX matrix approximation. No algorithmic result was presented, except for an exhaustive algorithm that ran in $\Omega(n^k)$ time.

To the best of our knowledge, the first nontrivial *algorithmic result* for relative-error low-rank matrix approximation was provided by a preliminary version of this paper [8]. In particular, an earlier version of Theorem 1 provided the first known relative-error column-based low-rank approximation in polynomial time [8]. The major difference between our Theorem 1 and our result in [8] is that the sampling probabilities in [8] are more complicated. The algorithm of [8] runs in $O(SVD(A_k))$ time (although it was originally reported to run in $O(SVD(A))$ time), and it has a sampling complexity of $O(k^2 \log(1/\delta)/\epsilon^2)$ columns.

Subsequent to the completion of the preliminary version of this paper [8], several developments have been made on relative-error low-rank matrix approximation algorithms. First, Har-Peled reported an algorithm that in roughly $O(mnk^2 \log k)$ time returns as output a rank-k matrix A' with a relative-error approximation guarantee [13]. His algorithm uses geometric ideas and involves sampling and merging approximately-optimal k-flats; it is not clear if this approximation can be expressed in terms of a small number of columns of A. Then, Deshpande and Vempala [3] reported an algorithm that also returns a relative-error approximation guarantee. Their algorithm extends ideas from [19, 2] and it leads to a CX matrix approximation consisting of $O(k \log k)$ columns of A. The complexity of their algorithm is $O(Mk^2 \log k)$, where M is the number of nonzero elements of A, and their algorithm can be implemented with $O(k \log k)$ passes over the data. In light of these developments, we simplified and generalized our preliminary results [8], and we performed a more refined analysis to improve our sampling complexity to $O(k \log k)$.

2   Proof of Theorem 1

Regardless of whether the columns are chosen with the EXACTLY(c) algorithm or EXPECTED(c) algorithm, we can construct a *column sampling matrix* S, such that $C = AS$. Similarly, we may introduce a *diagonal rescaling matrix* D in this expression, which rescales each sampled column by $1/\sqrt{cp_j}$ for the EXACTLY(c) algorithm and $1/\min\{1, \sqrt{cp_j}\}$ for the EXPECTED(c) algorithm. For details on this formalism, see [9]. Since scaling the columns of a matrix does not change the subspace spanned by its columns, $A - CC^+A = A - ASD\,(ASD)^+\,A$. Our careful choice for S and D will allow us to apply matrix perturbation results from [5, 21] to bound this latter expression. For simplicity, we assume that $\epsilon \in (0, 1]$.

2.1 Constructing C with the Exactly(c) Algorithm

The first claim of Theorem 1 considers the situation when the columns of A are sampled with the EXACTLY(c) algorithm. In this subsection, we provide its proof. The proof of the second claim is similar, and we outline the differences in the next subsection.

To prove our main result, we must "disentangle" the "top" singular subspace of A from the "bottom" singular subspace. To do so, first note that using the unitary invariance of the Frobenius norm, and since $\left(U_A \Sigma_A V_A^T SD\right)^+ = \left(\Sigma_A V_A^T SD\right)^+ U_A^T$, it follows that

$$\left\|A - CC^+ A\right\|_F^2 = \left\|\Sigma_A - \left(\Sigma_A V_A^T SD\right)\left(\Sigma_A V_A^T SD\right)^+ \Sigma_A\right\|_F^2 \tag{4}$$

$$= \left\|\begin{bmatrix} \Sigma_k \\ \mathbf{0} \end{bmatrix} - \left(\Sigma_A V_A^T SD\right)\left(\Sigma_A V_A^T SD\right)^+ \begin{bmatrix} \Sigma_k \\ \mathbf{0} \end{bmatrix}\right\|_F^2$$

$$+ \left\|\begin{bmatrix} \mathbf{0} \\ \Sigma_{\rho-k} \end{bmatrix} - \left(\Sigma_A V_A^T SD\right)\left(\Sigma_A V_A^T SD\right)^+ \begin{bmatrix} \mathbf{0} \\ \Sigma_{\rho-k} \end{bmatrix}\right\|_F^2. \tag{5}$$

Next, to upper bound the second term on the right hand side of (5), recall that since $I - \left(\Sigma_A V_A^T SD\right)\left(\Sigma_A V_A^T SD\right)^+$ is a projector matrix, it may be dropped without increasing a unitarily invariant norm, and thus

$$\left\|\left(I - \left(\Sigma_A V_A^T SD\right)\left(\Sigma_A V_A^T SD\right)^+\right)\begin{bmatrix} \mathbf{0} \\ \Sigma_{\rho-k} \end{bmatrix}\right\|_F^2 \leq \|A - A_k\|_F^2. \tag{6}$$

Finally, to establish the first claim of Theorem 1, we seek to upper bound the first term on the right hand side of (5) by $\epsilon \|A - A_k\|_F^2$. That is, we seek an upper bound that does not depend at all on *any* of the top k singular values of A. To this end, note that

$$\left\|\begin{bmatrix} \Sigma_k \\ \mathbf{0} \end{bmatrix} - \left(\Sigma_A V_A^T SD\right)\left(\Sigma_A V_A^T SD\right)^+ \begin{bmatrix} \Sigma_k \\ \mathbf{0} \end{bmatrix}\right\|_F^2$$

$$= \min_{X \in \mathbb{R}^{c \times k}} \left\|\begin{bmatrix} \Sigma_k \\ \mathbf{0} \end{bmatrix} - \left(\Sigma_A V_A^T SD\right) X\right\|_F^2 \tag{7}$$

$$\leq \left\|\begin{bmatrix} \Sigma_k \\ \mathbf{0} \end{bmatrix} - \left(\Sigma_A V_A^T SD\right)\left(\Sigma_k V_k^T SD\right)^+ \Sigma_k\right\|_F^2. \tag{8}$$

Equations (7) and (8) follow from least-squares approximation theory: (7) follows since $\left(\Sigma_A V_A^T SD\right)\left(\Sigma_A V_A^T SD\right)^+ \begin{bmatrix} \Sigma_k \\ \mathbf{0} \end{bmatrix}$ is the exact projection of the matrix $\begin{bmatrix} \Sigma_k \\ \mathbf{0} \end{bmatrix}$ on the subspace spanned by the columns of $\Sigma_A V_A^T SD$; and (8) follows since $X = \left(\Sigma_k V_k^T SD\right)^+ \Sigma_k \in \mathbb{R}^{c \times k}$ is a suboptimal – but as we will see below very convenient – choice for X in (7).

To see that (8) provides the bound we seek, let the rank of the $k \times c$ matrix $V_k^T SD$ be $\tilde{k}$, and let its SVD be $V_k^T SD = U_{V_k^T SD} \Sigma_{V_k^T SD} V_{V_k^T SD}^T$. Clearly $\tilde{k} \leq k$. Among other things, the following lemma states that, given our construction of S and D, all the singular values of $V_k^T SD$ are close to 1 and thus that the rank of $V_k^T SD$ is equal to k.

Lemma 1. *If $c \geq 40k^2/\beta\epsilon^2$, then with probability at least 0.9:*

- $\tilde{k} = k$, *i.e.*, $rank(V_k^T SD) = rank(V_k)$,
- $\left\| (V_k^T SD)^+ - (V_k^T SD)^T \right\|_2 = \left\| \Sigma_{V_k^T SD}^{-1} - \Sigma_{V_k^T SD} \right\|_2$,
- $(\Sigma_k V_k^T SD)^+ = (V_k^T SD)^+ \Sigma_k^{-1}$, *and*
- $\left\| \Sigma_{V_k^T SD} - \Sigma_{V_k^T SD}^{-1} \right\|_2 \leq \epsilon/\sqrt{2}$.

Proof: Note that for all $i \in [\tilde{k}]$,

$$\left| 1 - \sigma_i^2 \left(V_k^T SD \right) \right| = \left| \sigma_i \left(V_k^T V_k \right) - \sigma_i \left(V_k^T SDDS^T V_k \right) \right|$$
$$\leq \left\| V_k^T V_k - V_k^T SDDS^T V_k \right\|_2. \tag{9}$$

Since the probabilities of (2) satisfy the condition of Theorem 1 of [5]

$$\mathbf{E} \left[\left\| V_k^T V_k - V_k^T SDDS^T V_k \right\|_F^2 \right] \leq \frac{1}{\beta c} \|V_k\|_F^4 = \frac{k^2}{\beta c}, \tag{10}$$

where the equality follows since $\|V_k\|_F^2 = k$. By applying Markov's inequality to (10), taking square roots of both sides, combining it with (9), and using $\|\cdot\|_2 \leq \|\cdot\|_F$ and the assumed choice of c, it follows that $\left| 1 - \sigma_i^2 \left(V_k^T SD \right) \right| \leq \epsilon/2 \leq 1/2$, since $\epsilon \leq 1$. This implies that all singular values of $V_k^T SD$ are strictly positive, and thus that $\tilde{k} = k$. The remainder of the proof is similar to that of Lemma 4.1 of [9].

$\diamond$

Using Lemma 1, we manipulate the right hand side of (8) as follows:

$$\left\| \begin{bmatrix} \Sigma_k \\ 0 \end{bmatrix} - \left(\Sigma_A V_A^T SD \right) \left(\Sigma_k V_k^T SD \right)^+ \Sigma_k \right\|_F^2$$

$$= \left\| \begin{bmatrix} \Sigma_k \\ 0 \end{bmatrix} - \begin{bmatrix} \Sigma_k & 0 \\ 0 & \Sigma_{\rho-k} \end{bmatrix} \begin{bmatrix} V_k^T \\ V_{\rho-k}^T \end{bmatrix} SD \left(V_k^T SD \right)^+ \right\|_F^2$$

$$= \left\| \begin{bmatrix} \Sigma_k \\ 0 \end{bmatrix} - \begin{bmatrix} \Sigma_k V_k^T \\ \Sigma_{\rho-k} V_{\rho-k}^T \end{bmatrix} SD \left(V_k^T SD \right)^+ \right\|_F^2$$

$$= \left\| \Sigma_k - \Sigma_k \underbrace{V_k^T SD \left(V_k^T SD \right)^+}_{=I_k} \right\|_F^2 + \left\| \Sigma_{\rho-k} V_{\rho-k}^T SD \left(V_k^T SD \right)^+ \right\|_F^2 \tag{11}$$

$$= \left\| \Sigma_{\rho-k} V_{\rho-k}^T SD \left(V_k^T SD \right)^+ \right\|_F^2. \tag{12}$$

The first term of (11) is the most important point of the proof. The sampling probabilities $\{p_i\}$ are carefully constructed to guarantee that the $k \times c$ matrix $V_k^T SD$ has full rank; thus its columns – which are k-dimensional vectors – span $\mathbb{R}^k$. As a result, the projection of Σ_k on the subspace spanned by the columns of $V_k^T SD$ is equal to Σ_k. Thus, since Σ_k does not appear in (12), at this point in the proof, we have removed any dependency of the error on the top k singular values of A.

We can combine (5), (6), (8), and (12), and take the square root of both sides to get

$$\left\| A - CC^+ A \right\|_F \leq \left\| A - A_k \right\|_F + \left\| \Sigma_{\rho-k} V_{\rho-k}^T SD \left(V_k^T SD \right)^+ \right\|_F . \quad (13)$$

From this, the triangle inequality, and the fact that for any two matrices A and B, $\|AB\|_F \leq \|B\|_2 \|A\|_F$, we have that

$$\left\| \Sigma_{\rho-k} V_{\rho-k}^T SD \left(V_k^T SD \right)^+ \right\|_F$$
$$\leq \left\| X SD \left(V_k^T SD \right)^T \right\|_F + \left\| X SD \left(\left(V_k^T SD \right)^+ - \left(V_k^T SD \right)^T \right) \right\|_F$$
$$\leq \left\| X SDDS^T V_k \right\|_F + \left\| \Sigma_{V_k^T SD}^{-1} - \Sigma_{V_k^T SD} \right\|_2 \left\| X SD \right\|_F , \quad (14)$$

where we have let $X = \Sigma_{\rho-k} V_{\rho-k}^T$. The following lemma will be used to bound (14); the proof is omitted.

Lemma 2. *For any probabilities* $\{p_i\}$, $\left\| \Sigma_{\rho-k} V_{\rho-k}^T SD \right\|_F \leq 10 \left\| A - A_k \right\|_F$, *with probability at least 0.9.*

The following lemma will also be used to bound (14).

Lemma 3. *If* $c \geq 10k/\beta\epsilon^2$, *then* $\left\| \Sigma_{\rho-k} V_{\rho-k}^T SDDS^T V_k \right\|_F \leq \epsilon \left\| A - A_k \right\|_F$, *with probability at least 0.9.*

Proof: Note that $\Sigma_{\rho-k} V_{\rho-k}^T V_k = \mathbf{0}$, and we will view $\Sigma_{\rho-k} V_{\rho-k}^T SDDS^T V_k$ as approximating this matrix product. We apply Lemma 4 of [5] (see also Figure 5 of [5]) to get

$$\mathbf{E}\left[\left\| \Sigma_{\rho-k} V_{\rho-k}^T SDDS^T V_k - \Sigma_{\rho-k} V_{\rho-k}^T V_k \right\|_F^2 \right] \leq \frac{1}{\beta c} \left\| A - A_k \right\|_F^2 \left\| V_k \right\|_F^2$$
$$= \frac{k}{\beta c} \left\| A - A_k \right\|_F^2 .$$

The lemma follows by applying Markov's inequality and taking the square roots of both sides of the resulting inequality.

$\diamond$

If $c \geq 40k^2/\beta\epsilon^2$, then Lemmas 1, 2, and 3 hold simultaneously with probability at least $1 - 3(0.1) = 0.7$. We condition on this event. Then, from (14), using Lemmas 1, 2, and 3, we get

$$\left\| \Sigma_{\rho-k} V_{\rho-k}^T SD \left(V_k^T SD \right)^+ \right\|_F \leq 9\epsilon \left\| A - A_k \right\|_F .$$

By combining this with (13), it follows that

$$\left\|A - CC^+ A\right\|_F \leq (1 + 9\epsilon) \left\|A - A_k\right\|_F.$$

The first claim of Theorem 1 follows with probability at least 0.7 by letting $\epsilon' = \epsilon/9$ and adjusting c to $O(k^2/\beta\epsilon'^2)$; it follows with probability at least $1 - \delta$ by running $O(\log(1/\delta))$ trials and using standard boosint procedures.

Note that setting $c = O(k^2/\epsilon^2)$ was required by Lemma 1, but that Lemmas 2 and 3 hold with $c = O(k/\epsilon^2)$. In particular, (10) of Lemma 1 required setting $c = O(k^2/\epsilon^2)$ in order to bound the error by $\epsilon/2$. We conjecture that the same bound holds if $c = O(k \log k/\epsilon^2)$. This result would follow from a stronger spectral norm bound than that provided by the Frobenius norm bound of Theorem 1 of [5]. Instead, in the next section, we will reduce c to $O(k \log k/\epsilon^2)$ by slightly modifying our sampling technique and using Theorem 3.1 of [21].

2.2   Constructing C with the Expected(c) Algorithm

The second claim of Theorem 1 considers the situation when the columns of A are sampled with the EXPECTED(c) algorithm. In this subsection, we outline its proof.

If the columns of A are sampled with the EXPECTED(c) algorithm, then the number of columns of S, and thus the number of rows and columns of D, is a random variable with expectation at most c. On the other hand, with this sampling procedure we can directly bound the spectral norm of (9), as opposed to bounding it indirectly via the Frobenius norm. To do so, consider the following theorem, which is a small extension of Theorem 3.1 in [21] to include the β factor; see also [20].

Theorem 2. *Let $X \in \mathbb{R}^{m \times n}$ and let $c \leq n$ be a positive integer. If S and D are constructed with the EXPECTED(c) algorithm using sampling probabilities $p_i, i \in [n]$ such that $\sum_i p_i = 1$ and $p_i \geq \beta \left|X^{(i)}\right|^2 / \|X\|_F^2$, then*

$$\mathbf{E}\left[\left\|XX^T - XSDDS^T X^T\right\|_2\right] \leq O\left(\sqrt{\frac{\log c}{\beta c}}\right) \|X\|_F \|X\|_2.$$

All of the derivations of Section 2.1 up to Lemma 1 hold for this modified sampling procedure. The following lemma is the analog of Lemma 1 with this new sampling prodecure, and it leads to an improved dependency of c on k.

Lemma 4. (Analog of Lemma 1) *If $c = O\left(k \log k/\beta\epsilon^2\right)$, then each of the claims of Lemma 1 holds with probability at least 0.9.*

Proof: From Theorem 2 and since $\|V_k\|_F = \sqrt{k}$ and $\|V_k\|_2 = 1$, it follows that

$$\mathbf{E}\left[\left\|V_k^T V_k - V_k^T SDDS^T V_k\right\|_2\right] \leq O\left(\sqrt{\log c/\beta c}\,\|V_k\|_F\,\|V_k\|_2\right)$$
$$= O\left(\sqrt{k \log c/\beta c}\right).$$

Using the assumed value of c, by Markov's inequality, and since $\epsilon \leq 1$, $\left| 1 - \sigma_i^2 \left(V_k^T SD \right) \right| \leq \epsilon/2 \leq 1/2$ with probability at least 0.9, which implies that $\tilde{k} = k$. The rest of the proof is the same as in Lemma 1.

◇

The remainder of the proof parallels the proof of Section 2.1.

3   Concluding Remarks

We conclude with three open problems.

- To what extent do the results of the present paper generalize to other matrix norms?
- What hardness results can be established for the optimal choice of columns?
- Does there exist a deterministic (any factor) approximation algorithm to the problem we consider?

Acknowledgements. We would like to thank Ravi Kannan for numerous useful discussions, Sariel Har-Peled for writing up his results amidst travel in India [13], and Amit Deshpande and Santosh Vempala for graciously providing a copy of [3].

References

1. R. Bhatia. *Matrix Analysis*. Springer-Verlag, New York, 1997.
2. A. Deshpande, L. Rademacher, S. Vempala, and G. Wang. Matrix approximation and projective clustering via volume sampling. In *Proceedings of the 17th Annual ACM-SIAM Symposium on Discrete Algorithms*, pages 1117–1126, 2006.
3. A. Deshpande and S. Vempala. Adaptive sampling and fast low-rank matrix approximation. Technical Report TR06-042, Electronic Colloquium on Computational Complexity, March 2006.
4. P. Drineas, A. Frieze, R. Kannan, S. Vempala, and V. Vinay. Clustering in large graphs and matrices. In *Proceedings of the 10th Annual ACM-SIAM Symposium on Discrete Algorithms*, pages 291–299, 1999.
5. P. Drineas, R. Kannan, and M.W. Mahoney. Fast Monte Carlo algorithms for matrices I: Approximating matrix multiplication. *To appear in: SIAM Journal on Computing*.
6. P. Drineas, R. Kannan, and M.W. Mahoney. Fast Monte Carlo algorithms for matrices II: Computing a low-rank approximation to a matrix. *To appear in: SIAM Journal on Computing*.
7. P. Drineas, R. Kannan, and M.W. Mahoney. Fast Monte Carlo algorithms for matrices III: Computing a compressed approximate matrix decomposition. *To appear in: SIAM Journal on Computing*.
8. P. Drineas, M.W. Mahoney, and S. Muthukrishnan. Polynomial time algorithm for column-row based relative-error low-rank matrix approximation. Technical Report 2006-04, DIMACS, March 2006.
9. P. Drineas, M.W. Mahoney, and S. Muthukrishnan. Sampling algorithms for ℓ_2 regression and applications. In *Proceedings of the 17th Annual ACM-SIAM Symposium on Discrete Algorithms*, pages 1127–1136, 2006.

10. A. Frieze, R. Kannan, and S. Vempala. Fast Monte-Carlo algorithms for finding low-rank approximations. In *Proceedings of the 39th Annual IEEE Symposium on Foundations of Computer Science*, pages 370–378, 1998.
11. A. Frieze, R. Kannan, and S. Vempala. Fast Monte-Carlo algorithms for finding low-rank approximations. *Journal of the ACM*, 51(6):1025–1041, 2004.
12. G.H. Golub and C.F. Van Loan. *Matrix Computations*. Johns Hopkins University Press, Baltimore, 1989.
13. S. Har-Peled. Low rank matrix approximation in linear time. *Manuscript. January 2006*.
14. R.A. Horn and C.R. Johnson. *Matrix Analysis*. Cambridge University Press, New York, 1985.
15. F.G. Kuruvilla, P.J. Park, and S.L. Schreiber. Vector algebra in the analysis of genome-wide expression data. *Genome Biology*, 3:research0011.1–0011.11, 2002.
16. Z. Lin and R.B. Altman. Finding haplotype tagging SNPs by use of principal components analysis. *American Journal of Human Genetics*, 75:850–861, 2004.
17. M.Z. Nashed, editor. *Generalized Inverses and Applications*. Academic Press, New York, 1976.
18. P. Paschou, M.W. Mahoney, J.R. Kidd, A.J. Pakstis, S. Gu, K.K. Kidd, and P. Drineas. Intra- and inter-population genotype reconstruction from tagging SNPs. *Manuscript submitted for publication*.
19. L. Rademacher, S. Vempala, and G. Wang. Matrix approximation and projective clustering via iterative sampling. Technical Report MIT-LCS-TR-983, Massachusetts Institute of Technology, Cambridge, MA, March 2005.
20. M. Rudelson and R. Vershynin. Approximation of matrices. *Manuscript*.
21. R. Vershynin. Coordinate restrictions of linear operators in l_2^n. *Manuscript*.

Dobrushin Conditions and Systematic Scan[*]

Martin Dyer[1], Leslie Ann Goldberg[2], and Mark Jerrum[3]

[1] School of Computing
University of Leeds, Leeds LS2 9JT, UK
[2] Department of Computer Science, University of Warwick, Coventry CV4 7AL, UK
[3] School of Informatics, University of Edinburgh, Edinburgh EH9 3JZ, UK

Abstract. We consider Glauber dynamics on finite spin systems. The mixing time of Glauber dynamics can be bounded in terms of the influences of sites on each other. We consider three parameters bounding these influences — α, the total influence on a site, as studied by Dobrushin; α', the total influence of a site, as studied by Dobrushin and Shlosman; and α'', the total influence of a site in any given context, which is related to the path-coupling method of Bubley and Dyer. It is known that if any of these parameters is less than 1 then random-update Glauber dynamics (in which a randomly-chosen site is updated at each step) is rapidly mixing. It is also known that the Dobrushin condition $\alpha < 1$ implies that systematic-scan Glauber dynamics (in which sites are updated in a deterministic order) is rapidly mixing. This paper studies two related issues, primarily in the context of systematic scan: (1) the relationship between the parameters α, α' and α'', and (2) the relationship between proofs of rapid mixing using Dobrushin uniqueness (which typically use analysis techniques) and proofs of rapid mixing using path coupling. We use matrix-balancing to show that the Dobrushin-Shlosman condition $\alpha' < 1$ implies rapid mixing of systematic scan. An interesting question is whether the rapid mixing results for scan can be extended to the $\alpha = 1$ or $\alpha' = 1$ case. We give positive results for the rapid mixing of systematic scan for certain $\alpha = 1$ cases. As an application, we show rapid mixing of systematic scan (for any scan order) for heat-bath Glauber dynamics for proper q-colourings of a degree-Δ graph G when $q \geq 2\Delta$.

1 Introduction

A *spin system* consists of a set of *sites* and a set of spins. In this paper, both sets will be finite. We use $[n] = \{1, \ldots, n\}$ to denote the set of sites, and C to denote the set of spins. A *configuration* is an assignment of spins to sites, and Ω^+ denotes the set of all configurations. Sites interact locally, and these interactions specify the relative likelihood of possible (local) sub-configurations.

[*] Partially supported by the EPSRC grant Discontinuous Behaviour in the Complexity of Randomized Algorithms. Some of the work was done while the authors were visiting the Mathematical Sciences Research Institute in Berkeley. A full version, with all proofs, appears at http://www.eccc.uni-trier.de/eccc-reports/2005/TR05-075/index.html

J. Diaz et al. (Eds.): APPROX and RANDOM 2006, LNCS 4110, pp. 327–338, 2006.

Taken together, these give a well-defined probability distribution, π, on the set of configurations Ω^+. *Glauber dynamics* is a random walk on configurations that updates spins one site at a time, and converges to π. Before giving some examples, we formalise these concepts in a way that will be useful for this paper.

We use the following notation. If x is a configuration and j is a site then x_j denotes the spin at site j in x. For each site j, S_j denotes the set of pairs of configurations that agree off of site j. That is, S_j is the set of pairs $(x, y) \in \Omega^+ \times \Omega^+$ such that, for all $i \neq j$, $x_i = y_i$. For each site j, we will have a transition matrix $P^{[j]}$ on the state space Ω^+ which satisfies two properties: $P^{[j]}$ moves from one configuration to another by updating site j. That is, if $P^{[j]}(x, y) > 0$, then $(x, y) \in S_j$. Also, π is invariant with respect to $P^{[j]}$.

Most theoretical results about Glauber dynamics consider *random updates*, in which the random walk on configurations proceeds as follows. At each step, a site j is chosen uniformly at random. The configuration is then updated according to transition matrix $P^{[j]}$. Formally, random-update Glauber dynamics corresponds to a Markov chain $\mathcal{M}$ with state space Ω^+ and transition matrix $P = (1/n) \sum_{j=1}^{n} P^{[j]}$.

For example, consider the spin system corresponding to proper q-colourings of an n-vertex graph G with maximum degree $\Delta \leq q - 2$. The sites are the vertices $1, \ldots, n$. C is the set of colours $C = \{1, \ldots, q\}$. The distribution π assigns equal probability to all proper colourings (colourings with no monochromatic edges) and it assigns zero probability to all improper colourings. In so-called "heat-bath" Glauber dynamics, the transition matrix $P^{[j]}$ makes the following transition from a configuration x. Let C_x denote the set of colours that are not assigned to neighbours of site j in x. Let $x \to^j c$ denote the configuration obtained from x by changing the spin at site j to c. $P^{[j]}$ makes a transition to a uniformly-chosen configuration in $\{x \to^j c \mid c \in C_x\}$. Another example is the "Metropolis" Glauber dynamics for proper q-colourings of G. In this case, the transition matrix $P^{[j]}$ makes the following transition from x. A colour $c \in C$ is chosen uniformly at random. If $c \in C_x$ then a transition is made to $x \to^j c$ otherwise the new configuration is x. Further examples include corresponding dynamics for the Potts model and for the hard-core lattice gas model.

There has been much work on analyzing the *mixing time* of random-update Glauber dynamics. The mixing time from a specified initial configuration x (as a function of the deviation ε from stationarity) is $\tau_x(\mathcal{M}, \varepsilon) = \min\{t > 0 : d_{\mathrm{TV}}(P^t(x, \cdot), \pi(\cdot)) \leq \varepsilon\}$, where d_{TV} denotes total variation distance. The mixing time of $\mathcal{M}$ is $\tau(\mathcal{M}, \varepsilon) = \max_{x \in \Omega^+} \tau_x(\mathcal{M}, \varepsilon)$. $\mathcal{M}$ is said to be "rapidly mixing" if $\tau(\mathcal{M}, \varepsilon)$ is at most a polynomial in n and $\log(\varepsilon^{-1})$.

It is well-known that the mixing time can be bounded in terms of the influences of sites on each other. To be more precise, let $\mu_j(x, \cdot)$ be the distribution on spins at site j induced by $P^{[j]}(x, \cdot)$ and let $\rho_{i,j}$ be the influence of site i on site j which is given by $\rho_{i,j} = \max_{(x,y) \in S_i} d_{\mathrm{TV}}(\mu_j(x, \cdot), \mu_j(y, \cdot))$. We will be interested in three quantities. Let α be the total influence *on* a site, defined by $\alpha = \max_{j \in [n]} \sum_{i \in [n]} \rho_{i,j}$. Let α' be the total influence *of* a site, defined by $\alpha' =$

$\max_{i \in [n]} \sum_{j \in [n]} \rho_{i,j}$. Finally, let α'' be the total influence of a site *in any given context,* defined by $\alpha'' = \max_{i \in [n]} \max_{(x,y) \in S_i} \sum_{j \in [n]} d_{TV}(\mu_j(x, \cdot), \mu_j(y, \cdot))$.

The *Dobrushin condition* $\alpha < 1$, which says that the total influence on a site is small, implies that $\mathcal{M}$ is rapidly mixing. In particular, $\tau(\mathcal{M}, \varepsilon) = O(\frac{n}{1-\alpha} \log(n\varepsilon^{-1}))$. Dobrushin's original result [7] was not stated in terms of rapid mixing — instead, he was concerned with a closely related issue — uniqueness of the Gibbs measure for countable (not finite) spin systems. For a proof that the condition implies rapid mixing see, for example, Weitz's paper [20].

An easy application of the *path coupling* method of Bubley and Dyer [2] shows that $\alpha'' < 1$ implies rapid mixing. In particular, $\tau(\mathcal{M}, \varepsilon) = O(\frac{n}{1-\alpha''} \log(n\varepsilon^{-1}))$. Path coupling can also be used to show rapid mixing for the case $\alpha'' = 1$ provided the change in path length has enough variance. For details, see Dyer and Greenhill's survey paper [11].

An inspection of the definition of α' reveals that

$$\alpha' = \max_i \sum_j \max_{(x,y) \in S_i} d_{TV}(\mu_j(x, \cdot), \mu_j(y, \cdot)) \geq \alpha''.$$

Therefore, $\alpha' < 1$ implies $\tau(\mathcal{M}, \varepsilon) = O(\frac{n}{1-\alpha'} \log(n\varepsilon^{-1}))$. Dobrushin and Shlosman [8] were the first to derive uniqueness from the condition $\alpha' < 1$, which says that the total influence of a site is small. (As Weitz points out [20], Dobrushin and Shlosman stated their result in terms of the total influence *on* a site but they worked in a translation-invariant setting and what they used is that the total influence *of* a site is small. In fact, Dobrushin and Shlosman worked in a more general block-dynamics setting. This will be discussed below.)

While theoretical results about Glauber dynamics typically consider random updates, experimental work is often carried out using systematic strategies that cycle through sites in a deterministic manner, a dynamics we refer to as *systematic scan.* Formally, systematic scan corresponds to a Markov chain $\mathcal{M}_\rightarrow$ with state space Ω^+ and transition matrix $P_\rightarrow = \prod_{j=1}^n P^{[j]}$.

The Dobrushin condition implies that systematic scan is rapidly mixing. In particular, $\tau(\mathcal{M}_\rightarrow, \varepsilon) = O(\frac{1}{1-\alpha} \log(n\varepsilon^{-1}))$. A proof follows easily from the account of Dobrushin uniqueness in Simon's book [18], some of which is derived from the account of Föllmer [12].

This paper explores two related issues, primarily in the context of systematic scan. **(1.)** What is the relationship between conditions bounding the influence *on* a site (upper bounds on α) and conditions bounding the influence *of* a site (upper bounds on α' or α'')? Do they imply rapid mixing in the same circumstances? **(2.)** What is the relationship between proofs of rapid mixing using Dobrushin uniqueness, which typically use analysis techniques and proofs of rapid mixing using path coupling? Can they be used to prove rapid mixing in the same circumstances?

The second of these issues was first raised by Sokal [19], who shows how to translate a proof based on a Markovian coupling to analytic language. The first of these issues was highlighted by Weitz [20] who elucidated the dual nature of the Dobrushin condition $\alpha < 1$ and the Dobrushin-Shlosman condition $\alpha' < 1$.

A preliminary issue is the relationship between the parameters α, α' and α''. We have observed above that the two "influence of a site" parameters, α' and α'' are related by $\alpha'' \leq \alpha'$. Weitz [20, Section 5.3] gives an example in a block-dynamics context where the parameter analogous to α'' is less than one (allowing one to infer rapid mixing), but the other parameters are greater than one. In the full version we give a similar example in our (single-site) context. The example is a spin system for which $\alpha'' < 1$ but $\alpha > 1$ and $\alpha' > 1$. In fact, α'' can be made arbitrarily smaller than the other parameters. We also give an example where the "influence of a site" parameters α' and α'' are less than one but the "influence on a site" parameter α is greater than one. Also, we give an example where the "influence on a site" parameter α is less than one but the other parameters are greater than one.

The primary motivation for our work was the observation that Dobrushin's condition (the influence on a site is small) implies rapid mixing for systematic scan. There are some known coupling results for systematic scan using the fact that the influence *of* a site is small, but only in very special circumstances, for example, proper colourings of a path, where the scan is left-to-right along the path [9]. When Dobrushin's condition is satisfied then systematic scan is rapidly mixing regardless of the order in which vertices are scanned. In this paper we show that rapid mixing also occurs (for every scan order) if the influence *of* a site is small. In particular, we prove the following theorem.

Theorem 1. *Suppose $\alpha' < 1$. Then $\tau(\mathcal{M}_\rightarrow, \varepsilon) \leq \frac{2}{1-\alpha'} \log(4n^2(1-\alpha')^{-1}\varepsilon^{-1})$.*

Theorem 1 is proved in Section 2. The main ingredient in the proof is matrix balancing. This enables us to translate the Dobrushin-Shlosman condition (the influence of a site is small) into the Dobrushin condition (the influence on a site is small). The matrix-balancing can be viewed as a generic way of deriving weights. Simon's version of Dobrushin uniqueness corresponds to showing convergence in the L_1 norm of a particular vector. This convergence occurs for $\alpha < 1$. The matrix-balancing approach corresponds to showing convergence in a weighted L_1 norm. Weights have been used previously in mixing results, for example [2, 9, 20]. The point here is that the weights are derived automatically by the balancing. Then convergence holds for $\alpha' < 1$.

The statement of Theorem 1 can be generalised. In particular, the proof does not require each of the sites to be updated the same number of times. All that is really required is that the sequence of updates contains a *subsequence* of at least $\frac{2}{1-\alpha'} \log(4n^2(1-\alpha')^{-1}\varepsilon^{-1})$ scans. The same generalisation holds for the Dobrushin case. It will be useful to record the following corollary. Let $P_\leftarrow = \prod_{k=1}^{n} P^{[n-k+1]}$ and let $P_{\rightarrow\leftarrow} = P_\rightarrow P_\leftarrow$. Let $\mathcal{M}_{\rightarrow\leftarrow}$ be the Markov chain with state space Ω^+ and transition matrix $P_{\rightarrow\leftarrow}$.

Corollary 1. *Suppose $\alpha' < 1$. Then $\tau(\mathcal{M}_{\rightarrow\leftarrow}, \varepsilon) \leq \frac{2}{1-\alpha'} \log(4n^2(1-\alpha')^{-1}\varepsilon^{-1})$.*

The mixing time of a Markov chain is closely related to its spectral gap — the gap between the largest eigenvalue of its transition matrix and the second-largest eigenvalue. In the full version, we prove the following theorem, which applies when the eigenvalues of the transition matrix are real.

Theorem 2. *Suppose $\alpha' < 1$. Let β_1 be the second-largest eigenvalue of $P_{\rightarrow}$. If β_1 and its associated eigenvector are real then $1 - \beta_1 \geq (1 - \alpha')/2$.*

Having seen that the Dobrushin condition and the Dobrushin-Shlosman condition imply rapid mixing for systematic scan, an interesting question is whether these results can be extended to the $\alpha = 1$ or $\alpha' = 1$ case. In the full version we provide an example where $\alpha = 1$ and random-update Glauber dynamics is rapidly mixing, but systematic scan is not even ergodic. The example indicates that perhaps Dobrushin-like arguments cannot be extended to the $\alpha = 1$ case. The reason for this is that Dobrushin-like arguments imply rapid mixing for systematic scan, as well as for random updates. But we have an example where systematic scan is clearly not rapidly mixing. Nevertheless, we give positive results for the rapid-mixing of systematic scan for certain $\alpha = 1$ cases, particularly some cases for which there are *symmetric* upper bounds on the dependencies corresponding to a bound of 1, both on the total influence of a site, and on the total influence on a site.

A *dependency matrix* for a spin system is an $n \times n$ matrix R in which $R_{i,j} \geq \rho_{i,j}$ (so $R_{i,j}$ is an upper bound on the influence of site i on site j. We will be particularly interested in the case in which R is symmetric. In this case we can view R as a weighted adjacency matrix, and we refer to the resulting (undirected) graph on sites as the *dependency graph* of R. We say that R is *connected* if the resulting dependency graph is connected in the sense that there is a positive-weight path from every site to every other. In Section 3.1 we prove the following result, which says that systematic scan is rapidly mixing if there is a dependency matrix which (1) is symmetric, (2) has row and column sums at most 1 (corresponding to total influence at most 1 for every site), and (3) every connected component has a site with a row sum less than 1 (corresponding to total influence less than 1).

The mixing bound given in the theorem is a function of n, the number of sites, and also of N, the "precision" of R. We say that a dependency matrix R has precision N (for a positive integer N) if every entry $R_{i,j}$ can be expressed as a fraction of integers with denominator N.

Theorem 3. *Suppose that a spin system has a precision-N symmetric dependency matrix R with row sums and column sums at most 1. Suppose that every connected component has a site whose row sum is less than 1. Then $\tau(\mathcal{M}_{\rightarrow}, \varepsilon) = O(n^3 N \log(n^2 \varepsilon^{-1}))$ and $\tau(\mathcal{M}_{\rightarrow\leftarrow}, \varepsilon) = O(n^3 N \log(n^2 \varepsilon^{-1}))$.*

We now apply Theorem 3 to an example considered previously. Suppose that G is an n-vertex connected graph with maximum degree $\Delta \geq 2$. Let $q = 2\Delta$ and consider the spin system corresponding to heat-bath Glauber dynamics for proper q-colourings of G. If G is not Δ-regular then there is a connected symmetric dependency matrix R in which some vertex s of degree less than Δ has small total influence so its row sum is less than 1. As we will see in Section 3, Theorem 3 implies that systematic scan is rapidly mixing. Using the decomposition method of Martin and Randall, we can extend the result to the case in which G is Δ-regular (and every vertex has total influence 1), proving the following.

Theorem 4. *Let G be an n-vertex connected graph with maximum degree $\Delta \geq 2$ and let $q = 2\Delta$. Consider the spin system corresponding to heat-bath updates on proper colourings and let x be any proper colouring. Then $\tau_x(\mathcal{M}_{\rightarrow\leftarrow}, \varepsilon) = O(n^3 \log(n) \log((\varepsilon\pi(x))^{-1}))$.*

Theorem 4 is proved in Section 3.2 and a generalisation (to more general spin systems) is stated and proved in Section 4.

It would be nice to have a full characterisation of the situations in which $\alpha = 1$ (or $\alpha' = 1$) implies rapid mixing for systematic scan. Another interesting open question is whether bounds on the path-coupling parameter α'' imply rapid mixing for systematic scan.

All of our mixing results build upon the methods used in Simon's account of Dobrushin uniqueness (using analysis techniques). Thus, we might ask the question whether path coupling is a less useful technique for proving rapid mixing for systematic scan. In Section 5 we show that this may not be the case. In particular, we provide an alternative proof that the Dobrushin condition implies rapid mixing of systematic scan. This proof uses path coupling.

An issue that is not treated in this paper is how these methods generalise to dynamics other than Glauber. Dobrushin and Shlosman's result [8] actually applied to block-dynamics rather than just to single-site dynamics, though only when the underlying dependency graph is Z^d. For random-update Glauber dynamics, Weitz [20] shows rapid mixing both when the influence of a site is small and when the influence on a site is small. Weitz's results apply to "block dynamics" and to an arbitrary dependency graph. Both results are proved using coupling. Our results about systematic scan can be similarly generalised. This work, by Pedersen [15], is in preparation.

2   Rapid Mixing for $\alpha' < 1$

In this section, we prove rapid mixing of systematic scan for the case $\alpha' < 1$. We prove Theorem 1, Corollary 1, and Theorem 2. Consider a spin system with dependency matrix R. Suppose that some entry of R is non-zero and that the row sums of R are less than 1. Let $\gamma = 1 - \max_{i \in [n]} \sum_{j \in [n]} R_{i,j}$. Note that $\gamma \in (0, 1]$. Also, if a spin system is non-trivial in the sense that it has sites i and j with $\rho_{i,j} > 0$ and it has $\alpha' < 1$ then it has such a dependency matrix with $\gamma = 1 - \alpha'$.

2.1   Matrix Balancing

In this section we prove the following lemma.

Lemma 1. *There is an $n \times n$ diagonal matrix W such that, for every $i \in [n]$, $(\gamma/4n) \leq W_{i,i} \leq 1$ and every column sum of WRW^{-1} is at most $1 - \gamma/2$.*

Proof. Let R' be the matrix which is the same as R except that we add a new column $n + 1$ and a new row $n + 1$. For $i \in [n]$, let $R'_{i,n+1} = 1 - r_i$ where r_i is the row sum of R. Note that $R'_{i,n+1} \geq \gamma$ and that the row sums of R' are 1. For $j \in [n]$, let $R'_{n+1,j}$ be a quantity $x_j \in (1/(2n), 1]$ which will be chosen later so that $\sum_{j=1}^{n} x_j = 1$. Let $R'[n+1, n+1] = 0$. Note that R' is the transition matrix of a Markov chain which is ergodic and every state (including state $n+1$) has positive probability. Let π' be its stationary distribution.

Now let $W = \text{diag}(\pi'_1, \ldots, \pi'_{n+1})$ and let $R'' = WR'W^{-1}$ (this is multiplying row i by π'_i and dividing column j by π'_j). Note by stationarity of π' that the columns of R'' have sum 1. That is $\sum_{i=1}^{n+1} R'_{i,j} \frac{\pi'_i}{\pi'_j} = \frac{1}{\pi'_j} \sum_{i=1}^{n+1} \pi'_i R'_{i,j} = 1$. We will choose $x_j = \frac{1}{2n} + \frac{1}{2} \frac{\pi'_j}{(1-\pi'_{n+1})}$. Now $R'_{n+1,j} \frac{\pi'_{n+1}}{\pi'_j} = x_j \frac{\pi'_{n+1}}{\pi'_j} \geq \frac{1}{2} \frac{\pi'_{n+1}}{(1-\pi'_{n+1})}$ and $\pi'_{n+1} \geq \sum_{i=1}^{n} \pi'_i(1 - r_i) \geq \gamma(1 - \pi'_{n+1})$, so $R'_{n+1,j}\pi'_{n+1}/\pi'_j \geq \gamma/2$ and (if we remove the extra row and column from W) the column sums of WRW^{-1} are at most $1 - \gamma/2$. Finally, $W_{j,j} = \pi'_j \geq \pi'_{n+1} x_j \geq \frac{\gamma}{1+\gamma} \frac{1}{2n} \geq \gamma \frac{1}{4n}$.   □

2.2   The Effect of a Applying $P^{[j]}$ or $P_\rightarrow$

Let $M^{[j]}$ be the matrix constructed from the identity matrix by replacing column j with the jth column of R. For any function f from Ω^+ to $\mathbb{R}$, let $\delta_i(f) = \max_{(x,y)\in S_i} |f(x) - f(y)|$ and let $\delta(f)$ be the column vector $\delta(f) = (\delta_1(f), \ldots, \delta_n(f))$. Let $P^{[j]}f$ be the function from Ω^+ to $\mathbb{R}$ given by $P^{[j]}f(x) = \sum_y P^{[j]}(x,y)f(y)$. The following lemma is (a slight generalisation of one) in Simon's book [18].

Lemma 2. *The vector* $\delta(P^{[j]}f)$ *is component-wise less than or equal to the vector* $M^{[j]}\delta(f)$.

Now let W be the matrix from Lemma 1. Let $Q^{[j]} = WM^{[j]}W^{-1}$ and note that $Q^{[j]}$ is the matrix constructed from the identity matrix by replacing column j with the vector $(q_{1,j}, \ldots, q_{n,j})$, where $q_{i,j} = R_{i,j}w_i/w_j$. From Lemma 1, $\sum_{i=1}^{n} q_{i,j} \leq 1 - \gamma/2$. Now let $k_i(f) = w_i\delta_i(f)$ and let $k(f)$ be the column vector $k(f) = (k_1(f), \ldots, k_n(f))$. Let $K(f) = \sum_{i=1}^{n} k_i(f)$. The following lemma is a weighted version of a lemma in Simon's book (using the weights from the matrix balancing). It is proved in the full version of the paper.

Lemma 3. *For any* $m \in \{0, \ldots, n\}$,
$$K(P^{[1]} \cdots P^{[m]}f) \leq \sum_{i=1}^{m} (1 - \gamma/2)k_i(f) + \sum_{i=m+1}^{n} k_i(f).$$

Corollary 2. $K(P_\rightarrow f) \leq (1 - \gamma/2)K(f)$.

2.3   The Bound on the Mixing Time - Proof of Theorem 1 and Corollary 1

For a test function f, let $f_t(x) = \sum_z P_\rightarrow^t(x,z)f(z)$. Thus, Corollary 2 gives $K(f_t) \leq (1-\gamma/2)K(f_{t-1})$. Now let f be the indicator variable for being in some subset A of Ω^+. Then

$$\max_x f_t(x) - \min_y f_t(y) \leq \sum_i \frac{k_i(f_t)}{w_i} \leq \left(\frac{1}{\min_i W_{i,i}}\right) K(f_t) \leq \frac{4n}{\gamma} K(f_t)$$

$$\leq \frac{4n}{\gamma}(1 - \gamma/2)^t K(f_0) \leq \frac{4n^2}{\gamma}(1 - \gamma/2)^t,$$

which is at most ε for $t \geq (2/\gamma) \log(4n^2\gamma^{-1}\varepsilon^{-1})$. Also, $\min_y f_t(y) \leq E_\pi f_t \leq \max_x f_t(x)$ and $E_\pi f_t = \pi(A)$, which gives us $\tau(\mathcal{M}_\rightarrow, \varepsilon) \leq \frac{2}{\gamma} \log(4n^2\gamma^{-1}\varepsilon^{-1})$. This proves Theorem 1. Corollary 1 comes from the fact that the right-hand side of the expression in Lemma 3 is at most $K(f)$. Thus, extra updates do no harm.

Remark 1. The argument in Sections 2.2 and 2.3 gives an upper bound on the mixing tome of $\mathcal{M}_\rightarrow$ using the assumption that there is a weighting W so that the weighted column sums of R are bounded below 1. In particular, the assumption is that the column sums of WRW^{-1} are at most $1 - \gamma'$ for some $\gamma' \in (0, 1]$ (here $\gamma' = \gamma/2$) where W is an $n \times n$ diagonal matrix with $1/w \leq W_{i,i} \leq 1$ for some w (here $w = \gamma/(4n)$). The mixing time bound is $(1/\gamma') \log(wn\varepsilon^{-1})$. Lemma 1 shows that a suitable W can be found provided only that the system has a dependency matrix in which the maximum row sum is bounded below 1 (in particular, $1 - 2\gamma'$). It is easy to see that the same argument would apply if the *weighted* row sums of R are bounded below 1, where of course the resulting mixing time depends on the weights as above. Thus, the argument essentially shows how to automatically translate any weighting in which row sums are bounded below 1 into one in which column sums are bounded below 1.

2.4   Remarks - Contraction in Various Norms

Lemma 2 tells us that $\delta(P^{[j]}f) \leq M^{[j]}\delta(f)$. We are interested in the effect of M on the vector $\delta(f)$ where $M = \prod_{j=1}^n M^{[j]}$. Lemma 3 shows that $\delta(f)$ is contracting in the weighted L_1 distance $K(f)$. The contraction comes from the fact that the column sums of the weighted matrix $Q^{[j]}$ are less than 1. Simon's proof of Dobrushin's result for $\alpha < 1$ corresponds to taking W to be the identity matrix (so it is L_1 contraction). Our proof uses contraction in weighted L_1 distance, but there are other possibilities. For example, in the full version we revisit the random-updates Markov chain $\mathcal{M}$ and we prove mixing by observing a contraction in the L_∞ norm. This is possible when the row sums of the appropriate weighted matrix are less than 1.

3   Positive Results for $\alpha = 1$

3.1   Symmetric Dependency Matrices with a Row Sum Less Than 1

In this section we prove Theorem 3, which says that systematic scan is rapidly mixing if there is a dependency matrix which (1) is symmetric, (2) has row and column sums at most 1 (corresponding to total influence at most 1 for every site), and (3) every connected component has a site with a row sum less than 1 (corresponding to total influence less than 1). We start with the connected case. Recall that a matrix has precision N if every entry in it can be expressed as a fraction of integers with denominator N.

Theorem 5. *Suppose that a spin system has a precision-N symmetric connected dependency matrix R with row sums and column sums at most 1. Suppose that* *there is a site s with row sum less than* 1. *Then* $\tau(\mathcal{M}_\rightarrow, \varepsilon) = O(n^3 N \log(n^2 \varepsilon^{-1}))$ *and* $\tau(\mathcal{M}_{\rightarrow\leftarrow}, \varepsilon) = O(n^3 N \log(n^2 \varepsilon^{-1}))$.

Remark 2. The dependence of the running time on the precision N is one way to express the condition that the dependency matrix R needs to be sufficiently "mixing". There are other possible choices. See [4].

Proof. Without loss of generality assume that for $i \neq s$ we have $\sum_j R_{i,j} = 1$. If this is not the case then $R_{i,i}$ can be increased until it is. Note that increasing $R_{i,i}$ to make the row sum 1 does not increase the precision of the matrix. Let $\gamma = 1 - \sum_{j \in [n]} R_{s,j}$. Since the sum of row s is less than 1, γ is positive. Also, γ is less than 1 (otherwise sites in $[n]$ do not depend upon s, contradicting connectivity). Construct $R^{[2]}$ from R by adding an extra row and column. Set $R^{[2]}_{s,n+1} = R^{[2]}_{n+1,s} = \gamma$ and $R^{[2]}_{n+1,n+1} = 1-\gamma$ and make the rest of the entries in the new row and column equal to zero. Note that the row and column sums of $R^{[2]}$ are 1. It is the transition matrix of an ergodic Markov chain. Its stationary distribution, $\pi^{[2]}$, is uniform (by symmetry). Let $H(i,j)$ denote the hitting time of j from i and let H denote the maximum hitting time. Note that $H \leq (n+1)^2 N$. (To see this, view the Markov chain with transition matrix $R^{[2]}$ as a random walk on a N-regular undirected graph (with multiple edges and loops allowed). Then $H(G)$ is at most the number of vertices times the number of (directed) edges [13].) Note also that $H(i,i) = 1/\pi^{[2]}_i = (n+1)$. Now let $R^{[3]} = R^{[2]} - E$, where E is the all-zero matrix except that row $n+1$ of E is $(-\xi, \ldots, -\xi, n\xi)$ where $\xi = \frac{1}{n(n+1)^2 N}$. Note that γ can be expressed as a fraction of integers with denominator N. Since it is less than 1, $\gamma \leq \frac{N-1}{N} < 1 - n\xi$, so $R^{[3]}_{n+1,n+1}$ is non-negative. It is clear that the other entries of $R^{[3]}$ are also non-negative. Furthermore, $R^{[3]}$ is the transition matrix of an ergodic Markov chain. Let $\pi^{[3]}$ be its stationary distribution. We now use Theorem 2.1 of [3]. This says that for all j, $|\pi^{[2]}_j - \pi^{[3]}_j| \leq \frac{\|E\|_\infty}{2} \frac{\max_{i \neq j} H(i,j)}{H_{j,j}} \leq \frac{1}{2(n+1)}$. Now let $W = \text{diag}(\pi^{[3]}_1, \ldots, \pi^{[3]}_{n+1})$ and let $R^{[4]} = W R^{[3]} W^{-1}$ (this is multiplying row i by $\pi^{[3]}_i$ and dividing column j by $\pi^{[3]}_j$). Note by stationarity of $\pi^{[3]}$ that the columns of $R^{[4]}$ have sum 1. That is, $\sum_{i=1}^{n+1} R^{[3]}_{i,j} \frac{\pi^{[3]}_i}{\pi^{[3]}_j} = \frac{1}{\pi^{[3]}_j} \sum_{i=1}^{n+1} \pi^{[3]}_i R^{[3]}_{i,j} = 1$. Also for $j \in [n]$, we have

$$R^{[4]}_{n+1,j} = R^{[3]}_{n+1,j} \frac{\pi^{[3]}_{n+1}}{\pi^{[3]}_j} \geq \tfrac{1}{3} R^{[3]}_{n+1,j} \geq \tfrac{\xi}{3}, \text{ so } \sum_{i=1}^n R^{[4]}_{i,j} \leq 1 - \xi/3. \text{ Finally, observe}$$

that for i and j in $[n]$, $R^{[4]}_{i,j} = R^{[3]}_{i,j} \frac{\pi^{[3]}_i}{\pi^{[3]}_j} = R_{i,j} \frac{\pi^{[3]}_i}{\pi^{[3]}_j}$, so W (with row and column $n+1$

deleted) can be used as a weight matrix for R. Also $W_{j,j} = \pi^{[3]}_j \geq 1/(2(n+1))$. Now we proceed as in Section 2 and to get the variation distance down to ε, this takes $\frac{3}{\xi} \log(2n(n+1)\varepsilon^{-1}) = 3n(n+1)^2 N \log(2n(n+1)\varepsilon^{-1})$ scans. This proves the bound on $\tau(\mathcal{M}_\rightarrow, \varepsilon)$ in the theorem. The bound on $\tau(\mathcal{M}_{\rightarrow\leftarrow}, \varepsilon)$ is established in the same way as Corollary 1.

To prove Theorem 3, apply Theorem 5 to each connected component individually. Suppose that t is the maximum, over the components, of $\tau(\mathcal{M}_\rightarrow, \varepsilon/n)$. Then $d_{TV}(P^t_\rightarrow(x, \cdot), \pi) \leq \varepsilon$. The same argument applies to $\tau(\mathcal{M}_{\rightarrow\leftarrow}, \varepsilon)$.

3.2 Heat-Bath Updates and Proper Colourings with 2Δ Colours

Let G be a connected graph with maximum degree $\Delta \geq 2$ and let $q = |C| = 2\Delta$. Let Ω be the set of proper q-colourings of G. Let π be the uniform distribution on Ω (so $\pi(x) = 0$ for all configurations $x \in \Omega^+ - \Omega$). Let $P^{[j]}$ be the transition matrix for a "heat-bath" update on site j. To be precise, $\mu_j(x, \cdot)$ is the uniform distribution on colours that are not used at neighbours of site j in configuration x. For all edges (i, j) of G, $\rho_{i,j} = 1/(q - d(j))$, where $d(j)$ is the degree of site j, so we take $R_{i,j} = 1/(q - \Delta) = 1/\Delta$. If (i, j) is not an edge then $R_{i,j} = 0$ ([16]).

In the full version, we prove Theorem 4. If G is not Δ-regular, then the theorem follows from the observations that we have just made about $R_{i,j}$ and from Theorem 5. Some vertex has degree less than Δ, so has row sum less than 1. The difficult case is when G is Δ-regular.

The proof uses the decomposition theorem of Martin and Randall [14]. The "restriction" chain leaves the spin of a particular vertex fixed. The "projection chain" chains the spin of this vertex. The key point is that the mixing time of the restriction time can be bounded, using comparison, by The analysis in Section 3.1. See the full version for details.

4   A Generalised $\alpha = 1$ Case

The following is a generalisation of Theorem 4 which is proved in the full version. Consider a general spin system. Let $\Omega = \{x \in \Omega^+ \mid \pi(x) > 0\}$. As in Section 3.2 let $\Omega_c = \{x \in \Omega \mid x_n = c\}$ and let $P_{\rightarrow\leftarrow c}$ be the Markov chain on Ω_c defined by $P_{\rightarrow\leftarrow c}(x, y) = P_{\rightarrow\leftarrow}(x, y)$ for distinct x and y in Ω_c with $P_{\rightarrow\leftarrow c}(x, x) = 1 - \sum_{y \in \Omega_c : y \neq x} P_{\rightarrow\leftarrow c}(x, y)$.

Theorem 6. *Suppose that a spin system has a precision-N symmetric connected dependency matrix R with row sums equal to 1. Suppose there is a positive real ξ such that*

[(1)] For every site j, $P^{[j]}$ is reversible with respect to π. For every spin c, $P_{\rightarrow\leftarrow c}$ is irreducible. For every configuration x and every site j, $P^{[j]}(x, x_j) \geq \xi$. For every configuration x and every colour c, $\Pr(\tau_n = c) \geq \xi$ when τ is drawn from $P_{\rightarrow\leftarrow}(x, \cdot)$.

Then $\tau_x(\mathcal{M}_{\rightarrow\leftarrow}, \varepsilon) \leq O(\xi^{-3} n^3 N \log(n)) \log(1/(\varepsilon \pi(x)))$.

Remark 3. It is easy to see that Theorem 6 implies Theorem 4, though the implicit constants are slightly worse since Condition 4 gives a slightly worse analysis than the analysis given in Section 3.2. The theorem also applies to Glauber-dynamics on spin systems such as the Potts model or the hard-core lattice gas model.

Remark 4. The connectivity requirement in Theorem 6 can be removed by considering the connected components separately as in the proof of Theorem 3.

5 Proving Rapid Mixing for Systematic Scan Using Path Coupling

All of the mixing results so far have built upon Dobrushin uniqueness. We conclude by sketching an alternate proof, based on path coupling [2], that the Dobrushin condition $\alpha < 1$ implies rapid mixing of systematic scan. Similar ideas are implicit in [20], though this does not explicitly consider systematic scan.

We consider coupled chains X_t, Y_t. Let the (path) coupling be given by choosing the same vertex w_t in both chains, and then coupling the choice of spin maximally. Suppose the initial states X_0, Y_0 have shortest path P_0. The length of P_0 is the Hamming distance $H(X_0, Y_0)$. Consider the evolution of this path at time t to $P = (Z_0, Z_1, \ldots, Z_{\ell-1}, Z_\ell)$, with length $\ell \geq H(X_t, Y_t)$. (Note that we do not optimise the path length after each step, but assume instead that the path evolves naturally.) We will call any edge of P (Z_{r-1}, Z_r) $(r \in [\ell])$ an edge in S_i if $(Z_{r-1}, Z_r) \in S_i$. Note that P_0 has at most one edge in S_i for each $i = 1, 2, \ldots, n$. Suppose ν_i is the total number of edges of P in S_i. Clearly $\ell = \sum_{i=1}^{n} \nu_i$, so $\mathrm{E}[\ell] \leq n \max_i \mathrm{E}[\nu_i]$.

Suppose $w_t = j$. For every edge in S_i $(i \neq j)$, an edge in S_i will persist, and a new edge in S_j will appear with probability at most $\rho_{i,j}$. Every edge in S_j will either disappear, or persist with probability at most $\rho_{j,j}$. Thus, denoting the quantities at time $t + 1$ with primes, $\mathrm{E}[\nu_i'] = \mathrm{E}[\nu_i]$ $(i \neq j)$ and $\mathrm{E}[\nu_j'] \leq \sum_{i=1}^{n} \rho_{i,j} \mathrm{E}[\nu_i] \leq \alpha \max_i \mathrm{E}[\nu_i]$. In a complete scan we have $w_t = j$ for every j and some t. Hence the resulting values $\overrightarrow{\nu}_i$ after the scan will satisfy $\max_i \mathrm{E}[\overrightarrow{\nu}_i] \leq \alpha \max_i \mathrm{E}[\nu_i]$. Since $\max_i \mathrm{E}[\nu_i] \leq 1$ initially, after s complete scans $\max_i \mathrm{E}[\nu_i] \leq \alpha^s$, and thus $\mathrm{d}_{\mathrm{TV}}(X_{ns}, Y_{ns}) \leq \Pr(X_{ns} \neq Y_{ns}) \leq \mathrm{E}[H(X_{ns}, Y_{ns})] \leq \mathrm{E}[\ell] \leq n\alpha^s$.

Acknowledgements

We are very grateful to Alan Sokal for pointing out to us that Dobrushin's condition implies mixing of systematic scan, and for referring us to [18], and to Christian Borgs for bringing Föllmer's work to our attention.

References

1. D. Aldous, Some inequalities for reversible Markov chains, *J. London Math Society (2)* **25** (1982), pp. 564–576.
2. R. Bubley and M. Dyer, Path coupling: a technique for proving rapid mixing in Markov chains, FOCS 38, 1997, 223–231.
3. G.E. Cho and C.D. Meyer, Markov chain sensitivity measured by mean first passage times, *Linear Algebra and its Applications* vol 316 number 1–3, pages 21–28, 2000.
4. G.E. Cho and C.D. Meyer, Comparison of perturbation bounds for the stationary distribution of a Markov chain, *Linear Algebra and its Applications* vol 335 number 1–3, pages 137–150, 2001.
5. P. Diaconis and L. Saloff-Coste, Comparison theorems for reversible Markov chains, *Annals of Applied Probability* **3** (1993), pp. 696–730.

6. P. Diaconis and D. Stroock, Geometric bounds for eigenvalues of Markov chains, *Annals of Applied Probability* **1** (1991), pp. 36–61.
7. R.L. Dobrushin, Prescribing a system of random variables by conditional distributions, Theory Prob. and its Appl. 15 (1970) 458–486.
8. R.L. Dobrushin and S.B. Shlosman, Constructive criterion for the uniqueness of a Biggs field, in J. Fritz, A. Jaffe, D. Szasz, Statistical mechanics and dynamical systems, Birkhauser, Boston (1985) 347–370.
9. M. Dyer, L.A. Goldberg and M. Jerrum, Systematic scan for sampling colourings, To appear, Annals of Applied Probability.
10. M. Dyer, L.A. Goldberg, M. Jerrum and R. Martin, Markov chain comparison, Pre-print. 2004.
11. M. Dyer and C. Greenhill. Random walks on combinatorial objects. In J.D. Lamb and D.A. Preece, editors, *Surveys in Combinatorics*, volume 267 of *London Mathematical Society Lecture Note Series*, pages 101–136. Cambridge University Press, 1999.
12. H. Föllmer, A covariance estimate for Gibbs measures, J. Funct. Analys. 46 (1982) 387–395.
13. L. Lovśz and P. Winkler, Mixing of random walks and other diffusions on a graph, Surveys in Combinatorics, 1995, P. Rowlinson (editor), pp. 119-154, London Math. Soc. Lecture Note Series 218.
14. R. Martin and D. Randall, Sampling adsorbing staircase walks using a new Markov chain decomposition method, Proc. of the 41st Annual IEEE Symposium on Foundations of Computer Science (FOCS 2000), pp. 492–502.
15. K. Pedersen, personal communication.
16. J. Salas and A.D. Sokal, Absence of phase transition for antiferromagnetic Potts models via the Dobrushin uniqueness theorem, *J. Statistical Physics* **86** (1997), pp. 551–579.
17. A. Sinclair, Improved bounds for mixing rates of Markov chains and multicommodity flow, *Combinatorics, Probability and Computing* **1** (1992), pp. 351–370.
18. B. Simon, The Statistical Mechanics of Lattice Gases, Princeton University Press 1993
19. A. Sokal, A personal list of unsolved problems concerning lattice gases and antiferromagnetic potts models, to appear in *Markov Processes and Related Fields*.
20. D. Weitz, Combinatorial Criteria for Uniqueness of Gibbs Measurs, Random Structures and Algorithms, to appear, 2005.

Complete Convergence of Message Passing Algorithms for Some Satisfiability Problems

Uriel Feige[1], Elchanan Mossel[2], and Dan Vilenchik[3]

[1] Micorosoft Research and The Weizmann Institute
urifeige@microsoft.com
[2] U.C. Berkeley
mossel@stat.berkeley.edu
[3] Tel Aviv University
vilenchi@post.tau.ac.il

Abstract. Experimental results show that certain message passing algorithms, namely, *survey propagation*, are very effective in finding satisfying assignments in random satisfiable 3CNF formulas. In this paper we make a modest step towards providing rigorous analysis that proves the effectiveness of message passing algorithms for random 3SAT. We analyze the performance of Warning Propagation, a popular message passing algorithm that is simpler than survey propagation. We show that for 3CNF formulas generated under the planted assignment distribution, running warning propagation in the standard way works when the clause-to-variable ratio is a sufficiently large constant. We are not aware of previous rigorous analysis of message passing algorithms for satisfiability instances, though such analysis performed for decoding of Low Density Parity Check (LDPC) Codes. We discuss some of the differences between results for the LDPC setting and our results.

1 Introduction and Results

The effectiveness of some message passing algorithms, in particular Survey Propagation, was experimentally shown for "hard" formulas with clause-variable ratio below (yet rather close to) the conjectured satisfiability threshold, ~ 4.2 [3]. In this paper we analyze the performance of Warning Propagation (WP for brevity), a simple popular message passing algorithm, when applied to random satisfiable formulas generated under the planted distribution with a constant clause-variable ratio. We show that the standard way of running message passing algorithms – run message passing until convergence, simplify the formula according to the resulting assignment, and satisfy the remaining subformula (if nonempty), if possible, using a simple "off the shelf" heuristic – works for planted random satisfiable formulas with a sufficiently large yet constant clause-variable ratio. We are not aware of previous rigorous analysis of message passing algorithms for non-trivial SAT distributions.

1.1 Different SAT Distributions

A CNF formula over the variables $x_1, x_2, ..., x_n$ is a conjunction of clauses C_1, $C_2, ..., C_m$ where each clause is a disjunction of one or more literals. Each literal

J. Diaz et al. (Eds.): APPROX and RANDOM 2006, LNCS 4110, pp. 339–350, 2006.

is either a variable or its negation. A formula is said to be in k-CNF form if every clause contains exactly k literals. A CNF formula is satisfiable if there is a boolean assignment to the variables s.t. every clause contains at least one literal which evaluates to true. 3SAT is the language of all satisfiable 3CNF formulas. Although 2SAT is known to be in P, 3SAT is one of the most famous NP-complete problems. In [12] it is proved that it is NP-hard to approximate MAX-3SAT (the problem of finding an assignment that satisfies as many clauses as possible) within a ratio better than 7/8. Given the difficulty of designing algorithms that work well in the worst case, we consider the average case performance of algorithms. One possibility for rigorously modeling average-case instances is to use random models.

Algorithmic theory of random structures has been the focus of extensive research in recent years (see [10] for a survey). As part of this trend, uniformly random 3CNFs (generated by selecting at random $m = m(n)$ clauses over the variables $\{x_1, ..., x_n\}$) have attracted much attention. Random 3SAT is known to have a sharp satisfiability threshold in the clause-to-variable ratio [9]. Namely, a random 3CNF with clause-to-variable ratio below the threshold is satisfiable **whp** (with high probability, meaning with probability tending to 1 as n goes to infinity) and one with ratio above the threshold is unsatisfiable **whp**. This threshold is not known exactly (and not even known to be independent of n). The threshold is known to be at least 3.52 [14] and at most 4.506 [5].

In this work we mainly consider formulas with large clause-variable ratio. At such ratios almost all 3CNF formulas are not satisfiable, therefore more refined definitions are due. We consider three distributions on 3SAT instances. The first, analogous to the well known random graph model $G_{n,p}$, is the distribution in which every clause, out of $2^3 \binom{n}{3}$ possible clauses, is included with probability $p = p(n)$. We denote this distribution by $\mathcal{P}_{n,p}$. The second distribution is obtained from $\mathcal{P}_{n,p}$ by conditioning on satisfiability, namely $\mathcal{P}_{n,p}^{\text{sat}}[F] = \mathcal{P}_{n,p}[F|S]$ where S is the event that F is satisfiable. Lastly, we consider the planted distribution, $\mathcal{P}_{n,p}^{\text{plant}}$, which is obtained form $\mathcal{P}_{n,p}$ by conditioning on satisfiability by a specific "planted" assignment φ. Equivalently, in $\mathcal{P}_{n,p}^{\text{plant}}$, first an assignment φ to the variables is picked uniformly at random. Then, every clause satisfied by φ is included with probability $p = p(n)$. Throughout, we use φ to denote the planted assignment when the relevant instance is clear from context.

In the context of satisfiability algorithms, $\mathcal{P}_{n,p}^{\text{sat}}$ is arguably the most interesting and natural distribution to study. However, as pointed out frequently, $\mathcal{P}_{n,p}^{\text{sat}}$ seems hard to tackle rigorously and experimentally. The planted 3SAT distribution is an intermediate step towards analyzing $\mathcal{P}_{n,p}^{\text{sat}}$, and is an interesting, quite natural distribution on its own right, the analog the of planted clique, planted bisection, planted coloring, and planted bipartite hypergraphs studied e.g. in [2, 13, 6] The planted 3SAT distribution is also discussed e.g. in [8, 7]. Our main result (Theorem 2) relates to the planted 3SAT model, but some of our other results (such as Proposition 1 and Corollary 1) are relevant to the $\mathcal{P}_{n,p}^{\text{sat}}$ setting as well.

1.2   3SAT and Factor Graphs

Let $\mathcal{F}$ be a 3CNF formula on n variables and m clauses. The *factor graph* (e.g. [16]) of $\mathcal{F}$, denoted by $FG(\mathcal{F})$, is the following graph representation of $\mathcal{F}$. The factor graph is a bipartite multigraph, $FG(\mathcal{F}) = (V_1 \cup V_2, E)$ where $V_1 = \{x_1, x_2, ..., x_n\}$ (the set of variables) and $V_2 = \{C_1, C_2, ..., C_m\}$ (the set of clauses). $(x_i, C_j) \in E$ iff x_i appears in C_j. For a 3CNF $\mathcal{F}$ with m clauses it holds that $\#E = 3m$ (as every clause contains exactly 3 variables). To make presentation clearer, we denote by $\#A$ the size of a set A and by $|a|$ the absolute value of a real number a. For simplicity we assume that every clause contains three distinct variables, therefore FG is a graph (no parallel edges).

1.3   Warning Propagation

Warning Propagation (WP) is a simple iterative message passing algorithm, and serves as an excellent intuitive introduction to more involved message passing algorithms such as Belief Propagation [19] and Survey Propagation [3]. These algorithms are based on the *cavity method* in which the messages that a clause (or a variable) receives are meant to reflect a situation in which a "cavity" is formed, namely, the receiving clause (or variable) is no longer part of the formula. Messages in the WP algorithm can be interpreted as "warnings", telling a clause the values that variables will have if the clause "keeps quiet" and does not announce its wishes, and telling a variable which clauses will not be satisfied if the variable does not commit to satisfying them. We now present the algorithm in a formal way.

Let $\mathcal{F}$ be a CNF formula. For a variable x, let $N^+(x)$ be the set of clauses in $\mathcal{F}$ in which x appears positively (namely, as the literal x), and $N^-(x)$ be the set of clauses in which x appears negatively. For a clause C, let $N^+(C)$ be the set of positively appearing variables and respectively $N^-(C)$ the set of negatively appearing ones. There are two types of messages involved in the WP algorithm. Messages sent from a **variable** x_i to a **clause** C_j in which it appears:

$$x_i \rightarrow C_j = \sum_{C_k \in N^+(x_i), k \neq j} C_k \rightarrow x_i - \sum_{C_k \in N^-(x_i), k \neq j} C_k \rightarrow x_i$$

If x_i appears only in C_j then we set the message to 0. The intuitive interpretation of this message should be x_i signals C_j what is currently its favorable assignment by the other clauses it appears in (a positive message means TRUE, negative one means FALSE and a 0 message means undecided). The second type are messages sent from a **clause** C_j to a **variable** x_i appearing in C_j:

$$C_j \rightarrow x_i = \prod_{x_k \in N^+(C_j), k \neq i} I_{<0}(x_k \rightarrow C_j) \prod_{x_k \in N^-(C_j), k \neq i} I_{>0}(x_k \rightarrow C_j)$$

where $I_{<0}(b)$ equals 1 if $b < 0$ and 0 otherwise (and symmetrically $I_{>0}(b)$ for the case $b > 0$). If C_j contains only x_i (which cannot be the case in 3CNF formulas) then the message is set to 1. $C_j \rightarrow x_i = 1$ can be intuitively interpreted as C_j

sending a *warning* to x_i asking it to satisfy C_j (as all other literals signaled C_j that currently they evaluate to FALSE). Lastly, we define the current assignment of a variable x_i to be

$$B_i = \sum_{C_j \in N^+(x_i)} C_j \to x_i - \sum_{C_j \in N^-(x_i)} C_j \to x_i$$

If $B_i > 0$ then x is assigned TRUE, if $B_i < 0$ then x_i is assigned FALSE, otherwise x_i is UNASSIGNED. Assume some order on the clause-variable messages (e.g. the lexicographical order on pairs of the form (j, i) representing the message $C_j \to x_i$). Given a vector $\alpha \in \{0, 1\}^{3m}$ in which every entry is the value of the corresponding $C_j \to x_i$ message, a partial assignment $\psi \in \{TRUE, FALSE, UNASSIGNED\}^n$ can be generated according to the corresponding B_i values (as previously explained).

It would be convenient to think of the messages in terms of the corresponding factor graph. Every undirected edge (x_i, C_j) of the factor graph is replaced with two anti-parallel directed edges, $(x_i \to C_j)$ associated with the message $x_i \to C_j$ and respectively the edge $(C_j \to x_i)$.

```
Warning Propagation(CNF formula F) :
1. construct the corresponding factor graph FG(F).
2. randomly initialize the clause-variable messages to 0 or 1.
3. repeat until no clause-variable message changed from the
   previous iteration:
   3.a randomly order the edges of FG(F).
   3.b update all clause-variable messages Cj → xi according
       to the random edge order.
4. compute a partial assignment ψ according to the Bi messages.
5. return  ψ.
```

Note that in line 3.b. above, when evaluating the clause-variable message along the edge $C \to x$, $C = (x \vee y \vee z)$, the variable-clause messages concerning this calculation $(z, y \to C)$ are evaluated on-the-fly using the last updated values $C_i \to y$, $C_j \to z$ (allowing feedback from the same iteration). We allow the algorithm not to terminate (the clause-variable messages may keep changing every iteration). If the algorithm does return an assignment ψ then we say that it converged. In practice it is common to limit in advance the number of iterations, and if the algorithm didn't converge by then, return a failure.

1.4   Related Work and Techniques

The Survey Propagation algorithm [3] experimentally outperforms all known algorithms in finding satisfying assignments to $\mathcal{P}_{n,p}$ formulas with clause-variable ratio ρ close to the satisfiability threshold ($4 \leq \rho \leq 4.25$). However, theoretical understanding of Survey Propagation and other message passing algorithm for random SAT problems is still lacking. This should be compared with the success of message passing algorithms for decoding low-density-parity-check (LDPC) codes [11]. Here, the experimental success of message passing algorithms [11] was recently complemented rigourously by a large body of theoretical work, see e.g. [17, 20, 18]. Some important insights emerge from this theoretical work. In particular, it is shown that the quality of decoding improves exponentially with the number of iterations – thus all but a small constant fraction of the received codeword can be decoded correctly using a constant number of iterations. Our analysis of WP shows that much of the coding picture is valid also for $\mathcal{P}_{n,p}^{\text{plant}}$ thus providing important insights as to the success of message passing algorithms for random satisfiability problems. The planted 3SAT model is similar to LDPC in many ways. Both constructions are based on random factor graphs. In codes, the received corrupted codeword provides noisy information on a single bit or on the parity of a small number of bits of the original codeword. In $\mathcal{P}_{n,p}^{\text{plant}}$, φ being the planted assignment, the clauses containing a variable x_i contain noisy information on the polarity of $\varphi(x_i)$ in the following sense – each clause contains x_i in a polarity coinciding with $\varphi(x_i)$ with probability 4/7. Our results are similar in flavor to the coding results. However, the combinatorial analysis provided here allows to recover an assignment satisfying *all* clauses, whereas in the random LDPC codes setting, message passing allows to recover only $1 - o(1)$ fraction of the codeword correctly. In [18] it is shown that for the erasure channel, all bits may be recovered correctly using a message passing algorithm, however in this case the LDPC code is designed so that message passing works for it. We on the other hand take a well known SAT distribution and analyze the performance of a message passing algorithm on it. Moreover, the SAT setting is more involved, as there are many assignments satisfying the formula, while for the erasure channel there is a unique codeword satisfying the combinatorial constraints given by the message.

As for relevant results in random graph theory, the seminal work of [2] paved the road towards dealing with large-constant-degree planted distributions. [2] present an algorithm that **whp** k-colors planted k-colorable graphs with a sufficiently large constant expected degree. Building upon the techniques introduced by [2], [13] present an algorithm that 2-colors sparse planted 3-uniform bipartite hypergraphs and [8], solving an open question posed in [15], presents an algorithm for satisfying large constant degree planted 3SAT instances. Though in our analysis we use similar techniques to the aforementioned works, our result is conceptually different in the following sense. In [2, 13, 8] the starting point is the planted distribution, and then one designs an algorithm that works well under this distribution. The algorithm may be designed in such a way that makes its analysis easier. In contrast, our starting point is a given message passing algorithm (WP), and then we ask for which input distributions it works well. We cannot change the algorithm in ways that would simplify the analysis.

Another difference between our work and that of [2, 13, 8] is that unlike the algorithms analyzed in those other papers, WP is a randomized algorithm which makes its analysis more difficult. We could have simplified our analysis had we changed WP to be deterministic (for example, by initializing all clause-variable messages to 1 in step 2 of the algorithm), but there are good reasons why WP is randomized. For example, it can be shown that (the randomized version) WP converges with probability 1 on 2CNF formulas that form one cycle of implications, but might not converge if step 4 does not introduce fresh randomness in every iteration of the algorithm (details omitted).

1.5   Notation

Given a 3CNF $\mathcal{F}$, **simplify** $\mathcal{F}$ according to ψ, when ψ is a partial assignment, means: in every clause substitute every assigned variable with the value given to it by ψ. Then remove all clauses containing literals which evaluate to true. In all remaining clauses, remove all literals which evaluate to false (the resulting instance is not necessarily in 3CNF form). Denote by $\mathcal{F}|_\psi$ the 3CNF $\mathcal{F}$ simplified according to ψ. For a set of variables $A \subseteq V$, denote by $\mathcal{F}[A]$ the set of clauses in which all variables belong to A.

Given a 3CNF formula $\mathcal{F}$, we say that a variable x is **pure** in $\mathcal{F}$ if it appears only in one polarity (namely, always appears as the literal x or always as the literal $\bar{x}$). Let P_0 be the set of pure variables in $\mathcal{F}$, and C_0 be the set of clauses containing a pure variable. Let $L_0 = \mathcal{F}$, and $L_1 = L_0 \setminus C_0$. Let P_1 be the pure variables in L_1, namely the variables that become pure after setting the pure variables in a satisfying manner and simplifying $\mathcal{F}$. Similarly, define C_1 to be the set of clauses in L_1 containing a variable from P_1. Generally, define $L_i = L_{i-1} \setminus C_{i-1}$, P_i to be the pure variables in L_i, and C_i to be the clauses in L_i containing a variable from P_i. We say that a 3CNF $\mathcal{F}$ is r-pure if $L_r = \emptyset$. The following theorem is implicitly proved in [4].

Theorem 1. *Let $\mathcal{F}$ be randomly sampled according to $\mathcal{P}_{n,p}$, $p = d/n^2$, $d <$ 1.225, then **whp** $\mathcal{F}$ is $O(n)$-pure.*

Note that if there exists an r s.t. $\mathcal{F}$ is r-pure then in particular $\mathcal{F}$ is satisfiable. To better understand our results it would be convenient to have the somewhat informal notion of a *simple formula* in mind. We call a CNF formula simple, if it can be satisfied using simple well-known heuristics (examples include formulas whose factor graph is tree-like and r-pure formulas – both solvable using the pure-literal heuristic [4], formulas with small weight terminators – to use the terminology of [1] – efficiently solvable **whp** using a RWalkSat, etc).

1.6   Our Results

Theorem 2. *Let $\mathcal{F}$ be a 3CNF formula randomly sampled according to $\mathcal{P}_{n,p}^{\text{plant}}$, where $p \geq d/n^2$, d a sufficiently large constant, and let φ be its planted assignment. Then the following holds with probability $1 - e^{-\Theta(d)}$ (the probability taken over the choice of $\mathcal{F}$, the random choices in line 2 of the WP algorithm, and the random order in the first time line 4 executes):*

1. *WP$(\mathcal{F})$ converges after at most $O(\log n)$ iterations.*
2. *Let ψ be the partial assignment returned by WP$(\mathcal{F})$, let V_A denote the variables assigned to either TRUE or FALSE in ψ, and V_U the variables left*

UNASSIGNED. Then for every variable $x \in V_A$, $\psi(x) = \varphi(x)$. Moreover, $\#V_A \geq (1 - e^{-\Theta(d)})n$.

3. $\mathcal{F}|_\psi[V_U]$ is a simple formula which can be satisfied in time $O(n)$.

Remark 1. We also have a proof of Theorem 2 with '$1 - o(1)$' instead of '$1 - e^{-\Theta(d)}$'. This however involves a somewhat more complicated analysis exceeding the scope of this abstract (further discussion in Section 3). For the full details the reader is referred to the journal version.

Proposition 1. *Let $\mathcal{F}$ be a 3CNF formula randomly sampled according to $\mathcal{P}_{n,p}^{\text{plant}}$, where $p \geq c \log n / n^2$, with c a sufficiently large constant, and let φ be its planted assignment. Then **whp** after at most 2 iterations WP($\mathcal{F}$) converges, and the returned ψ equals φ. (This result can be extended to $\mathcal{P}_{n,p}^{\text{sat}}$, see below.)*

Proposition 2. *Let $\mathcal{F}$ be an r-pure CNF formula. Then after at most $O(r)$ iterations of WP($\mathcal{F}$), regardless of the initial messages and the order of execution, the following holds:*

1. *WP($\mathcal{F}$) converges.*
2. *Let ψ be the assignment returned by WP($\mathcal{F}$). If $\psi(x) \neq UNASSIGNED$, then in every satisfying assignment x is assigned according to $\psi(x)$.*
3. *If $\mathcal{F}$ contains no unit clauses then ψ is the all-UNASSIGNED vector.*

Corollary 1. *In the setting of Theorem 1, **whp** WP($\mathcal{F}$) converges after at most $O(n)$ iterations and the returned ψ is the all-UNASSIGNED vector.*

The corollary follows immediately from Theorem 1 and Proposition 2.

Proposition 3. *Let $\mathcal{F}$ be a satisfiable CNF formula whose corresponding factor graph contains no cycles. Then $\mathcal{F}$ is $O(n)$-pure.*

The main idea behind the proof of Theorem 2 is to show that the formula is dense enough so that **whp** there exists a large subformula forcing WP to point in the correct direction. The rest of the formula induces a factor graph containing only trees, which are also "easy" for WP. We note that formulas in $\mathcal{P}_{n,p}^{\text{plant}}$, with $n^2 p$ some large constant, are not known to be simple (in the sense that we defined above). On the contrary, "hardness" evidence can be found in works such as [1], showing that RWalkSat is very unlikely to hit a satisfying assignment in polynomial time when running on a random $\mathcal{P}_{n,p}^{\text{plant}}$ instance in the setting of Theorem 2. In the setting of Proposition 1, the formula is already dense enough so that **whp** it forces entirely WP to point to the planted assignment.

Proposition 2 combined with Proposition 3 provide a proof to the convergence of WP on trees. Our proof of this known result gives an explicit characterization of the fixed point to which WP converges (which is implicit for trees in [3]).

The remainder of the paper is structured as follows. In Section 2 we discuss some properties that a typical instance in $\mathcal{P}_{n,p}^{\text{plant}}$ possesses, and outline the proof of Theorem 2 and Proposition 1. In Section 3 we discuss a stronger version of Theorem 2. Most details of the proofs are omitted and can be found in the journal version.

2   Properties of a Random $\mathcal{P}_{n,p}^{\text{plant}}$ Instance

In this section we discuss relevant properties of a random $\mathcal{P}_{n,p}^{\text{plant}}$ instance. To simplify presentation, we assume w.l.o.g. (due to symmetry) that the planted assignment φ is the all-one vector.

2.1   Stable Variables

Definition 1. *A variable x **supports** a clause C with respect to a partial assignment ψ, if it is the only variable to satisfy C under ψ, and the other two variables are assigned by ψ.*

Proposition 4. *Let $\mathcal{F}$ be as in the setting of Theorem 2 and let F_{SUPP} be a random variable counting the number of variables in $\mathcal{F}$ whose support w.r.t. φ is less than $d/3$. Then, $E[F_{SUPP}] \leq e^{-\Theta(d)}n$.*

This follows from concentration arguments as every variable is expected to support $\frac{d}{n^2} \cdot \binom{n}{2} = \frac{d}{2} + O(\frac{1}{n})$ clauses.

Following the definitions in Section 1.3, given a CNF $\mathcal{F}$ and a variable x, we let $N^{++}(x)$ be the set of clauses in $\mathcal{F}$ in which x appears positively but doesn't support w.r.t. φ. Let $N^s(x)$ be the set of clause in $\mathcal{F}$ which x supports w.r.t. φ. Let $\pi = \pi(\mathcal{F})$ be some ordering of the clause-variable message edges in the factor graph of $\mathcal{F}$. For an index i and a literal ℓ_x (by ℓ_x we denote a literal over the variable x) let $\pi^{-i}(\ell_x)$ be the set of clause-variable edges $(C \to x)$ that appear before index i in the order π and in which x appears in C as ℓ_x. For a set of clause-variable edges $\mathcal{E}$ and a set of clauses $\mathcal{C}$ we denote by $\mathcal{E} \cap \mathcal{C}$ the subset of edges containing a clause from $\mathcal{C}$ as one endpoint.

Definition 2. *A variable x is **stable** in $\mathcal{F}$ w.r.t. an edge order π if the following holds for every clause-variable edge $C \to x$ (w.l.o.g. assume $C = (\ell_x \vee \ell_y \vee \ell_z)$, $C \to x$ is the i'th message in π):*

1. *$|\#\pi^{-i}(y) \cap N^{++}(y) - \#\pi^{-i}(\bar{y}) \cap N^-(y)| \leq d/30$.*
2. *$|\#N^{++}(y) - \#N^-(y)| \leq d/30$.*
3. *$\#N^s(y) \geq d/3$*

and the same holds for z.

Proposition 5. *Let $\mathcal{F}$ be as in the setting of Theorem 2, and let π be a random ordering of the clause-variable messages. Let F_{UNSTAB} be a random variable counting the number of variables in $\mathcal{F}$ which are not stable. Then, $E[F_{UNSTAB}] \leq e^{-\Theta(d)}n$.*

This follows from concentration arguments since $E[\#\pi^{-i}(y) \cap N^{++}(y) - \#\pi^{-i}(\bar{y}) \cap N^-(y)] = 0$, $E[\#N^{++}(y) - \#N^-(y)] = 0$, and since every variable is expected to appear in at most $O(d)$ clauses.

Let $\alpha \in \{0,1\}^{3\#\mathcal{F}}$ be a clause-variable message vector. For a set of clause-variable message edges $\mathcal{E}$ let $\mathbf{1}_\alpha(\mathcal{E})$ be the set of edges along which the value is 1 according to α. For a set of clauses $\mathcal{C}$, $\mathbf{1}_\alpha(\mathcal{C})$ denotes the set of clause-variable message edges in the factor graph of $\mathcal{F}$ containing a clause from $\mathcal{C}$ as one endpoint and along which the value is 1 in α.

Definition 3. *A variable x is **violated** by α in π if there exists a message $C \to x$, $C = (\ell_x \vee \ell_y \vee \ell_z)$, in place i in π s.t. one of the following holds:*

1. $|\#\mathbf{1}_\alpha(\pi^{-i}(y) \cap N^{++}(y)) - \#\mathbf{1}_\alpha(\pi^{-i}(\bar{y}) \cap N^-(y))| > d/30$
2. $|\#\mathbf{1}_\alpha(N^{++}(y)) - \#\mathbf{1}_\alpha(N^-(y))| > d/30$
3. $\#\mathbf{1}_\alpha(N^s(y)) < d/7$.

Or one of the above holds for z.

Proposition 6. *Let $\mathcal{F}$ be as in the setting of Theorem 2, and let X be a set of stable variables w.r.t. an arbitrary ordering π. Let α be a random clause-variable message vector. Let F_{VIO} be a random variable counting the number of violated variables in X. Then, $E[F_{VIO}] \leq e^{-\Theta(d)}\#X$.*

The proof again uses concentration arguments.

2.2 Dense Subformulas

The next property we discuss is analogous to a property proved in [2] for random graphs. Loosely speaking, [2] prove that **whp** a random graph doesn't contain a small induced subgraph with a large average degree. Using first moment calculations we show:

Proposition 7. *Let $c > 1$ be an arbitrary constant. Let $p \geq d/n^2$, $d = d(c)$ a sufficiently large constant. Then **whp** over $\mathcal{F} \in \mathcal{P}_{n,p}^{\text{plant}}$ there exists no subset of variables U, s.t. $\#U \leq e^{-\Theta(d)}n$ and there are at least $c\#U$ clauses in $\mathcal{F}$ containing two variables from U.*

2.3 The Core Variables

We describe a subset of the variables, denoted throughout by $\mathcal{H}$ and referred to as the *core variables*, which plays a crucial role in the analysis. Loosely speaking, a variable is considered "safe" if it is stable w.r.t. the initial random order π, and it is not violated by the initial clause-variable message assignments α. If in addition, a safe variable x_i supports many clauses w.r.t. φ (whose corresponding message is '1' in α), then its corresponding B_i value will agree with $\varphi(x_i)$ after the first iteration. This invariant needs to be preserved however in later iterations. The set $\mathcal{H}$ captures the notion of such variables with a self-preserving quality. There are several ways to obtain these desired properties. Formally, $\mathcal{H} = \mathcal{H}(\mathcal{F}, \varphi, \alpha, \pi)$ is constructed using the following iterative procedure:

```
Let A1 be the set of variables whose support w.r.t φ is at most d/3.
Let A2 be the set of non-stable variables w.r.t. π.
Let A3 be the set of stable variables w.r.t. π violated by α.
```
1. set $H_0 = V \setminus (A_1 \cup A_2 \cup A_3)$.
2. while $\exists a_i \in H_i$ supporting less than $d/4$ clauses in $\mathcal{F}[H_i]$ OR appearing in $d/30$ or more clauses not in $\mathcal{F}[H_i]$: let $H_{i+1} \leftarrow H_i \setminus \{a_i\}$.
3. define $\mathcal{H} = H_{m+1}$ where a_m is the last variable removed in step 2.

Proposition 8. *If both α and π are chosen uniformly at random then* **whp** $\#\mathcal{H} \geq (1 - e^{-\Theta(d)})n$.

The main idea of the proof is to observe that to begin with we eliminate very few variables (using the discussion in Section 2.1 to bound $\#A_1 \cup A_2 \cup A_3$). If too many variables were removed in the iterative step then a small but dense subformula exists. Proposition 7 bounds the probability of the latter occurring.

2.4   The Factor Graph of the Non-core Variables

Proposition 8 implies that for $p = c \log n/n^2$, c a sufficiently large constant, **whp** $\mathcal{H}$ contains already all variables. The following analysis is needed for the setting of Theorem 2. The *non-core* factor graph is the factor graph of the formula $\mathcal{F}$ simplified according to the partial assignment that assigns all core variables their planted assignment.

Proposition 9. **Whp** *every connected component in the non-core factor graph contains $O(\log n)$ variables.*

Proposition 9 will not suffice to prove Theorem 2, and we need a further characterization of the non-core factor graph.

Proposition 10. *With probability $1 - e^{-\Theta(d)}$, there exists no cycle in the non-core factor graph.*

2.5   Outline of Proof of Theorem 2 and Proposition 1

We start with Theorem 2 and derive Proposition 1 as an easy corollary of the analysis. The outline of the proof is as follows. We assume that the formula $\mathcal{F}$ and the run of WP are *typical* in the sense that Propositions 8, 9 and 10 hold. First we prove that after one iteration WP sets the core variables $\mathcal{H}$ correctly (B_i agrees with φ in sign) and this assignment does not change in later iterations. Therefore from iteration 2 and onwards WP is basically running on $\mathcal{F}$ in which variables belonging to $\mathcal{H}$ are substituted with their planted assignment. This subformula is satisfiable and its factor graph is a forest (namely, composed of disjoint trees). Therefore, convergence is guaranteed. The set V_A of Theorem 2 is composed of all variables from $\mathcal{H}$ and those variables from the forest that get assigned. The set V_U is composed of the UNASSIGNED variables from the forest.

We say that a message $C \to x$, $C = (\ell_x \vee \ell_y \vee \ell_z)$, is *correct* if its value is the same as it is when $y \to C$ and $z \to C$ agree in sign with their planted assignment (in other words, $C \to x$ is 1 iff x supports C w.r.t. φ).

Proposition 11. *If $x_i \in \mathcal{H}$ and all messages $C \to x_i$, $C \in \mathcal{F}[\mathcal{H}]$ are correct at the beginning of an iteration (line 3 in the WP algorithm), then this invariant is kept by the end of that iteration.*

Proposition 12. *If $x_i \in \mathcal{H}$ and all messages $C \to x_i$, $C \in \mathcal{F}[\mathcal{H}]$ are correct by the end of a WP iteration, then B_i agrees in sign with $\varphi(x_i)$ by the end of that iteration.*

Proposition 12 follows immediately from the definition of $\mathcal{H}$ and the message B_i. It remains to show then that after the first iteration all messages $C \to x_i$, $C \in \mathcal{F}[\mathcal{H}]$ are correct.

Proposition 13. *If $\mathcal{F}$ is a typical instance in the setting of Theorem 2, then after one iteration of WP(F), for every variable $x_i \in \mathcal{H}$, every message $C \to x_i$, $C \in \mathcal{F}[\mathcal{H}]$ is correct.*

Proposition 14. *Let $\mathcal{F}$ be a typical instance in the setting of Theorem 2, then for every variable $x_j \in V \setminus \mathcal{H}$, after $O(\log n)$ iterations either $B_j = 0$ or B_j agrees in sign with $\varphi(x_j)$.*

As for satisfying the set of unassigned variables in time $O(n)$, Propositions 3 and 10 imply that the pure-literal procedure [4] solves the subformula induced by the unassigned variables in linear time. Theorem 2 then follows.

To prove Proposition 1, observe that when $p = c \log n / n^2$, with c a sufficiently large constant, Proposition 8 implies $\mathcal{H} = V$. Combing this with Proposition 13, Proposition 1 readily follows.

3   Discussion

Theorem 2 establishes correct convergence of WP with high-*constant* probability. It is desirable to replace this with convergence **whp**. The constant probability in Theorem 2 follows from the fact that only with constant probability the non-core factor graph is a forest (Proposition 10 provides a lower bound on this probability, and one can prove that with constant probability the non-core factor graph indeed contains a cycle). Nevertheless, using similar arguments one can prove that **whp** every connected component in the non-core factor graph contains at most one cycle. Proving convergence of WP then boils down to proving converges on factor graphs with at most one cycle. This is more involved a task than proving convergence on tree-like factor graphs. In fact, convergence of WP on tree-like factor graphs is guaranteed with probability 1. However, the convergence of WP on cycles crucially relies on the random ordering of the messages (line 3.a. in WP). Moreover, the number of iterations to convergence is an unbounded random variable whose expectation is bounded by the square of the component's size, and whose tails decay exponentially. The stronger version of Theorem 2 with a complete proof is available in the journal version of this paper.

Acknowledgements. We thank Eran Ofek for many useful discussions. This work was done while the authors were visiting Microsoft Research, Redmond, Washington. E.M is supported by a Sloan fellowship in Mathematics, by NSF Career award DMS-0548249 and NSF grants DMS-0528488 and DMS-0504245.

References

1. M. Alekhnovich and E. Ben-Sasson. Linear upper bounds for random walk on small density random 3-cnf. In *Proc. 44th IEEE Symp. on Found. of Comp. Science*, pages 352–361, 2003.
2. N. Alon and N. Kahale. A spectral technique for coloring random 3-colorable graphs. *SIAM J. on Comput.*, 26(6):1733–1748, 1997.
3. A. Braunstein, M. Mezard, and R. Zecchina. Survey propagation: an algorithm for satisfiability. *Random Structures and Algorithms*, 27:201–226, 2005.
4. A. Z. Broder, A. M. Frieze, and E. Upfal. On the satisfiability and maximum satisfiability of random 3-cnf formulas. In *Proc. 4th ACM-SIAM Symp. on Discrete Algorithms*, pages 322–330, 1993.
5. O. Dubois, Y. Boufkhad, and J. Mandler. Typical random 3-sat formulae and the satisfiability threshold. In *Proc. 11th ACM-SIAM Symp. on Discrete Algorithms*, pages 126–127, 2000.
6. U. Feige and R. Krauthgamer. Finding and certifying a large hidden clique in a semirandom graph. *Random Structures and Algorithms*, 16(2):195–208, 2000.
7. U. Feige and D. Vilenchik. A local search algorithm for 3SAT. Technical report, The Weizmann Institute of Science, 2004.
8. A. Flaxman. A spectral technique for random satisfiable 3CNF formulas. In *Proc. 14th ACM-SIAM Symp. on Discrete Algorithms*, pages 357–363, 2003.
9. E. Friedgut. Sharp thresholds of graph properties, and the k-sat problem. *J. Amer. Math. Soc.*, 12(4):1017–1054, 1999.
10. A. M. Frieze and C. McDiarmid. Algorithmic theory of random graphs. *Random Structures and Algorithms*, 10(1-2):5–42, 1997.
11. T. G. Gallager. Low-density parity-check codes. *IRE. Trans. Info. Theory*, IT-8:21–28, January 1962.
12. J. Håstad. Some optimal inapproximability results. *J. ACM*, 48(4):798–859, 2001.
13. C. Hui and A. M. Frieze. Coloring bipartite hypergraphs. In *Proceedings of the 5th International Conference on Integer Programming and Combinatorial Optimization*, pages 345–358, 1996.
14. A. C. Kaporis, L. M. Kirousis, and E. G. Lalas. The probabilistic analysis of a greedy satisfiability algorithm. In *Proc. 10th Annual European Symposium on Algorithms*, volume 2461 of *Lecture Notes in Comput. Sci.*, pages 574–585. Springer, Berlin, 2002.
15. E. Koutsoupias and C. H. Papadimitriou. On the greedy algorithm for satisfiability. *Info. Process. Letters*, 43(1):53–55, 1992.
16. F. R. Kschischang, B. J. Frey, and H. A. Loeliger. Factor graphs and the sum-product algorithm. *IEEE Transactions on Information Theory*, 47(2):498–519, 2001.
17. M. Luby, M. Mitzenmacher, M. A. Shokrollahi, and D. Spielman. Analysis of low density parity check codes and improved designs using irregular graphs. In *Proceedings of the 30th ACM Symposium on Theory of Computing*, pages 249–258, 1998.
18. M. Luby, M. Mitzenmacher, M. A. Shokrollahi, and D. Spielman. Efficient erasure correcting codes. *IEEE Trans. Info. Theory*, 47:569–584, February 2001.
19. J. Pearl. *Probabilistic reasoning in intelligent systems: networks of plausible inference*. Morgan Kaufmann Publishers Inc., San Francisco, CA, USA, 1988.
20. T. Richardson, A. Shokrollahi, and R. Urbanke. Design of capacity-approaching irregular low-density parity check codes. *IEEE Trans. Info. Theory*, 47:619–637, February 2001.

Robust Mixing

Murali K. Ganapathy

University of Chicago, Chicago, IL 60637, USA
gmkrishn@cs.uchicago.edu

Abstract. In this paper, we develop a new "robust mixing" framework for reasoning about adversarially modified Markov Chains (AMMC). Let $\mathbb{P}$ be the transition matrix of an irreducible Markov Chain with stationary distribution π. An adversary announces a sequence of stochastic matrices $\{A_t\}_{t>0}$ satisfying $\pi A_t = \pi$. An AMMC process involves an application of $\mathbb{P}$ followed by A_t at time t. The robust mixing time of an irreducible Markov Chain $\mathbb{P}$ is the supremum over all adversarial strategies of the mixing time of the corresponding AMMC process. Applications include estimating the mixing times for certain non-Markovian processes and for reversible liftings of Markov Chains.

Non-Markovian card shuffling processes: The random-to-cyclic transposition process is a *non-Markovian* card shuffling process, which at time t, exchanges the card at position t (mod n) with a random card. Mossel, Peres and Sinclair (2004) showed that the mixing time of this process lies between $(0.0345 + o(1))n \log n$ and $Cn \log n + O(n)$ (with $C \approx 4 \times 10^5$). We reduce the constant C to 1 by showing that the random-to-top transposition chain (*a Markov Chain*) has robust mixing time $\leq n \log n + O(n)$ when the adversarial strategies are limited to those which preserve the symmetry of the underlying Markov Chain.

Reversible liftings: Chen, Lovász and Pak showed that for a reversible ergodic Markov Chain $\mathbb{P}$, any reversible lifting $\mathbb{Q}$ of $\mathbb{P}$ must satisfy $\mathcal{T}(\mathbb{P}) \leq \mathcal{T}(\mathbb{Q}) \log(1/\pi_*)$ where π_* is the minimum stationary probability. Looking at a specific adversarial strategy allows us to show that $\mathcal{T}(\mathbb{Q}) \geq r(\mathbb{P})$ where $r(\mathbb{P})$ is the relaxation time of $\mathbb{P}$. This helps identify cases where reversible liftings cannot improve the mixing time by more than a constant factor.

1 Introduction

In this paper, we develop a "robust mixing" framework which allows us to reason about adversarially modified Markov Chains (AMMC). This framework can be used to bound mixing times of some *non-Markovian processes* in terms of the robust mixing time of related Markov Chains. Another type of application is to estimate mixing times of complex Markov Chains in terms of that of simpler ones. We also use this framework to give an alternate proof of a reversible lifting result due to Chen et al. [4].

J. Diaz et al. (Eds.): APPROX and RANDOM 2006, LNCS 4110, pp. 351–362, 2006.

1.1   Robust Mixing

All stochastic processes considered here are discrete time processes with a finite state space. Markov Chains we consider here are not assumed to be reversible, unless otherwise specified. All logarithms are natural logarithms unless otherwise specified.

Let $\mathbb{P}$ be the transition probability matrix of an irreducible Markov chain on $\mathcal{X}$ and stationary distribution π. By abuse of notation we identify $\mathbb{P}$ with the associated Markov Chain.

Definition 1. *Let $\mathbb{P}$ be an irreducible Markov Chain with stationary distribution π. A stochastic matrix $\mathbb{A}$ (not necessarily irreducible) is said to be* compatible *with $\mathbb{P}$ if $\pi\mathbb{A} = \pi$. Notation: $\pi_* = \min_x \pi(x)$.*

Definition 2. *An* adversarially modified Markov Chain *(AMMC) $\mathcal{P}$ is a pair $(\mathbb{P}, \{\mathbb{A}_t\}_{t>0})$, where $\mathbb{P}$ is an irreducible Markov Chain and $\mathbb{A}_t$ are stochastic matrices compatible with $\mathbb{P}$. Given an AMMC and an initial distribution μ_0, the AMMC process evolves as follows:*

- *At time $t = 0$, pick $X_0 \in \mathcal{X}$ according to μ_0,*
- *Given X_t, pick Y_t according to the distribution $\mathbb{P}(X_t, \cdot)$,*
- *Given Y_t, pick X_{t+1} according to the distribution $\mathbb{A}_t(Y_t, \cdot)$*

An application of $\mathbb{P}$ followed by $\mathbb{A}_t$ is called a *round*. We use μ_t and ν_t to denote the distribution of X_t and Y_t respectively. Note that μ_t is the distribution after t-rounds.

Definition 3. *Let $\mathcal{P}$ be an AMMC. Its* mixing time *and* L_2-mixing time *are defined by the equations*

$$\mathcal{T}(\mathcal{P}, \epsilon) = \max_{\mu_0} \min_t \{\|\mu_t - \pi\|_{\mathrm{TV}} \le \epsilon\} \quad and \quad \mathcal{T}_2(\mathcal{P}, \epsilon) = \max_{\mu_0} \min_t \{\|\mu_t - \pi\|_{2,\pi} \le \epsilon\}$$
(1)

respectively. Here $\|\mu - \pi\|_{\mathrm{TV}} = \sum_x |\mu(x) - \pi(x)|/2$ is the total variation norm and $\|\mu - \pi\|_{2,\pi}^2 = \sum_x (\mu(x) - \pi(x))^2/\pi(x)$ is the $L_2(\pi)$ norm. When ϵ is not specified, we take it to be $1/4$ for $\mathcal{T}$ and $1/2$ for $\mathcal{T}_2$.

The standard mixing times $\mathcal{T}(\mathbb{P}, \epsilon)$ and $\mathcal{T}_2(\mathbb{P}, \epsilon)$ are defined as the mixing times of the AMMC where $\mathbb{A}_t = I$ for all $t > 0$.

Definition 4. *Let $\mathbb{P}$ be an irreducible Markov Chain. An* adversarially modified version *of $\mathbb{P}$ is an AMMC $(\mathbb{P}, \{\mathbb{A}_t\}_{t>0})$.*

Definition 5. *Let $\mathbb{P}$ be an ergodic Markov Chain. The* robust mixing time *and* robust L_2-mixing time *of $\mathbb{P}$ are defined by the equations*

$$R(\mathbb{P}, \epsilon) = \sup_{\mathcal{P}} \mathcal{T}(\mathcal{P}, \epsilon) \quad and \quad R_2(\mathbb{P}, \epsilon) = \sup_{\mathcal{P}} \mathcal{T}_2(\mathcal{P}, \epsilon)$$
(2)

respecitively, where the suprema are taken over adversarially modified versions $\mathcal{P}$ of $\mathbb{P}$. When $\mathbb{P}$ is clear from context, we drop it and when ϵ is not specified we take it to be $1/4$ for R and $1/2$ for R_2.

Since the set of stochastic matrices compatible with $\mathbb{P}$ is a bounded polytope it follows that the worst case for robust mixing time is achieved when each $\mathbb{A}_t$ is a vertex of the polytope.

When we need to distinguish between the standard notion of mixing time and robust mixing time, we refer to the standard notion as "standard mixing time." Note that our adversary is *oblivious*.

1.2   Properties of Robust Mixing Time

Like standard mixing time, robust mixing time is also sub-multiplicative. The proof for standard mixing time can be adapted to the Robust setting.

Theorem 1. *(Submultiplicativity) Let $\mathbb{P}$ be an ergodic Markov Chain. For $\epsilon, \delta > 0$,*

$$R(\mathbb{P}, \epsilon\delta/2) \leq R(\mathbb{P}, \epsilon/2) + R(\mathbb{P}, \delta/2) \quad and \quad R_2(\mathbb{P}, \epsilon\delta) \leq R_2(\mathbb{P}, \epsilon) + R_2(\mathbb{P}, \delta) \quad (3)$$

A useful property enjoyed by Robust mixing time not shared by the standard mixing time is the following convexity property.

Theorem 2. *(Convexity) Let $\mathbb{P}$ be an ergodic Markov Chain with stationary distribution π and $\mathbb{Q}$ any Markov Chain compatible with $\mathbb{P}$. Let $0 < a = 1-b < 1$. Then for $R' \in \{R, R_2\}$, we have $R'(a\mathbb{P} + b\mathbb{Q}) \leq R'(\mathbb{P}) + R'(\mathbb{Q}) - 1$. Moreover,*

- *if $R(\mathbb{P}) \geq 11$, then $R(a\mathbb{P} + b\mathbb{Q}) \leq 3R(\mathbb{P})/a$*
- *If $\pi_* \leq 1/16$ and $R_2(\mathbb{P}) \geq \log(1/\pi_*)/2$ then $R_2(a\mathbb{P} + b\mathbb{Q}) \leq 7R_2(\mathbb{P})/a$*

This result is proved in Section 2. Convex combinations of Markov Chains are considered by [2]. For reversible chains $\mathbb{P}$ and $\mathbb{Q}$, using standard results, one can show that $(\mathbb{P} + \mathbb{Q})/2$ mixes in time $O(\min(\mathcal{T}(\mathbb{P}), \mathcal{T}(\mathbb{Q})) \log(1/\pi_*))$. Our result allows us to eliminate the $\log(1/\pi_*)$ factor under some conditions. See Theorem 8 for one such instance.

1.3   Relation to Classical Parameters of Markov Chains

We now relate the robust mixing time of Markov chains to classical mixing parameters.

Definition 6. *Let $\mathbb{P}$ be an ergodic Markov Chain with stationary distribution π.*

- $\mathbb{S}(\mathbb{P}) = \sqrt{\Pi^{-1}}\mathbb{P}\sqrt{\Pi}$, *where Π is the diagonal matrix with $\Pi(x,x) = \pi(x)$,*
- $\overleftarrow{\mathbb{P}} = \Pi^{-1}\mathbb{P}^{\mathrm{T}}\Pi$ *denotes the reverse of $\mathbb{P}$, where $\mathbb{P}^{\mathrm{T}}$ is the transpose of $\mathbb{P}$.*

Definition 7. *Let $\mathbb{A}$ be any $N \times N$ real matrix. By a singular value decomposition of $\mathbb{A}$, we mean two orthonormal bases $\{x_0, \ldots, x_{N-1}\}$ and $\{y_0, \ldots, y_{N-1}\}$ and scalars $\sigma_0 \geq \sigma_1 \geq \cdots \geq \sigma_{N-1} \geq 0$ which satisfy $x_i\mathbb{A} = \sigma_i y_i$ and $y_i\mathbb{A}^{\mathrm{T}} = \sigma_i x_i$ for all $0 \leq i \leq N - 1$. The σ_i are called the singular values of $\mathbb{P}$.*

See Horn and Johnson [7, Chapter 3] for basic results about singular values. If $\mathbb{A}$ is the transition matrix of a reversible chain or a real symmetric matrix, let $\lambda_0(\mathbb{A}) \geq \lambda_1(\mathbb{A}) \geq \cdots \geq \lambda_{N-1}(\mathbb{A})$ denote its eigenvalues and put $\lambda_*(\mathbb{A}) = \max(\lambda_1(\mathbb{A}), |\lambda_{N-1}(\mathbb{A})|)$.

Definition 8. *Let $\mathbb{P}$ be an ergodic reversible Markov Chain. Its* relaxation time *is defined by $r(\mathbb{P}) = \frac{-1}{\log \lambda_*(\mathbb{P})}$.*

For reversible ergodic Markov Chains $\mathbb{P}$, $r(\mathbb{P}) \leq \mathcal{T}(\mathbb{P}) \leq \mathcal{T}_2(\mathbb{P}) \leq r(\mathbb{P})(\log(1/\pi_*)/2 + 1)$.

Like mixing time for reversible chains, the robust mixing time of a Markov Chain (not necessarily reversible) is determined by the second largest singular value of $\mathbb{S}(\mathbb{P})$ up to a $\log(1/\pi_*)$ factor. More specifically, we have

Theorem 3. *Let $\mathbb{P}$ be an ergodic Markov Chain with stationary distribution π. Then*

$$\max(\mathcal{T}(\mathbb{P}\overleftarrow{\mathbb{P}}), \mathcal{T}(\mathbb{P})) \leq R(\mathbb{P}) \leq 2r(\mathbb{P}\overleftarrow{\mathbb{P}})(\log(1/\pi_*)/2 + 1) \qquad (4)$$

In particular if $\mathbb{P}$ is reversible, we have $r(\mathbb{P}) \leq \mathcal{T}(\mathbb{P}) \leq R(\mathbb{P}) \leq r(\mathbb{P})\left(\frac{\log(1/\pi_*)}{2} + 1\right)$.

Proof. Considering the adversarial strategy $\mathbb{A}_t = I$ and $\mathbb{A}_t = \overleftarrow{\mathbb{P}}$, gives the lower bounds. For the upper bound, observe that for non-constant function f on $\mathcal{X}$, $\|\mathbb{P}f\|_{2,\pi} \leq \sigma_2(\mathbb{S}(\mathbb{P}))\|f\|_{2,\pi}$ and for any $\mathbb{A}$ compatible with $\mathbb{P}$, $\|\mathbb{A}f\|_{2,\pi} \leq \|f\|_{2,\pi}$.

Many techniques used to establish upper bounds on mixing time show that an appropriate "distance" measure between the t-step distribution and π falls by a factor $\gamma < 1$ for each application of $\mathbb{P}$. For many such "distance" measures one can also show that the adversary cannot increase the distance. Hence the upper bound obtained on the standard mixing time holds for the robust mixing time also.

Thus upper bounds on mixing time established using spectral gap estimation, conductance methods, congestion bounds [10], spectral profiling [5] as well as log-Sobolev inequalities for $\mathbb{P}\overleftarrow{\mathbb{P}}$ [8] immediately yield the same upper bounds on Robust mixing time as well. The most notable exceptions are coupling and entropy constant [3].

1.4 Cayley Walks with Restricted Adversaries

We now turn to a much-studied class of walks on groups.

Definition 9. *Let G be a finite group and P a probability distribution over G. By a* Cayley walk *on G induced by P we mean a Markov Chain on G with transition probability matrix $\mathbb{P}$ given by $\mathbb{P}(h, h \cdot s) = P(s)$ for all $h, s \in G$. By a* Cayley walk *on G, we mean a Cayley walk on G induced by P for some probability distribution P over G.*

In case of a *Cayley walk*, one can look at the robust mixing time when the adversary's strategies are restricted to preserving the symmetries of the group. We consider two natural restrictions.

Definition 10. *Let G be a group. A* Cayley strategy *is the transition probability matrix of some Cayley walk (not necessarily irreducible) on G. Denote by $\mathcal{C}$ the set of all Cayley strategies for the group G. A* Cayley adversary *is an adversary whose strategies are limited to Cayley strategies.*

Definition 11. *Let G be a group. A permutation J on G is said to be a* holomorphism *if it is the composition of a right translation and an automorphism of G. Note that holomorphisms are closed under composition and left translations are also holomorphisms. A* holomorphic strategy *is a convex combination of holomorphisms of G. Denote by $\mathcal{H}$ the set of all holomorphic strategies of G (G will be clear from the context). A* holomorphic adversary *is one who is limited to holomorphic strategies.*

We now turn to defining the robust mixing time against restricted adversaries.

Definition 12. *Let $\mathbb{P}$ be an irreducible Markov Chain. A set $\mathcal{S}$ of stochastic matrices is said to be* a valid set of strategies against $\mathbb{P}$ *if all elements of $\mathcal{S}$ are compatible with $\mathbb{P}$, $I \in \mathcal{S}$ and $\mathcal{S}$ is closed under products and convex combinations.*

Definition 13. *Let $\mathbb{P}$ be an irreducible Markov Chain and $\mathcal{S}$ a valid set of strategies against $\mathbb{P}$. The* $\mathcal{S}$-robust mixing time *and* $\mathcal{S}$-robust L_2-mixing time *are defined by the equations*

$$R^{\mathcal{S}}(\mathbb{P}, \epsilon) = \sup_{\mathcal{P}} \mathcal{T}(\mathcal{P}, \epsilon) \quad and \quad R_2^{\mathcal{S}}(\mathbb{P}, \epsilon) = \sup_{\mathcal{P}} \mathcal{T}_2(\mathcal{P}, \epsilon) \tag{5}$$

where $\mathcal{P} = (\mathbb{P}, \{\mathbb{A}_t\}_{t>0})$ ranges over adversarially modified versions of $\mathbb{P}$ where $\mathbb{A}_t \in \mathcal{S}$ for all t.

In case $\mathbb{P}$ is a Cayley walk on a group G, define the holomorphic robust mixing time *and* holomorphic robust L_2-mixing time *by taking $\mathcal{S} = \mathcal{H}$. Similarly taking $\mathcal{S} = \mathcal{C}$ define* Cayley robust mixing time *and* Cayley robust L_2-mixing time.

Theorem 1 as well as Theorem 2 can be extended to work with any valid set of strategies against $\mathbb{P}$. Hence we also have the following

Theorem 4. *(Submultiplicativity for Cayley walks) Let $\mathbb{P}$ be an ergodic random walk on a group G and $\mathbb{Q}$ any random walk on G. All conclusions of Theorem 1 hold when R is replaced by $R^{\mathcal{H}}$ and $R^{\mathcal{C}}$ and R_2 replaced by $R_2^{\mathcal{H}}$ and $R_2^{\mathcal{C}}$.*

Theorem 5. *(Convexity for Cayley walks) Let $\mathbb{P}$ be an ergodic random walk on a group G and $\mathbb{Q}$ any random walk on G. All conclusions of Theorem 2 hold when R is replaced by $R^{\mathcal{H}}$ and $R^{\mathcal{C}}$ and R_2 replaced by $R_2^{\mathcal{H}}$ and $R_2^{\mathcal{C}}$.*

Theorem 3 shows that $R(\mathbb{P})$ and $R_2(\mathbb{P})$ are determined up to a $\log(1/\pi_*)$ factor by the second largest singular value of $\mathbb{S}(\mathbb{P})$. However, it turns out that $R_2^{\mathcal{C}}$ and $R_2^{\mathcal{H}}$ are within a factor of 2 of $\mathcal{T}_2(\mathbb{P}\overleftarrow{\mathbb{P}})$. In fact we have,

Theorem 6. *Let $\mathbb{P}$ denote an irreducible Cayley walk on a group G. Then*

$$\max(\mathcal{T}_2(\mathbb{P}), \mathcal{T}_2(\mathbb{P}\overleftarrow{\mathbb{P}})) \leq R_2^{\mathcal{C}}(\mathbb{P}) \leq R_2^{\mathcal{H}}(\mathbb{P}) \leq 2\mathcal{T}_2(\mathbb{P}\overleftarrow{\mathbb{P}}) \qquad (6)$$

In particular if $\mathbb{P}$ is reversible and ergodic, $\mathcal{T}_2(\mathbb{P}) \leq R_2^{\mathcal{C}}(\mathbb{P}) \leq R_2^{\mathcal{H}}(\mathbb{P}) \leq 2\mathcal{T}_2(\mathbb{P}^2) \leq \mathcal{T}_2(\mathbb{P}) + 1$.

The proof of Theorem 6 will be given in Section 3. One consequence of Theorem 6 is that for a reversible ergodic Cayley walk, a holomorphic adversary cannot change the L_2-mixing time.

1.5 Applications: Card Shuffling

In this section we give some applications of the foregoing results in this paper.

Let $\mathbb{P}_{\mathrm{RC}}$ denote the (non-Markovian) random-to-cyclic transposition process. The problem of estimating the mixing time of $\mathbb{P}_{\mathrm{RC}}$ was raised by Aldous and Diaconis [1] in 1986. Mironov [9] used this process to analyze a cryptographic system known as RC4 and showed that $\mathbb{P}_{\mathrm{RC}}$ mixes in time $O(n \log n)$. Mossel et al. [11] showed that $\mathcal{T}(\mathbb{P}_{\mathrm{RC}}) = \Theta(n \log n)$. They showed a lower bound of $(0.0345 + o(1))n \log n$ and an upper bound of $Cn \log n + O(n)$ where $C \approx 4 \times 10^5$. We are able to reduce C to 1.

Theorem 7. *Let $\mathbb{P}_{\mathrm{RC}}$ denote the (non-Markovian) random-to-cyclic transposition process and $\mathbb{P}_{\mathrm{RT}}$ denote the (Markovian) random-to-top transposition chain, where we exchange a random card with the top card. Then $\mathcal{T}_2(\mathbb{P}_{\mathrm{RC}}) \leq \mathcal{T}_2(\mathbb{P}_{\mathrm{RT}}) + 1 \leq n \log n + O(n)$.*

Proof. Let $\alpha_t := t \pmod n$. For $k, r \in \{1, \ldots, n\}$, $(k\,r) = (1\,k)(1\,r)(1\,k)$. Hence if we let $\mathbb{A}_t$ correspond to right multiplication by $(1\,\alpha_t)(1\,\alpha_{t+1})$, it follows that the given adversarial modification of $\mathbb{P}_{\mathrm{RT}}$ simulates $\mathbb{P}_{\mathrm{RC}}$. Since the simulation was done by a Cayley adversary, we have $\mathcal{T}_2(\mathbb{P}_{\mathrm{RC}}) \leq R_2^{\mathcal{C}}(\mathbb{P}_{\mathrm{RT}}) \leq R_2^{\mathcal{H}}(\mathbb{P}_{\mathrm{RT}})$. From Theorem 6 and reversibility of $\mathbb{P}_{\mathrm{RT}}$ it follows that $R_2^{\mathcal{H}}(\mathbb{P}_{\mathrm{RT}}) \leq \mathcal{T}_2(\mathbb{P}_{\mathrm{RT}}) + 1$. But $\mathcal{T}_2(\mathbb{P}_{\mathrm{RT}}) \leq n \log n + O(n)$ ([12, Theorem 9.9]).

Another application of Theorem 6 is in estimating the mixing time of a convex combination of reversible Cayley walks on a group G. This allows us to eliminate the $\log |G|$ factor which would arise if we only use the second largest singular value of the chains.

Theorem 8. *Let $\mathbb{P}_1$ and $\mathbb{P}_2$ be two reversible ergodic Cayley walks on a group G and put $\mathbb{Q} = a_1\mathbb{P}_1 + a_2\mathbb{P}_2$ where $0 < a_1 = 1 - a_2 < 1$. Then assuming $\mathcal{T}_2(\mathbb{P}_i) \geq \log(|G|)/2$ for $i = 1, 2$ and $|G| \geq 16$, we have $\mathcal{T}_2(\mathbb{Q}) \leq 1 + \min\left(\frac{7\mathcal{T}_2(\mathbb{P}_1)}{a_1}, \frac{7\mathcal{T}_2(\mathbb{P}_2)}{a_2}, \mathcal{T}_2(\mathbb{P}_1) + \mathcal{T}_2(\mathbb{P}_2)\right)$.*

This follows from Theorem 5 and Theorem 6.

1.6   Applications: Reversible Lifting

Definition 14. *Let $\mathbb{P}$ and $\mathbb{Q}$ be Markov Chains on state spaces $\mathcal{X}$ and $\mathcal{Y}$ with stationary distributions π and μ respectively. $\mathbb{P}$ is said to be a* collapsing *of $\mathbb{Q}$ if there is a mapping $f : \mathcal{Y} \to \mathcal{X}$ such that*

- *$\pi(x) = \mu(\mathcal{Y}_x)$ for all $x \in \mathcal{X}$ where $\mathcal{Y}_x = f^{-1}(x)$, and*
- *For all $x_1, x_2 \in \mathcal{X}, \mathbb{P}(x_1, x_2) = \sum_{y_1 \in \mathcal{Y}_{x_1}} \sum_{y_2 \in \mathcal{Y}_{x_2}} \mu^{x_1}(y_1) \mathbb{Q}(y_1, y_2)$ where μ^x is the conditional distribution of $y \in \mathcal{Y}$ given $f(y) = x$, i.e. $\mu^x(y) = \mu(y)/\pi(x)$.*

A lifting *of $\mathbb{P}$ is a chain $\mathbb{Q}$ such that $\mathbb{P}$ is the collapsing of $\mathbb{Q}$.*

Chen et al. [4] showed that if $\mathbb{Q}$ is a reversible lifting of a Markov chain $\mathbb{P}$, then $T(\mathbb{Q}) \geq T(\mathbb{P})/\log(1/\pi_*)$. We give an alternate proof of the same result which is motivated by adversarial strategies. The crucial observation is the following

Theorem 9. *Let $\mathbb{Q}$ be a lifting of $\mathbb{P}$. Then $R(\mathbb{Q}) \geq T(\mathbb{P})$.*

If $\mathbb{Q}$ is reversible, using Theorem 3 we immediately have $T(\mathbb{Q}) \log(1/\mu_*) \geq T(\mathbb{P})$ where $\mu_* = \min_y \mu(y)$. When μ_* is only polynomially smaller than π_*, we have an alternate proof of the reversible lifting result. To fine tune the result, we use the adversarial strategy used in Theorem 9 to show

Theorem 10. *Let $\mathbb{Q}$ be a reversible lifting of $\mathbb{P}$. Then $T(\mathbb{Q}) \geq r(\mathbb{P})$.*

As a consequence, when $T(\mathbb{P}) = O(r(\mathbb{P}))$ no reversible lifting $\mathbb{Q}$ of $\mathbb{P}$ can mix faster than $\mathbb{P}$ (ignoring constant factors). Theorem 9 and Theorem 10 are proved in Section 4.

2   Convexity

We now prove Theorem 2 and Theorem 5. We start with a consequence of Hoeffding's inequality [6, Theorem 1].

Lemma 1. *Let $S = CT/p$ for $C > 1$ and $0 < p < 1$. Let $Z_1, \ldots, Z_S$ be independent Bernoulli random variables with $\Pr\{Z_i = 1\} = p$. Let $Z = \sum_i Z_i$. Then*

$$\Pr\{Z < T\} \leq \exp\{-T((C-1) - \log C)\}$$

Lemma 2. *Let $\mathbb{P}$ be an ergodic Markov Chain and $\mathbb{Q}$ be compatible with $\mathbb{P}$. Let S be a valid set of strategies against $\mathbb{P}$. Assume $\mathbb{Q} \in S$. Let $0 < a = 1 - b < 1$. Then $R^S(a\mathbb{P} + b\mathbb{Q}) \leq 2(1+\delta) R^S(\mathbb{P})/a$, as long as $2R^S(\mathbb{P})(\delta - \log(1+\delta)) \geq \log 8$. If $R^S(\mathbb{P}) \geq 11$, then δ may be taken to be $1/2$.*

Proof. Let $S = 2(1+\delta)T/a$, where $T = R^S(\mathbb{P})$ and $\delta > 0$ to be determined later. Put $\mathbb{P}' = a\mathbb{P} + b\mathbb{Q}$. Fix an adversarial strategy $\mathbb{A}_t$ and an initial distribution μ_0. The S-round distribution the adversarially modified $\mathbb{P}'$ is then given by

$$\mu_S = \mu_0 \mathbb{R}_1 \mathbb{A}_1 \mathbb{R}_2 \mathbb{A}_2 \ldots \mathbb{R}_S \mathbb{A}_S$$

where each $\mathbb{R}_i$ is $\mathbb{P}$ with probability a, $\mathbb{Q}$ with probability b and the choices of $\mathbb{R}_i$ for different i are independent. Let $Z = |\{i : \mathbb{R}_i = \mathbb{P}\}|$. Then $\mathbb{E}[Z] = Sa = 2(1+\delta)T$.

If $Z \geq 2T$, $||\mu_S - \pi||_{\mathrm{TV}} \leq 1/8$ since $R^{\mathcal{S}}(\mathbb{P}, 1/8) \leq 2T$ (by sub-multiplicativity) and all stochastic matrices unequal to $\mathbb{P}$ can be considered as an adversarial choice. By Lemma 1,

$$\Pr\{Z < 2T\} \leq \exp\left(-2T(\delta - \log(1+\delta))\right)$$

Thus if $2T(\delta - \log(1+\delta)) \geq \log 8$, we have

$$||\mu_S - \pi||_{\mathrm{TV}} \leq \Pr\{Z \geq 2T\} * 1/8 + 1/8 * 1 \leq 1/4$$

where we used the fact that $||\mu - \pi||_{\mathrm{TV}} \leq 1$ always.

Lemma 3. *Let $\mathbb{P}$ be an ergodic Markov Chain and $\mathcal{S}$ a valid set of strategies against $\mathbb{P}$. Let $\mathbb{Q}$ be compatible with $\mathbb{P}$ and $\mathbb{Q} \in \mathcal{S}$. Let $0 < a = 1-b < 1$. Assume that $R_2^{\mathcal{S}}(\mathbb{P}) \geq \log(1/\pi_*)/2$ and $\pi_* \leq 1/16$. Then $R_2^{\mathcal{S}}(a\mathbb{P}+b\mathbb{Q}) \leq 2(1+\delta)R_2^{\mathcal{S}}(\mathbb{P})/a$, as long as $R_2^{\mathcal{S}}(\mathbb{P})(\delta - \log(1+\delta)) \geq \log(1/\pi_*)/2$. In particular δ may be taken to be $5/2$.*

Proof. This proof is similar to Lemma 2. Put $T = R_2^{\mathcal{S}}(\mathbb{P})$ and $S = 2T(1+\delta)/a$ for $\delta > 0$ to be determined later. Fix an adversarial strategy $\mathbb{A}_t$ and an initial distribution μ_0. The S-round distribution is then given by

$$\mu_S = \mu_0 \mathbb{R}_1 \mathbb{A}_1 \mathbb{R}_2 \mathbb{A}_2 \ldots \mathbb{R}_S \mathbb{A}_S$$

where the $\mathbb{R}_i$ are independent and equal to $\mathbb{P}$ with probability a and $\mathbb{Q}$ with probability b. Let $Z = |\{i : \mathbb{R}_i = \mathbb{P}\}|$ so that $\mathbb{E}[Z] = Sa = 2(1+\delta)T$.

Going along the same lines as Lemma 2, we have

$$||\mu_S - \pi||_{2,\pi} \leq \Pr\{Z \geq 2T\} * 1/4 + \exp\left(-2T(\delta - \log(1+\delta))\right) \cdot \frac{1}{\sqrt{\pi_*}} \quad (7)$$

since the worst value for $||\mu - \pi||_{2,\pi} = 1/\sqrt{\pi_*}$. Substituting $T = \alpha \log(1/\pi_*)/2$ for $\alpha \geq 1$, we have

$$||\mu_S - \pi||_{2,\pi} \leq 1/4 + \sqrt{\pi_*}^{-(2\alpha(\delta - \log(1+\delta)))-1} \quad (8)$$

Since $\pi_* \leq 1/16$, we have $||\mu_S - \pi||_{2,\pi} \leq 1/2$ if $\alpha(\delta - \log(1+\delta)) \geq 1$.

The final bound $R^{\mathcal{S}}(a\mathbb{P} + (1-a)\mathbb{Q}) \leq R^{\mathcal{S}}(\mathbb{P}) + R^{\mathcal{S}}(\mathbb{Q}) - 1$ follows from the fact that after $R^{\mathcal{S}}(\mathbb{P}) + R^{\mathcal{S}}(\mathbb{Q}) - 1$, either $\mathbb{P}$ occurs $\geq R^{\mathcal{S}}(\mathbb{P})$ times or $\mathbb{Q}$ occurs $\geq R^{\mathcal{S}}(\mathbb{Q})$ times.

3   Cayley Walks on Groups

In this section, we consider Cayley walks driven by a probability measure P over a group G. The chain is irreducible iff the support of P generates G and aperiodic if $P(id) > 0$ where id is the identity element of G.

It is well known that the knowledge of all the singular values of the transition matrix $\mathbb{P}$ can give good bounds on the standard mixing time. In this section we show that the same conclusion holds for the robust mixing time against holomorphic adversaries.

Definition 15. *For a group G, the* holomorph *of G, denoted $\mathrm{Hol}(G)$ is the semi-direct product of G and $\mathrm{Aut}(G)$, where $\mathrm{Aut}(G)$ acts naturally on G. Elements of $\mathrm{Hol}(G)$ are called* holomorphisms.

Holomorphisms of G can be identified with permutations of G as follows: Elements of G act by right translation and those of $\mathrm{Aut}(G)$ act naturally. Since this permutation representation of $\mathrm{Hol}(G)$ is faithful, we can identify holomorphisms of G by the permutation they induce on G.

Definition 16. *A permutation $J : G \to G$ is said to be* G-respecting *if for some permutation $K : G \to G$, and all $g, h \in G$, $J(h)^{-1}J(g) = K(h^{-1}g)$.*

Lemma 4. *A permutation J of G is G-respecting iff it is a holomorphism of G.*

Holomorphic strategies $\mathcal{H}$ are precisely convex combinations of $\mathrm{Hol}(G)$ (viewed as permutations on G) while Cayley strategies $\mathcal{C}$ are convex combinations of $G \le \mathrm{Hol}(G)$.

We now look at the holomorphic robust mixing time of a Cayley walk on G. We identify a permutation on G with the $|G| \times |G|$ permutation matrix representing it.

Let $\mathbb{P}$ be the transition matrix of a Cayley walk on G and fix a holomorphic strategy $\{\mathbb{A}_t\}_{t>0}$. Define $\mathbb{Q}_0 = I$ and for $k > 0$, put $\mathbb{Q}_{k+1} = \mathbb{Q}_k \mathbb{P} \mathbb{A}_{k+1}$. If μ_t denotes the distribution after t rounds we have $\mu_t = \mu_0 \mathbb{Q}_t$, where μ_0 is the initial distribution.

Lemma 5. *If μ_0 is supported only at $g \in G$ then $||\mu_t - \pi||_2^2 = (\mathbb{Q}_t \mathbb{Q}_t^{\mathrm{T}})(g,g) - 1/N$, where $N = |G|$ and $||\cdot||_2$ denotes the Euclidean norm.*

Definition 17. *A matrix $\mathbb{B}$ whose rows and columns are indexed by elements of G is said to be a* G-circulant *if $\mathbb{B}(g,h) = P(g^{-1}h)$ for some function $P : G \to \mathbb{R}$.*

Lemma 6. *If J is a holomorphism of G and $\mathbb{B}$ is G-circulant, then so is $J^{-1}\mathbb{B}J$.*

Theorem 11. *Let $\mathbb{P}$ be the transition probability matrix of an ergodic Cayley walk on a finite group G. Let $1 = \sigma_0 \ge \sigma_1 \ge \ldots \sigma_{N-1} \ge 0$ denote the singular values of $\mathbb{P}$. Let $\{\mathbb{A}_t\}_{t>0}$ denote the moves of a holomorphic adversary. Then $||\mu_t - \pi||_{2,\pi}^2 \le \sum_{i=1}^{N-1} \sigma_i^{2t}$ where μ_t denotes the distribution after t-rounds and μ_0 is any initial distribution.*

Proof. Assume w.l.o.g. that all the $\mathbb{A}_t$ are holomorphisms of G and also that μ_0 is supported on one element g of G.

Let $\mathbb{Q}_t = \mathbb{P}\mathbb{A}_1 \mathbb{P}\mathbb{A}_2 \ldots \mathbb{P}\mathbb{A}_t$. By Lemma 5, and the relation $||\cdot||_{2,\pi}^2 = |G| ||\cdot||_2$, we have $||\mu_t - \pi||_{2,\pi}^2 \le N \cdot \mathbb{Q}_t \mathbb{Q}_t^{\mathrm{T}}(g,g) - 1$. Consider evaluating $\mathbb{Q}_t \mathbb{Q}_t^{\mathrm{T}}$ inside out, i.e. put

$$\mathbb{C}_{t+1} = I \quad \text{and for } k \le t \quad \mathbb{C}_k = \mathbb{P}(\mathbb{A}_k \mathbb{C}_{k+1} \mathbb{A}_k^{\mathrm{T}})\mathbb{P}^{\mathrm{T}} \tag{9}$$

From Lemma 6 and closure properties of G-circulant matrices it follows that $\mathbb{C}_1 = \mathbb{Q}_t \mathbb{Q}_t^{\mathrm{T}}$ is G-circulant and $N\mathbb{Q}_t \mathbb{Q}_t^{\mathrm{T}}(g,g) = \mathrm{tr}(\mathbb{Q}_t \mathbb{Q}_t^{\mathrm{T}}) = \sum_{i\geq 0} \sigma_i(\mathbb{Q}_t)^2$. Hence we have

$$||\mu_t - \pi||_{2,\pi}^2 = \sum_{i=1}^{N-1} \sigma_i(\mathbb{Q}_t)^2 \leq \sum_{i=1}^{N-1} \prod_{j=1}^{t} \sigma_i(\mathbb{P}\mathbb{A}_j)^2 \leq \sum_{i=1}^{N-1} \sigma_i(\mathbb{P})^{2t} \qquad (10)$$

where the first equality follows from $\sigma_0(\mathbb{Q}_t) = 1$, the second inequality from [7, Chapter 3] and the third inequality from $\sigma_i(\mathbb{P}\mathbb{A}_j) \leq \sigma_i(\mathbb{P})\sigma_1(\mathbb{A}_j) = \sigma_i(\mathbb{P})$.

Now we prove Theorem 6 and show that that holomorphic robust L_2-mixing time of $\mathbb{P}$ is within a factor 2 of the standard mixing time of $\mathbb{P}\overleftarrow{\mathbb{P}}$.

Proof (of Theorem 6). Let $\mathbb{P}$ be the transition matrix of a Cayley walk on G. Considering the adversarial strategy where $\mathbb{A}_t = I$ and the one where $\mathbb{A}_t = \overleftarrow{\mathbb{P}}$, we have $\max(\mathcal{T}_2(\mathbb{P}), \mathcal{T}_2(\mathbb{P}\overleftarrow{\mathbb{P}})) \leq R_2^{\mathcal{C}}(\mathbb{P}) \leq R_2^{\mathcal{H}}(\mathbb{P})$.

Let σ_i denote the singular values of $\mathbb{P}$. Let $v_t = v_0 \mathbb{Q}^t$ denote the t-step distribution of the Markov chain $\mathbb{Q} = \mathbb{P}\overleftarrow{\mathbb{P}}$ starting from v_0. From Lemma 5 and the relation $|| \cdot ||_{2,\pi}^2 = |G| ||| \cdot ||_2^2$, we have

$$||v_t - \pi||_{2,\pi}^2 = |G|\mathbb{Q}^{2t}(g,g) - 1 = \mathrm{tr}(\mathbb{Q}^{2t}) - 1 = \sum_{i=1}^{N-1} \sigma_i^{4t} \qquad (11)$$

Now consider a run of an adversarially modified version of $\mathbb{P}$ for $2t$-steps. Let μ_{2t} be the distribution after $2t$-rounds starting from μ_0. Theorem 11 together with (11) implies $||\mu_{2t} - \pi||_{2,\pi} \leq ||v_t - \pi||_{2,\pi}$. Hence $R_2^{\mathcal{H}}(\mathbb{P}) \leq 2\mathcal{T}_2(\mathbb{P}\overleftarrow{\mathbb{P}})$.

4   Reversible Liftings

In this section we reprove a result due to Chen et al. [4] on reversible liftings of Markov Chains. The proof is inspired by considering the Robust mixing time of a Markov Chain and looking at a particular adversarial strategy. We start with a proof of Theorem 9.

Proof (of Theorem 9). We prove $R(\mathbb{Q}) \geq T(\mathbb{P})$ by exhibiting an adversarial strategy which allows the adversary to simulate the evolution of $\mathbb{P}$.

Consider the following adversarial strategy $\mathbb{A}$: Given $y \in \mathcal{Y}$, the adversary picks a state $y' \in \mathcal{Y}$ according to the distribution μ^x where $x = f(y)$. Recall that μ^x is the conditional distribution of μ given that $f(y) = x$.

Since $\mu = \sum_{x \in \mathcal{X}} \pi(x)\mu^x$, it follows that this strategy fixes the stationary distribution μ of $\mathbb{Q}$. We now claim that with this strategy the adversary can simulate the evolution of $\mathbb{P}$ on $\mathcal{Y}$.

For a distribution v on $\mathcal{X}$, consider the distribution $F(v) = \sum_{x \in \mathcal{X}} v(x)\mu^x$ on $\mathcal{Y}$. One can verify that $F(v)\mathbb{Q}\mathbb{A} = F(v\mathbb{P})$ and in particular it follows that $||F(v)(\mathbb{Q}\mathbb{A})^t - \mu||_{\mathrm{TV}} = ||v\mathbb{P}^t - \pi||_{\mathrm{TV}}$ for any $t > 0$. Hence $R(\mathbb{Q}) \geq T(\mathbb{P})$.

Now if $\mathbb{Q}$ were reversible, Theorem 3 implies $R(\mathbb{Q}) \leq \mathcal{T}(\mathbb{Q})(1 + \log(1/\mu_*)/2)$. Hence

$$\mathcal{T}(\mathbb{Q}) \geq \frac{\mathcal{T}(\mathbb{P})}{\log(1/\mu_*)} \tag{12}$$

When μ_* is only polynomially smaller than π_*, this gives our result. We now improve the result by looking at the adversarial strategy in more detail and show $\mathcal{T}(\mathbb{Q}) \geq r(\mathbb{P})$.

Proof (of Theorem 10). Let $\mathbb{A}$ denote the stochastic matrix representing the adversarial strategy. Note that $\mathbb{A}$ is reducible and reversible.

Let $\boldsymbol{\alpha}$ denote the eigenvector (of length $|\mathcal{X}|$) corresponding to $\lambda_*(\mathbb{P})$ and define $\beta = \sum_x \boldsymbol{\alpha}(x)\mu^x(y)$ (of length $|\mathcal{Y}|$). One can easily check that $\beta(\mathbb{Q}\mathbb{A}) = \lambda_*(\mathbb{P})\beta$. Thus $\lambda_*(\mathbb{Q}\mathbb{A}) \geq \lambda_*(\mathbb{P})$.

Since $\mathbb{A}$ is a contraction (it is stochastic), we have

$$\lambda_*(\mathbb{Q}) \geq \lambda_*(\mathbb{Q}\mathbb{A}) \geq \lambda_*(\mathbb{P}) \tag{13}$$

hence $\mathcal{T}(\mathbb{Q}) \geq r(\mathbb{Q}) \geq r(\mathbb{Q}\mathbb{A}) \geq r(\mathbb{P})$.

[4] gives an example $\mathbb{Q}$ of a reversible random walk on a tree (with π_* exponentially small) and its collapsing $\mathbb{P}$ for which

$$\mathcal{T}(\mathbb{Q}) = \Theta(\mathcal{T}(\mathbb{P}))\frac{\log\log(1/\pi_*)}{\log(1/\pi_*)} \tag{14}$$

Since we know that $R(\mathbb{Q}) \geq \mathcal{T}(\mathbb{P})$ it shows that Theorem 3 is almost tight, even for reversible chains.

5 Questions

Question 1. Is is true that for reversible $\mathbb{P}$, with uniform stationary distribution $R(\mathbb{P}) = O(\mathcal{T}(\mathbb{P}))$?

Question 2. Is the robust mixing time of random-to-top $O(n \log n)$? More generally, if $\mathbb{P}$ is a Cayley walk on a group G, is it true that $R(\mathbb{P}) = O(\mathcal{T}(\mathbb{P}\overleftarrow{\mathbb{P}}))$?

Using log-Sobolev constants one can show $R((\mathbb{P}_{\mathrm{RT}} + I)/2) = O(n \log^2 n)$.

In all the examples we have seen, the adversarial strategy which achieves the robust mixing time can be taken to be homogenous. Is this always the case?

Question 3. Is it true that $R(\mathbb{P}) = \max_{\mathbb{A}} \mathcal{T}(\mathbb{P}\mathbb{A})$ where the maximum is taken over all $\mathbb{A}$ compatible with $\mathbb{P}$?

Acknowledgements

I would like to thank László Babai for introducing me to the fascinating area of Markov Chains and for very helpful discussions and suggestions.

References

[1] David Aldous and Persi Diaconis. Shuffling cards and stopping times. *The American Mathematical Monthly*, 93(5):333–348, May 1986.

[2] Ivona Bezakova and Daniel Stefankovic. Convex combinations of markov chains and sampling linear orderings. In preperation.

[3] Sergey Bobkov and Prasad Tetali. Modified log-sobolev inequalities, mixing and hypercontractivity. In *STOC '03: Proceedings of the thirty-fifth annual ACM symposium on Theory of computing*, pages 287–296, New York, NY, USA, 2003. ACM Press. ISBN 1-58113-674-9. doi: http://doi.acm.org/10.1145/780542.780586.

[4] Fang Chen, László Lovász, and Igor Pak. Lifting markov chains to speed up mixing. In *Proceedings of the thirty-first annual ACM symposium on Theory of computing*, pages 275–281. ACM Press, 1999.

[5] Montenegro R. Tetali P. Goel, S. Mixing time bounds via the spectral profile. *Electronic Journal of Probability*, 11:1–26, 2006. URL http://www.math.washington.edu/~ejpecp/EjpVol11/paper1.abs.html.

[6] Wassily Hoeffding. Probability inequalities for sums of bounded random variables. *American Statistical Association Journal*, pages 13–30, 1963.

[7] Roger Horn and Charles Johnson. *Topics in Matrix Analysis*. Cambridge University Press, 1991.

[8] Laurent Miclo. Remarques sur l'hypercontractivité et l'évolution de l'entropie pour des chanes de markov finies. *Séminaire de probabilités de Strasbourg*, 31: 136–167, 1997.

[9] Ilya Mironov. (Not so) random shuffles of RC4. In *Crypto '02*, pages 304–319, 2002.

[10] Ravi Montenegro. Duality and evolving set bounds on mixing times. preprint. URL http://www.ravimontenegro.com/research/evosets.pdf.

[11] Elchanan Mossel, Yuval Peres, and Alistair Sinclair. Shuffling by semi-random transpositions, 2004.

[12] Laurent Saloff-Coste. Random walks on finite groups. URL www-stat.stanford.edu/~cgates/PERSI/papers/rwfg.pdf.

Approximating Average Parameters of Graphs[*]
In Memory of Shimon Even (1935–2004)

Oded Goldreich[1,**] and Dana Ron[2,***]

[1] Department of Computer Science, Weizmann Institute of Science, Rehovot, Israel
oded.goldreich@weizmann.ac.il
[2] Department of Electrical Engineering-Systems, Tel Aviv University, Tel Aviv Israel

Abstract. Inspired by Feige (*36th STOC*, 2004), we initiate a study of sublinear randomized algorithms for approximating average parameters of a graph. Specifically, we consider the average degree of a graph and the average distance between pairs of vertices in a graph. Since our focus is on sublinear algorithms, these algorithms access the input graph via queries to an adequate oracle.

We consider two types of queries. The first type is standard neighborhood queries (i.e., *what is the ith neighbor of vertex v?*), whereas the second type are queries regarding the quantities that we need to find the average of (i.e., *what is the degree of vertex v?* and *what is the distance between u and v?*, respectively).

Loosely speaking, our results indicate a difference between the two problems: For approximating the average degree, the standard neighbor queries suffice and in fact are preferable to degree queries. In contrast, for approximating average distances, the standard neighbor queries are of little help whereas distance queries are crucial.

1 Introduction

In a recent work [8], Feige investigated the problem of estimating the average degree of a graph *when given direct access to the list of degrees* (of individual vertices). He observed two interesting ("phase transition") phenomena. Firstly, in contrast to the problem of estimating the average value of an arbitrary function $d : [n] \rightarrow [n-1]$ (where $[n] \stackrel{\text{def}}{=} \{1, \ldots, n\}$), sublinear-time approximations can be obtained when the function d represents the degree sequence of a simple graph over n vertices.[1] Secondly, whereas a $(2 + \epsilon)$-approximation can be obtained in $O(\sqrt{n})$-time, for every constant $\epsilon > 0$, a better approximation factor cannot be achieved in sublinear time (i.e., a $(2 - o(1))$-approximation requires time $\Omega(n)$).

Feige's work views the problem of estimating the average degree of a graph as a special case of estimating the average value of an arbitrary function $d : [n] \rightarrow [n-1]$.

---

[*] The research was supported in part by the Israel Internet Association (ISOC-IL).
[**] Part of this work was done while being a fellow of the Radcliffe Institute for Advanced Study, Harvard University.
[***] Part of this work was done while being a fellow of the Radcliffe Institute for Advanced Study, Harvard University.
[1] Here we also assume that there are no isolated vertices in the graph (i.e., each vertex has degree at least 1).

J. Diaz et al. (Eds.): APPROX and RANDOM 2006, LNCS 4110, pp. 363–374, 2006.
© Springer-Verlag Berlin Heidelberg 2006

Our perspective is different: We view Feige's work as a sublinear algorithm for a natural graph theoretic problem, which brings up two (open-ended) questions:

1. What type of operations (i.e., direct access queries to the input graph) are natural to consider for such an algorithm?
2. What other natural "average graph parameters" (i.e., averages of vertex-based quantities) are of interest?

In the following two subsections we briefly address these questions, and afterwards we present our results that refer to various combinations of "answers" to these questions.

1.1   Types of Direct Access Queries

When viewing the problem of estimating the average degree in a graph as a special case of the problem of estimating the average value of an arbitrary function $d : [n] \rightarrow [n-1]$, it seems natural to restrict the algorithm to "degree queries". However, from the point of view of sublinear-time algorithms for graphs (cf., e.g., [10, 11, 15, 3, 14]), it is natural to allow also other types of queries to the graph. The most natural queries are neighbor queries; that is, queries of the form (v, i) that are answered by the ith neighbor of v (or by a special symbol that indicates that v has less than i neighbors). In case of relatively dense graphs, it is also natural to consider adjacency queries (i.e., are vertices u and v adjacent in the graph). Thus, we consider two basic types of queries:

1. Standard neighbor (and adjacency) queries, which are natural in any algorithmic problem regarding graphs.
2. Problem-specific queries that associate values to vertices (or to sets of vertices), where our aim is to compute the average of these values. For example, in the case of approximating the average degree we consider degree queries.

We comment that degree queries can be emulated by a logarithmic number of neighbor queries (i.e., via binary search).

1.2   Other Natural Averaging Problems

In addition to the average degree of a graph, we consider two problems regarding distances in a graph. The first is approximating the all-pairs average distance in the graph, and the second is approximating the average distance of a fixed vertex to all the graph vertices. We refer to these problem by the terms all-pairs and single-source, respectively.

In addition to the standard neighbor queries, for the average distance approximation problems, we will also consider distance queries. That is, in both cases, we will consider queries of the form (u, v) that are answered by the distance between u and v in the graph.

1.3   Our Results

Our results indicate that for one problem (i.e., approximating the average degree) augmenting the problem-specific oracle with neighbor queries helps, whereas for the other problems (i.e., approximating average distances) such an augmentation does not help. Moreover, as noted above, degree queries are not of great help (for approximating the average degree), whereas distance queries are crucial to approximating average distances in sublinear-time. In both cases, our algorithms do not use adjacency queries (and our lower bounds show that these queries do not help).

Approximating the Average Degree of a Graph. We present a sublinear algorithm that obtains an arbitrarily good approximation of the average degree, *while making only neighbor queries.*[2] Specifically, for every constant $\epsilon > 0$, we obtain a $(1 + \epsilon)$-approximation to the average degree of a simple graph $G = (V, E)$ in $\widetilde{O}(\sqrt{|V|})$-time, where the O-notation hides a polynomial dependence on ϵ.

Our result should be contrasted with Feige's results [8]: Recall that Feige showed that, when using only degree queries, a $(2 - o(1))$-approximation (of the average degree of $G = (V, E)$) requires time $\Omega(|V|)$. Thus, neighbor queries are essential for sublinear-time algorithms that provide a $(2 - o(1))$-approximation. On the other hand, he showed that (for every constant $\epsilon > 0$) a $(2 + \epsilon)$-approximation can be obtained in $O(\sqrt{|V|})$-time (using only degree queries).

The running-time of our algorithm is essentially optimal: any constant-factor approximation of the average degree requires making $\Omega(\sqrt{|V|})$ queries of some graph $G = (V, E)$, even when allowed both neighbor and degree queries. Furthermore, a $(1 + \epsilon)$-approximation requires $\Omega(\sqrt{|V|/\epsilon})$ queries.

The above represents a simplified account of the results. We recall that Feige [8] provides his algorithm with a lower bound on the average degree of the input graph. This auxiliary input allows also to handle graphs that have isolated vertices (rather than assuming that each vertex has degree at least 1) and yields an improvement whenever the lower bound is better (than the obvious value of 1). Specifically, given a lower bound of ℓ (on the average degree), the complexity of Feige's algorithm is related to $\sqrt{|V|/\ell}$ rather than to $\sqrt{|V|}$. The same improvement holds also for our algorithms. Furthermore, we observe that our algorithms (as well as Feige's) can be adapted to work without this lower bound. Specifically, the complexity of the modified algorithm, which obtains no a priori information about the average degree, is related to $(|V|/\bar{d})^{1/2}$, where $\bar{d}$ denotes the actual average degree (which is, of course, not given to the algorithm). Thus, we get:

Theorem 1. *There exists an algorithm that makes only neighbor queries to the input graph and satisfies the following condition. On input $G = (V, E)$ and $\epsilon \in (0, 1)$, with probability at least $2/3$, the algorithm halts within $\widetilde{O}((|V|/\bar{d})^{1/2} \cdot \mathrm{poly}(1/\epsilon))$ steps and outputs a value in $[\bar{d}, (1 + \epsilon) \cdot \bar{d}]$, where $\bar{d} = 2|E|/|V|$.*

Again, this running-time is essentially optimal in the sense that a $(1+\epsilon)$-approximation requires $\Omega((|V|/(\epsilon\bar{d}))^{1/2})$ queries, for every value of $|V|$ and $\bar{d} \in [2, o(|V|)]$ and $\epsilon \in [\omega(|V|^{-1/4}), o(|V|/\bar{d})]$.

Approximating Average Distances. We present a sublinear algorithm that obtains an arbitrarily good approximation of the average (all-pairs and single-source) distances,

---

[2] Note that a degree query can be emulated using $O(\log |V|)$ neighbor queries, by performing a kind of binary search.

while making (only) *distance queries*. Specifically, we obtain a $(1 + \epsilon)$-approximation of the (relevant) average distance of a simple unweighted graph $G = (V, E)$ in time $O(\sqrt{|V|}) \cdot \text{poly}(1/\epsilon)$. Actually, as in the case of approximating the average degree, we obtained an improved performance as a function of the actual average distance.

Theorem 2. *There exists an algorithm that makes only distance queries to the input graph and satisfies the following condition. On input $G = (V, E)$ and $\epsilon \in (0, 1)$, with probability at least 2/3, the algorithm halts within $O((|V|/\overline{d}_G)^{1/2} \cdot \text{poly}(1/\epsilon))$ steps and outputs a value in $[\overline{d}_G, (1 + \epsilon) \cdot \overline{d}_G]$, where $\overline{d}_G$ is the average of the all-pairs distances in G. A corresponding algorithm exists for the average distances to a given vertex $s \in V$.*

This running time is essentially optimal: any constant-factor approximation of the average distance in $G = (V, E)$ requires making $\Omega((|V|/\overline{d}_G)^{1/2})$ queries, even when allowed both distance and neighbor queries. Furthermore, a $(1 + \epsilon)$-approximation requires $\Omega((|V|/(\epsilon \overline{d}_G))^{1/2})$ queries, for every value of $|V|$ and $\overline{d}_G = o(|V|)$ and $\epsilon = \omega(|V|^{-1})$.

We show that distance queries are essential for sublinear-time algorithms that provide any constant-factor approximation of the average distances. Specifically, *when using only neighbor queries*, a k-approximation of the average distance in $G = (V, E)$ requires making $\Omega(|E|/k^2 \log k)$ queries. In the case of the single-source problem, this means that (when using only neighbor queries) a constant-factor approximation is as hard to obtain as the exact value. In the case of the all-pairs problem, by emulating distance queries in a straightforward manner, we can obtain a $(1 + \epsilon)$-approximation in time $O(\sqrt{|V|} \cdot |E|) \cdot \text{poly}(1/\epsilon)$ *when using only neighbor queries*. For moderately sparse graphs, this yields an improvement over the straightforward approach of computing (or approximating) all pair-distances and computing the average of these $|V|^2$ values. Details follow.

If $|E| \ll |V|^{7/2}$ then our $O(\sqrt{|V|} \cdot |E|) \cdot \text{poly}(1/\epsilon)$-time $(1 + \epsilon)$-approximation is definitely preferable to computing the average of $|V|^2$ approximate values regardless of how the latter are obtained. On the other hand, if $|E| > |V|^{e_{mm}-0.5}$, where $e_{mm} \in [2, 2.376)$ is the matrix multiplication exponent (cf. [4]), then one can find all pair-distances as well as their average faster than the time it takes our algorithm approximates the latter (cf. [9, 16]). In the intermediate range[3] (of $|V|^{3/2} \gg |E| \gg |V|^{e_{mm}-0.5}$, where $e_{mm} - 0.5 < 1.876$), our algorithm should be compared against a host of all-pairs approximate distance algorithms and the preference may depend on additional parameters (e.g., the approximation sought and a priori bounds on the average all-pairs distance). Specific algorithms that may be relevant include those of [6, 5]. (The interested reader is referred to Zwick's survey [17] of algorithms for finding exact and approximate distances in graphs.)

1.4   Related Work

In addition to the work of Feige [8], we are aware of two other related results on estimating average parameters of graphs. Indyk [13] considers the problem of estimating

---

[3] Indeed, the intermediate range exists provided $e_{mm} > 2$ (or rather, that $e_{mm} = 2$ is not known).

the average distance in a distance metric over n points. In particular, such a metric is defined by the shortest distances in a connected weighted graph. Indyk gives a $(1 + \epsilon)$-approximation algorithm that runs in time $O(n/\epsilon^{7/2})$. This algorithm is linear in the number of points, but sublinear in the size of the input, which is an $n \times n$ matrix.

Bădoiu *et. al.* [2] consider the problem of computing the optimal cost of the metric facility location problem in sublinear time. It follows from their analysis that it is possible to obtain a $(1 + \epsilon)$-approximation of the average degree of a graph in time $\tilde{O}(n/\epsilon^2)$ in the following model: The algorithm does not have access to degree queries nor to neighbor queries, but rather is only allowed to traverse the incidence list of a vertex according to a fixed order. By definition, in this model it takes $\Theta(d(v))$ time to compute the degree $d(v)$ of a vertex v. This algorithm is sublinear in the size of the input when the graph is not sparse.

2   Preliminaries

Throughout the work, all algorithms are probabilistic and have direct access to their input. That is, such algorithms are actually probabilistic oracle machines that have access to one or more oracles. These oracles will typically represent a graph in a way to be understood from the context. For example, we consider oracles that answer queries such as neighbor queries and degree queries. The explicit input to these algorithms will consist of relevant parameters that always include the number of vertices in the graph, which in turn determines the vertex set (i.e., for simplicity, we assume that all n-vertex graphs have $[n] \stackrel{\text{def}}{=} \{1, \ldots, n\}$ as their vertex set). As the basic definition of approximation algorithms, we use the following standard one.

Definition 1. *For $\epsilon > 0$, a $(1 + \epsilon)$-approximation of a quantity $q : \{0, 1\}^* \rightarrow (0, \infty)$ is an algorithm that on input x, with probability at least $2/3$, outputs a value in the interval $[q(x), (1 + \epsilon) \cdot q(x)]$.*

The error probability can be decreased to 2^{-k} by invoking the basic algorithm for $O(k)$ times and outputting the median value. At times, when $\epsilon \ll 1$, for simplicity of presentation we allow the algorithm to output a value in the interval $[(1-\epsilon) \cdot q(x), (1+\epsilon) \cdot q(x)]$. (Indeed, the output can be "normalized" by division (by $1 - \epsilon$).) Our algorithms will all be uniform in the sense that we actually present an algorithm that takes ϵ as a parameter.

When stating lower bounds that depend on several parameters, we mean that these bounds hold uniformly for all choices of these parameters (or all choices satisfying explicitly stated conditions). That is, when we say that a $(1 + \epsilon)$-approximation of q requires $\Omega(f(n, \epsilon, p))$ queries, we means that there exists a constant $c > 0$ such for any possible value of the parameters n, ϵ and p and any $(1 + \epsilon)$-approximation algorithm A of the quantity q, there exists an n-vertex graph G with $q(G) = p$ such that A makes at least $c \cdot f(n, \epsilon, p)$ queries. (Since all our lower bounds refer to the query complexity of algorithms, linear speed-up phenomena do not arise.)

Throughout this work, we assume that the neighbors of each vertex are listed in arbitrary order. This reasonable assumption facilitates the proofs of the lower bound, which can be modified to handle also the case where the said lists are sorted.

In all that follows, when we say "with high probability" we mean with probability at least $1 - \delta$ for some small constant $\delta > 0$.

3 Approximating the Average Degree of a Graph

Let $G = (V, E)$ be a *simple* graph (i.e., having no parallel edges and no self-loops), where $|V| = n$, and let $d(v)$ denote the degree of vertex $v \in V$ in G. We denote by $\overline{d} \overset{\text{def}}{=} \frac{1}{n} \sum_{v \in V} d(v)$ the average degree in G. An algorithm for estimating $\overline{d}$ is allowed to perform two types of queries: *degree queries* and *neighbor queries*. Namely, for any vertex v of its choice the algorithm can obtain $d(v)$, and for any v and $j \leq d(v)$, the algorithm can obtain the jth neighbor of v. Actually, when degree queries are allowed then it suffices to allow the algorithm to obtain a random neighbor of any desired (i.e., queried) vertex.

We start by describing an algorithm that is provided with an *a priori known lower bound* ℓ on the value of $\overline{d}$. We later eliminate the need for this a priori knowledge. We close this section by establishing that our algorithm has almost optimal running-time (when referring to its dependence on the size of the graph). All missing proofs can be found in the full version of this paper [12]

3.1 The Algorithm

Our algorithm is inspired by the work of Kaufman *et. al.* [14], and more specifically, by a subroutine presented in [14] for sampling edges "almost uniformly". The basic idea of our algorithm is to sample vertices and to put them into "buckets" according to their degrees such that in bucket B_i we have vertices with degree between $(1 + \beta)^{i-1}$ and $(1 + \beta)^i$ (where $\beta = \epsilon/c$ for some constant $c > 1$). If S is the sample, then we denote by S_i the subset of sampled vertices that belong to B_i. We will focus on the sets S_i that are *sufficiently large*, because we want $|S_i|/|S|$ to be a good approximation of $|B_i|/n$. Let us denote the set of the corresponding i's by L.

Suppose we take $(1/|S|) \sum_{i \in L} |S_i|(1 + \beta)^{i-1}$ as our estimate for the average degree of the graph. Note that the expected value of $|S_i|/|S|$ is $|B_i|/n$ and that $(1/n) \sum_i |B_i|(1 + \beta)^{i-1} \leq \overline{d}$. Hence, with high probability, for a sufficiently large sample S, we would be overestimating the average degree by a factor of at most $(1 + \epsilon)$. The source of the overestimation is only the error in approximating $|B_i|/n$ by $|S_i|/|S|$. However, we may underestimate $\overline{d}$ by a factor of roughly 2. The reason is that the edges between large buckets and small buckets are only counted once, rather than twice, and the edges with both endpoints in small buckets are not counted at all. The "threshold of largeness" is set such that the number of vertices in small buckets is so small that we can discard all possible edges that have both end-points in small buckets. (This calls for taking a sample of size $O(\sqrt{n})$, setting the threshold at $\text{poly}(\log n)$, and concluding that the number of vertices in small buckets is at most $\sqrt{n}$.)

So far we have described a procedure that approximates $\overline{d}$ up to a factor of $2 + \epsilon$ while using only degree queries (i.e., we obtain Feige's result [8] using a different analysis). To get beyond the "factor 2 barrier" we observe that the main source of approximation error is due to edges with one endpoint in a large bucket and the other endpoint in a small bucket. These edges were counted once (in our estimate for $\overline{d}$), whereas they need to be counted twice. Thus, it suffices to estimate the number of such edges, which can be done by estimating, for each large bucket, the fraction of edges that are incident to vertices in the bucket and whose other endpoint is in a small bucket.

This estimate cannot be obtained using degree queries, but it can be obtained using "random neighbor" queries. Specifically, for every vertex v in a large S_i, we select uniformly a neighbor of v and check whether this neighbor resides in a small bucket. Adding our estimate of the number of edges between large buckets and small buckets to $(n/|S|) \sum_{i \in L} |S_i| (1 + \beta)^{i-1}$ yields a $(1 + \epsilon)$-approximation of $2|E|$ (and hence a $(1 + \epsilon)$-approximation of $\overline{d} = 2|E|/n$).

We are now ready to present the algorithm in full detail. For $t = \lceil \log_{(1+\beta)} n \rceil + 1$, we define a partition of V into the following buckets:

$$B_i = \left\{ v : d(v) \in \left((1+\beta)^{i-1}, (1+\beta)^i \right] \right\}, \quad \text{for } i = 0, 1, \ldots, t-1. \quad (1)$$

The algorithm refers to an a priori lower bound ℓ on $\overline{d}$, and the reader may think of $\ell = 1$ as in the foregoing motivating discussion. We will consider B_i to be *large* (and put $i \in L$) if the sample S contains at least $\Omega \left(\frac{\sqrt{\epsilon}}{t} \cdot \frac{|S|}{\sqrt{n/\ell}} \right)$ representatives of B_i. Otherwise it is considered *small*. For a large B_i, we let $\widetilde{\alpha}_i$ denote our approximation of the fractions of edges incident at B_i that have their other endpoint in a small bucket.

Average Degree Approximation Algorithm

1. Uniformly and independently select $K = \tilde{\Theta} \left(\sqrt{n/\ell} \cdot \mathrm{poly}(1/\epsilon) \right)$ vertices from V, and let S denote the (multi-)set of selected vertices.
2. For $i = 0, 1, \ldots, \lceil \log_{(1+\beta)} n \rceil$, let $S_i = S \cap B_i$.
3. Let $L = \left\{ i : \frac{|S_i|}{|S|} \geq \frac{1}{t} \cdot \sqrt{\frac{\epsilon}{6} \cdot \frac{\ell}{n}} \right\}$, where $t \overset{\text{def}}{=} \lceil \log_{(1+\beta)} n \rceil + 1$.
4. For every $i \in L$ and every $v \in S_i$, select at random a neighbor u of v, and let $\chi(v) = 1$ if $u \in \bigcup_{j \notin L} B_j$, and $\chi(v) = 0$ otherwise. For every $i \in L$, let $\widetilde{\alpha}_i = |\{v \in S_i : \chi(v) = 1\}| / |S_i|$.
5. Output $\tilde{\mathsf{d}} = \frac{1}{K} \cdot \sum_{i \in L} (1 + \widetilde{\alpha}_i) \cdot |S_i| \cdot (1 + \beta)^i$.

Lemma 1. *For every $\epsilon < 1/2$ and $\beta \leq \epsilon/8$, the above algorithm outputs a value $\tilde{\mathsf{d}}$ such that, with probability at least $2/3$, it holds that $(1 - \epsilon) \cdot \overline{d} < \tilde{\mathsf{d}} < (1 + \epsilon) \cdot \overline{d}$.*

Working without a degree lower bound. For sake of simplicity, we start by modifying the algorithm so that when given a valid lower bound ℓ, it does not output an overestimation of the average degree (except with small probability). This is done by simply decreasing the output by a factor of $1 + \epsilon$. Thus, the output, $\tilde{\mathsf{d}}$, of the algorithm satisfies $\Pr[(1 - 2\epsilon)\overline{d} < \tilde{\mathsf{d}} < \overline{d}] \geq 2/3$. Furthermore, by $O(1) + \log \log n$ repetitions, we may reduce the probability of error to below $1/(6 \log n)$.

An interesting feature of our algorithm is that, with high probability, it does not output an overestimate of $\overline{d}$ even in case it is invoked with a parameter ℓ that is *higher* than the average degree $\overline{d}$ (i.e., is not a valid lower bound). To verify this feature, observe that the only place in the analysis where we rely on the assumption $\ell \leq \overline{d}$ is in bounding the underestimation error (i.e., when bounding the total number of edges with both endpoints in U). (We comment that also Feige's algorithm [8] has this feature, but for different reasons.)

This feature allows us to present a version of our algorithm that does not require an a priori lower bound on the average degree. Specifically, let our algorithm be denoted by A. Then, starting with $\ell = n/2$, we may proceed in at most $2 \log_2 n$ iterations as follows. We invoke A with the current value of ℓ, and let $\tilde{d}$ denote the output obtained. If $\tilde{d} \geq \ell$ then we halt and output $\tilde{d}$, otherwise we proceed to the next iteration while setting $\ell \leftarrow \ell/2$. In case all iterations were completed and still $\tilde{d} < \ell$ in the last iteration (i.e., $\tilde{d} < 1/2n$) then the graph must have no edges and we halt outputting $\tilde{d} = 0$.

Let $\ell_j = n/2^j$ be the parameter used in the j-th invocation of algorithm A, and let $\tilde{d}_j$ denote the corresponding output. Then, with probability at least $2/3$, for every iteration j that took place, it holds that $\tilde{d}_j \leq \overline{d}$ and if $\overline{d} \geq \ell_j$ then $\tilde{d}_j \geq (1 - 2\epsilon)\overline{d}$. In this case, assuming the graph contains any edges at all,[4] the algorithm will stop after at most $\log(n/\overline{d}) + O(1)$ iterations, and will output a value that is in the interval $[(1 - 2\epsilon)\overline{d}, \overline{d}]$.

Thus, the overall running-time of the algorithm is $\mathrm{poly}(\epsilon^{-1} \log n) \cdot \sqrt{n/\overline{d}}$. **Theorem 1 follows**.

3.2 A Lower Bound

We observe that any constant approximation algorithm must perform $\Omega(\sqrt{n})$ queries. A more general bound, which depends also on the approximation parameter $\epsilon > 0$ and on the actual degree of the graph, is stated next.

Theorem 3. *For any* n, $\overline{d} \in [2, o(n)]$ *and* $\epsilon \in (\omega(1/\overline{d}n), o(n/\overline{d}))$, *a* $(1 + \epsilon)$-*approximation of the average degree of* $G = (V, E)$ *requires* $\Omega((n/(\epsilon\overline{d}))^{1/2})$ *queries, where* $\overline{d} = 2|E|/n$. *This holds even if the algorithm is allowed neighbor and adjacency queries as well as degree queries.*

4   Approximating the Average Distance from a Single Source

Let $G = (V, E)$ be a simple undirected unweighted *connected* graph, where $n = |V|$ and $m = |E|$. For some given ("designated") vertex $s \in V$ we are interested in the average distance of s to the graph's vertices. That is, suppose we have access to an oracle that for any vertex $v \in V$ provides us with the distance, denoted $\mathrm{dist}_G(s, v)$, between s and v (in G). We would like to estimate the *average* distance, denoted $\overline{d}_G(s)$, of vertices in the graph from s; that is, $\overline{d}_G(s) = \frac{1}{n}\sum_{v \in V} \mathrm{dist}_G(s, v)$.

We first consider algorithms that make only distance queries. We present an algorithm (in Section 4.1) and a roughly matching lower bound (in Section 4.2). We later discuss the case in which the algorithm is also allowed neighbor queries (resp., only allowed neighbor queries); see Section 4.3 (resp., Section 4.4). All missing proofs can be found in the full version of this paper [12].

---

[4] In case the graph contains no edges, the algorithm will complete all iterations with no output (because $\overline{d} = 0 < \ell_j$ whereas $\tilde{d}_j = 0$ for each $j \leq 2 \log n$), and thus output the correct value (i.e., 0) at the last step. In this case, the overall running-time of the algorithm is $\mathrm{poly}(\epsilon^{-1} \log n) \cdot n$. Clearly one can modify the algorithm so that its complexity is never more that $O(n)$ (i.e., the complexity of computing the exact average degree), by stopping once ℓ_j goes below $\mathrm{poly}(\epsilon^{-1} \log n)/n$ for an appropriate polynomial in $\log n$ and ϵ^{-1}.

4.1   An Algorithm

We start with the basic version of our result.

Theorem 4. *There exists an algorithm that, for any given $\epsilon \in (0, 1)$, makes $O\left(\sqrt{n}/\epsilon^2\right)$ distance queries and provides a $(1+\epsilon)$-approximation of the average distance of a given vertex to all graph vertices.*

The algorithm selects uniformly and independently $q = \Theta\left(\sqrt{n}/\epsilon^2\right)$ vertices $v_1, \ldots, v_q$, performs the distance queries $\text{dist}_G(s, v_i)$ for $i = 1, \ldots, q$, and outputs the average of the answers received. We show that, with high probability, the algorithm's output is an $(1 + \epsilon)$-approximation of $\overline{d}_G$.

Let $d_{\max}$ be the maximum distance of any vertex v from s. For each value $i = 0, \ldots, d_{\max}$ let p_i denote the fraction of vertices at distance i from s. Let η be a random variable that takes value i with probability p_i, and let $\eta_1, \ldots, \eta_q$ be independent random variables that are distributed the same as η. By definition, $\text{Exp}[\eta] = \overline{d}_G(s)$, and the output of our algorithm is distributed as $\frac{1}{q}\sum_{j=1}^{q}\eta_j$. Hence, we are interested in upper bounding the probability that $\frac{1}{q}\sum_{j=1}^{q}\eta_j$ deviates from its expected value, $\overline{d}_G(s)$, by more than $\epsilon \cdot \overline{d}_G(s)$. By Chebyshev's inequality

$$\Pr\left[\left|\frac{1}{q}\sum_{j=1}^{q}\eta_j - \text{Exp}[\eta]\right| \geq \epsilon \cdot \text{Exp}[\eta]\right] \leq \frac{\text{Var}[\eta]}{q \cdot \epsilon^2 \cdot \text{Exp}[\eta]^2} \tag{2}$$

Since $q = \Theta\left(\sqrt{n}/\epsilon^2\right)$, it suffices to show that the ratio between $\text{Var}[\eta] = \text{Exp}[\eta^2] - \text{Exp}[\eta]^2$ and $\text{Exp}[\eta]^2$ is $O(\sqrt{n})$. This follows from the next lemma, by using $\ell = 1/2$.

Lemma 2. *For η and p_i as defined above, $\text{Exp}[\eta^2] \leq \sqrt{2n/\ell} \cdot \text{Exp}[\eta]^2$, for any $\ell \leq \text{Exp}[\eta]$.*

Since all distances are integers, and all are non-negative with the exception of $\text{dist}(s, s) = 0$, we know that $\text{Exp}[\eta] \geq \frac{n-1}{n} \geq 1/2$, which means that $\ell = 1/2$ can always be used. Thus, Theorem 4 follows from Lemma 2 (when specialized to the obvious case of $\ell = 1/2$), but we will use the more general statement of the lemma later.

Proof: By the definitions of η and $d_{\max}$,

$$\text{Exp}[\eta^2] = \sum_{i=0}^{d_{\max}} p_i \cdot i^2 \leq d_{\max} \cdot \text{Exp}[\eta] \tag{3}$$

We next observe that by definition of $d_{\max}$, for every $i \leq d_{\max}$ we have that $p_i \geq 1/n$, and so

$$\text{Exp}[\eta] = \sum_{i=0}^{d_{\max}} p_i \cdot i > \frac{d_{\max}^2}{2n} \tag{4}$$

By multiplying the bound $\text{Exp}[\eta] \geq \ell$ (provided in the lemma's hypothesis) by Eq. (4), we get that $\text{Exp}[\eta]^2 \geq \frac{\ell \cdot d_{\max}^2}{2n}$ and so

$$\frac{\sqrt{\ell} \cdot d_{\max}}{\sqrt{2n}} \leq \mathrm{Exp}[\eta] \tag{5}$$

Finally, we multiply Eq. (3) & (5) and get that

$$\mathrm{Exp}[\eta^2] \cdot \frac{\sqrt{\ell} \cdot d_{\max}}{\sqrt{2n}} \leq d_{\max} \cdot \mathrm{Exp}[\eta]^2 \tag{6}$$

and the lemma follows. $\square$

An improved algorithm. As in Section 3, a better algorithm can be obtained, provided we are given an a priori lower bound on the average distance. Denoting such a lower bound by ℓ, Lemma 2 implies that using a sample of size $q = \Theta(\epsilon^{-2} \cdot \sqrt{n/\ell})$ will do. Actually, as in Section 3, we do not actually need this lower bound, and the algorithm can function without it and perform as well. That is:

Theorem 5. *There exists an algorithm that, on input a graph $G = (V, E)$, a vertex s and parameter $\epsilon \in (0, 1)$, makes $O(\epsilon^{-2}(n/\overline{d}_G(s))^{1/2})$ distance queries and provides a $(1 + \epsilon)$-approximation of the average distance of vertices in G to s (i.e., $\overline{d}_G(s)$).*

4.2   A Lower Bound

In this subsection we establish the essential optimality of the algorithm presented in the previous subsection.

Theorem 6. *For any n, $\overline{d} \in (2, o(n))$ and $\epsilon \in (\omega(1/\overline{d}n), o(n/\overline{d}))$, any algorithm that performs only distance queries and provides a $(1 + \epsilon)$-approximation of the average distance of vertices in $G = (V, E)$ from $s \in V$, where $\overline{d}_G(s) = \overline{d}$, must ask $\Omega((n/(\epsilon\overline{d}))^{1/2})$ queries.*

Proof: For parameters n and $k \in (\omega(1), o(n))$, consider a (randomly labeled version of a) graph, denoted $G_{n,k}$, consisting of a star of $n - k$ vertices centered at s and a path of length k also starting at vertex s. (The reader may think of such a graph as a broom; see Figure 1.)

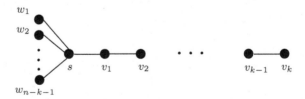

Fig. 1. An illustration of the "broom-like" graph $G_{n,k}$

By definition, the average distance of $G_{n,k}$ from s is

$$\overline{d}_{G_{n,k}}(s) = \frac{(n - k - 1) \cdot 1 + \sum_{i=1}^{k} i}{n} = 1 + \frac{k^2 - k - 2}{2n} = 1 + (1 - o(1)) \cdot \frac{k^2}{2n} \tag{7}$$

Given $\overline{d} \in (2, o(n))$ and $\epsilon \in (1/\sqrt{dn}, o(n/\overline{d}))$, we set k so that $1 + (k^2/2n) = \overline{d}$ (i.e., $k \approx (2(\overline{d} - 1)n)^{1/2})$ and $k' \approx (2((1 + \epsilon) \cdot \overline{d} - 1)n)^{1/2}$. Thus, $\overline{d}_{G_{n,k}}(s) = (1 - o(1)) \cdot \overline{d}$ and $\overline{d}_{G_{n,k'}}(s) = (1 + \epsilon) \cdot \overline{d}$. First, we observe that any $(1 + \epsilon)$-approximation algorithm must make $\Omega(n/k') = \Omega((n/(1 + \epsilon)\overline{d})^{1/2})$ queries in order to hit a vertex on the path (which is a necessary condition for distinguishing $G_{n,k}$ from $G_{n,k'}$). This establishes the claim for (say) $\epsilon > 1/10$. For the case of $\epsilon \leq 0.1$, we note that in order to distinguish $G_{n,k}$ from $G_{n,k'}$ the algorithm must hit one of the $k' - k$ vertices that are at distance greater than k from s in $G_{n,k'}$, which yields the lower bound of $\Omega(n/(k' - k)) = \Omega((n/\epsilon\overline{d})^{1/2})$. □

4.3 Adding Access to Neighbor and Adjacency Queries

A natural question is whether providing access to neighbor and adjacency queries, in addition to distance queries, can improve the query complexity of the average degree estimation problem. We answer this question negatively.

Theorem 7. *Let* $\overline{d} \in (2, o(n))$ *and* $\epsilon \in (\omega(1/\overline{d}n), o(n/\overline{d}))$, *and consider algorithms that are allowed distance queries, neighbor queries, adjacency queries and degree queries. Any such algorithm that provides a* $(1 + \epsilon)$-*approximation of the average distance of vertices in* $G = (V, E)$ *from* $s \in V$ *where* $\overline{d}_G(s) = \overline{d}$, *must perform* $\Omega((n/\epsilon\overline{d})^{1/2})$ *queries.*

4.4 Using Only Neighbor and Adjacency Queries

If we allow only neighbor and adjacency queries, then the problem becomes significantly harder.

Theorem 8. *Let* $k > 1$ *be a given approximation factor. Every algorithm that is allowed only neighbor, adjacency and degree queries must perform* $\Omega(m/(k \log k))$ *queries in order to obtain a* k-*approximation of* $\overline{d}_G(s)$ *in graphs* G *with* m *edges, provided* $m \in (\Omega(n), O(n^2/k \log k))$.

5 Approximating All-Pairs Average Distance

In continuation to Section 4, we now turn to the question of estimating the average distance between all pairs of vertices. That is, for any given graph G over n vertices, let $\overline{d}_G = \frac{1}{n^2} \sum_{u,v \in V} \text{dist}(u, v)$ denote the average distance between pairs of vertices in the graph.

Theorem 9. *There exists an algorithm that, on input a graph* $G = (V, E)$ *an a parameter* $\epsilon \in (0, 1)$, *makes* $O((n/\overline{d}_G)^{1/2}/\epsilon^2)$ *distance queries and provides a* $(1 + \epsilon)$-*approximation of* $\overline{d}_G$.

We also note that lower bounds analogous to the ones stated in Theorems 7 and 8 hold also for approximating the average of all-pairs distances (i.e., $\overline{d}_G$). That is, for a graph $G = (V, E)$, any $(1+\epsilon)$-approximation algorithm that uses distance queries must make $\Omega((n/\epsilon\overline{d}_G)^{1/2})$ queries, whereas any constant factor approximation algorithm that uses only standard queries must make $\Omega(|E|)$ such queries.

6   Extensions

The results of Sections 4 and 5 extend to the directed versions of these averaging problems: For the all-pairs problem, we require that the directed graph be strongly connected (so that all distances are defined). For the case of the single-source problem, it suffices to require that all vertices are reachable from the source.

Our algorithms for degree approximation have been recently extended to k-regular hypergraphs [1]. The complexity in this case is $\tilde{\Theta}\left(|V|^{\frac{k-1}{k}}\right)$.

Acknowledgments

We thank the anonymous referees of RANDOM for their comments. We also thank Kfir Barhum for pointing out an inaccuracy in a previous version.

References

1. K. Barhum. MSc. thesis, Weizmann Institute of Science. In preparation.
2. M. Bădoiu, A. Czumaj, P. Indyk, and C. Sohler. Facility Location in Sublinear Time. In *Proc. of the 32nd ICALP*, 2005.
3. B. Chazelle, R. Rubinfeld, and L. Trevisan. Approximating the Minimum Spanning Tree Weight in Sublinear Time. In *SIAM Journal on Computing*, Vol. 34, pages 1370–1379, 2005.
4. D. Coppersmith and S. Winograd. Matrix Multiplication via Arithmetic Progression. *Journal of Symbolic Computation*, Vol. 9, pages 251–280, 1990.
5. D. Dor, S. Halperin, and U. Zwick. All Pairs Almost Shortest Paths. *SIAM Journal on Computing*, Vol. 29, pages 1740–1759, 2000.
6. M. L. Elkin. Computing Almost Shortest Paths. Technical Report MCS01–03, Faculty of Mathematics and Computer Science, Weizmann Institute of Science, 2001.
7. S. Even. *Graph Algorithms*. Computer Science Press, 1979.
8. U. Feige. On sums of independent random variables with unbounded variance, and estimating the average degree in a graph. In *Proc. of the 36th STOC*, pages 594–603, 2004.
9. Z. Galil and O. Margalit. All Pairs Shortest Paths for Graphs with Small Integer Length Edges. *Information and Computation*, Vol. 54, pages = 243–254, 1997.
10. O. Goldreich and D. Ron. Property Testing in Bounded Degree Graphs. *Algorithmica*, Vol. 32 (2), pages 302–343, 2002.
11. O. Goldreich and D. Ron. A Sublinear Bipartitness Tester for Bounded Degree Graphs. *Combinatorica*, Vol. 19 (3), pages 335–373, 1999.
12. O. Goldreich and D. Ron. Approximating Average Parameters of Graphs. ECCC, TR05-073.
13. P. Indyk. Sublinear Time Algorithms for Metric Space Problems. in *Proc. of the 31st STOC*, pages 428–434, 1999.
14. T. Kaufman, M. Krivelevich, and D. Ron. Tight Bounds for Testing Bipartiteness in General Graphs. In *SIAM Journal on Computing*, Vol. 33, pages 1441–1483, 2004.
15. M. Parnas and D. Ron. Testing the diameter of graphs. *Random Structures and Algorithms*, Vol. 20 (2), pages 165–183, 2002.
16. R. Siedel. On the All-Pairs-Shortest-Path Problem in Unweighted Undirected Graphs. *Journal of Computer and System Sciences*, Vol. 51, pages 400–403, 1995.
17. U. Zwick. Exact and approximate distances in graphs - a survey. *Proceedings of the 9th Annual European Symposium on Algorithms (ESA)*, pages 33–48, 2001.

Local Decoding and Testing for Homomorphisms[*]

Elena Grigorescu, Swastik Kopparty, and Madhu Sudan

Massachusetts Institute of Technology, Cambridge, MA, USA
{elena_g, swastik, madhu}@mit.edu

Abstract. Locally decodable codes (LDCs) have played a central role in many recent results in theoretical computer science. The role of finite fields, and in particular, low-degree polynomials over finite fields, in the construction of these objects is well studied. However the role of group homomorphisms in the construction of such codes is not as widely studied. Here we initiate a systematic study of local decoding of codes based on group homomorphisms. We give an efficient list decoder for the class of homomorphisms from any abelian group G to a fixed abelian group H. The running time of this algorithm is bounded by a polynomial in $\log |G|$ and an agreement parameter, where the degree of the polynomial depends on H. Central to this algorithmic result is a combinatorial result bounding the number of homomorphisms that have large agreement with any function from G to H. Our results give a new generalization of the classical work of Goldreich and Levin, and give new abstractions of the list decoder of Sudan, Trevisan and Vadhan. As a by-product we also derive a simple(r) proof of the local testability (beyond the Blum-Luby-Rubinfeld bounds) of homomorphisms mapping $\mathbb{Z}_p^n$ to $\mathbb{Z}_p$, first shown by M. Kiwi.

1 Introduction

Given a pair of finite groups $G = (G, +)$ and $H = (H, \cdot)$, the class of homomorphisms between G and H forms an "error-correcting code". Namely, for any two distinct homomorphisms $\phi, \psi : G \to H$, the fraction of elements $\alpha \in G$ such that $\phi(\alpha) = \psi(\alpha)$ is at most $1/2$. This observation has implicitly driven the quest for many "homomorphism testers" [3, 2, 8, 1, 13], which test to see if a function $f : G \to H$ given as an oracle is close to being a homomorphism. In this paper, we investigate the complementary "decoding" question: Given oracle access to a function $f : G \to H$ find all homomorphisms $\phi : G \to H$ that are close to f.

To define the questions we study more precisely, let $\text{agree}(f, g)$ denote the agreement between $f, g : G \to H$, i.e., the quantity $\Pr_{x \leftarrow_U G}[f(x) = g(x)]$. Let $\text{Hom}(G, H) = \{\phi : G \to H \mid \phi(x + y) = \phi(x)\phi(y)\}$ denote the set of homomorphisms from G to H. We consider the *combinatorial* question: Given G, H and $\epsilon > 0$, what is the largest "list" of functions that can have ϵ-agreement with some

---

[*] Research supported in part by NSF Award CCR-0514915.

J. Diaz et al. (Eds.): APPROX and RANDOM 2006, LNCS 4110, pp. 375–385, 2006.

fixed function, i.e, what is $\max_{f:G\to H} |\{\phi : G \to H | \phi \in \text{Hom}(G, H), \text{agree}(f, g) \geq \epsilon\}|$?

We also consider the algorithmic question: Given G, H, $\epsilon > 0$ and oracle access to a function $f : G \to H$, (implicitly) compute a list of all homomorphisms $\phi : G \to H$ that have agreement ϵ with f. (A formal definition of implicit decoding will be given later. For now, we may think of this as trying to compute the value of ϕ on a set of generators of G.) We refer to this as the "local decoding" problem for homomorphisms.

Local decoding of homomorphisms for the special case of $G = \mathbb{Z}_2^n$ and $H = \mathbb{Z}_2$ was the central technical problem considered in the seminal work of Goldreich and Levin [4]. They gave combinatorial bounds showing that for $\epsilon = \frac{1}{2} + \delta$, the list size is bounded by $\text{poly}(1/\delta)$, and gave a local decoding algorithm with running time $\text{poly}(n/\delta)$.

The work of Goldreich and Levin was previously abstracted as decoding the class of degree one n-variate polynomials over the field of two elements. This led Goldreich, Rubinfeld, and Sudan [5] to generalize the decoding algorithm to the case of degree one polynomials over any finite field. (In particular, this implies a decoding algorithm for homomorphisms from $G = \mathbb{Z}_p^n$ to $H = \mathbb{Z}_p$, that decodes from $\frac{1}{p} + \epsilon$ agreement and runs in time $\text{poly}(n/\epsilon)$, where $\mathbb{Z}_p$ denotes the additive group of integers modulo a prime p.) Later Sudan, Trevisan, and Vadhan [11], generalized the earlier results to the case of higher degree polynomials over finite fields . This generalization, in turn led to some general reductions between worst-case complexity and average-case complexity.

Our work is motivated by the group-theoretic view of Goldreich and Levin, as an algorithm to decode group homomorphisms. While the group-theoretic view has been applied commonly to the complementary problem of "homomorphism testing", the decoding itself does not seem to have been examined formally before.

To motivate we start with a simple example.

Consider the case where $G = \mathbb{Z}_p^n$ and $H = \mathbb{Z}_p^m$. How many homomorphisms can have agreement $\frac{1}{p} + \delta$ with a fixed function $f : G \to H$? Most prior work in this setting used (versions) of the Johnson bound in coding theory. Unfortunately such a bound only works for agreement greater than $\frac{1}{\sqrt{p}}$ in this setting.[1] An ad-hoc counting argument gives a better bound on the list size of $\delta^{-O(m)}$. While better bounds ought to be possible, none are known, illustrating the need for further techniques. Our work exposes several such questions. It also sheds new light on some of the earlier algorithms.

Our results. Our results are restricted to the case of abelian groups G and H. Let $\Lambda = \Lambda_{G,H}$ denote the maximum possible agreement between two homomorphisms from G to H. Our main algorithmic result is an efficient algorithm, with running time $\text{poly}(\log |G|, \frac{1}{\epsilon})$ to decode all homomorphisms with agreement $\Lambda + \epsilon$

---

[1] For those familiar with the application of the Johnson bound in the setting of $m = 1$, we point out that it relied crucially on the fact that the agreement of any pair of homomorphisms was $\frac{1}{|H|}$ which is no longer true when $m \neq 1$.

with a function $f : G \to H$ given as an oracle, for any *fixed* group H. Note that in such a case the polynomial depends on H. See Theorem 2 for full details.

Crucial to our algorithmic result is a corresponding combinatorial one showing that there are at most $\text{poly}(\frac{1}{\epsilon})$ homomorphisms with agreement $\Lambda_{G,H} + \epsilon$ with any function $f : G \to H$, for any fixed group H. Once again, the polynomial in the bound depends on H. See Theorem 1 for details.

Finally, we also include a new proof of a result of Kiwi [8] on testing homomorphisms from $\mathbb{Z}_p^n$ to $\mathbb{Z}_p$. This is not related to our main quests, but we include it since some of the techniques we use to decode homomorphisms yield a simple proof of this result. See Theorem 3.

Organization of this paper. In Section 2 we present basic terminology and our main results. In Section 3 we exploit the decomposition theorem for abelian groups to reduce the proofs of the main theorems to the special case of p-groups. In Section 4 we tackle the combinatorial problem of the list-size for p-groups. In Section 5 we consider the corresponding algorithmic problem. Section 6 analyzes a homomorphism tester for functions from $\mathbb{Z}_p^n$ to $\mathbb{Z}_p$ using some techniques of the previous sections.

2   Definitions and Main Results

Let G, H be abelian groups, and let $\text{Hom}(G, H) = \{h : G \to H \mid h$ is a homomorphism$\}$. Note that $\text{Hom}(G, H)$ forms a *code*. Indeed, if $f, g \in \text{Hom}(G, H)$, then $G' = \{x \mid f(x) = g(x)\}$ is a subgroup of G. Since the largest subgroup of G has size at most $\frac{|G|}{2}$, it follows that f and g differ in at least $\frac{1}{2}$ of the domain. For two functions $f, g : G \to H$, define

$$\text{agree}(f, g) = Pr_{x \leftarrow_U G}[f(x) = g(x)],$$

and

$$\Lambda_{G,H} = \max_{f,g \in \text{Hom}(G,H), f \neq g} \{\text{agree}(f,\ g)\}.$$

In the case when $\text{Hom}(G, H)$ contains only the trivial homomorphism we define $\Lambda_{G,H} = 0$.

The notions of decodability and local list decoders are standard in the context of error correcting codes. Below we formulate them for the case of group homomorphisms.

Definition 1. *[11] (List decodability) The code* $\text{Hom}(G, H)$ *is* (δ, l)-*list decodable if for every function* $f : G \to H$, *there exist at most* l *homomorphisms* $h \in \text{Hom}(G, H)$ *such that* $\text{agree}(f, h) \geq \delta$.

Definition 2. *[14](Local list decoding) A probabilistic oracle algorithm $\mathcal{A}$ is a* (δ, T) *local list decoder for* $\text{Hom}(G, H)$ *if given oracle access to any function* $f : G \to H$, *(notation $\mathcal{A}^f$), the following hold:*

1. *With probability $\frac{3}{4}$ over the random choices of $\mathcal{A}^f$, $\mathcal{A}^f$ outputs a list of probabilistic oracle machines $M_1, \ldots, M_L$ s.t., for any homomorphism $h \in \mathrm{Hom}(G, H)$ with* $\mathrm{agree}(f, h) \geq \delta$,

$$\exists j \in [L], \forall x, \quad \Pr[M_j^f(x) = h(x)] \geq \frac{3}{4},$$

   where the probability is taken over the randomness of $M_j^f(x)$.
2. *$\mathcal{A}$ and each M_j^f run in time T.*

The model of computation with respect to groups is as follows. An abelian group G can be represented (see Sect. 3) by its cyclic decomposition $\mathbb{Z}_{p_1^{e_1}} \times \ldots \times \mathbb{Z}_{p_k^{e_k}}$, where p_i's are prime. An element of G is given by a vector $\alpha = (\alpha_1, \alpha_2, \ldots, \alpha_k)$, with $\alpha_i \in \mathbb{Z}_{p_i^{e_i}}$.

Our main results are the list decodability and local list decodability of group homomorphism codes.

Theorem 1. *Let H be a fixed finite abelian group. Then for all finite abelian groups G, $\mathrm{Hom}(G, H)$ is $\left(\Lambda_{G,H} + \epsilon, \mathrm{poly}_{|H|}(\frac{1}{\epsilon})\right)$ list decodable.*

Remark: The exact polynomial bound on the list size that our proof gives, in general, depends on the structure of the groups in an intricate way, but can nevertheless be uniformly bounded by $O\left(\frac{1}{\epsilon^{4 \log |H|}} |H|^5\right)$. Still, the precise bounds obtained by the proof are not optimal. For example, our proof gives that $\mathrm{Hom}(\mathbb{Z}_2^n, \mathbb{Z}_2^2)$ is $(\frac{1}{2} + \epsilon, O(\frac{1}{\epsilon^4}))$ list decodable, while it can be shown (via alternate means) that it is $(\frac{1}{2} + \epsilon, O(\frac{1}{\epsilon^2}))$ list decodable.

Theorem 2. *Let H be a fixed finite abelian group. Then for all finite abelian groups G there is a $(\Lambda_{G,H} + \epsilon, \mathrm{poly}_{|H|}(\log|G|, \frac{1}{\epsilon}))$ local list decoder for $\mathrm{Hom}(G, H)$.*

3   Decomposition and Reduction

We will embark on our quest by first decomposing the groups involved into slightly smaller but better-behaved groups. In this section we will see how these decompositions can be done and thereby reduce our main theorems to statements about list decoding on "p-groups". These statements will be proved in the following two sections by some Fourier analytic machinery and by generalizing the STV-style list decoders.

The structure theorem for finite abelian groups states that every abelian group G is of the form $\prod_{i=1}^{k} \mathbb{Z}_{p_i^{e_i}}$, where the p_i's are primes and the e_i's are positive integers. A *p-group* is a group of order p^r, for some positive integer r. The structure theorem implies that for any prime p, any finite abelian group G can be written as $G_p \times G'$, where G_p is a p-group and $\gcd(p, |G'|) = 1$ (take $G_p = \prod_{p_i = p} \mathbb{Z}_{p_i^{e_i}}$). This decomposition will play a crucial role in what follows.

Remark 1. $\Lambda_{G,H}$ behaves well under decomposition of G and H:

1. If $\gcd(|G|,|H|) = 1$ then $\mathrm{Hom}(G,H)$ contains only the trivial homomorphism and therefore, $\Lambda_{G,H} = 0$.
2. Otherwise, let p be the smallest prime s.t. $p \mid \gcd(|G|,|H|)$. Then $\Lambda_{G,H} = \frac{1}{p}$. Indeed, it is enough to bound $\mathrm{agree}(h,\mathbf{0})$, for any nontrivial homomorphism $h : G \to H$. Let $d = |\mathrm{image}\ (h)|$ and note that $d \mid |H|$, since $\mathrm{image}(h)$ is a subgroup of H. Since $G/\ker(h) \cong \mathrm{image}(h)$, it follows that $|\ker(h)|/|G| = 1/d \le 1/p$, and thus $\Lambda_{G,H} \le \frac{1}{p}$.
   Finally, if $G = \mathbb{Z}_{p^t} \times G'$, and $H = \mathbb{Z}_{p^r} \times H'$, then the homomorphism $h : G \to H$ definde by $h(a,b) = (ap^{r-1},0)$ satisfies $\mathrm{agree}(h,\mathbf{0}) = \frac{1}{p}$. Hence, $\Lambda_{G,H} = \frac{1}{p}$.
3. The above observations imply $\Lambda_{G_1 \times G_2,H} = \max\{\Lambda_{G_1,H},\Lambda_{G_2,H}\}$ and $\Lambda_{G,H_1 \times H_2} = \max\{\Lambda_{G,H_1},\Lambda_{G,H_2}\}$.

3.1   The Decompositions $G \to H_1 \times H_2$ and $G_1 \times G_2 \to H$

The following two propositions (whose proofs are omitted from this version) say that list decoding questions for $Hom(G,H)$ can be reduced to list decoding questions on summands of G or H.

Proposition 1. *Let G, H_1, H_2 be abelian groups. Let $a_i = \Lambda_{G,H_i}$. Suppose for all $\epsilon > 0$, $\mathrm{Hom}(G,H_i)$ is $(a_i + \epsilon, \ell_i(\epsilon))$-list decodable, with $(a_i + \epsilon, T_i(\epsilon))$ local list decoders, for $i = 1,2$. Then $\mathrm{Hom}(G,H_1 \times H_2)$ is $(\max\{a_1,a_2\} + \epsilon, \ell_1(\epsilon)\ell_2(\epsilon))$ list decodable and has a $(\max\{a_1,a_2\} + \epsilon, O\left((T_1(\epsilon)T_2(\epsilon))\right)$ local list decoder, for all $\epsilon > 0$.*

Proposition 2. *Let G_1, G_2, H be abelian groups. Let $a_i = \Lambda_{G_i,H}$. Suppose for all $\epsilon > 0$, $\mathrm{Hom}(G_i,H)$ is $(a_i + \epsilon, \ell_i(\epsilon))$-list decodable, with a $(a_i + \epsilon, T_i(\epsilon))$ local list decoder, for $i = 1,2$. Then $\mathrm{Hom}(G_1 \times G_2,H)$ is $(\max\{a_1,a_2\} + \epsilon, O(\frac{1}{\epsilon^2} \ell_1(\epsilon)\ell_2(\epsilon) |H|^2))$ list decodable, and has a $(\max\{a_1,a_2\}+\epsilon, O(\frac{|H|}{\epsilon^2} (T_1(\epsilon) + T_2(\epsilon)) + \ell_1(\epsilon)\ell_2(\epsilon) |H|^2)$ local list decoder, for all $\epsilon > 0$.*

3.2   Proof of the Main Theorems

Using the propositions proved in the previous section, our theorems will reduce to the main lemma given below. A proof is sketched in Section 4.

Lemma 1. *Let p be a fixed prime and $r > 0$ be a fixed integer. Then for any abelian p-group G, $\mathrm{Hom}(G,\mathbb{Z}_{p^r})$ is $\left(\frac{1}{p} + \epsilon, (2p)^{3r}\frac{1}{\epsilon^2}\right)$ list decodable.*

In Section 5, we shall use it to prove the corresponding algorithmic version.

Lemma 2. *Let p be a fixed prime and $r > 0$ be a fixed integer. Then for any abelian p-group G, $\mathrm{Hom}(G,\mathbb{Z}_{p^r})$ is $\left(\frac{1}{p} + \epsilon, \mathrm{poly}(\log |G|, \frac{1}{\epsilon})\right)$ locally list decodable.*

Proof (of Theorem 1). If $|G|, |H|$ are relatively prime then the result is obvious. Otherwise, let $p(= \frac{1}{A_{G,H}})$ be the smallest prime dividing both $|G|$ and $|H|$. Let $H = \prod_{i=1}^{r} \mathbb{Z}_{p_i^{\beta_i}}$. Let $i \in \{1, \ldots, r\}$. If $\gcd(p_i, |G|) = 1$, then $\mathrm{Hom}(G, \mathbb{Z}_{p_i^{\beta_i}})$ is $(\epsilon, 1)$ list decodable. Otherwise, write G as $G_{p_i} \times G'$, where G_{p_i} is a p_i-group and $\gcd(p_i, |G'|) = 1$. Then by Lemma 1 and Proposition 2, $\mathrm{Hom}(G, \mathbb{Z}_{p_i^{\beta_i}})$ is $\left(\frac{1}{p_i} + \epsilon, O(\frac{1}{\epsilon^4}(2p_i)^{3\beta_i} p^{2\beta_i}) \right)$ list decodable, and hence is also $\left(\frac{1}{p} + \epsilon, \frac{1}{\epsilon^4} p_i^{5\beta_i} \right)$ list decodable (since if $p_i || G|$, then $p \leq p_i$). Combining these for all $i \in \{1, \ldots, r\}$ by Proposition 1, $\mathrm{Hom}(G, H)$ is $\left(\frac{1}{p} + \epsilon, \prod_{p_i || G|} \frac{1}{\epsilon^4}(2p_i)^{5\beta_i} \right)$ list decodable, as required.

Proof (of Theorem 2). The proof of this theorem is directly analogous to the previous proof, using Lemma 2 instead of Lemma 1.

4   Combinatorial Bounds for p-Groups

In this section we will briefly touch upon how our main lemma (Lemma 1) is proved. Recall that we wish to obtain a combinatorial upper bound on the number of homomorphisms having agreement $\frac{1}{p} + \epsilon$ with a function $f : G \to \mathbb{Z}_{p^r}$, where G is a p-group. The starting point for our proof is the observation that $\mathbb{Z}_{p^r}$ is isomorphic to the multiplicative group μ_{p^r}, a subgroup of the complex numbers consisting of the p^rth roots of unity. This makes the tools of Fourier analysis available to us.

4.1   Sketch of the Argument

In this version we only give a sketch of the proof at a very high level. We are given a function $f : G \to \mathbb{Z}_{p^r}$. We begin by giving a formula that expresses the agreement between our function and any given homomorphism in terms of Fourier coefficients of some functions related to f. This will imply that every homomorphism having high agreement with f "corresponds" to some large Fourier coefficient. Now Parseval's identity tells us that there can only be few large Fourier coefficients, and the end of the proof looks near. Unfortunately, it is possible that many distinct homomorphisms "correspond" to the same Fourier coefficients. Nevertheless, we will be able to bound the number of occurences of the above pathology in terms of the number of homomorphisms in $\mathrm{Hom}(G, \mathbb{Z}_{p^l})$ that have high agreement with a related function $f' : G \to \mathbb{Z}_{p^l}$, for some $l < r$. Thus, inducting on r, we will arrive at the result.

In the proof we use the following version of the Johnson bound, which is the base case for the induction, and is also useful in Section 6.

Proposition 3. *Let G be a p-group. Then*

1. $\mathrm{Hom}(G, \mu_p)$ *is $(\frac{1}{p} + \epsilon, \frac{1}{\epsilon^2})$ list decodable, for any $\epsilon > 0$.*
2. *Let $f : G \to \mu_p$ and $\rho_t = \mathrm{agree}(f, \chi_t)$ for $\chi_t \in \mathrm{Hom}(G, \mu_p)$, then*

$$\sum_{\chi_t \in \mathrm{Hom}(G, \mu_p)} \left(\rho_t - \frac{1}{p-1}(1 - \rho_t) \right)^2 \leq 1.$$

5   Algorithmic Results for p-Groups

In this section we will turn our attention to the algorithmic decoding question suggested by the combinatorial results of the previous section. Here we will show Lemma 2 stated in Section 3.

Lemma 2. *Let p be a fixed prime and $r > 0$ be a fixed integer. Then for any abelian p-group G, $Hom(G, \mathbb{Z}_{p^r})$ is $\left(\frac{1}{p} + \epsilon, \text{poly}(\log |G|, \frac{1}{\epsilon})\right)$ locally list decodable.*

We will provide an algorithm which, given access to a function $f : G \to \mathbb{Z}_{p^r}$, with G a p-group, outputs an implicit representation of the homomorphisms that agree in a $\frac{1}{p} + \epsilon$ with f. Intuitively, to get the value of such a homomorphism $h \in Hom(G, \mathbb{Z}_{p^r})$ at a point x, we restrict our attention to a random coset of a random subgroup of G that contains x. Provided that h restricted to this coset has agreement at least $\frac{1}{p} + \epsilon/2$ with f, we can deduce the value of $h(x)$. Along the way we prove a lemma that says that random cosets of a random subgroup of a p-group "sample well", which is shown using the second moment method.

5.1   Cosets of Subgroups Generated by Enough Elements Sample Well

Definition 3. *Let G be an abelian group, and let $z_1, \ldots, z_k \in G$. Define $S_{z_1, \ldots, z_k}$ to be the subgroup of G generated by $z_1, \ldots, z_k$.*

Before giving our decoding algorithms, we state a useful lemma (whose proof is omitted in this version).

Lemma 3. *Let G be an abelian p-group, let $A \subseteq G$, with $\mu = \frac{|A|}{|G|}$ and let $x, z_1, \ldots z_k \in G$ be picked uniformly at random. Then*

$$Pr_{x, z_1, \ldots, z_k} \left[\left| \frac{|A \cap (x + S_{z_1, \ldots, z_k})|}{|S_{z_1, \ldots, z_k}|} - \mu \right| > \epsilon \right] \leq \frac{1}{\epsilon^2 p^k}.$$

5.2   The Generalized STV Algorithm

We begin with a simple but useful observation [3]: homomorphisms have simple and efficient self-correctors, i.e., for $g : G \to H$, there is a randomized procedure $Corr^g : G \to H$ running in time $\text{poly}(\log |G|)$ satisfying the following property

- *Self-corrector:* If $g : G \to H$ is such that there is some homomorphism $h : G \to H$ with $\text{agree}(g, h) > 7/8$, then with for all $x \in G$, $Corr^g(x) = h(x)$ with probability $> 3/4$.

Let $R_{x, z_1, \ldots, z_k}$ be the set $x + S_{z_1 - x, \ldots, z_k - x}$, i.e., the "affine subspace" passing through $x, z_1, \ldots, z_k$. Let $r_{x, z_1, \ldots, z_k} : [T]^k \to (x + S_{z_1 - x, \ldots, z_k - x})$ be the parametrization of $R_{x, z_1, \ldots, z_k}$ given by:

$$r_{x, z_1, \ldots, z_k}(\bar{\alpha}) = x + \sum_i \alpha_i (z_i - x).$$

For a function $g : G \to H$, define the restriction $g|_{R_{x,z_1,\dots,z_k}} : [T]^k \to H$ by $g|_{R_{x,z_1,\dots,z_k}}(\bar{\alpha}) = g(r_{x,z_1,\dots,z_k}(\bar{\alpha}))$. Notice that when we restrict homomorphisms to a set of the form $R_{x,z_1,\dots,z_k}$, we get an *affine homomorphism*, i.e., a function of the form $h + b$ where h is a homomorphism and $b \in H$.

The oracle $M^f_{z_1,\dots,z_k,a_1,\dots,a_k}(x)$:
For $b \in H$, define $h_b : [T]^k \to H$ by $h_b(\bar{\alpha}) = b + \sum \alpha_i(z_i - x)$.
1: For each b in H, estimate (by random sampling)
$l_b = \mathrm{agree}(f|_{R_{x,z_1,\dots,z_k}}, h_b)$.
2: If there is exactly one b with $l_b > \frac{1}{p} + \frac{\epsilon}{4}$ then output b, else fail.

The local list decoder:
Repeat $O(1)$ times:
1: Pick $z_1,\dots,z_k \in G$ uniformly and independently at random, where $k = c_1 \log_p \frac{1}{\epsilon}$.
2: For each $(a_1,\dots,a_k) \in H^k$, output $Corr^{M^f_{z_1,\dots,z_k,a_1,\dots,a_k}}$.

The analysis of the list-decoding algorithm is similar to that of [11] and we omit it in this version. It leads to the following lemma.

Lemma 4. *If h is a homomorphism s.t. $agree(h, f) \geq \frac{1}{p} + \epsilon$ then*

$$Pr_x[M^f_{z_1,\dots,z_k,h(z_1),\dots,h(z_k)}(x) = h(x)] \geq 7/8,$$

with probability $\frac{1}{2}$ over the choice of $z_1,\dots,z_k \in G$.

Proof of Lemma 2
Let h be a homomorphism that agrees with f on a $\frac{1}{p} + \epsilon$ fraction of points. Consider the oracle $M^f_{z_1,\dots,z_k,h(z_1),\dots,h(z_k)}$ (where the a_i are "consistent" with h). By Lemma 4, $M^f_{z_1,\dots,z_k,h(z_1),\dots,h(z_k)}(x)$ is correct on at least $\frac{15}{16} > \frac{7}{8}$ of the $x \in G$, and thus $Corr^{M^f_{z_1,\dots,z_k,h(z_1),\dots,h(z_k)}}$ computes h on all of G with probability at least $\frac{3}{4}$. It follows that each high-agreement homomorphism will appear w.h.p in the final list if the execution of the algorithm is repeated a constant number of times. This completes the proof of the lemma.

6   Homomorphism Tester

In this section we will prove a result of Kiwi using techniques related to Section 4. The result says that the 3 query linearity tester given below for homomorphisms in $\mathrm{Hom}(\mathbb{Z}_p^n, \mu_p)$ has very good acceptance probability/maximum agreement trade-offs. In particular, its performance is far better than that of the BLR [3] test for $p > 2$.

Given $f : \mathbb{Z}_p^n \to \mu_p$.

We are analyzing the following linearity test:

- Pick $x, y \in \mathbb{Z}_p^n$, $\alpha, \beta \in \mathbb{Z}_p^*$ uniformly at random
- Accept if $f(\alpha x + \beta y) = f(x)^\alpha f(y)^\beta$, else reject.

Kiwi [8] analyzed this test to get the following theorem.

Theorem 3. *Suppose f passes the above test with probability δ, then f has agreement at least δ with some homomorphism in $\mathrm{Hom}(\mathbb{Z}_p^n, \mu_p)$.*

In fact, [8] proved a more general result for testing vector-space homomorphisms over any finite field $\mathbb{F}_q^n \to \mathbb{F}_q$, not necessarily over prime fields. His proof uses the MacWilliams identities and properties of the Krawchouk polynomials. Here we give a simple proof of the above theorem using elementary Fourier analysis. Our proof also generalizes to the case of vector-space homomorphisms (using Trace functions) though we don't include the proof in this version.

Proof. The proof will use Fourier analysis, and modeled along the general lines of the argument in [2] (i.e., expressing agreement and acceptance probabilities in terms of Fourier coefficients).

For $\eta \in \mu_p$, define $\mathcal{S}(\eta) = \mathbb{E}_{c \in \mathbb{Z}_p^*}[\eta^c]$. It is easily seen that

$$\mathcal{S}(\eta) = \begin{cases} 1, & \text{if } \eta = 1 \\ \frac{-1}{p-1}, & \text{otherwise} \end{cases}$$

Recall that every homomorphism from $\mathbb{Z}_p^n \to \mu_p$ is a character χ_t for some $t \in \mathbb{Z}_p^n$, where $\chi_t(x) = e^{2\pi i (t \cdot x)/p}$. For $f : \mathbb{Z}_p^n \to \mathbb{C}$, the Fourier coefficient $\hat{f}(t)$ is defined to be $\mathbb{E}_{x \in \mathbb{Z}_p^n} f(x) \overline{\chi_t}(x)$. We will assume some familiarity with basic properties of characters and Fourier coefficients in this version of the paper.

For $t \in \mathbb{Z}_p^n$ let ρ_t be the agreement of f with χ_t. We shall prove that $\delta \le \max_{t \in \mathbb{Z}_p^n} \rho_t$. This will prove the result.

We begin by finding an explicit formula for ρ_t in terms of the Fourier coefficients.

$$\rho_t - \frac{1}{p-1}(1 - \rho_t) = \mathbb{E}_{x \in \mathbb{Z}_p^n}[\mathcal{S}(f(x)\overline{\chi_t}(x))] = \mathbb{E}_{x \in \mathbb{Z}_p^n, c \in \mathbb{Z}_p^*}[f(x)^c \overline{\chi_t}(x)^c] \quad (1)$$

$$= \mathbb{E}_{c \in \mathbb{Z}_p^*} \mathbb{E}_{x \in \mathbb{Z}_p^n}[f(x)^c \overline{\chi_{ct}}(x)] = \mathbb{E}_{c \in F_p^*}[\hat{f^c}(ct)] \quad (2)$$

We now find a similar formula for δ and perform some manipulations that allow us to relate it to our formula for ρ_t.

$$\delta - \frac{1}{p-1}(1 - \delta) = \mathbb{E}_{x,y \in \mathbb{Z}_p^n} \mathbb{E}_{\alpha,\beta \in \mathbb{Z}_p^*} \left[\mathcal{S}\left(f(x)^\alpha f(y)^\beta f(\alpha x + \beta y)^{-1} \right) \right] \quad (3)$$

$$= \mathbb{E}_{x,y \in \mathbb{Z}_p^n} \mathbb{E}_{\alpha,\beta \in \mathbb{Z}_p^*} \left[\mathbb{E}_{c \in \mathbb{Z}_p^*}[f(x)^{c\alpha} f(y)^{c\beta} f(\alpha x + \beta y)^{-c}] \right] \quad (4)$$

$$= p^n \mathbb{E}_{x,y,z} \mathbb{E}_{\alpha',\beta',\gamma'} \left[f(x)^{\alpha'} f(y)^{\beta'} f(z)^{\gamma'} \mathbf{1}(\alpha' x + \beta' y + \gamma' z = 0) \right] \quad (5)$$

where we substituted $\alpha' = c\alpha, \beta' = c\beta, \gamma' = -c, z = \alpha x + \beta y$ (and one verifies that $z = \alpha x + \beta y$ is equivalent to $\alpha' x + \beta' y + \gamma' z = 0$). Note that since $\gamma' \in \mathbb{Z}_p^*$, the probability that a random $z \in \mathbb{Z}_p^n$ is such that $\alpha' x + \beta' y + \gamma' z = 0$ is $\frac{1}{p^n}$.

$$(5) = p^n \mathbb{E}_{x,y,z} \mathbb{E}_{\alpha',\beta',\gamma'} \left[f(x)^{\alpha'} f(y)^{\beta'} f(z)^{\gamma'} \mathbb{E}_{t \in \mathbb{Z}_p^n} [\overline{\chi_t}(\alpha' x + \beta' y + \gamma' z)] \right]$$

$$= p^n \mathbb{E}_t \left[\mathbb{E}_{\alpha',\beta',\gamma'} \mathbb{E}_x \left[f(x)^{\alpha'} \overline{\chi_{\alpha' t}}(x) \right] \mathbb{E}_y \left[f(y)^{\beta'} \overline{\chi_{\beta' t}}(y) \right] \mathbb{E}_z \left[f(z)^{\gamma'} \overline{\chi_{\gamma' t}}(z) \right] \right]$$

$$= \sum_t \left[\mathbb{E}_{\alpha',\beta',\gamma'} \left[\hat{f^{\alpha'}}(\alpha' t) \hat{f^{\beta'}}(\beta' t) \hat{f^{\gamma'}}(\gamma' t) \right] \right]$$

$$= \sum_t \left(\mathbb{E}_{\alpha' \in \mathbb{Z}_p^*} [\hat{f^{\alpha'}}(\alpha' t)] \right)^3$$

$$= \sum_t \left(\rho_t - \frac{1}{p-1}(1 - \rho_t) \right)^3 \quad \text{(By (2))}$$

Simplifying the last expression and using Proposition 3 we get $\delta \leq \max_t \rho_t$.

Acknowledgments

Thanks to Amir Shpilka for many valuable discussions.

References

1. Michael Ben-Or, Don Coppersmith, Michael Luby, Ronitt Rubinfeld, Non-Abelian Homomorphism Testing, and Distributions Close to their Self-Convolutions. RANDOM 2004.
2. Mihir Bellare and Don Coppersmith and Johan Håstad and Marcos Kiwi and Madhu Sudan. Linearity testing over characteristic two. *IEEE Transactions on Information Theory*, 42(6), 1781-1795, 1996.
3. Manuel Blum and Michael Luby and Ronitt Rubinfeld. Self-Testing/Correcting with Applications to Numerical Problems. *Journal of Computer and System Sciences*, 47(3), 549-595, 1993.
4. Oded Goldreich and Leonid Levin. A hard-core predicate for all one-way functions. *Proceedings of the 21st Annual ACM Symposium on Theory of Computing*, 25–32, 1989
5. Oded Goldreich and Ronitt Rubinfeld and Madhu Sudan. Learning polynomials with queries: The highly noisy case. *SIAM Journal on Discrete Mathematics*, 13(4):535-570, 2000.
6. Venkatesan Guruswami and Madhu Sudan. List decoding algorithms for certain concatenated codes. *Proceedings of the 32nd Annual ACM Symposium on Theory of Computing*, 181-190, 2000.
7. Marcos Kiwi , Frédéric Magniez , Miklos Santha. Exact and approximate testing/correcting of algebraic functions: A survey. *Theoretical Aspects of Computer Science*, Teheran, Iran, Springer-Verlag, LNCS 2292, 30-83, 2002.
8. Marcos Kiwi. Testing and weight distributions of dual codes. *Theoretical Computer Science*, 299(1–3):81-106, 2003.

9. Eyal Kushilevitz and Yishay Mansour. Learning decision trees using the Fourier spectrum. SIAM Journal on Computing 22(6):1331-1348, 1993.
10. Dana Moshkovitz, Ran Raz. Sub-Constant Error Low Degree Test of Almost Linear Size, STOC 2006.
11. Madhu Sudan and Luca Trevisan and Salil Vadhan. Pseudorandom generators without the XOR lemma, *Proceedings of the 31st Annual ACM Symposium on Theory of Computing* 537-546, 1999.
12. Madhu Sudan. Algorithmic Introduction to Coding Theory. Lecture Notes, 2001.
13. Amir Shpilka and Avi Wigderson. Derandomizing Homomorphism Testing in General Groups. *Proceedings of the 36th Annual ACM Symposium on Theory of Computing (STOC)*, pp. 427-435, 2004.
14. L. Trevisan. Some Applications of Coding Theory in Computational Complexity. Survey Paper. *Quaderni di Matematica* 13:347-424, 2004

Worst-Case Vs. Algorithmic Average-Case Complexity in the Polynomial-Time Hierarchy

Dan Gutfreund[*]

Division of Engineering and Applied Sciences,
Harvard University, Cambridge, MA 02138
danny@eecs.harvard.edu

Abstract. We show that for every integer $k > 1$, if Σ_k, the k'th level of the polynomial-time hierarchy, is *worst-case* hard for probabilistic polynomial-time algorithms, then there is a language $L \in \Sigma_k$ such that for every probabilistic polynomial-time algorithm that attempts to decide it, there is a *samplable* distribution over the instances of L, on which the algorithm errs with probability at least $1/2 - 1/poly(n)$ (where the probability is over the choice of instances and the randomness of the algorithm). In other words, on this distribution the algorithm essentially does not perform any better than the algorithm that simply decides according to the outcome of an unbiased coin toss.

1 Introduction

Suppose that NP is worst-case hard. This means that every efficient algorithm A fails to solve SAT correctly on an infinite sequence of instances . A very natural question is the following: given the description of such an algorithm A, how hard is it to generate instances from this sequence? I.e. given an input length n, what is the complexity of finding an instance of length n on which A errs (and assume for now that for every n such an instance exist). Clearly, by exhaustive search one can do that in exponential time. Surprisingly, Gutfreund, Shaltiel and Ta-Shma [GSTS05] showed that it can actually be done in probabilistic polynomial-time with a constant probability of success.

Let A be a probabilistic polynomial-time algorithm trying to decide some language L. We say that a distribution $\mathcal{D}_A$ over instances of L is δ-hard for A, if with probability at least δ, A fails to decide correctly whether an instance x drawn from $\mathcal{D}_A$ is in L or not (where the probability is over the choice of x and the randomness of A).

Informally, the result of [GSTS05] says that if NP $\not\subseteq$ BPP, then there exist a language $L \in$ NP such that for every probabilistic polynomial-time algorithm A that tries to decide L, there exists a polynomial-time samplable distribution $\mathcal{D}_A$ that is δ-hard for A, where $0 < \delta < 1$ is some universal constant. By standard techniques, the result of [GSTS05] extends to any level of the polynomial-time

---

[*] Research supported by ONR grant N00014-04-1-0478. Part of this research was done while the author was at the Hebrew University.

hierarchy. I.e. for every integer $k > 0$, $\Sigma_k \not\subseteq \mathrm{BPP}$ implies that there is a language $L \in \Sigma_k$ such that for every efficient algorithm that attempts to decide L, there exists a δ-hard samplable distribution (with the same constant δ).

While the proof of [GSTS05] do not try to optimize δ, their technique imposes a barrier of $1/3$ on this constant. We will explain later the origin of this barrier (see Section 3.2). The authors of [GSTS05] ask whether this constant δ can be improved. In particular, whether it can be arbitrarily close to $1/2$. Note that $1/2$ is the best one can hope for since an algorithm that decides according to an unbiased coin toss will always give a correct answer (on every instance) with probability $1/2$.

In this paper we partially solve this open question. While we cannot improve the constant δ for NP, we can do so for every other level of the polynomial-time hierarchy (from the second and up). We show,[1]

Theorem 1. *For every integer $k > 1$ and a constant $c > 0$, there exist a language $L \in \Sigma_k$, such that if $\Sigma_k \not\subseteq BPP$ then for every probabilistic polynomial-time algorithm A that tries to decide L, there exists a polynomial-time samplable distribution $\mathcal{D}_A$ that is $(1/2 - n^{-c})$-hard for A.*

In other words, we show that if Σ_k is worst-case hard for efficient probabilistic algorithms, then for every such algorithm that tries to decide some language $L \in \Sigma_k$, there exist a samplable distribution on which the algorithm essentially does not preform any better than the simplest algorithm one can think of. I.e. the one that decides according to the outcome of an unbiased coin toss!

In fact, we show that the statement of Theorem 1 already holds for the class $\mathrm{P}_{||}^{\mathrm{NP}}$ of languages that can be decided by a deterministic polynomial-time Turing machine that has a non-adaptive access to an NP-oracle (see the statement of Theorem 5).

1.1   Motivation

Understanding the connections between worst-case and average-case complexities is fundamental to the fields of computational complexity and the foundations of cryptography. Worst-case complexity is a convenient and standard measure. On the other hand, average-case complexity may be a more realistic approach to measure the complexity of problems on instances that actually appear in practice. Furthermore, having access to an average-case hard problem, or in other words, being able to efficiently produce instances that are hard to solve, can be used for cryptography or derandomization.

It is well known that for complexity classes that contain the polynomial-time hierarchy, such as EXP, PSPACE and $\sharp$P, their worst-case hardness (against small circuits, or efficient algorithms) is equivalent to their average-case hardness (with respect to the uniform distribution) [Lip91, BFNW93, IW97, STV99, TV02]. However, such connections are not known to hold for NP or any other level of the polynomial-time hierarchy. The general techniques that are used to

---

[1] In Section 3 we give a more formal statement using the notations of [GSTS05].

prove worst-case/average-case connections for EXP and PSPACE do not apply to these seemingly lower classes. Furthermore it is known that certain proof techniques are unlikely to prove worst-case/average-case connections within the polynomial-time hierarchy [FF93, BT03, Vio03, GTS06].

The average-case complexity that we study here (and in [GSTS05]) is different to the standard notion that is used in, e.g., cryptography. In the standard setting that was defined by Levin [Lev86] (see also [Imp95]), a language is average-case hard if there exist *a single* samplable distribution that is hard for *every* efficient algorithm. Here we study the notion that was formulated by Kabanets [Kab01], in which the order of quantifiers is swapped. That is, now a language is hard on the average if for every efficient algorithm there exist a (possibly different) hard distribution. This notion seems to better capture "easiness" on the average, and we therefore refer to it as algorithmic average-case complexity (see [GSTS05] for a more detailed discussion about the two notions).

The result of [GSTS05] can be seen as a weak form of worst-case to average-case reduction within NP, and it is the first such connection under a natural notion of average-case complexity. Hopefully, it is a first step towards establishing worst-case/average-case connections in the usual sense of average-case complexity. Indeed very recently, Gutfreund and Ta-Shma [GTS06] used [GSTS05] to obtain new connections between derandomization, worst-case hardness and average-case hardness (under the standard notion) in a setting in which no such connections were known previously. We therefore believe that further investigating this notion of average-case complexity is an important task.

The applications of average-case hardness to cryptography and derandomization require extremely hard on the average languages. I.e. languages for which there is a samplable distribution (typically the uniform one) that is $(1/2 - 1/p(n))$-hard for an arbitrary large polynomial $p(n)$. Thus obtaining optimal hardness on the average from worst-case hardness, as we do in this paper, may hopefully find applications in cryptography or complexity theory.

Finally we mention that the average-case notion studied here is most suitable to understanding the power of heuristics for hard problems. As explained in [GSTS05], their result says that if NP is worst-case hard then no heuristic for SAT can do too well on every samplable distribution. In other words, every heuristic that solves SAT well in practice, must be bound to a specific samplable distribution (or a family of distributions). Still, their result left open the possibility that there is a heuristic that does quite well on every samplable distribution, i.e. succeeds with probability 2/3. In this paper we show that for complete problems in higher levels of the hierarchy even this is impossible.

2   Preliminaries

We assume that the reader is familiar with standard complexity classes such as NP and BPP. We define inductively the levels of the polynomial-time hierarchy: $\Sigma_1 = \text{NP}$, and $\Sigma_{k+1} = \text{NP}^{\Sigma_k}$. Where NP^A represents the class of languages that can be decided by a polynomial-time nondeterministic oracle machine with access to an oracle that solves some problem in the class A. The polynomial-time hierarchy is defined to be $PH = \bigcup_k \Sigma_k$. P^{NP} is the class of languages that can be decided by a deterministic polynomial-time oracle machine that has access to an NP-oracle. $P_{||}^{NP}$ is the same class with the restriction that the machine can only make non-adaptive queries to the NP-oracle.

For a language $L \subseteq \{0,1\}^*$, $L(x)$ is the characteristic function of L, i.e. $L(x) = 1$ if $x \in L$ and 0 otherwise. For a string $y \in \{0,1\}^*$, we denote by y_i the i'th bit in the string. $[n]$ denotes the set $\{1, \ldots, n\}$. For a set S we denote by $x \in_R S$ that x is chosen uniformly from S. A distribution $\mathcal{D}$ over $\{0,1\}^*$ is an ensemble of distributions $\{\mathcal{D}_n\}_{n \in \mathbb{N}}$, where $\mathcal{D}_n$ is a distribution over $\{0,1\}^n$. For a distribution $\mathcal{D}$ we denote by $x \leftarrow \mathcal{D}$, that x is a sample from $\mathcal{D}$. Let $A(\cdot;\cdot)$ be a probabilistic TM, using $m(n)$ bits of randomness on inputs of length n. We say that A is a sampler for the distribution $\mathcal{D} = \{\mathcal{D}_n\}_{n \in \mathbb{N}}$, if for every n, the random variable $A(1^n; y)$ is distributed identically to $\mathcal{D}_n$, where the distribution is over the random string $y \in_R \{0,1\}^{m(n)}$. In particular, A always outputs strings of length n on input 1^n. If A runs in time $t(n)$ we say that $\mathcal{D}$ is samplable in time $t(n)$. If $t(n)$ is a fixed polynomial, we simply say that $\mathcal{D}$ is samplable.

2.1   The Class Pseudo BPP

We now define the average-case notion used in [GSTS05].

Definition 1. *Let $L \subseteq \{0,1\}^*$ be a language, $\mathcal{D} = \{\mathcal{D}_n\}_{n \in \mathbb{N}}$ a distribution over $\{0,1\}^*$, $A(\cdot;\cdot)$ a probabilistic algorithm, and $\delta(n) : \mathbb{N} \rightarrow [0,1]$. We say that $\mathcal{D}$ is $(\delta(n), L)$-hard for A if for every large enough n,*

$$\Pr_{x \leftarrow \mathcal{D}_n, r} [A(x;r) \neq L(x)] \geq \delta(n)$$

If the above holds only for infinitely many n's we say that $\mathcal{D}$ is infinitely often $(\delta(n), L)$-hard for A.

We now define average-case classes under this notion. We use the definition and notations of Kabanets [Kab01].

Definition 2. *(Pseudo BPP) Let $\delta(n) : \mathbb{N} \rightarrow [0,1]$. We say that $L \in Pseudo_{1-\delta(n)}$ BPP if there exists a probabilistic polynomial-time algorithm A, such that no samplable distribution is infinitely often $(\delta(n), L)$-hard for A.*

In this notation, the result of [GSTS05] (generalized to the polynomial time hierarchy) is,

Theorem 2. *For some universal constant $0 < \delta < 1/3$, and for every integer $k \geq 1$,*

$$\Sigma_k \nsubseteq BPP \implies \Sigma_k \nsubseteq Pseudo_{1-\delta} BPP$$

2.2   List-Decodable Codes

Definition 3. *A binary code $C = \{C_n\}$ is a family of bijections $C_n : \{0,1\}^n \to \{0,1\}^{m(n)}$.*

A binary code C is (ϵ, ℓ)-list-decodable if for every n and every $x \in \{0,1\}^m$,

$$|\{y \in \{0,1\}^n : \Delta(x, C_n(y)) \leq 1/2 - \epsilon\}| \leq \ell$$

where Δ is the relative Hamming distance.

A binary (ϵ, ℓ)-list-decodable code C is computationally efficient if there is a pair of algorithms (Enc, Dec), such that,

1. *Enc computes the function C_n for every n, and its running time is polynomial in n and $\frac{1}{\epsilon}$. (In particular, $m(n, \epsilon) = poly(n, \frac{1}{\epsilon})$.)*
2. *Dec, on input $x \in \{0,1\}^m$, outputs all the strings $y \in \{0,1\}^n$ such that $\Delta(x, C_n(y)) \leq 1/2 - \epsilon$. Its running time is polynomial in n and $\frac{1}{\epsilon}$. (In particular, $\ell = \ell(n, \epsilon) = poly(n, \frac{1}{\epsilon})$.)*

Theorem 3. *[STV99] A computationally efficient binary (ϵ, ℓ)-list-decodable code exists.*

3   The Main Theorem and an Overview of the Techniques

3.1   The Main Theorem

We now state our main theorem (Theorem 1) in a formal way, using the notations from Section 2.1.

Theorem 4. *For every integer $k > 1$ and constant $c > 0$,*

$$\Sigma_k \not\subseteq BPP \;\Rightarrow\; \Sigma_k \not\subseteq Pseudo_{1/2+n^{-c}} BPP$$

The theorem follows directly from the following theorem.

Theorem 5. *For every constant $c > 0$,*

$$NP \not\subseteq BPP \;\Rightarrow\; P_{||}^{NP} \not\subseteq Pseudo_{1/2+n^{-c}} BPP$$

The proof of Theorem 5 appears in Section 4. To see why Theorem 4 follows from Theorem 5, note that for every integer $k > 1$, $P_{||}^{NP} \subseteq \Sigma_k$. Also it is well known that $NP \not\subseteq BPP$ if and only if $\Sigma_k \not\subseteq BPP$ (again for every $k > 1$). Therefore $\Sigma_k \not\subseteq BPP \Rightarrow NP \not\subseteq BPP \Rightarrow P_{||}^{NP} \not\subseteq Pseudo_{1/2+n^{-c}} BPP \Rightarrow \Sigma_k \not\subseteq Pseudo_{1/2+n^{-c}} BPP$. In fact, this shows that the statement of Theorem 4 holds for every class that is contained in the polynomial-time hierarchy and contains the class $P_{||}^{NP}$.

The rest of this section is devoted to a discussion about the techniques. We start by explaining why previous techniques fail to prove Theorem 5, and in particular we explain the origin of the $1/3$ barrier in [GSTS05]. We then describe our techniques.

3.2   [GSTS05] and the 1/3 Barrier

Let us briefly recall what is shown in [GSTS05]. They show how to construct, given a description of an algorithm A, a samplable distribution $\mathcal{D}_A$ such that if A does too well on $\mathcal{D}_A$ then a modification of A actually solves SAT on the worst-case. Thus they obtain a contradiction to the worst-case hardness of NP.

The proof shows how to efficiently generate a distribution over triplets of Boolean formulas of the form $\psi = \phi(\alpha_1, \ldots, \alpha_i, x_{i+1}, \ldots, x_n)$, $\psi^0 = \phi(\alpha_1, \ldots, \alpha_i, 0, x_{i+2}, \ldots, x_n)$, and $\psi^1 = \phi(\alpha_1, \ldots, \alpha_i, 1, x_{i+1}, \ldots, x_n)$. Where ϕ is a formula on n Boolean variables and $\alpha_1, \ldots, \alpha_i$ is a partial assignment to these variables. The analysis shows that A must give, with high probability, contradicting answers regarding the satisfiability of these formulas (i.e. claiming that ψ is satisfiable while ψ^0 and ψ^1 are not). In other words at this point we know that A must err on at least one of these formulas. The problem is that the formulas are obtained probabilistically. So it is possible that the probability over the coin tosses of the procedure that generates ψ, ψ^0, ψ^1, that the first, the second, or the third formula is the hard one for A, is the same probability $1/3$, while A gives the correct answer on the other two. Thus the best we can do is to pick one of the three at random. Therefore with probability at most $1/3$ we obtain an instance on which A makes an error.

The natural approach to obtain an even harder language on the average, is to apply standard hardness amplification techniques such as Yao's XOR Lemma [Yao82], and more generally direct product theorems. It turns out that this approach does not work for our algorithmic average-case notion (i.e. for pseudo classes). Roughly speaking, the reason is that in our setting, sampling from a hard distribution for a specific algorithm may take more time than running the algorithm. In particular the algorithm cannot run subroutines that sample from the distribution that is hard for it. In all the direct product proofs, the algorithm is required to run a reduction from a mildly hard on the average language to an extremely hard on the average language, and this reduction samples from the hard distribution for the algorithm.

3.3   Our Technique

Our proof is inspired by the argument of Sudan, Trevisan and Vadhan [STV99] that shows the equivalence between the worst-case complexity and average-case complexity of EXP. Similar to their proof, we will encode the truth-tables of a langauge (SAT in our case) at different lengths, to obtain truth-tables of a new language. We will argue that unless SAT is easy on the worst-case, the new language must be extremely hard on the average (in the algorithmic notion of average-case complexity).

As in [STV99] our encoding will be a concatenation of two codes, however, for our first encoding we will not use an error correcting code as they do,[2] but

---

[2] Using an error-correcting code results in a language that is computable in exponential-time, while we want this language to be computable within the polynomial-time hierarchy.

rather an encoding that encodes satisfying assignments. Roughly speaking, it will have the property that if an oracle can compute correctly an ϵ fraction of the entries in the encoded word, then we can use this oracle to solve the *search problem* of SAT on an ϵ fraction of SAT instances (solving the search problem on a given Boolean formula means to find a satisfying assignment if it exists and to answer "unsatisfiable" otherwise).

To justify the usefulness of this type of encoding, we make the following observation: when applying the argument of [GSTS05] directly on algorithms that are trying to solve the *search problem* of SAT (rather than the decision problem), the 1/3 barrier does not show up. That is, if an efficient algorithm solves the search problem with probability at least ϵ on every samplable distribution, then the algorithm actually solves SAT on the worst-case.

To obtain such an encoding, we use a "parallel" search to decision reduction due to Ben-David et. al. [BDCGL90], that is based on the reduction to unique solutions of Valiant and Vazirani [VV86]. This (randomized) reduction shows how to solve the search problem of SAT given an oracle that solves the decision problem. For every SAT instance, the encoding will contain a (non-Boolean) entry of the answers an NP-oracle would give to the queries that the reduction makes on that instance. Since the reduction is randomized, our encoding will contain a different entry for every possible random string for the reduction.

Finally, we will concatenate the first encoding with a good binary list-decodable code. That is, we will use such a code to encode each entry in the first code. This step, as in [STV99], follows the reasoning of Goldreich and Levin [GL89].

The final encoding will be the truth table of our new language. We will show that if an efficient algorithm solves this language with probability at least $1/2+\epsilon$ with respect to every samplable distribution, then we can solve the search problem of SAT with probability (approximately) ϵ with respect to every samplable distribution. This means, by our first observation, that we can solve SAT on every input. However, this contradicts the hypothesis that NP is worst-case hard.

4   The Proof of Theorem 5

4.1   Finding Incorrect Instances for Search Algorithms

Our first observation is that the argument of [GSTS05] gives a better conclusion when applied directly to search algorithms. That is, if NP is worst-case hard, we can efficiently sample (with high probability over the the coins of the sampler) a formula on which the algorithm errs (with high probability over the coins of the algorithm). We make this statement precise in Lemma 1 below. The proof of the lemma, which is omitted due to space limitations,[3] follows the outline and the ideas of [GSTS05]. In our case, though, we need to be much more accurate with the analysis of the success probability.

---

[3] The proof appears in a longer version of this paper that can be obtained from http://www.eecs.harvard.edu/~danny.

Lemma 1. *If NP $\not\subseteq$ BPP then for every constant $b > 0$, and every probabilistic polynomial-time algorithm SSAT that tries to solve the search problem of SAT, there is a constant $\delta > 0$ and a samplable distribution that for infinitely many input lengths samples with probability at least $1 - 2^{-n^{\delta}}$, a formula for which SSAT gives the correct answer (a satisfying assignment or 'no') with probability at most n^{-b} (over its random coin tosses).*

4.2 A "Parallel" Search to Decision Reduction

The next ingredient in our proof is a (randomized) search to decision reduction due to Ben-David et. al. [BDCGL90]. The reduction is based on the reduction to unique solutions of Valiant and Vazirani [VV86]. Here we will use an alternative reduction to unique solutions that is based on the isolation lemma of Mulmuley, Vazirani and Vazirani [MVV87].[4]

In Figure 1 we present this reduction. Namely, we describe an algorithm that solves the search problem of SAT given access to an oracle that solves the decision problem of SAT.

Input: A Boolean formula ϕ on n variables. An oracle B that decides SAT correctly.
Output: A satisfying assignment if ϕ is satisfiable, and 'no' otherwise.
The algorithm: 1. Choose uniformly a (weight) function $w : [n] \to [n^e]$ (for now e will be a parameter, we will set it later). We look at strings $y \in \{0,1\}^n$ as characteristic vectors of subsets from the universe $[n]$. We define $w(y)$ to be the sum of the weights that w assigns to the elements of the set characterized by y.
   2. For every $k \in [n^{e+1}]$ and $i \in [n]$, reduce the following NP-statement: "there exists a satisfying assignment y to ϕ, such that $w(y) = k$, and $y_i = 1$", to a Boolean formula $x_{k,i}$. Run B on all the formulas thus generated.
   3. For every $k \in [n^{e+1}]$, generate the following assignment α_k to ϕ: set the i'th bit of α_k to 1, if B said 'yes' on $x_{k,i}$, otherwise set it to 0.
   4. If for some k, α_k satisfies ϕ, output α_k. Otherwise output 'no'.

Fig. 1. An algorithm for the search problem of SAT

Claim. For every ϕ, with probability at least $1 - n^{-e+1}$, the algorithm from Figure 1 outputs the correct answer (a satisfying assignment if it exists and otherwise 'no').

Proof. [Sketch] If ϕ is not satisfiable then the algorithm never errs. Now assume that ϕ is satisfiable. By the isolation lemma [MVV87], with probability at least $1 - n^{-e+1}$ over the choice of w, there exists a $k \in [n^{e+1}]$ such that there is exactly one assignment y that satisfies ϕ and $w(y) = k$. In this case $\alpha_k = y$ and the algorithm finds a satisfying assignment. $\square$

---

[4] Our proof also works with the reduction of [VV86], but it is somewhat more convenient to use [MVV87].

4.3 The Proof of the Theorem

We can now prove Theorem 5.

Proof. Let us assume for contradiction that NP $\not\subseteq$ BPP, but $\mathrm{P}_{||}^{NP} \subseteq$ Pseudo$_{1/2+n^{-c}}$ BPP for some constant $c > 0$. Set $e = 4c + 1$.

Let ϕ be a Boolean formula over n variables (we will assume w.l.o.g. that $|\phi| = n$). Let $w : [n] \to [n^e]$ be a (weight) function. Define the function $f_{\phi,w} : [n^{e+1}] \times [n] \to \{0,1\}$ as follows, $f_{\phi,w}(k, i) = 1$ if and only if there exists a satisfying assignment y to ϕ, such that $w(y) = k$, and $y_i = 1$. We also denote by $f_{\phi,w}$ the truth-table (of length n^{e+2}) of the function. Let $\epsilon = \epsilon(n) = n^{-2c}/8$, and let $\mathcal{C} = \{C_n\}$ be the computationally efficient (ϵ, ℓ)-list-decodable code from Theorem 3. Define the language L whose instances are triples (ϕ, w, j) where ϕ, w are as above, and $j \in [m]$, where $m = m(n, \epsilon)$ is the length of encoded words in the code $C_{n^{e+2}}$. We define $L(\phi, w, j) = C_{n^{e+2}}(f_{\phi,w})_j$ (i.e. the j'th bit of the encoded word).

First we observe that $L \in \mathrm{P}_{||}^{NP}$. On inputs ϕ, w, j we generate in polynomial-time the formulas $x_{k,i}$ from the search to decision reduction in Figure 1. We then ask the NP oracle in parallel all the queries $x_{k,i}$ for $1 \le k \le n^{e+1}$ and $1 \le i \le n$. This gives us the truth-table of $f_{\phi,w}$. We then compute $C_{n^{e+2}}(f_{\phi,w})$ in polynomial-time using the algorithm Enc from Definition 3, and output the j'th bit of the encoded word.

Next we want to show that our assumption that $L \in$ Pseudo$_{1/2+n^{-c}}$ BPP, contradicts our assumption that NP $\not\subseteq$ BPP. Let $A(\cdot; \cdot)$ be a probabilistic polynomial-time algorithm that for every samplable distribution, for every large enough n, decides L correctly with probability at least $1/2+n^{-c}$ (over the choice of instances and the random coins of the algorithm). We denote by r the random coins of A (on input of length n, $|r| = poly(n)$). For a Boolean formula ϕ of length n, and w, j as above, we define $n' = |(\phi, w, j)|$. note that for large enough n, $n' \le n^2$.

In Figure 2 we define an algorithm SSAT that attempts to solve the search problem of SAT by using the algorithm A.

Since A runs in time $poly(n') = poly(n)$, and $\mathcal{C}$ is an efficiently computable list-decodable code, the running time of SSAT is polynomial in n.

By our assumption, NP $\not\subseteq$ BPP. Therefore, by Lemma 1, there exists a constant $\delta > 0$, and a samplable distribution $\mathcal{D} = \{\mathcal{D}_n\}$, that for infinitely many n's outputs with probability greater than $1 - 2^{-n^\delta}$ a formula ϕ on which SSAT answers the correct answer with probability at most $n^{-2c}/2$.

Define the samplable distribution $\mathcal{D}'$ on instances of L as follows: sample a formula ϕ of length n from $\mathcal{D}$ (and assume w.l.o.g. that ϕ is over n variables). Then sample uniformly a weight function $w : [n] \to [n^e]$, and $j \in [m]$. Output (ϕ, w, j). By our assumption, for every large enough n', A decides L correctly with probability at least $1/2 + n'^{-c} \ge 1/2 + n^{-2c}$ over an instance sampled from $\mathcal{D}'$ and the random coins of A. We say that ϕ is good if,

$$\Pr_{w,j,r} [A((\phi, w, j); r) = L(\phi, w, j)] > 1/2 + n^{-2c}/2$$

Input: A Boolean formula ϕ on n variables.
Output: A satisfying assignment if ϕ is satisfiable, and 'no' otherwise.
The algorithm: Repeat the following n^{2c+2} times:
  1. Choose uniformly a (weight) function $w : [n] \to [n^{e+1}]$.
  2. Repeat the following n^{2c+2} times:
      (a) Choose random coins r for A, appropriate for input length n' (recall that $n' = |(\phi, w, j)|$).
      (b) For every $1 \le j \le m$ query A on input (ϕ, w, j) and randomness r. Let $\bar{a} = (a_1, \dots, a_m)$ be the answers of A.
      (c) Run the decoding algorithm Dec from Definition 3 on $\bar{a}$ to obtain a list of $\ell = poly(1/\epsilon) = poly(n)$ strings, $\beta_1, \dots \beta_\ell$, in $\{0,1\}^{n^{e+2}}$.
      (d) We look at each $\beta_s \in \{0,1\}^{n^{e+2}}$ as if it is the vector of answers given by the oracle B, from the search to decision reduction in Figure 1, on the queries $x_{k,i}$ (Step 2 of the reduction). For every $1 \le s \le \ell$ and $1 \le k \le n^{e+1}$, we obtain the assignment α_k^s as it is done in Step 3 of the reduction.
      (e) If for some s and k, α_k^s satisfies ϕ then output this assignment and halt.
If we haven't halted until now then output 'no'.

Fig. 2. The algorithm $SSAT$

By triangle inequality,

$$\Pr_{\phi \leftarrow D_n} [\phi \text{ is good}] \ge n^{-2c}/2$$

Next we show that with high probability SSAT does well on good ϕ's. Clearly, when ϕ is not satisfiable, SSAT always output 'no' (regardless if ϕ is good or not). Fix a good satisfiable formula ϕ. We say that w is good for ϕ if the following holds,

1. $\Pr_{j,r}[A((\phi, w, j); r) = L(\phi, w, j)] > 1/2 + n^{-2c}/4$, and,
2. w defines a unique satisfying assignment for ϕ (in the sense described in the proof of Claim 4.2).

By the above and the isolation lemma (see the proof of Claim 4.2),

$$\Pr_{w}[w \text{ is good for } \phi] \ge n^{-2c}/4 - n^{-e+1} > n^{-2c}/8$$

(Recall that $e = 4c + 1$.) Since SSAT runs n^{2c+2} iterations with independent choices of w, with probability at least $1 - 2^{-n}$ it chooses a good w for ϕ in at least one of the iterations. Let us concentrate on this iteration, and fix a good w for ϕ. We say that r is good for ϕ and w if,

$$\Pr_{j}[A((\phi, w, j); r) = L(\phi, w, j)] > 1/2 + n^{-2c}/8$$

Since ϕ and w are good, it holds that,

$$\Pr_r[r \text{ is good for } \phi \text{ and } w] \geq n^{-2c}/8$$

SSAT runs n^{2c+2} iterations with independent choices of r, so with probability at least $1-2^{-n}$ it chooses a good r for ϕ and w in at least one of the iterations. Let us concentrate on this iteration. Whenever ϕ, w and r are good, the vector $\bar{a}$ (of A's answers) in Stage 2b of the algorithm SSAT, agrees with the language L on at least $1/2 + n^{-2c}/8$ points. When this happens, one of the strings $\beta_1, \ldots, \beta_\ell$ is the truth-table of $f_{\phi,w}$. This is because $\bar{a}$ is $(1/2 + n^{-2c}/8)$-close to $C_{n^{c+2}}(f_{\phi,w})$. Since w is good for ϕ, it defines a unique assignment that satisfies ϕ, and thus, as in the proof of Claim 4.2, at least one of the assignments α_k^s will satisfy ϕ, and SSAT will give the correct answer on ϕ. We conclude that on good ϕ's, SSAT gives the correct answer with very high probability, certainly greater than $n^{-2c}/2$. However, D_n samples a good ϕ with probability greater than $n^{-2c}/2 > 2^{-n^\delta}$ and this contradicts Lemma 1.   □

Acknowledgements

I would like to thank Ronen Shaltiel and Amnon Ta-Shma for numerous discussions about worst-case complexity, average-case complexity and beyond.

References

[BDCGL90] S. Ben-David, B. Chor, O. Goldreich, and M. Luby. On the theory of average case complexity. In *Proceedings of the 22nd Annual ACM Symposium on Theory of Computing*, pages 379–386, 1990.

[BFNW93] L. Babai, L. Fortnow, N. Nisan, and A. Wigderson. BPP has subexponential simulation unless Exptime has publishable proofs. *Computational Complexity*, 3:307–318, 1993.

[BT03] A. Bogdanov and L. Trevisan. On worst-case to average-case reductions for NP problems. In *Proceedings of the 44th Annual IEEE Symposium on Foundations of Computer Science*, pages 308–317, 2003.

[FF93] J. Feigenbaum and L. Fortnow. Random-self-reducibility of complete sets. *SIAM Journal on Computing*, 22:994–1005, 1993.

[GL89] O. Goldreich and L. A. Levin. A hard-core predicate for all one-way functions. In *Proceedings of the 21st Annual ACM Symposium on Theory of Computing*, pages 25–32, 1989.

[GSTS05] D. Gutfreund, R. Shaltiel, and A. Ta-Shma. if NP languages are hard in the worst-case then it is easy to find their hard instances. In *Proceedings of the 20th Annual IEEE Conference on Computational Complexity*, pages 243–257, 2005.

[GTS06] D. Gutfreund and A. Ta-Shma. New connections between derandomization, worst-case complexity and average-case complexity. Submitted for publication, 2006.

[Imp95]   R. Impagliazzo. A personal view of average-case complexity. In *Proceedings of the 10th Annual Conference on Structure in Complexity Theory*, pages 134–147, 1995.

[IW97]   R. Impagliazzo and A. Wigderson. P = BPP if E requires exponential circuits: Derandomizing the XOR lemma. In *Proceedings of the 29th Annual ACM Symposium on Theory of Computing*, pages 220–229, 1997.

[Kab01]   V. Kabanets. Easiness assumptions and hardness tests: Trading time for zero error. *Journal of Computer and System Sciences*, 63 (2):236–252, 2001.

[Lev86]   L. Levin. Average case complete problems. *SIAM Journal on Computing*, 15 (1):285–286, 1986.

[Lip91]   R. Lipton. New directions in testing. *Proceedings of DIMACS workshop on distributed computing and cryptography*, 2:191–202, 1991.

[MVV87]   K. Mulmuley, U. V. Vazirani, and V. V. Vazirani. Matching is as easy as matrix inversion. *Combinatorica*, 7(1):105–113, 1987.

[STV99]   M. Sudan, L. Trevisan, and S. Vadhan. Pseudorandom generators without the XOR Lemma. In *Proceedings of the 31st Annual ACM Symposium on Theory of Computing*, pages 537–546, 1999.

[TV02]   L. Trevisan and S. Vadhan. Pseudorandomness and average-case complexity via uniform reductions. In *Proceedings of the 17th Annual IEEE Conference on Computational Complexity*, pages 129–138, 2002.

[Vio03]   E. Viola. Hardness vs. randomness within alternating time. In *Proceedings of the 18th Annual IEEE Conference on Computational Complexity*, pages 53–62, 2003.

[VV86]   L. G. Valiant and V. V. Vazirani. NP is as easy as detecting unique solutions. *Theoretical Computer Science*, 47(1):85–93, 1986.

[Yao82]   A. C. Yao. Theory and applications of trapdoor functions. In *Proceedings of the 23rd Annual IEEE Symposium on Foundations of Computer Science*, pages 80–91, 1982.

Randomness-Efficient Sampling Within NC^1

Alexander Healy*

Division of Engineering and Applied Sciences, Harvard University,
Cambridge, MA 02138
ahealy@fas.harvard.edu

Abstract. We construct a randomness-efficient *averaging sampler* that is computable by uniform constant-depth circuits with parity gates (i.e., in uniform $AC^0[\oplus]$). Our sampler matches the parameters achieved by random walks on constant-degree expander graphs, allowing us to apply a variety expander-based techniques within NC^1. For example, we obtain the following results:

- Randomness-efficient error-reduction for uniform probabilistic NC^1, TC^0, $AC^0[\oplus]$ and AC^0: Any function computable by uniform probabilistic circuits with error $1/3$ using r random bits is computable by uniform probabilistic circuits with error δ using $r + O(\log(1/\delta))$ random bits.
- Optimal explicit ϵ-biased generator in $AC^0[\oplus]$: There is a $1/2^{\Omega(n)}$-biased generator $G : \{0,1\}^{O(n)} \to \{0,1\}^{2^n}$ for which poly(n)-size uniform $AC^0[\oplus]$ circuits can compute $G(s)_i$ given $(s,i) \in \{0,1\}^{O(n)} \times \{0,1\}^n$. This resolves a question raised by Gutfreund & Viola (*Random 2004*).
- uniform $BP \cdot AC^0 \subseteq$ uniform $AC^0/O(n)$.

Our sampler is based on the *zig-zag graph product* of Reingold, Vadhan and Wigderson (*Annals of Math 2002*) and as part of our analysis we give an elementary proof of a generalization of Gillman's *Chernoff Bound for Expander Walks* (*FOCS 1994*).

1   Introduction

Over the last three decades, *expander graphs* have found a wide variety of applications in Theoretical Computer Science. They have been used in designing novel algorithms, in the study of circuit complexity and to derandomize probabilistic computation, just to name a few notable examples from this vast literature.

Many of these applications involve a *random walk* on an expander. That is, we choose a random starting node v in an expander graph G, take a k-step random walk and use the k nodes visited by this walk in some way – often as a substitute for k independently-chosen nodes. Despite its simplicity, this process has some remarkable sampling properties which we discuss shortly. For the moment, we address the computational efficiency of expanders walks.

---

* Research supported by NSF grant CCR-0205423 and a Sandia Fellowship.

J. Diaz et al. (Eds.): APPROX and RANDOM 2006, LNCS 4110, pp. 398–409, 2006.

In applications, one often requires an expander graph that is exponentially large, say on 2^n nodes. In this case, a random walk on the graph is performed using an efficient *explicit* representation – that is, each node is identified with an n-bit string and it is possible to efficiently (e.g., in time $\text{poly}(n)$) find all the neighbors of a given node $v \in G$. Various constructions [Mar73, GG81, LPS88, RVW02] are known of such explicit constant-degree expander graphs of exponential size.

At first glance, a random walk on an expander graph seems like an inherently *sequential* process – indeed, each step of the walk seems to rely on the previous step in an essential way. A natural question, therefore, is whether expander-based techniques can be applied within highly-*parallel* models of computation, such as log-depth circuits (i.e., NC^1) or even constant depth circuits.

The main technical contribution of this work is a *sampler* which is just as good as a random walk on an expander graph (this is made precise in the next section), but which is computable in parallel time $O(\log n)$, i.e. computable by uniform NC^1 circuits. In fact, our sampler is computable by uniform constant-depth circuits with parity gates (i.e. $AC^0[\oplus]$), a class which is strictly weaker than NC^1 as it cannot even compute the majority of n bits [Raz87].

We now discuss the important sampling properties of random walks on expander graphs in order to better understand what properties we require of our sampler. A more formal definition of expander graphs is given in Section 3, but for the moment the reader may simply think of an expander graph as a constant-degree undirected graph, G, that is "highly-connected".

The following was first shown by Ajtai, Komlós & Szemerédi [AKS87]:

The Hitting Property: For any subset S of half the nodes of G, the probability that a k-step random walk never visits a node in S is at most $2^{-\Omega(k)}$.

This hitting property is quite useful (e.g. to reduce the error of RP algorithms), but some applications require an even stronger property, which we call the *strong hitting* property:

The Strong Hitting Property: For any sequence of subsets $S_1, \ldots, S_k$, each consisting of half the nodes of G, the probability that, for all i, a k-step random walk does not pass through S_i on the i-th step is at most $2^{-\Omega(k)}$.

The strong hitting property is what is necessary for the error reduction techniques of [CW89] and [IZ89] and for the derandomized XOR Lemma of [IW97], as well a variety of other applications.

Clearly, the strong hitting property generalizes the hitting property. Another natural generalization of the hitting property was first proved by Gillman [Gil94]:

The Chernoff Bound for Expander Walks: For any subset S of half the nodes of G, the fraction of time that a k-step random walk spends in S is $1/2 \pm \epsilon$ with probability $1 - 2^{-\Omega(\epsilon^2 k)}$.

It is not clear, however, that this Chernoff Bound subsumes the *strong* hitting property. The following Strong Chernoff Bound does:

The Strong Chernoff Bound for Expander Walks: Fix a sequence of subsets $S_1, \ldots, S_k$, each consisting of half the nodes of G. Then for a k-step random walk on G, the fraction of indices i such that the i-th step of the walk lands in S_i is $1/2 \pm \epsilon$ with probability $1 - 2^{-\Omega(\epsilon^2 k)}$.

Thus, the Strong Chernoff Bound subsumes all the other sampling properties and seems to represent the essential property of random walks on expanders that is necessary for most natural applications. This bound has only been proved recently – it follows from a more general result of Wigderson and Xiao [WX05].

In this work, we give a direct and elementary proof of the Strong Chernoff Bound for Expander Walks (Theorem 1). In contrast to most of the proofs in this area, our proof uses only basic linear algebra and, in particular, does not require any perturbation theory or complex analysis in order to obtain a bound that matches the parameters of Gillman's (non-strong) Chernoff bound.[1]

Theorem 1 (Implicit in [WX05]). *Let G be a regular λ-expander[2] on V. Fix a sequence of subsets $S_i \subseteq V$ each of density $\rho_i = |S_i|/|V|$, and for a random walk $v_1, \ldots, v_k$ on G, let T be the random variable that counts the number of steps i such that $v_i \in S_i$. Then for all $\epsilon > 0$,*

$$\Pr\left[\left|T - \sum_i \rho_i\right| \geq \epsilon k\right] \leq 2e^{-\epsilon^2(1-\lambda)k/36}.$$

Any omitted proofs/details can be found in the full version [Hea06].

2   Our Results

Our main result is the construction of a *sampler* that is computable by $AC^0[\oplus]$ circuits (see Section 3 for a definition) and possesses all the "sampling properties" of a random walk on a constant-degree expander graphs of size 2^n. To make this notion precise, we recall the following definition (essentially from [Zuc97]):

Definition 1. *A function $\Gamma : \{0,1\}^m \to (\{0,1\}^n)^k$ is said to be a strong (γ, ϵ)-averaging sampler if: for any sequence of functions $f_i : \{0,1\}^n \to \{0,1\}$ each with mean $\mu_i = \Pr_x[f_i(x) = 1]$,*

$$\Pr_s\left[\left|\sum_i (f_i(\Gamma(s)_i) - \mu_i)\right| \leq \epsilon k\right] \geq 1 - \gamma.$$

m is called the seed length *of the sampler, and k is its* sample complexity.

Theorem 1 immediately implies that a random walk on a constant-degree expander (with $\lambda = 1 - \Omega(1)$) of size 2^n is a strong averaging sampler with seed length $m = n + O(\log(1/\gamma)/\epsilon^2)$ and sample complexity $k = O(\log(1/\gamma)/\epsilon^2)$. Our main theorem is that $AC^0[\oplus]$ can compute a sampler that is just as good:

---

[1] [WX05] also gives a proof of a (strong) Chernoff bound using no perturbation theory, but their bound does not match Gillman's. It should be noted, however, that [WX05] considers the more general setting of matrix-valued functions.

[2] I.e., a regular graph whose normalized second-largest eigenvalue (in absolute value) is at most λ – see Section 3.

Theorem 2. *There is a strong (γ, ϵ)-averaging sampler $\Gamma : \{0,1\}^m \to (\{0,1\}^n)^k$, with $m = n + O(\log(1/\gamma)/\epsilon^2)$ and $k = O(\log(1/\gamma)/\epsilon^2)$, that is computable by uniform $AC^0[\oplus]$ circuits of size $\mathrm{poly}(n, 1/\epsilon, \log(1/\gamma))$.*

At this point, the reader may wish to disregard the exact parameters of our construction, and instead think of our construction as computing (intuitively) a walk of length k on a constant-degree expander graph of size 2^n. Indeed, in most applications that employ random walks on expander graphs, one can safely substitute a sampler with the above parameters in place of the expander walk.

Gutfreund & Viola [GV04] show that walks on the Gabber-Galil expander graph [Mar73, GG81] with 2^n nodes are computable in space $O(\log n)$ (and therefore that logspace has strong samplers). To the best of our knowledge, ours is the first work giving strong samplers within the class NC^1 of log-depth circuits; in fact, our construction is in the strictly-weaker class $AC^0[\oplus] \subsetneq TC^0 \subseteq NC^1 \subseteq L$.

Since expander walks are a widely-applicable tool, it is not surprising that our sampler should have a variety of applications. We mention three:

Randomness-Efficient Error Reduction within NC^1: One important application of random walks on expander graphs is in reducing the error of probabilistic algorithms. Error reduction was achieved for BPP by Cohen & Wigderson [CW89] and Impagliazzo & Zuckerman [IZ89]. Bar-Yosef et al. [BYGW99] show how to achieve modest-but-optimal error reduction for randomized logspace, and the expander walks of [GV04] imply randomness-efficient error reduction for the class $BP \cdot L$.[3] Our sampler implies error-reduction for classes below logspace:

Lemma 1. *Let $f : \{0,1\}^n \to \{0,1\}$ be in $\mathrm{poly}(n)$-size uniform $BP \cdot AC^0[\oplus]$ (resp., $BP \cdot TC^0$ or $BP \cdot NC^1$) with error $\leq 1/3$ using $r = r(n)$ random bits. Then for any $\delta \geq 1/2^{O(\mathrm{poly}(n))}$, f is in $\mathrm{poly}(n)$-size uniform $BP \cdot AC^0[\oplus]$ (resp. $BP \cdot TC^0$ or $BP \cdot NC^1$) with error $\leq \delta$ using $r + O(\log(1/\delta))$ random bits.*

Combining our sampler with Nisan's unconditional pseudorandom generators for constant depth circuits [Nis91], we obtain an even stronger result for AC^0:

Lemma 2. *Let $f : \{0,1\}^n \to \{0,1\}$ be in $\mathrm{poly}(n)$-size uniform $BP \cdot AC^0$ with error $\leq 1/3$ using $r = r(n)$ random bits. Then for any $\delta \geq 1/2^{O(\mathrm{poly}(n))}$, f is in $\mathrm{poly}(n)$-size uniform $BP \cdot AC^0$ with error $\leq \delta$ using $\min\{r, \mathrm{polylog}(n)\} + O(\log(1/\delta))$ random bits.*

Derandomization with Linear Advice: Recently, Fortnow & Klivans [FK06] have proved that $RL \subseteq L/O(n)$ – i.e., one can derandomize randomized logspace computation using only a linear amount of non-uniform advice. Their approach relies on a clever combination of Nisan's pseudorandom generator for small space [Nis92] and the logspace expander walks of [GV04]. Our techniques yield an analogous result for uniform probabilistic constant-depth circuits:

---

[3] $BP \cdot L$ refers to randomized logspace computations that are allowed *two-way* access to the random bits, whereas the result of Bar-Yosef et al. refers to algorithms that have only *one-way* access to the random bits. See the survey of Saks [Sak96] for a discussion of different notions of randomized space-bounded computation.

Corollary 1. uniform $BP \cdot AC^0 \subseteq$ uniform $AC^0/O(n)$.

Ajtai & Ben-Or [ABO84] show that nonuniform $BP \cdot AC^0 =$ nonuniform AC^0; Theorem 1 quantifies the amount of nonuniformity that is necessary for this derandomization, and thus can be viewed as a refinement of their result.

A similar approach, using a generator of Viola [Vio05], can be used to show that $BP \cdot AC^0[\oplus_{\log}] \subseteq AC^0[\oplus]/O(n)$ and $BP \cdot AC^0[\text{SYM}_{\log}] \subseteq TC^0/O(n)$, where $AC^0[\oplus_{\log}]$ is the class of poly(n)-size AC^0 circuits having $O(\log n)$ parity gates, and $AC^0[\text{SYM}_{\log}]$ is the class of poly(n)-size AC^0 circuits having $O(\log n)$ arbitrary symmetric gates (e.g., parity and majority gates).

An Optimal Explicit ϵ-Biased Generator in $AC^0[\oplus]$: Gutfreund & Viola [GV04] study the complexity of constructing *explicit* ϵ-biased generators (see Definition 2). They give a construction in $AC^0[\oplus]$ whose seed length is optimal for $\epsilon = \Omega(1/\text{poly} \log \log(m))$ (where m is the number of output bits) and sub-optimal for smaller ϵ. Healy and Viola [HV06] give an optimal construction in TC^0 and a sub-optimal construction in $AC^0[\oplus]$ whose parameters are incomparable to those of [GV04]. In this work, we resolve this question entirely: using our sampler, we construct an *optimal* explicit ϵ-biased generator in $AC^0[\oplus]$:

Corollary 2 ([NN90] + [GV04] + Thm. 2). *For any $\epsilon > 0$ and m, there is an ϵ-biased generator $G : \{0,1\}^n \to \{0,1\}^m$, $n = O(\log m + \log(1/\epsilon))$ for which* poly(n)-*size uniform $AC^0[\oplus]$ can compute $G(s)_i$ given $(s,i) \in \{0,1\}^n \times [m]$.*

Proof (Outline). Following [GV04], we implement the generator of [NN90] which requires a 7-wise independent generator (which can be constructed in $AC^0[\oplus]$ [GV04, HV06]) and an expander walk that we replace by our sampler.   □

Corollary 2 is tight in terms of seed length and complexity – see [GV04].

3   Preliminaries

ϵ-*Biased Sets and Generators:* Small-biased spaces appear in two ways in this work: (i) poly-size ϵ-biased sets are used to construct expander graphs on which our sampler is based (Lemma 3), (ii) one application of our sampler is to build exponential-size ϵ-biased sets that are computable *explicitly* (Corollary 2).

Definition 2. *A multi-set $S \subseteq \mathbb{F}_2^m$ is ϵ-biased if for all $0 \neq y \in \mathbb{F}_2^m$ we have* $\Pr_{x \in S}[\langle x, y \rangle \equiv 0 \bmod 2] \in [1/2 - \epsilon, 1/2 + \epsilon]$. *An ϵ-biased generator is a function* $\Gamma : \{0,1\}^\ell \to \{0,1\}^m$ *whose range is an ϵ-biased multi-set. An explicit ϵ-biased generator is a function $\Gamma : \{0,1\}^\ell \times [m] \to \{0,1\}$ such that the function $\Gamma'(s) = (\Gamma(s,1), \Gamma(s,1), \ldots, \Gamma(s,m))$ is an ϵ-biased generator.*

Expander Graphs. Informally, expander graphs are sparse-yet-highly-connected graphs. Of the various equivalent notions of graph expansion (see, e.g., [Gol99]), we choose to work with the spectral definition.

Definition 3. *A regular graph G of degree d is a λ-expander if the second-largest eigenvalue (in absolute value) of its probability transition matrix (i.e., $1/d$ times its adjacency matrix) is at most λ.*

When we refer to a "λ-expander", we really mean a "family of $\lambda(n)$-expanders of size $s(n)$" for some function $s(n)$, and when we refer to an "expander graph", without mention of λ, it is understood that we mean a "$(1 - \Omega(1))$-expander".

By a *random walk* $v_1, \ldots, v_k$ on an d-regular graph G, we mean the following process: Choose a random starting vertex $v_0 \in G$, and for $i = 1, \ldots, k$, let v_i be a random neighbor of v_{i-1} in G. Note that we implicitly discard the start vertex v_0 – although the distribution is unchanged if we keep v_0, this convention simplifies our presentation. Note that such a walk is described by a tuple $(v_0, s_1, \ldots, s_k) \in [|G|] \times [d] \times \cdots \times [d]$, and hence by a string of $O(\log |G| + k \log d)$ bits.

Constant-Depth Circuits. We consider three classes of unbounded fan-in constant-depth circuits: circuits over the bases $\{\wedge, \vee \neg\}$ (i.e., AC^0), $\{\wedge, \vee, Parity, \neg\}$ (i.e., $AC^0[\oplus]$), and $\{\wedge, \vee, Majority, \neg\}$ (i.e., TC^0). All circuits are of polynomial size and *uniform* – specifically, we adopt the standard of *Dlogtime*-uniformity, which is even more restrictive than logspace-uniformity and which has become the generally-accepted convention for uniformity in constant-depth circuits [BIS90]. Informally, a circuit is *Dlogtime*-uniform if, given indices of two gates, one can determine the types of the gates and whether they are connected in linear time in the length of the indices (which is logarithmic in the size of the circuit).

We indicate non-uniform circuits explicitly using *slash* notation: for example, $AC^0/O(n)$ is the class of boolean functions f that are computable by *Dlogtime*-uniform AC^0 circuits $C_n : \{0,1\}^n \times \{0,1\}^{O(n)} \to \{0,1\}$ for which there is an advice string a_n of length $O(n)$ such that $C_n(x, a_n) = f(x)$ for all $x \in \{0,1\}^n$.

The probabilistic classes $BP \cdot AC^0, BP \cdot AC^0[\oplus], BP \cdot TC^0$ and $BP \cdot NC^1$ are all defined in the natural way: the circuit takes two inputs, one of n bits and one of $r(n)$ random bits for some polynomially-bounded function $r(n)$, and for any fixed input $x \in \{0,1\}^n$, the circuit should correctly compute the function with probability at least $2/3$ over the $r(n)$ random bits.

Recall that $AC^0 \subsetneq AC^0[\oplus] \subsetneq TC^0 \subseteq NC^1 \subseteq$ logspace, where the last inclusion holds under logspace uniformity and the separations follow from works by Furst et al. [FSS84] and Razborov [Raz87], respectively (and hold even for non-uniform circuits). See, e.g., [Vol99] for background on constant-depth circuits.

4   The Sampler Construction

In this section, we describe our sampler and prove Theorem 2. Recall that our goal is to construct a sampler $\Gamma : \{0,1\}^m \to (\{0,1\}^n)^k$ that matches the parameters of random walks on expander graphs. Naturally, one way to achieve this would be to exhibit a family of constant-degree expander graphs on 2^n nodes and show that walks of length k on these expanders can be computed in $AC^0[\oplus]$ of size poly(n, k). Unfortunately, we do not know of any such family of expanders. Instead, we begin with a family of expander graphs of degree poly(n) where walks are computable in $AC^0[\oplus]$ – note that a walk of length k on such a graph is described by a seed of length $n + O(k \cdot \log n)$ – and then we *derandomize* the walk on this graph to achieve the optimal seed length $n + O(k)$. This derandomization uses random walks on a smaller expander graph, and its analysis is based on the *zig-zag graph product* of [RVW02]. By [GV04], it is known that AC^0 circuits can compute walks of length $\log n$ on a Gabber-Galil graph of size 2^n, so in the sequel we shall focus on the case where $k = \Omega(\log n)$.

Our first graph, G, is a Cayley graph on the group $\mathbb{F}_2^n$. Specifically, we construct a $1/n$-biased set $S \subset \mathbb{F}_2^n$ of size $\text{poly}(n)$ (see Definition 2) and let $\{v, w\}$ be an edge if and only if $v - w \in S$. The following well-known fact guarantees that G has second-largest eigenvalue at most $2/n$ (e.g., see [AR94]).

Lemma 3. *A Cayley graph on $\mathbb{F}_2^n$ with generators $S \subset \mathbb{F}_2^n$ is a 2ϵ-expander if and only if S is ϵ-biased.*

Before continuing, let us see how walks on G can be computed in $AC^0[\oplus]$. First, we note that a $1/n$-biased set S of size $\text{poly}(n)$ can be constructed in AC^0. For instance, we may use the "Powering Construction" of an ϵ-biased generator from [AGHP92] together with the results on field arithmetic of [HV06].[4] (For a nonuniform construction, one could simply hardwire S into the circuit.)

Thus, given a walk $(v, s_1, \ldots, s_k) \in \{0, 1\}^n \times \{0, 1\}^{O(\log n)} \times \cdots \times \{0, 1\}^{O(\log n)}$, to determine the i-th vertex visited by the walk, the circuit need only compute from each s_j (in parallel) the appropriate vector $v_{s_j} \in S$ and then compute the sum $v + \sum_{j \leq i} v_{s_j} \mod 2$, which is clearly computable in $AC^0[\oplus]$ of size $\text{poly}(n, k)$.

Our approach to derandomizing this walk is motivated by the zig-zag product of Reingold, Vadhan & Wigderson [RVW02]. Roughly speaking, one may interpret their results as saying: to derandomize a walk on a graph G of degree d, one may choose the steps according to a random walk on a constant-degree expander graph H of size d. (For technical reasons, we requires H to be the *square* of an expander, but we ignore this for the moment.) Specifically,

1. Choose a random starting vertex $v_0 \in G$
2. Choose a random $w_0 \in H$; take a random walk of length k, visiting $w_1, \ldots, w_k$
3. View $w_1, \ldots, w_k$ as indices in $[d] = [|H|]$
4. Use $w_1, \ldots, w_k$ as the steps of a walk (starting at v_0) in G
5. Output the nodes $v_1, \ldots, v_k \in G$ visited by this walk

Note that the seed length of such a sampler is $|v_0| + (|w_0| + O(k)) = n + \log |H| + O(k) = n + O(k)$ (since we assume $k = \Omega(\log n)$), as desired. Moreover, one can show that the above construction is a strong averaging sampler. What is not clear, however, is how to compute this sampler in $AC^0[\oplus]$, because it requires a long walk on H – while H is small (only $\text{poly}(n)$ nodes) compared to G (which has 2^n nodes), we do not know how to take such a long walk on any constant-degree expander family in $AC^0[\oplus]$ (or even in NC^1). In order to circumvent this obstacle, we use many short walks on H, rather than a single long walk.

---

[4] Specifically, let $m = \log n$ (assuming that $\log n$ is an integer for simplicity) and consider the finite field $\mathbb{F}_{2^{2m}}$ with 2^{2m} elements (viewed as the ring of polynomials over $\mathbb{F}_2$ modulo an irreducible polynomial of degree $2m$). The generator outputs $2^{4m} = n^4$ vectors $v_{\alpha, \beta}$ of dimension $2^m = n$, indexed by pairs of elements $\alpha, \beta \in \mathbb{F}_{2^{2m}}$, where the i-th bit of $v_{\alpha, \beta}$ is given by $\langle \alpha^i, \beta \rangle$ (mod 2). [AGHP92] shows that such a generator has bias less than $2^m/2^{2m} = 1/n$, and [HV06] shows that all the necessary field arithmetic can be carried out in uniform AC^0 of size $\text{poly}(n)$.

Construction 3.

1. *Choose a random starting vertex $v_0 \in G$*
2. *Take $k/\log n$ random walks $w_1^{(i)}, \ldots, w_{\log n}^{(i)} \in H$ for $i = 1, \ldots, k/\log n$.*
3. *View $w_1^{(1)}, \ldots, w_{\log n}^{(1)}, \ldots, w_1^{(k/\log n)}, \ldots, w_{\log n}^{(k/\log n)}$ as indices in $[d] = [|H|]$*
4. *Use $w_1^{(1)}, \ldots, w_{\log n}^{(1)}, \ldots, w_1^{(k/\log n)}, \ldots, w_{\log n}^{(k/\log n)}$ as the steps of a walk in G*
5. *Output the nodes $v_1, \ldots, v_k \in G$ visited by this walk*

This sampler has seed length $|v_0| + (k/\log n)(\log |H| + O(\log n)) = n + O(k)$ (again, since we assume that $k = \Omega(\log n)$). Furthermore, we show below that this construction is a strong averaging sampler, achieving essentially the same parameters as a random walk on an expander graph. Before proving this, however, we observe that it is computable in $AC^0[\oplus]$. Indeed, it is known how to compute walks of length $O(\log n)$ on poly-sized explicit expanders of constant degree in AC^0 [Ajt93, GV04],[5] and thus each of the five steps above is computable in constant depth. Theorem 2 is a consequence of the following lemma:

Lemma 4. *Let $H = \tilde{H}^2$ where $\tilde{H}$ is a constant-degree expander graph on $\mathrm{poly}(n)$ nodes. Then Construction 3 is a strong averaging boolean sampler with seed length $n + O(\log(1/\gamma)/\epsilon^2)$ and sample complexity $O(\log(1/\gamma)/\epsilon^2)$.*

Proof. We employ the zig-zag product of [RVW02], which we briefly recall.

Zig-Zag Product. Let G be a regular graph of degree d on vertices V_G whose edges are labeled with the names $1, \ldots, d$ in such a way that no two incident edges share the same label.[6] (Thus, if w is the "i-th neighbor of v", then v is the "i-th neighbor of w" – G, defined above, clearly has this property.) Then if g is a regular graph on vertices V_g where $|V_g| = d$, we may form the *zig-zag product* graph $G\textcircled{z}g$ where:

- $G\textcircled{z}g$ has vertices $V_G \times V_g$
- $\{(v, w), (v', w')\}$ is an edge if there is an $x \in g$ such that v' is the x-th neighbor of v in G and (w, x, w') is a path in g. (Note that the labeling condition on G ensures this is symmetric.)

Thus, to step from $(v, w) \in G\textcircled{z}g$, to a random neighbor (v', w'): (i) Choose a random neighbor x of w in g. (ii) Set v' to be the x-th neighbor of v in G. (iii) Choose a random neighbor w' of x in g.

In particular, if we consider the V_G-coordinate of an ℓ-step random walk in $G\textcircled{z}g$, it has the same distribution as the following: (i) Choose a random start vertex $v_0 \in V_G$. (ii) Take a random walk $w_1, w_2 \ldots, w_\ell$ in g^2. (iii) For $i > 0$, let v_i to be the w_i-th neighbor of v_{i-1} in G. (iv) Output $v_1, v_2, \ldots, v_\ell$.

---

[5] As with the $1/n$-biased set S above, the delicate issue here is the uniformity of the circuits; if we only wish to give a nonuniform construction we could simply hard-wire all the possible walks of length $\log n$ into the circuit.

[6] The zig-zag product of [RVW02] actually holds in much greater generality; however, this simplification suffices for our application.

Thus, each of of the segments of length $k/\log n$ in our sampler corresponds to a random walk on $G\textcircled{z}\tilde{H}$, projected onto the V_G-coordinate. For the boundaries between these segments, Construction 3 says we choose a new, entirely-random node of $\tilde{H}$ and then continue the walk on G. This is equivalent to taking a step on $G\textcircled{z}K_d$, i.e. the zig-zag product of G with a complete graph (with self-loops) on d nodes. Therefore, the output of our sampler is the projection onto the V_G-coordinate of a random walk on a time-varying graph that is usually $G\textcircled{z}\tilde{H}$, and $G\textcircled{z}K_d$ once every $\log n$ steps. We now show that this satisfies Definition 1.

First we note for any function $f : V_G \to \{0,1\}$ there is a natural *lift* of f to $\hat{f} : V_G \times V_{\tilde{H}} \to \{0,1\}$, defined by $\hat{f}(v,w) = f(v)$. It is clear that the lift $\hat{f}$ has the same average as f. Therefore, to conclude that the projection of a random walk yields a strong averaging sampler, it suffices to show that a random walk on the time-varying graph is a strong averaging sampler. By the remark after the proof of Theorem 1, it does not matter if the graph is varying as long as it is a λ-expander at every step. Thus, we are left with the task of showing that $G\textcircled{z}\tilde{H}$ and $G\textcircled{z}K_d$ are expanders. For this, we use the following result of [RVW02]:

Lemma 5 ([RVW02], Corollary to Theorem 4.3). *Let G be a regular graph of degree d whose edges are labeled with $1,\ldots,d$ in such a way that no two incident edges share the same label, and let g be a regular graph on d nodes. If G is a λ_G-expander and g is a λ_g-expander, then $G\textcircled{z}g$ is a $(\lambda_G + \lambda_g)$-expander.*

By Lemma 3, G is a $2/n$-expander, and by assumption $\tilde{H}$ is a $(1 - \Omega(1))$-expander, and so $G\textcircled{z}\tilde{H}$ is a $(1 - \Omega(1))$-expander. K_d, the complete graph (with self-loops) on d nodes, is a 0-expander, and therefore $G\textcircled{z}K_d$ is a $2/n$-expander.

Thus our sampler stretches $n + O(k)$ bits into k n-bit samples satisfying the bound from Theorem 1 with $\lambda = 1 - \Omega(1)$. Specifically, the sampler achieves error ϵ with confidence $1 - \gamma = 1 - e^{-\Omega(\epsilon^2 k)}$; i.e., the seed length is $n + O(k) = n + O(\log(1/\gamma)/\epsilon^2)$ and the sample complexity is $k = O(\log(1/\gamma)/\epsilon^2)$. $\qquad\square$

5   The Proof of Theorem 1

Theorem 1. *Let G be a regular λ-expander[7] on V. Fix a sequence of subsets $S_i \subseteq V$ each of density $\rho_i = |S_i|/|V|$, and for a random walk $v_1,\ldots,v_k$ on G, let T be the random variable that counts the number of steps i such that $v_i \in S_i$. Then for all $\epsilon > 0$,*

$$\Pr\left[\left|T - \sum\nolimits_i \rho_i\right| \geq \epsilon k\right] \leq 2e^{-\epsilon^2(1-\lambda)k/36}.$$

Wigderson and Xiao [WX05] have recently established the same bound (up to constants). As with Gillman's proof (which treats the case $S_1 = \cdots = S_k$), they employ results from perturbation theory to obtain their bound. In contrast, the proof presented here has modest prerequisites, summarized below.

---

[7] I.e., a regular graph whose normalized second-largest eigenvalue (in absolute value) is at most λ – see Section 3.

Background. Let G be a regular undirected graph N nodes. G's probability transition matrix, P, is clearly real and symmetric, and hence the eigenvectors of P form an orthogonal basis of $\mathbb{R}^N$. Since G is regular, $\mathbf{1} = (1, \ldots, 1)$ is an eigenvector with eigenvalue $\lambda_1 = 1$. By the Perron-Frobenius Theorem, all other eigenvalues $\lambda_2 \geq \ldots \geq \lambda_n$ are between 1 and -1. We write $\lambda = \max\{|\lambda_2|, |\lambda_n|\}$. For any $\boldsymbol{v} \in \mathbb{R}^N$, $\boldsymbol{v}^{\|}$ denotes the component of $\boldsymbol{v}$ in the direction of $\mathbf{1}$ and $\boldsymbol{v}^{\perp}$ denotes the component of $\boldsymbol{v}$ that lies in the orthogonal complement of $\mathbf{1}$; i.e., $\boldsymbol{v}^{\|} = \langle \mathbf{1}, \boldsymbol{v} \rangle \boldsymbol{u}$ and $\boldsymbol{v}^{\perp} = \boldsymbol{v} - \boldsymbol{v}^{\|} = \boldsymbol{v} - \langle \mathbf{1}, \boldsymbol{v} \rangle \boldsymbol{u}$, where $\boldsymbol{u} = (1/N, \ldots, 1/N)$. It is not hard to check that $\|P\boldsymbol{v}^{\perp}\| \leq \lambda \|\boldsymbol{v}^{\perp}\|$ for any $\boldsymbol{v} \in \mathbb{R}^N$.

Proof (Theorem 1). We shall bound $\Pr[T - \sum_i \rho_i \geq \epsilon k]$ and the same bound will follow for $\Pr[T - \sum_i \rho_i \leq -\epsilon k]$ by replacing the sets S_i with their complements. Let $r \leq \min\{1, \log(1/\lambda)/2\}$ be a positive parameter to be specified later.

$$\Pr\left[T \geq \epsilon k + \sum_i \rho_i\right] = \Pr\left[e^{rT} \geq e^{r(\epsilon k + \sum_i \rho_i)}\right] \leq \frac{\mathrm{E}\left[e^{rT}\right]}{e^{r(\epsilon k + \sum_i \rho_i)}} \qquad (1)$$

where the last step follows by applying Markov's inequality.

We now bound $\mathrm{E}\left[e^{rT}\right]$. Let P be the probability transition matrix for G, and for each set S_i let E_i be a diagonal matrix with $e_{j,j} = e^r$ if $j \in S_i$ and $e_{j,j} = 1$ otherwise. It is not hard to see that

$$\mathrm{E}\left[e^{rT}\right] = \mathbf{1}^T E_k P E_{k-1} P \cdots E_1 P \boldsymbol{u}. \qquad (2)$$

To bound this quantity, we require the following lemma whose proof we omit.

Lemma 6. *Let P be as above, and assume that $r \leq \log(1/\lambda)/2$. Let $S \subseteq V$ be of density $\rho = |S|/|V|$, and let E be the diagonal matrix with $e_{j,j} = e^r$ for $j \in S$ and $e_{j,j} = 1$ otherwise. Then for any $\boldsymbol{v} \in \mathbb{R}^N$:*

- $\|(EP\boldsymbol{v})^{\|}\| \leq (1 + \rho(e^r - 1)) \cdot \|\boldsymbol{v}^{\|}\| + (e^r - 1) \cdot \|\boldsymbol{v}^{\perp}\|$
- $\|(EP\boldsymbol{v})^{\perp}\| \leq (e^r - 1) \cdot \|\boldsymbol{v}^{\|}\| + \sqrt{\lambda} \cdot \|\boldsymbol{v}^{\perp}\|.$

We define a sequence $\boldsymbol{v}_0 = \boldsymbol{u}$ and $\boldsymbol{v}_i = E_i P \boldsymbol{v}_{i-1}$ for $i > 0$, noting that

$$\mathrm{E}\left[e^{rT}\right] = \mathbf{1}^T E_k P \cdots E_1 P \boldsymbol{u} = \langle \mathbf{1}, \boldsymbol{v}_k \rangle = \langle \mathbf{1}, \boldsymbol{v}_k^{\|} \rangle \leq \|\mathbf{1}\| \cdot \|\boldsymbol{v}_k^{\|}\| = \sqrt{N} \cdot \|\boldsymbol{v}_k^{\|}\|. \quad (3)$$

Lemma 6 yields a system of two simultaneous recurrences in the variables $\|\boldsymbol{v}_i^{\|}\|$ and $\|\boldsymbol{v}_i^{\perp}\|$. By solving these recurrences (details omitted), one can show that that $\mathrm{E}\left[e^{rT}\right] \leq \sqrt{N} \cdot \|\boldsymbol{v}_k^{\|}\| \leq \prod_{i=1}^{k}\left(1 + r\rho_i + 9r^2/(1 - \lambda)\right)$.

Taking logarithms and using the fact that $\log(1 + x) \leq x$ for all $x \geq 0$, we have $\log \mathrm{E}\left[e^{rT}\right] \leq k \cdot 9r^2/(1 - \lambda) + r \cdot \sum_i \rho_i$, and thus, by Equation (1), $\log \Pr[T - \sum_i \rho_i \geq \epsilon k] \leq \log\left(\mathrm{E}\left[e^{rT}\right]\right) - r\left(\epsilon k + \sum_i \rho_i\right) \leq k\left(\frac{9r^2}{1-\lambda} - \epsilon r\right)$. We minimize the right-hand side by setting $r = \epsilon(1 - \lambda)/18$, noting that r is indeed at most $\min\{1, \log(1/\lambda)/2\}$ simply because $1 - \lambda \leq \log(1/\lambda)$ for all $\lambda \in [0, 1]$. Finally, we have that $\log \Pr[T - \sum_i \rho_i \geq \epsilon k] \leq -\epsilon^2(1 - \lambda)k/36$. $\qquad \square$

Remark 1. One can readily see that the same proof works even if the graph is different for each of the k steps, as long as it is a λ-expander at each step, as is required in the proof of Theorem 2. This property has been exploited before, most notably in the hardness amplification result of Goldreich et al. [GIL+90] (although there, they only require the hitting property of expander walks).

6   Open Questions

Expander walks are known to yield optimal samplers (up to constants) for $\epsilon = \Omega(1)$, but not for smaller ϵ. Can $AC^0[\oplus]$ compute an optimal sampler?

We suspect that AC^0 cannot compute samplers that match the parameters of our $AC^0[\oplus]$ construction. One approach to showing this is to use the equivalence of samplers and extractors from [Zuc97] and show that AC^0 cannot compute an extractor for sources of high constant min-entropy. Viola [Vio04] has shown that AC^0 cannot compute an extractor for sources of low min-entropy; however, his techniques do not seem to apply directly in this setting.

Acknowledgements

Thanks to Danny Gutfreund, Salil Vadhan and Emanuele Viola for helpful discussions, comments, suggestions and for their encouragement. Thanks also to the anonymous Random 2006 reviewers for their comments.

References

[ABO84]   Miklós Ajtai and Michael Ben-Or. A theorem on probabilistic constant depth computation. In *Proceedings of STOC 1984*, pp. 471–474, 1984.

[AGHP92]   Noga Alon, Oded Goldreich, Johan Håstad, and René Peralta. Simple constructions of almost k-wise independent random variables. *Random Structures & Algorithms*, 3(3):289–304, 1992.

[Ajt93]   Miklós Ajtai. Approximate counting with uniform constant-depth circuits. In *Advances in computational complexity theory*, pp. 1–20. AMS, 1993.

[AKS87]   M. Ajtai, J. Komlos, and E. Szemeredi. Deterministic simulation in LOGSPACE. In *Proceedings of STOC 1987*, pp. 132–140, 1987.

[AR94]   Noga Alon and Yuval Roichman. Random cayley graphs and expanders. *Random Structures & Algorithms*, 5:271–284, 1994.

[BIS90]   David A. Mix Barrington, Neil Immerman, and Howard Straubing. On uniformity within NC^1. *J. of Comp. & Sys. Sci.*, 41(3):274–306, 1990.

[BYGW99]   Z. Bar-Yossef, O. Goldreich, and A. Wigderson. Deterministic amplification of space-bounded probabilistic algorithms. In *Proceedings of the 14th Conference on Computational Complexity*, pp. 188–198, 1999.

[CW89]   Aviad Cohen and Avi Wigderson. Dispersers, deterministic amplification, and weak random sources. In *Proceedings of FOCS 1989*, pp. 14–19, 1989.

[FK06]   Lance Fortnow and Adam Klivans. Linear advice for randomized logarithmic space. In *Proceedings of the 23rd STACS*, pp. 469 – 476, 2006.

[FSS84]   Merrick L. Furst, James B. Saxe, and Michael Sipser. Parity, circuits, and the polynomial-time hierarchy. *Math. Systems Theory*, 17(1):13–27, 1984.

[GG81]   O. Gabber and Z. Galil. Explicit construction of linear size superconcentrators. *Journal of Computer and System Sciences*, 22:407–420, 1981.

[GIL⁺90]   Oded Goldreich, Russell Impagliazzo, Leonid A. Levin, Ramarathnam Venkatesan, and David Zuckerman. Security preserving amplification of hardness. In *Proceedings of FOCS 1990*, pp. 318–326, 1990.

[Gil94]   David Gillman. A Chernoff bound for random walks on expander graphs. In *Proceedings of FOCS 1994*, pp. 680–691, 1994.

[Gol97]   Oded Goldreich. A sample of samplers - a computational perspective on sampling. *Elec. Colloquium on Computational Complexity*, 4(020), 1997.

[Gol99]   Oded Goldreich. *Modern cryptography, probabilistic proofs and pseudorandomness*. Springer-Verlag, Berlin, 1999.

[GV04]   Dan Gutfreund and Emanuele Viola. Fooling parity tests with parity gates. In *Proceedings of Random 2004*, pp. 381–392, 2004.

[Hea06]   Alexander Healy Randomness-efficient sampling within NC^1. *Elec. Col. on Comp. Complexity*, TR06-058, 2006. http://eccc.hpi-web.de/eccc/

[HV06]   A. Healy and E. Viola. Constant-depth circuits for arithmetic in finite fields of characteristic two. In *Proceedings of STACS 2006*, pp. 672 – 683, 2006.

[IW97]   R. Impagliazzo and A. Wigderson. $P = BPP$ if E requires exponential circuits: Derandomizing the XOR lemma. In *Proc. of STOC 1997*.

[IZ89]   Russell Impagliazzo and David Zuckerman. How to recycle random bits. In *Proceedings of FOCS 1989*, pp. 248–253, 1989.

[LPS88]   A. Lubotzky, R. Phillips, and P. Sarnak. Ramanujan graphs. *Combinatroica*, 8(3):261–277, 1988.

[Mar73]   G. A. Margulis. Explicit constructions of expanders. *Problemy Peredachi Informatssi; Problems of Information Transmission*, 9(4):71–80, 1973.

[Nis91]   Noam Nisan. Pseudorandom bits for constant depth circuits. *Combinatorica*, 11(1):63–70, 1991.

[Nis92]   Noam Nisan. Pseudorandom generators for space-bounded computation. *Combinatorica*, 12, 1992.

[NN90]   J. Naor and M. Naor. Small-bias probability spaces: efficient constructions and applications. In *Proceedings of STOC 1990*, pp. 213–223, 1990.

[Raz87]   Alexander A. Razborov. Lower bounds on the dimension of schemes of bounded depth in a complete basis containing the logical addition function. *Akademiya Nauk SSSR. Mat. Zametki*, 41(4):598–607, 623, 1987.

[RVW02]   Omer Reingold, Salil Vadhan, and Avi Wigderson. Entropy waves, the zig-zag graph product and new constant-degree expanders. *Annals of Mathematics*, 155(1):157–187, January 2002.

[Sak96]   M. Saks. Randomization and derandomization in space-bounded computation. In *Proc. of the 11th Conference on Computational Complexity*, 1996.

[Vio04]   E. Viola. The complexity of constructing pseudorandom generators from hard functions. *Computational Complexity*, 13(3-4):147–188, 2004.

[Vio05]   E. Viola. Pseudorandom bits for constant-depth circuits with few arbitrary symmetric gates. In *Proc. of 20th Conf. on Comp. Complexity*, 2005.

[Vol99]   Heribert Vollmer. *Introduction to circuit complexity*. Springer-Verlag, 1999.

[WX05]   Avi Wigderson and David Xiao. A randomness-efficient sampler for matrix-valued functions and applications. In *Proceedings of FOCS 2005*, 2005. See also ECCC Technical Report TR05-107, http://eccc.hpi-web.de/eccc/.

[Zuc97]   David Zuckerman. Randomness-optimal oblivious sampling. *Random Structures & Algorithms*, 11(4):345–367, 1997.

Monotone Circuits for the Majority Function

Shlomo Hoory[1], Avner Magen[2], and Toniann Pitassi[2]

[1] IBM Research Laboratory in Haifa, Israel
shlomoh@il.ibm.com
[2] Department of Computer Science, University of Toronto
{avner, toni}@cs.toronto.edu

Abstract. We present a simple randomized construction of size $O(n^3)$ and depth $5.3 \log n + O(1)$ monotone circuits for the majority function on n variables. This result can be viewed as a reduction in the size and a partial derandomization of Valiant's construction of an $O(n^{5.3})$ monotone formula, [15]. On the other hand, compared with the deterministic monotone circuit obtained from the sorting network of Ajtai, Komlós, and Szemerédi [1], our circuit is much simpler and has depth $O(\log n)$ with a small constant. The techniques used in our construction incorporate fairly recent results showing that expansion yields performance guarantee for the belief propagation message passing algorithms for decoding low-density parity-check (LDPC) codes, [3]. As part of the construction, we obtain optimal-depth linear-size monotone circuits for the promise version of the problem, where the number of 1's in the input is promised to be either less than one third, or greater than two thirds. We also extend these improvements to general threshold functions. At last, we show that the size can be further reduced at the expense of increased depth, and obtain a circuit for the majority of size and depth about $n^{1+\sqrt{2}}$ and $9.9 \log n$.

1 Introduction

The complexity of monotone formulas/circuits for the majority function is a fascinating, albeit perplexing, problem in theoretical computer science. Without the monotonicity restriction, majority can be solved with simple linear-size circuits of depth $O(\log n)$, where the best known depth (over binary AND, OR, NOT gates) is $4.95 \log n + O(1)$ [12]. There are two fundamental algorithms for the majority function that achieve logarithmic depth. The first is a beautiful construction obtained by Valiant in 1984 [15] that achieves monotone formulas of depth $5.3 \log n + O(1)$ and size $O(n^{5.3})$. The second algorithm is obtained from the celebrated sorting network constructed in 1983 by Ajtai, Komlós, and Szemerédi [1]. Restricting to binary inputs and taking the middle output bit (median), reduces this network to a monotone circuit for the majority function of depth $K \log n$ and size $O(n \log n)$. The advantage of the AKS sorting network for majority is that it is a completely uniform construction of small size. On the negative side, its proof is quite complicated and more importantly, the constant K is huge: the best known constant K is about 5000 [11], and as observed by

J. Diaz et al. (Eds.): APPROX and RANDOM 2006, LNCS 4110, pp. 410–425, 2006.

Paterson, Pippenger, and Zwick [12], this constant is important. Further converting the circuit to a formula yields a monotone formula of size $O(n^K)$, which is roughly n^{5000}.

In order to argue about a quality of a solution to the problem, one should be precise about the different resources and the tradeoffs between them. We care about the depth, the size, the number of random bits for a randomized construction, and formula vs circuit question. Finally, the conceptual simplicity of both the algorithm and the correctness proof is also an important goal. Getting the best depth-size tradeoffs is perhaps the most sought after goal around this classical question, while achieving uniformity comes next.

An interesting aspect of the problem is the natural way it splits into two subproblems, the solution to which gives a solution to the original problem. Problem I takes as input an arbitrary n-bit binary vector, and outputs an m-bit vector. If the input vector has a majority of 1's, then the output vector has at least a 2/3 fraction of 1's, and if the input vector does not have a majority of 1's, then the output vector has at most a 1/3 fraction of 1's. Problem II is a promise problem that takes the m-bit output of problem I as its input. The output of Problem II is a single bit that is 1 if the input has at least a 2/3 fraction of 1's, and is a 0 if the input has at most a 1/3 fraction of 1's. Obviously the composition of these two functions solves the original majority problem.

There are several reasons to consider monotone circuits that are constructed via this two-phase approach. First, Valiant's analysis uses this viewpoint. Boppana's later work [2] actually lower bounds each of these subproblems separately (although failing to provide lower bound for the entire problem). Finally, the second subproblem is of interest in its own right. Problem II can be viewed as an approximate counting problem, and thus plays an important role in many areas of theoretical computer science. Non monotone circuits for this promise problem have been widely studied.

Results: The contribution of the current work is primarily in obtaining a new and simple construction of monotone circuits for the majority function of depth $5.3 \log n$ and size $O(n^3)$, hence significantly reducing the size of Valiant's formula while not compromising at all the depth parameter.

Further, for subproblem II as defined above, we supply a construction of a circuit size that is of a *linear size*, but does not compromise the depth compared to Valiant's solution. A very appealing feature of this construction is that it is uniform, conditioned on a reasonable assumption about the existence of good enough expander graphs. To this end we introduce a connection between this circuit complexity question and another domain, namely *message passing algorithms*. The depth we achieve for the promise problem nearly matches the 1954 lower bound of Moore and Shannon [10].

Finally, we show how to generalize our solution to a general threshold function, and explore the tradeoffs between the different resources we use; specifically, we show that by allowing for a depth of roughly twice that of Valiant's construction, we may get a circuit of size $O(n^{1+\sqrt{2}+o(1)}) = O(n^{2.42})$.

Techniques: In obtaining our result we introduce the concept of *deterministic amplification*, replacing the *probabilistic amplification* used by Valiant. In probabilistic amplification, given a monotone boolean function $f : \{0,1\}^n \to \{0,1\}$, one considers the probability $A_f(p)$ that f is one when the n input variables are independently one with probability p. We say that f probabilistically amplifies (p_l, p_h) to (q_l, q_h) if $A_f(p_l) \leq q_l$ and $A_f(p_h) \geq q_h$. We say that a monotone function $f : \{0,1\}^n \to \{0,1\}^m$ deterministically amplifies (p_l, p_h) to (q_l, q_h) if for *every* input with up to $p_l n$ (at least $p_h n$) ones the proportion of ones in the output is at most q_l (at least q_h).

With this terminology splitting the problem into the two subproblems mentioned above can be easily described. We seek two function f_1 and f_2 so that $f_1 : \{0,1\}^n \to \{0,1\}^m$ deterministically amplifies $(1/2 - 1/n, 1/2)$ to $(\delta, 1 - \delta)$ for some small constant $\delta > 0$, and $f_2 : \{0,1\}^m \to \{0,1\}$ deterministically amplifies $(\delta, 1 - \delta)$ to $(0, 1)$. In the sequel, we will call the problem of constructing f_1 *phase I* and that of constructing f_2 *phase II*.

Our circuit for phase I is quite simple. Starting with the n input variables at level zero, we have alternating layers of AND/OR gates, where each gate independently chooses its two inputs from the previous layer. We prove that such a circuit satisfies the requirements if the number of layers is $3.3 \log n$, and if the layers are sufficiently large (width decreasing with depth from $O(n^3)$ to $O(n)$).

We give two constructions of circuits for phase II. Both constructions yield circuits for $f_2 : \{0,1\}^m \to \{0,1\}$ of size $O(m)$ and depth $(2 + \epsilon) \cdot \log m + O(1)$, for arbitrarily small $\epsilon > 0$, almost matching the depth lower bound of $2 \log \delta m$ of Moore-Shannon [10]. The first construction is a probabilistic argument similar to our phase I construction but with different parameters. In it we explore the somewhat surprising benefits gained when changing the fanin of the gates to a large enough parameter d.

In the second construction we derandomize our construction using good expander graphs. The construction is an application of a well-known message-passing belief-propagation algorithm on an expander graph. To compute the promise problem, we simulate a logarithmic number of rounds of the message-passing algorithm on a d-regular bipartite graph that is a sufficiently good expander. The message passing algorithm is similar to the belief propagation algorithm used to decode LDPC codes on the erasure channel, and the analysis is based on adaptation of a result of Burshtein and Miller [3] to our setting. For the construction to be completely uniform, we must assume the existence of an explicit construction of sufficiently good expanders. While not known to date, finding such expanders is the focus of a rapidly developing research area, which hopefully will produce the required good expanders.

One crucial parameter used in our analysis, is the number of different inputs the circuit must handle. It is appealing from a computational point of view, as it gives a progress measure toward the final goal of the circuit. One interesting aspect of our probabilistic construction is that it can translate an improvement in this parameter into a reduction in the circuit size. We obtain a variant to our construction by exploiting this property. This variant, has a preprocessing stage that partially sorts its input, and consequently has a smaller size, at the expense of an increased depth.

The organization of the rest of the paper is as follows. In Section 2, we define the two notions of amplification that we will be considering, and review Valiant's argument. In Section 3, we present our new monotone circuits for majority. In Section 4, we adapt our construction to obtain efficient monotone circuits for all threshold functions. In Section 5, we obtain smaller size monotone circuits for the majority, at the expense of increasing the depth. In the last section, we discuss the known lower bounds, and open problems.

2   Notions of Amplification

For a monotone boolean function H on k inputs, we define its *amplification function* $A_H : [0, 1] \to [0, 1]$ as $A_H(p) = \Pr[H(X_1, \ldots, X_k) = 1]$, where X_i are independent boolean random variables that are one with probability p. Valiant [15], considered the function H on four variables, which is the OR of two AND gates, $H(x_1, x_2, x_3, x_4) = (x_1 \wedge x_2) \vee (x_3 \wedge x_4)$. The amplification function of H, depicted in Figure 1, is $A_H(p) = 1 - (1 - p^2)^2$, and has a non-trivial fixed point at $\beta = (\sqrt{5} - 1)/2 \simeq 0.61$.

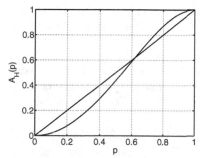

Fig. 1. $A_H(p)$ for $H(x_1, x_2, x_3, x_4) = (x_1 \wedge x_2) \vee (x_3 \wedge x_4)$

We say that a monotone function $F : \{0, 1\}^n \to \{0, 1\}$ *probabilistically amplifies* (p_l, p_h) to (q_l, q_h), if $q_l \geq A_F(p_l)$ and $q_h \leq A_F(p_h)$. In other words, applying F to independent boolean random variables that are one with probability p will amplify a promise that p is less than p_l or more than p_h, to a promise that F's output is one with probability less than q_l or more than q_h respectively. Since A'_H is continuous, for any $\epsilon > 0$ there exists $\Delta_0 > 0$ such that H probabilistically amplifies $(\beta - \Delta, \beta + \Delta)$ to $(\beta - (\gamma - \epsilon)\Delta, \beta + (\gamma - \epsilon)\Delta)$ for all $\Delta < \Delta_0$, where $\gamma = A_H(\beta)' = (\sqrt{5} - 1)^2 \simeq 1.52$. Let H_k be the depth $2k$ binary tree with alternating layers of AND and OR gates, where the root is labeled OR. Valiant's construction uses the fact that A_{H_k} is the composition of A_H with itself k times. Therefore, H_k probabilistically amplifies $(\beta - \Delta, \beta + \Delta)$ to $(\beta - (\gamma - \epsilon)^k \Delta, \beta + (\gamma - \epsilon)^k \Delta)$, as long as $(\gamma - \epsilon)^k \Delta < \Delta_0$. This implies that for any constant $\epsilon > 0$ we can take $2k = 3.3 \log n + O(1)$ to probabilistically amplify $(\beta - \Omega(1/n), \beta + \Omega(1/n))$ to $(\epsilon, 1 - \epsilon)$, where 3.3 is any constant bigger than

$\alpha = \log_{\sqrt{5}-1} 2 \simeq 3.27$. Further analysis shows that for $2k = 5.3 \log n + O(1)$, the tree H_k probabilistically amplifies $(\beta - \Omega(1/n), \beta + \Omega(1/n))$ to $(2^{-n-1}, 1 - 2^{-n-1})$, implying the existence a formula of depth $5.3 \log n + O(1)$ and size $O(n^{5.3})$ for the $\lceil \beta n \rceil$-th threshold function. Results of Boppana [2] and Dubiner and Zwick [5] show that no smaller formula can produce such an amplification.

One aspect of Valiant's construction that we are going to exploit, is that the use of a binary tree in the last $2 \log n$ layers is rather arbitrary. Similar analysis shows that replacing those layers by $2 \log_r n$ layers of an r-ary tree result with the similar probabilistic amplification. Replacing the r-ary AND, OR gates by formulas using binary gates results in a depth blowup of factor $\lceil \log r \rceil$. Therefore, the same depth as Valiant's construction can be obtained when r is a power of two, and taking any large value of r results in an arbitrarily small degradation in the constant before the log.

The approach of this paper, is to follow the same general scheme suggested by Valiant. However, instead of an $O(n^{5.3})$ formula, we produce an $O(n^3)$ circuit of similar depth. Because of the smaller size we cannot use a tree and maintain complete independence between the results computed at a certain layer, as is done in Valiant's tree. Instead we define a random circuit such that the values in a layer are completely independent, given the number of 1's of the previous a layer. In order that the portion of ones in each layer behaves as we would like, we need to make layer sizes sufficiently large. The crucial simple observation that enables us to keep layer sizes small, is that the circuit need only handle 2^n scenarios.

Definition 1. *Let F be a boolean function $F : \{0,1\}^n \to \{0,1\}^m$, and let $\mathcal{S} \subseteq \{0,1\}^n$ be some subset of the inputs. We say that F deterministically amplifies (p_l, p_h) to (q_l, q_h) with respect to $\mathcal{S}$, if for all inputs $x \in \mathcal{S}$, the following promise is satisfied (we denote by $|x|$ the number of ones in the vector x):*

$$|F(x)| \le q_l m \quad \text{if} \quad |x| \le p_l n$$
$$|F(x)| \ge q_h m \quad \text{if} \quad |x| \ge p_h n.$$

Note that unlike the probabilistic amplification, deterministic amplification has to work for all inputs or scenarios in the given set $\mathcal{S}$. From here on, whenever we simply say "amplification" we mean deterministic amplification.

For an arbitrary small constant $\epsilon > 0$, the construction we give is composed of two independent phases that may be of independent interest.

- A circuit $C_1 : \{0,1\}^n \to \{0,1\}^m$ for $m = O(n)$ that deterministically amplifies $(\beta - \Omega(1/n), \beta + \Omega(1/n))$ to $(\delta, 1 - \delta)$ for an arbitrarily small constant $\delta > 0$. This circuit has size $O(n^3)$ and depth $(\alpha + \epsilon) \cdot \log n + O(1)$.
- A circuit $C_2 : \{0,1\}^m \to \{0,1\}$, such that $C_2(x) = 0$ if $|x| < \delta m$ and $C_2(x) = 1$ if $|x| > (1 - \delta)m$, where $\delta > 0$ is a sufficiently small constant. This circuit has size $O(m)$ and depth $(2 + \epsilon) \cdot \log m + O(1)$.

The first circuit C_1 is achieved by a simple probabilistic construction that resembles Valiant's construction. We present two constructions for the second circuit, C_2. The first construction is probabilistic; the second construction is a simulation of a logarithmic number of rounds of a certain message passing algorithm on a good bipartite expander graph. The correctness is based on the analysis of a similar algorithm used to decode a low density parity check code (LDPC) on the erasure channel [3].

Combining the two circuits together yields a circuit $C : \{0,1\}^n \to \{0,1\}$ for the $\lceil \beta n \rceil$-th threshold function. The circuit is of size $O(n^3)$ and depth $(\alpha + 2 + 2\epsilon) \log n + O(1)$.

3   Monotone Circuits for Majority

In this section we give a randomized construction of the circuit $C : \{0,1\}^n \to \{0,1\}$ such that $C(x)$ is one if the portion of ones in x is at least βn and zero otherwise. The circuit C has size $O(n^3)$ and depth $(2 + \alpha + \epsilon) \cdot \log n + O(1)$ for an arbitrary small constant $\epsilon > 0$. As we described before, we will describe C as the compositions of the circuits C_1 and C_2 whose parameters are given by the following two theorems:

Theorem 1. *For every $\epsilon, \epsilon', c > 0$, there exists a circuit $C_1 : \{0,1\}^n \to \{0,1\}^m$ for $m = O(n)$, of size $O(n^3)$ and depth $(\alpha + \epsilon) \cdot \log n + O(1)$ that deterministically amplifies all inputs from $(\beta - c/n, \beta + c/n)$ to $(\epsilon', 1 - \epsilon')$.*

Theorem 2. *For every $\epsilon > 0$ there exists $\epsilon' > 0$ and a circuit $C_2 : \{0,1\}^n \to \{0,1\}$, of size $O(n)$ and depth $(2+\epsilon) \cdot \log n + O(1)$ that deterministically amplifies all inputs from $(\epsilon', 1 - \epsilon')$ to $(0,1)$.*

The two circuits use a generalization of the four input function H used in Valiant's construction. For any integer $d \geq 2$, we define the function $H^{(d)}$ on d^2 inputs as the d-ary OR of d d-ary AND gates, i.e $\vee_{i=1}^{d} \wedge_{j=1}^{d} x_{ij}$. Note that Valiant's function H is just $H^{(2)}$.

Each of the circuits C_1 and C_2 is a layered circuit, where layer zero is the input, and each value at the i-th layer is obtained by applying $H^{(d)}$ to d^2 independently chosen inputs from layer $i - 1$. However, the values of d we choose for C_1 and C_2 are different. For C_1 we have $d = 2$, while for C_2 we choose sufficiently large $d = d(\epsilon)$ to meet the depth requirement of the circuit. We let $\mathcal{F}_{n,m,F}$ denote a random circuit mapping n inputs to m outputs, where F is a fixed monotone boolean circuit with k inputs, and each of the m output bits is calculated by applying F to k independently chosen random inputs.

We start with a simple lemma that relates the deterministic amplification properties of $\mathcal{F}$ to the probabilistic amplification function A_F. [1]

Lemma 1. *For any $\epsilon, \delta > 0$, the random function $\mathcal{F}$ deterministically amplifies (p_l, p_h) to $(A_F(p_l) \cdot (1 + \delta), A_F(p_h) \cdot (1 - \delta))$ with respect to $\mathcal{S} \subseteq \{0,1\}^n$ with probability at least $1 - \epsilon$, if:*

$$m = \Omega \left(\frac{\log(|\mathcal{S}|) + \log(1/\epsilon)}{A_F(p_l) \cdot \delta^2} \right).$$

---

[1] Note that we talk about the deterministic amplification properties of a random function.

Proof. It is sufficient to prove that for any input $x \in \mathcal{S}$, the probability of failure of F is bounded by $\epsilon/|\mathcal{S}|$. By definition, for any application of $\mathcal{F}$, the probability to get 1 is $A_F(p)$, where $p = |x|/n$ is the portion of ones in x. By monotonicity, we may assume that $p = p_l$ or $p = p_h$. A straightforward application of the Chernoff bound is all we need here. For the case $p = p_l$, we have $\Pr[||\mathcal{F}(x)| > A_F(p_l)(1+\delta)m] < \exp(-mA_F(p)\delta^2/3)$, which is less than $\epsilon/|\mathcal{S}|$ for $m \geq 3(\log|\mathcal{S}| + \log(1/\epsilon))/(A_F(p_l) \cdot \delta^2)$. The case $p = p_h$ is handled similarly.

Proof (Proof of Theorem 1).

The circuit C_1 is a composition of $\mathcal{F}_{n,m_1,H}, \mathcal{F}_{m_1,m_2,H}, \ldots, \mathcal{F}_{m_{t-1},m_t,H}$, where the parameters $n = m_0, m_1, \ldots, m_t = m$ are positive integers to be fixed later, and are the sizes of the layers of the circuit. Since $\mathcal{F}_{\cdot,\cdot,H}$ is a random function, this describes a random construction of a circuit. We prove that with high probability such a circuit deterministically amplifies all inputs from $(\beta - c/n, \beta + c/n)$ to $(\epsilon', 1 - \epsilon')$. For simplicity, we only prove that with high probability for all inputs with portion of ones smaller than $\beta - c/n$ the output has fewer than $\epsilon'm$ ones. The proof of the other case is similar. For convenience of notation, we say that some circuit (deterministically or probabilistically) amplifies p to q as a short hand for amplifying $(p, \cdot)$ to $(q, \cdot)$ where the dot stands for the unspecified upper bounds.

The basic idea is that if layers have large size, we expect this circuit to have similar behavior to Valiant's tree. As observed before, for every constant $\epsilon > 0$ there is a constant $\Delta_0 > 0$ such that for any $p = \beta - \Delta$ with $0 < \Delta < \Delta_0$, we have $A_H(p) < \beta - (\gamma - \epsilon)\Delta$, where $\gamma = A_H(\beta)'$. This implies that if the portion of ones at some level i is p, then the expected portion of ones at level $i+1$ is $A_H(p) < \beta - (\gamma - \epsilon)\Delta$. We will set δ in Lemma 1 so that the deterministic amplification of $\mathcal{F}_{m_i,m_{i+1}}$ guarantees that the portion of ones at level $i+1$ will be at most $\beta - (\gamma - 2\epsilon)\Delta$. The details follow.

Let G_i be the be the circuit truncated to the first i layers. We prove that with high probability G_i deterministically amplifies the initial promise $\beta - c/n$ to $\beta - (\gamma - 2\epsilon)^i \cdot (c/n)$, as long as $(\gamma - 2\epsilon)^i \cdot (c/n) < \Delta_0$. The proof proceeds by inductions on i, where the basis $i = 0$ trivially holds. So, assume that $i > 0$, and that $(\gamma - 2\epsilon)^i \cdot (c/n) < \Delta_0$. Furthermore, assume that the first $i - 1$ layers are some *fixed circuit* G_{i-1} satisfying the hypothesis. Namely, deterministically amplifies $\beta - c/n$ to $\beta - (\gamma - 2\epsilon)^{i-1} \cdot (c/n)$, for all inputs. Let G_i be obtained by composing the fixed circuit G_{i-1} with the random circuit $\mathcal{F}_{m_{i-1},m_i,H}$. Then, as G_{i-1} is fixed, it has at most 2^n possible outputs. The crucial observation, is that it suffices for $\mathcal{F}_{m_j-1,m_j,H}$ to deterministically amplify $\beta - (\gamma - 2\epsilon)^{i-1} \cdot (c/n)$ to $\beta - (\gamma - 2\epsilon)^i \cdot (c/n)$, *only with respect to the 2^n outputs of G_{i-1}.*

Then, it suffice to choose the values δ in Lemma 1, as

$$\frac{\epsilon \cdot (\gamma - 2\epsilon)^{i-1} \cdot (c/n)}{\beta - (\gamma - \epsilon) \cdot (\gamma - 2\epsilon)^{i-1} \cdot (c/n)} = \Theta\left((\gamma - 2\epsilon)^{i-1} \cdot (c/n)\right).$$

That is, we can choose δ as an increasing geometric sequence, starting from $\Theta(1/n)$ for $i = 1$, up to $\Theta(1)$ for $i = \log_{\gamma-2\epsilon} n$. The implied layer size for error probability 2^{-n} (which is much better than we need), is $\Theta(n/\delta^2)$. Therefore, it decreases geometrically from $\Theta(n^3)$ down to $\Theta(n)$.

It is not difficult to see that after achieving the desired amplification from $\beta - c/n$ to $\beta - \Delta_0$, only a constant number of layers is needed to get down to ϵ'. The corresponding value of δ in these last steps is a constant (that depends on ϵ'), and therefore, the required layer sizes are all $\Theta(n)$.

Proof (Proof of Theorem 2).

The circuit C_2 is a composition of $\mathcal{F}_{n,m_1,H^{(d)}}, \mathcal{F}_{m_1,m_2,H^{(d)}}, \ldots, \mathcal{F}_{m_{t-1},m_t,H^{(d)}}$, where d and the layer sizes $n = m_0, m_1, \ldots, m_t = 1$ are suitably chosen parameters depending on ϵ. We prove that with high probability such a circuit deterministically amplifies all inputs from $(\epsilon', 1 - \epsilon')$ to $(0, 1)$. As before, we restrict our attention to the lower end of the promise problem and prove that C_2 outputs zero on all inputs with portion of ones smaller than ϵ'.

As in the circuit C_1, the layer sizes must be sufficiently large to allow accurate computation. However, for the circuit C_2, accurate computation does not mean that the portion of ones in each layer is close to its expected value. Rather, our aim is to keep the portion of ones bounded by a fixed constant ϵ', while making each layer smaller than the preceding one by approximately a factor of d. We continue this process until the layer size is constant, and then use a constant size circuit to finish the computation. Therefore, since the number of layers of such a circuit is about $\log n / \log d$, and the depth of the circuit for $H^{(d)}$ is $2\lceil \log d \rceil$, the total depth is about $2 \log n$ for large d.

By the above discussion, it suffices to prove the following: For every $\epsilon > 0$ there exists a real number $\delta > 0$ and two integers d, n_0, such that for all $n \geq n_0$ the random circuit $\mathcal{F}_{n,m,H^{(d)}}$ with $m = (1 + \epsilon) \cdot n/d$, deterministically amplifies δ to δ with respect to all inputs, with failure probability at most $1/n$. Since $A_H(\delta) = 1 - (1 - \delta^d)^d \leq d \cdot \delta^d$, the probability of failure for any specific input with portion of ones at most δ, is bounded by:

$$\binom{m}{\delta m} \cdot (A_H(\delta))^{\delta m} \leq \left(\frac{e}{\delta} \cdot d \cdot \delta^d \right)^{\delta m} = \left(de \cdot \delta^{d-1} \right)^{\delta m}.$$

Therefore, by a union bound the probability that $\mathcal{F}_{n,m,H^{(d)}}$ fails is bounded by:

$$\left(de \cdot \delta^{d-1} \right)^{\delta m} \cdot \binom{n}{\delta n} \leq \left[(de \cdot \delta^{d-1})^{(1+\epsilon)/d} \cdot (e/\delta) \right]^{\delta n} = \left[c(d, \epsilon) \cdot \delta^{(1+\epsilon) \cdot (d-1)/d - 1} \right]^{\delta n},$$

where $c(d, \epsilon)$ is some function of d and ϵ. Given, ϵ, we choose a sufficiently large d so that $(1 + \epsilon) \cdot (d - 1)/d - 1$ is positive. Then we take sufficiently small δ, so that the expression in the square brackets is smaller than one. Finally, we take a sufficiently large n_0 to guarantee that the exponentially small upper bound on the error probability, is smaller than $1/n$.

3.1   Derandomizing the Construction of Phase II

In this subsection we present a second construction of a small monotone circuit C that deterministically amplifies $(a, 1 - a)$ to $(0, 1)$ with respect to $\{0, 1\}^n$.

Our construction uses recent ideas and algorithms from belief propagation decoding, applied to solving majority. Underlying both belief propagation and algorithms for majority is the concept of amplification, first introduced in the classical 1954 paper of Moore and Shannon. Since then, the amplification method has been generalized and used in a variety of contexts. Luby, Mitzenmacher and Shokrollahi [8] used the amplification method to analyze the performance of a belief propagation message passing algorithm for decoding low density parity check (LDPC) codes. Today the use of belief propagation for decoding LDPC codes is one of the hottest topics in error correcting codes [9, 14, 13].

Let $G = (V_L, V_R; E)$ be a d regular bipartite graph with n vertices on each side, $V_L = V_R = [n]$. Consider the following message passing algorithm, where we think of the left and right as two players. The left player "plays AND" and the right player "plays OR". At time zero the left player starts by sending one boolean message through each left to right edge, where the value of the message m_{uv} from $u \in V_L$ to $v \in V_R$ is the input bit x_u. Subsequently, the messages at time $t > 0$ are calculated from the messages at time $t-1$. At odd times, given the left to right messages m_{uv}, the right player calculates the right to left messages m'_{vw}, from $v \in V_R$ to $w \in V_L$ by the formula $m'_{vw} = \vee_{u \in N(v) \setminus w} m_{uv}$. That is, the right player sends a 1 along the edge from $v \in V_R$ to $w \in V_L$ if and only if *at least one* of the incoming messages/values (not including the incoming message from w) is 1. Similarly, at even times the algorithm calculates the left to right messages m'_{vw}, $v \in V_L$, $w \in V_R$, from the right to left messages m_{uv}, by the formula $m'_{vw} = \wedge_{u \in N(v) \setminus w} m_{uv}$. That is, the left player sends a 1 along the edge from $v \in V_L$ to $w \in V_R$ if and only if *all* of the incoming messages/values (not including the incoming message from w) are 1. We further need the following definitions. We call a *left* vertex *bad* at even time t if it transmits at least one message of value one at time t. Similarly, a *right* vertex is *bad* at odd time t if it is a right vertex that transmits at least one message of value zero at time t. We let $b(t)$ be the number of bad vertices at time t. These definitions will be instrumental in providing a potential function measuring the progress of the message passing algorithm which is expressed in Lemma 2.

We say that a bipartite graph $G = (V_L, V_R; E)$ is (λ, e)-expanding, if for any vertex set $S \subseteq V_L$ (or $S \subseteq V_R$) of size at most λn, $|N(S)| \geq e|S|$. It will be convenient to denote the expansion of the set S by $e_S = |N(S)|/|S|$.

Lemma 2. *Consider the message passing algorithm using a $d \geq 4$ regular expander graph with $d - 1 > e \geq (d+1)/2$. If $b(t) \leq \lambda n/d^2$ then $b(t+2) \leq b(t)/\eta$, where $\eta = \frac{d-1}{2(d-e)}$.*

We postpone the proof of the lemma, and show its use for constructing the circuit C_2. First, we show that, for any $\epsilon > 0$, the algorithm provides a circuit of depth $(2 + \epsilon) \log n$ and of size $O(n \log n)$ for the promise problem with $a \leq \lambda(d-1)/d^3$. Suppose that there are at most an ones. Then $b(0) \leq an$. Therefore $b(2t) \leq an/\eta^t$, and so $b(2t) = 0$ for $t > \log(an)/\log \eta$ and all outputs are zero at that time. If there are at most an zeros, we analyze the number of bad *right* vertices as follows. Since each bad right vertex must be connected to at least

$d - 1$ left vertices associated with zero input bits, and since there are at most an left vertices transmitting zero, it follows that $b(1) \leq an \cdot d/(d-1) \leq \lambda/d^2$ whence the conditions of Lemma 2 are satisfied and $b(2t+1) \leq \frac{and}{d-1}/\eta^t$ and so $b(2t+1) = 0$ for $t > \log(a\frac{d}{d-1}n)/\log\eta$.

The better the expanders we use, the bigger $\eta = \frac{d-1}{2(d-e)}$ gets, and the better the time guarantee above gets. How good are the expanders that we may use? One can show the existence of such expanders for sufficiently large d large, and $e > d - c$ for an absolute constant c.

The best known explicit construction that gets close to what we need, is the result of [4]. However, that result does not suffice here for two reasons. The first is that it only achieves expansion $(1 - \epsilon)d$ for any $\epsilon > 0$ and sufficiently large d depending on ϵ. The second is that it only guarantees left-to-right expansion, while our construction needs both left-to-right and right-to-left expansion. We refer the reader to the survey [6] for further reading and background.

For such expanders, $\eta \geq \frac{d-1}{2c}$, and therefore, after $2\log(\frac{and}{d-1})/\log\frac{d-1}{2c} = (2 + \epsilon)\frac{\log n}{\log d-1}$ iterations, all messages contain the right answer, where ϵ can be made arbitrarily small by choosing sufficiently large d. It remains to convert the algorithm into a monotone circuit, which introduces a depth-blowup of $\log\lceil d - 1\rceil$ owing to the depth of a binary tree simulating a $(d - 1)$-ary gate. Thus we get a $(2 + \epsilon)\log n$-depth circuit for arbitrarily small $\epsilon > 0$. The size is obviously $dn \cdot \text{depth} = O(n\log n)$.

To get a linear circuit, further work is needed, which we now describe. The idea is to use a sequence of graphs $G_0 = G, G_1, \ldots$, where each graph is half the size of its preceding graph, but has the same degree and expansion parameters. We start the message passing algorithm using the graph $G = G_0$, and every t_0 rounds (each round consists of OR and then AND), we switch to the next graph in the sequence. Without the switch, the portion of bad vertices should decrease by a factor of η^{t_0}, every t_0 rounds. We argue that each switch can be performed, while losing at most a constant factor. To describe the switch from G_i to G_{i+1}, we identify $V_L(G_{i+1})$ with an arbitrary half of the vertices $V_L(G_i)$, and start the message passing algorithm on G_{i+1} with the left to right messages from each vertex in $V_L(G_{i+1})$, being the same as at the last round of the algorithm on G_i. As the number of bad left vertices cannot increase at a switch, their portion, at most doubles. For the right vertices, the exact argument is slightly more involved, but it is clear that the portion of bad right vertices in the first round in G_{i+1}, increases by at most a constant factor c, compared with what it should have been, had there been no switch. (Precise calculation, yields $c = 2d\eta$.) Therefore, to summarize, as the circuit consists of a geometrically decreasing sequence of blocks starting with a linear size block, the total size is linear as well. As for the depth, the amortized reduction in the portion of bad vertices per round, is by a factor of $\eta' = \eta/c^{1/t_0}$. Therefore, the resulting circuit is only deeper than the one described in the previous paragraph, by a factor of $\log\eta/\log\eta'$. By choosing a sufficiently large value for t_0, we obtain:

Theorem 3. *For any $\epsilon > 0$, there exists $a > 0$ such that for any n there exists a monotone circuit of depth $(2+\epsilon)\log n + O(1)$ and size $O(n)$ that solves a-promise problem.*

We note here that $O(\log n)$ depth monotone circuits for the a-promise problem can also be obtained from ϵ-halvers. These are building blocks used in the AKS network. However, our monotone circuits for the a-promise problem have two advantages. First, our algorithm relates this classical problem in circuit complexity to recent popular message passing algorithms. And second, the depth that we obtain is nearly tight. Namely, Moore and Shannon [10] prove that any monotone formula/circuit for majority requires depth $2\log n - O(1)$, and the lower bound holds for the a-promise problem as well.

Proof (Proof of Lemma 2). (builds on [3])

We consider only the case of bad left vertices. The proof for bad right vertices follows from the same proof, after exchanging ones with zeroes, ANDs with ORs, and lefts with rights. Let $B \subseteq V_L$ be the set of bad left vertices, and assume $|B| \leq \lambda n/d^2$ at some even time t and B' the set of bad vertices at time $t + 2$. We bound the size of B' by considering separately $B' \setminus B$ and $B' \cap B$. Note that all sets considered in the proof have size at most λn, and therefore expansion at least e.

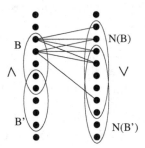

To bound $B' \setminus B$, consider the set $Q = N(B' \setminus B) \setminus N(B) = N(B' \cup B) \setminus N(B)$. Since vertices in Q are not adjacent to B, then at time $t + 1$ they send right to left messages valued zero. On the other hand, any vertex in $B' \setminus B$ can receive at most one such zero message (otherwise all its messages at time $t + 2$ will be valued zero and it cannot be in B'). Therefore, since each vertex in Q must have at least one neighbour in $B' \setminus B$, it follows that $|Q| \leq |B' \setminus B|$. Therefore, we have:

$$|N(B' \cup B)| = |N(B)| + |Q| \leq |N(B)| + |B' \setminus B| = e_B \cdot |B| + |B' \setminus B|.$$

On the other hand, $|N(B' \cup B)| \geq e \cdot |B' \cup B| = e \cdot (|B| + |B' \setminus B|)$.
Combining the above two inequalities, we obtain:

$$|B' \setminus B| \leq \frac{e_B - e}{e - 1} \cdot |B|. \tag{1}$$

To bound $B' \cap B$, consider the set $T = N(B' \cap B) \setminus N(B \setminus B') = N(B) \setminus N(B \setminus B')$. Let N_0 (resp. N_1) be the number of zero (resp. one) messages received by vertices in $B' \cap B$ at time $t + 1$. Then obviously, $N_0 + N_1 = d \cdot |B' \cap B|$. As before, a vertex in $B' \cap B$ can receive at most one zero message and therefore $N_0 \leq |B' \cap B|$. Also, let T_0 be the vertices of T that transmit at least one zero message at time $t+1$ to $B' \cap B$, and $T_1 = T \setminus T_0$. Clearly $|T_0| \leq N_0$. On the other hand, each vertex in T_1 transmits a one to some vertex in $B' \cap B$, and therefore must have at least two neighbors in $B' \cap B$, implying that $|T_1| \leq N_1/2$. Hence

$$|T| \leq N_0 + N_1/2 = (N_0 + N_1)/2 + N_0/2$$
$$\leq (d/2)|B' \cap B| + (1/2)|B' \cap B| = |B' \cap B| \cdot (d+1)/2.$$

Therefore

$$e_B \cdot |B| = |N(B)| = |N(B \setminus B')| + |T| \leq d \cdot |B \setminus B'| + |B' \cap B| \cdot (d+1)/2$$
$$= d \cdot |B| - |B' \cap B| \cdot (d-1)/2.$$

This implies:

$$|B' \cap B| \leq \frac{d - e_B}{(d-1)/2} \cdot |B|. \tag{2}$$

Combining inequalities (1) and (2) we get that:

$$|B'|/|B| \leq \frac{e_B - e}{e - 1} + \frac{d - e_B}{(d-1)/2}.$$

Since $e \geq (d+1)/2$, and $e_B \geq e$, this yields the required bound:

$$|B'|/|B| \leq 2(d - e)/(d - 1).$$

As noted before in Section 2, replacing the last $2 \log n$ layers of Valiant's tree with $2 \log_r n$ layers of r-ary AND/OR gates, results in an arbitrarily small increase in the depth of the corresponding formula for a large value of r. It is interesting to compare the expected behavior of the suggested belief-propagation algorithm to the behavior of the $(d - 1)$-ary tree. Assume that the graph G is chosen at random (in the configuration model), and that the number of rounds k is sufficiently small, $(d - 1)^{2k} \ll n$. Then, almost surely the computation of all but $o(1)$ fraction of the k-th round messages is performed by evaluating a $(d - 1)$-ary depth k trees. Moreover, introducing an additional $o(1)$ error, one may assume that the leaves are independently chosen boolean random variables that are one with probability p, where p is the portion of ones in the input. This observation sheds some light on the performance of the belief propagation algorithm. However, our analysis proceeds far beyond the number of rounds for which a cycle free analysis is applicable.

4 Monotone Formulas for Threshold-k Functions

Consider the case of the k-th threshold function, $T_{k,n}$, i.e. a function that is one on $x \in \{0,1\}^n$ if $|x| \geq (k + 1)$ and zero otherwise. We show that, by essentially the same techniques of Section 3, we can construct monotone circuits to this more general problem. We assume henceforth that $k < n/2$, since otherwise, we construct the circuit $T_{n-1-k,n}$ and switch AND with OR gates. For $k/n = \Theta(1)$, the construction yields circuits of depth $5.3 \log n + O(1)$ and size $O(n^3)$. However, when $k = o(n)$, circuits are shallower and smaller (this not surprising fact is also discussed in [2] in the context of formulas).

The construction goes as follows: (i) Amplify $(k/n, (k+1)/n)$ to $(\beta - \Omega(1/k), \beta + \Omega(1/k))$ by randomly applying to the input a sufficiently large number of OR gates with arity $\Theta(n/k)$ (ii) Amplify $(\beta - \Omega(1/k), \beta + \Omega(1/k))$ to $(O(1), 1 - O(1))$ using a variation of phase I, and (iii) Amplify $(O(1), 1 - O(1))$ to $(0, 1)$ using phase II.

We now give a detailed description. For the sake of the section to follow, we require the following lemma which is more general than is needed for the results of this section. The proof is omitted for lack of space.

Lemma 3. Let $\mathcal{S} \subseteq \{0,1\}^n$, and $\epsilon > 0$. Then, for any k, there is a randomized construction of a monotone circuit that evaluates $T_{k,n}$ correctly on all inputs from $\mathcal{S}$ and has

$$depth \leq \log(n) + 2.3 \log(k') + (2 + \epsilon) \cdot \log \log |\mathcal{S}| + O(1),$$
$$size \leq O(\log |\mathcal{S}| \cdot k'n).$$

Here $k' = \min(k, n - 1 - k)$, and the constants of the O depend only on ϵ.

To guarantee the correctness of a monotone circuit for $T_{n,k}$, it suffices to check its output on inputs of weight $k, k + 1$ (as the circuit is monotone). Plugging $\log |\mathcal{S}| = \log(\binom{n}{k} + \binom{n}{k+1}) = O(k \log(n/k))$ into the lemma yields:

Theorem 4. $T_{k,n}$ has a randomized construction of a monotone circuit with

$$depth \leq \log(n) + 4.3 \log(k') + O(\log \log(n/k)),$$
$$size \leq O((k')^2 n \log(n/k')),$$

where $k' = \min(k, n - 1 - k)$, and the constants of the O are absolute.

5   Reducing the Circuit Size

The result obtained so far for the majority, is a monotone circuit of depth $5.3 \log n + O(1)$ and size $O(n^3)$. In this section, we would like to obtain smaller circuit size, at the expense of increasing the depth somewhat. The crucial observation is that the size of our circuit depends linearly on the logarithm of the number of scenarios it has to handle. Therefore, applying a preprocessing stage to reduce the wealth of scenarios may save up to a factor of n in the circuit size. We propose a recursive construction that reduces the circuit size to about $n^{1+\sqrt{2}}$.

Initially, by Theorem 4, we have monotone circuits $C_{k,n}^0$ for the threshold functions $T_{k,n}$ with size $s_0(n) = O(n^3)$ and depth $d_0(n) = 5.3 \log n + O(1)$.

Given circuits $C_{k,n}^{(i)}$ for $T_{k,n}$ of size and depth bounded by $s_i(n)$, $d_i(n)$, one can calculate all threshold functions in parallel and obtain a sorting circuit $C_n^{(i)}$:

$\{0,1\}^n \rightarrow \{0,1\}^n$ of size and depth bounded by $ns_i(n)$, $d_i(n)$. The circuit $C_{k,n}^{(i+1)}$ is built of two stages. First, the n-bit input is partitioned into n/a_i blocks of size a_i, and each block is sorted in parallel using the circuit $C_a^{(i)}$. Second, the kth threshold function is calculated on the partially sorted n-bit string by the family of circuits with parameters given by Lemma 3. When the n/a blocks are sorted, there are only $(a_i + 1)^{n/a_i}$ possible inputs, as the number of ones in each blocks completely specifies the input. Therefore, the first stage reduces the number of scenarios to $(a_i + 1)^{n/a_i} \leq n^{n/a_i}$ and we have

$$s_{i+1}(n) = (n/a_i) \cdot (a_i + 1) \cdot s_i(a_i) + O\left((n/a_i) \cdot \log n \cdot n^2\right) = n \cdot s_i(a_i) + n^{3+o(1)}/a_i,$$
$$d_{i+1}(n) = d_i(a_i) + 5.3 \log n - 2 \log(a_i) + O(\log \log n).$$

Let $a_i = n^{\alpha_i}$ for some constants α_i, and assume that $s_i(n) = n^{\sigma_i + o(1)}$, and that $d_i(n) = \delta_i \log n + O(\log \log n)$. Then we obtain the following recurrence:

$$\sigma_{i+1} = \max(1 + \alpha_i \sigma_i, 3 - \alpha_i), \quad \delta_{i+1} = \alpha_i \delta_i + 5.3 \log n - 2\alpha_i.$$

We choose $\alpha_i = 2/(\sigma_i + 1)$ to equate $1 + \alpha_i \sigma_i$ with $3 - \alpha_i$. Consequently, $\sigma_{i+1} = 3 - 2/(\sigma_i + 1)$ and $\delta_{i+1} = 5.3 + (\delta_i - 2) \cdot 2/(\sigma_i + 1)$, yielding the following sequence:

i	0	1	2	3	4	5	6	7	8	9	10
α_i	0.500	0.571	0.583	0.585	0.586	0.586	0.586	0.586	0.586	0.586	0.586
σ_i	3.000	2.500	2.429	2.417	2.415	2.414	2.414	2.414	2.414	2.414	2.414
δ_i	5.271	6.906	8.074	8.814	9.259	9.522	9.677	9.768	9.821	9.852	9.870

The sequence α_i tends to $1 + \sqrt{2}$ which is the positive solution of $x = 3 - 2/(x + 1)$, and δ_i tends to $(1 + \sqrt{2})(\alpha - 2 + 2\sqrt{2}) \simeq 9.896$. Therefore:

Theorem 5. *There is a randomized construction of a monotone circuit for the majority of size $n^{1+\sqrt{2}+o(1)}$, and depth $9.9 \log n + O(1)$.*

6   Related Work and Open Problems

It is of great interest to improve upon the size or amount of randomness required by our construction. One approach, is to reduce the number of scenarios by preprocessing. The best result we have here is stated in Theorem 5. A second approach, is to improve the original bound ($n^3 \log n$ random bits, n^3 size). The obvious obstacle are the first few layers of the phase I circuit. The current n^3 size upper bound follows from a union bound, which we do not know to be tight. In fact, we do not even know how to save on the size or amount of randomness required to construct the first layer! This very problem can be cast as a discrepancy problem on hypergraphs. Indeed, if we restrict ourselves to repeated applications of $H(x_1, x_2, x_3, x_4) = (x_1 \wedge x_2) \vee (x_3 \wedge x_4)$ so that each application is associated with the two pairs $\{x_1, x_2\}$ and $\{x_3, x_4\}$, we have the following discrepancy problems on 4-uniform hypergraphs. Find a hypergraph on n vertices with each edge composed of two size-two sets $e = e_1 \cup e_2$. The graph should have as few edges as possible while satisfying that for every vertex subset $S \subset [n]$, the portion of edges that have at least one of their halves inside S is close to $A_H(|S|/n)$. This problem seems to generalize a similar problem for graphs: constructing a graph where for every vertex subset S, the portion of edges with both end points in S is close to $(|S|/n)^2$. Not surprisingly, this is equivalent to expansion.

In seminal work, Karchmer and Wigderson [7] gave a precise characterization of both monotone and monotone formula/circuit size based on the complexity of related communication search problems. For the majority function, the monotone search problem is as follows. Let mMaj-search be the following two-player communication complexity problem. Player I is given a subset $A \subset [n]$ of size $n/2 + 1$. Player II is given a subset $B \subset [n]$ of size $n/2$. They want to determine an element $i \in [n]$ such that i is in their intersection. In the non-monotone version of the problem, Maj-search, the input is the same, but now they are allowed to find either an element i in their intersection, or an element j lying outside of both sets. By the main theorem of [7], the minimal monotone formula/circuit size for majority is equal to the communication complexity of mMaj-search, and the minimal formula/circuit size for majority is equal to the communication complexity of Maj-search. The monotone communication complexity problem for the promise problem is as above except that now Players I and II are given subsets A and B each of size $2n/3$ and again they want to find some element in their common intersection. Likewise, for the monotone version, they want to compute either an element in the common intersection, or an element j lying outside of both sets. We find it useful to consider the upper and lower bounds for majority, as well as for the promise version of majority within this communication complexity setting.

There are two central open problems related to this work. First, is the promise version really simpler than majority? A lower bound greater than $2 \log n$ on the communication complexity of mMaj-search would settle this question. Boppana [2] and more recent work [5] show lower bounds on a particular method for obtaining monotone formulas for majority. However we are asking instead for lower bounds on the size/depth of unrestricted monotone formulas/circuits. Secondly, the original question remains unresolved. Namely, we would like to obtain explicit uniform formulas for majority of optimal or near optimal size. A related problem is to come up with a natural (top-down) communication complexity protocol for mMaj-Search that uses $O(\log n)$ many bits.

References

1. M. Ajtai, J. Komlós, and E. Szemerédi. Sorting in $c \log n$ parallel steps. *Combinatorica*, 3(1):1–19, 1983.
2. R. B. Boppana. Amplification of probabilistic boolean formulas. *IEEE Symposium on Foundations of Computer Science (FOCS)*, pages 20–29, 1985.
3. D. Burshtein and G. Miller. Expander graph arguments for message-passing algorithms. *IEEE Trans. Inform. Theory*, 47(2):782–790, 2001.

4. M. Caplbo,O. Reingold, S. Vadhan, and A. Wingderson. Randomness conductors and constant-degree expansion beyond the degree 2 barrier. In *Proceedings 34th Symposium on Theory of Computing*, Pages 659–668, 2002.
5. M. Dubiner and U. Zwick. Amplification by read-once formulas. *SIAM J. Comput.*, 26(1):15–38, 1997.
6. S. Hoory, N. Linial, and A. Wigderson. Expander graphs and their applications. to appear at the Bulletin of the AMS.
7. Mauricio Karchmer and Avi Wigderson. Monotone circuits for connectivity require super-logarithmic depth. In *Proceedings of the Twentieth Annual ACM Symposium on Theory of Computing*, Pages 539–550, Chicago, IL, May 1988
8. M. Luby, M. Mitzenmacher, and A. Shokrollahi. Analysis of random processes via and-or tree evaluation. In *ACM-SIAM Symp. on Discrete Algorithms (SODA)*, 1998.
9. M. Luby, M. Mitzenmacher, A. Shokrollahi, and D. A. Spielman. Analysis of low density codes and improved designs using irregular graphs. *ACM Symposium on Theory of Computing (STOC)*, 1998.
10. E. F. Moore and C. E. Shannon. Reliable circuits using less reliable relays. I, II. *J. Franklin Inst.*, 262:191–208, 281–297, 1956.
11. M. S. Paterson. Improved sorting networks with $O(\log N)$ depth. *Algorithmica*, 5(1):75–92, 1990.
12. M. S. Paterson, N. Pippenger, and U. Zwick. Optimal carry save networks. In *Boolean function complexity (Durham, 1990)*, volume 169 of *London Math. Soc. Lecture Note Ser.*, pages 174–201. Cambridge Univ. Press, Cambridge, 1992.
13. T. Richardson and R. Urbanke. Modern coding theory. Draft of a book.
14. T. Richardson and R. Urbanke. The capacity of low-density parity-check codes under message-passing decoding. *IEEE Trans. Inform. Theory*, 47(2):599–618, 2001.
15. L. G. Valiant. Short monotone formulae for the majority function. *J. Algorithms*, 5(3):363–366, 1984.

Space Complexity vs. Query Complexity

Oded Lachish[1], Ilan Newman[2,*], and Asaf Shapira[3,**]

[1] University of Haifa, Haifa, Israel
loded@cs.haifa.ac.il
[2] University of Haifa, Haifa, Israel
ilan@cs.haifa.ac.il
[3] School of Computer Science, Raymond and Beverly Sackler Faculty of Exact
Sciences, Tel Aviv University, Tel Aviv, Israel
asafico@tau.ac.il

Abstract. Combinatorial property testing deals with the following relaxation of decision problems: Given a fixed property and an input x, one wants to decide whether x satisfies the property or is "far" from satisfying it. The main focus of property testing is in identifying large families of properties that can be tested with a certain number of queries to the input. Unfortunately, there are nearly no general results connecting standard complexity measures of languages with the hardness of testing them. In this paper we study the relation between the space complexity of a language and its query complexity. Our main result is that for any space complexity $s(n) \leq \log n$ there is a language with space complexity $O(s(n))$ and query complexity $2^{\Omega(s(n))}$. We conjecture that this exponential lower bound is best possible, namely that the query complexity of a languages is at most exponential in its space complexity.

Our result has implications with respect to testing languages accepted by certain restricted machines. Alon et al. [FOCS 1999] have shown that any regular language is testable with a constant number of queries. It is well known that any language in space $o(\log \log n)$ is regular, thus implying that such languages can be so tested. It was previously known that there are languages in space $O(\log n)$ which are not testable with a constant number of queries and Newman [FOCS 2000] raised the question of closing the exponential gap between these two results. A special case of our main result resolves this problem as it implies that there is a language in space $O(\log \log n)$ that is not testable with a constant number of queries, thus showing that the $o(\log \log n)$ bound is best possible. It was also previously known that the class of testable properties cannot be extended to all context-free languages. We further show that one cannot even extend the family of testable languages to the class of languages accepted by single counter machines which is perhaps the weakest (uniform) computational model that is strictly stronger than finite automata.

1 Introduction

Basic Definitions: Combinatorial property testing deals with the following relaxation of decision problems: given a fixed property $\mathcal{P}$ and an input x, one wants

---

* Research was supported by the Israel Science Foundation (grant number 55/03).
** Research supported in part by a Charles Clore Foundation Fellowship.

to decide whether x satisfies $\mathcal{P}$ or is "far" from satisfying the property. This notion was first introduced in the work of Blum, Luby and Rubinfeld [5], and was explicitly formulated for the first time by Rubinfeld and Sudan [16]. Goldreich, Goldwasser and Ron [8] have started a rigorous study of what later became known as "combinatorial property testing". Since then much work has been done, both on designing efficient algorithms for specific properties, and on identifying natural classes of properties that are efficiently testable. For detailed surveys on the subject see [6, 15].

In this paper we focus on testing properties of strings, or equivalently languages [1]. In this case a string of length n is ϵ-far from satisfying a property $\mathcal{P}$ if at least ϵn of the string's entries should be modified in order to get a string satisfying $\mathcal{P}$. An ϵ-tester for $\mathcal{P}$ is a randomized algorithm that given ϵ and the ability to query the entries of an input string, can distinguish with high probability (say 2/3) between strings satisfying $\mathcal{P}$ and those that are ϵ-far from satisfying it. The query complexity $q(\epsilon, n)$ is the maximum number of queries the algorithm makes on any input of length n. Property $\mathcal{P}$ is said to be testable with a *constant number of queries* if $q(\epsilon, n)$ can be bounded from above by a function of ϵ only. For the sake of brevity, we will sometimes say that a language is *easily testable* if it can be tested with a constant number of queries [2].

If a tester accepts with probability 1 inputs satisfying $\mathcal{P}$ then it is said to have a 1-sided error. If it may err in both directions then it is said to have 2-*sided error*. A tester may be *adaptive*, in the sense that its queries may depend on the answers to previous queries, or *non-adaptive*, in the sense that it first makes all the queries, and then proceeds to compute using the answers to these queries. All the lower bounds we prove in this paper hold for the most general testers, namely, 2-sided error adaptive testers.

Background: One of the most important questions in the field of property testing is to prove general testability results, and more ambitiously to classify the languages that are testable with a certain number of queries. While in the case of (dense) graph properties, many general results are known (see [2] and [3]) there are not too many general results for testing languages that can be decided in certain computational models. Our investigation is more related to the connection between certain classical complexity measures of languages and the hardness of testing them, which is measured by their query complexity as defined above. A notable result in this direction was obtained by Alon et al. [1] were it was shown that any regular language is easily testable. In fact, it was shown in [1] that any regular language can be tested with an optimal constant number of queries $\Theta(1/\epsilon)$ (the hidden constant depends on the language). It has been long known (see exercise 2.8.12 in [14]) that any language that can

---

[1] It will sometimes be convenient to refer to properties $\mathcal{P}$ of strings as languages L, as well as the other way around, where the language associated with the property is simply the family of strings that satisfy the property.

[2] We note that some papers use the term *easily testable* to indicate that a language is testable with $poly(1/\epsilon)$ queries.

be recognized in space [3] $o(\log \log n)$ is in fact regular. By the result of [1] this means that any such language is easily testable. A natural question is whether it is possible to extend the family of easily testable languages beyond those with space complexity $o(\log \log n)$. It was (implicitly) proved in [1] that there are properties in space $O(\log n)$ that are not easily testable, and Newman [13] raised the question of closing the exponential gap between the $o(\log \log n)$ positive result and the $\Omega(\log n)$ negative result. Another natural question is whether the family of easily testable languages can be extended beyond those of regular languages by considering stronger machines. Newman [13] has considered *non-uniform* extensions of regular languages and showed that any language that can be accepted by read-once branching programs of constant width is easily testable. Fischer and Newman [7] showed that this can not be further extended even to read twice branching programs of constant width. For the case of *uniform* extensions, it has been proved in [1] that there are context-free languages that are not easily testable.

In this paper we study a relation between the space complexity and the query complexity of languages. As a special case of this relation we resolve the open problem of Newman [13] concerning the space complexity of the easily testable languages. We also show that the family of easily testable languages cannot be extended to essentially any family of languages accepted by (uniform) machines stronger than finite state automata.

Main Results: As we have discussed above there are very few known connections between standard complexity measures and query complexity. Our first and main investigation in this paper is about the relation between the space complexity of a language and the query complexity of testing it. Our main result shows that in some cases the relation between space complexity and query complexity may be at least exponential. As we show in Theorem 2 below, it can be shown that there are languages, whose space complexity is $O(\log n)$ and whose query complexity is $\Omega(n)$. Also, as we have previously noted, languages whose space complexity is $o(\log \log n)$ can be tested with $\Theta(1/\epsilon)$ queries. Therefore, the interesting space complexities $s(n)$ that are left to deal with are in the "interval" $[\Omega(\log \log n), O(\log n)]$. For ease of presentation it will be easier to assume that $s(n) = f(\log \log n)$ for some integer function $x \le f(x) \le 2^x$. As in many cases, we would like to rule out very "strange" complexity functions $s(n)$. We will thus say that $s(n) = f(\log \log n)$ is *space constructible* if the function f is space constructible, that is, if given the *unary* representation of a number x it is possible to generate the *binary* representation of $f(x)$ using space $O(f(x))$. Note that natural functions, such as $s(n) = (\log \log n)^2$ and $s(n) = \sqrt{\log n}$ are space constructible [4].

---

[3] Throughout this paper we consider only deterministic space complexity. Our model for measuring the space complexity of the algorithm is the standard Turing Machine model, where there is a read only input tape, and a work tape where the machine can write. We only count the space used by the work tape. See [14] for the precise definitions. For concreteness we only consider the alphabet $\{0, 1\}$.

[4] We use the standard notion of space constructibility, see e.g. [14]. Note that when $s(n) = (\log \log n)^2$ we have $f(x) = x^2$ and when $s(n) = \sqrt{\log n}$ we have $f(x) = 2^{x/2}$.

Theorem 1 (Main Result). *For any (space constructible) function $s(n)$ there is a language in space $O(s(n))$, whose query complexity is $2^{\Omega(s(n))}$.*

We believe it will be interesting to further study the relation between these two measures. Specifically, we raise the following conjecture claiming that the lower bound of Theorem 1 is best possible:

Conjecture 1. Any language in space $s(n)$ can be tested with query complexity $2^{O(s(n))}$.

As we have mentioned above, one of the steps in the proof of Theorem 1 is the following result that may be of independent interest.

Theorem 2. *There is a language in space $O(\log n)$, whose query complexity is $\Omega(n)$.*

To the best of our knowledge, the lowest complexity class that was previous known to contain a language, whose query complexity is $\Omega(n)$, is **P** (see [9]). If Conjecture 1 is indeed true then Theorem 2 is essentially best possible.

As an immediate application of Theorem 1 we deduce the following corollary, showing that the class of easily testable languages cannot be extended from the family of regular languages even to the family of languages with space complexity $O(\log \log n)$ thus answering the problem raised by Newman in [13] concerning the space complexity of easily testable languages.

Corollary 1. *For any $k > 0$, there is a language in space $O(\log \log n)$, whose query complexity is $\Omega(\log^k n)$.*

Corollary 1 rules out the possibility of extending the family of easily testable languages from regular languages, to the entire family of languages, whose space complexity is $O(\log \log n)$.

We turn to address another result, ruling out another possible extension of regular languages. As we have mentioned before, it has been shown in [1] that there are context-free languages that are not testable. Hence, a natural question is whether there exists a uniform computational model stronger than finite state machines and weaker than stack machines such that all the languages that are accepted by machines in this model are testable. Perhaps the weakest uniform model within the class of context-free languages is that of a *deterministic single-counter automaton* (also known as one-symbol push-down automaton). A deterministic single-counter automaton is a finite state automaton equipped with a counter. The possible counter operations are increment, decrement and do nothing, and the only feedback from the counter is whether it is currently 0 or positive (larger than 0). Thus, such an automaton, running on a string ω reads an input character at a time, and based on its current state and whether the counter is 0, jumps to the next state and increments/decrements the counter or leaves it unchanged. Such an automaton accepts a string ω if starting with a counter holding the value 0 it reads all the input characters and ends with the counter holding the value 0. It is quite obvious that such an automaton is equivalent to a deterministic push-down automaton with one symbol stack (and a read-only bottom symbol to indicate empty stack). This model of computation can recognize a very restricted subset of context free languages. Still, some interesting languages are recognized by such an automaton, e.g. D_1 the first Dyck language, which is the language of balanced parentheses. Formal definition and discussion on variants of counter automata can be found in [18].

In this paper we also prove the following theorem showing that the family of testable properties cannot be extended even to those accepted by single-counter automata.

Theorem 3. *There is a language that can be accepted by a deterministic single-counter automaton and whose query complexity is $\Omega(\log \log n)$ even for 2-sided error tests.*

Combining Theorem 3 and Corollary 1 we see that the family of testable properties cannot be extended beyond that of the regular languages in two natural senses.

Organization: The rest of the paper is organized as follows. In Section 2 we prove the exponential relation between space complexity and query complexity of Theorem 1. An important step in the proof is Theorem 2 that we also prove in this section. Section 3 contains the proof of Theorem 3 showing that there are languages accepted by counter machines that are not easily testable. Section 4 contains some concluding remarks and open problems. Due to space limitations several proofs are omitted and will appear in the full version.

2   Space Complexity vs. Query Complexity

In this section we prove that languages in space $s(n)$ may have query complexity exponential in $s(n)$. We start with an overview containing the important details of the proof of Theorem 2 stating that there are languages in space $O(\log n)$ that have query complexity $\Omega(n)$. We then show how to use Theorem 2 in order to prove the general lower bound of Theorem 1.

Overview of the Proof of Theorem 2: The construction of the language L in Theorem 2 is based on dual-codes of asymptotically good linear codes over $GF(2)$, which are based on Justesen's construction [10]. We begin with some brief background from Coding Theory (see [12] for a comprehensive background). A *linear code* C over $GF(2)$ is just a subset of $\{0,1\}^n$ that forms a linear subspace. The (Hamming) *distance* between two words $x, y \in C$, denoted $d(x, y)$, is the number of indices $i \in [n]$ for which $x_i \neq y_i$. The *distance* of the code, denoted $d(C)$ is the minimum distance over all pairs of distinct words of C, that is $d(C) = \min_{x \neq y \in C} d(x, y)$. The *size* of a code, denoted $|C|$ is the number of words in C. The *dual-code* of C, denoted $C^\perp$ is the linear subspace orthogonal to C, that is $C^\perp = \{y : \langle x, y \rangle = 0 \text{ for all } x \in C\}$, where $\langle x, y \rangle = \sum_{i=1}^{n} x_i y_i \pmod 2$ is the dot product of x and y over $GF(2)$. The *generator matrix* of a code C is a matrix G whose rows span the subspace of C. Note, that a code is a family of strings of fixed size n and our interest is languages containing strings of unbounded size.

We will thus have to consider families of codes of increasing size. The following notion will be central in the proof of Theorem 2:

Definition 1. *An infinite family of codes* $\mathcal{C} = \{C_1, C_2, \ldots\}$, *where* $C_n \subseteq \{0,1\}^n$, *is said to be* asymptotically good *if there exist positive reals* d *and* r *such that* $\liminf_{n \to \infty} \frac{d(C_n)}{n} \geq d$ *and* $\liminf_{n \to \infty} \frac{\log(|C_n|)}{n} \geq r$.

We turn to discuss the main two Lemmas needed to prove Theorem 2. The first is the following:

Lemma 1. *Suppose* $\mathcal{C} = \{C_1, C_2, \ldots\}$ *is an asymptotically good family of linear codes. Then, for any infinite* $S \subseteq \mathcal{N}$, *the language* $L = \bigcup_{n \in S} C_n^{\perp}$ *has query complexity* $\Omega(n)$.

Lemma 1 is essentially a folklore result. Its (simple) proof relies on the known fact that if C is a code with distance t then $C^{\perp}$ is a *t-wise independent* family, that is, if one uniformly samples a string from $C^{\perp}$ then the distribution induced on any t coordinates is the uniform distribution. Such families are sometimes called in the coding literature *orthogonal array of strength* t, see [12]. The fact that the codes in $\mathcal{C}$ satisfy $\frac{\log(|C_{n_i}|)}{n_i} \geq r$ implies that a random string is with high probability far from belonging to $C_{n_i}^{\perp}$. These two facts allow us to apply Yao's principle to prove that even *adaptive* testers must use at least $\Omega(n)$ queries in order to test L for some fixed ϵ_0. As pointed to us by Eli Ben-Sasson, Lemma 1 can also be proved by applying a general non-trivial result about testers for membership in linear codes (see Theorem 3.3 in [4] for more details).

A well known construction of Justesen [10] gives an asymptotically good family of codes. By exploiting the fact that for appropriate prime powers n, one can perform arithmetic operations over $GF(n)$ in space $O(\log n)$, one can use the main idea of [10] in order to prove the following:

Lemma 2. *There is an asymptotically good family of linear codes* $\mathcal{C} = \{C_1, C_2, \ldots\}$ *and a space* $O(\log n)$ *algorithm, with the following property: Given integers* n, i *and* j, *the algorithm generates entry* i, j *of the generator matrix of* C_n.

Apparently this result does not appear in any published paper. However, most details of the construction appear in Madhu Sudan's lecture notes [17]. Lemma 2 immediately implies that the language $C^{\perp} = \cup C_i^{\perp}$ is recognizable in $O(\log n)$ space. Theorem 2 will follow by applying the above two lemmas. The proofs of the above Lemmas will appear in the full version of the paper.

Proof of Theorem 1: In this subsection we apply Theorem 2 in order to prove Theorem 1. To gain intuition for the construction, let us consider the case $s(n) = \log \log n$. Consider the following language L_s: a string $x \in \{0, 1, \#\}^n$ is in L_s if it is composed of $n/\log n$ blocks of size $\log n$ each, separated by the $\#$ symbol, such that each block is a word of the language of Theorem 2. It can be shown that the query complexity of testing L_s is $\Omega(\log n)$. As the language of Theorem 2 is in space $O(\log n)$ it is clear that if the blocks of an input are indeed or length $O(\log n)$, then we can recognize L_s using space $O(\log \log n)$;

we just run the space $O(\log n)$ algorithm on each of the blocks, whose length is $O(\log n)$. Of course, the problem is that if the blocks are not of the right length then we may be "tricked" into using too much space. We thus have to add to the language some "mechanism" that will allow us to check if the blocks are of the right length. This seems to be difficult as we need to initiate a counter that will hold the value n, but we need to do so without using more than $O(\log \log n)$ space, and just holding the value n requires $\Theta(\log n)$ bits.

The following language comes to the rescue: consider the language $\mathcal{B}$ over the alphabet $\{0, 1, *\}$, which is defined as follows: for every integer $r \geq 1$, the language $\mathcal{B}$ contains the string $s_r = \mathrm{bin}(0) * \mathrm{bin}(1) * \ldots * \mathrm{bin}(2^r - 1)*$, where $\mathrm{bin}(i)$ is the binary representation of the integer i as a word of length r (that is, including leading 0's). Therefore, for every r there is precisely one string in $\mathcal{B}$ of length $(r + 1)2^r$. This language is the standard example for showing that there are languages in space $O(\log \log n)$ that are not regular (see [14] exercise 2.8.11).

Note that after verifying that a string $x \in \mathcal{B}$ we have an implicit representation of a number very close to $\log(|x|)$: this is just the number of entries before the first $*$ symbol. This also gives us a value close to $\log \log n$, which we needed in the previous example. The main idea for the proof of Theorem 1 is to "interleave" the language $\mathcal{B}$ with a language consisting of blocks of length $2^{s(n)}$ of strings from the language of Theorem 2. For ease of presentation the language we construct to prove Theorem 1 is over the alphabet $\{0, 1, \#, *\}$. It can easily be converted into a language over $\{0, 1\}$ with the same asymptotic properties by encoding each of the 4 symbols using 2 bits. The details follow.

Let L_2 be the language of Theorem 2 and let $s(n)$ satisfy $s(n) = f(\log \log n)$ for some space constructible function $n \leq f(n) \leq 2^n$ (recall the discussion before the statement of Theorem 1). In what follows we will use the notation L^k to denote the strings of some language L whose lengths is k, that is $L^k = L \cap \{0, 1, \#, *\}^k$. Given the function f, we define a language L_f that we need in order to prove Theorem 1 as the union of families of strings X_r of length $n(r)$, where for any $r \geq 1$ we define

$$n(r) = 2(r + 1)2^r.$$

A string $x \in \{0, 1, \#, *\}^{n(r)}$ belongs to X_r if it has the following two properties:

1. The odd entries of x form a string from $\mathcal{B}$ (thus the odd entries are over $\{0, 1, *\}$).
2. In the even entries of x, substrings between consecutive $\#$ symbols [5] form a string from L_2^k, where $k = 2^{f(\lfloor \log r \rfloor)}$. The only exception is the last block for which the only requirement is that it would be of length at most k (thus the even entries are over $\{0, 1, \#\}$).

Note that the words from L_2, which appear in the even entries of strings belonging to X_r all have length $2^{f(\lfloor \log r \rfloor)}$. We now define

$$L_f = \bigcup_{r=1}^{\infty} X_r \, . \tag{1}$$

---

[5] The first $\#$ symbol is between the first block and the second block.

and

$$K_f = \{2^{f(\lfloor \log r \rfloor)} : r \in \mathcal{N}\}. \tag{2}$$

Observe that the words from L_2, which appear in the even entries of strings belonging to L_f, all have lengths that belong to the set K_f. With a slight abuse of notation we now define the language L_2^f as the subset of L_2 consisting of words with lengths from K_f. By Theorem 2, when taking K_f as the set S in the statement of the theorem, we get the following claim:

Claim 1. *For some $\epsilon_0 > 0$, every ϵ_0-tester of L_2^f has query complexity $\Omega(n)$.*

We now turn to prove the main claims needed to obtain Theorem 1.

Claim 2. *The language L_f has space complexity $O(s(n)) = O(f(\log \log n))$.*

Proof. To show that L_f is in space $O(f(\log \log n))$ we consider the following algorithm for deciding if an input x belongs to L_f. We first consider only the odd entries of x and use the $O(\log \log n)$ space algorithm for deciding if these entries form a string from $\mathcal{B}$. If they do not we reject and if they do we move to the second step. Note, that at this step we know that the input's length n is $2(r+1)2^r$ for some $r \leq \log n$. In the second step we initiate a binary counter that stores the number $\lfloor \log r \rfloor \leq \log \log n$. Observe, that the algorithm can obtain r by counting the number of odd entries between consecutive $*$ symbols, and that we need $O(\log \log n)$ bits to hold r. We then construct a counter that holds the value $k = 2^{f(\lfloor \log r \rfloor)}$, using space $O(f(\lfloor \log r \rfloor))$ by exploiting the fact that f is space constructible [6]. We then verify that the number of even entries between consecutive $\#$ symbols is k, besides the last block for which we check that the length is at most k. Finally, we run the space $O(\log n)$ algorithm of L_2 in order to verify that the even entries between consecutive $\#$ symbols form a string from L_2 (besides the last block).

The algorithm clearly accepts a string if and only if it belongs to L_f. Regarding the algorithm's space complexity, recall that we use an $O(\log \log n)$ space algorithm in the first step (this algorithm was sketched at the beginning of this section). Note, that after verifying that the odd entries form a string from the language $\mathcal{B}$, we are guaranteed that $r \leq \log n$. The number of bits needed to store the counter we use in order to hold the number $k = 2^{f(\lfloor \log r \rfloor)}$ is $f(\lfloor \log r \rfloor) \leq f(\log \log n)$ as needed. Finally, as each block is guaranteed to be of length $2^{f(\lfloor \log r \rfloor)}$, the $O(\log n)$ algorithm that we run on each of the blocks uses space $O(\log(2^{f(\lfloor \log r \rfloor)})) = O(f(\lfloor \log r \rfloor)) = O(f(\log \log n))$ as needed. ∎

---

[6] More precisely, given the binary encoding of $\lfloor \log r \rfloor$ we form an unary representation of $\lfloor \log r \rfloor$. Such a representation requires $O(\log \log n)$ bits. We then use the space constructibility of f to generate a binary representation of $f(\lfloor \log r \rfloor)$ using space $O(f(\lfloor \log r \rfloor))$. Finally, given the binary representation of $f(\lfloor \log r \rfloor)$ it is easy to generate the binary representation of $2^{f(\lfloor \log r \rfloor)}$ using space $O(f(\lfloor \log r \rfloor))$.

Claim 3. *The language L_f has query complexity $2^{\Omega(f(\log\log n))} = 2^{\Omega(s(n))}$.*

Proof. By Claim 1, for some fixed ϵ_0 every ϵ_0-tester for L_2^f has query complexity $\Omega(n)$. We claim that this implies that every $\frac{\epsilon_0}{3}$-tester for L_f has query complexity $2^{\Omega(f(\log\log n))}$. Consider any $\frac{\epsilon_0}{3}$-tester T_f for L_f and consider the following ϵ_0-tester T_2 for L_2^f: Given an input x, the tester T_2 immediately rejects x in case there is no integer r for which $|x| = 2^{f(\lfloor\log r\rfloor)}$. Recall that the strings of L_2^f are all taken from K_f as defined in (2). In case such an integer r exists, set $n = 2(r+1)2^r$. The tester T_2 now *implicitly* constructs the following string x' of length n. The odd entries of x' will contain the unique string of $\mathcal{B}$ of length $(r+1)2^r$. The even entries of x' will contain repeated copies of x separated by the # symbol (the last block may contain some prefix of x). Note that if $x \in L_2^f$ then $x' \in L_f$. On the other hand, observe that if x is ϵ-far from L_2^f then x' is $(\frac{\epsilon}{2} - o(1))$-far from L_f, because in the even entries of x', one needs to change an ϵ-fraction of the entries in the substring between consecutive # symbols, in order to get a words from L_2^f (the $o(1)$ term is due to the fraction of the string occupied by the # symbols that need not be changed). This means that it is enough for T_2 to simulate T_f on x' with error parameter $\frac{\epsilon_0}{3}$ and thus return the correct answer with high probability. Of course, T_2 cannot construct x' "for free" because to do so T_2 must query all entries of x. Instead, T_2 only answers the oracle queries that T_f makes as follows: given a query of T_f to entry $2i - 1$ of x', the tester T_2 will supply T_f with the ith entry of the unique string of $\mathcal{B}$ of length $(r+1)2^r$. Given a query of T_f to entry $2i$ of x', the tester T_2 will supply T_f with the jth entry of x, where $j = i \pmod{|x| + 1}$. To this end, T_2 will have to perform a query to the entries of x.

We thus get that if L_f has an $\frac{\epsilon_0}{3}$-tester making t queries on inputs of length $2(r+1)2^r$, then L_2^f has an ϵ_0-tester making t queries on inputs of length $2^{f(\lfloor\log r\rfloor)}$. We know by Claim 1 that the query complexity of any ϵ_0-tester of L_2^f on inputs of length $2^{f(\lfloor\log r\rfloor)}$ is $\Omega(2^{f(\lfloor\log r\rfloor)})$. This means that the query complexity of T_2 on the inputs x' we described must also be $\Omega(2^{f(\lfloor\log r\rfloor)})$. The lengths of these inputs is $n = 2(r+1)2^r$. This means that $r = \log n - \Theta(\log\log n)$ and therefore the query complexity on these inputs is $\Omega(2^{f(\lfloor\log r\rfloor)})) = \Omega(2^{f(\log\log n-2)}) = 2^{\Omega(f(\log\log n))}$, where in the last equality we used the fact that $f(x) \leq 2^x$. ∎

Proof of Theorem 1. Take the language L_f and apply Claims 3 and 2. ∎

3   Testing Counter Machine Languages May Be Hard

In this section we define a language $\mathcal{L}$ that is decidable by a deterministic single-counter machine and sketch an $\Omega(\log\log n)$ lower bound on the query complexity of *adaptive*, 2-sided error testers for testing membership in $\mathcal{L}$. We start with defining the language $\mathcal{L}$.

Definition 2. *$\mathcal{L}$ is the family of strings $s \in \{0,1\}^*$ such that $s = 0^{k_1}1^{k_1}\ldots0^{k_i}1^{k_i}$ (The integers k_i are arbitrary). For every integer n we set $\mathcal{L}_n = \mathcal{L} \cap \{0,1\}^n$.*

We proceed with the proof of Theorem 3. First note that one can easily see that $\mathcal{L}$ can be accepted by a deterministic counter automaton as defined in Subsection 1. What we are left with is thus to prove the claimed lower bound on testing $\mathcal{L}$. Note that any adaptive tester of a language $L \subseteq \{0,1\}^*$ with query complexity $q(\epsilon, n)$ can be simulated by a non-adaptive tester with query complexity $2^{q(\epsilon,n)}$. Therefore, in order to prove our $\Omega(\log \log n)$ lower bound, we may and will prove an $\Omega(\log n / \log \log n)$ lower bound that holds for *non-adaptive* testers. To this end we apply Yao's minmax principle, which implies that in order to prove a lower bound of $\Omega(\log n / \log \log n)$ for non-adaptive testers it is enough to show that there is a distribution $\mathcal{D}$ over legitimate inputs (that is, inputs from $\mathcal{L}_n$ and inputs that are $\frac{1}{120}$-far from $\mathcal{L}_n$), such that for any non-adaptive *deterministic* algorithm Alg, which makes $o(\log n / \log \log n)$ queries, the probability that Alg errs on inputs generated by $\mathcal{D}$ is at least $1/3$.

One of the key ingredients needed to construct $\mathcal{D}$ are the following two pairs of strings:

$$BAD_\ell = \left\{ \begin{array}{l} 0^\ell\ 1^\ell\ 0^\ell\ 1^\ell\ 0^\ell\ 1^\ell\ 1^\ell\ 1^\ell\ 0^\ell\ 0^\ell\ 0^\ell\ 1^\ell, \\ 0^\ell\ 0^\ell\ 1^\ell\ 1^\ell\ 0^\ell\ 0^\ell\ 0^\ell\ 1^\ell\ 0^\ell\ 1^\ell\ 1^\ell\ 1^\ell \end{array} \right\}$$

$$GOOD_\ell = \left\{ \begin{array}{l} 0^\ell\ 0^\ell\ 1^\ell\ 1^\ell\ 0^\ell\ 0^\ell\ 1^\ell\ 1^\ell\ 0^\ell\ 0^\ell\ 1^\ell\ 1^\ell, \\ 0^\ell\ 1^\ell\ 0^\ell\ 1^\ell\ 0^\ell\ 1^\ell\ 0^\ell\ 1^\ell\ 0^\ell\ 1^\ell\ 0^\ell\ 1^\ell \end{array} \right\}$$

where ℓ is a positive integer. Note, that each of the 4 strings is of length 12ℓ. We refer to strings selected from these sets as 'phrase strings'. We view the phrase strings as being composed of 12 disjoint intervals of length ℓ, which we refer to as 'phrase segments'. By the definition of the 'phrase strings' each 'phrase segment' is an homogeneous substring (that is, all its symbols are the same).

Note that for any ℓ, the 4 strings in BAD_ℓ and $GOOD_\ell$ have the following two important properties: (i) The 4 strings have the same (boolean) value in phrase segments $1, 4, 5, 8, 9$ and 12. (ii) In the other phrase segments, one of the strings in BAD_ℓ has the value 0 and the other has value 1, and the same applies to $GOOD_\ell$. The idea behind the construction of $\mathcal{D}$ and the intuition of the lower bound is that in order to distinguish between a string chosen from BAD_ℓ and a string chosen from $GOOD_\ell$ one must make queries into 2 distinct phrase segments. The reason is that by the above observation, if all the queries belong to segment $i \in [12]$, then either the answers are all identical and are known in advance (in case $i \in \{1, 4, 5, 8, 9, 12\}$), or they are identical and have probability 0.5 to be either 0 or 1, *regardless* of the set from which the string was chosen.

In the construction of the distribution $\mathcal{D}$ we select with probability $1/2$ whether the string we choose will be a positive instance or a negative instance. We select a positive instance by concatenating a set of strings uniformly and independently selected from $GOOD_\ell$ with strings of the form $0^t 1^t$. We construct negative instance in the same manner except that we replace the selection of strings from $GOOD_\ell$, by selecting strings from BAD_ℓ. Thus, the only way to distinguish between a positive instance and a negative instance is if at least two queries are located in the same phrase string, but in different phrase segments. The distribution $\mathcal{D}$ will be such that if the number of queries that is used is $o(\log n/\log\log n)$, then with high probability there will be no two queries in two different phrase segments that belong to the same phrase string. As each phrase string is selected independently this makes it impossible for the tester to know whether the string is a positive instance or a negative one.

We assume in what follows that $n \geq 16$. Let $\mathcal{D}_N$ be a distribution over $\{0,1\}^n$ that is defined by the following process of generating a string $\alpha \in \{0,1\}^n$:

1. Uniformly select an integer $s \in [1, \lfloor \log n \rfloor - 3]$ and set $\ell = 2^s$.
2. Independently and uniformly select integers $b \in [6\ell]$, until the first time that the integers $b_1, \ldots, b_r$ selected satisfy $\sum_{i=1}^r (2b_i + 12\ell) \geq n - 24\ell$.
3. Independently and uniformly select r strings $\beta_1, \ldots, \beta_r \in BAD_\ell$.
4. For each $i \in [r]$ set $B_i = 0^{b_i} 1^{b_i} \beta_i$. We refer to B_i as the ith 'block string'. We refer to the substring $0^{b_i} 1^{b_i}$ as the 'buffer string' and β_i as the 'phrase'.
5. Set $\alpha = B_1 \cdots B_r 0^t 1^t$, where $t = (n - \sum_{i=1}^r |B_i|)/2$.

Let $\mathcal{D}_P$ be a distribution over $\{0,1\}^n$ that is defined in the same manner as $\mathcal{D}_N$ with the exception that in the third stage we select independently and uniformly r strings $\beta_1, \ldots, \beta_r \in GOOD_\ell$. In the full version of the paper we use these two distributions to prove the required lower bound on testing $\mathcal{L}$.

4   Concluding Remarks and Open Problems

Our main result in this paper gives a relation between the space complexity and the query complexity of a language, showing that the later may be exponential in the former. We also raise the conjecture that this relation is tight, namely that the query complexity of a language is at most exponential in its space complexity. The results of this paper further show that the family of easily testable languages cannot be extended beyond that of the regular languages in terms of two natural senses; the space complexity of the accepting machine or the minimal computational model in which it can be recognized.

An intriguing related question is to understand the testability of languages with sublinear number of queries, that is $poly(\log n)$ or even just $o(n)$ queries. In particular, an intriguing open problem is whether all the context free languages can be tested with a sublinear number of queries. Currently, the lower bounds for testing context-free languages are of type $\Omega(n^\alpha)$ for some $0 < \alpha < 1$. It seems that as an intermediate step towards understanding the testability of context-free languages, it will be interesting to investigate whether all the languages acceptable by single-counter automata can be tested with $o(n)$ queries. We note that the language we constructed in order to prove Theorem 3 can be tested with $poly(\log n, \epsilon)$ queries. See [11] for the full details.

Acknowledgments. The authors would like to thank Noga Alon, Madhu Sudan and Eli Ben-Sasson for helpful discussions.

References

1. N. Alon, M. Krivelevich, I. Newman and M. Szegedy, Regular languages are testable with a constant number of queries, SIAM J. on Computing 30 (2001), 1842-1862.
2. N. Alon and A. Shapira, A characterization of the (natural) graph properties testable with one-sided error, Proc. of FOCS 2005, 429-438.
3. N. Alon, E, Fischer, I. Newman and A. Shapira, A combinatorial characterization of the testable graph properties: it's all about regularity, Proc. of STOC 2006, 251-260.
4. E. Ben-Sasson, P. Harsha and S. Raskhodnikova, Some 3-CNF properties are hard to test, Proc. of STOC 2003, 345-354.
5. M. Blum, M. Luby and R. Rubinfeld, Self-testing/correcting with applications to numerical problems, JCSS 47 (1993), 549-595.
6. E. Fischer, The art of uninformed decisions: A primer to property testing, The Computational Complexity Column of The Bulletin of the European Association for Theoretical Computer Science 75 (2001), 97-126.
7. E. Fischer, I. Newman and J. Sgall, Functions that have read-twice constant width branching programs are not necessarily testable, Random Struct. and Alg., in press.
8. O. Goldreich, S. Goldwasser and D. Ron, Property testing and its connection to learning and approximation, JACM 45(4): 653-750 (1998).
9. O. Goldreich and L. Trevisan, Three theorems regarding testing graph properties, Random Structures and Algorithms, 23(1):23-57, 2003.
10. J. Justesen, A class of constructive asymptotically good algebraic codes, IEEE Transcations on Information, 18:652-656, 1972.
11. O. Lachish and I. Newman, Languages that are Recognized by Simple Counter Automata are not necessarily Testable, ECCC report TR05-152.
12. F. MacWilliams and N. Sloane, **The Theory of Error-Correcting Codes**, North-Holland, Amsterdam, 1997.
13. I. Newman, Testing of functions that have small width branching programs, Proc. of 41th FOCS (2000), 251-258.
14. C. Papadimitriou, **Computational Complexity**, Addison Wesley, 1994.
15. D. Ron, Property testing, in: *Handbook of Randomized Computing*, Vol. II, Kluwer Academic Publishers, 2001, 597–649.
16. R. Rubinfeld and M. Sudan, Robust characterization of polynomials with applications to program testing, *SIAM J. on Computing* 25 (1996), 252–271.
17. M. Sudan, Lecture Notes on Algorithmic Introduction to Coding Theory, available at http://theory.lcs.mit.edu/~madhu/FT01/scribe/lect6.ps.
18. L.G. Valiant, M. Paterson, Deterministic one-counter automata, Journal of Computer and System Sciences, 10 (1975), 340–350.

Consistency of Local Density Matrices Is QMA-Complete

Yi-Kai Liu

Computer Science and Engineering
University of California, San Diego
y9liu@cs.ucsd.edu

Abstract. Suppose we have an n-qubit system, and we are given a collection of local density matrices $\rho_1, \ldots, \rho_m$, where each ρ_i describes a subset C_i of the qubits. We say that the ρ_i are "consistent" if there exists some global state σ (on all n qubits) that matches each of the ρ_i on the subsets C_i. This generalizes the classical notion of the consistency of marginal probability distributions.

We show that deciding the consistency of local density matrices is QMA-complete (where QMA is the quantum analogue of NP). This gives an interesting example of a hard problem in QMA. Our proof is somewhat unusual: we give a Turing reduction from Local Hamiltonian, using a convex optimization algorithm by Bertsimas and Vempala, which is based on random sampling. Unlike in the classical case, simple mapping reductions do not seem to work here.

1 Introduction

Quantum mechanical systems exhibit many unusual phenomena, such as coherent superpositions and nonlocal entanglement. It is interesting to compare this with the behavior of classical probabilistic systems. In a classical system, such as a Markov chain or a graphical model, one may have correlations or dependencies among different parts of the system; in particular, local properties can affect the joint probability distribution of the entire system. Many quantum systems have a similar flavor, though their behavior is more complicated. In this paper, we investigate one problem of this kind, and its relationship to the complexity class QMA.

First, consider a classical problem. Suppose we have random variables $X_1, \ldots, X_n$, with some unknown joint distribution D, and we are given marginal distributions $D_1, \ldots, D_m$, where each D_i describes a subset C_i of the variables. (We assume that the random variables X_j take on values in some fixed finite set, and the subsets C_i have size at most some constant k.) Does there exist a joint distribution D that matches each of the marginals D_i on the subsets C_i? If so, we say that the marginals D_i are "consistent."

Deciding the consistency of marginal distributions is NP-hard, by a straightforward reduction from 3-coloring. (We are given a graph $G = (V, E)$. For each vertex $v \in V$, construct a random variable X_v which takes on values in $\{r, g, b\}$.

J. Diaz et al. (Eds.): APPROX and RANDOM 2006, LNCS 4110, pp. 438–449, 2006.

For each edge $(u, v) \in E$, specify that the marginal distribution of X_u and X_v must be uniform over the set $\{r, g, b\}^2 \setminus \{rr, gg, bb\}$. These marginals are consistent iff G is 3-colorable.)

Now consider the generalization of this problem to quantum states. Suppose we have an n-qubit system, and we are given local density matrices $\rho_1, \ldots, \rho_m$, where each ρ_i describes a subset C_i of the qubits. Does there exist a global state σ on all n qubits that matches each of the local states ρ_i on the subsets C_i? If so, we say that the local states ρ_i are "consistent." (This problem was first suggested to me by Dorit Aharonov, in connection with the class QCMA [1].)

We will show that this problem is QMA-complete, where QMA is the quantum analogue of NP. QMA is the class of languages that have poly-time quantum verifiers, where the witness is allowed to be a quantum state. QMA arises naturally in the study of quantum computation, and it also has a complete problem, Local Hamiltonian, which is a generalization of k-SAT [2, 3].

Our result is interesting, because we only know of a few QMA-complete problems, and most of them look like universal models of quantum computation. For instance, the fact that Local Hamiltonian is QMA-complete [2, 3, 4, 5, 6] is closely related to the fact that adiabatic quantum computation is equivalent to the standard quantum circuit model [7]. Other QMA-complete problems such as Identity Check involve properties of quantum circuits [8]. The Consistency problem, however, does not seem to embody any particular model of quantum computation; this will become clearer when we present our reduction from Local Hamiltonian.

Why are there so few QMA-complete problems, when there is such an astonishing variety of NP-complete problems? The reason seems to be that the techniques used to show NP-hardness, such as mapping reductions using combinatorial gadgets, break down when we apply them to a "quantum" problem like Local Hamiltonian. For instance, to reduce Local Hamiltonian to the Consistency problem, we would try to use local density matrices to "simulate" local Hamiltonians. But we run into problems due to the presence of non-commuting matrices. (In cases where quantum gadgets do work, such as [5, 6], they are much more subtle than classical gadgets.)

Instead, our proof that the consistency problem is QMA-hard uses a randomized Turing reduction from Local Hamiltonian. The basic idea is that Local Hamiltonian can be expressed as a convex program in polynomially many variables, which can be solved using convex optimization algorithms, given an oracle for the Consistency problem. In particular, we use a class of convex optimization algorithms [9, 10, 11] which are based on random walks, and only require a membership oracle, rather than a separation oracle. We also use a nifty representation of local density matrices in terms of the expectation values of Pauli matrices.

Note that the Consistency problem has a rather different structure from Local Hamiltonian. For instance, a local density matrix contains complete information about the local state of the system, whereas in many cases a local Hamiltonian only constrains the local state of the system to lie within a certain subspace.

Finally, we remark that our reduction from Local Hamiltonian to Consistency preserves the "neighborhood structure" of the problem, in that the local density matrices act on the same subsets of qubits as the local Hamiltonians. So, using the QMA-hardness results for 2-Local Hamiltonian [5] and Local Hamiltonian on a 2-D square lattice [6], we can immediately get QMA-hardness results for the corresponding special versions of the Consistency problem.

We also mention some related work. In [13], one considers the Common Eigenspace Problem, verifying the consistency of a set of eigenvalue equations $H_i|\psi\rangle = \lambda_i|\psi\rangle$, where the operators H_i commute. We do something similar, translating each local density matrix into constraints on the expectation values of Pauli matrices, though in our case the Pauli matrices do not commute. Also, in [14], one considers a quantum analogue of 2-SAT, where we seek a state $|\psi\rangle$ whose local density matrices have support on prescribed subspaces. However, this problem is more closely related to Local Hamiltonian than to Consistency, since the constraints can be written in the form $\Pi_i|\psi\rangle = 0$ where the Π_i are local projectors.

2   Preliminaries

2.1   Density Matrices

A quantum state of an n-qubit system is represented by a density matrix, which is a $2^n \times 2^n$ positive semidefinite matrix with trace 1. A classical joint probability distribution on n bits is a special case, where the density matrix is diagonal, and the diagonal entries are the probabilities of the 2^n possible outcomes. A subset of qubits C is described by a reduced density matrix, which is obtained by taking the partial trace over the qubits not in C. This is analogous to a marginal distribution, which is obtained by summing over some of the variables.

We measure the difference between two quantum states using the L_1 matrix norm, $\|\rho - \sigma\|_1 = \operatorname{tr}|\rho - \sigma|$. Note that this is also called the trace or statistical distance (when normalized by a factor of $1/2$).

Let X, Y and Z denote the Pauli matrices for a single qubit, and define $\mathcal{P} = \{I, X, Y, Z\}$. We can construct n-qubit Pauli matrices by taking tensor products $P = P_1 \otimes \cdots \otimes P_n \in \mathcal{P}^{\otimes n}$. Any 2^n-dimensional Hermitian matrix can be written as a real linear combination of n-qubit Pauli matrices. Furthermore, the n-qubit Pauli matrices are orthogonal with respect to the Hilbert-Schmidt inner product: $\operatorname{tr}(P^\dagger Q) = 2^n$ if $P = Q$, and 0 otherwise. So, if σ is an n-qubit state, we can write it in the form

$$\sigma = \frac{1}{2^n} \sum_{P \in \mathcal{P}^{\otimes n}} \alpha_P P,$$

where the coefficients are uniquely determined by $\alpha_P = \operatorname{tr}(P\sigma)$; note that these are the expectation values of the Pauli matrices P. This application of the Pauli matrices is closely related to quantum state tomography.

2.2   QMA and the Local Hamiltonian Problem

The class QMA, or "Quantum Merlin-Arthur," is defined as follows [2, 3]: a language L is in QMA if there exists a poly-time quantum verifier V and a polynomial p such that

- If $x \in L$, then there exists a quantum state ρ on $p(|x|)$ qubits such that $V(x, \rho)$ accepts with probability $\geq 2/3$.
- If $x \notin L$, then for all quantum states ρ on $p(|x|)$ qubits, $V(x, \rho)$ accepts with probability $\leq 1/3$.

(Here, $|x|$ denotes the length of the string x.) This is similar to the definition of NP, except that the witness is allowed to be a quantum state, and the verifier is a quantum circuit with bounded error probability.

The Local Hamiltonian problem is defined as follows:

Consider a system of n qubits. We are given a Hamiltonian $H = H_1 + \cdots + H_m$, where each H_i acts on a subset of qubits $C_i \subseteq \{1, \ldots, n\}$. The H_i are Hermitian matrices, with eigenvalues in some fixed interval (for instance $[0, 1]$), and each matrix entry is specified with $\mathrm{poly}(n)$ bits of precision. Also, $m \leq \mathrm{poly}(n)$, and each subset C_i has size $|C_i| \leq k$, for some constant k.

In addition, we are given two real numbers a and b (specified with $\mathrm{poly}(n)$ bits of precision) such that $b - a \geq 1/\mathrm{poly}(n)$.

The problem is to distinguish between the following two cases:
- If H has an eigenvalue that is $\leq a$, output "YES."
- If all the eigenvalues of H are $\geq b$, output "NO."

Kitaev showed that Local Hamiltonian is in QMA, and the case of $k = 5$ is QMA-hard [2, 3]. With greater effort, one can show that Local Hamiltonian with $k = 2$ is also QMA-hard [4, 5].

2.3   Convex Programming

Consider the following version of convex programming:

Let $K \subseteq \mathbb{R}^n$ be a convex set, specified by a membership oracle O_K.
Let $f : K \to \mathbb{R}$ be a convex function, which is efficiently computable.
Find some $x \in K$ that minimizes $f(x)$.

Note that the membership oracle O_K is not as powerful as a separation oracle. We would like to solve this problem with precision ε; that is, we want to find some x that lies within distance ε of an optimal solution x^*.

We can solve this problem in time $\mathrm{poly}(n, \log(1/\varepsilon))$, using an algorithm by Bertsimas and Vempala which is based on random sampling [9, 11]. Actually, for our purposes we only need to solve the special case where f is a linear function; for this case, we can use a slightly faster simulated annealing algorithm [10], or an algorithm based on the shallow-cut Ellipsoid method [12]. But for simplicity we will stick with the Bertsimas and Vempala algorithm.

Theorem 1. *(Bertsimas and Vempala). Consider the convex program described above. Suppose K is contained in a ball of radius R centered at the origin. Also, suppose we are given a point y, such that the ball of radius r around y is contained in K. Then this problem can be solved in time* $\mathrm{poly}(n, L)$, *where* $L = \log(R/r)$.

3   Consistency of Local Density Matrices

We define the Consistency problem as follows [1]:

> Consider a system of n qubits. We are given a collection of local density matrices $\rho_1, \ldots, \rho_m$, where each ρ_i acts on a subset of qubits $C_i \subseteq \{1, \ldots, n\}$. Each matrix entry is specified with $\mathrm{poly}(n)$ bits of precision. Also, $m \leq \mathrm{poly}(n)$, and each subset C_i has size $|C_i| \leq k$, for some constant k.
>
> In addition, we are given a real number β (specified with $\mathrm{poly}(n)$ bits of precision) such that $\beta \geq 1/\mathrm{poly}(n)$.
>
> The problem is to distinguish between the following two cases:
> - There exists an n-qubit state σ such that, for all i, $\|\mathrm{tr}_{\{1,\ldots,n\}-C_i}(\sigma) - \rho_i\|_1 = 0$. In this case, output "YES."
> - For all n-qubit states σ, there exists some i such that $\|\mathrm{tr}_{\{1,\ldots,n\}-C_i}(\sigma) - \rho_i\|_1 \geq \beta$. In this case, output "NO."

Theorem 2. *Consistency is in QMA.*

Proof sketch: The basic idea is as follows. Given a witness state σ, the verifier will pick a subset C_i, and perform measurements to compare σ (on the subset C_i) to ρ_i. There is a complication, however, because the verifier requires many independent copies of the witness σ, and the prover might try to cheat using entanglement among the different copies. One can deal with this problem using a Markov argument. For details, see the discussion of QMA+ in [15]. $\square$

4   Consistency is QMA-Hard

Theorem 3. *Consistency is QMA-hard, via a poly-time randomized Turing reduction from Local Hamiltonian. Furthermore, the reduction uses the same value of k for both problems, so we get that Consistency with $k = 2$ is QMA-hard.*

We begin by discussing the basic idea of the proof, and the complications that arise. We then describe the actual reduction from Local Hamiltonian to Consistency, and finally we deal with some issues of numerical precision.

4.1   The Basic Idea

Say we are given a local Hamiltonian $H = H_1 + \cdots + H_m$, where H_i acts on the subset C_i. Consider the following convex program:

Let ρ be any $2^n \times 2^n$ complex matrix.
Find some ρ that minimizes $\mathrm{tr}(H\rho)$,
such that $\rho \succeq 0$ and $\mathrm{tr}(\rho) = 1$.

It is easy to see that H has an eigenvalue $\leq \gamma$ if and only if the convex program achieves $\mathrm{tr}(H\rho) \leq \gamma$ for some ρ. (Note that, although the convex program allows mixed states ρ, the optimal ρ can always be chosen to be a pure state.) Unfortunately, this convex program has 4^n variables, which makes it unwieldy.

We now construct another convex program, which is equivalent to the previous one, but has only a polynomial number of variables:

Let $\rho_1, \ldots, \rho_m$ be complex matrices, where ρ_i has size $2^{|C_i|} \times 2^{|C_i|}$.
(We interpret each ρ_i as the reduced density matrix for the subset C_i.)
Find some $\rho_1, \ldots, \rho_m$ that minimize $\mathrm{tr}(H_1\rho_1) + \cdots + \mathrm{tr}(H_m\rho_m)$,
such that each ρ_i satisfies $\rho_i \succeq 0$ and $\mathrm{tr}(\rho_i) = 1$,
and $\rho_1, \ldots, \rho_m$ are consistent.

Note that the set of feasible solutions is indeed convex: if $\rho_1, \ldots, \rho_m$ are consistent, and $\rho'_1, \ldots, \rho'_m$ are consistent, then any convex combination $\rho''_i = q\rho_i + (1-q)\rho'_i$ $(i = 1, \ldots, m)$ is also consistent.

Observe that the optimal value of this convex program is equal to the optimal value of the previous convex program; this is because, if $\rho_1, \ldots, \rho_m$ are consistent with some n-qubit state σ, then $\mathrm{tr}(H\sigma) = \mathrm{tr}(H_1\rho_1) + \cdots + \mathrm{tr}(H_m\rho_m)$. Also, note that the number of variables in this convex program is $\sum_{i=1}^{m} 4^{|C_i|} \leq 4^k m \leq \mathrm{poly}(n)$.

This convex program has a "consistency" constraint, which we do not know how to evaluate. But if we have an oracle for the Consistency problem, then we can solve this convex program, using the algorithm of Bertsimas and Vempala. To make this work, we will have to find a suitable representation for the set of feasible solutions,

$$K = \{(\rho_1, \ldots, \rho_m) \text{ which are consistent}\}.$$

Also, we will have to address some questions about the accuracy of the Consistency oracle, i.e., how well does it approximate K, and how does this affect the Bertsimas-Vempala algorithm.

We could represent each element $(\rho_1, \ldots, \rho_m) \in K$ by writing down the matrix entries for the ρ_i; then we could view K as a subset of $\mathbb{C}^d$, where $d = \sum_{i=1}^{m} 4^{|C_i|}$. But this straightforward approach runs into some trouble. Observe that the matrix entries must satisfy some algebraic constraints: each ρ_i must be Hermitian, $(\rho_i)^\dagger = \rho_i$; and ρ_i and ρ_j must agree on their intersection $C_i \cap C_j$, that is, $\mathrm{tr}_{C_i - (C_i \cap C_j)}(\rho_i) = \mathrm{tr}_{C_j - (C_i \cap C_j)}(\rho_j)$. Because of these constraints, the set K actually lies in a lower-dimensional subspace of $\mathbb{C}^d$. We would need to characterize this subspace, before we can apply the Bertsimas-Vempala algorithm. We can avoid this problem by switching to a different representation for the set K.

4.2   The Actual Reduction

We will represent each element of K using the expectation values of the "local" Pauli matrices on the subsets $C_1, \ldots, C_m$. These local Pauli matrices form a basis for the space of all local Hamiltonians (acting on the subsets C_i). For an n-qubit state σ, knowing the expectation values of these Pauli matrices is equivalent to knowing the projection of σ onto this subspace; and this is equivalent to knowing the local density matrices of σ.

First, some notation. Let P be an n-qubit Pauli matrix, and define the "support" of P be the set of qubits on which P acts nontrivially; that is, $\text{supp}(P) = \{i \mid P_i \neq I\}$. Also, for any subset of qubits C, define the "restriction" of P to C, $P|C = \bigotimes_{i \in C} P_i$.

Define $\mathcal{S}$ to be the set of "local" Pauli matrices:

$$\mathcal{S} = \{P \in \mathcal{P}^{\otimes n} \mid \text{supp}(P) \subseteq C_i \text{ for some } i\} - \{I\},$$

where we excluded the identity matrix I because its expectation value is always 1. Also let $d = |\mathcal{S}|$, and note that $d \leq 4^k m - 1 \leq \text{poly}(n)$.

For each $P \in \mathcal{S}$, let α_P be the corresponding expectation value; and let $(\alpha_P)_{P \in \mathcal{S}}$ denote the collection of these α_P. Also, let $\alpha_I = 1$. We define the set K' to be

$$K' = \{(\alpha_P)_{P \in \mathcal{S}} \text{ which are consistent}\},$$

where we say the α_P are "consistent" if there exists an n-qubit state σ such that for all $P \in \mathcal{S}$, $\alpha_P = \text{tr}(P\sigma)$. Note that K' is a subset of $\mathbb{R}^d$. Also, clearly K' is convex.

Lemma 4. *There is a linear bijection between K and K'.*

Proof: Given some $(\rho_1, \ldots, \rho_m) \in K$, we can construct $(\alpha_P)_{P \in \mathcal{S}} \in K'$ as follows:

> For each $P \in \mathcal{S}$: We know that $\text{supp}(P) \subseteq C_i$ for some i. So we can write P in the form $P = (P|C_i) \otimes I$. Then we set $\alpha_P = \text{tr}((P|C_i)\rho_i)$.

If the ρ_i are consistent with some n-qubit state σ, then the α_P are also consistent with σ; to see this, write $\alpha_P = \text{tr}((P|C_i)\rho_i) = \text{tr}(P\sigma)$. (Note that in the case where $\text{supp}(P) \subseteq C_i \cap C_j$, it makes no difference whether we pick i or j in the above procedure, because ρ_i and ρ_j yield the same reduced density matrix on $C_i \cap C_j$.)

Going in the opposite direction, given some $(\alpha_P)_{P \in \mathcal{S}} \in K'$, we can construct $(\rho_1, \ldots, \rho_m) \in K$ as follows:

> For $i = 1, \ldots, m$: We construct ρ_i by using the α_P for all P with $\text{supp}(P) \subseteq C_i$. Note that we can write P in the form $P = (P|C_i) \otimes I$. We set
> $$\rho_i = \frac{1}{2^{|C_i|}} \sum_{P : \text{supp}(P) \subseteq C_i} \alpha_P (P|C_i).$$

If the α_P are consistent with some n-qubit state σ, then the ρ_i are also consistent with σ; to see this, write σ in terms of the α_P, where we now include the expectation values $\alpha_P = \mathrm{tr}(P\sigma)$ for all $P \in \mathcal{P}^{\otimes n}$,

$$\sigma = \frac{1}{2^n} \sum_{P \in \mathcal{P}^{\otimes n}} \alpha_P P;$$

note that when we trace out the qubits not in C_i, we get that $\mathrm{tr}_{\{1,...,n\}-C_i}(P)$ equals $2^{n-|C_i|}(P|C_i)$ if $\mathrm{supp}(P) \subseteq C_i$, and 0 otherwise; thus we have

$$\mathrm{tr}_{\{1,...,n\}-C_i}(\sigma) = \frac{1}{2^{|C_i|}} \sum_{P\,:\,\mathrm{supp}(P) \subseteq C_i} \alpha_P (P|C_i) = \rho_i.$$

Finally, observe that these maps (between K and K') are linear, and they are inverses of each other. $\qquad\square$

So we can restate our convex program, using the set K' instead:

Let α_P (for $P \in \mathcal{S}$) be real numbers.
Find some α_P that minimize

$$\sum_{i=1}^{m} \frac{1}{2^{|C_i|}} \sum_{P\,:\,\mathrm{supp}(P) \subseteq C_i} \alpha_P \, \mathrm{tr}(H_i(P|C_i)),$$

such that $(\alpha_P)_{P \in \mathcal{S}} \in K'$ (i.e., the α_P are consistent).

Lemma 5. *The optimal value of this convex program is equal to the smallest eigenvalue of the local Hamiltonian $H = H_1 + \cdots + H_m$.*

Proof: This follows from the remarks in the previous section, and Lemma 4. $\quad\square$

Next, we prove some bounds on the geometry of the set $K' \subseteq \mathbb{R}^d$.

Lemma 6. *K' is contained in a ball of radius $R = \sqrt{d}$ centered at the origin.*

Proof: Suppose $(\alpha_P)_{P \in \mathcal{S}} \in K'$, and say it is consistent with some state σ. Since $\alpha_P = \mathrm{tr}(P\sigma)$, it follows that $-1 \leq \alpha_P \leq 1$, which implies the result. $\quad\square$

Lemma 7. *The ball of radius $r = 1/\sqrt{d}$ around the origin is contained in K'.*

Proof: Let $(\alpha_P)_{P \in \mathcal{S}}$ be any vector in $\mathbb{R}^d$ of length at most $1/\sqrt{d}$. By the Cauchy-Schwartz inequality, $\sum_{P \in \mathcal{S}} |\alpha_P| \leq 1$; let $p = \sum_{P \in \mathcal{S}} |\alpha_P|$. Now define $\sigma = (1/2^n)(I + \sum_{P \in \mathcal{S}} \alpha_P P)$. This is a legal density matrix, because it can be written as

$$\sigma = \frac{1}{2^n}\left((1-p)I + \sum_{P \in \mathcal{S}}(|\alpha_P|I + \alpha_P P)\right)$$

$$= (1-p)\frac{I}{2^n} + \sum_{P \in \mathcal{S}} |\alpha_P| \frac{I + \mathrm{sign}(\alpha_P)P}{2^n},$$

which is (with probability $1 - p$) the fully mixed state, and (with probability $|\alpha_P|$, for $P \in \mathcal{S}$) the mixture of all eigenstates of P with eigenvalue $\text{sign}(\alpha_P)$. Furthermore, the α_P are consistent with σ; thus we conclude that $(\alpha_P)_{P \in \mathcal{S}} \in K'$. $\qquad\square$

4.3   Numerical Precision

First, we will show that the Consistency oracle gives a good approximation to the set K'. We start by defining a new problem, Consistency', using the expectation values α_P of the local Pauli matrices $P \in \mathcal{S}$ (similar to the definition of K'):

> As in the original Consistency problem, we have an n-qubit system, and subsets $C_1, \ldots, C_m$, with $|C_i| \leq k$. But instead of the local density matrices $\rho_1, \ldots, \rho_m$, we are given real numbers α_P for all $P \in \mathcal{S}$. Each α_P is specified with $\text{poly}(n)$ bits of precision.
>
> In addition, we are given a real number β' (specified with $\text{poly}(n)$ bits of precision) such that $\beta' \geq 1/\text{poly}(n)$.
>
> The problem is to distinguish between the following two cases:
> - There exists an n-qubit state σ such that, for all $P \in \mathcal{S}$, $\text{tr}(P\sigma) = \alpha_P$. In this case, output "YES."
> - For all n-qubit states σ, $\left(\sum_{P \in \mathcal{S}} (\text{tr}(P\sigma) - \alpha_P)^2\right)^{1/2} \geq \beta'$. In this case, output "NO."

Lemma 8. *There is a poly-time mapping reduction from Consistency' to Consistency.*

Proof sketch: The reduction is as follows: Use the α_P to construct $\rho_1, \ldots, \rho_m$ as described in Lemma 4. Set $\beta = \beta'/\sqrt{d}$, where $d = |\mathcal{S}|$. Details omitted. $\qquad\square$

Next, we will show that the Bertsimas-Vempala algorithm succeeds in solving our convex program, even when the oracle for the set K' is slightly inaccurate. (The shallow-cut ellipsoid method would also work, see [12] for details.) We will make some general remarks about the algorithm, and then show that it works for our specific problem.

For our purposes, we only need to solve a simpler problem, deciding the feasibility of a convex program:

> As before, let K be a convex set, and let f be a convex function.
> Given some $t \in \mathbb{R}$, does there exist a point $x \in K$ such that $f(x) \leq t$?

The Bertsimas-Vempala algorithm is built around a subroutine that solves the feasibility problem [9]. The basic idea is as follows:

> Let P be the set K.
> Randomly sample some points from P, and compute an approximate centroid of P; call this point z.
> If $f(z) \leq t$, stop and return true.

Compute $\nabla f(z)$, and use this to cut out a portion of the set P.[1]

Repeat the procedure starting from line 2. If P gets too small, stop and return false.

The critical step is to sample random points from the set P. (Note that P is convex, and we have a membership oracle for P.) One way is to do a random walk known as the "ball walk":

Pick a point y uniformly at random in the ball of radius δ centered at the current position x. If $y \in P$, then move to y, otherwise stay at x. Repeat.

The points where the membership oracle makes mistakes all lie close to the boundary of P; call this the "boundary layer" P_b. Intuitively, if the boundary layer is thin, it should not have much effect on the random walk. Indeed, using an argument by Lovász and Simonovits [16], one can show the following:

Lemma 9. *For any polynomial p, there exists a polynomial q such that, if we run the ball walk for at most $p(n)$ steps, and $\mathrm{vol}(P_b)/\mathrm{vol}(P) \leq 1/q(n)$, then with probability $2/3$ we will never enter the region P_b.*

So, if we can show that the boundary layer is small compared to the total volume of P, then our algorithm will work fine. (As long as the random walk does not enter the boundary layer, the algorithm will perform exactly as if it had access to a perfect membership oracle.)

Finally, there may still be errors due to finite numerical precision—using n bits of precision, we have errors of size 2^{-n}. This will not be a problem for us, since we only need accuracy of $1/\mathrm{poly}(n)$.

We are now ready to prove that the Bertsimas-Vempala algorithm works for our specific problem:

Proof of Theorem 3: Given an instance of Local Hamiltonian, use Lemma 5 to express it as a convex program. Let f denote the objective function, and set $t = (a+b)/2$. By Lemma 8, we can assume we have an oracle for Consistency', which approximates the set K' with error β', for any $\beta' \geq 1/\mathrm{poly}(n)$. Then use the Bertsimas-Vempala algorithm to solve the following problem: does there exist a solution $\alpha \in K'$ such that $f(\alpha) \leq t$?

Recall that the objective function

$$f(\alpha) = \sum_{i=1}^{m} \frac{1}{2^{|C_i|}} \sum_{P\,:\,\mathrm{supp}(P)\subseteq C_i} \alpha_P \,\mathrm{tr}(H_i P)$$

is linear. We claim that its derivatives in all directions are at most $4^k m \leq \mathrm{poly}(n)$. To see this, note that there are at most $4^k m$ terms in the sum, and for each term, we have

$$|\mathrm{tr}(H_i P)| \leq \mathrm{tr}(|H_i P|) \leq \|H_i\|_2 \|P\|_2 \leq 2^{|C_i|},$$

---

[1] Specifically, we can deduce a hyperplane that separates z from the set $\{x \,|\, f(x) \leq t\}$. Then we take the intersection of P with the half-space that does not contain z.

using the Cauchy-Schwartz inequality for the L_2 matrix norm [17], and the fact that the eigenvalues of H_i lie in the interval $[0, 1]$, while the eigenvalues of P are ± 1.

Now suppose we have a "YES" instance of Local Hamiltonian. Then there exists some $\alpha^* \in K'$ such that $f(\alpha^*) \leq a$. We claim that the set $\{\alpha \in K' \,|\, f(\alpha) \leq t\}$ contains a ball of radius $\delta \geq 1/\mathrm{poly}(n)$. To see this, let σ^* be the density matrix which corresponds to α^*. Perturb σ^* by mixing it with the state $I/2^n$, then add a small contribution of each of the Pauli matrices $P \in \mathcal{S}$. This generates a ball contained in K'. Moreover, this ball can have radius $\delta \geq 1/\mathrm{poly}(n)$ and still satisfy the condition $f(\alpha) \leq t$; this is because f does not vary too quickly, and there is a gap between a and t.

So, in the Bertsimas-Vempala algorithm, the set P always contains a ball of radius δ. Now set the error threshold for the membership oracle to be $\beta' \leq \delta/n^d$. We will show that the boundary layer P_b is small compared to the total volume of P. Define P^+ to be the set P expanded by an amount β', that is, $P^+ = P + \beta' B$, where B is the unit ball. We have that

$$P^+ \subseteq P + (\beta'/\delta)P = (1 + \beta'/\delta)P,$$

where the equality holds because P is convex. This implies that

$$\mathrm{vol}(P^+) \leq (1 + \beta'/\delta)^n \, \mathrm{vol}(P) \leq (1 + 2/n^{d-1}) \, \mathrm{vol}(P).$$

So we can conclude that $\mathrm{vol}(P_b) \leq \mathrm{vol}(P^+) - \mathrm{vol}(P) \leq (2/n^{d-1}) \, \mathrm{vol}(P)$. Therefore, by Lemma 9, the Bertsimas-Vempala algorithm will work correctly in this case.

Now suppose we have a "NO" instance of Local Hamiltonian. Then for all $\alpha \in K'$, $f(\alpha) \geq b$. In addition, there is some $\delta \geq 1/\mathrm{poly}(n)$ such that, for all α within distance δ of K', $f(\alpha) > t$; this is because f does not vary too quickly, and there is a gap between b and t.

Set the error threshold for the membership oracle to be $\beta' \leq \delta$. Then the set $\{\alpha \in K' \,|\, f(\alpha) \leq t\}$ is empty, even when the membership oracle makes mistakes. So the Bertsimas-Vempala algorithm will work correctly in this case.

Finally, we claim that the Bertsimas-Vempala algorithm runs in time polynomial in n. This follows from Theorem 1 and Lemmas 6 and 7; note that $L = \log(R/r) = \log(\mathrm{poly}(n)) = O(\log n)$. □

5   Discussion

Consistency of local density matrices is an interesting problem that gives some new insight into the class QMA. The reduction from Local Hamiltonian is nontrivial, and in that sense, Consistency seems to be an easier problem to deal with. One direction for future work is to try to find additional QMA-complete problems by giving reductions from Consistency (rather than from Local Hamiltonian).

Another question is whether Consistency remains QMA-hard under mapping reductions. We mention that we can build zero-knowledge proof systems for

Consistency [18], using techniques developed by Watrous [19]. If we could show that Consistency is QMA-hard under mapping reductions, then we could get zero-knowledge proof systems for any language in QMA.

Acknowledgements. Thanks to Dorit Aharonov for suggesting this problem and pointing out an error in a previous version of the paper; thanks also to Russell Impagliazzo and the anonymous reviewers for their helpful comments. Supported by an ARO/NSA Quantum Computing Graduate Research Fellowship.

References

1. D. Aharonov, private communication, 2004.
2. A.Yu. Kitaev, A.H. Shen and M.N. Vyalyi, *Classical and Quantum Computation*, AMS, 2002.
3. D. Aharonov and T. Naveh, "Quantum NP - A Survey," Arxiv: quant-ph/0210077.
4. J. Kempe and O. Regev, "3-Local Hamiltonian is QMA-complete," Quantum Info. and Comput., Vol.3(3), pp.258-264, 2003, Arxiv: quant-ph/0302079.
5. J. Kempe, A. Kitaev and O. Regev, "The Complexity of the Local Hamiltonian Problem," FSTTCS 2004, pp.372-383, Arxiv: quant-ph/0406180.
6. R. Oliveira and B.M. Terhal, "The complexity of quantum spin systems on a two-dimensional square lattice," Arxiv: quant-ph/0504050.
7. D. Aharonov, W. van Dam, J. Kempe, Z. Landau, S. Lloyd and O. Regev, "Adiabatic Quantum Computation is Equivalent to Standard Quantum Computation," FOCS 2004, pp.42-51, Arxiv: quant-ph/0405098.
8. D. Janzing, P. Wocjan and T. Beth, "Identity check is QMA-complete," Arxiv: quant-ph/0305050.
9. D. Bertsimas and S. Vempala, "Solving Convex Programs by Random Walks," Journal of the ACM 51 (4) pp.540-556 (2004).
10. A. Kalai and S. Vempala, "Convex Optimization by Simulated Annealing," preprint, 2004.
11. S. Vempala, "Geometric Random Walks: A Survey," MSRI volume on Combinatorial and Computational Geometry, 2005.
12. M. Grötschel, L. Lovász and A. Schrijver, *Geometric Algorithms and Combinatorial Optimization*, Springer, 1988.
13. S. Bravyi and M. Vyalyi, "Commutative version of the local Hamiltonian problem and common eigenspace problem," Quantum Info. and Comput., Vol.5, No.3 (2005), pp.187-215, Arxiv: quant-ph/0308021.
14. S. Bravyi, "Efficient algorithm for a quantum analogue of 2-SAT," Arxiv: quant-ph/0602108.
15. D. Aharonov and O. Regev, "A Lattice Problem in Quantum NP," FOCS 2003, pp.210-219, Arxiv: quant-ph/0307220.
16. L. Lovász and M. Simonovits, "Random Walks in a Convex Body and an Improved Volume Algorithm," Random Structures and Algorithms, Vol.4, No.4 (1993).
17. R. Bhatia, *Matrix Analysis*, Springer, 1997.
18. Y.-K. Liu, in preparation.
19. J. Watrous, "Zero-knowledge against quantum attacks," Arxiv: quant-ph/0511020.

On Bounded Distance Decoding for General Lattices

Yi-Kai Liu*, Vadim Lyubashevsky**, and Daniele Micciancio**

University of California, San Diego
9500 Gilman Drive, La Jolla, CA 92093-0404, USA
{y9liu, vlyubash, daniele}@cs.ucsd.edu

Abstract. A central problem in the algorithmic study of lattices is the *closest vector problem*: given a lattice $\mathcal{L}$ represented by some basis, and a target point y, find the lattice point closest to y. *Bounded Distance Decoding* is a variant of this problem in which the target is guaranteed to be close to the lattice, relative to the minimum distance $\lambda_1(\mathcal{L})$ of the lattice. Specifically, in the α-Bounded Distance Decoding problem (α-**BDD**), we are given a lattice $\mathcal{L}$ and a vector y (within distance $\alpha \cdot \lambda_1(\mathcal{L})$ from the lattice), and we are asked to find a lattice point $x \in \mathcal{L}$ within distance $\alpha \cdot \lambda_1(\mathcal{L})$ from the target. In coding theory, the lattice points correspond to codewords, and the target points correspond to lattice points being perturbed by noise vectors. Since in coding theory the lattice is usually fixed, we may "pre-process" it before receiving any targets, to make the subsequent decoding faster. This leads us to consider α-**BDD** with pre-processing. We show how a recent technique of Aharonov and Regev [2] can be used to solve α-**BDD** with pre-processing in polynomial time for $\alpha = O\left(\sqrt{(\log n)/n}\right)$. This improves upon the previously best known algorithm due to Klein [13] which solved the problem for $\alpha = O(1/n)$. We also establish hardness results for α-**BDD** and α-**BDD** with pre-processing, as well as generalize our results to other ℓ_p norms.

1 Introduction

A lattice is the set of intersection points of a regular (but not necessarily orthogonal) n-dimensional grid. One of the most central problems in the algorithmic study of lattices is the *closest vector problem*: given a lattice $\mathcal{L}$ (typically represented by a basis, see Section 2 for details), and a target point y, find the lattice point closest to y. Beside having numerous applications in theoretical computer science, lattices are a central object in coding theory [1, 8]. In this setting, lattice points represent codewords, and the target point y represents a perturbed codeword (encoding a message being transmitted). In this scenario, the closest vector problem corresponds exactly to the maximum likelyhood decoding problem for white Gaussian noise channels. The closest vector problem is NP-hard to solve

---

  * Supported by an ARO/NSA Quantum Computing Graduate Research Fellowship.
 ** Supported by NSF CAREER 0093029 and NSF ITR 0313241.

even approximately for any constant [5] and some sub-polynomial [9] approximation factors. On the positive side, the best general approximation algorithm to solve the closest vector problem in (random) polynomial time achieves only approximation factors almost exponential in the dimension of the lattice [3]. We remark that there are two fundamental differences between the closest vector problem as typically studied in complexity theory and coding theory:

1. In complexity theory, the lattice is considered as part of the input to the problem, while in coding theory the lattice (defining the error correcting code) is usually fixed once and for all.
2. In complexity theory the target point can be arbitrarily far from the lattice, while in coding theory it is usually assumed that the distance of the target from the lattice is less than half the minimum distance between lattice points.

The first issue has been recently addressed [15, 11, 18, 4], considering a version of the **CVP** *with pre-processing* (**CVPP**). In **CVPP**, the lattice is fixed and can be arbitrarily preprocessed, and the complexity of the **CVP** algorithm is measured without taking pre-processing time into account. In the sequence of papers [15, 11, 18, 4] it is shown that there are lattices such that **CVPP** is NP-hard to solve exactly, or even approximate within any constant factor.[1]

The second issue is equally important, but has so far received far less attention. The relevance of the second issue stems from the fact that the amount of error (i.e., the distance of the target from the lattice) depends on the properties of the communication channel, and it is usually known to the code designer. This allows the code designer to choose a lattice code whose minimum distance is, in some respect, "large" compared to the maximum error.

The variant of the closest vector problem where the target is guaranteed to be close to the lattice relative to the minimum distance, $\lambda_1(\mathcal{L})$, of the lattice is called the *Bounded Distance Decoding* problem (**BDD**) [23]. Specifically, in the α-Bounded Distance Decoding problem (α-**BDD**), we are given a lattice $\mathcal{L}$ and a vector $\boldsymbol{y}$ (within distance $\alpha \cdot \lambda_1(\mathcal{L})$ from the lattice), and are asked to find a lattice point $\boldsymbol{x} \in \mathcal{L}$ within distance $\alpha \cdot \lambda_1(\mathcal{L})$ from the target. Typically $\alpha = 1/2$, but other values of α can be interesting as well, as some decoding algorithms may not work up to the unique decoding radius $\lambda_1/2$, while in other cases even for $\alpha > 1/2$ it may be possible to come up with a relatively short (i.e., polynomially long) list of candidate lattice points. (The latter is called the "list decoding problem", and it has received an great deal of attention lately in the context of codes over finite fields. We are not aware of any result of the same kind for lattice codes, but the problem is certainly very interesting and natural.)

Our contribution. In this paper we investigate the bounded distance decoding problem α-**BDD** (with pre-processing), and prove both algorithmic and computational hardness results about this problem. On the algorithmic side, we

---

[1] The factor achieved by a **CVP** approximation algorithm is defined as the ratio between the distance (from the target) of lattice point output by the algorithm, over the distance of the optimal solution.

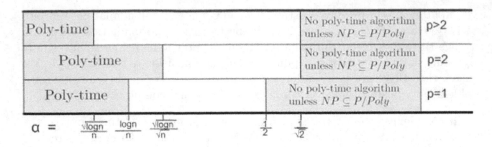

Fig. 1. Complexity of α-BDD with pre-processing for various l_p norms

show that α-**BDD** with pre-processing can be solved in polynomial time for $\alpha = O(\sqrt{\log n/n})$. Previously, the problem was known to be polynomial time solvable only for factors $O(1/n)$ [6, 13]. On the computational hardness side, we show (under standard complexity assumptions) that the α-**BDD** problem cannot be solved in polynomial time (even in its pre-processing variant) for any constant factor $\alpha > 1/\sqrt{2}$. Specifically, for the α-**BDD** problem with pre-processing, any polynomial time solution would imply that NP is contained in P/poly. We also adapt some of our results to other ℓ_p norms (see figure 1).

Related work. Our work is closely related, and builds upon, previous work of Aharonov and Regev [2] (for the algorithmic results) and Micciancio [16] (for the hardness results), as well as previous algorithms [6, 13] and NP-hardness results [4] for the **CVP** problem with pre-processing.

The well known *nearest plane algorithm* [6] (together with the bounds in [14]) yields a polynomial time solution to α-**BDD** with pre-processing for $\alpha = 2/n$. That solution had been subsequently improved to any $\alpha = O(1/n)$ in [13], which was still the best general solution prior to our work. Our algorithm uses a technique developed by Aharonov and Regev [2], and it is closely related to their work. In [2], Aharonov and Regev show how to compute (with pre-processing) a function $f(x)$ which approximates the distance of a target point x from a lattice within a factor of $O(\sqrt{n/\log n})$. Unfortunately, being able to approximate the distance of a target point from the lattice does not, in general, allow us to find lattice points which are close to the target.[2] So, the Aharonov and Regev's **CVP** pre-processing algorithm does not directly yield a solution to α-**BDD**. In this paper we observe that, for a certain region of $\mathbb{R}^n$ close to the lattice, the function f of Aharonov and Regev allows us to distinguish which of two points in $\mathbb{R}^n$ is closer to the lattice. So, by adding a noise vector to our target point, we can

---

[2] Technically, there is no known approximation preserving reduction from the problem of approximating **CVP** in its search version (i.e., finding an approximately closest lattice point), to approximating **CVP** in its decision or distance estimation version (i.e., determining if a target point is close or far away from a lattice). (See [17, Chapter 3].) Such a reduction trivially exists for the exact versions of **CVP** and for small (subpolynomial) approximation factors by NP-hardness [5, 9], but finding such a reduction for polynomial approximation factors is an open problem.

generate a nearby point which is closer to the lattice and verify that it actually is closer. Our α-**BDD** algorithm performs a "guided" walk starting from the target and moving closer and closer to the lattice, until we get within distance $O(\lambda_1/n)$ from it, at which point some other known algorithm for α-**BDD** (i.e. [6, 13]) can be used.

On the complexity front, no hardness result was known prior to our work because, interestingly, all known NP-hardness results for **CVP** (with or without pre-processing) [5, 9, 15, 11, 18, 4] employed lattices with very small minimum distance. We prove our hardness results using a technique of Micciancio [16] to embed a lattice $\mathcal{L}$ and target $\boldsymbol{y}$ into a higher dimensional space in such a way that the minimum distance of the lattice increases, without at the same time substantially increasing the distance of the target from the lattice. The hardness results for ℓ_p norms for $p > 2$ are obtained by an application of a recent technique of Regev and Rosen [19] to our result for the ℓ_2 norm.

Questions regarding the complexity of α-**BDD** (and related problems) had been previously considered in the setting of linear codes over finite fields. For example, Vardy [23] conjectured the problem to be NP-hard for $\alpha = 1/2$, and for the closely related *relatively near codeword* problem (RNC) NP-hardness results for any $\alpha > 1/2$ were proven by Dumer, Micciancio and Sudan [10], and later adapted by Regev [18] to the pre-processing variant of the problem.

2 Preliminaries

2.1 Lattices

The set of all integer combinations of vectors $B = (\boldsymbol{b}_1, \ldots, \boldsymbol{b}_n)$ defines a *lattice* $\mathcal{L}(B)$ in $\mathbb{R}^n$. B is said to form a basis of $\mathcal{L}(B)$ (when the basis is clear from context, we may write $\mathcal{L}$ instead of $\mathcal{L}(B)$). For any basis $\boldsymbol{b}_1, \ldots, \boldsymbol{b}_n$, the *Gram-Schmidt* basis is denoted by $\boldsymbol{b}_1^*, \ldots, \boldsymbol{b}_n^*$ where $\boldsymbol{b}_i^*$ is the component of $\boldsymbol{b}_i$ which is orthogonal to the vector space formed by $\boldsymbol{b}_1, \ldots, \boldsymbol{b}_{i-1}$. We denote by $\lambda_1(\mathcal{L})$ the length of the shortest vector of $\mathcal{L}$ (equivalently, the minimum distance of $\mathcal{L}$).

Lemma 1. *For every n-dimensional lattice $\mathcal{L}$, there exists a basis $\boldsymbol{b}_1, \ldots, \boldsymbol{b}_n$ such that $\min_i \|b_i^*\| \geq \lambda_1(\mathcal{L})/n$.*

Lattice Problems. Given a lattice basis $\mathbf{B}$, vector $\boldsymbol{x}$ and a real t, the decisional version of the closest vector problem (**CVP**) asks whether $dist(\mathcal{L}(B), \boldsymbol{x}) \leq t$. The approximate version of decisional **CVP** can be formulated as a promise problem $\mathbf{GapCVP}_\gamma$. Given a lattice basis $\mathbf{B}$, vector $\boldsymbol{x}$ and a real t, an algorithm for $\mathbf{GapCVP}_\gamma$ should answer "YES" if $dist(\mathcal{L}(B), \boldsymbol{x}) \leq t$ and "NO" if $dist(\mathcal{L}(B), \boldsymbol{x}) > \gamma t$. For all values in between, any answer is acceptable.

An algorithm that solves $\mathbf{GapCVP}_\gamma$ with pre-processing works in two steps. First, it is given a basis $\boldsymbol{B}$. The algorithm then outputs an advice string A. The time that the algorithm expends in obtaining A does not count towards its running time. Then, it is given a vector $\boldsymbol{x}$ and a real t. With the ability to use the advice string A, it should answer "YES" if $dist(\mathcal{L}(\boldsymbol{B}), \boldsymbol{x}) \leq t$ and "NO" if $dist(\mathcal{L}(\boldsymbol{B}), \boldsymbol{x}) > \gamma t$. For all values in between, any answer is acceptable. Alekhnovich, et. al. [4] showed that the $\mathbf{GapCVP}_\gamma$ with pre-processing problem is NP-hard for any constant γ. Aharonov and Regev [2] showed a polynomial algorithm for this problem for $\gamma = O\left(\sqrt{n/\log n}\right)$.

In the alpha-Bounded Distance Decoding problem (α-**BDD**), we are given a lattice basis $\mathbf{B}$ and a vector $\boldsymbol{x}$ and are asked to find a lattice vector $\mathbf{y} \in \mathcal{L}(\mathbf{B})$ such that $dist(\mathbf{x}, \mathbf{y}) \leq \alpha \cdot \lambda_1(\mathcal{L}(\mathbf{B}))$ (if such a vector exists). In Section 4 we show that this problem is NP-hard for $\alpha > 1/\sqrt{2}$.

As for $\mathbf{GapCVP}_\gamma$ with pre-processing, an algorithm for α-**BDD** with pre-processing works in two steps. First, it is given a basis $\boldsymbol{B}$. The algorithm then outputs an advice string A. The time that the algorithm expends in obtaining A does not count towards its running time. Then, it is given a vector $\boldsymbol{x}$. With the ability to use the advice string A, it should find a lattice vector $\mathbf{y} \in \mathcal{L}(\mathbf{B})$ (if one exists) such that $dist(\mathbf{x}, \mathbf{y}) \leq \alpha \cdot \lambda_1(\mathcal{L}(\mathbf{B}))$. In Section 3, we provide a polynomial algorithm for values of $\alpha = O\left(\sqrt{\log n/n}\right)$ and in Section 4, we show that if a polynomial algorithm exists for $\alpha > 1/\sqrt{2}$, then $NP \subseteq P/poly$.

We define one last lattice problem. This problem will be useful in Section 4 to prove the hardness results. Given a full rank n-dimensional lattice basis $\mathbf{B}$, vector $\boldsymbol{x}$ and a real t, an algorithm for $\mathbf{GapCVP}'_\gamma$ should answer "YES" if $\exists z \in \{0,1\}^n$ such that $dist(\boldsymbol{B}z, \boldsymbol{x}) \leq t$ and "NO" if $dist(\mathcal{L}(\boldsymbol{B}), s\boldsymbol{x}) > \gamma t$ for all $s \in \mathbb{Z} \setminus \{0\}$. In all other cases, any answer is acceptable. Arora, et. al. [5] showed this problem to be NP-hard for all constants $\gamma \geq 1$.

2.2   Gaussian Functions on Lattices

In [2], Aharonov and Regev considered the following function f on any $\boldsymbol{x} \in \mathbb{R}^n$,

$$f(\boldsymbol{x}) = \frac{\sum_{\mathbf{y} \in \mathcal{L}} e^{-\pi \|\boldsymbol{x}-\mathbf{y}\|^2}}{\sum_{\mathbf{y} \in \mathcal{L}} e^{-\pi \|\mathbf{y}\|^2}}$$

and showed that with polynomial advice, one can estimate its value on exponentially many points in the quotient group $\mathbb{R}^n/\mathcal{L}$.

Lemma 2. ([2, Lemma 1.3]) Let $\mathcal{L}$ be an n-dimensional lattice. For any set S consisting of $2^{poly(n)}$ points in the group $\mathbb{R}^n/\mathcal{L}$ and any constant $c > 0$, there exists an advice string of size $poly(n)$ that allows one to evaluate the function f in polynomial time with error at most n^{-c} on every point in S.

A property of the function f is that if $\boldsymbol{x} \in \mathbb{R}^n$ has only one lattice point close to it, then the value of f will almost entirely determined by the distance of $\boldsymbol{x}$ from this point.

Lemma 3. Let $\mathcal{L}$ be an n-dimensional lattice whose shortest vector has length greater than $\sqrt{\frac{n}{2\pi}}$ and $\boldsymbol{x} \in \mathbb{R}^n$. If all points in $\mathcal{L}$ other than $\mathbf{y}'$ are at a distance more than $\sqrt{\frac{n}{2\pi}}$ from $\boldsymbol{x}$, then $f(\boldsymbol{x}) = e^{-\pi\|\boldsymbol{x}-\mathbf{y}'\|^2} \pm 2^{-\Omega(n)}$.

The next lemma shows that the function f is very sensitive at a distance less than $\sqrt{\log n}$ away from the lattice when the length of the shortest vector of the lattice is greater than $\sqrt{\frac{n}{2\pi}}$. Thus, if we are at a point $x \in \mathbb{R}^n$ and move closer to the lattice, there will be a noticeable change in the value of the function.

Lemma 4. *Let $\mathcal{L}$ be an n-dimensional lattice whose shortest vector has length greater than $\sqrt{\frac{n}{2\pi}}$, and let $y \in \mathcal{L}$. Suppose $x, x' \in \mathbb{R}^n$ are points such that $\|x - y\| = D$, $\|x' - y\| \leq (D + n^{-4})\sqrt{1 - \frac{1}{n}}$ where $\frac{1}{n} \leq D \leq \sqrt{\log n}$, and for all $y' \in \mathcal{L} \setminus \{y\}$, $\|x - y'\|, \|x' - y'\| > \sqrt{\frac{n}{2\pi}}$. Then $f(x') - f(x) > n^{-6.5}$.*

The below lemma is a partial converse of lemma 4.

Lemma 5. *Let $\mathcal{L}$ be an n-dimensional lattice whose shortest vector has length greater than $\sqrt{\frac{n}{2\pi}}$, and let $y \in \mathcal{L}$. Suppose $x, x' \in \mathbb{R}^n$ are points such that $\|x - y\|, \|x' - y\| \leq \sqrt{\log n}$, and for all $y' \in \mathcal{L} \setminus \{y\}$, $\|x - y'\|, \|x' - y'\| > \sqrt{\frac{n}{2\pi}}$, and $f(x') - f(x) > n^{-7.1}$. Then $\|x' - y\| \leq \sqrt{1 - 1/n^8}\|x - y\|$.*

3   Finding the Closest Lattice Vector

The following theorem was proved by Klein in [13]:

Theorem 1. *There is an algorithm that, when given an n-dimensional lattice $\mathcal{L}$ represented by basis vectors $b_1, \ldots, b_n$, and a target $x \in \mathbb{R}^n$ that's at distance D away from $\mathcal{L}$, will find the closest lattice vector to x, in time $n^{D^2 / \min_i \|b_i^*\|^2}$.*

Combining Theorem 1 with Lemma 1, implies that whenever the target point is within $O(\lambda_1(\mathcal{L})/n)$ of the lattice, there is a basis that can be used as advice to find the nearest lattice point in polynomial time.

Without loss of generality, we may assume that our lattice is scaled such that $\lambda_1(\mathcal{L}) > \sqrt{n}$. If the target that we're given is within distance $1/\sqrt{n}$ of the lattice (and thus within distance $\lambda_1(\mathcal{L})/n$ of the lattice), we can just find the closest vector by applying Theorem 1. If the target point is not that close to the lattice but is still within $\sqrt{\log n}$ of it, we will find another point closer to the lattice that is also close to the target point. We will proceed in like manner by finding points closer to the lattice until we get within $1/\sqrt{n}$ of the lattice at which time we will apply the algorithm in Theorem 1.

Theorem 2. *Let $\mathcal{L}$ be a lattice with shortest vector at least $\sqrt{n}$ and let A be the polynomial size advice string as in Lemma 2 that allows us to approximate the function f with error at most n^{-8}. Then there is a polynomial time algorithm using advice A that, when given a point $x \in \mathbb{R}^n$ that is within distance $\sqrt{\log n}$ of a lattice point $y \in \mathcal{L}$, will find a point x' that is within distance $1/\sqrt{n}$ of y.*

Proof. Let f_A be the function that uses advice A and approximates f to within n^{-8} on exponentially many points (as in Lemma 2). To be precise, we would need to put a grid on $\mathbb{R}^n$ everywhere within $\sqrt{\log n}$ of the lattice, and only be able to approximate the function f at the intersection points of the grid. Since both f

and f_A are symmetric with respect to the lattice, it will suffice to consider only grid points within distance $\sqrt{\log n}$ of the origin. Within this region we can make the grid very fine (i.e the diagonal of a grid square can be n^{-c} for any constant c), which is good enough for our purposes. For simplicity, we will assume that for all $x \in \mathbb{R}^n$, where x is within $\sqrt{\log n}$ of the lattice, $|f_A(x) - f(x)| < n^{-8}$. Consider the following algorithm: (in the algorithm u_i is the ith standard unit vector)

GetCloser(x, f_A)
    while$(f_A(x) < (e^{-\pi/n} + n^{-8}))$
        compute $D_A = \sqrt{\dfrac{-\log f_A(x)}{\pi}}$
        construct set $S = \{x - j(D_A/\sqrt{n})u_i \ : i \in \{1, \ldots, n\}, j \in \{-1, 1\}\}$
        set $x \leftarrow \underset{x' \in S}{\operatorname{argmax}} f_A(x')$

    return x

First we will show that at each iteration of the while loop, D_A is very close to the correct distance D between x and the lattice point y. Then we will show that one of the $2n$ elements of the set S is a vector that is within distance $(D + n^{-4})\sqrt{1 - 1/n}$ of the lattice. This will imply that the element $x' \in S$ for which the value $f_A(x')$ is the largest is within distance $D\sqrt{1 - 1/n^8}$ of the lattice. And by continuing to loop, we will eventually get within $1/\sqrt{n}$ of the lattice point.

Lemma 6. *At every step of the algorithm, $|D_A - \|x - y\|| \leq n^{-4}$.*

Lemma 7. *Let x and y be points in $\mathbb{R}^n$ such that $\|x - y\| = D$ and $n > 6$. For any $c > 0$ and $D_A \in [D - c, D + c]$, the set*

$$S = \{x - j(D_A/\sqrt{n})u_i \ : i \in \{1, \ldots, n\}, j \in \{-1, 1\}\}$$

contains a vector x' such that $\|x' - y\| < (D + c)\sqrt{1 - 1/n}$.

Proof. Without loss of generality, assume that $y = 0$. Then $D^2 = \|x\|^2 = \sum_i x_i^2 > (D - c)^2$. Thus, there must exist an i, such that $|x_i| \geq (D - c)/\sqrt{n}$. Let $j \in \{-1, 1\}$ have the same as the sign of x_i. Then the vector $x' = x - j\frac{D_A}{\sqrt{n}}u_i$ is in S and

$$\|x'\|^2 = \|(x_1, \ldots, x_{i-1}, x_i - j\tfrac{D_A}{\sqrt{n}}, x_{i+1}, \ldots, x_n)\|^2 = D^2 - 2|x_i|\tfrac{D_A}{\sqrt{n}} + \tfrac{D_A^2}{n}$$

Since $|x_i| \geq (D - c)/\sqrt{n}$ and $(D - c) \leq D_A \leq (D + c)$, we have

$$\|x'\|^2 \leq D^2 - 2\tfrac{(D-c)^2}{n} + \tfrac{(D+c)^2}{n} < (D + c)^2(1 - 1/n)$$

where the last inequality holds for $n > 6$.

Lemma 6 says that the value of D_A that we calculate using f_A is within n^{-4} of the actual distance D. Lemma 7 says that the set S contains a point that is a distance at most $(D + n^{-4})\sqrt{1 - 1/n}$ away from the lattice. By Lemma 4, we know that if $\|\boldsymbol{x}' - \boldsymbol{y}\| \leq (\|\boldsymbol{x} - \boldsymbol{y}\| + n^{-4})\sqrt{1 - 1/n}$, then $f(\boldsymbol{x}') - f(\boldsymbol{x}) > n^{-6.5}$, which by the triangular inequality implies that $f_A(\boldsymbol{x}') - f_A(\boldsymbol{x}) > n^{-6.5} - 2n^{-8} > n^{-7}$. Thus, there is a point $\boldsymbol{x}' \in S$ such that $f_A(\boldsymbol{x}') - f_A(\boldsymbol{x}) > n^{-7}$. By the triangular inequality, this implies that $f(\boldsymbol{x}') - f(\boldsymbol{x}) > n^{-7} - 2n^{-8} > n^{-7.1}$, which by Lemma 5 means that $\|\boldsymbol{x}' - \boldsymbol{y}\| \leq \sqrt{1 - 1/n^8}\|\boldsymbol{x} - \boldsymbol{y}\|$. So at every step, we are guaranteed to be getting closer to the lattice by a factor of $\sqrt{1 - 1/n^8}$. The loop ends once $f_A(\boldsymbol{x}) > (e^{-\pi/n} + n^{-8})$, which means that $f(\boldsymbol{x}) > e^{-\pi/n}$. Since at every step of the loop we made sure we were getting closer to the lattice, it must be that the point $\boldsymbol{x}$ is within distance $\sqrt{\log n}$ of the lattice, and thus by Lemma 3, $f(\boldsymbol{x}) = e^{-\pi\|\boldsymbol{x} - \boldsymbol{y}'\|} \pm 2^{-\Omega(n)}$. Since $f(\boldsymbol{x}) > e^{-\pi/n}$, it must be that $\|\boldsymbol{x} - \boldsymbol{y}\| < 1/\sqrt{n} + 2^{-\Omega(n)}$. To calculate the running time of the algorithm, we note that at every step of the loop, the value of the function f_A increases by n^{-7} with constant probability. Since the starting value of f_A is greater than $e^{-\pi\log n} - n^{-8} \approx n^{-\pi}$, the running time of the algorithm will be $O(n^5)$ multiplied by the time it takes to get a closer point $\boldsymbol{x}'$ (which is $O(n)$), multiplied by the time it takes to evaluate f_A (which is $poly(n)$).

3.1   Other ℓ_p Norms

Our algorithm can be adapted to solve α-**BDD** for the ℓ_1 norm, with $\alpha = \log n/n$. (The naive approach, using the ℓ_2 algorithm to solve the problem, works for $\alpha = \sqrt{\log n}/n$.) We believe our algorithm should also work for ℓ_p norms, $1 < p < 2$, with $\alpha = \sqrt[p]{\log n}/n$; but we are unable to give a rigorous proof in this case. When $p > 2$, however, the algorithm no longer works, and new ideas are probably necessary.

First, we recall how the technique of Aharonov and Regev [2] works. For any lattice $\mathcal{L}$, we define a periodic function $F(\boldsymbol{x}) = \sum_{\boldsymbol{y} \in \mathcal{L}} f(\boldsymbol{x} - \boldsymbol{y})$. $F(\boldsymbol{x})$ can be written in terms of its Fourier coefficients $\hat{F}(\boldsymbol{w})$, $\boldsymbol{w} \in \mathcal{L}^*$, as follows: $F(\boldsymbol{x}) = \sum_{\boldsymbol{w} \in \mathcal{L}^*} \hat{F}(\boldsymbol{w}) e^{2\pi i \langle \boldsymbol{w}, \boldsymbol{x}\rangle}$. Note that $\hat{F}(\boldsymbol{w}) = (1/\det(\mathcal{L}))\hat{f}(\boldsymbol{w})$, where $\hat{f}$ is the Fourier transform of f. Provided that the Fourier coefficients are non-negative, we can view them as a probability distribution over the dual lattice; and given some "advice" consisting of points in the dual lattice sampled according to this distribution, we can approximately compute $F(\boldsymbol{x})$.

For general ℓ_p norms, it is natural to use the function $f(\boldsymbol{x}) = e^{-\|\boldsymbol{x}\|_p^p}$. Note that f is a product of one-dimensional functions, $f(\boldsymbol{x}) = g(x_1) \cdots g(x_n)$, where $g(x) = e^{-|x|^p}$. Hence its Fourier transform is $\hat{f}(\boldsymbol{k}) = \hat{g}(k_1) \cdots \hat{g}(k_n)$. In order to use the technique of Aharonov and Regev, we would like to show that $\hat{f}(\boldsymbol{k}) \geq 0$; functions f with this property are called "positive definite" [22].

In addition, both our work and [2] make use of a technical lemma due to Banaszczyk [7] (see also [21]). We need to prove a generalization of this result for ℓ_p norms. (It is this step that determines the ratio $\alpha = \sqrt[p]{\log n}/n$.) It turns out that this requires us to show that $\hat{f}(\boldsymbol{k})$ is a non-increasing function of $\|\boldsymbol{k}\|$.

Algorithm for the ℓ_1 norm. In this case, $g(x) = e^{-|x|}$, and $\hat{g}(k) = 2/(1 + (2\pi k)^2)$. So it is easy to see that $\hat{f}(\boldsymbol{k}) \geq 0$ and is a decreasing function of $\|\boldsymbol{k}\|$. Our algorithm then works with minor modifications.

Algorithm for the ℓ_p norms with $1 < p < 2$. It is known that $\hat{g}(k) \geq 0$ [20]. Numerical investigation suggests that $\hat{g}(k)$ is a decreasing function of $|k|$, but we have not been able to prove this analytically. If this is true, then $\hat{f}(k)$ has the required properties, and our algorithm works.

What goes wrong for ℓ_p norms with $p > 2$. In this case, it is not true that $\hat{g}(k) \geq 0$ [20]. Hence $\hat{f}(k)$ can be negative for some k. We can still interpret $|\hat{F}(w)|$ as a probability distribution; however, note that $F(0) < \sum_w |\hat{F}(w)|$. Heuristic arguments suggest that when we normalize the probability distribution, $F(0)$ will be exponentially small, so our algorithm breaks down. (For instance, it is easy to see that $f(0) = 2^{-\Omega(n)} \int_{\mathbb{R}^n} |\hat{f}(k)| dk$.)

4   Hardness Results

In this section we prove the hardness of the α-**BDD** and α-**BDD** with pre-processing problems. The proofs are by reduction from a version of the closest vector problem **GapCVP**$'_\gamma$, and are based on techniques developed by Micciancio [16] to prove that the shortest vector problem is NP-hard to approximate within any constant factor smaller than $\sqrt{2}$. In fact, the $\alpha > 1/\sqrt{2}$ requirement in our proof comes from exactly the same limiting factor that makes Micciancio's proof [16] work only for approximation ratios bounded by $\sqrt{2}$. It is an interesting open question whether employing techniques used in the proof of stronger in-approximability results for **SVP** [12] it is possible to improve our NP-hardness result for α-**BDD** to $\alpha = 1/2$ or maybe even $\alpha = \Omega(1)$ or $\alpha = o(1)$.

Theorem 3. *For any ℓ_p norm $(p \geq 1)$ and $\alpha > 1/\sqrt[p]{2}$ and $\gamma > 1/\sqrt[p]{1 - \alpha^{-p}/2}$, there is a probabilistic polynomial time reduction from **GapCVP**$'_\gamma$ to α-**BDD**. Moreover, the reduction has the following two additional properties:*

1. *On input **GapCVP**$'_\gamma$ instance $(\mathbf{B}, \mathbf{y}, t)$, the reduction makes a single call to the α-**BDD** oracle on a lattice that depends only on $\mathbf{B}$ and t (but not $\mathbf{y}$).*
2. *Randomness is only used in the construction of the α-**BDD** lattice, and with high probability the constructed lattice is good for all target points $\mathbf{y}$.*

*In particular, there is a probabilistic polynomial time reduction from **GapCVP**$'_\gamma$ with pre-processing, to α-**BDD** with pre-processing such that randomness is only used during the pre-processing stage (and therefore can be equivalently be replaced by a non-uniform advice).*

Proof: Throughout the proof, $\| \cdot \|$ always denotes the ℓ_p norm. Let $(\mathbf{B}, \mathbf{y}, t)$ be a **GapCVP**$'_\gamma$ instance. We want to determine if $\mathbf{y}$ is close or far from the lattice. The idea is to use the α-**BDD** oracle to find a lattice vector close to $\mathbf{y}$. If the oracle fails or returns a vector far away from the target $\mathbf{y}$, we would like to conclude that $\mathbf{y}$ is indeed far from the lattice. The problem is that the α-**BDD** oracle is guaranteed to work only when the distance of $\mathbf{y}$ is small, relative to the minimum distance of the lattice $\lambda_1(\mathbf{B})$. Following [16], we address this problem by embedding $\mathbf{B}$ and $\boldsymbol{y}$ in a higher dimensional space, with the effect of increasing the minimum distance of the lattice, without substantially increasing the distance of the target from the lattice. The proof in [16] is based on the construction of a lattice basis $\mathbf{L}$ with some very special properties as described in the following lemma.

Lemma 8. *For any ℓ_p norm $(p \geq 1)$ and any constant $\sigma \in [1, \sqrt[p]{2})$ there exists a (probabilistic) algorithm that on input $k \in \mathbb{Z}^+$ outputs, in $k^{O(1)}$ time, two positive integers $m, r \in \mathbb{Z}^+$, a lattice basis $\mathbf{L} \in \mathbb{Z}^{(m+1) \times m}$, a vector $\boldsymbol{s} \in \mathbb{Z}^{m+1}$, and a linear integer transformation $\mathbf{T} \in \mathbb{Z}^{k \times m}$ such that*

1. $\lambda(\mathcal{L}(\mathbf{L})) > \sigma \cdot r$,
2. *with probability at least $1 - 1/n^{O(k)}$ for all $\boldsymbol{x} \in \{0,1\}^k$ there exists a $\boldsymbol{z} \in \mathbb{Z}^m$ such that $\mathbf{T}\boldsymbol{z} = \boldsymbol{x}$ and $\|\mathbf{L}\boldsymbol{z} - \boldsymbol{s}\| \leq r$.*

Informally the lemma states that lattice $\mathcal{L}(\mathbf{L})$ contains an unusually dense cluster of lattice points around the center $\boldsymbol{s}$.

Our reduction first invokes Lemma 8 with $\sigma = 1/(\alpha \sqrt[p]{1 - \gamma^{-p}})$ and $k = n$ to obtain $\mathbf{L}$, $\boldsymbol{s}$ and r. (Notice that under the assumptions $\alpha > 1/\sqrt[p]{2}$ and $\gamma > 1/\sqrt[p]{1 - \alpha^{-p}/2}$, we get $\sigma < \sqrt[p]{2}$ as required by Lemma 8.) Then, it combines $(\mathbf{B}, \boldsymbol{y})$ and $(\mathbf{L}, \boldsymbol{s})$ to define a α-**BDD** instance

$$\mathbf{B}' = \begin{bmatrix} \mathbf{BT} & 0 \\ \beta \mathbf{L} & 0 \\ 0 & \beta \sigma r \end{bmatrix} \qquad \boldsymbol{y}' = \begin{bmatrix} \boldsymbol{y} \\ \beta \boldsymbol{s} \\ 0 \end{bmatrix},$$

where β is an appropriate normalization factor to be chosen. Finally, the α-**BDD** oracle is invoked on input $(\mathbf{B}', \boldsymbol{y}')$. The **GapCVP**$'_\gamma$ instance is accepted if and only if the α-**BDD** oracle returns a lattice point $\boldsymbol{v} = [(\mathbf{BT}\boldsymbol{z})^T, (\mathbf{L}\boldsymbol{z})^T, (\beta \sigma r) \cdot z]^T$ (where $\boldsymbol{z} \in \mathbb{Z}^m$ and $z \in \mathbb{Z}$) such that $\mathbf{BT}\boldsymbol{z}$ is within distance γt from target $\boldsymbol{y}$.

We now prove that the reduction is correct. Notice that if the reduction accepts, then there is a lattice vector $\mathbf{BT}\boldsymbol{z}$ within distance γt from the target $\boldsymbol{y}$, so $(\mathbf{B}, \boldsymbol{y}, t)$ is certainly not a NO instance of **GapCVP**$'_\gamma$, and YES is a valid answer for the reduction. (Remember that when an instance does not satisfy the promise, any answer is acceptable.) All that remains to be shown is that when $(\mathbf{B}, \boldsymbol{y}, t)$ is a YES instance, then the reduction is guaranteed to accept (provided Lemma 8 was invoked successfully). So, assume the input to the reduction is a YES instance, i.e., there exists a binary vector $\boldsymbol{x} \in \{0,1\}^k$ such that $\|\mathbf{B}\boldsymbol{x} - \boldsymbol{y}\| \leq t$. First of all, we bound the minimum distance of the lattice $\mathbf{B}'$, so we can argue that the target is within the decoding radius of the α-**BDD** oracle. Considering only the second block of coordinates, we see that any lattice vector that is not a multiple of the last column has norm at least $\beta \lambda_1(\mathbf{L}) > \beta \sigma r$. It follows that the last column in the basis matrix is the (unique, up to sign change) shortest vector in the lattice and $\lambda_1(\mathbf{B}) = \beta \sigma r$. Now, consider the distance of $\boldsymbol{y}'$ from the lattice $\mathbf{B}'$. We know from Lemma 8 that there exists an integer vector $\boldsymbol{z} \in \mathbb{Z}^m$ such that $\mathbf{T}\boldsymbol{z} = \boldsymbol{x}$ and $\|\mathbf{L}\boldsymbol{z} - \boldsymbol{s}\| \leq r$. Multiplying the basis matrix $\mathbf{B}'$ by $[\boldsymbol{z}^T, 0]^T$, we get a lattice vector within distance

$$\sqrt[p]{\|\mathbf{B(Tz)} - \boldsymbol{y}\|^p + \beta^p \|\mathbf{Lz} - \boldsymbol{s}\|^p} \leq \sqrt[p]{t^p + \beta^p r^p}$$

from the target $\boldsymbol{y}'$. So, the α-**BDD** promise $\mathrm{dist}(\mathbf{B}', \boldsymbol{y}') \leq \alpha\lambda_1(\mathbf{B}') = \alpha\beta\sigma r$ is satisfied whenever

$$\sqrt[p]{t^p + \beta^p r^p} \leq \alpha\beta\sigma r. \tag{1}$$

Moreover, if (1) is satisfied, then the α-**BDD** oracle returns a vector $\boldsymbol{v} = [(\mathbf{BTz})^T, (\mathbf{Lz})^T, (\beta\sigma r) \cdot z]^T$ within distance $\alpha\lambda_1(\mathbf{B}') \leq \alpha\beta\sigma r$ from the target. Since the distance of $\boldsymbol{v}$ from $\boldsymbol{y}$ is at least $\|\mathbf{BTz} - \boldsymbol{y}\|$, we conclude that the reduction accepts, provided conditions (1) and

$$\alpha\beta\sigma r \leq \gamma t \tag{2}$$

are satisfied. Both (1) and (2) are easily verified setting $\beta = t\gamma/\alpha\sigma r$ and substituting $\sigma/(\alpha\sqrt[p]{1 - \gamma^{-p}})$. □

Since the **GapCVP**$'_\gamma$ problem is NP-hard for all constants $\gamma \geq 1$, we immediately get the following corollary:

Corollary 1. *For all $\alpha > 1/\sqrt[p]{2}$, α-**BDD** in the ℓ_p norm is NP-hard.*

In addition, the reduction in the proof of Theorem 3 only depends on $\mathbf{B}$ and t, and not on the actual vector $\boldsymbol{y}$. Thus it should be possible to use the hardness result of Alekhnovich, et. al. [4] to show the hardness of α-**BDD** with pre-processing. Two minor details arise, though. First, the reduction in [4], proves the hardness of **GapCVP**$_\gamma$ with pre-processing, not **GapCVP**$'_\gamma$ with pre-processing. But it can be extracted from the proof in [4] that the special version **GapCVP**$'_\gamma$ with pre-processing is NP-hard as well. Another point is that in the proof of [4], the algorithm is allowed to pre-process only the basis $\mathbf{B}$, while in our reduction, the algorithm would get both $\mathbf{B}$ and t for pre-processing before getting the vector $\boldsymbol{y}$. This problem can be solved by observing that in [4] the NP-hardness of **GapCVP**$_\gamma$ is proved by a reduction from the coding problem **NCPP**. As a result, there are only polynomially many values for t, and thus the advice string for **GapCVP**$_\gamma$ can contain advice for all possible values of t.

Corollary 2. *For any constant $\alpha > 1/\sqrt[p]{2}$, if there exists a polynomial time algorithm that can solve α-**BDD** with pre-processing in the ℓ_p norm , then $NP \subseteq P/poly$.*

In [19], Regev and Rosen showed reductions from problems in the ℓ_2 norm to problems in the ℓ_p norm by using the fact that for any p, there exist embedding functions $f : \mathbb{R}^n \to \mathbb{R}^m$ (where m is $poly(n)$) such that for any $\boldsymbol{x} \in \mathbb{R}^n$ $\|\boldsymbol{x}\|_2 \approx \|f(\boldsymbol{x})\|_p$. Using the same idea, we can obtain the following corollary:

Corollary 3. *For any $p > 2$ and constant $\alpha > 1/\sqrt{2}$, if there exists a polynomial time algorithm that can solve α-**BDD** with pre-processing in the ℓ_p norm, then $NP \subseteq P/poly$.*

References

1. E. Agrell, T. Eriksson, A. Vardy, and K. Zeger. Closest point search in lattices. *IEEE Trans. on Inf. Theory*, 48(8):2201–2214, 2002.
2. D. Aharonov and O. Regev. Lattice problems in NP ∩ coNP. *Journal of the ACM*, 52(5):749–765, 2005.
3. M. Ajtai, R. Kumar, and D. Sivakumar. Sampling short lattice vectors and the closest lattice vector problem. In *CCC*, pages 53–57, 2002.
4. M. Alekhnovich, S. Khot, G. Kindler, and N. Vishnoi. Hardness of approximating the closest vector problem with pre-processing. In *FOCS*, 2005.
5. S. Arora, L. Babai, J. Stern, and Z. Sweedyk. The hardness of approximate optima in lattices, codes, and systems of linear equations. *Journal of Computer and System Sciences*, 54(2):317–331, 1997.
6. L. Babai. On Lovász' lattice reduction and the nearest lattice point problem. *Combinatorica*, 6(1):1–13, 1986.
7. W. Banaszczyk. New bounds in some transference theorems in the geometry of numbers. *Mathematische Annalen*, 296(4):625–635, 1993.
8. A. H. Banihashemi and A. K. Khandani. On the complexity of decoding lattices using the Korkin-Zolotarev reduced basis. *IEEE Trans. on Inf. Theory*, 44(1):162–171, January 1998.
9. I. Dinur, G. Kindler, R. Raz, and S. Safra. Approximating CVP to within almost-polynomial factors is NP-hard. *Combinatorica*, 23(2):205–243, 2003.
10. I. Dumer, D. Micciancio, and M. Sudan. Hardness of approximating the minimum distance of a linear code. *IEEE Trans. on Inf. Theory*, 49(1):22–37, January 2003.
11. U. Feige and D. Micciancio. The inapproximability of lattice and coding problems with preprocessing. *Journal of Computer and System Sciences*, 69(1):45–67.
12. S. Khot. Hardness of approximating the shortest vector problem in lattices. In *FOCS*, pages 126–135, 2004.
13. P. Klein. Finding the closest lattice vector when it's unusually close. In *SODA*, pages 937–941, 2000.
14. J. C. Lagarias, H. W. Lenstra Jr., and C. P. Schnorr. Korkin-zolotarev bases and successive minima of a lattice and its reciprocal lattice. *Combinatorica*, 10(4):333–348, 1990.
15. D. Micciancio. The hardness of the closest vector problem with preprocessing. *IEEE Trans. on Inf. Theory*, 47(3):1212–1215, 2001.
16. D. Micciancio. The shortest vector problem is NP-hard to approximate to within some constant. *SIAM J. on Computing*, 30(6):2008–2035, 2001.
17. D. Micciancio and S. Goldwasser. *Complexity Of Lattice Problems: A Cryptographic Perspective*. Kluwer Academic Publishers, 2002.
18. O. Regev. Improved inapproximability of lattice and coding problems with pre-processing. *IEEE Trans. on Inf. Theory*, 50(9):2031–2037, 2004.
19. O. Regev and R. Rosen. Lattice problems and norm embeddings. In *STOC*, 2006.
20. I.J. Schoenberg. Metric spaces and positive definite functions. *Trans. Amer. Math. Soc.*, 44(3):522–536, 1938.
21. D. Stefankovic. Fourier transforms in computer science. Master's thesis, University of Chicago, 2000.
22. J. Stewart. Positive definite functions and generalizations, an historical survey. *Rocky Mountain J. Math*, 6(3), 1976.
23. A. Vardy. Algorithmic complexity in coding theory and the minimum distance problem. In *STOC*, pages 92–109, 1997.

Threshold Functions for Asymmetric Ramsey Properties Involving Cliques

Martin Marciniszyn[1,*], Jozef Skokan[2,**],
Reto Spöhel[1,***], and Angelika Steger[1]

[1] Institute of Theoretical Computer Science, ETH Zurich, 8092 Zurich, Switzerland
{mmarcini, rspoehel, steger}@inf.ethz.ch
[2] Instituto de Matemática e Estatística, Universidade de São Paulo,
05508-090 São Paulo, SP, Brazil
jozef@member.ams.org

Abstract. Consider the following problem: For given graphs G and $F_1, \ldots, F_k$, find a coloring of the edges of G with k colors such that G does not contain F_i in color i. For example, if every F_i is the path P_3 on 3 vertices, then we are looking for a proper k-edge-coloring of G, i.e., a coloring of the edges of G with no pair of edges of the same color incident to the same vertex.

Rödl and Ruciński studied this problem for the random graph $G_{n,p}$ in the symmetric case when k is fixed and $F_1 = \cdots = F_k = F$. They proved that such a coloring exists asymptotically almost surely (a.a.s.) provided that $p \leq bn^{-\beta}$ for some constants $b = b(F, k)$ and $\beta = \beta(F)$. Their proof was, however, non-constructive. This result is essentially best possible because for $p \geq Bn^{-\beta}$, where $B = B(F, k)$ is a large constant, such an edge-coloring does not exist. For this reason we refer to $n^{-\beta}$ as a *threshold function*.

In this paper we address the case when $F_1, \ldots, F_k$ are cliques of different sizes and propose an algorithm that a.a.s. finds a valid k-edge-coloring of $G_{n,p}$ with $p \leq bn^{-\beta}$ for some constants $b = b(F_1, \ldots, F_k, k)$ and $\beta = \beta(F_1, \ldots, F_k)$. Kohayakawa and Kreuter conjectured that $n^{-\beta(F_1, \ldots, F_k)}$ is a threshold function in this case. This algorithm can be also adjusted to produce a valid k-coloring in the symmetric case.

1   Introduction

The edge-chromatic number $\chi'(G)$ is one of the classical and well studied graph parameters. It is defined as the minimum number of colors k such that G allows for a k-edge-coloring with no pair of adjacent edges of the same color. Viewed

* The author was supported by FAPESP (Proj. Temático–ProNEx Proc. FAPESP 2003/09925-5).
** The author was supported by NSF grant INT-0305793, by NSA grant H98230-04-1-0035, and by FAPESP (Proj. Temático–ProNEx Proc. FAPESP 2003/09925-5 and Proc. FAPESP 2004/15397-4).
*** The author was supported by SNF grant 200021-108158.

J. Diaz et al. (Eds.): APPROX and RANDOM 2006, LNCS 4110, pp. 462–474, 2006.
© Springer-Verlag Berlin Heidelberg 2006

from a slightly different perspective, one can equivalently define $\chi'(G)$ as the minimum number of colors k such that G admits a k-edge-coloring avoiding monochromatic paths of length 2. This definition has led to a fruitful and well-studied area in deterministic graph theory. For given graphs G and F, is there an edge-coloring with k colors of G that avoids a monochromatic copy of F?

It follows from Ramsey's celebrated result [1] that *every* k-coloring of the edges of the complete graph on n vertices contains a monochromatic copy of F if n is sufficiently large. While this seems to rely on the fact that K_n is a very dense graph, Folkman [2] and, in a more general setting, Nešetřil and Rödl [3] showed that there also exist locally sparse graphs $G = G(F)$ with the property that every k-coloring of the edges of G contains a monochromatic copy of F. By transferring the problem into a random setting, Rödl and Ruciński [4] showed that in fact such graphs G are quite frequent. More preciwsely, they proved the following result. Let

$$G \to (F)^e_k$$

denote the property that *every* edge-coloring of G with k colors contains a monochromatic copy of F. Recall that in the binomial random graph $G_{n,p}$ on n vertices, every edge is present with probability $0 \le p = p(n) \le 1$ independently of all other edges. Then the theorem of Rödl and Ruciński reads as follows.

Theorem 1 ([4], [5], [6]). *Let $k \ge 2$ and F be a non-empty graph that is not a forest. Then there exist constants $b, B > 0$ such that*

$$\lim_{n \to \infty} \mathbb{P}\left[G_{n,p} \to (F)^e_k\right] = \begin{cases} 0 & \text{if } p \le bn^{-1/m_2(F)} \\ 1 & \text{if } p \ge Bn^{-1/m_2(F)} \end{cases},$$

where

$$m_2(F) := \max\left\{\frac{|E(H)| - 1}{|V(H)| - 2} : H \subseteq F \land |V(H)| \ge 3\right\}.$$

A function $p_0 = p_0(n)$ like the function $n^{-1/m_2(F)}$ in Theorem 1 is called threshold or threshold function. In Theorem 1, this function can be motivated as follows. For the sake of simplicity, suppose that $m_2(F) = (|E(F)| - 1)/(|V(F)| - 2)$. Then, for $p = cn^{-1/m_2(F)}$, the expected number of copies of F containing a given edge of $G_{n,p}$ is a constant depending on c. If this constant is close to zero, the copies of F in $G_{n,p}$ are loosely scattered and a valid coloring should thus exist. On the other hand, if this constant is large, the copies of F in $G_{n,p}$ highly intersect with each other, and the existence of a valid coloring becomes unlikely.

In Theorem 1 the same graph F is forbidden in every color class. We can generalize this setup by allowing for k different forbidden graphs, one per color. Within classical Ramsey theory the study of these so-called asymmetric Ramsey properties led to many interesting questions (see e.g. [7]) and results, most notably the celebrated paper of Kim [8] where he established an asymptotically sharp bound on the Ramsey number $R(3, t)$, that is, the minimum number n such that every 2-edge-coloring of the complete graph on n vertices contains either a red triangle or a blue clique of size t.

Within the random setting only very little is known about asymmetric Ramsey properties. Let

$$G \to (F_1, \ldots, F_k)^e$$

denote the property that in *every* edge-coloring of G with k colors, there exists a color i such that F_i is contained in the subgraph of G spanned by the edges which are assigned to i. In [9] Kohayakawa and Kreuter proved the following result for cycles C_ℓ of fixed length ℓ.

Theorem 2 ([9]). *Let $k \geq 2$ and $3 \leq \ell_1 \leq \cdots \leq \ell_k$ be integers. Then there exist constants $b, B > 0$ such that*

$$\lim_{n \to \infty} \mathbb{P}\left[G_{n,p} \to (C_{\ell_1}, \ldots, C_{\ell_k})^e\right] = \begin{cases} 0 & \text{if } p \leq bn^{-1/m_2(C_{\ell_2}, C_{\ell_1})} \\ 1 & \text{if } p \geq Bn^{-1/m_2(C_{\ell_2}, C_{\ell_1})} \end{cases},$$

where

$$m_2(C_{\ell_2}, C_{\ell_1}) := \frac{\ell_1}{\ell_1 - 2 + 1/m_2(C_{\ell_2})}.$$

On the basis of their results in [9], Kohayakawa and Kreuter formulated the following conjecture.

Conjecture 3 ([9]). Let F_1, F_2 be graphs with $1 < m_2(F_1) \leq m_2(F_2)$. Then there exists a constant $b > 0$ such that for all $\varepsilon > 0$, we have

$$\lim_{n \to \infty} \mathbb{P}\left[G_{n,p} \to (F_1, F_2)^e\right] = \begin{cases} 0 & \text{if } p \leq (1-\varepsilon)bn^{-1/m_2(F_1, F_2)} \\ 1 & \text{if } p \geq (1+\varepsilon)bn^{-1/m_2(F_1, F_2)} \end{cases},$$

where

$$m_2(F_1, F_2) := \max\left\{\frac{|E(H)|}{|V(H)| - 2 + 1/m_2(F_1)} : H \subseteq F_2 \wedge |V(H)| \geq 2\right\}.$$

The threshold function from Conjecture 3 is supported by the following observation. The expected number of copies of F_2 in $G_{n,p}$ with $p = \Theta\left(n^{-1/m_2(F_1, F_2)}\right)$ is

$$\Theta\left(n^{|V(F_2)|}p^{|E(F_2)|}\right) = \Omega\left(n^{2 - 1/m_2(F_1)}\right).$$

Since every edge-coloring of $G_{n,p}$ must avoid monochromatic copies of F_2 in color 2, there is at least one edge of color 1 in every subgraph of $G_{n,p}$ isomorphic to F_2. Select one such edge from each copy of F_2 arbitrarily. It is plausible that these edges span a graph G' with edge density $\Omega\left(n^{-1/m_2(F_1)}\right)$ that satisfies certain pseudo-random properties. As it turns out, that seems just about the right density in order to embed a copy of F_1 into G', no matter which edges were selected from the original graph.

In this paper, we consider the threshold function p_0 for cliques $K_{\ell_1}, \ldots, K_{\ell_k}$. A threshold phenomenon consists of two separate statements, the so-called 0-statement and the 1-statement, which are usually proved in entirely different ways. In our setting, the two statements are as follows. For the 1-statement one has to show that if $p \geq Bp_0$, a random graph $G_{n,p}$ asymptotically almost surely (a.a.s.) satisfies $G_{n,p} \to (K_{\ell_1}, \ldots, K_{\ell_k})^e$, i.e., *every* k-edge-coloring of $G_{n,p}$ contains at least one of the forbidden monochromatic cliques. For the 0-statement we suppose that $p \leq bp_0$ for some sufficiently small constant $b > 0$ and need to provide a k-edge-coloring of a random graph $G_{n,p}$ that avoids every forbidden clique K_{ℓ_i}, $1 \leq i \leq k$, in the corresponding color class i.

A standard way of attacking the 1-statement, which was also pursued in [9], is via the sparse version of Szemerédi's regularity lemma, which was independently developed by Kohayakawa [10] and Rödl (unpublished). Using properties of regularity, one can find a monochromatic copy of a forbidden subgraph in the colored graph $G_{n,p}$. Unfortunately, generalizing this argument from cycles to cliques requires a proof of Conjecture 23 in [11] of Kohayakawa, Łuczak, and Rödl. This so-called KLR-Conjecture formulates a probabilistic version of the classical embedding lemma for dense graphs. It implies many interesting extremal results for random graphs. In their monograph on random graphs [12], Janson, Łuczak, and Ruciński consider the verification of this conjecture one of the most important open questions in the theory of random graphs. Despite recent progress [13], the conjecture is, in its full generality, still wide open. However, assuming that it is true, a proof of the 1-statement is routinely obtained. We omit the proof in this extended abstract due to space restrictions.

From an algorithmic or constructive point of view, the 0-statement is much more interesting. The way of proving it that was pursued in [5] and [9] is by contradiction. This approach shows the existence of a coloring, but provides no efficient way of obtaining the coloring from the proof. Our approach is constructive. We provide a (polynomial-time) algorithm that computes a valid coloring for graphs that satisfy certain properties. We employ techniques similar to those in [5] and [9] in order to prove that these properties a.a.s. hold in $G_{n,p}$ with p sufficiently small. Indeed, the results in [5] yield that our algorithm also computes valid colorings of $G_{n,p}$ in the symmetric case, unless the forbidden graph is one of a few special cases, e.g., a triangle. In fact, the symmetric case of triangles was solved in [6] by different methods.

We prove the threshold from Conjecture 3 for cliques. As in Theorems 1 and 2, the threshold function is slightly weaker than conjectured, allowing for distinct constants in the 0- and the 1-statement.

Theorem 4 (Main Result). *Let $k \geq 2$ and $\ell_1 \geq \cdots \geq \ell_k \geq 3$ be integers. Then there exist constants $b, B > 0$ such that*

$$\lim_{n \to \infty} \mathbb{P}\left[G_{n,p} \to (K_{\ell_1}, \ldots, K_{\ell_k})^e\right] = \begin{cases} 0 & \text{if } p \leq bn^{-1/m_2(K_{\ell_2}, K_{\ell_1})} \\ 1 & \text{if } p \geq Bn^{-1/m_2(K_{\ell_2}, K_{\ell_1})} \end{cases},$$

where

$$m_2(K_{\ell_2}, K_{\ell_1}) := \frac{\binom{\ell_1}{2}}{\ell_1 - 2 + 1/m_2(K_{\ell_2})},$$

and the 1-statement holds provided Conjecture 23 in [11] is true for K_{ℓ_2}.

In this extended abstract, we will outline the proof of the 0-statement of Theorem 4 under the additional assumption that $\ell_2 > 3$. For $\ell_2 = 3$, we face additional difficulties. Due to space restrictions, we focus on the main case and sketch how to deal with triangles in Section 3.

1.1 Notation

Our notation is mostly adopted from [12]. All graphs are simple and undirected. We abbreviate the number of vertices $|V(G)|$ of a graph G by $v(G)$ and similarly the number of edges $|E(G)|$ by $e(G)$. We say that a property $\mathcal{P}$ holds in $G_{n,p}$ *asymptotically almost surely* (a.a.s.) if we have

$$\lim_{n \to \infty} \mathbb{P}\left[G_{n,p} \text{ satisfies } \mathcal{P}\right] = 1 \ .$$

2 An Algorithm for Computing Valid Edge Colorings

Suppose $G = G_{n,p}$ with $p \leq bn^{-1/m_2(K_{\ell_2}, K_{\ell_1})}$ is given. In order to provide a valid coloring of G, it suffices to compute a 2-coloring of $E(G)$ such that there is no copy of K_{ℓ_1} in color 1 and no copy of K_{ℓ_2} in color 2. That implies the 0-statement of Theorem 4 also for k-colorings. Hence, we focus on 2-colorings and abbreviate ℓ_1 by r and ℓ_2 by ℓ in the following. For the rest of this section, $r > \ell > 3$ shall remain fixed.

We describe an algorithm that finds a valid edge-coloring of G a.a.s. The basic idea of the algorithm is to remove edges from the graph successively. An edge e is deleted from G if there are no two cliques of size ℓ and r respectively that intersect exactly on e. Assuming that all edges of G can be removed in this way, it is easy to create a valid coloring by inserting them in the reverse order one by one, always assigning a valid color instantly. The actual algorithm is more complex since sometimes one has to *forget* about the existence of certain small cliques in order to remove really all edges from G. As we shall see, we can easily deal with those cliques later.

In order to simplify notation, we define, for any graph G, the families

$$\mathcal{L}_G := \{L \subseteq G : L \cong K_\ell\} \quad \text{and} \quad \mathcal{R}_G := \{R \subseteq G : R \cong K_r\}$$

of all ℓ-cliques and r-cliques in G respectively. Furthermore, we introduce the family

$$\mathcal{L}_G^* := \left\{L \in \mathcal{L}_G : \forall e \in E(L) \ \exists R \in \mathcal{R}_G \text{ s.t. } E(L) \cap E(R) = \{e\}\right\} \subseteq \mathcal{L}_G \ .$$

The algorithm is given in Figure 1. Note that edges are removed from and inserted into a working copy $G' = (V, E')$ of G. The local variable $\mathcal{L}$ contains the same elements as $\mathcal{L}_{G'}$ up to the first execution of lines 12-13. In general, we have $\mathcal{L} \subseteq \mathcal{L}_{G'}$.

Lemma 5. *If algorithm* ASYM-EDGE-COL *terminates without error, then it has indeed found a valid coloring of* G.

ASYM-EDGE-COL($G = (V, E)$)

```
 1   s ← EMPTY-STACK()
 2   E' ← E
 3   L ← L_G
 4   while E' ≠ ∅
 5   do if ∃ e ∈ E' s.t. ∄(L, R) ∈ L × R_{G'=(V,E')} : E(L) ∩ E(R) = {e}
 6      then for each L ∈ L : e ∈ E(L)
 7           do s.PUSH(L)
 8               L.REMOVE(L)
 9         s.PUSH(e)
10         E'.REMOVE(e)
11      else if ∃L ∈ L \ L*_{G'=(V,E')}
12           then s.PUSH(L)
13                L.REMOVE(L)
14      else error "stuck"
15   while s ≠ ∅
16   do if s.TOP() is an edge
17      then e ← s.POP()
18           e.SET-COLOR( blue )
19           E'.ADD(e)
20      else  L ← s.POP()
21           if L is entirely blue
22              then f ← any e ∈ E(L) s.t. ∄R ∈ R_{G'=(V,E')} : E(L) ∩ E(R) = {e}
23                   f.SET-COLOR( red )
```

Fig. 1. The implementation of algorithm ASYM-EDGE-COL

Proof. First, we argue that the algorithm never creates a blue copy of K_ℓ. Observe that *every* copy of K_ℓ that exists in G' is pushed on the stack in the first loop. Therefore, in the execution of the second loop, the algorithm must check the coloring of every such copy. Due to the order of the elements on the stack, each check is performed only after all edges of the corresponding clique were inserted and colored. For every blue copy of K_ℓ, one particular edge is recolored to red. Since red edges are never flipped back to blue, no blue copy of K_ℓ can occur in the coloring found by the algorithm.

It remains to prove that the assignment of color red to some edge by the algorithm can never create an entirely red copy of K_r. By the condition on f in line 22 of the algorithm, at the very moment there exists no copy of K_r in G' that intersects with L exactly in f. So there is either no K_r containing f at all, or every such copy contains also another edge from L. In the latter case, those copies cannot become entirely red since L is entirely blue.

Our last step is to show that the edge f in line 22 always exists. Since the second loop inserts edges into G' in the reverse order in which they were deleted during the first loop, when we select f in line 22, G' has the same structure as at the moment when L was pushed on the stack. This could have happened either in line 7, when there exists no r-clique in G' that intersects with L on some particular edge $e \in E(L)$, or in line 12, when L satisfies the condition of the if-clause in line 11. In both cases we have $L \notin \mathcal{L}_{G'}^*$, and, therefore, there exists an edge $e \in E(L)$ such that all currently existing copies of K_r do not intersect with L exactly in e. □

It remains to prove the following lemma.

Lemma 6. *There exists a positive constant* $b = b(\ell, r)$ *such that the algorithm* ASYM-EDGE-COL *a.a.s. terminates on* $G_{n,p}$ *with* $p \leq bn^{-1/m_2(K_\ell, K_r)}$ *without error.*

2.1 Proof of Lemma 6

We prove Lemma 6 by providing an algorithm GROW that, if ASYM-EDGE-COL fails on an arbitrary graph G, explicitly computes a subgraph $F \subseteq G$ which is either too large or too dense to appear in $G_{n,p}$ with p as in the lemma. More precisely, we shall show that for any graph F that GROW may return, the probability that F appears in $G_{n,p}$ is small compared to the size of $\mathcal{F}$, the class of all graphs that GROW may return. It follows that $G_{n,p}$ a.a.s. does not contain any of these graphs, which implies Lemma 6 by contradiction. Note that we employ algorithm GROW only for proving the lemma. It does not contribute to the running time of algorithm ASYM-EDGE-COL.

In order to formulate algorithm GROW, we need some definitions. Let

$$\gamma = \gamma(\ell, r) := 1/m_2(K_\ell, K_r) - 2/(\ell + r - 3) \ .$$

Note that for $r > \ell > 3$, we have

$$\gamma(\ell, r) = \frac{2\big((\ell^2 - 3\ell - 2)r - 2\ell(\ell - 3)\big)}{r(r-1)(\ell+1)(\ell+r-3)} > 0 \ .$$

Remark 7. Observe that $\gamma(3, r)$ is negative for $r \geq 3$. This is why we have to modify our proof for the case $\ell = 3$, see Section 3. The proof we present here also covers the symmetric case for $\ell = r \geq 5$ since then $\gamma(\ell, \ell) > 0$.

For any graph F, let

$$\lambda(F) := v(F) - e(F)/m_2(K_\ell, K_r) \ .$$

The definition of $\lambda(F)$ is motivated by the fact that the number of copies of F in $G_{n,p}$ with $p = bn^{-1/m_2(K_\ell, K_r)}$ has order of magnitude

$$n^{v(F)}p^{e(F)} = b^{e(F)}n^{\lambda(F)} \ .$$

For any graph F, we call an edge $e \in E(F)$ *eligible for extension* if it satisfies

$$\nexists(L, R) \in \mathcal{L}_F \times \mathcal{R}_F \text{ s.t. } E(F) \cap E(F) = \{e\} \ .$$

The implementation of algorithm GROW is shown in Figure 2. The intended input is the graph $G' \subseteq G$ after ASYM-EDGE-COL got stuck. It proceeds as

```
GROW(G' = (V, E))
1   i ← 0
2   F_0 ← any R ∈ R_{G'}
3   while i < log(n) ∧ λ(F_i) > −γ
4   do if ∃R ∈ R_{G'} \ R_{F_i} s.t. |V(R) ∩ V(F_i)| ≥ 2
5          then F_{i+1} ← F_i ∪ R
6          else e ← ELIGIBLE-EDGE(F_i)
7                 F_{i+1} ← EXTEND-L(F_i, e, G')
8       i ← i + 1
9   return F_i
```

```
EXTEND-L(F, e, G')
1   L ← any L ∈ L*_{G'} : e ∈ E(L)
2   F' ← F ∪ L
3   for each e' in E(L) \ E(F)
4   do R_{e'} ← any R ∈ R_{G'} : E(L) ∩ E(R) = {e'}
5       F' ← F' ∪ R_{e'}
6   return F'
```

Fig. 2. The implementation of algorithm GROW

follows: the seed F_0 is any copy of K_r in G'. In every iteration i, it extends F_i to F_{i+1} by adding new vertices and edges to it. As long as there are copies of K_r in G' that intersect with F_i in at least two vertices but not in all edges, it greedily adds those to F_i. If there are no such copies, it calls a subroutine ELIGIBLE-EDGE that takes F_i as input and returns an edge $e \in E(F_i)$ eligible for extension that is *unique up to isomorphism of F_i*, i.e., in such a way that for any two isomorphic graphs F and F', there exists an isomorphism φ with $\varphi(F) = F'$ such that

$$\text{ELIGIBLE-EDGE}(F') = \varphi(\text{ELIGIBLE-EDGE}(F)) \ .$$

Note that this implies in particular that e depends only on the graph F_i and not on the surrounding graph G'. Clearly, one way to implement this procedure would be keeping a large table of representatives for all isomorphism classes of graphs with up to n vertices that maps to each entry one particular edge eligible for extension. Since we only want to show the existence of a certain structure in G' and do not care about complexity issues here, the actual implementation of that procedure is irrelevant. Procedure EXTEND-L then adds a graph $L \in L*_{G'}$ that contains the edge e returned by ELIGIBLE-EDGE to F_i. It glues to each new edge $e' \in E(L) \setminus E(F_i)$ a graph $R_{e'} \in R_{G'}$ that intersects with L only in e'. The algorithm stops and returns $F_i \subseteq G' \subseteq G$ as soon as $\lambda(F_i) \leq -\gamma$ or $i \geq \log(n)$.

We argue that GROW terminates without error, i.e., that ELIGIBLE-EDGE always finds an edge eligible for extension, and that EXTEND-L always finds suitable graphs L and $R_{e'}$, $e' \in E(L)$. Let us consider the properties of G' when ASYM-EDGE-COL gets stuck. As the condition in line 5 of ASYM-EDGE-COL fails, G' is in the family

$$\mathcal{C}(\ell, r) := \{G = (V, E) : \forall e \in E(G) \ \exists (L, R) \in L_G \times R_G \text{ s.t. } E(L) \cap E(R) = \{e\}\} \ .$$

In fact, every edge of G' is contained in a copy $L \in \mathcal{L}$, and as the condition in line 11 fails as well, G' is even in the smaller family

$$\mathcal{C}^*(\ell, r) := \{G = (V, E) : \forall e \in E(G) \; \exists L \in \mathcal{L}_G^* \text{ s.t. } e \in E(L)\} \subseteq \mathcal{C}(\ell, r) \ .$$

Claim 8. *Algorithm* GROW *terminates without error on any nonempty input graph $G' \in \mathcal{C}^*(\ell, r)$. Moreover, every iteration of the while-loop adds at least one edge to F.*

Proof. Suppose there is no edge in F_i that is eligible for extension. Then we have $F_i \in \mathcal{C}(\ell, r)$ by definition. This implies that every vertex $v \in V(F_i)$ has degree at least $(\ell - 1) + (r - 1) - 1 = \ell + r - 3$, i.e., $e(F_i)/v(F_i) \geq (\ell + r - 3)/2$. It follows that

$$\lambda(F_i) \leq e(F_i) \left(\frac{2}{\ell + r - 3} - \frac{1}{m_2(K_\ell, K_r)} \right) = -e(F_i)\gamma \leq -\gamma \ ,$$

where we used that $\gamma = \gamma(\ell, r)$ is positive. Consequently, GROW terminates in line 3 without calling ELIGIBLE-EDGE. Hence, ELIGIBLE-EDGE always returns an edge eligible for extension when called from GROW.

Property $\mathcal{C}^*(\ell, r)$ of G' guarantees the existence of suitable graphs L and $R_{e'}$, $e' \in E(L)$, when EXTEND-L is called. Moreover, by definition of $\mathcal{L}_{G'}^*$, there exists, in particular, $R_e \in \mathcal{R}_{G'}$ such that e is the intersection of R_e and L. When EXTEND-L(F, e, G') is called, R_e has already been added to F during a previous iteration in lines 4 and 5 of GROW. Hence, the L returned in line 1 of EXTEND-L is not contained in F, as otherwise e would not be eligible for extension. On the other hand, it is clear that an R found in line 4 adds at least one new edge to F. Together this proves that every iteration adds at least one edge to F. $\square$

Now, we will consider the evolution of F in more detail. We say that iteration i of the while-loop in procedure GROW is *non-degenerate* if we have the following assertions:

- The condition in line 4 evaluates to false and, hence, EXTEND-L is called.
- In line 2 of EXTEND-L, we have $V(F) \cap V(L) = e$.
- In every execution of line 5 of EXTEND-L, we have $V(F') \cap V(R_{e'}) = e'$.

Otherwise, we call iteration i *degenerate*. In non-degenerate iterations, F_{i+1} is uniquely defined up to isomorphism for a given F_i, depending only on the implementation of subroutine ELIGIBLE-EDGE, which determines the position where to attach the next K_ℓ. A graph F_2 that results from two non-degenerate iterations is depicted in Figure 3 for $r = 6$ and $\ell = 4$. The little dashed circle identifies F_0. The greater dotted circle circumscribes F_1. Observe that the structures which are added in every step are isomorphic.

Claim 9. *If iteration i of the while-loop in procedure* GROW *is non-degenerate, we have*

$$\lambda(F_{i+1}) = \lambda(F_i) \ .$$

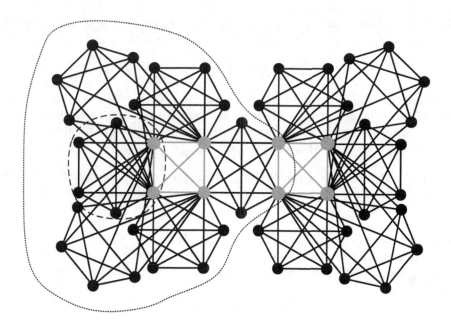

Fig. 3. A graph F_2 resulting from two non-degenerate iterations for $r = 6$ and $\ell = 4$. The two central copies of K_4 are shaded.

Proof. In a non-degenerate iteration, the L added in line 1 of EXTEND-L contributes $\ell - 2$ new vertices and $\binom{\ell}{2} - 1$ new edges to F. Each of these new edges then is replaced by a copy of K_r. Hence, we have

$$\lambda(F_{i+1}) - \lambda(F_i) = \ell - 2 + \left(\binom{\ell}{2} - 1\right)(r - 2) - \left(\binom{\ell}{2} - 1\right)\binom{r}{2}/m_2(K_\ell, K_r)$$

$$= \left(\binom{\ell}{2} - 1\right)\left(\frac{\ell - 2}{\binom{\ell}{2} - 1} + r - 2 - \left(r - 2 + \frac{1}{m_2(K_\ell)}\right)\right) = 0 .$$

$\square$

In a degenerate iteration i, the structure of F_{i+1} does not only depend on F_i, but varies with the structure of G'. Suppose that F_i is extended with an r-clique in line 5. This R can intersect with F_i in virtually every possible way. Moreover, there may be several copies of K_r which satisfy the condition in line 4. The same is true for the graphs added in lines 2 and 5 of EXTEND-L. Thus, degenerate iterations cause difficulties since they enlarge the family of graphs that algorithm GROW may potentially return. However, we will show that at most a *constant* number of degenerate iterations can occur before the algorithm terminates. This allows us to control the number of non-isomorphic graphs that can be the output of GROW. The key to proving this is the next claim.

Claim 10. *There exists a constant $\kappa = \kappa(\ell, r) > 0$ such that if iteration i of the while-loop in procedure* GROW *is degenerate, we have*

$$\lambda(F_{i+1}) \leq \lambda(F_i) - \kappa .$$

The proof of Claim 10 is the main technical part of our work and beyond the scope of this extended abstract. In combination with Claim 9, it yields the next claim, which in turn leads to a polylogarithmic bound on the number of non-isomorphic graphs that GROW can return.

Claim 11. *There exists a constant $m_0 = m_0(\ell, r)$ such that algorithm* GROW *performs at most m_0 degenerate iterations before it terminates, regardless of the input instance G'.*

Proof. An easy calculation yields that $\lambda(F_0) = \lambda(K_r) = 2 - 2/(\ell+1)$. The value of the function λ remains unchanged in every non-degenerate iteration due to Claim 9. However, Claim 10 yields a constant κ, which depends solely on ℓ and r, such that

$$\lambda(F_{i+1}) \leq \lambda(F_i) - \kappa$$

for every degenerate iteration i. Hence, after at most

$$m_0 := \frac{\lambda(F_0) + \gamma}{\kappa}$$

degenerate iterations, we have $\lambda(F_i) \leq -\gamma$, and the algorithm terminates.   □

Let $\mathcal{F}(\ell, r, n)$ denote a family of representatives for the isomorphism classes of all graphs that can be the output of GROW with parameters n and $\gamma(\ell, r)$ on *any* input instance G'.

Claim 12. *There exists $C = C(\ell, r)$ such that $|\mathcal{F}(\ell, r, n)| \leq \log(n)^C$.*

Proof. For $t \geq d \geq 0$, let $\mathcal{F}(t, d)$ denote a family of representatives for the isomorphism classes of all graphs F_t that algorithm GROW can generate after exactly t iterations if it performs exactly d degenerate iterations along the way, and let $f(t, d) := |\mathcal{F}(t, d)|$ denote its cardinality.

Observe that in every iteration, we add at most

$$K := \ell - 2 + \binom{\ell}{2}(r - 2)$$

new vertices to F, which is exactly the number of vertices added in a non-degenerate iteration. Hence, we have $v(F_t) \leq r + Kt$. It also follows that in every iteration, the new edges $E(F_{t+1}) \setminus E(F_t)$ span a graph from $\mathcal{G}_K$, where $\mathcal{G}_K$ denotes the set of all graphs on at most K vertices. F_{t+1} is uniquely defined if one specifies $G \in \mathcal{G}_K$, the number y of vertices in which G intersects F_t, and two ordered lists of vertices from G and F_t respectively of length y, which specify the mapping of the intersection vertices from G into F_t. Thus, the number of ways to extend F_t is bounded from above by

$$\sum_{G \in \mathcal{G}_K} \sum_{y=2}^{v(G)} v(G)^y (v(F_t))^y \leq C_1 (r + Kt)^K \leq t^{C_2} \leq \log(n)^{C_2} \ ,$$

where the constants C_1 and C_2 depend only on ℓ and r.

As the selection of the edge to be extended is unique up to isomorphism of F, the evolution of F is uniquely defined if there are no degenerate iterations along the way, regardless of the input instance G'. This implies in particular that $f(t, 0) = 1$ for all t, and more generally that for $t \geq d \geq 0$

$$f(t, d) \leq \binom{t}{d} \left(\log(n)^{C_2} \right)^d \leq \log(n)^{(C_2+1)d} \ .$$

Here the binomial coefficient corresponds to the choice of the d degenerate iterations. We conclude from Claim 11 that there exists a constant $C = C(\ell, r) > 0$ such that

$$|\mathcal{F}(\ell, r, n)| \leq \sum_{t=0}^{\log(n)} \sum_{d=0}^{m_0} f(t, d) \leq (\log(n) + 1)(m_0 + 1) \log(n)^{(C_2+1)m_0} \leq \log(n)^C$$

for n sufficiently large. □

Claim 13. *There exists a constant $b > 0$ such that for $p \leq bn^{-1/m_2(K_\ell, K_r)}$, $G_{n,p}$ does not contain any graph from $\mathcal{F}(\ell, r, n)$ a.a.s.*

Proof. Let $\mathcal{F}_1$ and $\mathcal{F}_2$ denote the classes of graphs that algorithm GROW can output if it terminates due to the first or the second condition in line 3, respectively. Owing to Claim 12 we have a polylogarithmic bound on the cardinality of $\mathcal{F} = \mathcal{F}(\ell, r, n) = \mathcal{F}_1 \cup \mathcal{F}_2$, and Claims 9 and 10 imply that $\lambda(F_i)$ is nonincreasing. It follows that for $b := e^{-\lambda(F_0)-\gamma}$, the expected number of copies of graphs from $\mathcal{F}$ in $G_{n,p}$ with $p \leq bn^{-1/m_2(K_\ell, K_r)}$ is bounded by

$$\sum_{F \in \mathcal{F}} n^{v(F)} p^{e(F)} = \sum_{F \in \mathcal{F}} b^{e(F)} n^{\lambda(F)} \leq \sum_{F \in \mathcal{F}_1} e^{(-\lambda(F_0)-\gamma)\log(n)} n^{\lambda(F)} + \sum_{F \in \mathcal{F}_2} n^{-\gamma}$$

$$\leq (\log(n))^C n^{-\gamma} = o(1) \ ,$$

which implies the claim due to Markov's inequality. Here we used again that γ is positive. Note that crucially, for all $F \in \mathcal{F}_1$, we have $e(F) \geq \log(n)$ since F was generated in $\lceil \log(n) \rceil$ iterations, each of which introduces at least one new edge. □

Suppose now that algorithm ASYM-EDGE-COL applied to $G_{n,p}$ with p as claimed gets stuck, and consider $G' \subseteq G$ at this moment. The call to GROW(G') returns a copy of a graph $F \in \mathcal{F}(\ell, r, n)$ that is contained in G'. But we just proved that a.a.s. we have $F \not\subseteq G_{n,p}$, which contradicts our assumption. This proves that ASYM-EDGE-COL finds a valid coloring of $G_{n,p}$ with $p \leq bn^{-1/m_2(K_\ell, K_r)}$ a.a.s.

3 Triangles

As stated in Remark 7, the proof presented in Section 2 does not cover the case $\ell = 3$ since $\gamma(3, r) = -1/(r^2 - r) < 0$. In particular, this implies that, for any $b > 0$, $G_{n,p}$ with $p = bn^{-1/m_2(K_3, K_r)}$ may contain copies of K_{r+1}. Since K_{r+1} is a member of the family $\mathcal{C}^*(3, r)$, ASYM-EDGE-COL will terminate with an error. Some rather technical work is required to show that, for $r \geq 6$, K_{r+1} is essentially the only graph in $\mathcal{C}^*(3, r)$ that is sparse enough to appear in $G_{n,p}$ and cause problems. Once this is established, it is not hard to prove that when ASYM-EDGE-COL gets stuck, G' is a.a.s. the union of edge-disjoint copies of K_{r+1} and can be colored easily. Some further complications arise in the cases $r = 4$ and $r = 5$, but the main line of argument is the same.

References

1. Ramsey, F.P.: On a problem of formal logic. Proceedings of the London Mathematical Society **30** (1930) 264–286
2. Folkman, J.: Graphs with monochromatic complete subgraphs in every edge coloring. SIAM J. Appl. Math. **18** (1970) 19–24
3. Nešetřil, J., Rödl, V.: The Ramsey property for graphs with forbidden complete subgraphs. J. Combinatorial Theory Ser. B **20** (1976) 243–249
4. Rödl, V., Ruciński, A.: Threshold functions for Ramsey properties. J. Amer. Math. Soc. **8** (1995) 917–942
5. Rödl, V., Ruciński, A.: Lower bounds on probability thresholds for Ramsey properties. In: Combinatorics, Paul Erdős is eighty, Vol. 1. Bolyai Soc. Math. Stud. János Bolyai Math. Soc., Budapest (1993) 317–346
6. Łuczak, T., Ruciński, A., Voigt, B.: Ramsey properties of random graphs. J. Combin. Theory Ser. B **56** (1992) 55–68
7. Chung, F., Graham, R.: Erdős on graphs. A K Peters Ltd., Wellesley, MA (1998) His legacy of unsolved problems.
8. Kim, J.H.: The Ramsey number $R(3, t)$ has order of magnitude $t^2 / \log t$. Random Structures Algorithms **7** (1995) 173–207
9. Kohayakawa, Y., Kreuter, B.: Threshold functions for asymmetric Ramsey properties involving cycles. Random Structures Algorithms **11** (1997) 245–276
10. Kohayakawa, Y.: Szemerédi's regularity lemma for sparse graphs. In: Foundations of computational mathematics (Rio de Janeiro, 1997). Springer, Berlin (1997) 216–230
11. Kohayakawa, Y., Łuczak, T., Rödl, V.: On K^4-free subgraphs of random graphs. Combinatorica **17** (1997) 173–213
12. Janson, S., Łuczak, T., Rucinski, A.: Random graphs. Wiley-Interscience Series in Discrete Mathematics and Optimization. Wiley-Interscience, New York (2000)
13. Gerke, S., Marciniszyn, M., Steger, A.: A probabilistic counting lemma for complete graphs. In Felsner, S., ed.: 2005 European Conference on Combinatorics, Graph Theory and Applications (EuroComb '05). Volume AE of DMTCS Proceedings., Discrete Mathematics and Theoretical Computer Science (2005) 309–316

Distance Approximation in Bounded-Degree and General Sparse Graphs

Sharon Marko[1],[*] and Dana Ron[2],[**]

[1] Department of Computer Science, Weizmann Institute of Science, Rehovot, Israel
sharon.marko@gmail.com
[2] Department of Electrical Engineering-Systems, Tel Aviv University, Tel Aviv 69978, Israel
danar@eng.tau.ac.il

Abstract. We address the problem of approximating the distance of bounded degree and general sparse graphs from having some predetermined graph property $\mathcal{P}$. Namely, we are interested in sublinear algorithms for estimating the fraction of edges that should be added to / removed from a graph so that it obtains $\mathcal{P}$. This fraction is taken with respect to a given upper bound m on the number of edges. In particular, for graphs with degree bound d over n vertices, $m = dn$. To perform such an approximation the algorithm may ask for the degree of any vertex of its choice, and may ask for the neighbors of any vertex.

The problem of estimating the distance to having a property was first explicitly addressed by Parnas et. al. (*ECCC 2004*). In the context of graphs this problem was studied by Fischer and Newman (*FOCS 2005*) in the dense-graphs model. In this model the fraction of edge modifications is taken with respect to n^2, and the algorithm may ask for the existence of an edge between any pair of vertices of its choice. Fischer and Newman showed that every graph property that has a testing algorithm in this model with query complexity that is independent of the size of the graph, also has a distance-approximation algorithm with query complexity that is independent of the size of the graph.

In this work we focus on bounded-degree and general sparse graphs, and give algorithms for all properties that were shown to have efficient testing algorithms by Goldreich and Ron (*Algorithmica, 2002*). Specifically, these properties are k-edge connectivity, subgraph-freeness (for constant size subgraphs), being a Eulerian graph, and cycle-freeness. A variant of our subgraph-freeness algorithm approximates the size of a minimum vertex cover of a graph in sublinear time. This approximation improves on a recent result of Parnas and Ron (*ECCC 2005*).

1 Introduction

Distance approximation is an extension of *Property Testing*. Property testing algorithms are required to distinguish between the case that an object (e.g., graph) has a predetermined property $\mathcal{P}$ and the case that it has a relatively large distance (i.e., greater than ϵ for a given distance parameter $\epsilon \in [0, 1]$) to having $\mathcal{P}$. Distance approximation algorithms are required to compute *an estimate of this distance* where the estimate may

---

[*] This work is part of the author's MSc thesis at the Computer Science Department, Weizmann Institute of Science, Rehovot, Israel.
[**] Research supported by a grant from the Israel Science Foundation.

J. Diaz et al. (Eds.): APPROX and RANDOM 2006, LNCS 4110, pp. 475–486, 2006.
© Springer-Verlag Berlin Heidelberg 2006

be up-to an additive error or up-to both an additive and a multiplicative error. Distance approximation and the closely related notion of *tolerant testing* (where the goal is to distinguish between being ϵ_1-close and ϵ_2-far to having the property) were first studied in [18]. Following that work, there have been several results on distance approximation, both positive [1, 10, 6] and negative [5]. These works considered properties of functions and strings [18, 1, 5, 10], ensembles of points [18], and (dense) graphs [6].

Distance Approximation in Dense Graphs.   In particular, Fischer and Newman [6] proved a general result on the relation between distance approximation and property testing in the *dense-graphs* model (introduced in [7]). In this model, the distance of a graph $G = (V, E)$ to having a property is defined as the fraction of edges that should be added/removed in order to obtain the property, where the fraction is with respect to $n^2 = |V|^2$. This model allows *vertex-pair* queries. That is, the algorithm may query whether there is an edge between any pair of vertices of its choice. Fischer and Newman [6] proved that *every* property that has a testing algorithm in the dense-graphs model whose complexity is only a function of the distance parameter ϵ, has a distance approximation algorithm with an additive error δ in this model, whose complexity is only a function of δ. The dependence on δ may be quite high (a tower of height polynomial in $1/\delta$), but there is *no* dependence on the size of the graph.

Bounded Degree and General Sparse Graphs.   The model in which Fischer and Newman obtained their result is clearly appropriate for dense graphs but not for sparse graphs. When studying property testing of sparse graphs, two models were considered (see [8] and [16]). In both models the testing algorithm may perform *degree queries* and *neighbor queries*.[1] That is, for any vertex v the algorithm may ask for the degree of v, and for any index i it may ask for the ith neighbor of v.[2] As in the dense-graphs model, the distance of a graph to having a property $\mathcal{P}$ is defined as the fraction of edges, normalized with respect to a relevant upper bound, that should be added/removed from a graph so that it obtains $\mathcal{P}$.

The difference between the models is the setting of the aforementioned upper bound. In the dense graphs model, the upper bound is n^2. When dealing with bounded-degree graphs, that is, graphs whose vertices all have degree at most d, this fraction is defined with respect to $d \cdot n$. In general, when the degree of the vertices in the graph is not bounded, then the fraction is taken with respect to the number of edges in the graph, or an upper bound on this number. We denote the (upper bound on the) number of edges by m. We assume that the algorithm is provided with m as input. Otherwise, it is possible to obtain an estimate of the number of edges [4, 9], but this comes at a cost of $\Theta(\sqrt{n})$ queries (in the case of sparse graphs).

Our Results.   Focusing on properties that have efficient testing algorithms for bounded-degree and general sparse graphs, we ask which of these also have efficient distance approximation algorithms. We establish that all properties shown to have efficient property testing algorithms in [8] also have efficient distance approximation algorithms. We leave open the interesting question of whether there exists a general transformation

---

[1] A third model, appropriate for testing properties of graphs that are neither dense nor sparse [12], also allows vertex-pair queries.

[2] If v has less than i neighbors then a special symbol is returned.

from property testing algorithms to distance approximation algorithms as in the case of dense graphs.

To state our results precisely, we introduce some notation. Recall that n denotes the number of vertices in the graph and m denotes (an upper bound on) the number of edges. In the case of bounded-degree graphs $m = dn$ where d is the maximum degree in the graph. Unless stated otherwise, the graphs we consider are not necessarily simple (i.e., they may have parallel edges), and the multiplicity of each edge is at most β. Let $\bar{d} \stackrel{\text{def}}{=} \frac{m}{n}$, so that $\bar{d}$ is roughly (an upper bound on) the average degree in the graph. For a property $\mathcal{P}$ and a graph G, we let $\Delta_{\mathcal{P}}(G)$ denote the relative *distance* of G to having the property $\mathcal{P}$. That is, $m \cdot \Delta_{\mathcal{P}}(G)$ is the minimum number of edge modification that should be performed on G so that it obtains $\mathcal{P}$. For $\alpha \geq 1$, we say that an algorithm is an α-*distance approximation algorithm* for a property $\mathcal{P}$ if, for any given $\delta \in (0, 1)$, it outputs an estimate $\widehat{\Delta}$ such that with probability at least $2/3$, $\Delta_{\mathcal{P}}(G) - \delta \leq \widehat{\Delta} \leq \alpha \cdot \Delta_{\mathcal{P}}(G) + \delta$.[3] We say that it is a *distance approximation algorithm* if it is a 1-distance approximation algorithm. Note that an α-distance approximation algorithm can be used to perform property testing simply by setting $\delta = \epsilon/2$ and accepting if and only if $\widehat{\Delta} \leq \epsilon/2$.

- k-**Edge-Connectivity.** We give a distance approximation algorithm for the k-edge-connectivity property in general sparse graphs. Its query complexity and running time are $\text{poly}(k\beta/(\delta\bar{d}))$.

- **Subgraph-Freeness.** We give a 3-distance approximation algorithm for the triangle-freeness property in bounded-degree graphs. The query and time complexity of the algorithm are $d^{O(\log(d/\delta))}$. The algorithm generalizes to subgraph-freeness for any fixed (constant size) subgraph H.

- **Eulerian.** We give a distance approximation algorithm for the Eulerian property in general sparse graphs. Its query complexity and running time are $O((\delta\bar{d})^{-4}\beta)$.

- **Cycle-Freeness.** We give a distance approximation algorithm for the cycle-freeness property in simple bounded-degree graphs. Its query complexity and running time are $O(\delta^{-3})$.

By adapting our subgraph-freeness distance-approximation algorithm we can get a sublinear approximation algorithm for the size of a minimum vertex cover. Specifically, an approximation with a multiplicative error of 2 and an additive error of δn, is achieved in time $d^{O(\log(d/\delta))}$. This algorithm improves on a recent result presented in [17] which achieve the same approximation in time $d^{O(\delta^{-3}\log d)}$.

A few notes are in place:

- With the exception of subgraph-freeness, the complexity of our algorithms is polynomially related to the corresponding complexity of the testing algorithms [8] (where our error parameter δ is replaced by the distance parameter ϵ, and $\bar{d}$ is replaced by d).

---

[3] We have chosen a non-symmetric definition in terms of the multiplicative factor α. It is of course possible to use a symmetric definition, in which case a factor C approximation according to our definition is equivalent to a factor $\sqrt{C}$ approximation in the symmetric definition. However, we find that it is less natural in our context.

- Approximating the distance to k-connectivity for the special case $k = 1$ was addressed in [3] as a central part of their algorithm for estimating the weight of a minimum spanning tree.

- Subgraph-freeness is the only result in which we have a multiplicative factor in addition to the additive error. In view of the work on dense graphs of [6], it is interesting to know whether or not the distance to subgraph-freeness can also be approximated up to any additive factor δ, using a number of queries that is independent of n.

- Testing subgraph-freeness in the *general sparse model* requires $\Omega(\sqrt{n})$ queries [2], and the same is true for cycle-freeness.

Techniques. Among the aforementioned results, the more interesting techniques are applied in distance approximation of subgraph freeness and k-connectivity.

In particular, the *testing* algorithm for subgraph freeness is a simple "brute-force" algorithm that uniformly selects vertices and checks, using a local search, whether they belong to a forbidden subgraph. The straightforward adaptation of this testing algorithm to the approximation task would give a multiplicative approximation factor that depends on the maximum degree d (in addition to the additive error δ). To get a constant factor approximation that does not depend on d, we take a different approach. Our approach can be viewed as following the paradigm (presented in [17]) for transforming local distributed algorithms into sublinear algorithms (e.g., for the minimum vertex cover).

Specifically, we first present an algorithm that inspects the whole graph, but it is essentially a local algorithm in which vertices perform local computations. We later transform this algorithm into a sublinear approximation algorithm. The (non-sublinear) algorithm is similar in spirit to the $O(\log n)$-rounds distributed approximation algorithm for the maximal independent set [13].

In the case of k-connectivity ($k > 1$), a relatively direct adaptation of the algorithm in [8] would give a multiplicative error of k (in addition to the additive error). To get a purely additive error we need to take a different approach. Specifically, we use different combinatorial representations of the connectivity structure of graphs (based on [15]), rather than those used in [8]. On top of this, we adapt and extend some of the ideas in [8]. We believe that the analysis we present for distance approximation is actually easier to follow than the analysis of the original testing algorithm.

Organization. Our result for k-connectivity is given in Section 2, and the result for subgraph freeness in Section 3. All missing proofs as well as the results for the Eulerian property and cycle freeness can be found in the full version of this paper [14].

2   Distance Approximation to k-Edge-Connectivity

In this section we consider the graph property of k-edge-connectivity. Recall that a graph is k-*edge-connected* or simply k-*connected* if there are k edge-disjoint paths between any pair of vertices in the graph. Equivalently, a graph is k-connected if the removal of any $k - 1$ edges from the graph results in a connected graph.

Goldreich and Ron [8] gave a testing algorithm for this property that works for bounded-degree graphs and runs in time[4] $\tilde{O}(k^3 \cdot \epsilon^{-3+2/k})$. They improve the running time to $\tilde{O}(1/\epsilon)$ for $k = 1, 2$ and to $\tilde{O}(\epsilon^{-2})$ for $k = 3$. Parnas and Ron [16] showed that these algorithms can be extended to the general sparse model.

We present a distance approximation algorithm for k-connectivity whose query complexity and running time are $O\left(\left(k/(\delta\bar{d}) \right)^6 \beta^{3/2} \log \left(k/(\delta\bar{d}) \right) \right)$. For the case $k = 1$, this problem was addressed by Chazelle, Rubinfeld and Trevisan [3] as part of their algorithm for approximating the weight of a minimum spanning tree of a graph. Using our terminology, they give a distance approximation algorithm for connectivity of general simple sparse graphs whose query complexity and running time are $O(1/(\delta^2 \cdot \bar{d}) \cdot \log(1/\delta))$.

The problem of approximating the distance of a graph from being k-connected for $k > 1$ is more complicated, but as we shall show, is still solvable in time that is independent of the size of the input graph. As noted earlier, the corresponding testing problem was solved by Goldreich and Ron in [8]. By trying to extend their approach to distance approximation, one can get a multiplicative factor of k, in addition to the additive factor. Here we partly build on their ideas, but use different graph structures. This allows us to obtain an additive approximation, without any multiplicative factor. Specifically, we build on the *extreme-sets tree* and the *extreme-sets partition*, introduced by Naor, Gusfield and Martel in [15] as part of their algorithm for optimally increasing the edge-connectivity of a graph. Since the algorithm in [15] works under the assumption that the graph is connected, we shall assume as well that the graph is connnected. This assumption can be easily removed, as we show in the full version of this paper [14].

2.1 Preliminaries

Let $G = (V, E)$ be a connected undirected graph. A *minimum cut* in the graph is a set of edges with minimal size whose removal from the graph disconnects it into two sets of vertices A and $\bar{A}$. We denote the cut by $(A, \bar{A})$. The *degree* of the set A, denoted by $d(A)$, is the number of edges with exactly one endpoint in A, thus it equals to the size of the cut $(A, \bar{A})$.

Definition 1. *We say that a set U is ℓ-extreme if it has degree ℓ and the degree of every proper subset of U is strictly larger than ℓ. That is, if $d(U) = \ell$ and for every $W \subset U$, $d(W) > \ell$.*

We note that extreme sets are different but related to the *connectivity classes* of the graph. An *ℓ-class* is a maximal set of vertices that cannot be disconnected by the removal of less than ℓ edges. By this definition and the definition of extreme sets, every $(\ell - 1)$-extreme set is also an ℓ-class but not vice versa since an ℓ-class might have degree greater than $\ell - 1$.

Extreme sets have the property that every two of them are either disjoint or one is contained in the other. More precisely, if U is ℓ-extreme and W is j-extreme for $\ell \geq j$ then either U and W are disjoint or $U \subseteq W$ (see [15]). This property is the key for

---

[4] The $\tilde{O}(\cdot)$ notation hides logarithmic factors, that is, $\tilde{O}(f(n))$ means $O(f(n) \cdot \text{polylog}(f(n)))$.

representing the collection of all the extreme sets of a graph in a tree called *extreme-sets tree*. The leaves of the tree are the vertices of the graph where each vertex of degree d is a d-extreme set. The parent of an extreme set U is the minimal extreme set W containing U. If U is an ℓ-extreme set and W is a j-extreme set than necessarily $\ell > j$. The root of the tree corresponds to V, which is a 0-extreme set. We shall use the notation $U \sqsubset W$ to denote that U is a child of W in the tree.

We next make a simple but important observation. Given any partition $\mathcal{P} = \{V_1, \ldots, V_q\}$ of the vertices of G, the minimum number of edges that should be added to G in order to make it connected is lower bounded by $\lceil \phi(\mathcal{P})/2 \rceil$ where $\phi(\mathcal{P})$ is the edge *demand* of the partition $\mathcal{P}$, defined by $\phi(\mathcal{P}) = \sum_{i=1}^{q} \max\{0, k - d(V_i)\}$. This is true since for every i, if $k - d(V_i) > 0$ then at least $k - d(V_i)$ endpoints of edges must be attached to vertices of V_i in order to increase the connectivity to k. Therefore, the number of edges that must be added to the graph is at least $\max_{\mathcal{P}}\{\lceil \phi(\mathcal{P})/2 \rceil\}$.

Naor, Gusfield and Martel [15] defined a partition called the *extreme-sets partition* (ES) and presented an algorithm for increasing the connectivity of G to k that adds exactly $\lceil \phi(ES)/2 \rceil$ edges. Given the aforementioned lower bound, we have that $\lceil \phi(ES)/2 \rceil$ equals m times the distance of G from k-connectivity, which is just the value that we would like to estimate. In what follows we describe this partition. For an extreme set U in the extreme-sets tree the *demand* of U is defined by $\phi(U) = \max\{0, k - d(U), \sum_{W \sqsubset U} \phi(W)\}$. Thus $\phi(U)$ is a lower bound on the number of endpoints of edges that must be attached to vertices in U in order to increase the edge-connectivity of the graph to k. The demand of the root V is defined by $\phi(V) = \sum_{U \sqsubset V} \phi(U)$. Using these notions, the *extreme-sets partition* is defined as follows.

Definition 2. *The* extreme-sets partition *of a graph G is the partition $ES = ES(G) = \{X_1, \ldots, X_q\}$ that satisfies the following conditions:*

1. *For every i, X_i is an extreme set with the property that either $\phi(X_i) = 0$ or $\phi(X_i) > \sum_{Y \sqsubset X_i} \phi(Y)$.*
2. *For every i, X_i is not contained in any other extreme set satisfying Condition 1.*

Given the extreme-sets tree, the partition ES of V can be constructed by recursively finding the partition of every child of V. Whenever an extreme set that satisfies Condition 1 from Definition 2 is found, it is added to the partition ES and the recursion ends. Since the leaves of the tree, i.e., the vertices of the graph, are all extreme sets that satisfy Condition 1, the partition ES is well defined. Observe that the demand of the partition ES satisfies $\phi(ES) = \sum_{i=1}^{q} \phi(X_i)$. In addition, this is exactly the demand of the root $\phi(V) = \sum_{U \sqsubset V} \phi(U)$ since for every $i \in \{1, \ldots, q\}$, the demand of every ancestor of X_i exactly equals the sum of the demands of its children. For an illustration of an extreme-sets tree and an extreme-sets partition see Figure 1 in the appendix.

For a graph G' and for any vertex v in G', we denote by $X_v(G')$ the set in $ES(G')$ that contains v. When $G' = G$ we use the shorthands X_v and ES, respectively.

2.2 The Algorithm

When approximating the distance to 1-connectivity the algorithm estimates, for every vertex selected, the size of its connected component. In an analogous way, in order to approximate the distance to k-connectivity, which equals to $\frac{1}{m}\lceil \phi(ES)/2 \rceil$ where

$\phi(ES) = \sum_{v \in V} \frac{\phi(X_v)}{|X_v|}$, we estimate for every vertex v the demand and the size of X_v. More precisely, since X_v may be large for some vertices, we introduce a certain refinement of ES that consists of subsets of a bounded size.

Definition 3. *Given a graph G and a size bound t, the t-bounded extreme-sets partition is the partition $ES^{(t)} = ES^{(t)}(G) = \{X_1^{(t)}, \ldots, X_q^{(t)}\}$ that satisfies the following conditions:*

1. *For every i, the size of $X_i^{(t)}$ is at most t.*
2. *For every i, $X_i^{(t)}$ is an extreme set with the property that either $\phi(X_i^{(t)}) = 0$ or*
   $$\phi(X_i^{(t)}) > \sum_{Y \sqsubset X_i^{(t)}} \phi(Y).$$
3. *For every i, $X_i^{(t)}$ is not contained in any other extreme set of size at most t satisfying Conditions 1 and 2.*

It is not hard to verify that $\phi(ES) \geq \phi(ES^{(t)})$.

For any graph G' and a vertex v in G', we denote the set in $ES^{(t)}(G')$ that contains v by $X_v^{(t)}(G')$. When $G' = G$ we use the shorthands $ES^{(t)}$ and $X_v^{(t)}$, respectively.

The following procedure searches for $X_v^{(t)}$ given a size bound t and a repetition parameter r, both of which will be set subsequently. It uses the *contraction* operation of a set A of vertices in which the vertices of A are merged into a single vertex a and for every edge (v, u) such that $v \in A$ and $u \notin A$, there is an edge between a and u.

Procedure 1 (*Extreme-set search from a given vertex v*).

1. *Repeat the following process for every $i = 1, \ldots, r$.*
   (a) *(Random Search Process) Start with $S_i = \{v\}$. As long as $|S_i| \leq t$ and the size of the cut $(S_i, \overline{S}_i)$ is less than $3t^2\beta$, assign a random cost in the range $[0, 1]$ to the edges of the cut $(S_i, \overline{S}_i)$ that were not yet assigned costs. Traverse the edge of lowest cost and add the new vertex reached to S_i.*
   (b) *(Extreme-Set Search) Let G_{S_i} be the graph obtained from G by contracting the set $\overline{S}_i$ to a single vertex $\overline{s}_i$. Construct the extreme-sets tree of G_{S_i} and let $X_v^{S_i}$ be the set $X_v(G_{S_i})$.*
2. *Let X_v^{max} be the maximal set among $\{X_v^{S_i}\}_{i=1}^r$. Declare X_v^{max} as the set in ES of G containing v i.e., as $X_v^{(t)}(G)$.*

Lemma 1. *For every v and size bound t, Procedure 1 finds $X_v^{(t)}(G)$ with probability at least $1 - e^{-2r/t^2}$. Its query complexity and running time are $O(t^4 r \beta^{3/2})$.*

Proof: To analyze the correctness of Procedure 1, assume first that at least one iteration of the random search process (Step 1.a) finds a set S that contains $X_v^{(t)}(G)$. In the following claim we establish that in this case, the procedure declares $X_v^{(t)}(G)$ as the required set.

Claim 2. *If at least one iteration of the random search process of Step 1.a finds a set S that contains $X_v^{(t)}(G)$, then Procedure 1 finds $X_v^{(t)}(G)$.*

What is left to analyze is the probability that the random search process of Step 1.a finds a set S containing $X_v^{(t)}$. To this end we lower bound the probability that all the vertices of $X_v^{(t)}$ are added to the growing set S before any other vertex is. But first we note (and it is not hard to verify) that for every $S \subseteq X_v^{(t)}$, the size of the cut $(S, \overline{S})$ is less than $3t^2\beta$. Therefore, if the algorithm adds to S only vertices from $X_v^{(t)}$, it won't stop before all the vertices of $X_v^{(t)}$ are in S.

Now, consider the graph G_X obtained from G by contracting the set $\overline{X_v^{(t)}}$ into a single vertex $\overline{x}$. Assume that the random search process of Step 1.a runs on G_X for $t' = |X_v^{(t)}|$ steps. The cut $(X_v^{(t)}, \overline{x})$ is a minimum cut of G_X since $X_v^{(t)}$ is an extreme set. Goldreich and Ron proved in [8] that in this case the probability that no cut edge is traversed before $X_v^{(t)}$ is found is at least $2t^{-2}$. Their analysis is based on Karger's analysis of his algorithm for finding minimum cut in a graph [11].

Lemma 3. *[8] For an undirected graph G, let L be a set of at most t vertices such that the cut $(L, \overline{L})$ is a minimum cut. Then, starting with some vertex $v \in L$, the random search process of Step 1.a succeeds in finding the cut $(L, \overline{L})$ with probability at least $2t^{-2}$.*

Corollary 4. *If we repeat the random search process r times, then, with probability at least $1 - (1 - 2t^{-2})^r > 1 - e^{-2t^{-2} \cdot r}$, at least one iteration finds a set containing $X_v^{(t)}$.*

Combining Corollary 4 with Claim 2, with probability at least $1 - e^{-2t^{-2}r}$, Procedure 1 finds $X_v^{(t)}$, thus proving Lemma 1.   ■

We now present the distance approximation algorithm that uses Procedure 1 to estimate the distance of a connected graph from being k-connected.

Algorithm 1 (Distance approximation to k-connectivity).

1. *Uniformly and independently sample $s = 32k^2/(\delta^2 \overline{d}^2)$ vertices from G. Let $S = \{u_1, \dots, u_s\}$ be the multiset of the sampled vertices.*

2. *For every sampled vertex u_j, run Procedure 1 using the size bound $t = 4k/\delta\overline{d}$ and the repetition constant $r = t^2 \ln(\frac{32k^2}{\delta^2 \overline{d}^2})$. Let X be the extreme set found and let $\widehat{n}_j = |X|$.*

3. *Calculate the demand of X and denote it by $\widehat{\phi}_j$.*

4. *Let $\widehat{\phi} = \frac{n}{s} \sum_{i=1}^{s} \frac{\widehat{\phi}_j}{\widehat{n}_j}$, let $\widehat{C} = \left\lceil \frac{\widehat{\phi}}{2} \right\rceil$ and output $\frac{1}{m}\widehat{C}$.*

Theorem 1. *For every $k > 1$, Algorithm 1 is a distance approximation algorithm for k-connectivity of connected graphs. The query complexity and running time of the algorithm are $O\left((k/(\delta\overline{d}))^6 \beta^{3/2} \log(k/(\delta\overline{d}))\right)$.*

As noted previously, in the full version of this paper [14] we show how to deal with the case of unconnected graphs. The only difference is a slight modification in Procedure 1.

3   Distance Approximation to Subgraph-Freeness

For a fixed graph H, we say that G is H-free if it contains no subgraph isomorphic to H. In this section we consider the problem of approximating the distance of a graph from being H-free for some fixed subgraph H in the bounded-degree model. We note that testing subgraph-freeness in the general sparse model requires $\Omega(\sqrt{n})$ queries [2].

In what follows we focus on triangles and then generalize the result to arbitrary subgraphs (in Subsection 3.2). We first present a non-sublinear algorithm for approximating the minimum number of edges that should be removed in order to obtain a triangle-free graph. Later we show how to transform it into a distance approximation algorithm whose running time is independent of n.

3.1   Triangle-Freeness

Let G be an undirected graph with degree at most d and let $m = dn$. We say that two triangles in G are *neighbors* if they share a common edge. For a triangle t, the set of its neighboring triangles is denoted by $\Gamma(t)$. The *degree* of a triangle, denoted by $d(t)$, is defined as the size of $\Gamma(t)$. The *distance* between two triangles t and t' is the minimum number of triangles minus 1 in a sequence $t_1, \ldots, t_\ell$ of triangles for which $t_1 = t$ and $t_\ell = t'$ and for every $i \in \{1, \ldots, \ell - 1\}$, the triangles t_i and t_{i+1} are neighbors. The *k-neighborhood* of a triangle t is defined as the set of triangles whose distance from t is at most k. In an analogous way, the *k-neighborhood* of a vertex v is the set of vertices whose distance from v is at most k. For a set S of edges, we say that S is a *triangle cover* if its removal from the graph results in a triangle-free graph and denote by C_{OPT} the minimum size of a triangle cover of G.

The following algorithm gets as input a graph G with degree at most d and a parameter δ, and approximates C_{OPT}.

Algorithm 2 (*Minimum triangle cover approximation*).

1. *Let T be the set of all the triangles in G and let $TC = \emptyset$ be the initial triangle cover.*
2. *From $i = 1$ to $r = \Theta(\log(d/\delta))$*
   (a) *Select each triangle $t \in T$ with probability $\frac{1}{c \cdot d(t)}$, where c is some constant that will be defined later. If $d(t) = 0$ then t is selected with probability 1.*
   (b) *Un-select every two neighboring triangles that were selected.*
   (c) *Add all the edges of the selected triangles to TC.*
   (d) *Remove from T all the selected triangles and their neighbors and update the degrees of the remaining triangles accordingly.*
3. *Add to TC one edge of every remaining triangle in T.*
4. *Output TC.*

Theorem 2. *For every δ, Algorithm 2 constructs a triangle cover TC of size C such that with probability at least $5/6$, $C_{OPT} \leq C \leq 3 \cdot C_{OPT} + \frac{\delta m}{2}$.*

Proof: First it is clear that $\mathcal{TC}$ is indeed a triangle cover and therefore $C \geq C_{OPT}$.

To show that $C \leq 3 \cdot C_{OPT} + \frac{1}{2}\delta m$ consider first the triangles that the algorithm adds to the cover during the loop of Step 2. Observe that these triangles are all edge-disjoint since whenever the algorithm selects neighboring triangles in Step 2.a, it un-selects them in Step 2.b. Also, any neighbor of a selected triangle is removed from $\mathcal{T}$ and cannot be selected on the following iterations. Therefore, any other triangle cover must contain at least one edge of every triangle from $\mathcal{TC}$ so the number of edges added to $\mathcal{TC}$ during the loop is at most $3 \cdot C_{OPT}$.

In order to upper bound the number of triangles left in $\mathcal{T}$ at the end of the loop of Step 2 we apply the following lemma.

Lemma 5. *For every* $i \in \{1, \dots, r\}$ *let* T_i *be the number of triangles left in* $\mathcal{T}$ *at the end of the i'th iteration of Step 2. For $i = 0$ let* $T_i = |\mathcal{T}|$. *Then for every $i > 0$,* $\mathrm{Exp}\,[T_i \mid T_{i-1}] \leq \left(1 - \frac{1}{c_1}\right) T_{i-1}$ *where* $c_1 = 3c^2/(c-3)$ *and c is the constant used in Step 2.a of Algorithm 2.*

The next corollary follows from Lemma 5.

Corollary 6. *By taking $c = 6$, after $r = 36(\log(\frac{d}{\delta})+3)$ iterations,* $\mathrm{Exp}[T_r] \leq \frac{\delta}{12d}|\mathcal{T}|$.

Using Markov's inequality, the probability that $T_r > \frac{\delta}{2d}|\mathcal{T}|$ is less than $1/6$. Now, since every edge belongs to at most d triangles, we have $|\mathcal{T}| \leq dm$ and so with probability at least $5/6$ the number of edges added to the cover $\mathcal{TC}$ in Step 3 is at most $\frac{1}{2}\delta m$. We conclude that in this case, the size of the cover is upper bounded by $3 \cdot C_{OPT} + \frac{1}{2}\delta m$, which completes the proof of Theorem 2. ∎

Next we show how to modify Algorithm 2 in order to achieve a 3-distance approximation algorithm for triangle-freeness whose running time is independent of n. Specifically, the algorithm uniformly and independently selects $\Theta(1/\delta^2)$ vertices and then for each triangle attached to a sampled vertex, determines whether or not it would have been added to $\mathcal{TC}$ by Algorithm 2. This can be determined by examining only the $\Theta(\log(d/\delta))$-neighborhood of every sampled vertex.

Algorithm 3 (*Distance approximation to triangle-freeness*).

1. *Uniformly and independently sample $s = 2/\delta^2$ vertices from G. Let $S = \{u_1, \dots, u_s\}$ be the multiset of the sampled vertices.*
2. *For every $j \in \{1, \dots, s\}$ observe the subgraph $G_r(u_j)$ induced by the $(r+1)$-neighborhood of u_j, where $r = \Theta(\log\frac{d}{\delta})$ is as in Algorithm 2.*
3. *Run Algorithm 2 on $\bigcup_{j=1}^{s} G_r(u_j)$. For every $u_j \in S$, let χ_j be the number of edges incident to u_j that the algorithm adds to the cover.*
4. *Let $\widehat{C} = \frac{n}{2s}\sum_{j=1}^{s}\chi_j$ and output $\frac{1}{dn}\widehat{C}$.*

Theorem 3. *Algorithm 3 is a 3-distance approximation algorithm for triangle-freeness. The query complexity and running time of the algorithm are $d^{O(\log(d/\delta))}$.*

3.2   Generalizing the Result to Arbitrary Subgraphs

The result for triangle-freeness can be generalized to arbitrary subgraphs using some subgraph specific parameters. Assume that H consists of m_H edges and its diameter is ρ_H. Also, let d_H be the maximal number of subgraphs a single edge can belong to.

Theorem 4. *For every constant size subgraph H, there exists a variant of Algorithm 3 that is an m_H-distance approximation algorithm for H-freeness. The query and time complexity of the algorithm is $d^{O(\rho_H \cdot m_H^2 \cdot \log(d_H/\delta))}$.*

3.3   A Sublinear Approximation for the Size of a Minimum Vertex Cover

As discussed in the introduction, the distance approximation algorithm for triangle-freeness can be transformed into a sublinear algorithm for approximating the size of a minimum vertex cover, where vertices replace edges and edges replace triangles.

The following theorem states the result. Its proof is very similar to the proof of Theorem 3. We remark that the same modifications of the algorithms in [17] can be applied here to achieve a dependence on $\Theta(\bar{d}/\delta)$ instead of d in the running time and query complexity.

Theorem 5. *For every $\delta > 0$ and every graph G with degree-bound d, it is possible to obtain, with probability at least $2/3$, an estimate of the size if a minimum vertex cover, which has a multiplicative error of at most 2 and an additive error of at most δn. The query complexity and running time are $d^{O(\log(d/\delta))}$.*

Acknowledgments. We would like to thank Ran Raz for many helpful discussions.

References

1. N. Ailon, B. Chazelle, S. Comandur, and D. Liue. Estimating the distance to a monotone function. In *Proceedings of the 8th RANDOM*, pages 229–236, 2004. To appear in *Random Structures and Algorithms*.
2. N. Alon, T. Kaufman, M. Krivilevich, and D. Ron. Testing triangle-freeness in general graphs. In *Proceedings of the 17th SODA*, pages 279–288, 2006.
3. B. Chazelle, R. Rubinfeld, and L. Trevisan. Approximating the minimum spanning tree weight in sublinear time. *SICOMP*, 34(6):1370–1379, 2005.
4. U. Feige. On sums of independent random variables with unbounded variance and estimating the average degree in a graph. In *Proceedings of the 36th STOC*, pages 594–603, 2004.
5. E. Fischer and L. Fortnow. Tolerant versus intolerant testing for boolean properties. In *Proceedings of the 20th IEEE Conference on Computational Complexity*, pages 135–140, 2005.
6. E. Fischer and I. Newman. Testing versus estimation of graph properties. In *Proceedings of the 37th STOC*, pages 138–146, 2005.
7. O. Goldreich, S. Goldwasser, and D. Ron. Property testing and its connection to learning and approximation. *JACM*, 45(4):653–750, 1998.
8. O. Goldreich and D. Ron. Property testing in bounded degree graphs. *Algorithmica*, 32(2):302–343, 2002.

9. O. Goldreich and D. Ron. Approximating average parameters of graphs. In these proceedings., 2006.

10. V. Guruswami and A. Rudra. Tolerant locally testable codes. In *Proceedings of the 9th RANDOM*, pages 306–317, 2005.

11. D. Karger. Global min-cuts in RNC and other ramifications of a simple mincut algorithm. In *Proceedings of the 4th SODA*, pages 21–30, 1993.

12. T. Kaufman, M. Krivelevich, and D. Ron. Tight bounds for testing bipartiteness in general graphs. *SICOMP*, 33(6):1441–1483, 2004.

13. M. Luby. A simple parallel algorithm for the maximal independent set problem. *SICOMP*, 15(2):1036–1055, 1986.

14. S. Marko and D. Ron. Distance approximation in bounded-degree and general sparse graphs. Manuscript. Available from: www.eng.tau.ac.il/~danar., 2006.

15. D. Naor, D. Gusfield, and C. Martel. A fast algorithm for optimally increasing the edge connectivity. *SICOMP*, 26(4):1139–1165, 1997.

16. M. Parnas and D. Ron. Testing the diameter of graphs. *Random Structures and Algorithms*, 20(2):165–183, 2002.

17. M. Parnas and D. Ron. On approximating the minimum vertex cover in sublinear time and the connection to distributed algorithms. ECCC Report TR05-094., 2005.

18. M. Parnas, D. Ron, and R. Rubinfeld. Tolerant property testing and distance approximation. ECCC Report TR04-010, 2004, to appear in JCSS.

A   An Example of an Extreme-Sets Partition

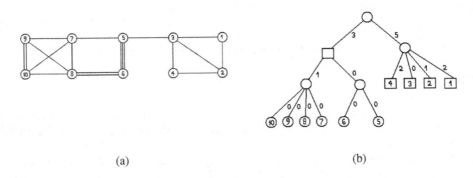

(a)   (b)

Fig. 1. (a) A graph and (b) its extreme-sets tree and extreme-sets partition. Each node represents an extreme set. The values on the edges are the demands of the corresponding extreme sets for $k = 4$. The squared nodes represent the sets of the extreme-sets partition.

Fractional Matching Via Balls-and-Bins

Rajeev Motwani[1,*], Rina Panigrahy[2,**], and Ying Xu[2,**]

Dept of Computer Science, Stanford University
{rajeev, rinap, xuying}@cs.stanford.edu

Abstract. We relate the problem of finding structures related to perfect matchings in bipartite graphs to a stochastic process similar to throwing balls into bins. We view each node on the left of a bipartite graph as having balls that it can throw into nodes on the right (bins) to which it is adjacent. We show that several simple algorithms based on throwing balls into bins deliver a near-perfect fractional matching, where a perfect fractional matching is a weighted subgraph on all nodes with nonnegative weights on edges so that the total weight incident at each node is 1.

1 Introduction

We study the problem of finding perfect (fractional) matchings in unweighted bipartite graphs using algorithms based on throwing balls into bins. While the problem of finding matchings in graphs is well-studied, as is the balls-and-bins formulation, this paper explore a novel connection between the two problems.

A *perfect matching* is a subgraph on all nodes where every node has degree exactly 1. The problem of finding perfect matchings in bipartite graphs is one fundamental problem in computing with applications in a wide variety of fields ranging from operations research, scheduling to load balancing. There are a variety of algorithms for computing maximum matchings in bipartite graphs and the related assignment problem [10, 11, 6, 5]. Most such algorithms are based on finding augmenting paths by reduction to maximum flow [7, 1]. The fastest algorithm for finding a perfect matching in a bipartite graph with m edges and n nodes runs in $O(m\sqrt{n})$ time [17].

A closely related notion is the well-known *perfect k-matching* [14], where the subgraph has non-negative integral weights on edges and the weighted degree of any node is exactly k. As k becomes large, in the limiting case this becomes a *perfect fractional matching* [14], which is a weighted subgraph on all nodes with nonnegative and possibly fractional weights on edges so that the total weight incident at each node is 1. The fractional matching (and k-matching) arises in any setting involving the matching problem where the nodes on the left side of the bipartite graph represent types and several entities of each type need to

---

* Supported in part by NSF Grants EIA-0137761 and ITR-0331640, and grants from Media-X and SNRC.
** Supported in part by a Stanford Graduate Fellowship and NSF Grants EIA-0137761 and ITR-0331640.

J. Diaz et al. (Eds.): APPROX and RANDOM 2006, LNCS 4110, pp. 487–498, 2006.
© Springer-Verlag Berlin Heidelberg 2006

be matched to the neighboring right nodes – for instance, each left node could represent several jobs of a type. Fractional matching is also meaningful if a single job may be split across multiple machines. More generally in a dynamic setting, a large number of jobs may arrive (and depart) over time, and need to be load balanced across the machines; this is equivalent to computing an online k-matching. This framework has been used to model many important applications [13, 12]. Recently [16] used this model to study Ad-Words problem where Internet queries (left nodes) on a search engine such as Google are matched to advertisements (right nodes) based on a set of *ad-words*; k-matching has been used as a crucial component in several auction design problems [20, 22]; fractional matchings were also used by Azar and Litichevskey [3] to model switch scheduling problem where packets arrive at *input ports* over time and need to be continuously matched to *output ports*.

The stochastic process of throwing balls into random bins is also a well-studied problem (see, for example, [2, 4, 21, 18]). It is well known if n balls are randomly inserted into n bins, then with high probability the bin with maximum load contains $(1 + o(1)) \log n / \log \log n$ balls; when $m > n \log n$ the maximum load is $m/n + O(\sqrt{m \log n/n})$ [21]. Azar et al [2] showed that instead of choosing one bin, if $d \geq 2$ bins are chosen at random and the ball is inserted into the least-loaded bin, the maximum load reduces dramatically to $\log \log n / \log d + O(1)$. Berenbrink et al [4] extended this result to the case when the number of balls m is greater than the number of bins n, showing that the maximum load is at most $\log \log n / \log d + O(1)$ above the average.

In this paper we relate the problem of finding structures related to perfect matchings in bipartite graphs to a variant of the balls-and-bins process. Given a bipartite graph with n nodes on each side, view each node on the left as having balls that it can throw into nodes on the right (bins) to which it is adjacent. Balls from different throwers are distinguishable; each thrower has a different color. Each ball thrown into a bin represents an edge between the thrower on the left and the bin on the right. The *load* of a bin on the right is the number of balls it receives and the load of a thrower on the left is the number of balls it throws. If each node on the left throws exactly one ball and each bin on the right gets exactly one ball, then the edges represented by the balls form a perfect matching. On the other hand, if each thrower is allowed to throw a large but equal number of balls, and each bin on the right receives an equal number of balls, then the set of ball placements corresponds to a perfect fractional matching. For example, if each node on the left throws exactly k balls and each bin on the right receives exactly k balls, then the set of ball placements corresponds to perfect k matching, where the degree of each node in the subgraph is exactly k; then, by assigning a weight $1/k$ to each instance of an edge in the subgraph, we obtain a perfect fractional matching. In this context, for a weighted subgraph, we define the *load* of a vertex to be its weighted degree.

Assuming the graph has a perfect matching, we show how to efficiently compute a *near perfect k-matching* where each node has degree close to k, or a *near-perfect fractional matching* where the load of any left vertex is 1, and load of any right vertex is at most $1 + \epsilon$. While the Hopcroft-Karp Algorithm [10] can be used to find a matching of size $n(1 - \epsilon)$, it is not clear that it can be used to find a fractional matching where every node is matched and has degree at most $1 + \epsilon$. Many of our algorithms are also applicable in an online setting. A key advantage of our matching algorithms is that most computation is local to each node, unlike other algorithms based on augmenting paths.

There are also extensive references on approximation algorithms for multi-commodity flow and generalized flow problems, see for example [8, 9, 19]. While some of those algorithms when applied to the matching problem may result in an algorithm similar to our first algorithm in spirit, we propose a much simpler framework which admits simpler and faster algorithms.

We show that simple algorithms based on throwing balls into bins find a near-perfect fractional matching. For example, consider the process where we iteratively pick a random node on the left and throw a ball into its least-loaded neighbor (a right vertex adjacent to the left node). We show that the distribution obtained from this algorithm is no worse than randomly throwing kn balls into n bins, implying that the maximum load on any node is at most $k + O(\sqrt{k \log n} + \log n)$. This gives a near perfect k-matching where the load of a node differs from k by at most $O(\sqrt{k \log n} + \log n)$, and an algorithm for finding a fractional matching in time $O(m \log n/\epsilon^2)$, where every node on the left has degree 1 and every node on the right has degree at most $1 + \epsilon$. This can also be viewed as an online algorithm for the problem of assigning jobs represented by nodes on the left to machines on the right, under the assumption of random job arrivals/departures. When the graph does not have a perfect matching, we can obtain a near-optimal fractional matching, where an optimal fractional matching is a subgraph containing all nodes in which all vertices on the left have load 1 and the maximum load on right vertices is minimized.

Another algorithm is based on the d-choice load-balancing of balls and bins. By picking a constant number of nodes on the left and inserting a ball into the least-loaded of their neighbors, we achieve a better distribution of load amongst the bins on the right — the maximum load is at most $\log \log n/\log d + O(1)$. However, this process ignores the load of vertices on the left which is the number of balls of each color. By appropriately choosing a lightly-loaded node on the left from the random choices and picking its least-loaded neighbor, we show how to find a subgraph in which the total load on both sides is exactly n and the maximum-loaded bin has load $\log \log n/\log d$. By increasing the number of choices from d to $\log n$, the maximum load can be reduced to 4.

By combining the load-balancing algorithms with the traditional augmenting path algorithms, we show how to find in time $O(m \log^2 n)$ a subgraph where every node on the left has load exactly one and every node on the right has load at most two. This can be generalized to finding a subgraph where every node has load either k, $k + 1$ or $k - 1$ in time $O(km \log n + m \log^2 n)$, implying that a near-perfect fractional matching where each right node has load within $1 \pm \epsilon$ can be computed in $O(m \log n/\epsilon)$ time.

2   Summary of Results

Given a bipartite graph with n vertices on each side, we associate a color with each vertex on the left. View the vertices on the left as *throwers* and those on the right as *bins*. Each left vertex can throw a ball of its color into any one of its *neighboring* bins. The objective is for each left vertex to throw k balls so that each bin on the right gets close to k balls. This gives a subgraph (with edge repetition) in which each left vertex has degree k and each right vertex has degree close to k; or, if we assign each edge a weight of $\frac{1}{k}$, it gives a near-perfect fractional matching. Define the *load* or the *height* of a vertex to be the number of balls in that bin; or for a weighted subgraph, the *load* of a vertex is its weighted degree. In what follows, when we say *subgraph*, it could be a weighted subgraph, or a multi-graph.

Throughout the paper, we assume that the graph has a perfect matching, unless otherwise specified.

We propose a set of "load balancing" algorithms and study their performance in Section 3.

- **Round-Robin Algorithm:** Perform iteratively for k rounds: in each round, go through the throwers in some given order, each thrower throwing its ball into the least-loaded neighboring bin.
- **Random-Color Algorithm:** Repeatedly choose a ball of a random color (or equivalently, choose a left vertex randomly) and throw into its least-loaded neighboring bin. Do this kn times.
- **Move-to-Low Algorithm:** Start with any assignment having k balls of each color. Perform iteratively: find any ball that can be moved to a bin adjacent to its color (adjacent to the vertex the ball comes from) whose load is at least 2 less than the current bin load, and move it.

We show that the distribution obtained from Random-Color is no worse than randomly throwing kn balls into n bins, implying that the maximum load of any bin is at most $k + O(\sqrt{k \log n} + \log n)$. When the graph does not have a perfect matching, this gives us a near-optimal fractional matching, where an optimal fractional matching corresponds to a subgraph where all vertices on the left have load 1 and the maximum load on a right vertex is minimized. We obtain the same upper bound for Move-to-Low. The bounds are tight for those two algorithms. The Round-Robin Algorithm has the weakest bound, which may not be tight, but the lower bound of $k + O(\sqrt{k \log n} + \log n)$ still holds. We prove that after k iterations of the Round-Robin Algorithm, the maximum height of the bins is at most $8k + O(\log n)$.

Another algorithm (see Section 4) is based on the d-choice load-balancing of balls and bins. It gives a bound of $\log \log n / \log d + O(1)$ on the maximum load of a bin, the average load being 1 on each side. By increasing the number of choices to $\Omega(\log n)$, the maximum load can be reduced to 4.

In Section 5, we combine the load-balancing algorithms with the traditional augmenting path algorithms. We show how to find in time $O(m \log n)$ a subgraph containing all the vertices where every node on the left has load exactly 1 and every node on the right has load at most 2. This can be generalized to finding a subgraph where every left node has load exactly k and every right node has load either k, $k+1$ or $k-1$ in time $O(km \log n)$.

Using Algorithm Random-Color or Move-to-Low, choosing k to be sufficiently large, we obtain in time $m \log n/\epsilon^2$ a fractional matching where every node on the left has load 1 and every node on the right has load at most $1 + \epsilon$. The augmenting path algorithms gives a near-perfect fractional matching where each right node has load within $1 \pm \epsilon$, with a better running time $O(m \log n/\epsilon)$.

3 Performance of Load Balancing Algorithms

Now we analyze the performance of the three load balancing algorithms introduced in Section 2.

3.1 Algorithm Random-Color

We show that the distribution of balls into bins obtained by Random Color algorithm is no worse than the distribution obtained when each ball is randomly thrown into a bin regardless of its color (we call it the **Pure-Random Algorithm**). It is known that if kn balls are randomly thrown into n bins, then the maximum bin size is $k + O(\sqrt{k \log n} + \log n)$ with high probability. (More precisely, when k is constant, the maximum load is $k + O(\log n/\log \log n)$ balls; when $k > \log n$ the maximum load is $k + O(\sqrt{k \log n})$ [21].) This bound is tight for Random-Color, because it is tight for Pure-Random, and when the graph has only n edges the two processes are equivalent.

The essential observation is that since there is a perfect matching, we can associate each color i with a unique bin b_i, that is, its matched neighbor in the perfect matching. Since this is a one-to-one mapping between colors and bins, picking a random color is the same as picking a random bin. When a ball of a certain color i is chosen, the algorithm always have the option to place the ball into b_i. It will end up placing the ball into a bin with height at most that of b_i. This amounts to picking a random bin b_i and placing the new ball into a bin of height at most b_i. In this sense, we are always doing better than the random process of assigning balls to random bins.

To formalize the argument, we use the notion of "majorization" and coupling (see also [4]).

Definition 1. *A* **load vector** *$u = (u_1, \ldots, u_n)$ specifies that the number of balls in the ith bin is u_i.*

Majorization: A load vector $u = (u_1, \ldots, u_n)$ is **majorized** *by a load vector v, written as $u \leq v$, if for all i, the total number of balls in the i most heavily loaded bins of u is at most that of v, that is, $\forall i, \sum_{1 \leq j \leq i} u_{\pi(j)} \leq \sum_{1 \leq j \leq i} v_{\sigma(j)}$, where π and σ are permutations of $1, \ldots, n$ such that $u_{\pi(1)} \geq u_{\pi(2)} \geq \ldots \geq u_{\pi(n)}$, and $v_{\sigma(1)} \geq v_{\sigma(2)} \geq \ldots \geq v_{\sigma(n)}$.*

The intuition is that if $u \leq v$, v can be converted into u by moving balls to lower heights. We are going to use the coupling technique to compare Random-Color and

Pure-Random. Here coupling means that the two considered stochastic processes are tied together (sharing same random bits) such that each process for itself looks exactly like the original process, but at any point of time the load vector of Random-Color is majorized by that of Pure-Random.

Theorem 1. *If the graph has a perfect matching, then there is a coupling between Random-Color and Pure-Random, such that the load vector obtained by Random-Color is majorized by the load vector obtained using Pure-Random.*

Proof. We prove by induction on the number of balls thrown. Let u be the load vector obtained after throwing some fixed number of balls using Random-Color, and v be the vector after throwing the same number of balls using Pure-Random. Since the order of bins does not matter in majorization, we can assume u is ordered, i.e. $u_1 \geq u_2 \geq \ldots \geq u_n$; so is v. By induction, we assume that $u \leq v$.

Let u' and v' denote the load vectors after throwing one more ball using the two algorithms respectively. We use the following coupling of Random-Color and Pure-Random. We choose uniformly at random a number i from 1 to n. In Random-Color, we choose the left vertex that is matched to bin i in the perfect matching, and put the ball into the least loaded bin adjacent to the vertex. In Pure-Random, we directly throw the ball into the ith bin. Note that the two bins may not be the same because bins in u and v are ordered according to the loads. It is easy to see that the probabilities for these assignments remains the same as those in the original processes. Thus coupling is well defined. We are going to prove that $u' \leq v'$.

Random-Color always put the ball in a bin with load at most u_i. Consider the load vector obtained by adding the new ball into bin i of u, denoted by u''. It is easy to see that $u' \leq u''$. We only need to show $u'' \leq v'$, which follows a known property of majorization: for any two ordered load vectors u and v, $u \leq v$ implies $u + e_i \leq v + e_i$, where e_i denotes the ith unit vector (Lemma 3.4 in [2]).

3.2   Algorithm Move-to-Low

Theorem 2. *If the graph has a perfect matching, and if algorithm Move-to-Low is allowed to run to a fixed point, the fraction of bins with load at least $k + j$ is at most $k^j \frac{k!}{(k+j)!}$; the maximum load on any bin is $O(\sqrt{k \log n} + \log n)$ above the average.*

Proof. We will compute a recurrence relation on the number of bins with load j. Observe that after equilibrium, if a ball of a certain color is in a bin of load j, all adjacent bins of that color are at least at height $j - 1$. Let p_j denote the number of bins at height j or more. The total number of balls in such bins is at least jp_j. Since there are only k balls of each color, the number of different colors in these balls is at least jp_j/k. If there is a perfect matching, these throwers (colors) have at least as many neighbors all of which have load at least $j - 1$.

So $\frac{jp_j}{k} \leq p_{j-1}$ or $p_j \leq \frac{k}{j} p_{j-1}$, or $p_{k+j} \leq \frac{k}{k+j} p_{k+j-1}$. After simplification since $p_k \leq n$, we get $p_{k+j} \leq k^j \frac{k!}{(k+j)!} n$. If $j = O(\sqrt{k \log n} + \log n)$ this becomes less than one.

The fraction $k^j \frac{k!}{(k+j)!}$ is remarkably close to the fraction of bins that receive $k + j$ balls when nk balls are randomly thrown into n bins, which is $\binom{kn}{k+j}(1/n)^{k+j}(1 - 1/n)^{nk-k-j} \approx k^{k+j}e^{-k}/(k + j)!k^j \frac{(k/e)^k}{(k+j)!}$. A tighter bound of the maximum bin size can be found along the lines of the bounds in [21] for random balls and bins.

The above analysis is tight. We can construct a subgraph where $\lceil \frac{jp_j}{k} \rceil = p_{j-1}$ and none of the balls can be moved to lower heights, by iteratively creating x new throwers for each bin of load x, and then creating x bins of load $x - 1$ for each new throwers.

The Random-Color and Move-to-Low algorithm can be generalized to graphs without perfect matchings, even with different numbers of vertices on left and right. Define the minimum expansion of a bipartite graph to be $min_{V \subset L} \frac{|N(V)|}{|V|}$. Suppose the minimum expansion of a bipartite graph G is c ($c \leq 1$; equality holds only when G has a perfect matching), then the distribution of bin heights obtained by Random-Color is no worse than that obtained by Pure-Random on cn bins. This is because the probability that a ball falls into one of the x heaviest loaded bins in Random Color is at most x/cn, while in Pure-Random, this probability is exactly x/cn. By a similar coupling argument as Theorem 1, the load vector of Random-Color is majorized by that of Pure-Random. The analysis for Move-to-Low can be generalized similarly. The detailed proof can be found in the complete version of our paper [15].

3.3   Algorithm Round-Robin

Theorem 3. *If the graph has a perfect matching, then after one iteration of algorithm Round-Robin, the maximum height of bins is at most $\log n + 1$. After k iterations, the maximum height of bins is at most $8k + O(\log n)$.*

We refer the readers to the complete version of our paper [15] for the proof.

4   Using d-Choice Load Balancing

The above algorithms produce a maximum load of at least $O(\log n/ \log \log n)$ with a total of n balls. We now show how ideas based on d-choice hashing can be used to find a better distribution. It is known that while throwing n balls into n bins if each ball picks $d \geq 2$ bins at random and is inserted into the least loaded of the d bins, then the maximum load of any bin is at most $\log \log n/ \log d + O(1)$ with high probability. We now show an algorithm that produces a subset of the edges (allowing repetitions) of the bipartite graph so that the total degree on each side is n and the maximum degree is $\log \log n/ \log d + O(1)$ in time $O(md)$ with high probability. In particular for $d = \Omega(\log n)$ the maximum load is 4.

Intuitively - if we pick $d \geq 2$ random vertices on the left side and throw a ball into the least loaded vertex among all their neighbors then this is similar to the d-choice load balancing. The color of the ball corresponds to one of the d vertices whose neighbor is chosen (breaking ties arbitrarily). If this is done n times the maximum load of any bin on the right side is at most $\log\log n/\log d + O(1)$. However, there is no guarantee of a low load on the vertices on the left; that is the maximum number of balls of a certain color. To remedy this we modify the algorithm slightly: instead of picking d vertices on the left we pick $2d-1$ and consider the d of these that have lowest loads. Then we look at the set of their neighbors on the right and add a ball into the least loaded bin. Again the color of the ball is the same as the vertex whose neighbor is chosen. Essentially if the $2d-1$ chosen left vertices have loads $u_1, u_2, .., u_{2d-1}$ and their matched neighbors in a perfect matching have loads $v_1, v_2, .., v_{2d-1}$, then we find an edge with load on the left at most the median value of u_i's on load and the right at most the median value of v_i's. This is like throwing n balls into n bins where each ball picks $2d-1$ bins at random and is inserted into one of the d least loaded of the $2d-1$ chosen bins.

Theorem 4. *If the graph has a perfect matching, and if n balls are inserted by the above process, we get a subset of the edges (allowing repetitions so it is a multiset) of the bipartite graph so that the total degree on each side is n and the maximum degree is $\log\log n/\log d + O(1)$ with high probability.*

Theorem 5. *If the graph has a perfect matching, and if n balls are inserted by above process except that we choose $d = \Omega(\log n)$ nodes for each ball, then we get a subset of the edges (allowing repetitions so it is a multiset) of the bipartite graph so that the total degree on each side is n and the maximum degree is 4 with high probability.*

Lemma 1. *Assume the graph has a perfect matching. If n balls are thrown into n bins where each ball picks $2d-1$ bins at random and is inserted arbitrarily into one of the d least loaded of the $2d-1$ chosen bins, then the maximum load of any bin is at most $\log\log n/\log d + O(1)$ with high probability. This is true even if instead of inserting a ball in one of the d least loaded of the $2d-1$ chosen bins, it is inserted into any other bin whose load does not exceed the median load of the $2d-1$ chosen bins.*

Proof. We use the layered induction technique similar to the one used to bound the load of bins in the two choice hashing [2]. Let p_i denote an upper bound on the fraction of bins with load at least i at the end of the process. We will derive a recurrence relation on p_i. Let us compute the probability that during an insertion the new ball lands at a height of $i+1$ or higher. For this to happen at least d of the bins chosen must have height i or greater; otherwise each of the d least loaded of the $2d-1$ bins has load less than i. Probability that this happens is at most $\binom{2d-1}{d}p_i^d \le 2^{2d}p_i^d = (4p_i)^d$. So the expected number of balls that fall at height $i+1$ or higher is at most $(4p_i)^d n$. This is also a bound on the expected number of bins with load at least $i+1$. So given p_i, the expected value of p_{i+1} is at most $(4p_i)^d$. This can be converted into a high probability bound – probability at least $1 - 1/n^c$ – by using Chernoff bounds as long as $(4p_i)^d n \ge c \log n$ where c is a large enough constant.

This implies, given p_i, if $(4p_i)^d n \ge c \log n$ then with high probability $p_{i+1} \le (4p_i)^d$, which implies $16p_{i+1} \le (16p_i)^d$. Now, $p_{32} \le 1/32$ as at most $1/32$ fraction of the bins can have load 32 or higher. Using the recurrence we get $p_{i+32} \leq 1/2^{d^i}$ as long as $1/2^{d^i} \geq c \log n/n$. Let k denote the highest value of $i + 32$ for which this holds; $k = \log \log n / \log d + c'$ where c' is some constant (possibly negative). Now we know that $(4p_k)^d \leq c \log n/n$. We will show that with high probability there is no bin with load $k + 2$. Again Chernoff bounds can be used to show that with high probability $p_{k+1} \leq 2c \log n/n$. Now we argue that the probability that a bin has load $k + 2$ is at most $n(8c \log n/n)^d$ using a simple union bound on the probability that any of the ball thrown lands at height $k + 2$. So the probability that there is a ball at height $32 + \log \log n / \log d + c' + 2$ is at most $n(8c \log n/n)^d + n/n^c \leq 1/n^{\Omega(1)}$. The conditioning arguments can be made more precise along the lines of the proof in [2]: Essentially let E_i denote the event that the fraction of bins with height at least $i + 32$ is at most the bound p_{i+32} computed above. Then E_0 is true; probability that E_i holds and E_{i+1} does not is negligible which recursively implies that each E_i holds with high probability except for the sum of the negligible probabilities.

Given Lemma 1 it is easy to argue that the loads of the vertices of the left side and right side grow according to the process described. For the right side vertices, since $2d - 1$ vertices are chosen at random from the left side, this is equivalent to choosing $2d - 1$ vertices at random on the right side in the optimal matching. After discarding $d - 1$ of them we pick a vertex with load that is no more than that of the least loaded of the d remaining right side bins. So again the arguments in proof of Lemma 1 go through; the only difference is instead of placing the ball in one of the d least loaded of the $2d - 1$ chosen bins, it is placed into some bin with possibly lower load but no higher.

Theorem 5 can also be proven similarly: The fraction of nodes on any side with load at least 3 is at most $1/3$ at any time. If $d = \Omega(\log n)$ nodes are chosen at random then, with high probability $(1 - 1/n^2)$, the median load is at most 3. Since we can always find an edge whose load is at most the median load on both the left and the right side, we can with high probability, add a ball along an edge so that after the addition the loads are at most 4.

5 Combining Load Balancing with Augmenting Paths

Classical perfect matching algorithms for bipartite graphs leverage max flow algorithms based on augmenting paths. Now we combine the above load balancing algorithms with augmenting paths, and get efficient algorithms to find near-perfect fractional matchings.

Definition 2. *An s-almost matching in a bipartite graph G is a multi-set of edges in G, such that the load of any left vertex in the matching is exactly s, while the load of any right vertex is in $[s - 1, s + 1]$.*

In this section, we show an algorithm finding a k-almost matching in time $O(m \log^2 n + mk \log n)$.

We start with 1-almost matchings. Recall that the *residual graph* [7] with respect to a subgraph is obtained by first directing all edges from left to right and then flipping the directions of all edges in the subgraph; *an augmenting path is a simple path in the residual graph*. Given a subgraph where the load of any left vertex is 1, its residual graph has the following property.

Lemma 2. *For every right vertex u of height $h \geq 3$, there exists an augmenting path of length at most $2 \log n$ from u to some right vertex v of height at most $h - 2$.*

Proof. Let d be the length of the shortest such augmenting path; N_p be the number of bins reachable from u within $2p$ steps ($2p < d$). The total number of balls in those bins is at least $(h - 1)N_p$, otherwise we have founded a bin of height $h - 2$ or less. Those balls correspond to $(h - 1)N_p$ distinct left vertices, the size of whose neighborhood is at least $(h - 1)N_p$ because there exists a perfect matching. This whole neighborhood is reachable from u in another 2 steps. Therefore, $N_{p+1} \geq (h-1)N_p \geq (h-1)^{p+1}$. For $h \geq 3$, $(h-1)^{d/2+1} \geq 2^{d/2}$. Because the total number of balls in the system is n, we have $2^{d/2} \leq n$, or $d \leq 2 \log n$.

Moving one ball from u along the augmenting path to v decreases the height of v by one. Given any subgraph, we can eliminate all highest bins by such moves until the maximum height is 2. In the end, the load of any left vertex in the subgraph remains 1, and the load of any right vertex is at most 2. As in the well known Edmonds-Karp algorithm (see Ch26.2 in [7]), augmenting paths are selected such that the lengths of augmenting paths increase with each phase. A phase consists of augmentations along a maximal set of edge-disjoint augmenting paths, which can be implemented by one breadth-first search. Thus we get a 1-almost matching.

1-almost matching Algorithm:
1. Run one round of Round-Robin Algorithm: throw n balls and get a maximum bin size of $\log n + 1$;
2. As long as the maximum bin height h is more than 2, use the Edmonds-Karp algorithm to eliminate bins of height h.

Edmonds-Karp algorithm:
1. Let any bin of height h be a source; any bin of height $h' \leq h - 2$ be a sink;
2. while there exists a source do
2-1. by performing a breadth-first search, find the length of the shortest augmenting path from any source to any sink; denoted this length by d;
2-2. find a maximal set of edge-disjoint source-sink paths of length d, under the constraint that any sink of height h' can appear in at most $h - h' - 1$ paths. For each augmenting path selected, flip the directions of the edges, i.e. perform a sequence of ball moves that decreases the height of source bin by 1 and increases the height of the sink by 1.

The Edmonds-Karp algorithm is based on the following property: if we always augment along the shortest path, then the shortest distance from source to any node in the residual graph increases monotonically. See Ch26.2 in [7] for the proof. This is summarized in the following lemma.

Lemma 3. *In the Edmonds-Karp algorithm, the length of the shortest augmenting path d increases monotonically. If a maximal set of augmenting paths are chosen in each phase, then d must strictly increase in each phase.*

We analyze the time complexity of the above algorithm. Using the standard implementation of Edmonds-Karp algorithm, each phase takes time $O(m)$: use a breadth-first search to decide the shortest distance d, which takes time m; then use depth-first search to find a maximal set of edge-disjoint augmenting paths of length d, which takes time $O(m)$ (see Ch26 in [7] for details).

To reduce the maximum bin height by at least 1, we need to run Edmonds-Karp algorithm for at most $2\log n$ phases as the length of the augmenting path strictly increases in each phase, and by Lemma 2 we need to look at augmenting paths of length at most $2\log n$. This needs to be done at most $\log n + 1$ times because the maximum bin height after one round of Round-Robin is at most $\log n + 1$. Since each phase takes time $O(m)$, the total running time is $O(m\log^2 n)$.

Theorem 6. *If the graph has a perfect matching, then a 1-almost matching can be found in time $O(m\log^2 n)$.*

We can use the above algorithm to find a k-almost matching in $O(mk\log n)$ time, where the load of each left vertex in the matching is k, and the load of each right vertex is between $k \pm 1$. The idea is to recursively double the balls in each bin, and move the balls by finding augmenting paths of length at most $k\log n$. We start with $k = 1$ and get a 1-almost matching using the above algorithm. Assume we have a $\frac{k}{2}$-almost matching. Now double each edge, and we get a subgraph where each left vertex has load k and any right vertex has load $k \pm 2$. To convert the matching into a k-almost matching, all we need to do is to eliminate bins of height $k + 2$ and $k - 2$ by finding augmenting paths to bins of height k. We can bound the length of the shortest augmenting path similar to Lemma 2. Please refer to [15] for the complete proof.

Theorem 7. *If the graph has a perfect matching, then a k-almost matching can be found in time $O(m\log^2 n + mk\log n)$.*

References

1. R.K. Ahuja, T.L. Magnanti, and J.B. Orlin. *Network Flows: Theory, Algorithms, and Applications.* Prentice-Hall, 1993.
2. Y. Azar, A.Z. Broder, A.R. Karlin, and E. Upfal. "Balanced allocations." *SIAM Journal on Computing*, 29:180–200, 1999.
3. Y. Azar and A. Litichevskey. "Maximizing throughput in multi-queue switches." In *Proceedings of 12th ESA Conference,* 2004.
4. P. Berenbrink, A. Czumaj, A. Steger, and B. Vöcking. "Balanced allocations: The heavily loaded case." In *Proceedings of the 32nd Annual ACM Symposium on Theory of Computing (STOC)*, pp. 745–754, 2000.
5. D.P. Bertsekas. "The Auction Algorithm: A Distributed Relaxation Method for the Assignment Problem." *Annals of Operations Research*, 14:105–123, 1988.

6. B.V. Cherkassky, A.V. Goldberg, P. Martin, J.C. Setubal, and J. Stolfi. "Augment or push: a computational study of bipartite matching and unit-capacity flow algorithms." *ACM J. Exp. Algorithmics*, 3, 1998.

7. T. Cormen, C. Leiserson, R. Rivest and C. Stein. *Introduction to Algorithms*. Second Edition. MIT Press, 2001.

8. N. Garg and J. Könemann. "Faster and Simpler Algorithms for Multicommodity Flow and other Fractional Packing Problems." In *Proceedings of the 39th Annual IEEE Symposium on Foundations of Computer Science*, pp.300-309, 1998.

9. A.V.Goldberg. "A natural randomization strategy for multicommodity flow and related algorithms." In *Information Processing Letters*, 42(5):249-256, 1992.

10. J. Hopcroft and R. Karp. "An $n^{5/2}$ algorithm for maximum matchings in bipartite graphs." *SIAM Journal on Computing*, 2:225-231, 1973.

11. H.W. Kuhn. "The Hungarian method for the assignment problem." *Naval Res. Logist. Quart.*, 2:83-97, 1955.

12. B. Kalyanasundaram and K.R. Pruhs. "An optimal deterministic algorithm for online b-matching." *Theoretical Computer Science*, 233:319-325, 2000.

13. R.M. Karp, U.V. Vazirani, and V.V. Vazirani. "An optimal algorithm for online bipartite matching." In *Proceedings of the 22nd Annual ACM Symposium on Theory of Computing*, 1990.

14. L. Lovasz and M.D. Plummer. *Matching Theory*. Annals of Discrete Mathematics. North Holland, 1986.

15. R. Motwani, R. Panigrahy and Y. Xu. "Fraction Matching via Balls-and-Bins." Technical Report, 2005.

16. A. Mehta, A. Saberi, U.V. Vazirani and V.V. Vazirani. "AdWords and Generalized On-line Matching." In *Proceedings of the 46th Annual IEEE Symposium on Foundations of Computer Science*, 2005.

17. S. Micali and V. Vazirani. "An $O(E\sqrt{V})$ algorithm for finding maximum matchings in general graphs." In *Proceedings of the 21st IEEE Symposium on the Foundations of Computer Science*, pp. 17-27, 1980.

18. R. Panigrahy. "Efficient Hashing with Lookups in Two Memory Accesses." In *Proceedings of SODA*, 2005.

19. S.A. Plotkin, D. Shmoys, and E. Tardos. "Fast approximation algorithms for fractional packing and covering problems". In *Proceedings of the 32nd Annual IEEE Symposium on the Foundations of Computer Science*, 1991.

20. M. Penn and M. Tennenholtz. "Constrained multi-object auctions and b-matching." *Information Processing Letters*, 75:29-34, 2000.

21. M. Raab and A. Steger. "Balls into bins – a simple and tight analysis." In *Proceedings of the 2nd International Workshop on Randomization and Approximation Techniques in Computer Science*, LNCS volume 1518, pp. 159-170, 1998.

22. M. Tennenholtz. "Tractable combinatorial auctions and b-matching." *Artif. Intell.*, 140:231-243, 2002.

A Randomized Solver for Linear Systems with Exponential Convergence

Thomas Strohmer and Roman Vershynin*

Department of Mathematics, University of California, Davis, CA 95616-8633, USA
strohmer@math.ucdavis.edu, vershynin@math.ucdavis.edu

Abstract. The Kaczmarz method for solving linear systems of equations $Ax = b$ is an iterative algorithm that has found many applications ranging from computer tomography to digital signal processing. Despite the popularity of this method, useful theoretical estimates for its rate of convergence are still scarce. We introduce a randomized version of the Kaczmarz method for overdetermined linear systems and we prove that it converges with expected exponential rate. Furthermore, this is the first solver whose *rate does not depend on the number of equations* in the system. The solver does not even need to know the whole system, but only its small random part. It thus outperforms all previously known methods on extremely overdetermined systems. Even for moderately overdetermined systems, numerical simulations reveal that our algorithm can converge faster than the celebrated conjugate gradient algorithm.

1 Introduction and State of the Art

We study a consistent linear system of equations

$$Ax = b, \tag{1}$$

where A is a full rank $m \times n$ matrix with $m \geq n$, and $b \in \mathbb{C}^m$. One of the most popular solvers for such overdetermined systems is *Kaczmarz's method* [12], which is a form of alternating projection method. This method is also known under the name *Algebraic Reconstruction Technique* (ART) in computer tomography [9, 13], and in fact, it was implemented in the very first medical scanner [11]. It can also be considered as a special case of the POCS (Projection onto Convex Sets) method, which is a prominent tool in signal and image processing [15, 1].

We denote the rows of A by $a_1^*, \ldots, a_m^*$, where $a_1, \ldots, a_m \in \mathbb{C}^n$, and let $b = (b_1, \ldots, b_m)^T$. The classical scheme of Kaczmarz's method sweeps through the rows of A in a cyclic manner, projecting in each substep the last iterate orthogonally onto the solution hyperplane of $\langle a_i, x \rangle = b_i$ and taking this as the next iterate. Given some initial approximation x_0, the algorithm takes the form

$$x_{k+1} = x_k + \frac{b_i - \langle a_i, x_k \rangle}{\|a_i\|} a_i, \tag{2}$$

* T.S. was supported by the NSF DMS grant 0511461. R.V. was supported by Alfred P. Sloan Foundation and by the NSF DMS grant 0401032.

J. Diaz et al. (Eds.): APPROX and RANDOM 2006, LNCS 4110, pp. 499–507, 2006.

where $i = k \mod m + 1$. Note that we refer to one projection as one iteration, thus one sweep in (2) through all m rows of A consists of m iterations. We will refer to this as *one cycle*.

While conditions for convergence of this method are readily established, useful theoretical estimates of the *rate of convergence* of the Kaczmarz method (or more generally of the alternating projection method for linear subspaces) are difficult to obtain, at least for $m > 2$. Known estimates for the rate of convergence are based on quantities of the matrix A that are hard to compute and difficult to compare with convergence estimates of other iterative methods (see e.g. [2, 3, 6] and the references therein). What numerical analysts would like to have is estimates of the convergence rate with respect to standard quantities such as $\|A\|$ and $\|A^{-1}\|$. The difficulty that no such estimates are known so far stems from the fact that the rate of convergence of (2) depends strongly on the *ordering* of the equations in (1), while quantities such as $\|A\|$, $\|A^{-1}\|$ are independent of the ordering of the rows of A.

It has been observed several times in the literature that using the rows of A in Kaczmarz's method in random order, rather than in their given order, can greatly improve the rate of convergence, see e.g. [13, 1, 10]. While this randomized Kaczmarz method is thus quite appealing for applications, no guarantees of its rate of convergence have been known.

In this paper, we propose the first randomized Kaczmarz method with *exponential expected rate of convergence*, cf. Section 2. Furthermore, this *rate does not depend on the number of equations* in the system. The solver does not even need to know the whole system, but only its small random part. Thus our solver outperforms all previously known methods on extremely overdetermined systems. We analyze the optimality of the proposed algorithm as well as of the derived estimate, cf. Section 3. Our numerical simulations demonstrate that even for moderately overdetermined systems, this random Kaczmarz method can outperform the celebrated conjugate gradient algorithm, see Section 4.

Notation: For a matrix A, $\|A\| := \|A\|_2$ denotes the spectral norm of A, $\|A\|_F$ is the Frobenious norm, i.e. the square root of the trace of A^*A, where the superscript * stands for the conjugate transpose of a vector or matrix. The left inverse of A (which we always assume to exist) is written as A^{-1}. Thus $\|A^{-1}\|$ is the smallest constant M such that the inequality $\|Ax\| \geq \frac{1}{M}\|x\|$ holds for all vectors x. As usual, $\kappa(A) := \|A\|\|A^{-1}\|$ is the condition number of A. The linear subspace spanned by a vector x is written as $\mathrm{lin}(x)$. Finally, $\mathbb{E}$ denotes expectation.

2   Randomized Kaczmarz Algorithm and Its Rate of Convergence

It has been observed [13, 1, 10] that the convergence rate of the Kaczmarz method can be significantly improved when the algorithm (2) sweeps through the rows of A in a random manner, rather than sequentially in the given order. Here we propose a specific version of this randomized Kaczmarz method, which chooses each row of A with probability proportional to its relevance – more precisely, proportional to the square of its Euclidean norm. This method of sampling from a matrix was proposed in [5] in the context of computing a low-rank approximation of A, see also [14] for subsequent work and references. Our algorithm thus takes the following form:

Algorithm 1 (Random Kaczmarz algorithm). *Let $Ax = b$ be a linear system of equations as in (1) and let x_0 be arbitrary initial approximation to the solution of (1). For $k = 0, 1, \ldots$ compute*

$$x_{k+1} = x_k + \frac{b_{r(i)} - \langle a_{r(i)}, x_k \rangle}{\|a_{r(i)}\|} a_{r(i)}, \tag{3}$$

where $r(i)$ is chosen from the set $\{1, 2, \ldots, m\}$ at random, with probability proportional to $\|a_{r(i)}\|^2$.

Our main result states that x_k converges exponentially fast to the solution of (1), and the rate of convergence depends *only* on the norms of the matrix and its inverse.

Theorem 2. *Let x be the solution of (1). Then Algorithm 1 converges to x in expectation, with the average error*

$$\mathbb{E}\|x_k - x\|^2 \leq \left(1 - \frac{1}{R}\right)^k \cdot \|x_0 - x\|^2, \tag{4}$$

where $R = \|A^{-1}\|^2 \, \|A\|_F^2$.

Proof. There holds

$$\sum_{j=1}^{m} |\langle z, a_j \rangle|^2 \geq \frac{\|z\|^2}{\|A^{-1}\|^2} \qquad \text{for all } z \in \mathbb{C}^n. \tag{5}$$

Using the fact that $\|A\|_F^2 = \sum_{j=1}^{m} \|a_j\|^2$ we can write (5) as

$$\sum_{j=1}^{m} \frac{\|a_j\|^2}{\|A\|_F^2} \left| \left\langle z, \frac{a_j}{\|a_j\|} \right\rangle \right|^2 \geq \frac{1}{R} \|z\|^2 \quad \text{for all } z \in \mathbb{C}^n. \tag{6}$$

The main point in the proof is to view the left hand side in (6) as an expectation of some random variable. Namely, recall that the solution space of the j-th equation of (1) is the hyperplane $\{y : \langle y, a_j \rangle = b\}$, whose normal is $\frac{a_j}{\|a_j\|}$. Define a random vector Z whose values are the normals to all the equations of (1), with probabilities as in our algorithm:

$$Z = \frac{a_j}{\|a_j\|} \quad \text{with probability} \quad \frac{\|a_j\|^2}{\|A\|_F^2}, \quad j = 1, \ldots, m. \tag{7}$$

Then (6) says that

$$\mathbb{E}|\langle z, Z \rangle|^2 \geq \frac{1}{R} \|z\|^2 \quad \text{for all } z \in \mathbb{C}^n. \tag{8}$$

The orthogonal projection P onto the solution space of a random equation of (1) is given by $Pz = z - \langle z - x, Z \rangle Z$.

Now we are ready to analyze our algorithm. We want to show that the error $\|x_k - x\|^2$ reduces at each step in average (conditioned on the previous steps) by at least the factor of $(1 - \frac{1}{R})$. The next approximation x_k is computed from x_{k-1} as $x_k = P_k x_{k-1}$, where $P_1, P_2, \ldots$ are independent realizations of the random projection P. The vector $x_{k-1} - x_k$ is in the kernel of P_k. It is orthogonal to the solution space of the equation onto which P_k projects, which contains the vector $x_k - x$ (recall that x is the solution to all equations). The orthogonality of these two vectors then yields

$$\|x_k - x\|^2 = \|x_{k-1} - x\|^2 - \|x_{k-1} - x_k\|^2.$$

To complete the proof, we have to bound $\|x_{k-1} - x_k\|^2$ from below. By the definition of x_k, we have

$$\|x_{k-1} - x_k\| = \langle x_{k-1} - x, Z_k \rangle$$

where $Z_1, Z_2, \ldots$ are independent realizations of the random vector Z. Thus

$$\|x_k - x\|^2 \leq \left(1 - \left|\left\langle \frac{x_{k-1} - x}{\|x_{k-1} - x\|}, Z_k \right\rangle\right|^2\right) \|x_{k-1} - x\|^2.$$

Now we take the expectation of both sides conditional upon the choice of the random vectors $Z_1, \ldots, Z_{k-1}$ (hence we fix the choice of the random projections $P_1, \ldots, P_{k-1}$ and thus the random vectors $x_1, \ldots, x_{k-1}$). Then

$$\mathbb{E}|_{\{Z_1,\ldots,Z_{k-1}\}}\|x_k - x\|^2 \leq \left(1 - \mathbb{E}_{\{Z_1,\ldots,Z_{k-1}\}}\left|\left\langle \frac{x_{k-1} - x}{\|x_{k-1} - x\|}, Z_k \right\rangle\right|^2\right) \|x_{k-1} - x\|^2.$$

By (8) and the independence,

$$\mathbb{E}|_{\{Z_1,\ldots,Z_{k-1}\}}\|x_k - x\|^2 \leq \left(1 - \frac{1}{R}\right) \|x_{k-1} - x\|^2.$$

Taking the full expectation of both sides, by induction we complete the proof. □

Remark (Dimension-free perspective, robustness). The rate of convergence in Theorem 2 does not depend on the number of equations nor the number of variables, and obviously also not on the order of the projections. It is only controlled by the intrinsic and stable quantity R of the matrix A. This continues the dimension free approach to operators on finite dimensional normed spaces, see [14].

2.1   Quadratic Time

Let n denote the number of variables in (1). Clearly, $n \leq R \leq \kappa(A)^2 n$, where $\kappa(A)$ is the condition number of A. Then as $k \to \infty$,

$$\mathbb{E}\|x_k - x\|^2 \leq \exp\left([1 - o(1)]\frac{k}{\kappa(A)^2 n}\right) \cdot \|x_0 - x\|^2. \tag{9}$$

Thus the algorithm converges exponentially fast to the solution in $O(n)$ iterations (projections). Each projection can be computed in $O(n)$ time; thus the algorithm takes $O(n^2)$ operations to converge to the solution. This should be compared to the Gaussian elimination, which takes $O(mn^2)$ time. (Strassen's algorithm and its improvements reduce the exponent in Gaussian elimination, but these algorithms are, as of now, of no practical use). Of course, we have to know the (approximate) Euclidean lengths of the rows of A before we start iterating; computing them takes $O(nm)$ time. But the lengths of the rows may in many cases be known a priori. For example, all of them may be equal to one (as is the case for Vandermonde matrices arising in trigonometric approximation) or be tightly concentrated around a constant value (as is the case for Gaussian random matrices).

The number m of equations is essentially irrelevant for our algorithm, as seen from (9). The algorithm does not even need to know the whole matrix, but only $O(n)$ random rows. Such Monte-Carlo methods have been successfully developed for many problems, even with precisely the same model of selecting a random submatrix of A (proportional to the squares of the lengths of the rows), see [5] for the original discovery and [14] for subsequent work and references.

3   Optimality

We discuss conditions under which our algorithm is optimal in a certain sense, as well as the optimality of the estimate on the expected rate of convergence.

3.1   General Lower Estimate

For any system of linear equations, our estimate can not be improved beyond a constant factor of R, as shown by the following theorem.

Theorem 3. *Consider the linear system of equations* (1) *and let* x *be its solution. Then there exists an initial approximation* x_0 *such that*

$$\mathbb{E}\|x_k - x\|^2 \geq \left(1 - \frac{2k}{R}\right) \cdot \|x_0 - x\|^2 \tag{10}$$

for all k, *where* $R = R(A) = \|A^{-1}\|^2 \|A\|_F^2$.

Proof. For this proof we can assume without loss of generality that the system (1) is homogeneous: $Ax = 0$. Let x_0 be a vector which realizes R, that is $R = \|A^{-1}x_0\|^2 \|A\|_F^2$ and $\|x_0\| = 1$. As in the proof of Theorem 2, we define the random normal Z associated with the rows of A by (7). Similar to (8), we have $\mathbb{E}|\langle x_0, Z\rangle|^2 = 1/R$. We thus see $\mathrm{lin}(x_0)$ as an "exceptional" direction, so we shall decompose $\mathbb{R}^n = \mathrm{lin}(x_0) \oplus (x_0)^\perp$, writing every vector $x \in \mathbb{R}^n$ as

$$x = x' \cdot x_0 + x'', \quad \text{where} \quad x' \in \mathbb{R}, \quad x'' \in (x_0)^\perp.$$

In particular,

$$\mathbb{E}|Z'|^2 = 1/R. \tag{11}$$

We shall first analyze the effect of one random projection in our algorithm. To this end, let $x \in \mathbb{R}^n$, $\|x\| \leq 1$, and let $z \in \mathbb{R}^n$, $\|z\| = 1$. (Later, x will be the running approximation x_{k-1}, and z will be the random normal Z). The projection of x onto the hyperplane whose normal is z equals

$$x_1 = x - \langle x, z \rangle z.$$

Since

$$\langle x, z \rangle = x'z' + \langle x'', z'' \rangle, \tag{12}$$

we have

$$|x_1' - x'| = |\langle x, z \rangle z'| \leq |x'||z'|^2 + |\langle x'', z'' \rangle z'| \leq |z'|^2 + |\langle x'', z'' \rangle z'| \tag{13}$$

because $|x'| \leq \|x\| \leq 1$. Next,

$$\|x_1''\|^2 - \|x''\|^2 = \|x'' - \langle x, z \rangle z''\|^2 - \|x''\|^2$$
$$= -2\langle x, z \rangle \langle x'', z'' \rangle + \langle x, z \rangle^2 \|z''\|^2 \leq -2\langle x, z \rangle \langle x'', z'' \rangle + \langle x, z \rangle^2$$

because $\|z''\| \leq \|z\| = 1$. Using (12), we decompose $\langle x, z \rangle$ as $a+b$, where $a = x'z'$ and $b = \langle x'', z'' \rangle$ and use the identity $-2(a+b)b + (a+b)^2 = a^2 - b^2$ to conclude that

$$\|x_1''\|^2 - \|x''\|^2 \leq |x'|^2|z'|^2 - \langle x'', z'' \rangle^2 \leq |z'|^2 - \langle x'', z'' \rangle^2 \tag{14}$$

because $|x'| \leq \|x\| \leq 1$.

Now we apply (13) and (14) to the running approximation $x = x_{k-1}$ and the next approximation $x_1 = x_k$ obtained with a random $z = Z_k$. Denoting $p_k = \langle x_k'', Z_k'' \rangle$, we have by (13) that $|x_k' - x_{k-1}'| \leq |Z_k'|^2 + |p_k Z_k'|$ and by (14) that $\|x_k''\|^2 - \|x_{k-1}''\|^2 \leq |Z_k'|^2 - |p_k|^2$. Since $x_0' = 1$ and $x_0'' = 0$, we have

$$|x_k' - 1| \leq \sum_{j=1}^{k} |x_j' - x_{j-1}'| \leq \sum_{j=1}^{k} |Z_j'|^2 + \sum_{j=1}^{k} |p_j Z_j'| \tag{15}$$

and

$$\|x_k''\|^2 = \sum_{j=1}^{k} \left(\|x_j''\|^2 - \|x_{j-1}''\|^2 \right) \leq \sum_{j=1}^{k} |Z_j'|^2 - \sum_{j=1}^{k} |p_j|^2.$$

Since $\|x_k''\|^2 \geq 0$, we conclude that $\sum_{j=1}^{k} |p_j|^2 \leq \sum_{j=1}^{k} |Z_j'|^2$. Using this, we apply Cauchy-Schwartz inequality in (15) to obtain

$$|x_k' - 1| \leq \sum_{j=1}^{k} |Z_j'|^2 + \left(\sum_{j=1}^{k} |Z_j'|^2 \right)^{1/2} \left(\sum_{j=1}^{k} |Z_j'|^2 \right)^{1/2} = 2 \sum_{j=1}^{k} |Z_j'|^2.$$

Since all Z_j are copies of the random vector Z, we conclude by (11) that $\mathbb{E}|x_k' - 1| \leq 2k\mathbb{E}|Z'|^2 \leq \frac{2k}{R}$. Thus $\mathbb{E}\|x_k\| \geq \mathbb{E}|x_k'| \geq 1 - \frac{2k}{R}$. This proves the theorem, actually with the stronger conclusion

$$\mathbb{E}\|x_k - x\| \geq \left(1 - \frac{2k}{R} \right) \cdot \|x_0 - x\|.$$

(the actual conclusion follows by Jensen's inequality). $\qquad\square$

3.2   The Upper Estimate Is Attained

If $\kappa(A) = 1$ then the estimate in Theorem 2 becomes an equality. This follows directly from the proof of Theorem 2.

Furthermore, there exist arbitrarily large systems and with arbitrarily large $\kappa(A)$ for which the estimate in Theorem 2 becomes an equality. More precisely, let n and $m \geq n$, $R \geq n$ be arbitrary positive numbers such that $\frac{1}{R}m$ is an integer. Then there exists a system (1) of m equations in n variables and with $R(A) = R$ for which the estimate in Theorem 2 becomes an equality.

To see this, we define the matrix A with the help of any orthogonal set $e_1, \ldots, e_n$ in $\mathbb{R}^n$. Let the first $\frac{1}{R}m$ rows of A be equal to e_1, the other rows of A be equal to one of the vectors e_j, $j > 1$, so that every vector from this set repeats at least $\frac{1}{R}m$ times as a row (this is possible because $R \geq n$). Then $R(A) = R$ (note that (5) is attained for $z = e_1$).

Let us test our algorithm on the system $Ax = 0$ with the initial approximation $x_0 = e_1$ to the solution $x = 0$. Every step of the algorithm brings the running approximation to 0 with probability $\frac{1}{R}$ (the probability of picking the row of A equal to e_1 in uniform sampling), and leaves the running approximation unchanged with probability $1 - \frac{1}{R}$. By the independence, for all k

$$\mathbb{E}\|x_k - x_0\|^2 = \left(1 - \frac{1}{R}\right)^k \cdot \|x_0 - x\|^2.$$

4   Numerical Experiments and Comparisons

In recent years conjugate gradient (CG) type methods have emerged as the leading iterative algorithms for solving large linear systems of equations, since they often exhibit remarkably fast convergence. How does the proposed random Kaczmarz method compare to CG algorithms?

It is not surprising, that one can easily construct examples for which CG (or its variations, such as CGLS or LSQR [8]) will clearly outperform the proposed method. For instance, take a matrix whose singular values, all but one, are equal to one, while the remaining singular value is ε, a number close to zero, say 10^{-8}. It follows from well known properties of the CG method (cf. [16]) that CGLS will converge in two steps, while the proposed Kaczmarz method will converge extremely slow, since $R \approx \varepsilon^{-2}$ and thus $1 - \frac{1}{R} \approx 1$ in this example.

On the other hand, the proposed algorithm outperforms CGLS in cases for which CGLS is actually quite well suited. We consider a Gaussian random matrix with $m \geq n$. While one iteration of CG requires $\mathcal{O}(mn)$ operations, one iteration (i.e., one projection) of Kaczmarz takes $\mathcal{O}(n)$ operations. Thus a cycle of m Kaczmarz iterations corresponds to one iteration of CG. Therefore, for a fair comparison, in the following we will compare the number of iteration cycles (1 iteration cycle for CGLS equals one standard CGLS iteration, and 1 iteration cycle for Kaczmarz equals m random projections). We let $m = 400, n = 100$ and construct 1000 random matrices. For each of them we run CGLS and the random Kaczmarz method described in Algorithm 1 (which does not require any preprocessing in this case since all rows of A have approximately the same norm). The resulting average rate of convergence for both methods is displayed in Figure 1.

Somewhat surprisingly, Algorithm 1 gives faster convergence than CGLS. Classical results about Gaussian random matrices [7, 4], combined with convergence estimates for the CG algorithm [8] and a little algebra yield that the (expected) convergence rate of CG for Gaussian $m \times n$ matrices is governed by $(\sqrt{\frac{n}{m}})^k$. Whereas for Algorithm 1 the expected convergence rate is bounded by $(1 - \frac{(\sqrt{m}-\sqrt{n})^2}{mn})^{\frac{k}{2}}$ which is inferior to the value computed for CG. Yet, numerical experiments clearly demonstrate the better performance of Algorithm 1. We will give a more thorough discussion of this performance gain compared to its theoretical prediction elsewhere.

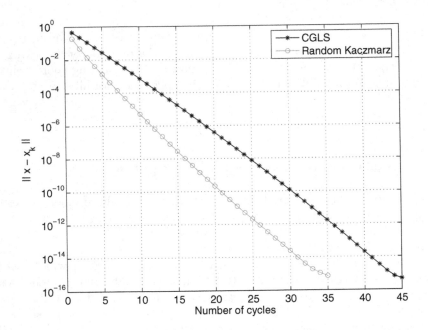

Fig. 1. Comparison of rate of convergence for the random Kaczmarz method described in Algorithm 1 and the conjugate gradient least squares algorithm

References

1. C. Cenker, H. G. Feichtinger, M. Mayer, H. Steier, and T. Strohmer. New variants of the POCS method using affine subspaces of finite codimension, with applications to irregular sampling. In *Proc. SPIE: Visual Communications and Image Processing*, pages 299–310, 1992.
2. F. Deutsch. Rate of convergence of the method of alternating projections. In *Parametric optimization and approximation (Oberwolfach, 1983)*, volume 72 of *Internat. Schriftenreihe Numer. Math.*, pages 96–107. Birkhäuser, Basel, 1985.

3. F. Deutsch and H. Hundal. The rate of convergence for the method of alternating projections. II. *J. Math. Anal. Appl.*, 205(2):381–405, 1997.
4. A. Edelman. Eigenvalues and condition numbers of random matrices. *SIAM J. Matrix Anal. Appl.*, 9(4):543–560, 1988.
5. A. Frieze, R. Kannan and S. Vempala, *Fast Monte-Carlo Algorithms for finding low-rank approximations*, Proceedings of the Foundations of Computer Science, 1998, pp. 378–390, journal version in Journal of the ACM 51 (2004), 1025-1041
6. A. Galántai. On the rate of convergence of the alternating projection method in finite dimensional spaces. *J. Math. Anal. Appl.*, 310(1):30–44, 2005.
7. S. Geman. A limit theorem for the norm of random matrices. *Ann. Probab.*, 8(2):252–261, 1980.
8. G.H. Golub and C.F. van Loan. *Matrix Computations.* Johns Hopkins, Baltimore, third edition, 1996.
9. G.T. Herman. *Image reconstruction from projections.* Academic Press Inc. [Harcourt Brace Jovanovich Publishers], New York, 1980. The fundamentals of computerized tomography, Computer Science and Applied Mathematics.
10. G.T. Herman and L.B. Meyer. Algebraic reconstruction techniques can be made computationally efficient. *IEEE Transactions on Medical Imaging*, 12(3):600–609, 1993.
11. G.N. Hounsfield. Computerized transverse axial scanning (tomography): Part I. description of the system. *British J. Radiol.*, 46:1016–1022, 1973.
12. S. Kaczmarz. Angenäherte Auflösung von Systemen linearer Gleichungen. *Bull. Internat. Acad. Polon.Sci. Lettres A*, pages 335–357, 1937.
13. F. Natterer. *The Mathematics of Computerized Tomography.* Wiley, New York, 1986.
14. M. Rudelson and R. Vershynin. Sampling from large matrices: an approach through geometric functional analysis, 2006. preprint.
15. K.M. Sezan and H. Stark. Applications of convex projection theory to image recovery in tomography and related areas. In H. Stark, editor, *Image Recovery: Theory and application*, pages 415–462. Acad. Press, 1987.
16. A. van der Sluis and H.A. van der Vorst. The rate of convergence of conjugate gradients. *Numer. Math.*, 48:543–560, 1986.

Maintaining External Memory
Efficient Hash Tables
(Extended Abstract)*

Philipp Woelfel

Univ. of Toronto, Dept. of Computer Science, Toronto, ON M5S3G4
philipp.woelfel@utoronto.ca

Abstract. In typical applications of hashing algorithms the amount of
data to be stored is often too large to fit into internal memory. In this case
it is desirable to find the data with as few as possible non-consecutive or
at least non-oblivious probes into external memory. Extending a static
scheme of Pagh [11] we obtain new randomized algorithms for maintaining hash tables, where a hash function can be evaluated in constant time
and by probing only one external memory cell or $O(1)$ consecutive external memory cells. We describe a dynamic version of Pagh's hashing
scheme achieving 100% table utilization but requiring $(2 + \epsilon) \cdot n \log n$
space for the hash function encoding as well as $(3 + \epsilon) \cdot n \log n$ space for
the auxiliary data structure. Update operations are possible in expected
constant amortized time. Then we show how to reduce the space for the
hash function encoding and the auxiliary data structure to $O(n \log \log n)$.
We achieve 100% utilization in the static version (and thus a minimal
perfect hash function) and $1 - \epsilon$ utilization in the dynamic case.

1 Introduction

In this paper, we devise randomized algorithms for efficiently maintaining hash
tables under circumstances typical for applications dealing with massive data.
Consider a set S of n keys from a finite universe U and assume that each key
$x \in S$ has some data D_x associated with it. A static dictionary for S supports
a query operation which returns the data D_x for a given key x. A dynamic
dictionary also support update operations which allow to insert new data into
the dictionary or to remove data from it. For many applications it is especially
important to be able to retrieve the data D_x as quickly as possible (examples
are databases used by web-servers, where a huge amount of queries have to be
answered in short time). A typical solution is to maintain a hash function h
mapping each key $x \in S$ to an entry of a hash table T. Such a hash function h
is called *perfect for* S if it is injective on S. If h has range $[n] := \{0, \dots, n-1\}$,
$n = |S|$, then h is called *minimal perfect*. If h is perfect on S, the data associated
with each key in S can be stored in $T[h(x)]$. We call such an implementation of
a dictionary a *hash table implementation based on perfect hashing*. An algorithm

---

* The research was supported by DFG grant Wo 1232/1-1.

J. Diaz et al. (Eds.): APPROX and RANDOM 2006, LNCS 4110, pp. 508–519, 2006.

dynamically maintaining a perfect hash function is called *stable* if $h(x)$ remains fixed for the duration that x is in S.

The following assumptions are typical for many dictionary and hashing applications: Firstly, efficiency is much more a concern for lookups than for update operations. For example, in database backends of webservers a huge number of queries have to be answered momentarily while updates of the database only rarely occur or sometimes even can be delayed to times of low load.

Secondly, the data set is so massive that even the description of the hash function does not fit into the internal memory. For example, the encoding of a minimal perfect hash function requires at least $\Omega(n) + \log \log |U| - O(\log n)$ bits, assuming that $|U| \geq n^{2+\epsilon}$ [8]. In this case, just in order to evaluate the hash function we have to access external memory. But then usually the number of non-consecutive accesses to external memory dominate the evaluation time of our hash function.

Finally, the data D_x associated with a key x requires much more space than its key x. Therefore it is especially important that a hash table implementation achieves a high utilization, since we have to reserve a fixed amount of space for each table entry (if we want to avoid another level of indirection). Assuming that the hash table is implemented by an extendible array $T[0], T[1], \ldots$, its utilization is given as $|S|/(\max\{h(S)\} + 1)$. In particular, even a small constant utilization seems infeasible, and a utilization as close to 100% as possible should be achieved. A minimal perfect hash function for the set S achieves 100% utilization.

Although general dictionary implementations (not necessarily based on perfect hashing) can be used to maintain minimal perfect hash functions by associating each key $x \in S$ with a unique value from $[n]$, such solutions require another level of indirection.

Previous and Related Work. Throughout this paper we assume that $|U| = n^{O(1)}$. It is well-known how to reduce the size of the universe by choosing a random hash function $\zeta : U \to [n^{O(1)}]$ such that ζ is injective on S with high probability. Moreover, we assume that the size of the internal memory is bounded by n^ϵ, $\epsilon < 1$.

Fredman, Komlós, and Szemerédi [9] were the first to devise an algorithm which constructs a perfect hash function (with $O(n \log n)$ bits) in expected linear time such that the hash function can be evaluated in constant time. The utilization is less than 0.2 in the case that only consecutive probes into external memory are allowed for hash function evaluation. Dietzfelbinger, Karlin, Mehlhorn, Meyer auf der Heide, Rohnert, and Tarjan [5] have devised a dynamic version of that scheme with essentially the same parameters, but which also supports updates in expected amortized constant time. Later improvements have either focused on reducing the space requirements or on obtaining a constant update time even with high probability. All schemes which do in fact achieve a constant update time with high probability are mostly of complexity theoretical interest (as opposed to practical). Demaine, Meyer auf der Heide, Pagh, and Pǎtraşcu [1] show an upper space bound of essentially $O(n \log \log(u/n) + n \log n - n \log t)$ for maintaining a perfect hash function with range $[n+t]$ and $O(n \log \log(u/n) + n \cdot r)$

for a dynamic dictionary where the data associated with each key comprises r bits. Update operations are supported with high probability in constant time and the algorithm is stable. For the static case, Hagerup and Tholey [10] hold the space record: They show how to construct a minimal perfect hash function in expected $O(n + \log\log|U|)$ time such that its encoding requires only almost optimal $(1 + o(1))(n \cdot \log e + \log\log|U|)$ space. Multiple non-oblivious probes into external memory are required for lookups in these space efficient dynamic or static schemes.

Dictionary algorithms such as Cuckoo-Hashing [12] and its extensions [7, 6] also allow the retrieval of data with few non-consecutive probes into external memory. Especially space and external memory efficient is the Cuckoo-Hashing variant of Dietzfelbinger and Weidling [6], where two hash functions h_1 and h_2 and two hash tables T_1 and T_2 are used. A table position consists of d consecutive memory cells, and the data D_x is stored in one of the $2 \cdot d$ memory cells from $T_1[h_1(x)]$ and $T_2[h_2(x)]$. For $d \geq 90 \cdot \ln(1/\epsilon)$ a utilization of $1 - \epsilon$ can be achieved and clearly the data can be retrieved with only two non-consecutive probes into external memory. Due to the large constant for d, it may be disadvantageous if the data associated with the keys is very large (now the time for finding D_x depends on the size of the data). Moreover, such dictionary solutions do not provide perfect hash functions without an additional level of indirection.

Pagh [11] showed how to construct a minimal perfect hash function in expected linear time which can be very efficiently evaluated with very simple arithmetics (essentially one or two multiplications) and by probing only one word from external memory. The hash function itself can be encoded in $(2 + \epsilon) \cdot n \cdot \log n$ bits. Dietzfelbinger and Hagerup [3] improved Pagh's scheme so that the resulting hash function can be encoded with $(1 + \epsilon) \cdot n \cdot \log n$ bits. Both schemes yield a static dictionary with 100% utilization.

In this paper we devise a dynamic variant of Pagh's scheme. Maintaining 100% utilization and using exactly the same hash functions, we show how to perform updates in expected amortized constant time. I.e., the hash functions can be evaluated very efficiently in constant time and with only one probe into external memory. In addition to the $(2+\epsilon) \cdot n \cdot \log n$ bits for encoding of the hash function we also need an auxiliary data structure comprising $(3 + \epsilon) \cdot n \cdot \log n$ bit. However, this auxiliary data structure is only needed for update operations and not for lookups. For many applications updates occur infrequently, e.g., at night time, so that the auxiliary data structure may be swapped out (or it can be removed and later be rebuild from scratch in expected linear time if needed). We believe that this scheme is quite practical if the main focus is on lookup performance, although the algorithm for updates is not very simple.

In Section 3 we investigate how much the space for the hash function description can be reduced under the constraint that evaluation requires only consecutive probes into external memory. We show that it is possible to reduce the encoding size of the hash functions and the space for the auxiliary data structure to $O(n \log\log n)$ bits. In the dynamic case we obtain a utilization of $1 - \epsilon$, for arbitrary small $\epsilon > 0$. In the static case we still achieve 100% utilization, hence we even have a minimal perfect hash function. For both implicit versions the corresponding hash functions can be evaluated in constant time and by probing $O(1)$ consecutive words from external memory. (Here $O(1)$ is a very small constant, e.g. 4). The hash functions itself are a little bit more complicated – their evaluation times are dominated by the arithmetics involved for evaluating two polynomials of small constant degree.

Our dynamic hashing algorithms are not stable. For updates we have to assume that the key x can be retrieved from the hash table entry $T[h(x)]$. But the hash function description itself is independent from the table contents.

2   The Displacement Scheme

As explained before, we assume throughout the paper that $U = [n^{O(1)}]$. Moreover, we assume a word-RAM model where every key in U fits into a memory word (i.e., we have a word-size of $\Omega(\log n)$).

Let $\mathcal{H}_a$ be a family of hash functions $h : U \to [a]$. $\mathcal{H}_a$ is c-universal if for any $x, x' \in U$, $x \neq x'$ and for randomly chosen $h \in \mathcal{H}_a$, it holds $Prob\big(h(x) = h(x')\big) \leq c/a$. If $\mathcal{H}_a$ is c-universal for some arbitrary constant c, then we call it approximately universal. $\mathcal{H}_a$ is uniform if $h(x)$ is uniformly distributed over $[a]$ for all $x \in U$.

Examples of efficient 1- and 2-universal and uniform hash families can be found in [2, 4, 13]. For our purposes it suffices to know that most hash functions from c-universal hash families can be evaluated in constant time with a few arithmetic operations (usually dominated by one multiplication and a division) and that they can be encoded in $O(\log |U|)$ or even $O(\log n + \log \log |U|)$ bits.

Pagh [11] showed how to construct minimal perfect hash functions $h_{g,f,d}$ defined in the following. Let a and b be positive integers and suppose that S is a set of n keys from the universe U. Let $f : U \to [a]$, $g : U \to [b]$ and $d = (d_0, \ldots, d_{b-1}) \in [a]^b$. Then $h_{g,f,d} : U \to [a]$ is defined by $x \mapsto (f(x) + d_{g(x)}(x)) \bmod a$.

One can visualize the hash function $h_{f,g,d}$ by a $(b \times a)$-matrix M, where the ith row, $i \in [b]$, is associated with the displacement value d_i. In order to evaluate h for an element $x \in U$, one first maps x into the matrix element in row $g(x)$ and column $f(x)$. Then the row is rotated cyclically $d_{g(x)}$ steps to the right, where $d_{g(x)}$ is the displacement associated with this row. We call two row displacements d_i, d_j, $i \neq j$, compatible (with respect to g, f and S), if for all $x \in g^{-1}(i) \cap S$ and $x' \in g^{-1}(j) \cap S$ it holds $f(x) + d_i \neq f(x') + d_j$. Clearly, $h_{f,g,d}$ is injective on S if and only if all row displacements are pairwise compatible. According to the informal description above, $S \cap g^{-1}(i)$ is the set of elements which are mapped into the ith row of the matrix M. We call $|S \cap g^{-1}(i)|$ the weight of row i. In order to construct the minimal perfect hash functions, Pagh used the following notion.

Definition 1. Let $S \subseteq U$ and $f : U \to [a]$, $g : U \to [b]$, and $w_i = |g^{-1}(i) \cap S|$ for $i \in [b]$. The pair (f,g) is δ-nice for S if the function $x \mapsto (f(x), g(x))$ is injective on S, and $\sum_{i, w_i > 1} w_i^2 \leq \delta \cdot a$.

Note that what we call δ-nice was originally denoted as r-good, where $r = \delta \cdot a/n$; we chose a different notion because it is more convenient for our purposes. Pagh has shown that a pair (f, g) being δ-nice for a set S, $\delta < 1$, suffices to find a displacement vector d such that $h_{f,g,d}$ is injective on S. On the other hand, δ-nice hash functions can be found easily using universal hash families.

Lemma 1 (Pagh [11]). *Let H_a be c_f-universal and H_b be c_g-universal. If $2 \cdot c_g \cdot n^2/(a \cdot b) \leq \delta \leq 4/c_f$, then for any n-element set $S \subseteq U$, the probability that a randomly chosen pair $(f, g) \in H_a \times H_b$ is δ-nice for S is more than $\left(1 - \delta \cdot c_f/4\right) \cdot \left(1 - 2 \cdot c_g \cdot n^2/(a \cdot \delta \cdot b)\right)$.*

For instance, let $a = n$ and $b = (2 + \epsilon_b)n$, $\epsilon_b > 0$. If H_a is 4-universal and H_b is 1-universal, then there exists a $\delta < 1$ such that (f, g) is δ-nice with positive constant probability. We refer the reader to [11] for more details on suitable hash families, or on constructions where $f(x)$ and $g(x)$ can be determined with essentially one multiplication if b is chosen only slightly larger.

Consider again the matrix M as described above. If (f, g) is δ-nice for S, all elements in S are mapped to disjoint matrix elements (for any displacement vector d). Pagh's algorithm finds in expected linear time a vector d such that all displacements are compatible. The row displacements are chosen randomly one after the other in an order with decreasing row weights. This order and the δ-niceness guarantee that a compatible displacement can be found for the next row to be processed. Clearly, such an ordering of the rows cannot be used for a dynamic algorithm. Our idea is the following: If an insertion yields incompatible displacements, then we randomly choose one such displacement anew. This new displacement may now be incompatible with other displacements, but after a constant number of tries, the total number of elements in rows with weight larger than one and with incompatible displacements decreases by a constant factor. Rows with weight are taken care of in the end – for them new compatible displacements can be found easily by keeping track of empty table cells.

Consider a hash table $T[0], \ldots, T[N-1]$ for a set $S \subseteq U$ of at most N elements. We first consider a fixed value of N and later show how to adapt if $|S|$ exceeds N. We store a perfect hash function $h := h_{f,g,d} : U \to [N]$ and an auxiliary data structure. Every element $x \in S$ is stored in $T[h(x)]$. If $|S| = N$, then h is minimal perfect. Since f and g can be stored in $O(\log n)$ bits, one probe into external memory (to retrieve $d_{g(x)}$) suffices for computing $h(x)$.

The functions f and g as well as the displacements can be chosen from the same sets H_a and H_b as in the static case. Fix some $\delta < 1$ such that (f, g) is δ-nice with constant probability. Throughout the description of the algorithm let $w_i = |S \cap g^{-1}(i)|$, if S and g are clear from the context.

For insertions and deletions we need an auxiliary data structure consisting of b linked lists $L_0, \ldots, L_{b-1}$. The list L_i contains all table positions $h(x)$ for which there is an element $x \in S \cap g^{-1}(i)$ (i.e., the columns j in the matrix to which all elements of the row i are mapped to). We don't need to store row weights w_i since we can compute them by searching through the lists L_i, but we store the sum $W = \sum_{i \in [b]} w_i^2$. Finally, we use a data structure for storing all empty table cells, i.e., the indices j such that $T[j] = \infty$. We use a function free_pos which returns in expected constant time the index of an arbitrary empty table cell (if there is one). The implementation of such a data structure comprising $\epsilon \cdot n \cdot \log n$ space is easy – a description can be found in the full version of the paper.

Update Operations. In order to delete an element it suffices to set $T[h(x)]$ to ∞ and to remove x from the list L_i, where $i = g(x)$. This requires $O(w_i)$ time. Since g is chosen from an approximately universal hash family H_b, the expectation of $w_{g(x)}$ is $O(n/b) = O(1)$.

Assume that $h_{f,g,d}$ is injective on S and that we want to insert a new element $x \notin S$. Let $S' = S \cup \{x\}$. Using the list L_i, it is easy to update the sum of squared row weights to $W' = W - w_i^2 + (w_i + 1)^2$ if $w_i > 0$ and $W' = W$ if $w_i = 0$. Using this list we can also check whether (f,g) is still δ-nice for S'. All this can be done in $O(w_i + 1)$ time. If (f,g) is not δ-nice for S', we have to perform a global rehash, i.e., we have to remove all keys and insert them again with a new randomly chosen pair (f,g).

Now assume that (f,g) is δ-nice for S'. If $T[j]$ is empty for $j = h_{g,f,d}(x)$, we simply store x in $T[j]$ and insert j into the lists L_i. If $T[j]$ is already occupied, then we have to determine new displacement values for some rows. For that we maintain a set Q of *possibly bad* rows such that all rows in $[b] - Q$ are compatible. With each row $i \in Q$ we also store the set $V_i := S' \cap g^{-1}(i)$ of all keys which are hashed to that row. Consequently we remove every element $x \in V_i$ from the hash table at the time we insert it in V_i. Initially, Q contains only row i and we collect the set V_i with the help of the list L_i. Let $w(Q) = \sum_{i \in Q, w_i > 1} w_i$.

We repeat the following procedure until $w(Q) = 0$. First we pick an arbitrary row i in Q where $w_i > 1$ and choose a new displacement d_i'. We now define a condition in which we accept this new displacement.

Definition 2. *Fix a set $S \subseteq U$ and a hash function $h_{f,g,d}$ and let d_i' be a new displacement of row $i \in Q$. The set $J(d_i', i)$ contains the indices of all rows $j \in [b] - Q$ such that row i and row j are not compatible. The displacement d_i' is acceptable if $\sum_{j \in J(d_i'), w_j > 1} w_j < w_i \cdot (1 + \delta)/2$.*

Clearly, we can determine the set $J(d_i', i)$ in time $O(w_i)$. By seeking (at least partly) through the lists L_j with $j \in J(d_i', i)$ we can also check in time $O(w_i)$ whether d_i' is acceptable. 1 below states that with constant probability a randomly chosen displacement is acceptable. Hence, in expected time $O(w_i)$ we can find an acceptable new displacement d_i'.

Then we change Q to $Q' = J(d_i', i) \cup Q - \{i\}$, create the sets V_j, $j \in J(d_i', i)$, and accordingly remove the elements in V_j from the hash table. Finally we store all elements $x \in V_i$ at their designted places $T[h_{f,g,d'}(x)]$ (since row i is now compatible with all rows not in Q', these table positions are not occupied).

Repeating this procedure eventually leads to a set Q^* with $w(Q^*) = 0$ (see the time analysis below). Hence, Q^* consists only of possibly bad rows with weight 1. For these rows it is easy to find new compatible displacement values using the function free_pos. After that, the resulting hash function h_{f,g,d^*} is injective on S' and all elements in S' are stored in their designated table positions.

Time Analysis. We first consider the case that the hash function pair (f, g) is still δ-nice for S' and that no global rehash is necessary. The following proposition shows that we can quickly find an acceptable displacement for each row. The (straight forward) proof has to be omitted due to space restrictions.

Proposition 1. *Let $(f, g) \in H_a \times H_b$ be δ-nice, $\delta < 1$, for some n-element set S. With positive constant probability a randomly chosen displacement is acceptable.*

Hence, the total expected time for finding an acceptable displacement for a row $i \in Q$ is $O(w_i)$. Then in expected time $O(w_i)$ we can decrease the value $w(Q)$ to a value of $w(Q') \leq w(Q) - w_i + w_i \cdot (1+\delta)/2 = w(Q) - \Omega(w_i)$ (using $\delta < 1$). It follows from the linearity of expectation that in order to obtain a set Q^* with $w(Q^*) = 0$ expected time $O(w(Q))$ suffices. Now recall that we started with $Q = \{i\}$, where $i = g(x)$ and where x was the element we inserted. Hence, the total expected time until the resulting set Q^* contains only rows with weight 1 is $O(w_i)$. Since we can collect only $O(t)$ rows with weight 1 in time t, the total number of rows in Q^* is also $O(t)$. By the assumption that the operation free_pos can be executed in expected constant time, we can redisplace these rows in expected time $O(t)$. To conclude, the total expected time for inserting an element x is $O(w_{g(x)})$. As argued in the section about deletions, $E(w_{g(x)}) = O(1)$.

We have shown so far that we obtain an expected constant insertion time as long as (f, g) remains δ-nice for the resulting set. By 1, a simple calculation shows that a randomly chosen set pair (f, g) is with constant probability δ-nice even for the $\lfloor \alpha N \rfloor$ sets obtained from some set S of size N during a sequence of $\lfloor \alpha N \rfloor$ update operations, for some sufficiently small $\alpha > 0$. Therefore, during $\lfloor \alpha \cdot N \rfloor$ update operations we expect only a constant number of global rehashes and thus the expectation of the amortized update time is constant.

A Dynamic Hash Table with 100% Utilization. So far we can insert and delete elements from a hash table of size N, as long as $|S| \leq N$. We now sketch an algorithm which maintains a dynamic hash table T where all n elements in S are stored in the table positions $T[0], \ldots, T[n-1]$ at all times. A complete description will be given in the full version of the paper.

The problem is mainly with deletions. If an arbitrary element is deleted, a "hole" is left behind in the middle of the hash table, say at position i. But in order to store all remaining $n-1$ elements in the table positions $T[0], \ldots, T[n-2]$, we have to move the element x from $T[n-1]$ to some other position. Since it is not clear how to bound the weight of the row $g(x)$ of that element, we don't know how to obtain an expected constant deletion time. The idea is now to ensure that the last $\gamma \cdot n$ entries of the table, $\gamma > 0$, are filled with elements from rows with weight one. Then we can easily choose a new displacement for the corresponding rows, so that the element in $T[n-1]$ moves into the hole $T[i]$.

We now interpret the displacements of the hash function $h_{f,g,d}$ differently: Let $i = g(x)$. Then $h_{f,g,d}(x) = (f(x) + d_i) \bmod a$ if $d_i < a$, and $h_{f,g,d}(x) = d_i$ if $d_i \geq a$. This way, the range of $h_{f,g,d}$ is not limited to $[a]$.

Consider a situation right before a rehash. Let S be the n-element set currently stored and let S_k, $k = 0, 1, 2, \ldots$ be the set obtained after the next k operations (i.e., $S_0 = S$). We let $a = (1 - \gamma)n$ for some sufficiently small $\gamma > 0$, and b, H_a, and H_b as before. Now we have to perform a global rehash also if the size of the set S drops below a.

In order to insert a new element x in a set S of size $n \geq a$, we redisplace rows exactly as before, but using only displacement values $d_i < a$ for rows with weight $w_i > 1$. As before we end up with a set Q^* with $w(Q^*) = 0$, i.e., $w_i = 1$ for all rows $i \in Q^*$. Now for one of the remaining rows in Q^* we choose the displacement value d_i such that the unique element x^* in that row obtains a hash value of n and we store x^* in $T[n]$. All other displacement values for rows with weight 1 can be determined using free_pos as before.

The insertion procedure guarantees that displacement values $d_i \geq a$ are only used for rows with weight $w_i = 1$. Hence, as long as $n = |S| > a$, the element x^* stored in $T[n-1]$ belongs to a row i^* with weight $w_{i^*} = 1$. Hence, in order to delete an element x from S, $|S| > a$, we simply change d_{i^*} to a value such that x^* moves into the table cell $T[h_{f,g,d}(x)]$, formerly occupied by x.

Theorem 1. *For any $\epsilon > 0$ a dynamic hash table with 100% utilization can be maintained with constant amortized update time and $(2 + \epsilon)n \log n$ space for the hash function encoding and $(3 + \epsilon)n \log n$ space for the auxiliary data structure. The hash function can be evaluated in constant time and with only one probe into external memory.*

Corrupted Hash Table Cells. For the following sections we need to consider a variant of the above scheme, which may also be of independent interest. Consider a hash table $T[0], \ldots, T[n + k - 1]$ with k *corrupted* cells. If a cell $T[i]$ is corrupted, then none of the keys in S may be stored there, but we assume that we can check in constant time whether a cell $T[i]$ is corrupted or not. Let $I \subseteq \{0, \ldots, n + k - 1\}$, $|I| = k$, be the set of indices of corrupted table cells. For $k = o(\sqrt{n})$, we can modify our data structure in such a way that an n-element set S is stored in the hash table T without using any corrupted cells. If a new element is inserted we use the same algorithm as above, except that when we choose a new displacement d_i' for a row i we have to ensure that none of the keys from that row are hashed to a corrupted cell. Thus, for every n-element set S we can maintain a hash function $h := h_{f,g,d}$ which is injective on S and where $h(S) = \{0, \ldots, n + k - 1\} - I$, and with the same time- and space-complexity as in 1.

3 Implicit Hash Functions

We now show how to reduce the space of our hash functions significantly. Recall that we assume a word-size of $\Omega(\log n)$. Similar as in [10] we use one additional hash function $\hat{h}$ in order to split the n-element set S into small groups.

For the following we need two functions $\mu(n) = (\log n)/K$ and $\lambda(n) = n/(\log n)^K$ for some large enough constant K. Let $\hat{h} : U \to [\hat{a}]$, $\hat{a} \in \mathbb{N}$, and let $S \subseteq U$ be an n-element set. We call a group $G_i := S \cap \hat{h}^{-1}(i)$, $i \in [\hat{a}]$, c-*small* if $|G_i| \le \log n/(c \cdot \log\log n)$. If G_i is not c-small, then it is c-*large*. The hash function $\hat{h}$ is c-*good* for S, if all groups have a size of at most $\mu(n)$ and if the total number of c-large groups is at most $\lambda(n)$

Let $\hat{b} = \lfloor n^\gamma \rfloor$ and $\hat{a} = \lceil Z \cdot n \cdot \log\log n/\log n \rceil$. We use the polynomial hash families $\mathcal{H}_s^k$ described in [10]. For a prime $p > |U|$ and $a \in [p]^{k+1}$ the hash function $r_a : U \to [s]$ is given as $x \mapsto \left(\sum_{i=0}^k a_i \cdot x^i \mod p\right) \mod s$. The hash family $\mathcal{H}_s^k$ consists of all hash functions r_a, $a \in [p]^{k+1}$.

Lemma 2. *For any n-element set S, any integer $c > 1$, any $\hat{b} = n^\gamma$, $\gamma > 0$, and any $\hat{a} = \lceil Z \cdot n \cdot \log\log n/\log n \rceil$, $Z > c$, there exist k_a, k_b such that for a randomly chosen pair $(f, g) \in \mathcal{H}_{\hat{a}}^{k_a} \times \mathcal{H}_{\hat{b}}^{k_b}$ and a randomly chosen vector $\hat{d} \in [\hat{a}]^{[\hat{b}]}$ the following is true:*

1. *With probability $1 - o(1)$ a random hash function $\hat{h} = \hat{h}_{\hat{f},\hat{g},\hat{d}}$ is c-good for S.*
2. *For every element $x \in S$, the probability that x is in a c-large group is $2^{-\Omega(\log n/\log\log n)}$.*

The idea of that proof, which we have to omit due to space restrictions, is very similar to a proof in [10], Lemma 3. The main difference is here that our expected group sizes are smaller by a $\log\log n$ factor and we therefore can only achieve that most groups instead of all groups deviate little from their expectation.

The Implicit Data Structure. We now sketch the dynamic scheme which achieves $1 - \epsilon$ utilization, $\epsilon > 0$, but requires only $O(n \cdot \log\log n)$ space for the hash function encoding and the auxiliary data structure. We choose $Z > c > 1$, $Z' = (1 + \alpha)Z$ for some arbitrary small $\alpha > 0$, and $\hat{a} = \lceil Z' \cdot n \cdot \log\log n/\log n \rceil$. As just described we use a hash function $\hat{h} : U \to [\hat{a}]$ in order to split the set S into groups. Consider a subsequence of operations between two global rehashes, i.e., during the time the hash function $\hat{h}$ remains c-good. The hash table T is split up in hash tables $T_0, \ldots, T_{\hat{a}-1}$ as well as one hash table T'. The tables T_i, $i \in [\hat{a}]$, are of size $t = \lfloor \log n/(c \cdot \log\log n) \rfloor$, and T' is of size $a' = O(n/\log n)$ (the constant factor can be chosen arbitrarily). In the following we call a group G_i *clean*, if it has been c-small since $\hat{h}$ was chosen the last time. At the moment a group G_i becomes c-large it is *dirty* and remains so until the next global rehash, even if it becomes c-small again before that. The idea is that all elements from a clean group G_i, $i \in [\hat{a}]$, are stored in the corresponding hash table T_i. For all bad groups the one larger hash table T' will suffice.

If after an insertion the function $\hat{h}$ is not c-good anymore, it has to be chosen anew (which triggers a global rehash). However, it can be shown that with constant probability $\hat{h}$ remains c-good during any sequence of $\lfloor \alpha n \rfloor$ update operations. Therefore we just discuss the update operations under the assumption that no global rehashes occur.

For each element $x \in S$ its group i is determined by $i = \hat{h}(x)$. With each group G_i we keep track of the number of its elements and store a bit indicating whether it is bad or not. If the group is clean, then it is also c-small and we can use the dynamic scheme as described in the previous section. It is easy to see that if we choose c as a large enough constant, then we can store all displacements and the auxiliary data structure in one word of size $\Omega(\log n)$. Thus, all the information for one group G_i can be stored in $O(1)$ words.

If the group G_i is bad, then we use instead one hash function h_i from an approximately universal and uniform hash family $H_{a'}$, where $a' = \lceil n/\log n \rceil$. An element $x \in G_i$ is now stored in the table position in $T'[h_i(x)]$. As $O(\log n)$ bits suffice for storing h_i, we can store the hash function information for each group in a constant number of words. We also maintain a list L_i' containing pointers to the table positions in T' for all elements in group G_i. This is the auxiliary data structure for a bad group G_i. Since all lists for bad groups contain only $O(n/\log n)$ elements altogether, linear space suffices for all of them.

It is not hard to see that the total space for storing all hash functions and the auxiliary data structures is $O(n \cdot \log \log n)$ and that in order to evaluate the hash function it suffices to read a constant number of consecutive memory cells from a data structure with more than n^ϵ space.

Insertions and Deletions. Between two global rehashes we know that in each clean group G_i there are at most t elements, and thus we can insert and delete just as described in Section 2. We now discuss updates for elements hashed into bad groups by $\hat{h}$.

Let S' be the set of elements in bad groups and assume that a newly inserted element x is mapped by $\hat{h}$ to a bad group G_i. Since $\hat{h}$ is c-good we know that $n' := |S'| \leq \lambda(n) = n/(\log n)^K$. The designated table entry for x is $T'[h_i(x)]$. Since h_i is chosen from an approximately universal and uniform hash family $H_{a'}$, the probability that this table position is already occupied by an element in S' is at most $n'/a' = O((\log n)^{1-K})$. If that table position is already occupied we randomly choose h_i from the universal hash family $H_{a'}$ anew. We use the list L_i' to collect all elements from G_i and rehash them again using the new hash function. The probability that one of the $O(\log n)$ elements in group G_i is mapped by h_i to one of the already occupied table cells in T' or that two of the elements in the group collide is at most $n'/a' + |G_i|^2/a' = O((\log n)^{1-K})$. Such a rehash requires $|G_i| = O(\log n)$ time if it is successfull, and thus x can be inserted in expected $O(\log n)$ time, given that a rehash is necessary. On the other hand, as we have seen, with probability $1 - O((\log n)^{1-K})$ the element x can be inserted without any rehash. Thus, for large enough K, x can be inserted in constant expected time given that it is hashed by $\hat{h}$ to a bad group.

We still have to discuss the transition from clean to dirty groups, though: If we insert a new element x into a clean group G_i, then this group may become dirty. In this case we have to move all elements from T_i to T' using a newly sampled hash function h_i (i.e., we rehash group G_i in expected $O(\log n)$ time). By the bound from part two of 2 on the probability that element x is in a bad group, the total expected time for inserting x is still constant in this case.

In order to delete an element x from a bad group we may simply set $T'[h'(x)]$ to ∞. Hence, we can delete elements in bad groups in worst-case constant time.

Theorem 2. *For any $\epsilon, \epsilon' > 0$ a dynamic hash table with $1 - \epsilon$ utilization can be maintained with constant amortized update time and $O(n \log \log n)$ space for the hash function encoding and the auxiliary data structure. The hash function can be evaluated in constant time and by probing $O(1)$ consecutive words from external memory (if the internal memory has size $n^{\epsilon'}$).*

Minimal Perfect Hashing with Implicit Hash Functions. We finally sketch an algorithm which constructs a minimal perfect hash function h in expected linear time such that the encoding of h requires only $O(n \log \log n)$ space and that h can be evaluated with only a few consecutive probes into external memory. The idea is again to use a hash function $\hat{h}$ to split the set S into $\hat{a}$ groups, but now we can use the fact that the group sizes do not change.

Let $S \subseteq U$ be a fixed n-element set. We will store all elements from S in a table $T = T[0], \ldots, T[n-1]$. As in the previous section we choose an integer c, a value $Z > c$, and let $\hat{a} = \lceil Z \cdot n \cdot \log \log n \rceil$. By 2 it is obvious how to find in $O(n)$ expected time a c-good hash function $\hat{h}$. Let $G_i = S \cap \hat{h}^{-1}(i)$, $i \in [\hat{a}]$.

We first process all c-large groups, one after the other. When we process a c-large group i, we create a hash function $h_i : U \to [n]$ mapping all elements in G_i to non-occupied table positions. The hash function h_i is a mapping $x \mapsto \lfloor \log n \rfloor \cdot h_i^*(x)$, where $h_i^* : U \to [\lfloor n/\log n \rfloor]$ is chosen from an approximately universal and uniform hash family $H_{\lfloor n/\log n \rfloor}$. We randomly sample such a h_i^* and then try to store each element $x \in G_i$ in the table position $T[h_i(x)]$. If that table cell is already occupied, we have to sample h_i anew. By arguments similar to those used in the dynamic case, the expected number of tries for each hash function h_i is only constant. Therefore, we can find in $O(n)$ expected time all hash functions h_i for c-large groups such that they map the elements from these groups to disjoint table positions.

Once we have found all hash functions h_i for the c-large groups, some of the table positions in T are occupied, which causes some interference with the c-small groups. That is where the notion of corrupted table cells (see Section 2) comes in handy. From now on we assume that every table cell $T[i]$ is corrupted, if one of the elements from a c-large group is stored there. Since we obtained each hash value $h_i(x)$ for an element x in a c-large group by multiplying a hash value $h_i^*(x)$ with $\lfloor \log n \rfloor$, we know that any $\lfloor \log n \rfloor$-sized interval of table cells, $T[i], \ldots, T[i + \lfloor \log n \rfloor]$, contains at most one corrupted cell.

We now process all c-small groups in increasing order. As in the dynamic case we find a hash function $h_i = h_{f_i, g_i, d_i}$ for each c-small group, mapping the elements of that group to a subtable T_i. We keep track of an offset o_i for each group i, indicating at which position in T the subtable T_i starts. Let a_i be the number of table cells we need for the ith group (this may be one more than the number of elements stored there, in the case that one table cell is corrupted). Then we can construct a hash function $h_i = o_i + h_{f_i, g_i, d_i}$ with the obvious random choices for f_i, g_i and d_i, which maps G_i injectively to the table positions $T[o_i], \ldots, T[o_i + a_i - 1]$ and spares out the corrupted table cell (if there is any). As we have seen in Section 2, h_i can be constructed even dynamically in expected constant time for each insertion and for $o(\sqrt{a_i})$ corrupted table cells.

Thus, we can compute all hash functions h_i, $1 \leq i \leq \hat{a}$, in expected time $O(n)$. The resulting mapping $h : S \rightarrow [n]$, $x \mapsto h_{\hat{h}(x)}(x)$, is a bijection. Each hash function h_i can be stored with $O(\log n)$ bits and thus the total space for storing h is $O(\hat{a} \cdot \log n) = O(n \cdot \log \log n)$.

Theorem 3. *For any n-element set $S \subseteq U$ a bijection $h : S \rightarrow [n]$ with encoding size $O(n \cdot \log \log n)$ can be constructed in expected time $O(n)$. The hash function can be evaluated in constant time and by probing $O(1)$ consecutive words from external memory (if the internal memory has size n^ϵ, $\epsilon > 0$).*

Acknowledgment

The author is grateful to Martin Dietzfelbinger and Rasmus Pagh for enlightening discussions on the subject of the paper. The anonymous referees provided very helpful comments.

References

1. E. D. Demaine, F. Meyer auf der Heide, R. Pagh, and M. Pătrascu. De dictionariis dynamicis pauco spatio utentibus (lat. on dynamic dictionaries using little space). In *Proc. of the 7th LATIN*, volume 3887 of *LNCS*, pp. 349–361. 2006.
2. M. Dietzfelbinger. Universal hashing and k-wise independent random variables via integer arithmetic without primes. In *Proc. of 13th STACS*, volume 1046 of *LNCS*, pp. 569–580. 1996.
3. M. Dietzfelbinger and T. Hagerup. Simple minimal perfect hashing in less space. In *Proc. of 9th ESA*, number 2161 in LNCS, pp. 109–120. 2001.
4. M. Dietzfelbinger, T. Hagerup, J. Katajainen, and M. Penttonen. A reliable randomized algorithm for the closest-pair problem. *J. of Alg.*, 25:19–51, 1997.
5. M. Dietzfelbinger, A. Karlin, K. Mehlhorn, F. Meyer auf der Heide, H. Rohnert, and R. E. Tarjan. Dynamic perfect hashing: Upper and lower bounds. *SIAM J. on Comp.*, 23:738–761, 1994.
6. M. Dietzfelbinger and C. Weidling. Balanced allocation and dictionaries with tightly packed constant size bins. In *Proc. of 32nd ICALP*, volume 3580 of *LNCS*, pp. 166–178. 2005.
7. D. Fotakis, R. Pagh, P. Sanders, and P. G. Spirakis. Space efficient hash tables with worst case constant access time. *Theory of Comp. Syst.*, 38:229–248, 2005.
8. M. L. Fredman and J. Komlós. On the size of separating systems and families of perfect hash functions. *SIAM Journal on Algebraic and Discrete Methods*, 5:61–68, 1984.
9. M. L. Fredman, J. Komlós, and E. Szemerédi. Storing a sparse table with $O(1)$ worst case access time. *J. of the ACM*, 31:538–544, 1984.
10. T. Hagerup and T. Tholey. Efficient minimal perfect hashing in nearly minimal space. In *Proc. of 18th STACS*, volume 2010 of *LNCS*, pp. 317–326. 2001.
11. R. Pagh. Hash and displace: Efficient evaluation of minimal perfect hash functions. In *Proc. of 6th WADS*, volume 1663 of *LNCS*, pp. 49–54. Berlin, 1999.
12. R. Pagh and F. F. Rodler. Cuckoo hashing. *J. of Alg.*, 51:122–144, 2004.
13. P. Woelfel. Efficient strongly universal and optimally universal hashing. In *Proc. of 24th MFCS*, volume 1672 of *LNCS*, pp. 262–272. 1999.

Author Index

Lecture Notes in Computer Science

For information about Vols. 1–4046

please contact your bookseller or Springer

Vol. 4096: E. Sha, S.-K. Han, C.-Z. Xu, M.H. Kim, L.T. Yang, B. Xiao (Eds.), Embedded and Ubiquitous Computing. XXIV, 1170 pages. 2006.

Vol. 4094: O. H. Ibarra, H.-C. Yen (Eds.), Implementation and Application of Automata. XIII, 291 pages. 2006.

Vol. 4093: X. Li, O.R. Zaïane, Z. Li (Eds.), Advanced Data Mining and Applications. XXI, 1110 pages. 2006. (Sublibrary LNAI).

Vol. 4092: J. Lang, F. Lin, J. Wang (Eds.), Knowledge Science, Engineering and Management. XV, 664 pages. 2006. (Sublibrary LNAI).

Vol. 4091: G.-Z. Yang, T. Jiang, D. Shen, L. Gu, J. Yang (Eds.), Medical Imaging and Augmented Reality. XIII, 399 pages. 2006.

Vol. 4090: S. Spaccapietra, K. Aberer, P. Cudré-Mauroux (Eds.), Journal on Data Semantics VI. XI, 211 pages. 2006.

Vol. 4089: W. Löwe, M. Südholt (Eds.), Software Composition. X, 339 pages. 2006.

Vol. 4088: Z.-Z. Shi, R. Sadananda (Eds.), Agent Computing and Multi-Agent Systems. XVII, 827 pages. 2006. (Sublibrary LNAI).

Vol. 4085: J. Misra, T. Nipkow, E. Sekerinski (Eds.), FM 2006: Formal Methods. XV, 620 pages. 2006.

Vol. 4083: S. Fischer-Hübner, S. Furnell, C. Lambrinoudakis (Eds.), Trust and Privacy in Digital Business. XIII, 243 pages. 2006.

Vol. 4082: K. Bauknecht, B. Pröll, H. Werthner (Eds.), E-Commerce and Web Technologies. XIII, 243 pages. 2006.

Vol. 4081: A. M. Tjoa, J. Trujillo (Eds.), Data Warehousing and Knowledge Discovery. XVII, 578 pages. 2006.

Vol. 4080: S. Bressan, J. Küng, R. Wagner (Eds.), Database and Expert Systems Applications. XXI, 959 pages. 2006.

Vol. 4079: S. Etalle, M. Truszczyński (Eds.), Logic Programming. XIV, 474 pages. 2006.

Vol. 4077: M.-S. Kim, K. Shimada (Eds.), Geometric Modeling and Processing - GMP 2006. XVI, 696 pages. 2006.

Vol. 4076: F. Hess, S. Pauli, M. Pohst (Eds.), Algorithmic Number Theory. X, 599 pages. 2006.

Vol. 4075: U. Leser, F. Naumann, B. Eckman (Eds.), Data Integration in the Life Sciences. XI, 298 pages. 2006. (Sublibrary LNBI).

Vol. 4074: M. Burmester, A. Yasinsac (Eds.), Secure Mobile Ad-hoc Networks and Sensors. X, 193 pages. 2006.

Vol. 4073: A. Butz, B. Fisher, A. Krüger, P. Olivier (Eds.), Smart Graphics. XI, 263 pages. 2006.

Vol. 4072: M. Harders, G. Székely (Eds.), Biomedical Simulation. XI, 216 pages. 2006.

Vol. 4071: H. Sundaram, M. Naphade, J.R. Smith, Y. Rui (Eds.), Image and Video Retrieval. XII, 547 pages. 2006.

Vol. 4070: C. Priami, X. Hu, Y. Pan, T.Y. Lin (Eds.), Transactions on Computational Systems Biology V. IX, 129 pages. 2006. (Sublibrary LNBI).

Vol. 4069: F.J. Perales, R.B. Fisher (Eds.), Articulated Motion and Deformable Objects. XV, 526 pages. 2006.

Vol. 4068: H. Schärfe, P. Hitzler, P. Øhrstrøm (Eds.), Conceptual Structures: Inspiration and Application. XI, 455 pages. 2006. (Sublibrary LNAI).

Vol. 4067: D. Thomas (Ed.), ECOOP 2006 – Object-Oriented Programming. XIV, 527 pages. 2006.

Vol. 4066: A. Rensink, J. Warmer (Eds.), Model Driven Architecture – Foundations and Applications. XII, 392 pages. 2006.

Vol. 4065: P. Perner (Ed.), Advances in Data Mining. XI, 592 pages. 2006. (Sublibrary LNAI).

Vol. 4064: R. Büschkes, P. Laskov (Eds.), Detection of Intrusions and Malware & Vulnerability Assessment. X, 195 pages. 2006.

Vol. 4063: I. Gorton, G.T. Heineman, I. Crnkovic, H.W. Schmidt, J.A. Stafford, C.A. Szyperski, K. Wallnau (Eds.), Component-Based Software Engineering. XI, 394 pages. 2006.

Vol. 4062: G. Wang, J.F. Peters, A. Skowron, Y. Yao (Eds.), Rough Sets and Knowledge Technology. XX, 810 pages. 2006. (Sublibrary LNAI).

Vol. 4061: K. Miesenberger, J. Klaus, W. Zagler, A.I. Karshmer (Eds.), Computers Helping People with Special Needs. XXIX, 1356 pages. 2006.

Vol. 4060: K. Futatsugi, J.-P. Jouannaud, J. Meseguer (Eds.), Algebra, Meaning, and Computation. XXXVIII, 643 pages. 2006.

Vol. 4059: L. Arge, R. Freivalds (Eds.), Algorithm Theory – SWAT 2006. XII, 436 pages. 2006.

Vol. 4058: L.M. Batten, R. Safavi-Naini (Eds.), Information Security and Privacy. XII, 446 pages. 2006.

Vol. 4057: J.P.W. Pluim, B. Likar, F.A. Gerritsen (Eds.), Biomedical Image Registration. XII, 324 pages. 2006.

Vol. 4056: P. Flocchini, L. Gąsieniec (Eds.), Structural Information and Communication Complexity. X, 357 pages. 2006.

Vol. 4055: J. Lee, J. Shim, S.-g. Lee, C. Bussler, S. Shim (Eds.), Data Engineering Issues in E-Commerce and Services. IX, 290 pages. 2006.

Vol. 4054: A. Horváth, M. Telek (Eds.), Formal Methods and Stochastic Models for Performance Evaluation. VIII, 239 pages. 2006.

Vol. 4053: M. Ikeda, K.D. Ashley, T.-W. Chan (Eds.), Intelligent Tutoring Systems. XXVI, 821 pages. 2006.

Vol. 4052: M. Bugliesi, B. Preneel, V. Sassone, I. Wegener (Eds.), Automata, Languages and Programming, Part II. XXIV, 603 pages. 2006.

Vol. 4051: M. Bugliesi, B. Preneel, V. Sassone, I. Wegener (Eds.), Automata, Languages and Programming, Part I. XXIII, 729 pages. 2006.

Vol. 4049: S. Parsons, N. Maudet, P. Moraitis, I. Rahwan (Eds.), Argumentation in Multi-Agent Systems. XIV, 313 pages. 2006. (Sublibrary LNAI).

Vol. 4048: L. Goble, J.-J.C.. Meyer (Eds.), Deontic Logic and Artificial Normative Systems. X, 273 pages. 2006. (Sublibrary LNAI).

Vol. 4047: M. Robshaw (Ed.), Fast Software Encryption. XI, 434 pages. 2006.